THE COMPLETE BOOK OF VEGETABLES, HERBS & FRUIT

THE COMPLETE BOOK OF VEGETABLES, HERBS & FRUIT

REVISED EDITION

MATTHEW BIGGS, JEKKA McVICAR & BOB FLOWERDEW

KYLE BOOKS

To Henry John William and Chloe Elizabeth
Matthew Biggs

To Mac, Hannah and Alistair
Jekka McVicar

To all those who helped make our glorious fruits from such humble beginnings
Bob Flowerdew

This revised and enlarged edition was published in Great Britain in 2016 by
Kyle Books, and imprint of Kyle Cathie Ltd.
192–198 Vauxhall Bridge Road
London SW1V 1DX

First published 1997, revised 2008

ISBN 978-0-85783-348-8

Interior Design Geoff Hayes
Editorial team Caroline Taggart, Hannah Coughlin, Claire Rogers and Tara O'Sullivan

Colour reproduction by Alta image, London
Printed in China by C&C Offset Printing Co., Ltd.

A Cataloguing in Publication record for this title is available from the British Library.

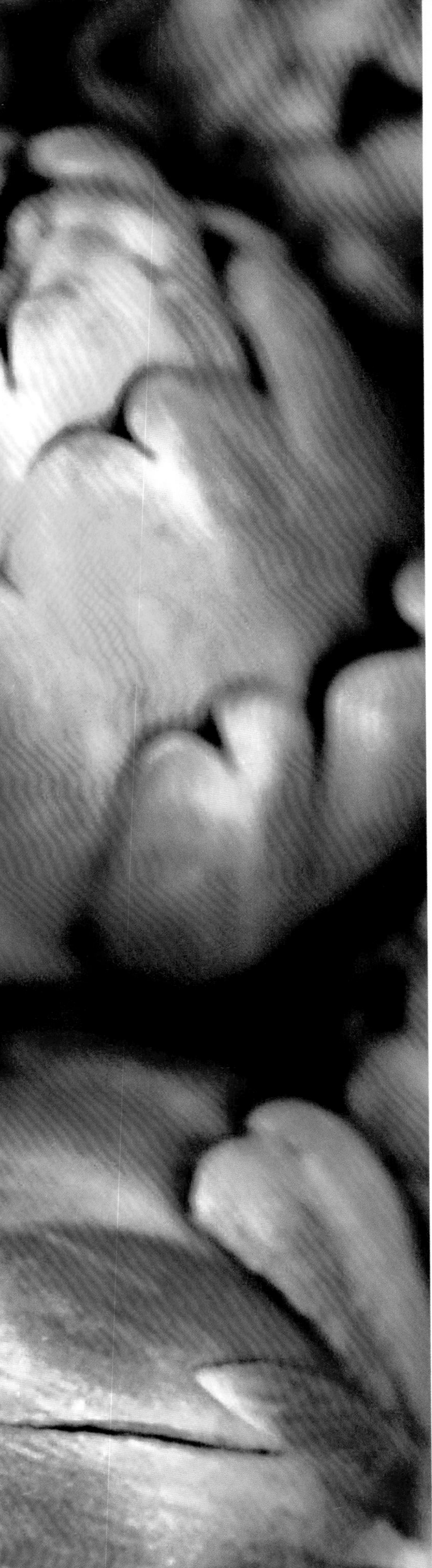

CONTENTS

INTRODUCTION

A flourishing, productive garden, containing vegetables, herbs and fruit plants, is a testament to diligent, imaginative gardening and a promise of a delicious harvest to come. The range of colour, texture, scent and flavour offered by these plants is unrivalled, and there is space in any garden – even in a window box – for a selection of edible and useful plants.

Vegetables, herbs and fruit have always been essential to humanity. They are the basis of the food chain – even for meat-eaters – and are a vital component in creating tempting, palatable meals, as well as providing unique flavouring and aromas. All provide essential vitamins and minerals for a balanced diet, and many herbs have the added dimension of being used medicinally.

Vegetables and herbs can be widely defined. Vegetables are those plants where a part, such as the leaf, stem or root, can be used for food. Herbs, similarly, are those plants that are used for food, medicine, scent or flavour. Fruits tend to be the sweet, juicy parts of the plants, containing the seed. There is considerable overlap between the three types of plant – one further distinction is that fruits are generally sweet, or used in sweet dishes, while vegetables are savoury, although this is by no means clear-cut.

For centuries throughout the world, productive gardens have been the focal point of family and community survival. Our earliest diet as hunter-gatherers must have included a wide range of seeds, fruits, nuts, roots, leaves and any moving thing we could catch. Gradually, over millennia, we learned which plants could be eaten and how to prepare them – as with the discovery that eddoes were edible only after being washed several times and cooked to remove the injurious calcium oxalate crystals. Fruit trees and bushes sprang up at the camp sites of nomadic people and were waiting for them when they returned, growing prolifically on their fertile waste heaps. Vegetables and herbs were collected from the surrounding countryside, and gradually were domesticated. Cultivated wheat and barley have been found dating from 8000 to 7000 BC, and peas from 6500 BC, while rice was recorded as a staple in China by 2800 BC.

With domestication came early selection of plants for beneficial characteristics such as yield, disease resistance and ease of germination. These were the first cultivated varieties, or 'cultivars'. This selection has continued extensively and by the eighteenth century in Europe, seed selection had become a fine art in the hands of skilled gardeners. Gregor Mendel's work with peas in 1855–64 in his monastery garden at Brno in Moravia yielded one of the most significant discoveries, leading to the development of hybrids and scientific selection. Most development has centred on the major food crops. Minor crops, such as seakale, have changed very little, apart from the selection of a few cultivars. Others, like many fruits, are similar to their wild relatives, but have fleshier, sweeter edible parts. Herbs have in general had less intensive work done on selection; many of the most popular and useful herbs are the same as or closely related to plants found in the wild.

The Vegetable Garden, Coombe by Paul Riley, 1988

Food plants have spread around the world in waves, from the Roman Empire, which brought fruits such as peaches, plums, grapes and figs from the Mediterranean and North Africa to northern Europe, to the influx of plants such as potatoes and maize from the New World in the fifteenth century. In between, monasteries guarded fruits, vegetables and herbs for their own use and for their medicinal value. During the famine and winter dearth of the Middle Ages and beyond, the commonplace scurvy and vitamin deficiencies would have seemed to many people almost miraculously cured by monks' potions containing little more than preserved fruits, vegetables or herbs full of nutrients and vitamin C. In 1597 John Gerard wrote his *Herball*, detailing numerous plants and their properties, and giving practical advice on how to use them.

Productive gardening developed on several levels. The rich became plant collectors and used the latest technology to overwinter exotic plants in hothouses and stovehouses. Doctors followed on in the traditions of the monasteries and had physic gardens of medicinal herbs. Villagers had cottage gardens filled with fruit trees and bushes, underplanted with vegetables and herbs.

In the twentieth century, the expense of labour and decrease in the amount of land available meant that productive gardening declined. Home food production revived during the Second World War, but the availability of ready-made foods afterwards again hit edible gardening at home. The later years of the twentieth century saw a reaction

St Paul de Vence by Margaret Loxton

against the blandness and cost of mass-produced food. There was also an increasing awareness of the infinite variety of herbs, and their use in herbalism, cosmetics and cooking all over the world.

The wider realization that we had polluted our environment and destroyed much of the ecology of our farms, countryside and gardens was to bring about a real revolution. A mass revulsion against chemical-based methods was mirrored in the rise of organic production and the slowly improving availability of better foods. Vegetarianism also increased as many people turned away from meat, in part because of factory farming. These trends have meant that there is now an increased demand for fruits and vegetables, often organically produced or with a fuller flavour, and supermarkets now offer a huge range all year round.

But there has also been a move towards people growing their own. The health benefits, ecology and economy of gardening appeal to a greener generation. An increased awareness of alternative medicine, including herbalism and aromatherapy, has revived interest in a range of herbs. With food processors, juicers and freezers, it is easier than ever to store and preserve what we harvest. In addition, the genetic richness represented by the huge range of food plants has been recognised and organizations such as the Henry Doubleday Research Association in the UK, Seed Savers in the USA and Seed Savers International are working to safeguard and make available the old and rare varieties.

The availability of different gardening techniques also offers great opportunities at home. Dwarfing fruit rootstocks, varieties that store well or resist disease, glass or plastic cover, and controlled heating in greenhouses give us scope to grow a huge variety of crops, even in a small garden. The earlier and later seasons, combined with gardening under cover, also mean that we can be planting and harvesting for a larger proportion of the year.

This book is intended to guide the reader in choosing which vegetables, herbs and fruit to grow, and then in producing a crop successfully. The vegetable and herb sections are arranged alphabetically by the botanical Latin name. The fruit section is grouped into five chapters covering different types of fruit plants – Orchard Fruits, Soft, Bush and Cane Fruits, Tender Fruits, Shrub and Flower Garden Fruits, and Nuts – according to how they are usually grown in temperate gardens.

Under each plant, after a brief introduction covering origins and history, the most useful and recommended varieties are given, followed by details of cultivation, including propagation, growing under glass and in containers, a maintenance calendar, pruning and training (if needed), dealing with pests and diseases, companion planting and harvesting and storing. Information and ideas are given for using the plant, including recipes and medicinal and cosmetic uses. If any part of the plant is toxic or harmful in any way, a detailed warning is given. If the plant is of particular ornamental or wildlife value in the garden, this is indicated. The fruit and vegetable sections also cover tropical and sub-tropical crops.

The end section of the book covers the practical aspects of making a productive garden, including planning your plot and preparing the soil, creating an ornamental edible garden, crop rotation, pollination, propagation and growing in containers, maintenance, companion planting and dealing with pests, diseases and weeds. A yearly calendar details the tasks in the productive garden month by month, although precise dates will vary according to conditions in different regions.

There is nothing more satisfying to the soul, eye and stomach than a garden well stocked with produce. This book will help you to grow what you want with confidence, and perhaps to experiment and try out new plants and flavours.

A Note on Botanical Names

While most gardeners refer to plants by their common names, one plant may have several common names, or the same common name may refer to different plants in different parts of the world. For this reason, botanical names are vital to ensure that plants are identified accurately. The system of botanical names used today is known as the Binomial System and was devised by the 18th-century Swedish botanist Carl Linnaeus (1707–78). In this system, each plant is classified by using two words in Latin form. The first word is the name of the genus (e.g. *Thymus*) and the second the specific epithet (e.g. *vulgaris*): together they provide a universally known name (e.g. *Thymus vulgaris*).

The Binomial System has been developed so that all living things are divided into a multi-branched family tree according to their characteristics. Plants are gathered into particular families according to the structure of their flowers, fruits or seeds. A family may contain just one or a few genera (the plural of genus) or it may contain many. The Cannabaceae family, for example, to which cannabis and hops belong, consists of only eleven genera, while the Asteraceae contains over 800 genera, including *Achillea* (mace and yarrow), *Arnica* and *Artemisia* (southernwood, wormwood, etc.) to name a few, and over 13,000 species.

Plants are cultivated for the garden from the wild to improve their leaf or their flower or their root. This can be done either by selection from seedlings or by spotting a mutation. Such plants are known as cultivars (a combination of 'cultivated varieties'). Propagation from these varieties is normally done by cuttings or division. Cultivars are given vernacular names, which are printed within quotes, e.g. *Thymus* 'Doone Valley', to distinguish them from wild varieties whose name are written in Latin form in italics, e.g. *Thymus pulegioides*. Sexual crosses between species, usually of the same genus, are known as hybrids and are indicated by a multiplication sign, e.g. *Thymus* x *citriodorus*.

Finally, a problem that seems to be getting worse. Many plants are undergoing reclassification and long-established names are being changed. This is the result of scientific studies and research whereby it is found either that a plant has been incorrectly identified or that its classification has changed. This book uses the latest information available, and where there has been a recent change in the botanical name, the previous one is shown in brackets.

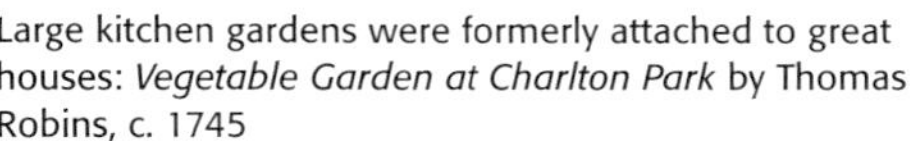

Large kitchen gardens were formerly attached to great houses: *Vegetable Garden at Charlton Park* by Thomas Robins, c. 1745

VEGETABLES
A-Z

***Agaricus bisporus* and others**

MUSHROOMS, EDIBLE FUNGI

Simple organisms growing on decaying substrate or symbiotically with living plants, some with edible fruiting bodies. Half hardy/tender. Value: low in calories, moderate potassium, linoleic and folic acid, carbohydrates, iron, niacin and B vitamins.

Fungi are extraordinary organisms: lacking both chlorophyll and root systems, they are more akin to moulds and yeasts than to traditional vegetable plants. The fleshy mushroom, or bracket, that you eat is a fruiting body, dispersing spores in order to reproduce in the same way that plant fruits disperse seeds. Instead of drawing nutrients through roots, however, a fungus is sustained by a network of fine – often microscopic – threads (known collectively as the *mycelium*). This can extend over vast distances into rotting wood, soil or some other preferred medium. To help identify a fungus it is important to know the particular substrate on which it depends, or the higher plant species with which it lives in symbiosis (certain fungi, for example, grow only in the vicinity of specific trees such as birch or oak); when attempting to cultivate any kind of mushroom, you must provide similarly congenial conditions.

However, few of the many thousand fungus genera are amenable to cultivation. Even in nature, fruiting is wildly unpredictable: the organism depends on precise moisture and temperature variables to produce fruiting bodies, and more generally is sensitive to environmental changes such as recent air pollution and high nitrate levels.

Their erratic behaviour and mysterious origins – allied with the deadly toxins some contain – have given rise to a 'love-hate' attitude towards fungi. Some have been collected as 'wild food' since ancient times. The Romans esteemed them as a delicacy and the rich employed collectors to find the most desirable species. However, Gerard, writing in his *Herball*, remained unimpressed: 'few…are good to be eaten and most of them do suffocate and strangle the eater'. John Evelyn advised that all types of mushrooms should be kept well out of the kitchen.

By the late seventeenth century, varieties of *Agaricus* began to be grown in underground caves in the Paris region, in which giant heaps of manure were impregnated with soil taken from areas where field and horse mushrooms grew naturally. For many centuries cultivated mushrooms were a delicacy enjoyed only by the wealthy, and from the eighteenth century most stableyards had a shady corner where there was a mushroom bed. Some garden owners had outhouses converted to provide ideal growing conditions: George IV had a large mushroom house at Kensington Palace in London. In seasons when wild or cultivated crops were plentiful, surplus mushrooms were conserved in the form of sauces and ketchups, and only recently has the role of mushroom sauce in the kitchen been usurped by tomato sauce.

Cultivated *Agaricus* species have remained popular in northern Europe and the English-speaking world, yet elsewhere they are eclipsed by other mushrooms. In Japan, velvet shank, nameko, oyster and shiitake mushrooms are established as the cultivated varieties, and some of them are slowly becoming popular in other countries.

A number of species is commercially available, some of them to amateur growers, and usually work by inoculating the growing medium with *mycelium* or 'spawn'. Home growing of many fungus species is still in its infancy, but it is gaining ground each year. As adventurous gardeners and mushroom eaters increasingly experiment, advances in mushroom cultivation will also help to conserve wild species.

Ensure fungi are correctly identified before eating

species

The following lists include several of the more common and better known edible fungi.

Commonly available for home cultivation

These fungi are grown commercially and can occasionally be bought from specialist suppliers as kits, 'spawn' or impregnated dowels for inoculating logs. A notable advantage of buying mushrooms in this form is that you are assured of their identity. You should follow suppliers' detailed cultivation instructions carefully to increase the chances of success. Some producers have their own selected strains of these fungi.

Agaricus bisporus (cultivated mushroom, champignon de Paris) has a smooth white to brownish cap and white stem and flesh. Excellent flavour raw and cooked. This accounts for 60 per cent of the world's mushroom production. Available as kits or spawn (see 'Cultivation' page 16). Commercial growers are developing new races with colour variation in the cap: watch out for them in the future as kits.

Flammulina velutipes (enokitake, velvet shank, velvet foot, winter mushroom) occurs naturally on dead wood and is grown commercially on sawdust, particularly in Japan. Small tan-yellow caps on dark brown stems. Do not eat stems and wipe or peel off any stickiness from the caps before cooking.

Hericium erinaceus (lion's mane, monkey head) produces large, rounded clusters of icicle-like growths that taste like lobster – delicious fried with butter and onions. Grows in the wild from wounds on living hardwoods such as beech, and can be cultivated on stumps or logs.

Cultivated mushrooms are easy to grow

Lentinula edodes (shiitake) is a small-to-medium size mushroom with light brown stem and pale to dark reddish-brown cap. Documents record this strongly flavoured gourmet mushroom being eaten in AD199 and it is the second most important cultivated fungus. In Japan it is grown commercially on logs of chestnut, oak or hornbeam. Different forms tolerate warmer or colder conditions, and include **'Snowcap'**, a thick-fleshed form with a long fruiting season, grown on large logs, and **'West Wind'**, which is ideal for inexperienced growers, yielding well over a long period.

Pholiota nameko (nameko, viscid mushroom) is among the four most important fungi cultivated in Japan. Grows in clusters on tree trunks or wood chips. Orange-brown caps atop paler stems are 5–6cm in diameter. It is pleasantly aromatic.

Pleurotus (or oyster mushroom) is a genus with a number of distinct species and strains. Popular in Japan and Central Europe, they are increasingly available in kit form, or can be grown from plugs of spawn. Eat when small, discarding the tough stem. Sauté in butter until tender, season, then add cream or yogurt. The first two are fairly common in the wild and are also offered in seed catalogues.

Pleurotus ostreatus (oyster mushroom) has large fan-shaped caps, slate-blue to white in colour and with white or pale straw-coloured gills; used coffee grounds (sterilised as the coffee is brewed) are becoming a popular medium for inoculating with the spawn.

P. cornucopiae (golden oyster) has a white stem with a cream cap turning to ochre-brown. Grows on the cut stumps of deciduous trees, usually elm or oak. Other species may be harder to find, and some need warmth to fruit.

P. ergyngii (king pleurotus) has a concave cap, whitish becoming grey-brown. It tastes sweet and meaty and grows in clusters on the decaying roots of plants in the carrot family. It can be grown on chopped straw.

P. flabellatus is an oyster mushroom with pink caps.
P. pulmonarius has brown or grey caps.
P. samoneus-tramineus from Asia is pink-capped.
P. sajor-caju has brown caps.

Stropharia rugosoannulata (king stropharia) is a brown-capped, violet-gilled fungus commonly cultivated

Chanterelles

in eastern Europe; it is claimed to be capable of growing in vegetable gardens. It requires a substrate of humus containing rotting hardwood or sawdust. Grow spawn from reputable sources: look-alikes include the deadly *Cortinarius* species.

Volvariella volvaceae (Chinese or straw mushroom, paddy straw, padi-straw) has a grey-brown cap, often marked with black, and a dull-brown stem. Grown on composted rice straw, it is regarded as an expensive delicacy in China and other Asian countries. Needs high temperatures and humidity to grow well. Harvest when it is immature.

Naturally occurring edible fungi

Many edible species of fungi may be found growing in your garden if it happens to provide the host trees or other conditions that form their natural habitat. In Continental Europe, fungi collected in the wild are often sold in markets; but local pharmacists or health inspectors are on hand to verify that those on sale are edible species.

NEVER EAT WILD FUNGI UNLESS YOU HAVE FIRST HAD THEIR IDENTITY CONFIRMED BY AN EXPERT

Agaricus arvensis (horse mushroom) and ***A. campestris*** (field mushroom) are cousins of the cultivated mushroom, found in clusters or rings in grazed or mown grassland. Dome-shaped white 'buttons' open to wide caps. Horse mushrooms can grow to soup-plate size with thick, firm flesh smelling of aniseed; the gills are pale greyish-pink darkening to chocolate-brown. Field mushrooms are smaller and rather more delicate in stature, with a 'mushroomy' smell; their deep-pink gills become dark brown to black at maturity.

Boletus edulis (cep, penny bun) grows on the ground near trees, favouring pine, beech, oak and birch woodlands. The rounded, bun-like brown cap (often covered with a white bloom when young) sits on a bulbous whitish stem – also edible. Tube-like pores (rather than gills) beneath the cap are white, turning dull yellow at maturity. This delicious, fleshy fungus is highly prized in Continental markets and can be eaten fresh, pickled or dried. Related species of *Boletus* are also edible.

Cantharellus cibarius (chanterelle) is funnel-shaped; egg yolk-yellow caps, fading with age, are thick and fleshy with gill-like wrinkles running down from the cap underside into the stem. Has a mild peppery aftertaste when eaten raw; and an excellent flavour when cooked. True chanterelles grow on soil in broad-leaved woodland: similar-looking species are highly toxic.

Hirneola auricula judae, syn. ***Auricularia auricula-judae*** (Jew's ear, wood-ear) looks like a human ear and is date-brown, drying to become small and hard. Found on living and dead elder, beech and sycamore. Can be dried and reconstituted with water. Popular in Taiwan and China.

Wild – and free!

Hypholoma capnoides, syn. ***Nematoloma capnoides***, is a gilled fungus found growing in clusters on conifer stumps. Caps 2–6cm diameter are pale ochre with a buff-coloured margin. Check identity carefully: other *Hypholoma* species are suspect.

If you can't grow your own, try the market for a wide range of varieties

MUSHROOMS, EDIBLE FUNGI

Oyster mushrooms

Laetiporus sulphureus, syn. ***Polyporus sulphureus*** (chicken of the woods) is a bracket fungus found on many hardwoods and softwoods, often on sweet chestnut, oak and beech. Has the flavour of chicken breast and is an orange to sulphur-yellow colour. Eat young, but only try a little the first time: it can cause nausea and dizziness in some people. The largest ever found weighed 45.4kg.

Langermannia gigantea (giant puffball) can be enormous: large and round with white skin, it sits on the ground like a giant soccer ball. The biggest ever recorded, according to the *Guinness Book of Records*, was 2.64m in circumference and weighed 22kg! It is found on soil in fields, hedgerows, woodlands and gardens, often near nettles. Eat when young, while the flesh is pure white; it tastes good sliced, dipped in breadcrumbs and fried. Other related (and smaller) species of puffball are also edible while they remain white all through.

Lepista nuda (blewit, wood blewit, blue-stalk) is medium to large with a light cinnamon-brown to tan cap; the gills and stems are violet to lavender. It is found in woodlands, parks and hedges. It can be grown in leaf debris around compost heaps. Better eaten young, it is well flavoured and particularly good in stews or fried. Never eat raw: cook thoroughly to remove traces of cyanic acid.

Marasmius oreades (fairy ring champignon) is found on lawns in a 'ring' of dark green grass with dying grass in the centre. Small with a bell-shaped light tan-coloured cap, matching gills and similar stem. Good in omelettes. Beware: similar-looking species are toxic.

Morchella esculenta (common morel) is a delicious fungus that emerges annually in spring, earlier than most autumn-fruiters. The hollow cap has a surface covered in honeycomb-like pits and varies from round to conical in shape. (The many crevices of the cap often harbour dirt and insects – rinsing in water is advised.) Found on well-drained soils under deciduous trees, particularly in ash and elm woods, in gardens and near old hedges. Other edible species include ***M. rotunda***, found on heavier soil, and ***M. vulgaris***, on richer soil. Some similar-looking mushroom species are highly toxic.

Sparassis crispa (cauliflower mushroom, brain fungus) has a folded, rounded fruiting body, creamy-white when young, which looks more like a cauliflower than a mushroom. It tastes nutty, with a spicy fragrance. Found at the base of pines and other conifers. As with morels, rinse to remove any debris.

Tuber melanosporum (the Perigord or black truffle), often found in oak woods, is highly desirable and the most valuable truffle. Pigs and trained dogs are used to sniff them out. Truffle-inoculated trees and hazel are now available to amateur gardeners but need specific growing conditions to flourish.

Pick fungi at their prime

Boletus edulis – delicious

cultivation

Propagation
Nameko mushrooms favour moist toilet rolls or arrive with packs of clean straw which the fungi grow on. It is also possible to cultivate mushrooms in the garden; the spawn of field mushrooms grows in rich soil and is an ideal candidate for cultivating around the compost heap, while blewit mushrooms thrive in a mixture of well-rotted leaf mould and pine needles. They are also grown using kits or on heaps of rotted horse manure.

It is also possible to buy mushroom growing kits or 'logs' supplied with dowels inoculated with spores of fungi like shiitake, which are hammered into holes drilled in the logs, or to buy the dowels and use logs of beech, birch, oak or similar wood that have been harvested at home. Put them in the shed, cellar or a shady corner of the garden, in conditions similar to those enjoyed by shade-loving plants and cover them with damp sacking.

Mushroom logs take six months to a year to become productive; flushes of growth appear when temperatures drop in autumn, though shiitake logs can be shocked into production in summer by plunging them into cold water.

Each log crops for about three years.

Growing
Humidity is essential for success, along with a plentiful supply of organic matter. Choose a shady position and, on a damp day in spring or autumn, 'plant' blocks of spawn, about the size of a golfball, 5cm below the soil and 30cm apart. A good crop of mushrooms often appears when spent mushroom compost is used as a mulch around other crops.

Maintenance
Indoor crops can be planted any time of the year. Plant spawn outdoors in spring or autumn.

Protected Cropping
Indoor crops grow well in an airy shed, cellar, greenhouse or cold frame. They do not need to be grown in the dark.

Container Growing
See 'Propagation'.

Harvesting and Storing
The first 'button' mushrooms are ready for harvesting 4–6 weeks after 'casing'; there may be another 2 weeks before the next 'flush'.

Harvesting lasts for about 6 weeks. To harvest, twist and pull mushrooms upwards, disturbing the compost as little as possible, removing broken stalks and filling holes with 'casing'.

Mushrooms last in a ventilated polythene bag in the salad drawer of a refrigerator for up to 3 days. Most species can be dried. Thread them on to a string and hang them over a radiator or in an airing cupboard, then store in a cool dry place. Reconstitute with water or wine.

After the final harvest, you can use the spent mixture as a mulch; you should never try to respawn for a second crop.

Pests and Diseases
Mushroom fly can be a problem; pick mushrooms when young.

medicinal

Edible fungi lower blood cholesterol, stimulate the immune system and deactivate viruses. Shiitake mushrooms are particularly effective. Jew's ear has been used in herbal medicine for treating sore throats.

warning

If you gather wild mushrooms, be certain of their identity before eating. Best of all, collect with an expert. Those that are highly toxic are often similar to edible species. Mistakes can be fatal.

Feeling hungry yet?

culinary

Fungi should always be eaten fresh as the flavour is soon lost and the quality deteriorates. Avoid washing most kinds: the fruit bodies absorb water, spoiling the texture and flavour. Simply clean the surface by wiping with a damp cloth or brushing off any dirt. Peel only when necessary.

Both caps and stems of *Agaricus* species can be eaten. With some other fungi, stems may be discarded as inedibly tough. Check for any special instructions on preparation: some fungi, for instance, are toxic unless cooked.

Harvested cultivated mushrooms at the 'button' stage can be eaten raw, added to salads; more mature caps can be baked, grilled or fried whole, or sliced and stir-fried, made into soups, pies or stuffings, added to stews or the stockpot, or used as a garnish. Other fungi can be prepared in many similar ways. Large fruit bodies can be stuffed.

Try frying in butter and a little lemon juice for 3–5 minutes. Brush with oil and seasoning and grill each side for 2–3 minutes. Add yogurt or cream before serving, or dip in breadcrumbs and fry. Garlic mushrooms are especially delicious.

Mushroom Soup
Serves 4

This is particularly satisfying on a cold winter's day. Open-capped cultivated mushrooms make a good alternative to the wild variety.

75g butter
4 shallots, finely chopped
1 clove garlic, crushed
500g field mushrooms, cleaned and sliced
1 litre chicken stock
1 tablespoon plain flour
Dash soy sauce
A little thick cream
Salt and freshly ground black pepper

Heat 50g of the butter in a heavy-bottomed pan and sauté the shallots until softened; add the garlic and cook for 1 minute more. Add the mushrooms and stir to coat well. Pour in the stock and bring to the boil. Season and cover, simmering for 10–15 minutes, until the mushrooms are cooked. Remove from the heat.

In a separate pan, heat the remaining butter and stir in the flour to make a roux. Cook for 2 minutes and remove from the stove. In a liquidiser, blend the roux with the soup (this may need to be done in batches). Add soy sauce, check seasoning and serve with cream.

Grilled Shiitake Mushrooms
Serves 4

Allow 2 mushrooms per person for a first course to be served with Italian bread.

8–12 shiitake mushrooms
6 tablespoons olive oil
1 sprig fresh rosemary
1 tablespoon fresh thyme leaves
2 tablespoons balsamic vinegar
2 tablespoons Barolo wine
1 small onion, finely chopped
4 slices Italian bread, toasted
1 tablespoon butter
1 clove garlic, crushed
Salt and freshly ground black pepper

Wash the shiitakes, ensuring the gills are free from dirt. Discard the stems and add to a stockpot. Dry the caps. Leave the mushrooms in a marinade of oil, rosemary, thyme, vinegar, wine, onion and seasoning for 30–45 minutes, turning them occasionally.

Grill the marinated caps under a preheated grill for 5 minutes each side, brushing with the marinade juices. Serve on slices of toasted Italian bread, buttered and rubbed with garlic. Pour over a little of the juices on each helping.

Mushroom Mélange

Mushroom Mélange
Serves 4

500g oyster mushrooms
500g shiitake mushrooms
4 tablespoons butter
2 cloves garlic, crushed
2 shallots, finely chopped
4 tablespoons white wine
150ml double cream
4 tablespoons grated Parmesan
Salt and freshly ground black pepper

Wash the mushrooms well and chop them, including the stalks. Melt the butter in a heavy saucepan, add the garlic and shallots and allow them to soften over a gentle heat. Add the mushrooms and stir well. Pour in the wine and the double cream and bring the mixture to simmering point. Then cover and leave to stew for 15 minutes.

Pour into a greased sauté pan, season with salt and pepper and sprinkle over the Parmesan. Put under a preheated grill on the highest setting for 3–5 minutes until the cheese is melted.

Serve immediately.

Allium cepa. Alliaceae

ONION

Biennial; grown as annual for swollen bulbs. Half hardy. Value: small amounts of most vitamins and minerals.

A vegetable of antiquity, the onion was cultivated by the Egyptians not only as food, but also to place in the thorax, pelvis or near the eyes during mummification. Pliny recorded six varieties in Ancient Rome. The onion was highly regarded for its antiseptic properties, but many other legends became attached to it. In parts of Ireland it was said to cure baldness: 'Rub the sap mixed with honey into a bald patch, keep on rubbing until the spot gets red. This concoction if properly applied would grow hair on a duck's egg.' Many varieties have been bred over the centuries; some, like 'The Kelsae', are famous for their size, while newer varieties have incorporated hardiness, disease resistance and colour.

varieties

Varieties of *Allium cepa* fall into several different groups according to their colour, shape and use. The bulb or common onion has brown, yellow or red skin and is round, elongated or spindle-shaped, or flattened. (Grouped with these are Japanese onions, a type of the perennial *Allium fistulosum*, which are grown as an annual for overwintering.) Spring or bunching onions are harvested small for salads, and pickling varieties (also known as 'silverskin', 'mini' or 'button' onions) are allowed to grow larger before harvesting.

Bulb or Common Onions

'Ailsa Craig', an old favourite, is a large variety; round and straw-coloured with a mild flavour. **'Albion'** is a round white bulb and ideal in salads or stir-fries. **'Buffalo'** is high-yielding and good for sowing in summer and harvesting the following year. The round, firm bulbs are well flavoured. **'Express Yellow O-X'** is a Japanese onion for sowing in summer and harvesting the following year. **'Marshalls Giant Fen Globe'** is an old, heavy-cropping variety with a mild flavour. **'Red Baron'** is a gorgeous dark, red-skinned onion with a strong flavour and red outer flesh to each ring. Good for storing. **'Rijnsburger'** is large, pale yellow and round, and an excellent keeper. **'Senshyu Yellow'** is a Japanese onion with a deep yellow skin and good taste. **'Sturon'**, an old, high-yielding variety, has straw-coloured skin and an excellent resistance to running to seed. **'Stuttgarter Giant'** is a reliable variety with flattened bulbs and a mild flavour. A good keeper and slow to bolt. **'The Kelsae'**, a large, round onion with

ONION

mild flesh, does not store well. **'Long Red Florence'** is spindle-shaped, sweet and mild-flavoured.

Bunching, Spring or Salad Onions
'Beltsville Bunching' is a vigorous, mild-tasting variety, tolerant of both winter cold and hot, dry weather. **'Ishikura'**, a cross between a leek and coarse chives, is prolific, tender and a rapid grower with upright, white stems and dark green leaves. It can be left in the ground to thicken and still retains its taste. **'Kyoto Market'** is mild, easy to germinate and excellent for early sowings. **'Redmate'** is a colourful variety with a red base, and ideal for livening up salads. It can be thinned to 7.5cm apart for mild bulb onions. **'Santa Claus'**, another red variety, is ready from about 6 weeks, keeps its taste well and can be harvested until the size of a leek. The colour is stronger during cold weather and when they are earthed up. **'White Lisbon'** is a tasty, popular and reliable variety. It is fast-growing and very hardy. **'Winter-Over'** is a well-flavoured, extremely hardy variety for sowing in autumn. **'Winter White Bunching'** has slim stalks, stiff leaves and a mild flavour. It is hardy and overwinters well.

Pickling Onions
'Brown Pickling SY300' is a pale brown-skinned early variety. It stores well and remains firm when pickled. **'Crystal White Wax'** grows to a consistent size and shape and produces perfect cocktail onions. **'Paris Silverskin'** is a popular, excellent 'cocktail' onion, which grows rapidly and thrives in poor soil. Sow from mid-spring and lift when the size of your thumbnail.

cultivation

Onions require an open, sunny site, fertile soil and free drainage. 'Sets' (immature bulbs that have been specifically grown for planting) are more tolerant than seedlings and do not need a fine soil or such high levels of fertility. Pickling onions tolerate poorer soil than other types. Rotate crops annually.

Propagation
For a constant supply, two or three plantings are needed, one in spring and another in summer when Japanese varieties are planted, or autumn when old, hardy types are used.

Onion sets have several advantages over seed. They are quick to mature, are better in cooler areas with shorter growing seasons, they grow well in poorer soils and are not attacked by onion fly or mildew. They are easy to grow and mature earlier, but are more expensive and prone to run to seed. (Buying modern varieties and heat-treated sets about 2cm in diameter reduces that risk.) There is a greater choice of varieties when growing from seed. If planting is delayed, spread out sets in a cool, well-lit place to prevent premature sprouting. It is possible to save your own sets from bulbs grown the previous year. Plant onion sets when the soil warms from late winter to mid-spring. Sets that have been heat-treated should not be planted until late spring. Plant in shallow drills or push them gently into the soil until only the tips are above the surface. For medium-sized onions, plant 5cm apart in rows 25cm apart; for larger onions space sets 10cm apart in rows.

Sow seed indoors in late winter at 10–16°C in seed trays, pots or modules (about 6 seeds in each module). Harden off the seedlings carefully by gradually increasing ventilation, then plant out in early spring when the seedlings have 2 true leaves. When transplanting those raised in modules and pots, ensure that the roots fall down into the planting hole and that the base of the bulb is about 1cm below the surface.

Onions can be sown outdoors in a seedbed in early to mid-spring in cool temperate zones. Use cloches or polythene to ensure the soil is warm, as cold, wet soil leads to poor germination and disease. Use treated seed to protect against fungal disease.

Spring onions add bite to salads

Onion flowers are beautifully ornamental

When the soil is moist and crumbly, rake in a general fertiliser about 2 weeks before sowing and walk over the plot to create a firm seedbed, then sow onions 12–20mm deep in rows 30cm apart. Once they germinate, thin to 4cm apart for medium-sized onions and 7.5–10cm for large onions. Thin when the soil is moist to deter onion fly. Plant multi-sown blocks 25–30cm apart. Plant firmly.

Sowing times for Japanese onions are critical; sown too early, they run to seed; sown too late, they are too weak to survive the winter. To cover for losses over winter, sow seeds about 2.5cm apart in rows 30cm apart. Top-dress with nitrogen in mid-winter. Sow pickling onions in spring, either broadcast or in drills the width of a hoe and about 10cm apart. Thin according to the size of onions required and harvest them when the leaves have died back.

Sow salad or bunching onions thinly, watering the drills before sowing in dry weather. Rows should be 10cm apart; thin to a final spacing of 1–2.5cm, when the seedlings are large enough to handle, for good-sized onions. For a regular supply sow at 2- to 3-week intervals through late spring and early summer, watering thoroughly during dry weather.

Growing

Dig thoroughly during early winter, incorporating liberal quantities of well-rotted manure or compost if needed. Do not grow on freshly manured ground. Lime acid soils. Before planting, rake the surface level, removing any debris and adding a general granular fertiliser to it at 60g/sq m. In summer pull back the earth or mulch from around the bulb to expose it to the sun.

Maintenance

Spring Plant sets or seeds. Keep weed-free, particularly in the early stages of growth.
Summer Mulch to reduce water loss and weeds. Watering is only vital during drought.
Autumn Lift early autumn.
Winter Push back any sets that have been lifted by frost or birds.

Protected Cropping

Onions do not need protection, although early sowings in cold weather and overwintering onions benefit from cloching or from horticultural fleece in exceptionally cold or wet weather.

Early sowings of salad or bunching onions can be made in late summer or early autumn and protected with cloches during severe weather for harvesting the following spring.

Onions drying on a metal rack

Rows of onion plants growing in the USA

Harvesting and Storing

Harvesting commences when the tops bend over naturally and the leaves begin to dry out. Do not bend the leaves over. Allow the bulbs and leaves to dry out while still in the ground during fine weather; wait until the dried foliage rustles before lifting. In adverse weather, spread out the bulbs on sacking or in trays in cold frames, cloches or a shed, turning them regularly. Handle bulbs carefully to avoid damage and disease. Before storing, be sure to remove any damaged, soft, spotted or thick-necked onions and use them immediately. Onions can be stored in trays, net bags or tights, or tied to a length of cord as onion ropes in a cool place.

Harvest salad or bunching onions before the bases swell. During dry weather, water before harvesting to make pulling easier.

Making an Onion Rope

Storing onions on a rope enables the air to circulate, reducing the possibility of diseases. It is attractive and a convenient method of storage. You can plait the stems to form a rope, as with garlic, but they are usually too short and are better tied to raffia or strong string. Firmly tie in 2 onions at the base, then wind the leaves of each onion firmly round the string, with each bulb just resting on the onions below. When you reach the top of the string, tie a firm knot around the bulbs at the top, then hang them up to dry. Cut onions from the rope as they are needed.

Pests and Diseases

If birds are a nuisance, protect plants with black thread or netting. The larvae of onion fly tunnel into bulbs, causing the stems to wilt and become yellow. Seedlings and small plants may die (August-sown crops are most vulnerable). Cultivate the ground thoroughly over winter. Remove and destroy affected plants, rotate crops, sow under mesh. White rot can be a problem, particularly on salad onions. White mould like cotton-wool, dotted with tiny, black spots, appears round the base. Leaves turn yellow and die. It is almost impossible to eradicate. Remove affected onions with as much of the soil round them as possible, dispose of plants and any debris – do not put them on the compost heap. Avoid spreading contaminated soil on tools or boots. Grow on a new site and from seed.

When attacked by stem eelworm, bulbs become distorted, crack, soften, then die. Grow plants from seed; rotate crops; in severe cases do not grow in the same place again. Dispose of plant debris thoroughly and remove any affected plants.

companion planting

Parsley sown with onions is said to keep onion fly away.

container growing

Bulb onions can be grown in containers, but yields will be small and not really worth the trouble.

medicinal

Used as an antiseptic and diuretic, the juice is good for coughs and colds. The bulbs and stems were formerly applied as poultices to carbuncles.

ONION

culinary

So indispensable are onions for flavouring sauces, stocks, stews and casseroles that there is hardly a recipe that does not start with some variant of 'fry (or sauté or sweat) the onion in the oil or fat until soft…' They also make a delicious vegetable or garnish in their own right: roasted or boiled whole, cut into rings, battered and deep-fried, or sliced and slowly softened into a meltingly sweet 'marmalade'. Finely chopped raw onion adds zing to dishes like rice salad; you can also use the thinnings to flavour salads.

Bunching onions are perfect for salads, pastas, soups and flans. In France they are chopped, sautéed in butter and added to chicken consommé with vermicelli. Pickled onions are an excellent accompaniment to bread, strong cheese and pickled beetroot – the traditional Ploughman's Lunch. Besides being pickled, pickling varieties can be used fresh in salads and stir-fries, added to stews or else threaded on to kebab skewers for barbecuing.

Onion Tart
Serves 4

This Alsatian dish is full of flavour and very filling. Enjoy it with a simple fresh green salad.

90g lard or olive oil
1kg onions, sliced into rings
90g smoked bacon, diced
240ml double cream
3 eggs, lightly beaten
Salt and freshly ground black pepper
Shortcrust pastry to line a 20–23cm tart tin

In a heavy pan, heat the lard or oil and sauté the onions until soft but not browned. Drain well on kitchen paper. Add the bacon to the pan and cook briskly for a couple of minutes, then drain off the fat. Next, mix the cream and the eggs and season well, then stir in the onions and the bacon and fill the pastry case.

Preheat the oven to 220°C/425°F/gas mark 7 and bake for 10–15 minutes, turning the heat down to 190°C/375°F/gas 5 for a further 15 minutes, or until the filling is set. Serve warm.

Onion and Walnut Muffins
Makes 20

This wonderful recipe comes from chef Wally Malouf's *Hudson Valley Cookbook*.

1 large onion, peeled and quartered
250g unsalted butter, melted
2 large eggs
6 tablespoons sugar
1 teaspoon sea salt
1 teaspoon baking powder
300g shelled walnuts, coarsely crushed
350g plain flour

Preheat the oven to 220°C/425°F/gas mark 7. Purée the onion finely in a food processor and measure it to achieve 250g. Beat together the butter, eggs and sugar and add the onion purée. Stir in the remaining ingredients one by one and mix thoroughly. Fill the muffin tins almost full. Bake them for 20 minutes, or until they are puffed and well browned. Serve warm.

Allium cepa (Aggregatum Group). *Alliaceae*

SHALLOT

Small onion, grown as an annual, forming several new bulbs. Hardy. Value: small amounts of most vitamins and minerals.

Shallots are hardy, mature rapidly, are good for colder climates, tolerate heat and will grow on poorer soils than common onions. Sets are more expensive than seed and are inclined to bolt unless they are heat-treated; buy virus-free stock which is higher-yielding and vigorous, or save healthy bulbs of your own for the following year.

Shallots are mild enough to be added whole to dishes

varieties

'Atlantic' can be sown early and produces heavy yields of moderate to large bulbs which are crisp, tasty and store well. **'Creation F1'**, a seed-grown variety, is delicious, highly resistant to bolting and stores well. **'Longor'** is a French variety with elongated bulbs and mild flesh. **'Giant Yellow Improved'** is well worth considering. The bulbs have yellow-brown skins and are consistently large and high-yielding. **'Golden Gourmet'** is a mild-tasting shallot for casseroles and salads. It is reliable and high-yielding, stores well and produces good edible shoots. **'Hative de Niort'** is an extremely attractive variety with elongated, pear-shaped bulbs, dark brown skins and white flesh. **'Pikant'** is prolific and resistant to bolting. Its skin is dark reddish-brown, the flesh strongly flavoured and firm. **'Mikor'** has large elliptical bulbs, crisp white flesh and a superb taste. **'Sante'** is large and round with brown skin and pinkish-white flesh, packed with flavour. Yields are high and it stores well. However, it is inclined to bolt and should only be planted from mid- to late spring when conditions improve. **'Red Sun'** is red with firm skin and a crisp and mild flavour that is ideal for salads.

Propagation

The ideal size for sets is about 2cm diameter, which will result in a high yield of good-sized shallots; larger sets will produce a greater number of smaller shallots. Plant from late winter or early spring, as soon as soil

SHALLOT

conditions are suitable. Shallots can also be planted from late autumn to mid-winter for early crops. Cover the soil with cloches, fleece or polythene 2 weeks before planting to warm the soil.

If the weather is unfavourable, bulbs can be planted in 10cm pots of compost and transplanted when conditions improve.

Space sets 23cm apart with 30–38cm between the rows. Make small holes with a trowel rather than pushing bulbs into the ground (the compaction this causes, particularly in heavier soils, can act as a barrier to young roots). Leave the tips of the bulbs just above the soil. Alternatively, plant in drills, 1cm deep, 18cm apart, then cover with soil.

F1 hybrids that are grown from seed produce one bulb, rather than several. From early to mid-spring, as soon as soil conditions allow, sow seed thinly 1cm deep in broad drills, the width of a hoe, thinning until there is 2.5–5cm around each plant. If spaced farther apart, clusters of bulbs are more likely to form.

Unlike onions, shallots develop in small clusters

Undersized shallots can be grown for their leaves, or you can pick a few leaves from those being grown for bulbs. Plant from autumn to spring under cloches for earlier crops, in seed trays or pots of compost under cover and outdoors when the soil becomes workable. Each bulb should be about 2.5cm apart.

Growing
Shallots flourish in a sheltered, sunny position on moist, free-draining soil, preferably one that has been manured for the previous crop.

Alternatively, double dig the area in early autumn, incorporating plenty of well-rotted organic matter into the lower spit.

Before planting, level the soil and rake in a general fertiliser at 110g/sq m. If you are sowing sets, a rough tilth will suffice, but seeds require a seedbed of a finer texture.

Water during dry periods and keep crops weed-free, particularly while becoming established. Use an onion hoe with care, as damaged bulbs cannot be stored.

Maintenance
Spring Plant sets when soil conditions allow. Sow seed when the soil warms up.
Summer Keep crops weed-free. Water during dry periods.
Autumn Dig in well-rotted organic matter if needed. Plant sets for early crops.
Winter Plant sets from late winter onwards.

Protected Cropping
Shallots are extremely hardy, but benefit from temporary protection under cloches or fleece in periods of severe winter weather, particularly if the soil is poorly drained.

Harvesting and Storing
From mid-summer onwards, as the leaves die back, carefully lift the bulbs and in dry weather leave them on the surface for about a week to dry out; otherwise dry them as onions. Do not cut off the green foliage as this may cause fungal infection, spoiling the bulbs for storage. Break up the bulbs in each clump, remove any soil and loose leaves, then store them in a dry, cool, well-ventilated place. Store on slatted trays, in net bags or in a pair of old tights. Harvest shallots grown for their foliage when the leaves are about 10cm high.

Pests and Diseases
Shallots are usually free of pests and diseases. Bolting may be a problem in early plantings or if temperatures fluctuate. Use resistant varieties for early plantings. Bulbs infected with virus are stunted and yields are poor. Use disease-free stock. If mildew is a problem, treat as for onions.

Birds can be a nuisance, pulling sets from the ground. Sprinkling a layer of fine soil over the tips can help; otherwise protect the crop with humming wire or similar bird scarers.

Bulbs lifted by frost should be carefully replanted immediately.

Eelworms and onion fly should be treated in the same way as for onions.

companion planting

Shallots make good companions for apples and strawberry plants; storing sulphur, they are believed to have a fungicidal effect.

container growing

Shallots will grow in large pots or containers of soil-based compost. Add slow-release fertiliser to the mix and put a good layer of broken crocks or polystyrene in the bottom of the pot, for drainage. Keep plants well watered in dry periods.

culinary

Shallots have a milder taste than onions; generally, the yellow-skinned varieties are larger and keep better, while red types are smaller and have the best flavour. The bulbs can be eaten raw or pickled and the leaves used like spring onions.

Shallots can be finely chopped and added to fried steak just before serving. Do not brown them, as it makes them bitter. Béarnaise sauce is made by reducing shallots and herbs in wine vinegar before thickening with egg and butter.

Allium porrum. Alliaceae

LEEK

Biennial grown as annual for blanched leaf bases. Hardy. Value: good source of potassium and iron, smaller amounts of beta carotene and vitamin C, particularly in green leaves.

The Bible mentions 'the cucumbers, and the melons, and the leeks, and the onions and the garlic' which grew in Egypt, where the leek was held as a sacred plant, and to swear by the leek was the equivalent to swearing by one of the gods.

Giant leek contests have been held in pubs and clubs throughout the north-east of England since the mid–1880s. At one show in 1895, W. Robson was awarded a second prize of £1 and a sheep's heart; now the world championship has a first prize of over £1,300. At it's peak, the European Community produced over 7 million tonnes per year.

varieties

Older varieties are divided into two main groups, long thin and short stout types. In many modern cultivars, such differences are less obvious. There are also early, mid-season and late varieties.

'Autumn Mammoth – Cobra' is a mid- to late harvest, medium-length variety with good bolting resistance. **'Autumn Mammoth 2 – Argenta'** and the similar **'Goliath'** mature in late autumn and can be harvested until mid-spring. A high-yielding leek with a medium shank length and thick stems. **'Bleu de Solaise'**, a French winter variety, can be harvested until spring. **'Bulgarian Giant'** is long, thin and of excellent quality for autumn harvest. **'King Richard'** is a high-yielding, mild-tasting, early variety with a long shank. Good for growing at close spacing for 'mini leeks'. **'Prelina'** is harvested in early autumn and has a moderate-length shank.

cultivation

Leeks flourish in a sunny, sheltered site on well-drained, neutral to slightly acid soil.

Propagation

Leeks need a minimum soil or compost temperature of at least 7°C to germinate, so you will achieve more consistent results, particularly with early crops, when they are sown under cover. For rapid germination, sow early varieties indoors during late winter at 13–16°C in trays, pots or modules of seed compost. Pot on those grown in trays or pots, spacing them about 5cm apart, when 2 true leaves are produced or when they begin to bend over. Harden off gradually before planting out in late spring.

Leeks can also be sown in unheated glasshouses, in cold frames or under cloches. Sow seeds from late winter to early spring, pot on, harden off and transplant in late spring.

Although all varieties are suitable, later sowings of mid-season and later types are particularly successful in seedbeds. Warm the soil using cloches, black polythene or horticultural fleece and rake the seedbed to a fine tilth. Sow thinly in rows 15cm apart and 2.5cm deep, providing protection during cold spells. They can also be sown directly in the vegetable plot, 2.5cm deep in rows 30cm apart, thinning when seedlings have 2 or 3 leaves. Sowing seeds in modules – either in pairs or singly – keeping the most vigorous of the two, and multi-sowing 3 to 5 per cell, avoids the necessity of 'pricking out' or thinning.

Growing

Fertile, moisture-retentive soil is essential, so dig in plenty of organic matter the winter before planting, particularly on light soils. On heavy soils, add organic matter and horticultural sand to improve drainage as crops are poor on heavy or waterlogged soil. Rake, level and firm the soil before planting in spring. As they are a long-term, high-nitrogen crop, apply a general fertiliser, fish, blood and bone or ammonium sulphate at 60–90g/sq m 1 or 2 weeks before planting.

Transplant leeks when they are 15–20cm tall. Trim the leaf tips back if they drag on the ground, but not the roots, as is often recommended. If the soil is dry, water the area thoroughly before planting. Planting 15cm apart in rows 30cm apart provides a high yield of moderately sized leeks. Planting them 7.5–10cm apart in rows gives a high yield of slim leeks. A spacing of 15–17.5cm each way provides a reasonable crop of medium-sized leeks. Leeks grown in modules should be planted 23cm apart each way.

Leeks growing in a vegetable bed

LEEK

There are two methods of planting to ensure well-blanched stems. I find the first method better, as deeply planted leeks are more drought-resistant and soil is less likely to fall down between the leaves. Make a hole 15–20cm deep with a dibber, drop the plant into it and fill the hole with water (this washes some soil into the bottom of the hole), but do not fill any further.

Alternatively, plant leeks 7.5cm deep and several times through the season pull the earth up around the stems, 5–7.5cm at a time, with a draw hoe. Stop earthing up when the plants reach maturity and make sure that the soil does not fall down between the leaves. Earthing up is easier on light soil.

Whichever method you use, after planting, water gently with a seaweed-based fertiliser. If there is a dry period after planting, water leeks daily until the plants are well established and thereafter only during drought conditions. Hand weed or hoe carefully to keep down weeds, using an onion hoe around younger plants to avoid any damage. In poorer soils, feed weekly in summer with a liquid seaweed or comfrey fertiliser.

Maintenance

Spring Pot on leeks grown under glass; sow seed outdoors.
Summer Transplant seedlings, water and feed. Keep crops weed-free by hoeing or mulching.
Autumn Harvest crops as required.
Winter Harvest mid- and late-season crops. Sow seed under glass in late winter.

Protected Cropping

Apart from early sowings in the greenhouse or cold frame or under cloches and horticultural fleece, leeks are an extremely hardy outdoor crop.

Harvesting and Storing

Early varieties are ready for lifting from early to mid-autumn, mid-season types from early to mid-winter and lates from early to mid-spring. Lift leeks carefully with a garden fork and dispose of any leaf debris to reduce the risk of disease in the future. Late varieties taking up space in the vegetable garden that is needed for spring planting can be stored for several weeks in a shallow, angled trench 15–20cm deep; cover them lightly with soil and leave the tops exposed. If inclement weather is likely to hinder harvest, they can be lifted and packed closely together in a cold frame. The top should be raised to provide ventilation on warmer days.

Pests and Diseases

Leeks share many diseases with their close relatives, onions. Leek rust appears as orange pustules on the leaves during summer and is worse in wet seasons. Foliage developing later in the season is healthy. Feed with high-potash fertiliser, remove infected plants and debris. Improve drainage; do not plant leeks on the site for 4–5 years; grow partially resistant varieties like **'Autumn Mammoth'**, **'Titan'** or **'Gennevilliers-Splendid'**. Slugs can be damaging. Collect them at night, use biological control, set traps or use aluminum sulfate. Mature leeks usually survive slug damage. Stem eelworm causes swelling at the base and distorted leaves. Destroy affected plants immediately and rotate crops. Leek moth – leaves develop whitish brown patches and tunnels are present in stems and bulbs where caterpillars have burrowed through. Squash caterpillars in leaves. Grow under fleece.

companion planting

Leeks grow well with celery. When planted with onions and carrots they discourage onion and carrot fly. Grow leeks in your rotation programme alongside garlic, onions and shallots. Leeks are a useful crop after early potatoes and if they are grown at a 30cm spacing; their upright growth makes them ideal for intercropping with lettuces like **'Tom Thumb'**, with land cress or winter purslane.

culinary

Leeks can be boiled or steamed, made into terrines, cooked in casseroles, added to pasta dishes, wrapped in suet pastry and baked. They are a useful addition to soups and an important ingredient in 'cock-a-leekie' soup and in French Vichyssoise. Braise in stock with a little wine added and bake in a moderate oven. Partially cook trimmed leeks in boiling water, drain well and roll in slices of good country ham and lay them in a dish; cover with a well-flavoured cheese sauce and bake in a hot oven until well browned. Slice or chop young leeks and use raw as a spring onion substitute in salads.

Leek and Ricotta Pie

Serves 4

4 largish leeks, trimmed and roughly chopped
2 tablespoons olive oil

2 cloves garlic, finely chopped
225g ricotta
2 tablespoons pine nuts
3 tablespoons raisins, softened in warm water
1 egg
Salt and freshly ground black pepper

For the Pastry:
90g butter
175g plain flour
3 tablespoons water
Pinch salt

Make the pastry by crumbling the butter into the flour and then adding water to make a dough. Add the salt and sprinkle with flour. Wrap in cling film and chill in the fridge for 30 minutes.

Steam the leeks gently for about 10 minutes and drain.

Preheat the oven to 190°C/375°F/gas mark 5. In a heavy frying pan heat the oil and gently fry the garlic. Then add the leeks and stir to coat well with oil; allow them to cook for about 5 minutes, stirring occasionally.

Remove from the heat. In a bowl mix the ricotta with the pine nuts and raisins, and bind with an egg. Add the leeks, mix well and season.

Gently roll out the pastry to fit a 20cm tart tin. Prick the base and bake blind for 10–15 minutes. Fill the tart with the leek and ricotta mixture and continue cooking for 30 minutes. Serve with a green salad.

Allium sativum. Alliaceae

GARLIC

Perennial grown as annual for strongly aromatic bulbs. Half hardy. Value: contains small quantities of vitamins and minerals.

Prized throughout the world for its culinary and medicinal properties, garlic, now known only as a cultivated plant, is thought to have originated in western Asia. It has been grown since Egyptian times and for centuries in China and India. Its reputation as a 'cure all' has been endorsed by modern science. The Egyptians placed it in their tombs and gave it to the slaves who built the pyramids to ward off infection, while Hippocrates prescribed it for uterine tumours. In medieval Europe it was hung outside doors to deter witches. Today almost 3 million tonnes per annum are produced globally.

varieties

There are two main categories of garlic; soft necks, which don't produce a flower stalk, and hard necks, which do. Hardnecks, also known as 'gourmet' garlics, have a greater colour range and taste but don't store well.

Softneck
'Germidor' crops early, producing large cloves and purple bulbs. **'Inchelium Red'** is a regular winner in taste tests. **'Long Keeper'** is well adapted to a cool, temperate climate. The bulbs are white-skinned and firm. **'Solent Wight'**, producing large cloves with a mild flavour, is heavy-cropping with an appealing bouquet.

Hardneck
'Chesnok Red' is robustly flavoured, **'Brown Tempest'** and **'Purple Moldovan'** are renowned for their taste and **'Spanish Rioja'** is sought after by enthusiasts.

cultivation

Propagation
Garlic is usually grown from healthy, plump bulb segments (cloves) saved from a previous crop. Where possible, buy nematode- and virus-resistant stock.

Plant cloves, a minimum of 13mm diameter, in late autumn or early spring, at a depth of 2.5cm and 10cm apart, with the rows 15–20cm apart.

Garlic is surprisingly hardy and needs a cold, dormant period of 1 or 2 months when temperatures are 0–10°C to yield decent-sized bulbs; for this reason it is generally better planted in late autumn. A long growing period is also beneficial for the ripening process.

A garlic field in Aomori, Japan

In areas with heavy soil, cloves can be planted any time over winter in pots or modules containing loam-based compost with added horticultural sand, and can be planted out as soon as soil conditions are favourable. Plant cloves vertically with the flattened base plate at the bottom, twice the depth of the clove with at least 2.5cm of soil above the tip. On good soils, planting up to 10cm deep increases the yield. When planting, you should handle the cloves lightly: do not press them into the soil as this reduces root development. The amount of leaf growth dictates the size of the mature bulb which develops during long summer days.

Growing
Garlic favours an open, sunny position on light, well-drained soil. On heavier soil, grow in ridges or improve the drainage by working horticultural sand or grit in to the topsoil.

Garlic is less successful in areas of heavy rainfall. On poor soils, it is beneficial to rake in a general fertiliser about 10 days before planting. Garlic can be grown on soil manured for the previous crop as well as limed acid soils.

Rotate the crop and do not grow in sites where onions have been planted the previous year. Keep the bulbs weed-free throughout the growing season.

Maintenance
Spring Mulch to suppress weeds. Water if necessary.
Summer Keep weed-free.
Autumn Plant cloves.
Winter Plant cloves in containers for planting out in the spring.

Protected Cropping
Garlic can be grown in an unheated greenhouse for an early crop.

GARLIC

Harvesting and Storing

From mid- to late summer, as soon as the leaves and stems begin to yellow, lift the bulbs carefully with a fork and leave them to dry off in the sun. Delaying harvest causes the bulbs to shrivel and increases the possibility of disease during storage. Handle them delicately as they are easily bruised.

In inclement weather, dry them under cover on trays. Store them in cool, dry conditions indoors or in a shed or garage. Hang them in bunches tied by the leaves, in string bags or plait the stems together. Plants that have gone to seed are still able to produce usable bulbs for the kitchen.

Pests and Diseases

Onion fly lay their eggs around the base of garlic: the larvae tunnel into the bulb and the plant turns yellow and dies. Rotate crops, dig the plot over winter, grow under horticultural fleece and cultivate the ground thoroughly in autumn.

Downy mildew is a common problem in wet seasons. Grey patches appear on the leaves. White rot, a grey fungus on the roots, turns the leaves yellow. Lift and destroy infected plants, and do not grow garlic, onions or shallots on the site for 8 years.

Stem and bulb eelworm seedlings become blunted, bloated and distorted, and stems rot. Lift and destroy affected plants. Rotate crops.

companion planting

Planted beside rose bushes, garlic controls greenfly. Good companions are lettuce, beetroot, summer savory, Swiss chard and strawberries. It should not be planted with peas and beans.

container growing

Garlic can be grown in pots, windowboxes or containers, in a moisture-retentive, free-draining compost. Water the plants regularly to produce decent-sized bulbs and place the container in a sunny position to allow the bulbs to develop.

medicinal

Garlic has powerful anti-viral, anti-bacterial and anti-fungal properties, and is effective for digestive complaints, bowel disorders and insect stings. It contains 2 chemicals which combine to form the bactericide allicin, which gives it the characteristic odour. Modern herbalists believe a cold will be cured by rubbing garlic on the soles of feet. What a combination of odours! Current research indicates its ability to reduce blood cholesterol levels and the chance of heart attack. There is also a lower incidence of colonic and other types of cancer where it is part of the daily diet.

culinary

Garlic is used as a seasoning in dishes from curries to pasta – whole plants can be added to salads! Harvest very young leaves of wild garlic, a few per plant as a milder substitute.

Jean-Christophe Novelli's Honey-glazed Lamb

Serves 4

4 lamb knuckles or shanks
Olive oil
Salt and freshly ground black pepper
2 carrots, cut into chunks
2 onions, cut into chunks
2 leeks, cut into chunks
225g celeriac, cut into chunks
1 head of garlic, broken into cloves
Handful of mixed fresh herbs, like rosemary, bay and thyme

For the Red Wine Sauce:
20g shallots, sliced
10g celery, sliced
20g button mushrooms, sliced
25g unsalted butter
1 tablespoon finely chopped mixed fresh herbs, like tarragon, parsley and basil
250ml red wine
Juice of ½ a lemon
2 teaspoons honey

Remove the excess fat from the knuckles and trim the meat away to expose the bone, reserving the meat trimmings. Heat a little oil in a roasting pan on the hob. Season the knuckles and place into the pan. Add the chunks of carrot, onion, leek and celeriac and brown the meat quickly all over. Meanwhile, bring a large pan of salted water to the boil. Lift the knuckles from the roasting pan, and plunge into the boiling water. Blanch for 1–2 minutes to help it retract around the bone and keep it in one piece during cooking.

Drain and put in a clean pan with the vegetables, garlic and herbs, together with enough water to cover the meat. Bring to the boil, lower the heat and simmer gently for about 1½ hours to the point where the meat is almost falling off the bone. Remove the knuckles from the stock and reserve the stock.

Preheat the oven to 160°C/325°F/gas mark 3. Sweat the sliced vegetables in a little of the butter to soften them without browning. Add the reserved meat trimmings and the chopped herbs and cook for a further 2–3 minutes. Pour in the red wine and bubble until reduced by half. Add 500ml of the reserved lamb stock and reduce again by half. Cut the remaining butter into small pieces and whisk these in, a few at a time. Season to taste and pass through a fine sieve. Pour half of the sauce into the roasting pan. Stir in the lemon juice and honey. Add the knuckles and coat with this glaze. Cook in the oven for about 20 minutes, removing every 5 minutes or so to baste the knuckles.

Just before the knuckles come out of the oven, heat through the remaining sauce. Place each knuckle or lamb shank on a warmed plate, pour on a little sauce and serve with mashed potatoes and seasonal vegetables of choice.

Pestou

This is a version of the classic garlic and basil butter, which is extremely good over pasta and which the Italians eat with fish. For 50g butter, have 2 plump cloves garlic, 5–6 sprigs of fresh basil, 2 tablespoons Parmesan and a pinch of salt. Pound the garlic in a mortar, tear the basil leaves roughly, then add to the garlic with the butter and cheese. Pound to mix.

Amaranthus spp. *Amaranthaceae*

AMARANTH

Also known as Kiwicha, Chinese Spinach, Calaloo, Bayam. Half hardy annual.

Value: extremely rich in protein; contains carotene, zinc, phosphorus and calcium, some of the vitamin B complex and vitamin E.

Amaranth was a staple grain of the Incas and Aztecs as long ago as 6000 BC until it was banned by the Spanish, who forced them to grow European crops in a form of 'botanical colonialism', and it retreated to the high Andes. The seeds remain viable for centuries – when archeologists sowed seeds found in an Aztec ruin, they germinated and grew successfully.

It is also an excellent ornamental plant, which was prized by Victorians in their bedding schemes and is known as 'Love Lies Bleeding'. With high levels of protein and minerals, it has been compared favourably with milk and has become an important nutrient source in tropical highlands.

Amaranth share an extremely efficient type of photosynthesis, similar to some cacti and succulents, which allow them to grow efficiently in high temperatures, intense sunlight and under dry conditions.

varieties

'A 452' has leaves and flowers in shades of gold, green, red and purple. **'Bolivia'** reaches up to 2m, the upright, red flower-heads followed by deep purple seeds. **'Golden'** is early maturing, with rich golden seeds. **'Plainsman'** is a short growing variety with maroon seed-heads followed by high volumes of dark purple seeds. **'Popping'** is a selection for use as a 'popping' grain.

Amaranthus caudatus is successful in cool conditions and the most important species for grain, reaching up to 2m or more at maturity, depending on the variety. Broad, lance-shaped leaves appear from side branches and main stem. Extravagant upright or hanging flower- and seed-heads, often called 'cat's tails', produce up to 500,000 seeds per plant.

Amaranthus gangeticus – the greatest variety of those grown for their leaves – is found in India. **'Asia Red'** is coloured bright red with tender and delicious leaves and shoot tips – ideal for hotter climates. There are several selections of fast-growing **'Bayam'** with dark green leaves and tender stems. The top-quality dwarf variety **'Tender leaf'** reaches 20cm tall, was developed in Taiwan and is ideal as a 'cut and come again' crop for harvesting at any stage.

Amaranth growing in Bhutan

AMARANTH

cultivation

Propagation
To extend the growing season, sow amaranth under glass, several per module, from mid-spring at 20°C, lightly but thoroughly, covering the seeds with compost or an opaque cover as they need darkness to germinate.

Germination is rapid, usually within 4 days and is aided by fluctuating day and night temperatures. When they are about 1cm tall, thin to leave the strongest seedling before planting outdoors once the danger of frost has passed, and protecting with cloches for 2–3 weeks to acclimatise.

Alternatively, in warmer climates, broadcast or sow seed in drills outdoors once the danger of frost has passed, into a seedbed raked to a fine tilth, mixing the fine poppy-like seeds with sharp sand to aid sowing, then thin to the final spacing.

Planting distance varies according to the climate. In cool temperate conditions, varieties grown for grain should be 20 in. apart, and up to 3¼ ft. apart in warmer climates or under the protection of a greenhouse or polythene tunnel where plants grow larger.

Cuttings from non-flowering side shoots of established or bought plants root easily. For a continuous supply of 'cut and come again' seedlings of leafy varieties, sow at 2-week intervals throughout the season.

Growing
Amaranth prefer a sunny position in light, moist, rich, free-draining soil similar to that enjoyed by pumpkins and squashes, but are tolerant of most soils, providing they are not waterlogged. Incorporate well-rotted organic matter, grit or sharp sand into the soil if necessary before planting. The plants are drought-tolerant once established and perform well in poor soils, though crops are reduced. Critical periods for watering are at germination and pollination, though regular watering during drought increases the size of the plant and consequently the crop. Amaranth are not frost-hardy – the most tolerant cultivars withstand temperatures down to 4°C, growing best at 21–28°C. Plant them with brassicas in your crop-rotation scheme. Hoe around young plants to break up the soil and keep them weed-free; they are easily overwhelmed by weed seedlings, though the leaves rapidly grow large enough to suppress weed growth.

Maintenance
Spring Sow seeds.
Summer Keep crops weed-free, water if necessary, harvest leaves. Harvest seeds from late summer.
Autumn Harvest seeds.
Winter Order seeds for next year.

Protected cropping
Sow seeds under cover to extend the growing season for 'leaf' or 'grain' varieties. Growing plants to maturity in greenhouses or polythene tunnels in cooler climates results in larger plants and increased production of leaves and seeds, particularly in poor summers.

Harvesting and storing
Plants take 16–24 weeks from sowing to harvest and yield well, even in poor summers. Harvest when the first seeds start to ripen and fall when rubbed between the fingers – usually from September; leave it too late and they will be lost during harvesting. Pull up the plants or cut off the seed-heads then strip the grains with your fingers and riddle into a bucket before drying the seeds. Store the seeds in dry conditions in jars or paper bags.

Harvest leaves from those being grown primarily for their seeds when plants are well established and the loss of a few leaves does not check their growth. Frequent harvesting delays flowering and encourages leaf and stem growth. Thinned seedlings can be used as transplants or pot herbs. Varieties grown for their leaves should be harvested before flowering, and are harvested at various stages as micro-greens, 'cut and come again' seedlings or young plants.

Pests and diseases
Seedlings can suffer from 'damping off' if they are sown too densely.

container growing

Varieties grown as 'cut and come again' crops for their leaves can be grown in growing bags or containers of moisture retentive, free-draining compost, with drainage holes in the base.

medicinal

Betalaina, a yellow or green dye is obtained from red-flowering varieties and used as a food colouring and for clothing. Medicinally, it has several uses; it is a diuretic, astringent and is applied to skin sores.

culinary

The leaves, stems and young shoots can be eaten raw, in soups, stir-fries or salads, steamed or lightly cooked like spinach and young leaves as micro-greens. The seeds have a nutty flavour when 'popped' like corn and can be added to rice, or ground as a wheat substitute in unleavened chapattis or tortillas; when used in baking bread it should be mixed with wheat as amaranth is gluten-free. Because of its high nutritional value, it is considered good for children, the ill and the elderly.

Apium graveolens var. *dulce*. *Umbelliferae*

CELERY

Biennial grown as annual for fleshy leaf stems and leaves. Hardy. Value: low in carbohydrate and calories, high in potassium.

The species is a biennial plant, native to Europe and Asia. It is usually found on marshy ground by rivers, particularly where the water is slightly saline. Its Latin generic name *Apium* is derived from the Celtic *apon*, water, referring to its favoured habitat, while *graveolens* means heavily scented, alluding to its aroma. The stems of the wild plant are very bitter, distinguishing it from var. *dulce* – meaning sweet or pleasant – from which the culinary varieties have been bred. Celery became popular in Italy in the seventeenth century and during the following 200 years spread throughout Europe to North America. 'Trench celery' (so called from the method used for blanching the stems) is very hardy and is harvested from late autumn to early spring, while the more recently developed self-blanching and American green types have a shorter growing season and are less hardy, cropping from mid-summer until mid-autumn. Less succulent, but full of flavour, is the smaller-stemmed 'cutting celery'.

The dense foliage of celery suppresses weed growth

varieties

Trench celery

This is grouped into white, pink and the hardier red varieties: **'Giant Pink'** is a hardy variety harvested from mid- to late winter. The crisp, pale pink stalks blanch easily. **'Giant Red'** is hardy and vigorous and the outer stalks turn shell-pink when blanched. **'Giant White'** is an old, tall, white celery variety with crisp stems and a solid, well-flavoured heart. It needs good growing conditions to flourish. **'Hopkins Fenlander'** is a late-maturing green celery with good flavour. Its sticks are of medium length and free from string. **'Standard Bearer'** red celery has the reputation of being the latest of all to reach maturity.

Self-blanching and American green celery

These include: **'Celebrity'**, an early-maturing variety with long, crisp stems and a nutty-flavoured heart. It has good bolting resistance and is one of the least stringy self-blanching varieties. **'Golden Self Blanching'** is compact with firm golden-yellow hearts which are crisp and tasty. Does not become stringy. **'Greensleeves'**, a green variety, produces tasty green sticks. **'Ivory Tower'** is 'stringless' and well flavoured. **'Lathom Self Blanching'** is a vigorous, well-flavoured early

CELERY

variety with crisp stems. **'Tall Utah Triumph'** has long, succulent, tender green stems. It crops from late summer to early autumn but the season can be extended by growing under cloches.

Leaf, cutting or soup celery
This produces leaves and stems over a long period and is very hardy. It is usually sold as seed mixes, but cultivars are available: **'French Dinant'** is excellent for drying and full of flavour. **'Soup Celery d'Amsterdam'** is aromatic and prolific, producing thin stems and lots of leaves. **'Thai Bai Khuen Chai'** is spicy and leafy with thin stems. Its strong flavour is good in Thai soup, salads, curries and stir-fries.

cultivation

Celery is a crop for cool temperate conditions, flourishing at 15–21°C on an open site. It requires rich, fertile soil which is constantly moist yet well drained and a pH of 6.5–7.5. Lime acid soils before planting if needed.

Propagation
Sow celery from mid- to late spring, in trays of moist seed compost, scattering the seed thinly over the surface; do not cover it with compost, as light is needed for germination. Keep the tray in a propagator or in a greenhouse at 13–16°C; germination can take several weeks, so be as patient as possible.
When two true leaves appear, transplant the seedlings into trays of moist seed compost about 6cm apart or individually into 7.5cm pots and allow them to establish. Harden off before planting outdoors from late spring to early summer when they have 5 to 7 true leaves.

Low temperatures after germination sometimes cause bolting later in life; temperatures should not fall below 10°C for longer than 12 hours until the seedlings have become established. They are particularly sensitive at transplanting size, so cover them with cloches and do not try to slow down the growth of advanced seedlings by putting them outdoors. It is much better to trim plants back to about 7.5cm with sharp scissors and keep them in the warm until outdoor temperatures are satisfactory. Cutting back also seems to lead to more successful transplanting. Planting in modules of seed compost lessens transplanting shock, which can also result in bolting. If you are unable to provide the necessary conditions, plantlets can always be bought.

Celery can also be sown *in situ* but germination is usually erratic and it is not worth the trouble. Celery's low germination rate can be improved by 'fluid sowing'. If possible, use treated seed to control celery leaf spot. Several sowings at 3-week intervals lengthens the harvesting season.

Sow cutting celery in trays of seed compost from late spring to late summer before hardening off and planting out 15cm apart each way. Alternatively, multi-sow in modules, about 6–8 seeds in each, and plant each module group 20cm apart. Leave a few plants to run to seed the following year, then transplant self-sown seedlings at the recommended spacing.

Growing
The planting method is different for 'trench celery' and self-blanching types. For the first, dig a trench 38–20 in. wide and 30cm deep in late autumn or early spring and incorporate as much well-rotted manure or compost as you can find. If more than one trench is needed, their centres should be 120cm apart. Trench celery can also be grown by filling in the trench to a depth of about 7.5–10cm and leaving the remaining soil alongside for earthing up. A week or 10 days before planting, rake a balanced general fertiliser at a rate of 60–90g/sq m into the bottom of the trench. Celery is easier to manage when planted in single rows with plants 30–45cm apart. If you plant in double rows, set the plants 23cm apart in pairs, rather than staggered. This makes blanching easier. Water thoroughly after planting.

Blanch by earthing up plants when they are about 30cm high. Before you start, tie the stems loosely, just below the leaves, using raffia or soft string and make sure the soil is moist, watering if necessary (or earth up after rain). Draw soil up the stems about 7.5cm at a time, repeating this two or three times at 3-week intervals until only the tops of plants are exposed. Do not earth up higher than the leaves, nor should you let soil fall into the heart of the plant. If heavy frosts are forecast in winter, place bracken, straw or other protective material over plants to keep in good condition for as long as possible.

Plants can also be blanched with 'collars'. Use 23–25cm strips of thick paper like newspaper, corrugated cardboard, brown wrapping paper or thick black polythene. (Ideally this should be lined with paper to prevent sweating.) I have also seen tile drainpipes and plastic guttering being used to good effect.

Celery being grown on a commercial scale

Celery blanched by wrapping cardboard around the stem

Begin blanching when plants are about 30cm high, tying the collar quite loosely around the plant to give it room to expand and leaving about one-third of the plant exposed. Further collars can be added every 2 to 3 weeks as the plants grow.

Remember to unwrap them periodically to remove any slugs hiding beneath. If collars are used in exposed sites, support them by staking with a cane. Cover the top of the cane with a flower pot, film case or ping-pong ball to avoid inflicting any eye damage.

Labour-saving self-blanching types do not need earthing up. They also tolerate a wider range of soils, and are particularly good where the ground is heavy and trenching or waterlogging would be a problem. They are, however, shallow-rooted and should be fed and watered regularly throughout the growing season. Self-blanching celery is planted at ground level. Dig in generous amounts of well-rotted organic matter in spring before planting. The spacing varies according to your requirements and plants should be arranged in a square pattern, not staggered rows. Spacing about 15cm apart gives a high yield of very tender, small-stemmed sticks; 27cm apart each way, the optimum spacing, gives high yields of longer, well-blanched sticks and 23cm apart each way gives moderate stem growth. Plant with the crown at soil level and put straw around the outer plants when they mature to help blanching.

For good-quality crops celery must be watered copiously throughout the growing season and the soil should not be allowed to dry out. Apply up to 22 litres/sq m per week during dry periods. Mulching with straw or compost once plants have established conserves moisture and suppresses weeds. Feed with a granular or liquid general fertiliser about 4 to 6 weeks after transplanting. Rotate crops, but do not plant next to parsnip, as both are attacked by celery fly. Avoid anything that checks plant growth throughout the season as this can cause bolting, so transplant the seedlings when the soil is warm, water and feed them regularly, and always mulch or hoe round the plants very carefully.

Maintenance

Spring Prepare the ground for planting. Sow seed and plant earlier crops out under cloches.
Summer Plant out in early summer, keep soil moist and weed regularly. Check for pests and diseases.
Autumn Harvest with care using a garden fork.
Winter Cover with straw, bracken or similar materials to allow harvest to continue during heavy frosts.

Protected Cropping

Protect newly transplanted plantlets with cloches or horticultural fleece for several weeks after planting, until they become established. This is particularly necessary in cooler conditions.

Harvesting and Storing

Lift celery carefully with a garden fork, easing the roots from the ground. Bracken, straw or other protective material placed over trenches assists lifting in frosty weather. Self-blanching celery can be harvested from mid-summer to early autumn. Before the first frosts, lift and store any remaining plants and put them in a cool, frost-free shed. They will keep for several weeks. Harvest cutting celery regularly from about 5 weeks after planting.

Pests and Diseases

Celery leaf miner or celery fly larvae tunnel through the leaves leaving brown blisters. Severe attacks check growth. Grow under horticultural fleece or protective mesh, pinch out affected leaves, do not plant seedlings which have affected leaves. Check plants in late spring and late summer. Do not plant near to parsnips as they can be affected.

Slugs are a major problem, particularly on heavy soil. Use biological control, traps, hand pick or use aluminum sulfate-based slug pellets. Carrot fly attack the roots and stem bases, stunting growth. Grow under fleece until harvest or put fine mesh netting barriers 45–75cm high around the crop before or straight after transplanting.

Celery leaf spot shows as brown spots on older leaves, spreading to younger ones. Severe attacks can stunt growth; use treated seed, rotate crops or use copper-based fungicide.

Celery pale leaf spot (early blight) appears as tiny yellow spots on the leaf surfaces with accompanying grey mould in damp conditions. This disease spreads rapidly. Spray with Bordeaux mixture, be vigilant and destroy all plant debris at the end of the season.

companion planting

Celery helps brassicas by deterring damaging butterflies. It grows well with beans, tomatoes and particularly leeks. If left to flower, celery attracts beneficial insects.

medicinal

Cultivated varieties are said to be beneficial in the treatment of rheumatism and as a diuretic.

CELERY

culinary

Usually eaten raw rather than cooked, celery adds welcome crunchiness to salads, particularly in winter months. It is a key ingredient of the famous Waldorf Salad, made with equal quantities of chopped red-skinned apples and celery, combined with walnuts and bound with mayonnaise.

Celery goes well with cheese – sticks filled with cream cheese or pâté are an appetizing 'nibble'. The 'heart' is particularly tasty. Cook celery in soups and stews, or stir-fry. Braise hearts by simmering in boiling water for 10 minutes, then cook in a covered dish for 45 minutes in a low oven to accompany roasts.

Add leaves to meat dishes, like parsley. Fresh or dried leaves flavour soups and stuffings. Cutting celery is a flavouring for salads, soups and stews; the seeds can also be used.

Celery will stay fresh in a polythene bag in the refrigerator for up to 3 days. Do not stand in water for long periods, or the freshness is lost.

Freeze celery by washing and cutting the sticks into 2.5cm lengths, blanch for 3 minutes, cool, drain and pack into polythene bags. Use frozen celery only in cooked dishes.

Celery and Courgette with Blue Cheese Dip
Serves 4

2 tablespoons olive oil
1 teaspoon chilli powder
½ teaspoon paprika
1 clove garlic, crushed
4 fresh basil leaves, roughly chopped
4 courgettes, sliced lengthwise into quarters
Fresh chives, to garnish
4 stalks celery, cut into 7.5cm lengths

For the Dip:
4 tablespoons cottage cheese
2 tablespoons crumbled Roquefort cheese
50g yogurt
Salt and freshly ground pepper

In a heavy frying pan over a gentle heat, mix the oil with the chilli powder, paprika, garlic and basil. Turn up the heat and fry the courgette slices, cut side down, until browned and turn them to brown the second side. In a small bowl mix together all the dip ingredients. On 4 small plates, arrange a fan of alternating courgette and celery sticks and fill the centre with the dip. Garnish with finely snipped chives and serve.

Apium graveolens var. *rapaceum*. *Umbelliferae*

CELERIAC

Also known as Celery Root. Biennial usually grown as annual for edible root. Hardy. Value: rich in potassium; moderate amounts of vitamin C.

This swollen-stemmed relative of celery has long been popular in Europe. It was introduced to Britain in the early eighteenth century by the writer and seedsman Stephen Switzer, who brought seed from Alexandria and wrote about the vegetable in his book, *Growing Foreign Kitchen Vegetables*. It is an excellent, versatile winter vegetable, hardier and more disease-resistant than celery, but with similar flavour and aroma.

varieties

The lowest part of the stem, known as the 'bulb', is eaten; the roots that grow below are removed. **'Alabaster'** is high-yielding with upright foliage, round bulbs and has good resistance to running to seed. **'Balder'** has round, medium-sized roots, which have excellent flavour when eaten cooked and raw. The 'bulb' of **'Brilliant'** is smooth with white flesh and does not discolour. **'Iram'** is a medium-sized 'bulb' with few side shoots. It stores well and the flesh remains white when cooked. **'Marble Ball'**, a well-known variety, is medium-sized, globular and strongly flavoured. It also stores well. **'Monarch'** is a popular variety with smooth skin and succulent flesh. **'Prinz'** has white-fleshed roots with an aromatic flavour and is also resistant to leaf disease and bolting. **'Tellus'** grows quickly and remains white after boiling. It has firm flesh and a smoother skin than many varieties.

cultivation

Propagation

Celeriac needs a long growing season. Sow in late winter to early spring in a propagator at 18°C or in mid- to late spring in a cold greenhouse, under cloches or in a cold frame. Plant seeds in peat substitute-based compost either several to a pot or in seed boxes or modules.

Germination is notoriously erratic. Pot on strong seedlings when they are about 1cm tall and large enough to handle. Plant them into single 7.5cm pots, modules or in seed trays at 6cm intervals, keeping the temperature at 13–16°C. Harden off when the weather becomes warm in late spring and plant outdoors once there is no danger of frost.

Celeriac is sensitive to cold at the transplanting stage; do not try to slow the growth of fast-growing seedlings by lowering the temperature, as this will encourage them to run to seed later in the season. Maintain the temperature and cut off the tops of the plants with sharp scissors to 8cm – the ideal size for transplanting.

Growing

Celeriac needs rich, fertile, moisture-retentive soil and is ideal for damper parts of the garden. In autumn, incorporate as much well-rotted manure or compost as possible. Space plants 30–38cm apart each way. Do not bury the crowns; they should be planted at ground level. Plant firmly and water thoroughly and continually. In mid-summer remove the outer leaves to expose the crown and encourage the bulb to develop, and remove side shoots if they appear.

Maintenance

Spring Plant out seedlings – harden off. Keep weed-free.
Summer Water in dry weather. Mulch to conserve moisture; feed weekly with a liquid manure, particularly in poorer soils.
Autumn Begin harvesting. Cover with straw.
Winter Prepare ground. Sow seeds under glass.

Protected Cropping

Celeriac only benefits from protection when it is at the seedling stage.

Harvesting and Storing

Celeriac can be harvested through the winter. Harvest when the plants are 7–13cm diameter, though they can be lifted when larger with no loss of flavour. Ideally they should remain in the ground until required. Before the onset of severe winter weather, protect plants with a layer of straw, bracken or with horticultural fleece to prevent the ground from freezing. If the soil is heavy, the site exposed or needed for another crop, lift and remove the outer leaves, keeping the central tuft attached; cut off the roots and store in a cool shed in boxes of damp peat substitute or sand.

Alternatively, lift and 'heel in' or transplant the crop in another part of the garden, laying them close together in a trench and covering the bulbs with soil. They last for several weeks when stored in this manner.

Bulbs can be frozen; cut into cubes, blanch for 3 minutes, dry, store in polythene bags and put in the freezer. They will keep for a week in the salad drawer of a refrigerator.

Pests and Diseases

Celeriac has the same problems as celery. Protect against slugs; use ferric phosphate-based slug pellets, pick off slugs at night and encourage natural predators. Slugs congregate under lettuce leaves or wet paper; pick off and destroy. Carrot fly is often a pest when established on carrots. Grow under fleece or place a barrier 75cm high of fine netting or polythene, erected before or just after sowing. Keep it in place until harvesting time. Celery fly is less of a problem than on celery. Pick off any brown, blistered leaflets or grow under horticultural fleece.

companion planting

Celeriac grows well where legumes have been planted the previous year and benefits from being placed alongside beans, brassicas, leeks, tomatoes and onions.

medicinal

Celeriac oil has a calming effect and is a traditional remedy for skin complaints and rheumatism. It is also said to restore sexual potency after illness! Celeriac is rich in calcium, phosphorus and vitamin C.

warning

Celeriac is a diuretic. Pregnant women and those with a kidney disorder should avoid eating it in large quantities.

'Prinz'

culinary

Containing only 14 calories per 100g, celeriac is excellent for anyone on a diet.

Scrub the bulb well to remove dirt before peeling. It discolours rapidly when cut; put immediately into acidulated water. Grated celeriac can be added raw to winter salads. Alternatively, blanch the slices or cubes in boiling water for a few seconds beforehand. In France it is cut into cubes and mixed with mayonnaise and Dijon mustard to make *céleri-rave rémoulade*.

The bulb adds flavour to soups or stews and is good with lamb or beef, puréed or seasoned with pepper, salt and butter, it is an ideal accompaniment for stronger-flavoured game.

The leaves are strongly flavoured and can be used sparingly to garnish salads or dried for use in cooking. The stems can be cooked and eaten like seakale. Celeriac can be made into delicious chips: boil a whole, peeled root in salted water until just tender and then cut into chips and fry in a mixture of butter and oil until lightly browned. These chips make an excellent accompaniment to game or plain grilled steaks.

Boiled and sliced, celeriac can be covered with a cheese sauce well flavoured with French mustard. It also makes an excellent soup.

Monkfish with Celeriac

Serves 4

750g monkfish, cut into chunks
50g celeriac, cut into julienne strips
1 large onion, finely sliced
1 carrot, peeled and cut into julienne strips
75g butter
1 tablespoon flour
2 teaspoons French mustard
2 tablespoons Greek yogurt
1 tablespoon double cream
Salt and freshly ground black pepper

Season the monkfish and prepare the vegetables. Heat half the butter in a heavy frying pan and cook the monkfish gently for about 7–8 minutes, turning it until just tender. Remove from the pan and keep warm. Using the rest of the butter, add the vegetables to the pan and sauté until soft. Stir in the flour and cook for a couple of minutes; then add the mustard, yogurt and cream. Stir well and heat through gently. Put the fish pieces in, stir to coat well and serve piping hot.

Arctium lappa. Compositae

BURDOCK

Also known as Edible Burdock, Gobo. Hardy biennial, grown as an annual. Value: moderate levels of dietary fibre plus vitamin B, potassium, calcium and inulin.

After walking his dog one day in the early 1940s, inventor George de Mestral noticed that burdock seeds had attached themselves to his woollen clothes and the fur of his dog; when analysing them under a microscope, he discovered that the sharply hooked 'burrs' were firmly linked to the loops of fur and material. After much research, he devised the famous fastening system Velcro®, which is widely used today. Burdock is naturalised throughout the world; the Chinese introduced it to Japan and neatly packed boxes are a feature of Japanese food markets. The famous French seedsmen Vilmorin-Andrieux, in their book, *The Vegetable Garden*, noted that, 'although it cannot be termed delicious, it is certainly not a bad vegetable and is, therefore, deserving of serious consideration'. It has an unusual but rather pleasing flavour – perhaps you should give it a try!

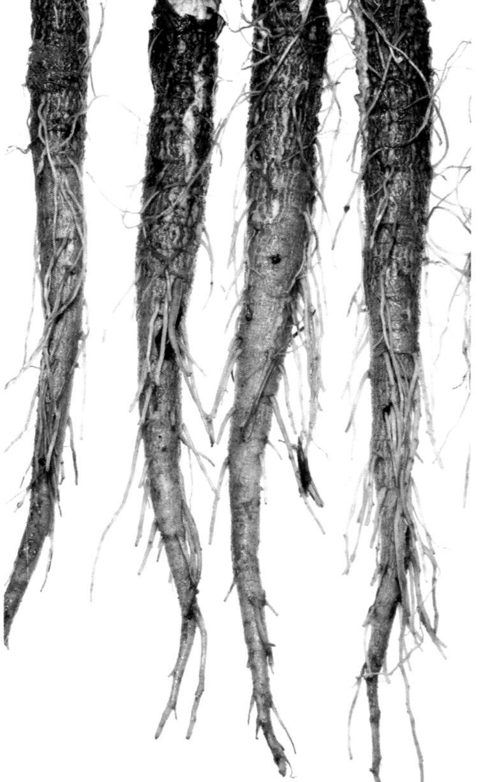

Burdock roots close up

varieties

Arctium lappa is a large biennial up to 1.2m tall, notable for its large, rough, heart-shaped leaves. Roots can be up to 2.4m long and 2.5cm thick. The small, thistle-like flower heads are followed by prickly burrs. However, avoid cultivating wild burdock as a crop as the roots are comparatively small. **'Mitoya Shirohada'** has roots that are white and smooth with a tender texture and just over 1m long. **'Takinogawa Long'** has small leaves and large roots. **'Watanabe Early'** is ideal for spring sowing, matures rapidly and has roots around 90cm long.

cultivation

Propagation

Soak seeds overnight or 'scarify' before sowing and place them on the surface as they need light to germinate. Sow in autumn but not too early in the season; smaller plants are more resistant to winter weather and larger over-wintering plants usually bolt in spring. Alternatively, sow in spring when temperatures are at a minimum of 20°C, placing three seeds 40cm apart, with 50cm between rows, then thin in late winter, leaving the strongest seedling. Another option is to germinate seeds under glass and transplant before a 'taproot' starts to form; modules are ideal for this.

Growing

Burdock grows well in temperate and warm, humid climates at temperatures from 20–25°C and is most successful on moisture-retentive, free-draining, light or sandy soils for increased rooting depth and easy lifting of the roots. It is less successful on heavier soils, though it can be grown successfully in raised beds or ridges. Clear the site before planting, incorporating copious amounts of well-rotted organic matter as deeply as possible and adding bone-meal to encourage root growth. Keep young plants weed-free until the foliage is large enough to prevent them from germinating. Water plants well during drought.

Burdock is vigorous and spreading, demanding high levels of nutrients and water and should be sited well away from other plants.

Maintenance

Spring Sow seeds under glass or outdoors if soil is warm.
Summer Water well and keep weed-free.
Autumn Lift roots.
Winter Protect over-wintering seedlings under cloches.

Protected cropping

Plants can be started off under glass before transplanting them outdoors.

Harvesting and Storing

Plants sown in autumn or spring are ready for harvest the following autumn. Lift the first roots when the foliage dies back after the first frosts; it may not be possible to dig out whole roots because of their length, but those that remain in the ground will not become a problem. Lifting roots when they are immature, up to 45cm long, ensures they are tender. Cover plants with bracken, straw or similar materials in winter; the roots can then still be lifted when the surrounding soil is frozen. Alternatively, lift all roots after the first frosts and store in boxes of damp sand or wrapped in paper in the salad drawer of a fridge.

Pests and Diseases

Burdock is pest- and disease-free.

container growing

Burdock is not suitable for most containers because of the size of the plant above and below ground, but raised beds should provide adequate space.

medicinal

Burdock contains a chemical that encourages lactation and is also used as a hair and scalp conditioner, blood purifier, diuretic and for the treatment of skin conditions.

Burdock growing in a field

culinary

Burdock is a major ingredient in the drinks 'Dandelion and Burdock' and 'Root Beer'. Young roots, which are of a better quality, can be eaten raw or stir-fried, whereas older roots are boiled until soft or added to casseroles. One seed catalogue even suggests it can be fried with red peppers and makes a good accompaniment to baked beans! The young leaves and stem can be lightly cooked; immature flower stems can also be harvested and eaten before flowering.

In Japan, the roots are pickled and served with sweet sauce, or shredded and braised with several ingredients, including carrot, sake and sesame oil, in a dish called *kinpira gobo*. Burdock *makizushi* is sushi filled with pickled burdock root instead of fish. Soak older roots in water for an hour before use to purge any bitterness, adding a dash of vinegar to prevent discoloration. Older roots taste earthy and become woody and less palatable with age. Buy burdock in Asian or health food stores – or get it for free when out in the countryside. This dish comes from southern Italy.

Bucatini with Burdock, Artichokes and Peas

Serves 4

1 burdock root, about 25cm long
Balsamic vinegar
500g bucatini pasta
1 tablespoon olive oil
1 shallot, finely chopped
1 small onion, finely chopped
3–4 cloves garlic, crushed
Zest of 1 lemon
4–6 marinated artichoke hearts
200g peas
1 large handful basil leaves
100ml chicken or vegetable stock
25g butter
Salt and freshly ground black pepper
Freshly grated pecorino cheese, to serve

Peel the burdock root if you prefer, but remember the vitamins are right under the skin and it has more flavour with the skin left on. Cut into matchstick-sized pieces, cover with water, stir in a couple of teaspoons of balsamic vinegar and leave to soak for 30 minutes.

Cook the pasta in boiling, salted water until *al dente*, then drain. Meanwhile, strain the burdock. Heat a large, heavy-bottomed frying pan over a medium heat and pour in the olive oil. Sauté the shallot and onion for 2–3 minutes, then add the garlic, lemon zest and burdock root and cook until the vegetables are softened and lightly coloured. Add the artichoke hearts, peas, basil and the butter, and sauté for a further 5 minutes. Combine the vegetables with the drained pasta, season with salt and pepper and serve sprinkled with cheese.

Armoracia rusticana. Cruciferae

HORSERADISH

Perennial herb sometimes grown as annual for its strong-flavoured fleshy roots. Hardy. Value: rich in vitamin C and calcium, moderate in carbohydrates.

Thought to be native of southern Russia and the eastern Ukraine, horseradish is now found throughout the temperate zones of the world. Cultivated since classical times, horseradish was probably carried round Europe by the Romans, who used it as a medicine and flavouring. Parkinson noted in 1640, 'it is too strong for tender and gentle stomaches', yet it was extensively eaten by country folk in Germany; it is known in France as *moutarde des Allemands*. Its hardiness and ability to regenerate from the smallest particle of root have ensured its success – and sometimes makes it a pernicious weed both in and outside the garden. The common name 'horseradish' distinguishes it from the salad radish, 'horse' signifying coarse.

varieties

The species ***Armoracia rusticana*** grows up to 90cm tall; the broad, oblong, dark green leaves have serrated margins. The thick tapering roots penetrate 60cm or more into the soil. ***Armoracia rusticana* 'Variegata'** has leaves splashed with cream and is of ornamental merit, though its flavour is not as good as the species.

cultivation

As horseradish is difficult to eradicate once established, lift all plants from late autumn to early winter, keeping some side roots for propagation the following spring. Alternatively, let it grow as a perennial for some years and divide in spring.

Propagation

Grow from root cuttings or young plants. In late autumn or winter lift crowns, remove side shoots of pencil thickness about 15–20cm long and store them in damp sand in a cool, frost-free shed. In mid-spring, make holes with a dibber 60cm apart, deep enough for the top of the root cutting to be covered with soil to a depth of 5–7cm; insert a piece of root into each hole with the thickest end uppermost and fill with soil.

Growing

Horseradish dislikes heavy shade but grows on any soil with reasonable drainage; roots flourish in deep, rich, well-drained soils. Dig in well-rotted organic matter the winter before planting; where crops are rotated, horseradish should follow a heavily manured crop like beans. After planting, apply a general fertiliser, water as needed, particularly during drought. Keep weed-free.

Roots are at their hottest when eaten fresh

HORSERADISH

Maintenance

Spring Plant root cuttings.
Summer Feed, water and keep the crop weed-free.
Autumn Lift whole crowns, keep large roots for use in the kitchen, use others for propagation.
Winter Prepare the ground for replanting.

Harvesting and Storing

Lift and harvest the crowns in autumn when the flavour is strongest and use stored roots over winter and spring. As exposure to light causes greening, store roots in the cool and dark. In summer, lift roots as required.

Pests and Diseases

Do not plant in soil affected by clubroot. Horseradish pale leaf spot – near-white spots with dark margins on the leaves – does not affect the roots and no treatment is needed.

companion planting

It is said to improve the disease resistance of potatoes.

container growing

Restrict the growth of horseradish as a perennial by planting in a bucket or dustbin (with drainage holes in the base) sunk into the ground. Use soil with plenty of rotted organic matter or a loam-based compost with a moderate fertiliser content. Water well. Divide in autumn and winter as necessary.

medicinal

Horseradish is a diaphoretic, digestive, diuretic and stimulant. Modern research has indicated anti-microbial activity against some micro-organisms. It is also said to staunch bleeding, prevent scarring and cure stomach cramp. In folk medicine the vapour from grated roots was inhaled to treat colds. It is rich in vitamin C, sulphur, potassium and calcium.

warning

Grate horseradish using the shredder attachment of a food processor to prevent your eyes watering.

culinary

The root stays fresh for about 2 weeks in the salad drawer of a refrigerator, but it is better to freeze grated root in polythene bags and use as required. Trim off the rootlets, scrub or scrape under the cold tap to remove soil. Finely grate or use a food processor, discarding the central core. Or store by grating into white vinegar (red wine or cider vinegar discolours the root).

Horseradish sauce, made with milk or oil and vinegar, is the traditional accompaniment for roast beef, asparagus and smoked fish like mackerel. Fold into whipping cream, yogurt or sour cream and season with salt, sugar and a little vinegar to make horseradish cream. Don't grate until just before serving or it will lose its flavour. My father made his horseradish sauce just before the roast beef was carved: a teaspoonful was absolutely lethal and sent us running for a glass of cold water! Serve with ham, on baked potatoes or cold meats. Grated horseradish can be used in steak tartare, or added to coleslaw, dips and sauces. Blended with butter and chilled it is an alternative to garlic butter to accompany grilled steak.

Apricot and Horseradish Sauce

Serve this more delicate sauce to accompany roast chicken. To serve it with fish, use fennel rather than tarragon.

500g apricots, stoned
Juice of ½ a lemon
Caster sugar
1–2 tablespoons freshly grated horseradish
1 tablespoon fresh tarragon, chopped
Salt and freshly ground white pepper

Soften the apricots in a little water and purée. To the purée, add the lemon juice, sugar to taste, horseradish and tarragon. Season to taste.

Asparagus officinalis. Asparagaceae

ASPARAGUS

Long-lived perennial grown for slender young shoots and ornamental foliage. Half hardy. Value: high in potassium and folic acid, moderate source of beta carotene and vitamin E.

The genus *Asparagus* provides us with a range of robust foliage houseplants and one of the world's most desirable vegetables. The delicious taste, succulent texture and suggestive shape of the emergent shoots combine to create an eating experience verging on the decadent which has been celebrated for over 2,000 years. Pliny the Elder describes cultivation methods used by the Romans for producing plants with blanched stems, and mentions a cultivar of which three 'spears' weighed a pound. These spears were once believed to arise from rams' horns buried in the soil. Wild asparagus grows in Europe, Asia and north-west Africa, in habitats including dry meadows, sand dunes, limestone cliffs and volcanic hillsides.

varieties

'Connover's Colossal', an early, heavy-cropping, old variety producing large, tasty spears, is suitable for light soils and freezes well. **'Lucullus'** crops heavily, and has long, slim, straight spears. **'Martha Washington'**, an established favourite in USA, crops heavily, has long spears and is rust-resistant. The dark purple spears of **'Purple Jumbo'** seem almost black if temperatures are cold. Tender spikes become green with boiling and makes delicious soup. **'Purple Passion'** also turns green when cooked, is vigorous and excellent for salads.

cultivation

Asparagus thrives in an open, sheltered position on well-drained soil. As a bed it can be productive for up to 20 years, thorough preparation is essential.

Propagation

Asparagus can be grown from seed, though it is easier and less time-consuming to plant crowns. Soak seed for 2 days before sowing in mid-spring, 2.5cm deep in drills 45cm apart. Thin seedlings when they are 7.5cm tall until they are 15cm apart. Alternatively, sow indoors in late winter at 13–16°C directly into modules, pots or trays. Pot on, harden off and plant outdoors in early summer. Male plants are the more productive, so the following year, remove any females (identifiable by their fruits) before they shed their fruits. Transplant the remaining male crowns into their permanent position in mid-spring the following year.

Growing

The autumn or winter before planting, dig in plenty of well-rotted organic matter; lime acid soil to create a pH of 6.5–7.5. It is vital to remove perennial weeds. Fork over the soil 1 or 2 weeks before planting and rake in a general fertiliser at approximately 90g/sq m.

One-year-old crowns establish quickly; 2- and 3-year-old crowns tend to suffer from a growth check after trans-planting. Plant in mid-spring, once the soil is warm. The roots desiccate quickly and are easily damaged, so cover with sacking until ready to plant, then handle with care.

Either plant in single rows with the crowns 30–45cm apart or in beds with 2 or 3 rows 30cm apart. For several beds, set them 90cm apart. Before planting dig a trench 30cm by 20cm and make a 4cm mound of soil in the base; plant crowns along the top, spreading out the roots, and cover them with 5cm of sifted soil. As the stems grow, gradually cover with soil; by autumn, the trench should be filled with soil. Keep beds weed-free by hand weeding or hoeing carefully to avoid damaging the shallow roots. On more exposed sites, support the 'ferns' when windy to avoid damage to the crown, and water during dry weather.

After harvesting, apply a general fertiliser to nurture stem growth and build up the plants for the following year. In autumn, when stems have turned yellow, cut

The autumn fruits of Connover's Colossal

ASPARAGUS

back to within 2.5–5cm of the surface and tidy up the bed. Ferns can be shredded and composted. Each spring apply a general fertiliser as growth begins. Mulching with manure has little value beyond suppressing weeds and conserving moisture.

Maintenance

Spring Sow seed and plant crowns. Harvest late spring.
Summer Keep weed-free and water as necessary. Stop harvesting by mid-summer.
Autumn Cut back yellowing ferns and tidy beds.
Winter Prepare new beds: mix in organic matter and remove perennial weeds.

Protected Cropping

Protect the crowns from late frosts with horticultural fleece or cloches.

Harvesting and Storing

However tempting, do not cut spears until the third year after planting (except possibly with **'Franklim'**). Harvesting lasts for 6 weeks in the first year and 8 weeks in subsequent years. Do not harvest after mid-summer: it can result in thin spears the following year. When spears are 10–17.5cm long, cut them obliquely about 2.5–5cm below the surface with a sharp knife or a serrated asparagus knife.

Pests and Diseases

The black and yellow adults and small greyish larvae of asparagus beetles appear from late summer, stripping stems and foliage. Control by removing dying foliage.

companion planting

Where growing conditions allow, asparagus is compatible with tomatoes, parsley and basil.

medicinal

Asparagus is used to treat rheumatism, gout and cystitis. Anyone who lacks the enzyme to break down asparagine produces urine with a strong odour – a disconcerting but harmless phenomenon.

warning

The berries are poisonous.

culinary

Asparagus spears should be used as fresh as possible, preferably within an hour of harvesting. They can be refrigerated in a polythene bag for up to 3 days. To freeze, tie into bundles and blanch thick spears for 4 minutes, thin for 2 minutes. Freeze in a plastic container.

Asparagus is best eaten steamed or boiled and served hot with butter. Also good cold with vinaigrette, Parmesan or mayonnaise. Asparagus tips can be added to salads and pizza toppings.

To boil, wash spears, peel away the skin below the tips, and soak in cold water until all have been prepared. Sort into stalks of even length (perhaps 20 stalks if thin varieties and 6–8 if thicker-stemmed), and tie with soft string or raffia, one close to the base and another just below the tip. Stand bundles upright in boiling salted water, with the tips above water level. Cover and boil gently for 10–15 minutes until al dente, then drain and serve. Don't overcook: the tips should be firm, and the spears should not bend when held at the base. The water can be used in soup.

Jean-Christophe Novelli's Steamed Asparagus

Serves 4

500g butter
4 egg yolks
1 tablespoon white wine vinegar
1 vanilla pod, split and seeds scraped out
20 asparagus spears
Salt and freshly ground black pepper
Toasted almonds

To make a hollandaise sauce, melt the butter and keep warm. Place a bowl over a pan of boiling water. Put the egg yolks, vinegar and vanilla seeds in the bowl and whisk until the mix is light and fluffy. Take off the heat and slowly whisk in the butter, a little at a time, until all is incorporated. Season to taste.

Steam the asparagus until still slightly crunchy, spoon a generous spoonful of hollandaise on top and garnish with some toasted almonds.

Barbarea verna. Brassicaceae

LAND CRESS

Also known as American Cress. Biennial or short-lived perennial grown as annual for young leaves. Hardy. Value: low in calories, good source of iron, calcium, beta carotene and vitamin C.

The genus *Barbarea* was known as *herba Sanctae Barbarae*, the 'herb of St Barbara', patron saint of miners and artillerymen and protectress from thunderstorms! Land cress is a fast-growing, hardy biennial with a rosette of deeply lobed, shiny leaves and yellow flowers. Native to south-western Europe, it has been grown as a salad crop since the seventeenth century; by the eighteenth century, extensive cultivation had died out in England, though plants became naturalised and are still common in the wild.

cultivation

Land cress grows in wet, shady conditions, but is best in moist fertile soil; in summer plant in light shade, under deciduous trees.

Propagation

Sow immediately the soil becomes workable, in early spring to early summer for a summer crop, and in mid- to late summer for autumn to spring crops. Sow in seed trays or modules for transplanting when large enough to handle, or in drills 1cm deep, thinning the seedlings to 15–20cm. Germination takes about 3 weeks in spring but half that time in mid-summer. If a few plants are left to run to seed in late spring, the following year they will seed freely; transplant seedlings into rows and water well.

Growing

Before sowing, dig in well-rotted manure or compost. Transplant seedlings sown in late summer under glass. In heat and drought they run to seed, so water often. Pick flower stalks as they appear.

Maintenance

Spring Sow seed, water well.
Summer Water as necessary so plants do not run to seed.
Autumn Sow winter crops and transplant plantlets under glass.
Winter Prepare beds for spring sowing; harvest protected crops.

Protected Cropping

Improve the quality of autumn and winter crops by growing in an unheated greenhouse or cold frame, or under cloches.

Harvesting

Harvest after 7 weeks when plants are 7–10cm long. Pick or cut the tender young leaves about 2.5cm above ground. Do not harvest heavily until the plant is established. Soak in water to loosen dirt, then wash it off.

Pests and Diseases

Flea beetle may affect plants; grow under crop covers.

companion planting

Makes a good edging plant for borders. Can be grown between taller crops, like sweetcorn and brassicas.

container growing

Grow in moisture-retentive, peat-substitute compost, water well and feed with a dilute general liquid fertiliser every 3 weeks.

culinary

Use its peppery tasting leaves as a watercress substitute – as a garnish, in salads and sandwiches. Or cook them like spinach and make into soup. Good in rice, pasta salads and stir-fries.

Cress, Anchovy and Barley Salad

Serves 4

300g pearl barley
4 tablespoons virgin olive oil
2 tablespoons white wine vinegar
2 tablespoons finely chopped fresh dill
4 anchovy fillets, roughly chopped
350g cress, washed and dried
1 cucumber, diced
Salt and freshly ground pepper

Cook the barley in boiling water until tender and drain. Set aside. Make the dressing: mix the oil and vinegar with the dill and seasoning. In a separate bowl, mix the barley with the anchovy, add the cress and cucumber, and pour over the dressing. Toss well.

Beta vulgaris subsp. *cicla*. *Chenopodiaceae*

SWISS CHARD

Also known as Silver Chard, Silver Beet, Seakale Beet. Biennial grown as annual for leaves and midribs. Hardy. Value: high in sodium, potassium, iron, and exceptional source of beta carotene, the precursor of vitamin A.

The umbrella name 'leaf beet' includes Swiss chard and also encompasses perpetual spinach or spinach beet. (The 'true' spinach and New Zealand spinach both belong to other genera.) A close relative of the beetroot, leaf beet is an ancient vegetable cultivated for its attractive, tasty leaves. Native to the Mediterranean, it was well known to the Greeks, who also ate its roots with mustard, lentils and beans. Aristotle wrote of red chard in the fourth century BC, and Theophrastus recorded both light and dark green varieties. The Romans introduced it to central and northern Europe and from there it slowly spread, reaching the Far East in the Middle Ages and China in the seventeenth century. The name 'chard' comes from the French *carde* and derives from the resemblance of the leaf stalks to those of globe artichokes and cardoons. In 1597 John Gerard wrote in his *Herball*, '... it grew with me to the height of eight cubits and did bring forth his rough seeds very plentifully.' If the measurement is correct, his Swiss chard would be approximately 4m tall. I wonder where that variety is today, was his yardstick wrongly calibrated, or had it simply bolted?

Both leaves and stems are tasty

varieties

Swiss chard has broad red or white leaf stems and midribs. **'Bright Yellow'** has golden yellow leaf stems and a mid-green puckered leaf. Delicious! **'Erbette'**, an Italian variety, is well flavoured and has an excellent texture. It is good as a 'cut and come again' crop. **'Fordhook Giant'** has huge, glossy green leaves with white veins and stems. It is tasty and high-yielding, producing bumper crops even at high temperatures. **'Lucullus'** is vigorous and crops heavily, producing pale yellow-green leaves with succulent midribs. Tolerant of high temperatures, it does not bolt.

'Perpetual Spinach', or spinach beet, is smaller, with narrower stems and dark, fleshy leaves, and is very resistant to bolting. Good for autumn and winter cropping. **'Rhubarb Chard'** (**'Ruby Chard'**) is noted for its magnificent bright crimson leaf stalks and dark green puckered leaves. Ideal for the ornamental border or 'potager', it needs growing with care, as it is prone to bolting.

'Fordhook Giant'

cultivation

Though they tolerate a wide range of soils, the best growing conditions are sunny or lightly shaded positions in rich, moisture-retentive, free-draining soil. In poor soil, bolting can be a problem, so dig in plenty of well-rotted organic matter the winter before planting. The ideal pH is 6.5–7.5 and acid soils should be limed.

The ideal growing temperature is 16–18°C, though the range of tolerance is remarkably broad. They survive in winter temperatures down to about −14°C and are more tolerant of higher summer temperatures than true spinach, which is inclined to bolt.

Propagation

For a constant supply throughout the year, make two sowings, one in mid-spring for a summer harvest and another in mid- to late summer. The later crop is usually lower-yielding.

Sow 3–4 seeds in 'stations' 23cm apart, in drills 1–2cm deep. Swiss chard needs 45cm between the rows, and perpetual spinach 38cm. Thin seedlings when large enough to handle to leave the strongest seedling.

SWISS CHARD

Alternatively, sow in modules or trays and transplant to their final spacing when they are large enough to handle. Swiss chard is particularly successful as a 'cut and come again' crop. Prepare the seedbed thoroughly and broadcast or sow seed in drills the width of a hoe.

Growing
Keep crops weed-free by hoeing or, preferably, mulching with well-rotted organic matter and keep the soil continually moist. In dry conditions, plants will need 9–13.5 litres per week, but are surprisingly drought-tolerant. A dressing of general granular or liquid fertiliser can be given to plants needing a boost.

Maintenance
Spring Sow the first crop in mid-spring in 'stations'; thin, leaving a strong seedling.
Summer Weed, water and feed as necessary. Sow a second crop in mid- to late summer. Harvest as needed.
Autumn Protect with cloches or fleece in late autumn for good-quality growth.
Winter Dig over the area where the following year's crop is to be planted. Harvest overwintering crops.

Protected Cropping
Though they are hardy enough to withstand winters outdoors, plants protected in cloches, cold frames, polythene tunnels or fleece produce better crops of higher-quality leaves.

Harvesting and Storing
Seeds sown in mid-spring are ready to harvest from early to midsummer. Harvest the outer leaves first, working towards the centre of the plant and cutting at the base of each stem: snapping them off is likely to disturb the roots. Choose firm leaves and discard any that are damaged or wilted. Pick regularly to ensure a constant supply of tender regrowth, so harvest even if you are unable to use them – they are certain to be welcomed by friends.

They can also be grown as 'cut and come again' crops from seedling stage through to maturity. Cut seedlings when about 5cm tall. After 2–3 crops have been harvested, allow them to regrow to about 7.5cm.

Semi-mature plants are harvested leaf by leaf and mature plants can be cut about 2.5cm above the ground; from this, new growth appears.

Swiss chard and perpetual spinach are best eaten straight from the plant. Leaves (minus the stalks) keep in a refrigerator in the salad compartment or in polythene bags for 2–3 days.

Pests and Diseases
They are relatively trouble-free, though beware of downy mildew when dense patches of seedlings are sown for 'cut and come again' crops. It appears as brown patches on leaves. Birds sometimes attack seedlings, so protect crops; growing plants under brassicas or beans also gives them some protection.

companion planting

They grow well with all beans except runners, and flourish alongside brassicas, onions and lettuce. Herbs like sage, thyme, mint, dill, hyssop, rosemary and garlic are also compatible.

container growing

Swiss chard can be grown in containers and makes a fine ornamental feature. Either transplant seedlings or sow directly into loam-based compost or garden soil, with added organic matter.

medicinal

Leaves are vitamin- and mineral-rich with high levels of iron and magnesium. In folk medicine the juice is used as a decongestant; the leaves are said to neutralise acid and have a purgative effect. Beware of eating it in large quantities!

culinary

Perpetual spinach can be lightly boiled, steamed or eaten raw. Swiss chard takes longer to cook: try it steamed and served with butter. Soup can be made from the leaves and its midribs cooked and served like asparagus, or added to pork pies. 'Rhubarb Chard' tastes milder than white-stemmed varieties.

Spinach Beet Fritters
Serves 4

These robust fritters go well with salmon or cod.

750g perpetual spinach, well washed
Knob of butter
2 large eggs, separated
1 tablespoon grated Parmesan
1 teaspoon grated lemon zest
Salt and freshly ground black pepper
Pinch of nutmeg
Olive oil, for frying

Prepare the leaves, removing the midribs, and chop roughly. Cook in the water that clings to the leaves until wilted – 2–3 minutes – and drain well. Chop finely and return to the pan with the butter, cooking until all the liquid evaporates. Leave to cool for 5 minutes and stir in the egg yolks, Parmesan and lemon zest. When almost cold, fold in the stiffly beaten egg whites and season with salt, pepper and nutmeg to taste.

Drop spoonfuls into hot fat, heated in a heavy frying pan, and cook for a minute or two, turning halfway.

Drain well and serve hot.

Beta vulgaris subsp. *vulgaris. Chenopodiaceae*

BEETROOT

Also known as Beet. Biennial grown as annual for swollen root and young leaves. Hardy. Value: slightly higher in carbohydrates than most vegetables, good source of folic acid and potassium.

Beetroot is a form of the maritime sea beet which has been selected over many centuries for its edible roots. From the same origin come mangold (a cattle fodder), the beet used for commercial sugar production and Swiss chard. Grown since Assyrian times, the vegetable was highly esteemed by the ancient Greeks and was used in offerings to Apollo. There were many Roman recipes for beetroot, which they regarded more highly than the greatly revered cabbage. It appeared in fourteenth-century English recipes and was first described as the beetroot we know today in Germany in 1558, though it was a rarity at that time in northern Europe. The typical red colouration comes from its cell sap, but there are also varieties in other colours.

Beets come in beautiful colours

varieties

Beetroots are grouped according to shape – round or globe-shaped, tapered or long, and flat or oval. To reduce the amount of thinning needed, breeders have introduced 'monogerm' varieties.

Globe

'Boltardy' is a delicious, well-textured, smooth-skinned variety and an excellent early cropper as it is very resistant to bolting. Good in containers. **'Bonel'** has deep red, succulent, tasty roots and is high-yielding. It crops over a long period and is resistant to bolting. **'Detroit 2 Little Ball'** produces deep red, smooth-skinned 'baby beet', which are ideal for pickling, bottling or freezing. A good crop for late sowing and for storing. **'Monogram'** is dark red, well-flavoured and vigorous, with smooth skin and rich red flesh. It is a 'monogerm' variety. **'Monopoly'** is also a 'monogerm' and is resistant to bolting, with a good colour and rough skin. **'Pablo'** is a tasty early with uniform roots and smooth skin. Also bolting resistant. **'Regala'** has very dark roots and is quite small, even at maturity. An excellent variety for containers, and resistant to bolting.

Tapered

'Cheltenham Green Top' is a tasty, old variety with rough skin and long roots. It stores well. **'Cheltenham Mono'** is a tasty, medium-sized 'monogerm' which is resistant to bolting, good for slicing and stores well.

Others

'Albina Vereduna' (**'Snowhite'**) is a wonderful globe-shaped white variety, with smooth skin and

BEETROOT

sweet flesh. It also has the advantage that it does not stain. The curly leaves can be used as 'greens' and are full of vitamins. It does not store well and is prone to bolting. **'Barbabietola di Chioggia'** is a mild, traditional Italian variety. Sliced, it reveals unusual white internal 'rings' and gives an exotic look to salads. When cooked, it becomes pale pink. Sow from mid-spring. **'Burpees Golden'** has beautiful orange skin and tasty yellow flesh. It looks great in salads, keeps its colour when cooked, does not bleed when cut and the leaves can be used as 'greens'. It stores well, has good bolting resistance and is better harvested when small. **'Cylindrica'** has sweet-tasting, dark, oval roots that have excellent keeping qualities and good flavour. Because of its shape, it is perfect for slicing and cooks well. Harvest when young. **'Egyptian Turnip Rooted'** (**'D'Egypte'**, **'Egyptian Flat'**) has smooth roots with deep red, delicious flesh. An American introduction, it was first grown around Boston about 1869. **'Forono'** is very tasty with large, cylindrical roots, smooth skin and good colour. Slow to go woody, it is ideal for summer salads and stores well. Susceptible to bolting, it should be sown from mid-spring.

Varieties of 'mini vegetables' include **'Pronto'**, **'Action'** and **'Monaco'**.

cultivation

Propagation

In most varieties, each 'seed' is a corky fruit containing 2 or 3 seeds, so a considerable amount of thinning is required. 'Monogerm' varieties, each containing a single seed, reduce such a work load. They also contain a natural inhibitor which slows or even prevents germination. Remove this by soaking seeds or washing them in running water for 30–60 minutes before sowing. At soil temperatures below 7°C germination is slow and erratic. To overcome this, sow early crops in modules, 'fluid sow' or sow in drills or stations after warming the soil with cloches. These can be left in place after sowing until the weather warms tip. Use bolting-resistant varieties until mid-spring; after that, any variety can be used.

Sow the first crops under cloches from late winter to early spring 12mm–2cm deep and 2.5cm apart with 23cm between rows. Thin to a final spacing of 10cm between plants. Alternatively, sow 2–3 seeds at 'stations' 10cm apart, thinning to leave the strongest seedling when the first true leaf appears. Round varieties can be 'multi-sown' in a cool greenhouse planting 3 seeds per module, thinning to 4–5 seedlings, then planting the modules 10cm apart when about 5cm high. Early crops can also be sown thinly in broad flat drills 12mm–2cm deep, in a similar way to peas.

Thin as soon as seedlings are touching and keep thinning as plants grow: those large enough can be used whole. If you grow beetroot under horticultural fleece or a similar cover (put in place once the seedlings have established), yields can be increased by up to 50 per cent. Remove protection 4–6 weeks after sowing. From mid-spring, if the weather is warm seeds can be sown without the protection of cloches, thinning to 7.5–10cm apart.

Beetroot grown for pickling need to be about 5cm in diameter. Sowing in rows 7.5cm apart and thinning plants to 6.5cm apart will give you the correct size.

From late spring to early summer sow the main crop, using any round or long variety. Harvest throughout the summer and for winter storing. Sow in drills or at 'stations', thinning to leave a final spacing of 7.5cm apart in rows 20cm apart or 12.5–15cm in and between the rows.

For a constant supply of beetroot, sow round cultivars under glass from late winter at 4-week intervals for mid-spring crops; and for a late autumn crop sow from early to mid-summer in mild areas (for lifting during winter thin to 10cm). For winter storage sow in late May, early June.

Growing

Beetroot needs an open site with fertile, well-drained light soil which has been manured for the previous crop. The pH should be 6.5–7.5, so acid soils will need liming. Autumn-maturing varieties tolerate heavier conditions and long-rooted varieties require a deeper soil. The best quality roots grow in moderate temperatures around 16°C.

Scatter a slow-release general fertiliser at 30–60g, 2–3 weeks before sowing, raking the seedbed to a fine tilth. For good-quality beetroot, it is important to avoid any check in growth; at the onset of drought, water at a rate of 11 litres/sq m every 2 weeks. Do not over water as this results in excessive leaf growth and small roots. If watering is neglected, yields are low, roots become woody and when it rains or you water suddenly, the roots will split. Keep weed-free and hoe with care as damage causes the roots to bleed: use an onion hoe or mulch round the plants. Mulching with a 5cm layer of well-rotted compost or spent mushroom compost will conserve moisture.

Maintenance

Spring Sow early crops under glass or cloches. Mid-spring crops can be sown without protection.
Summer Sow successionally every month, harvest earlier crops, keep the plot weed-free and water as required. Sow main crops.
Autumn Lift later crops and those for storage.
Winter In mild areas leave overwintering crops outdoors and protect with bracken, straw or similar materials. Alternatively, lift and store indoors.

Protected Cropping

Grow early crops under glass in modules and transplant under cloches. Alternatively, grow under cloches or crop covers and remove these about 6 weeks after sowing.

Harvesting and Storing

Beetroot takes 60–90 days to mature. It must always be harvested before it becomes woody and inedible. Harvest salad beetroot from late spring to mid-autumn and maincrop varieties from mid-summer onwards.

'Boltardy'

Early varieties are best harvested when the size of a golf ball; when later crops reach that size, lift every other plant and use for cooking, leaving the rest for lifting when they reach cricket ball-size.

Lift roots carefully with a fork, shake off soil and twist off the leaves. Do not cut off the leaves: it causes bleeding and makes a terrible mess! Use any damaged roots immediately. Lift beet for storage by mid-autumn and put in stout boxes of moist peat substitute, sand or sawdust, leaving a gap between each root. Store in a cool, frost-free shed or garage. Roots should keep until mid-spring the following year but check regularly and remove any that deteriorate. The long-rooted types are traditionally grown for storage, although most varieties store successfully.

In mild areas and on well-drained soil, they can be left over winter, but need a dense protective covering of straw or similar material before the frosts. This also makes lifting easier.

Pests and Diseases

Beetroot are generally trouble-free but may suffer from the following problems:

Black bean aphid forms dense colonies on the leaves. Yellow blotches between the veins, the symptom of manganese deficiency, appear on older leaves first and can be a problem on extremely alkaline soil.

Rough patches on the surface of the root and waterlogged brown patches and rings at its centre are a sign of boron deficiency. There may also be corky 'growths' on the shoots and leaf stalks.

Make sure that beetroot seedlings are protected against birds. Slugs make holes in leaves. The problem is worse in damp conditions.

companion planting

Beetroot flourish in the company of kohlrabi, carrots, cucumber, lettuce, onions, brassicas and most beans (not runners). Dill or Florence fennel planted nearby attracts predators. Because they combine well with so many other crops and small roots mature within 9–13 weeks, beetroots are good for intercropping and useful catch-crops.

culinary

The roots are eaten raw – try them grated as a crudité – or cooked and served fresh or pickled. Young 'tops' can be cooked like spinach and used as 'greens'. They add colour and flavour to salads, particularly the red, yellow, white and bi-coloured varieties. Bean and beetroot salad is particularly tasty. Wash in cold water, keeping root and stems intact: do not 'top and tail' or damage the skin, as bleeding causes loss of flavour and colour. Boil for up to 2 hours in saltwater, depending on the size, then carefully rub off the skin. It is delicious served hot as a vegetable, otherwise cool for pickling or for a fresh salad.

Beetroot can also be baked, and is the basis for borscht soup when cooked with white stock. It also makes excellent chutney and wine.

Freeze small beets which are no more than 5cm across. Wash and boil, skin and cool, then cut roots into slices or cubes and freeze in a rigid container. You should use within 6 months.

In a polythene bag or salad compartment of the fridge, they stay fresh for up to 2 weeks.

Spicy Beetroot Salad
Serves 4

750g beetroots, washed and trimmed
Juice of ½ a lemon
½ teaspoon ground cumin
½ teaspoon ground cinnamon
½ teaspoon paprika
1 tablespoon orange flower water
2 tablespoons olive oil
Salt and freshly ground black pepper
2 tablespoons chopped fresh parsley
Lettuce (coloured varieties mixed with green leaves such as lamb's lettuce)

Cook the beetroots in a steamer for 20 or 30 minutes until tender. Peel and slice them when cool, reserving the liquid that accumulates on the plate.

Toss them in lemon juice and coat with the spices, orange flower water and olive oil, together with the liquid. Season, cover and chill. To serve, toss with the parsley and arrange on individual plates on a bed of lettuce leaves.

container growing

Unless growing for exhibition, grow only globe varieties in containers – about 20cm deep – or troughs or growbags. Sow seed thinly 1.2–2cm deep from mid-spring to mid-summer, thinning to 10–12.5cm apart. Water regularly, harvest when the size of a tennis ball and keep weed-free.

other uses

The foliage is attractive and ideal for inclusion in an ornamental border or 'potager', particularly varieties like **'Bull's Blood'**. The leaf mineral content is 25 per cent magnesium, making it useful on the compost heap.

medicinal

Used in folk medicine as a blood tonic for gastritis, piles and constipation; mildly cardio-tonic. Recent research has shown that taking at least one glass of raw beetroot juice a day helps control cancer. It is regarded as a 'superfood'.

warning

The sap stains clothes. To remove, rub the stain with a slice of pear and wash as usual.

Brassica juncea. Brassicaceae

ORIENTAL MUSTARD

Also known as Mustard Greens. Hardy annual or biennial. Value: rich in vitamin A and C, moderate calcium, iron potash and phosphorous.

Commonly grown in Europe for mustard seed, mustard greens have their greatest diversity of shape and form in central Asia and the Himalayas which, along with the warm, central-Chinese province of Sichuan, India and the Caucasus, are thought to be one of the ancient areas of domestication. Sanskrit records show that mustard greens have been cultivated since 3000 BC and there have been an astonishing range of selections for their desirable characteristics. Among them are forms with tumescent swellings on the leaf stems, fleshy tap roots, large stems, multi-shoots, green stems and curled leaves plus heading varieties both large and small. This selection process is set to continue long into the future, as oriental mustard strengthens its position as an indispensable crop for oriental cuisine.

varieties

Commonly grown groups include:

Giant-leaved Mustard
This is very hardy. **'Green Chirimen'** has green, blistered leaves, while **'Miike Giant'** is broad-leaved and vigorous with mildly-flavoured, crinkled leaves and purple veins. **'Osaka Purple'** is fast growing, the large green leaves becoming purple in cold weather. It is easy to grow and cold-hardy.

Wrapped heart
These varieties grow better when there are high temperatures immediately after sowing. **'Amsoi'** produces tender greens with a mustard tang which are often used for pickling. **'Chicken heart'** is a semi-heading variety and well flavoured. **'Kekkyu Takana'** forms a small head and the mild flavour becomes stronger as it matures.

Leaf Mustard
Sow in summer and autumn. **'South Wind'** has purple-tinted, undulating leaves and is heat tolerant and good for stir-fries.

Green in the snow
This is fast growing and cold tolerant with frilly leaves. Young leaves are moderately spicy, older leaves should be cooked. Sow in autumn, can be grown as a 'cut and come again' crop.

Curled mustard
These are hardy but also tolerant of high temperatures. **'Green Wave'** is an All American Seed selection; it grows to 60cm tall with dark green, frilly-margined leaves. It is highly productive, slow to bolt and the flavour is hot.

Swollen stem mustard
These have unusually shaped or formed stems or leaf stems. **'Horned Mustard'** has bright green, frilly leaves with a distinct 'horn' in the centre of the stem. The leaf buds are also delicious. **'Tsa Tsai Round'** is unusual in that it forms thick, rounded, tumescent-like stems that are up to 15cm diameter and 225g in weight below the leaves. It is a cool-season crop.

ORIENTAL MUSTARD

'Osaka Purple'

cultivation

Propagation

Sow seeds thinly in situ and thin to the required spacing; summer and autumn sowings can also be made in modules for planting out. Sow according to the ultimate size of the variety or the stage of growth at harvest. Young plants should be 10–15cm apart; mature at 30cm apart in and between rows. Large varieties with spreading heads being grown to maturity should be 45cm apart.

Growing

Grow in an open, sheltered site in fertile soil of any kind, add well-rotted organic matter if necessary; they need plenty of moisture when in growth as they 'bolt' in dry conditions. All of the mustard varieties are cool season crops, though the resistance to cold and heat varies according to the variety. Sow from mid- to late summer for growing to maturity outdoors, or under cover, or in spring, for harvesting as 'cut and come again' crops. Sow successionally to extend the harvesting season.

Maintenance

Spring Sow outdoors when the soil is warm and friable.
Summer Sow crops for harvesting at maturity from mid-summer where frosts occur and late summer where frost free. Harvest 'curly leaved' and 'headed' types or 'cut and come again'.
Autumn Harvest.
Winter Harvest crops outdoors or under glass.

Protected Cropping

Oriental mustards make good autumn-sown crops for unheated glasshouses or polythene tunnels.

Harvesting and Storing

Cut and use fresh as required.

Pests and Diseases

They suffer from most of the usual brassica problems including flea beetle, cabbage root fly and slugs.

container growing

Some varieties can be grown as 'cut and come again' seedling crops in growing bags or large containers. They tend to be slow growing and are not as productive as other types of oriental vegetables using this method.

culinary

Oriental mustards are renowned for their spicy and peppery flavour which varies according to the plant's age and variety. They can be used raw or cooked in stir-fries and soups, combine well with crab, can be made into pickles and the leaves and shoots can be dried for winter eating. The inner leaves are milder and used in salads; the outer leaves are stronger flavoured and usually cooked.

Brassica napus (Napobrassica Group). *Brassicaceae*

SWEDE

Also known as Rutabaga, Swedish Turnip. Biennial grown as annual for swollen root and young leaves. Hardy. Value: small amounts of niacin (vitamin B) and vitamin C, low in calories and carbohydrates.

Swede is one of the hardiest of all root crops and is the perfect winter vegetable for cool temperate climates. An abbreviation of 'Swedish turnip', its name indicates its origins. Eaten in France and southern Europe in the sixteenth century, it came to Britain from Holland in 1755 and rapidly became popular as the 'turnip-rooted cabbage'. Along with the turnip, it was first used as winter fodder for sheep and cattle, improving milk production during a traditionally lean period. During times of famine, swedes were eaten by country folk and still have the reputation among many as peasant food.

To despise them is your loss; they are robust, undemanding and one of the easiest vegetables to grow. New varieties are disease-resistant, tasty and a wonderful accompaniment to sprouts as a winter vegetable – particularly when mashed with butter, cream and spices.

varieties

'Acme' has round roots with pale purple skin but its tops are prone to powdery mildew. **'Angela'** produces purple roots and is resistant to powdery mildew. **'Lizzy'** is a round variety with purple tops and yellow flesh, a soft texture and sweet, nutty flavour. **'Marian'** is purple with yellow flesh and is very tasty and quick-growing. **'Magres'** has deliciously flavoured yellow flesh and is mildew resistant. **'Virtue'** has red skin and sweet flesh.

cultivation

Propagation

Swedes need a long growing season and should be sown from early spring in cooler climates to early summer where temperatures are warmer and germination and growth are rapid. Sow in drills 2cm deep and 40–45cm apart, thinning seedlings to 23–30cm. Thin when they are no more than 2.5cm high, when the first true leaves appear, to ensure that the roots develop properly. Firm the soil after thinning.

Growing

Swedes prefer a sheltered and open site in fertile, well-drained but moisture-retentive soil. Good drainage is essential. Summer sowings can be made in moderate shade, provided they receive sufficient moisture. Swedes prefer a pH of 5.5–7.0, so very acid soil will need liming. If the ground has not been manured for the previous crop, double dig in autumn, incorporate plenty of well-rotted organic matter and allow the soil to 'weather' over winter.

About a week prior to sowing, remove any weeds or debris and rake general fertiliser into the soil at 60g/sq m. In common with other brassicas, swedes grow poorly on loose soil, so rake the soil to a fine tilth and firm it with the head of a rake or by carefully treading. If the soil is dry, water thoroughly before sowing and stand on a planting board to avoid compacting the soil. Mark each row with canes or twigs, and label and date the crop.

Keep crops weed-free by hand weeding and careful hoeing. Swedes need a constant supply of water throughout the growing season, otherwise they tend to run to seed or produce small, woody roots. Sudden watering or rain after a period of drought causes the roots

SWEDE

to split, so they will need up to 10 litres/sq m per week in dry periods. This improves the size and quality but usually reduces the flavour. Rotate swedes with other brassicas.

Swede tops can be blanched for eating raw as a winter salad vegetable. Lift a few roots in early to mid-winter, cut back the leaves and plant the roots under the greenhouse staging, or stand them upright in boxes or wooden trays filled with peat substitute, humus-rich garden soil or with a thick layer of straw. Cover with upturned boxes or black polythene to exclude the light and put them in a cellar, garage or shed. After 3–4 weeks shoots will appear and can be cut when they are 10–12.5cm long.

Maintenance

Spring Warm soil under cloches for early sowings.
Summer Water crops and keep weed-free.
Autumn Harvest early varieties.
Winter Lift and store crops before severe weather starts.

Protected Cropping

Cover ground with cloches, fleece or black polythene for 2–3 weeks in late winter to early spring to warm the ground before sowing early crops. Seedlings should be protected until they are well-established.

Harvesting and Storing

Harvest begins any time from early to mid-autumn. Swedes are extremely hardy and can be left in the ground until needed, though it is advisable not to leave them for too long or they will become woody. Lift when they are about the size of a grapefruit. They can be stored in boxes in a garage or cool shed. Twist off the leaves and place roots between layers of peat substitute, sawdust or sand in a stout box.

Smaller swedes are more tasty and succulent, so begin lifting as soon as the roots are large enough to use, before they reach their maximum size.

Pests and Diseases

Swedes are affected by the same problems as turnips.

Powdery mildew is common. It appears as a white powdery deposit over shoots, stems and leaves, causing stunted growth. In severe cases leaves become yellow and die. It is more of a problem when plants are dry at the roots and if it is cold at night and warm and dry in the day.

Swedes are susceptible to clubroot – to be avoided at all costs. Roots swell and distort, young plants wilt on hot days but recover overnight, growth is stunted and crops ruined. It is more of a problem on poorly drained, acid soils. Spores remain in the soil for up to 20 years.

Flea beetles are 3mm long and black with yellow stripes. They nibble leaves of seedlings, checking growth.

companion planting

Swedes grow well with peas.

medicinal

Swedes have been used in folk medicine for the treatment of coughs, kidney stones and whooping cough, though their efficacy has not been recorded.

A mature crop ready to be harvested

culinary

Swedes can be sliced or cubed and roasted like parsnips, or added to casseroles and stews. Otherwise, they are delicious mashed with potatoes and served with meat or fish.

To boil, peel off the outer 'skin', cut into slices or cubes and boil for 30 minutes. Drain thoroughly before serving.

Neeps

Serves 4

This is the traditional accompaniment for haggis and tatties (mashed potatoes), washed down with plenty of whisky, on Burns Night in Scotland.

750g swedes
Salt and black pepper
Pinch nutmeg
Butter or olive oil

Clean the swedes, cut them up and boil in enough water to prevent them from burning until tender. Process in a blender or put through a sieve, discarding the juices. Season, and reheat with either butter or olive oil.

Brassica oleracea* (Acephala Group). *Brassicaceae

KALE

Also known as Borecole, Collards, Colewort, Sprouts. Biennial grown as annual for young leaves and shoots. Hardy. Value: good source of calcium, iron, beta carotene, vitamins E and C.

Kales are exceptionally robust, and an ideal winter crop. Also, they are untroubled by common brassica problems. A primitive cabbage, kales are among the earliest cultivated brassicas (the Romans grew several types), with many similarities to the wild *Brassica oleracea* on the western coasts of Europe.

The Celtic 'kale' derives from *coles* or *caulis* used by the Greeks and Romans to describe brassicas; the German Kohl has the same origin. First recorded in North America by 1669, kales are thought to have been introduced much earlier.

varieties

Varieties are classified into groups, including the true kale, Siberian kale and 'Collards'. These are popular in the southern states of America and other warm climates. Kales vary in height from dwarf types, about 30–40cm high, to tall varieties growing to 90cm and spreading to 60cm. **'Nero de Toscana'** is famous for its ornamental value but the 'peppery' leaves are good steamed, stir-fried or in salads.

Siberian Kale, Rape Kale or Curled Kitchen Kale *(Brassica napus* Pabularia Group*)*
This is a relative of the swede or rutabaga, grown for the leaves, not the roots. It is variable in form and colour, with broader leaves than kale, which are sometimes curled or frilled. It must be sown *in situ*, not transplanted and crops when true kales have finished. **'Hungry Gap'** is a late cropper, robust and very reliable. **'Laciniato'**, an Italian variety, has deeply cut, flat leaves. **'Ragged Jack'** has pink-tinged leaves and midribs. **'Red Russian'** has tender and frilly leaves which have excellent flavour. **'True Siberian'** is fast-growing, with blue-green frilly leaves and harvests continually throughout winter.

Kale (Scotch Kale, Curly-leaved Kale or Borecole)
True kale usually has dark green or glaucous leaves with heavily frilled margins. **'Darkibor'** is exceptionally hardy with dark green, curled leaves. **'Dwarf Blue Curled Scotch'** (**'Dwarf Blue Curled Vates'**) is low-growing with glaucous leaves and is also hardy and slow to bolt. **'Dwarf Green Curled'** is compact, hardy and easy to grow; ideal for the small garden or windswept sites. **'Fribor'** with dark green leaves and **'Redbor'** winter well. **'Pentland Brig'**, a cross between curly and plain-leaved kale, is grown for the texture and flavour of its young leaves, side shoots and immature flower heads (which are cooked like broccoli). **'Showbor'** is a 'mini vegetable'. **'Spurt'** produces tender, deep green, curly leaves and is ready to harvest 6–8 weeks after sowing. It crops for a long period so grow as 'cut and come again'. **'Tall Green Curled'** (**'Tall Scotch Curled'**) is excellent for freezing, and shows good resistance to clubroot and cabbage root fly. **'Thousand Head'** is a plain-leaved, tall, old variety for harvesting through winter and spring. Exceptionally hardy.

Collards (or Greens)
These have smooth, thinner leaves than true kale that taste milder and are more heat-tolerant. **'Blue Max'** is tender and mild-flavoured. **'Green Glaze'** is attractive, delicious and quite frost hardy. **'Hevi-crop'** has deep, blue/green leaves and is very high-yielding. **'Top Bunch'** is upright and bolting resistant with blue-green savoyed leaves, good yields and good flavour.

cultivation

Propagation
Sow in early spring in trays of moist seed compost or modules, or outdoors in seedbeds in milder areas, for summer crops. Sow from late spring for autumn and winter crops. Alternatively, sow thinly in drills 1cm

Healthy plants packed with goodness

Kale can be highly ornamental

deep *in situ*, gradually thinning to final spacing, or sow 3 seeds in 'stations' at the final spacing, thinning to leave the strongest seedling. When 10–15cm high, water the rows the day before, then transplant in the early evening, keeping the lowest leaves just above the soil surface. Water well with liquid seaweed after planting and until they become established. The final spacing for dwarf varieties should be 45cm apart, with taller types 60–75cm apart.

Collards are sown in late spring in areas with cool summers and in late summer in hotter climates. Sow Siberian kale thinly in early summer, *in situ*, in drills 45cm apart, thinning when large enough to handle to a final spacing of 45cm apart.

Growing
Kales are very hardy: some survive temperatures down to −15°C; others tolerate high summer temperatures. They grow in poorer soils than most brassicas, but flourish in a sunny position on well-drained soil with moderate nitrogen levels. Excessive nitrogen encourages soft growth, making plants prone to damage. Lime acid soils.

Prepare the seedbed when the soil is moist. Lightly fork the surface, remove any weeds, then firm (but do not compact) the soil.

Overwintering crops need top-dressing with high-nitrogen liquid or granular fertiliser at 60g/sq m in spring to encourage side shoots.

Keep crops weed-free, water thoroughly before the onset of dry weather, mulch with a 5cm layer of organic matter and in autumn, firm or earth up round the base of taller plants to prevent wind rock. On more exposed sites, they may need staking.

Kales are good where peas, early potatoes or very early crops have been grown. Rotate with brassicas.

Maintenance
Spring Sow early crops under cover; in mid-spring sow in trays or modules, or outdoors in a seedbed.
Summer Transplant seedlings, feed and water well during dry spells. Keep weed-free. Harvest regularly.
Autumn Firm round plant bases; stake if necessary.
Winter Harvest regularly.

Protected Cropping
Make early sowings as a 'cut and come again' crop in the greenhouse border or in mild areas outdoors under cloches, from mid- to late winter. Sow in drills the width of a hoe, broadcast thinly, or sow sparingly in drills, thinning to 7.5cm apart.

Sow kale under cover in mid- to late winter for transplanting from mid-spring onwards when soil conditions are suitable.

Harvesting and Storing
Harvest by removing young leaves with a knife when they are 10–12.5cm long. Cut from several plants for an adequate picking.

Harvest regularly for constant young growth and a long cropping season; older shoots become bitter and tough. Harvest early sowings of 'cut and come again' crops when about 5–7.5cm high, or thin to 10cm and harvest when 12.5–15cm high.

Pests and Diseases
Kale is resistant to cabbage root fly and clubroot, and is usually ignored by pigeons. Take precautions against whitefly, cabbage caterpillar and flea beetle. Caterpillars eat irregular holes in the leaves: check regularly for eggs, squashing them. Beware of infestations of cabbage aphids, which check growth and may kill young plants.

companion planting

Kale can be planted with corn and peas.

culinary

Young leaves of later varieties taste better after being frosted.

Wash well and boil in 2.5cm of water for 8 minutes at most. Serve with butter or white sauce – an excellent accompaniment for poached eggs, fried fish, bacon and other fatty meats. Kale is also good with plain fish dishes such as salmon or cod. In parts of North America kale is served with hog jowls, and the juice is eaten with hot corn bread.

Kale can also be used in salads, soups and stews; it can be creamed, or braised with onions, parsley, spices and bacon or ham.

Before freezing, blanch young shoots for 1 minute, then cool, drain and chop. Kale stays fresh for about 3 days in the fridge or in a polythene bag.

Stir-fried Kale
Serves 4

750g young kale
3 tablespoons peanut or olive oil
Salt and freshly ground black pepper
Juice of ½ a lemon

Thoroughly wash and dry the kale in a salad spinner. Chop it roughly. Pour the oil in a wok or heavy frying pan over a high heat and, when the oil is steaming, toss in the kale and cook for 3 minutes, stirring constantly. Season and add the lemon juice, adding shavings of lemon peel for decoration if you wish. Serve piping hot with stews or roasted meats.

Brassica oleracea (Botrytis Group). *Brassicaceae*

CAULIFLOWER

Annual or perennial grown for immature flowerheads. Hardy or half hardy. Value: good source of vitamin C, traces of most other vitamins.

Cauliflowers are believed to have originated in Cyprus and the oldest record dates from the sixth century BC. One thousand years later they were still widely grown there, being known in England as 'Cyprus coleworts'. A Jewish-Italian traveller wrote from Cyprus in 1593 that cabbages and cauliflowers were to be found growing in profusion, and, 'For a quattrino one can get more almost than one can carry.' Gerard in his *Herball* of 1597 calls them 'Cole flowery'. Moorish scholars in twelfth-century Spain described three varieties as introductions from Syria, where cauliflowers had been grown for over a thousand years and were much developed by the Arabs. Even in 1699, John Evelyn suggested that the best seed came from Aleppo (now Halab, in northern Syria). Cultivation methods improved after 1700, and by the end of the eighteenth century the cauliflower was highly regarded throughout Europe. Dr Johnson is said to have remarked, 'Of all the flowers in the garden, I like the cauliflower.' But Mark Twain wrote disdainfully, 'Cauliflower is nothing but cabbage with a college education.' This, of course, is a matter of opinion.

Purple cauliflower – great for dips!

varieties

There are four main groups for spring, summer, autumn and winter, but many overlap the seasons.

'All the Year Round' is sown in late autumn or spring for spring or summer harvest. It produces good-quality white heads and is excellent for successional sowing. **'Alverda'** has yellow-green heads. Sow late spring to mid-summer for autumn cropping. **'Autumn Giant 3'** has beautiful white, firm heads. Excellent for late autumn and winter cropping. **'Cheddar'** (summer – autumn) has orange-yellow flowerheads. **'Clapton'** is clubroot resistant and takes 85 days to harvest. **'Dok Elgon'** has firm, snow-white heads and is a reliable variety for early and late autumn cropping. **'Emeraude'** (late summer – autumn) has bright green curds. **'Graffiti'** (summer – autumn) has deep purple curds. **'Mystique'** (spring) grows large white heads that are firm, crisp and have outstanding flavour. **'Purple Cape'** is a hardy, overwintering purple type for cropping from late winter to mid-spring. Good raw or cooked; the head turns green when boiled. **'Romanesco Veronica'**, also in catalogues under 'Broccoli', has a tender stem like asparagus – delicious! **'Snowball Self Blanching'** is high-yielding and the leaves naturally blanch the curds. **'Veitch's Autumn Giant'** is a huge plant with large leaves and massive heads that grow to 30cm in diameter; very tasty and also stores well. **'White Rock'** produces plenty of leaves to protect the curd and is a very versatile variety.

cultivation

Propagation
Start off crops as described in 'Protected Cropping' unless stated. Plants are ready for transplanting when they have 5–6 leaves; water before moving and retain as much soil as possible around the roots.

Sow successionally to ensure regular cropping all year round. Sow early summer crops in mid-autumn and leave them to overwinter under cover. Harden off in late winter, transplant from mid-spring as soon as the soil is workable and warm.

Space plants 50cm apart with 60cm between the rows or with 50cm between the rows and plants. Protect with crop covers until established.

Sow summer cauliflowers in early spring under cover or in a seedbed outdoors if the soil is warm and workable, for transplanting in mid-spring and harvesting from mid-summer to autumn. Spacing as above.

Sow early autumn cauliflowers in mid-spring for transplanting in early summer and harvesting from late summer to early autumn. Space 52cm apart in and between rows, or 50cm between plants and 60cm between rows.

A promising harvest

Sow autumn cauliflowers in late spring for transplanting in mid-summer and harvesting from mid- to late autumn. Space plants 60cm apart in and between rows.

If you have available space, sow overwintering cauliflowers, but remember that they can be in the ground for almost a year. Sow in seedbeds outdoors and transplant in mid-summer. Winter cauliflowers for harvesting from mid-winter to early spring should be 65cm apart each way; those for cutting from mid-spring to early summer should be spaced 60cm apart in and between the rows. Most varieties need frost protection to avoid damage to the heads.

Growing
Cauliflowers need a sheltered sunny site on deep, moisture-retentive, free-draining soil with a pH of 6.5–7.5. Dig in plenty of well-rotted organic matter in autumn before planting and lime acid soils where necessary. Avoid planting overwintering types in frost pockets. Rake over the area before planting: the ground should be firm, but not compacted. Keep plants well watered from germination to harvest, as checks in growth spoil the quality of the heads. They need at least 22 litres/sq m every 2 weeks.

Cauliflowers also need moderate nitrogen levels, though excessive amounts encourage soft leafy growth. Overwintering cauliflowers require low nitrogen levels, or they will be too soft to survive colder weather.

Keep crops weed-free with regular hoeing or mulching. Bend a few leaves over the heads of summer varieties to protect them from sunshine; do the same with winter crops to protect plants from frost and snow. Leaves can be easily tied in place with garden twine.

If the weather is hot and dry, mist plants every now and then to maintain humidity and cool temperatures. Unwrap heads occasionally to check for hiding pests. After cutting the head, feed with a general liquid fertiliser or scatter a granular general fertiliser at 15g/sq m to encourage the sideshoots to grow. Rotate crops.

Maintenance
Spring Sow early crops under glass and outdoors in seedbeds. Keep well watered and weed-free. Transplant.
Summer Sow indoors or outdoors in seedbeds. Transplant, harvest.
Autumn Sow overwintering crops. Harvest.
Winter Prepare ground for the following year. Harvest.

Foliage protects from sun or frost

Protected Cropping
Sow under a cold frame or under cloches and thin after germination to about 5cm, or sow 2–3 seeds in 'stations' at the required spacing and thin to leave the strongest seedling. Grow on in the seedbed or transplant into small pots or modules. Alternatively, sow directly in small pots or modules and germinate in a propagator or an unheated greenhouse according to the time of year. Pot on when they are large enough to handle. Ventilate on warm days, then harden off and transplant as required.

Harvesting and Storing
Harvest cauliflowers successionally while they are small, rather than waiting until they all mature. If the heads become brown or if florets start to separate, it is almost too late: cut them immediately. Harvest in the morning, except in frosty conditions when you should wait until midday. Cauliflowers can be stored for up to 3 weeks by lifting whole plants, shaking the soil off the roots and hanging upside down in a cool shed; mist the heads occasionally to maintain freshness.

Freeze tight heads only. To prepare, divide into sprigs, blanch for 3 minutes in water with a squeeze of lemon juice. Cool, drain, pack and freeze in polythene bags.

Cauliflowers will store, wrapped, in the salad drawer of a refrigerator for up to 1 week.

Pests and Diseases
Beware of clubroot, cabbage root fly, caterpillars and birds.

companion planting

Plant with rosemary, thyme, sage, onions, garlic, beet and chards.

medicinal

This is another vegetable reputedly good for reducing the risk of cancer, especially of the colon and stomach.

CAULIFLOWER

culinary

Separate into florets, boil or steam and serve with cheese or white sauce and grated nutmeg or flaked almonds. Alternatively, dip in batter, fry and eat as fritters.

Sprigs can be served raw with mayonnaise and other dips, or added to soups, soufflés and pickles.

Cauliflower with Chillies and Black Mustard Seeds

Cauliflower with Chillies and Black Mustard Seeds
Serves 4

In Indian cuisine, cauliflower can be cooked into curries but it is more often served dry, as in this recipe from Southern India, given to me by the chef at Chennai's Chola Hotel. For this dish, buy washed or white dahl.

5 tablespoons vegetable oil
½ teaspoon asafetida
1 teaspoon black mustard seeds
1 teaspoon urad dhal
2 dried hot red chillies, left whole
6 fresh hot green chillies, left whole
750g cauliflower, broken into bite-size florets
Salt
2 tablespoons fresh coconut, grated

Either use a karhari or a heavy frying pan. Heat the oil over high heat and add the asafetida, then the mustard seeds. When the seeds pop, add the dhal; this will turn red, at which point add the chillies and cook until the red ones start to darken. Then stir in the cauliflower and cook for a minute or so. Add a tablespoon of water, season with salt and keep stirring, adding more water as necessary; you will probably need to cook for 4–5 minutes, using 4–5 tablespoons of water. At this point, turn the heat right down and cook, covered, for a further 5 minutes, until all the liquid has evaporated. Take care not to let the cauliflower burn.

Stir in the coconut and serve, discarding the chillies unless you like really hot food. You have been warned!

Cauliflower Soufflé
Serves 4

500g cauliflower florets
1 tablespoon butter
1 tablespoon plain flour
150ml milk
3 large eggs, separated
4 tablespoons grated Cheddar
Pinch of nutmeg
Salt and freshly ground pepper

Steam or boil the cauliflower in salted water until just tender. Drain and dice finely. Keep warm. In a heavy saucepan, make a roux from the butter and flour and cook for 1 minute. Stir in the milk and bring to the boil, stirring constantly, to thicken the sauce. Then remove from the heat and stir in the egg yolks. Beat well and add the cheddar and cauliflower, seasoning well. Whisk the egg whites until stiff, then, using a metal spoon, fold into the cauliflower mixture.

Pour into a buttered soufflé dish and cook on the middle shelf of a preheated oven, 190°C/375°F/gas mark 5, until the top is gloriously browned.

This should take about 30–35 minutes. Serve the soufflé immediately.

Brassica oleracea (Capitata Group). *Brassicaceae.*

CABBAGE

Biennial grown as annual for leaves and hearts. Half hardy/hardy. Value: rich in beta carotene and vitamin C – especially green varieties and outer leaves; outer leaves contain vitamin E.

The Latin *brassica* comes from *bresic*, the Celtic word for cabbage, a plant cultivated for centuries in the eastern Mediterranean and Asia Minor. The Romans believed that cabbages rose from Jupiter's sweat as he laboured to explain two contradicting oracles – esteeming wild and cultivated cabbages as a cure-all as well as recommending them to prevent unseemly drunkenness.

Many varieties have been developed over the centuries. Heat-tolerant types were bred in southern Europe, while many hard-headed varieties were introduced by the Celts and Scandinavians. White cabbages appeared after AD814 and German literature records the cultivation of red cabbages in 1150; in the sixteenth century Estienne and Liébault believed these were made by watering cabbages with red wine or by growing them in hot places. By the thirteenth century 'headed cabbage' was well known, and three kinds of 'Savoy' were mentioned in a German herbal of 1543.

'January King'

varieties

Cabbages are usually grouped according to the season when they are harvested. They range from fairly loose-leaved heads of pointed or conical shape to rounded 'ball' shapes with varying degrees of densely packed leaves.

Spring

Spring cabbages traditionally have pointed heads, but there are now round-headed types. For 'spring greens', use their immature leaves or choose a leafy variety bred for this purpose.

'Duncan' is high yielding so grow for leaves or small solid heads in June. **'Offenham 1 Myatts Offenham Compacta'** is a tasty, very early spring cabbage and has dark green leaves. **'Pixie'** is another very early spring variety that has tight, compact hearts; ideal for small gardens.

Early summer/summer

'Derby Day' is an excellent 'ball-head' cabbage for harvesting from early summer. **'Hispi'** is reliable, with pointed heads of good quality and taste. It matures rapidly and is ideal for close spacing. **'First of June'** is a dark-leaved variety with a compact head, good for successional sowing in summer. **'Ruby Ball'** is an ornamental and colourful red cabbage, very reliable and particularly excellent for use in salads. **'Kalibos'**, also red, is slug-resistant and good for small gardens.

A heart is forming

Late summer/autumn

'Golden Acre', a ball-headed variety, is compact and sweet-tasting and ready in 12 weeks. It tolerates close spacing, is high-yielding and grows well on poorer soils. **'Kilaton'** is delicious, good quality and club root-resistant. **'Kilaxy'** produces compact, tasty heads, stands well and is also club root-resistant. **'Stonehead'** has a tightly packed head and is early-maturing. It is resistant to yellow and black rot.

Winter/winter storage

Winter cabbages are usually ball- or drum-headed. The white-leaved Dutch cabbage (used primarily for coleslaw) matures from mid- to late autumn and can be cut for storage or left to stand in mild conditions. The extremely hardy, tasty and attractive Savoy types with puckered green leaves mature from mid-autumn to late winter.

'January King 3' is an excellent drum-head Savoy type, which is extremely hardy and frost-resistant, maturing in mid-autumn to early winter. **'Multiton'** is a winter-storage cabbage and **'Savoy Express'** is small, sweet and ideal where space is limited.

cultivation

Propagation

Seeds sown in trays, modules or seedbeds will be ready for transplanting about 5 weeks after sowing. Spacings can be modified according to the size of 'head' required. Closer spacing means a smaller head while wider spacing produces slightly larger heads. Sow spring cabbage from mid- to late summer; transplant from early to mid-autumn. Space 25–30cm apart in and between the rows. Protect with cloches or crop covers over winter to encourage earlier cropping. For 'spring greens', grow suitable varieties with 25cm in and between the plants, or space spring cabbage plants 10–15cm apart and harvest when immature.

Sow cabbages for summer and autumn harvest successionally, with 'earlies' and 'lates' for an extended cropping season (see 'Varieties'). You should make the first sowings in a propagator or heated glasshouse at 13–16°C from late winter to early spring, pot on and transplant in mid- to late spring. Follow these with sowings in cold frames or a seedbed under cloches or crop covers. Make further sowings without protection until late spring.

Transplant from early to mid-summer, spacing plants 35–50cm apart depending on the size of head you are after.

Sow winter-maturing varieties successionally from mid- to late spring under cover or outdoors for transplanting from early to mid-summer. Space about 45cm apart.

Growing

Cabbages flourish at around 15–20°C and should not be transplanted at temperatures above 25°C, while some overwintering varieties survive temperatures down to −10°C.

Cabbage plants in a vegetable patch

They flourish in a rich, fertile, moisture-retentive soil with a pH of 5.5–7.0. Dig in plenty of well-rotted organic matter several weeks before planting and lime the soil if necessary. Rake the soil level before planting and ensure that it is firm but not compacted.

Do not fertilise spring cabbages after planting, as this encourages soft growth and the nutrients are washed away by winter rains. Wait until early to mid-spring and scatter a general granular fertiliser round the plants, or liquid feed.

Summer, autumn and winter cabbages need a dressing of fertiliser after transplanting and will benefit from a further granular or liquid feed in the growing season. To increase stability, earth up spring and winter cabbages as they grow.

Provided growing conditions are good and plants healthy, you can produce a second harvest from spring or early summer varieties. After cutting the head, cut a cross shape 13mm deep in the stump, which will sprout a cluster of smaller cabbages.

Keep cabbages moist and weed-free with regular hoeing, hand weeding or mulching. Rotate cabbages with other brassicas.

Maintenance

Spring Sow and transplant summer, autumn and winter cabbage. Harvest.
Summer Sow spring cabbage. Harvest.
Autumn Transplant spring cabbage. Harvest.
Winter Sow summer cabbages. Harvest.

Harvesting and Storing

Spring cabbages are ready to harvest from mid- to late spring. Summer and autumn varieties are ready to harvest from mid-summer to mid-autumn. Winter types can be harvested from late autumn to mid-spring. Spring and summer varieties are eaten immediately after harvest.

Dutch winter white cabbages and some red cabbages can be lifted for storing indoors. Choose those that are healthy and undamaged and dig them up before the first frosts for storage in a cool, slightly humid, frost-free place.

Remove the loose outer leaves and stand the heads on a slatted shelf or a layer of straw on the shed floor. Alternatively, suspend them in nets. They can also be stored in

CABBAGE

a spare cold frame if it is well ventilated to discourage rotting. They should store for up to 5 months.

Freeze only the best quality fresh crisp heads. Wash, shred coarsely, blanch for about 1 minute and pack into polythene bags or rigid plastic containers.

Wrapped in plastic cling film in a refrigerator, cabbages stay fresh for about a week.

Pests and Diseases

Cabbages suffer from the common brassica problems, including cabbage root fly, clubroot, aphids and birds.

companion planting

Cabbages thrive in the company of herbs like dill, mint, rosemary, sage, thyme and chamomile. They also grow well with many other vegetables including onions, garlic, peas, celery, potatoes, broad beans and beets.

Like all brassicas, they benefit from the nitrogen left in the soil after legumes have been grown. The belief that they do not grow well with vines, oregano and cyclamen stems from Classical times. In the sixteenth century, it was well known that 'Vineyards where Coleworts grow, doe yeeld the worser Wines'.

medicinal

Eating cabbage is said to reduce the risk of colonic cancer, stimulate the immune system and kill bacteria. Drinking the juice is alleged to prevent and heal ulcers. Some active principles are partly destroyed on cooking, so cabbage is much more nutritious eaten raw.

According to folklore, placing heated cabbage leaves on the soles of the feet reduces fever; placed on a septic wound, they draw out pus or a splinter.

culinary

Traditionally cabbage is cooked by boiling – preferably as briefly as possible – in a small amount of water, to preserve the nutrients. Add the cabbage to boiling water, which should not stop boiling while you place the younger leaves from the heart on top of the older leaves below. Cover, cook briefly for 3 minutes, then drain. Or steam for about 6–8 minutes.

Stir-frying, the oriental way, is almost as fast; alternatively, bake, braise or stuff. Use as a substitute for vine leaves in dolmades.

Eat shredded white or red cabbage raw in salads. Coleslaw is a mixture of shredded cabbage, carrot, apple and celery with French dressing or a mayonnaise/sour cream blend; its name derives from 'cole', the old name for cabbage, and the Dutch *slaw*, meaning salad.

Pickle red cabbage in vinegar and white cabbage in brine (as *sauerkraut*).

Czerwona Kapusta

Serves 6

The Polish and Czechs are extremely keen on red cabbage. This dish combines subtle flavours to make a refreshing change from our usual ways of cooking the vegetable.

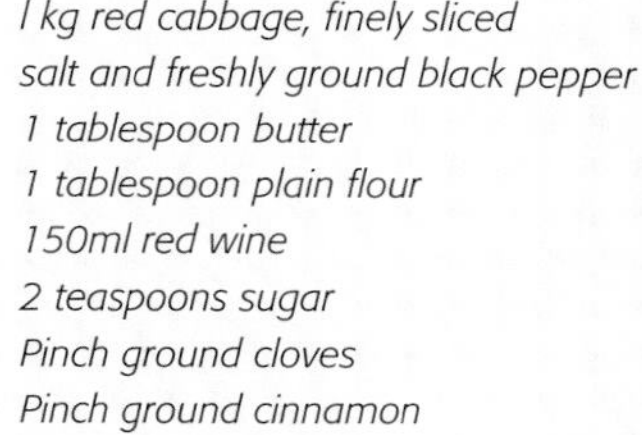

l kg red cabbage, finely sliced
salt and freshly ground black pepper
1 tablespoon butter
1 tablespoon plain flour
150ml red wine
2 teaspoons sugar
Pinch ground cloves
Pinch ground cinnamon

Put the cabbage in a colander and sprinkle with 1 teaspoon salt; leave for 15 minutes then rinse well under cold water. Transfer to a pan of boiling water and simmer gently until the cabbage is just cooked. Drain and keep warm. Reserve a little of the liquid.

Heat the butter in a saucepan over a medium heat and mix in the flour to make a roux. Cook for 2 minutes without burning. Dilute with the cooking liquid to make a thick sauce and stir in the cabbage. Season and add the red wine, sugar, cloves and cinnamon. Mix well and simmer for a further 5 minutes. Serve.

Czerwona Kapusta

Colcannon

Serves 4

Probably the most famous Irish dish, some believe this was traditionally made with kale but today it is commonly made with cabbage. Use a Savoy.

500g potatoes, peeled and cooked
1 leek, cleaned, sliced and cooked in a little cream or milk
500g cabbage, sliced and cooked
4 tablespoons butter
Salt and freshly ground black pepper

Mash the potatoes and season them before stirring in the slices of leek and juices in which they were cooked. Then add the cabbage and mix thoroughly over a low heat. Arrange on a warmed serving dish and make a hole in the centre. Keep warm. Partly melt the butter, season, and pour it into the cavity. Serve immediately, piping hot.

Brassica oleracea (Gemmifera Group). *Brassicaceae*

BRUSSELS SPROUT

Biennial grown as annual for leafy buds and 'tops'. Hardy. Value: excellent source of vitamin C, rich in beta carotene, folic acid, vitamin E and potassium.

First recorded as a spontaneous sport from a cabbage plant found in the Brussels region of Belgium around 1750, this vegetable had reached England and France by 1800. The Brussels version may not have been the first occurrence: a plant described as *brassica capitata polycephalos* (a many-headed brassica with knob-like heads) was illustrated in D'Alechaps's *Historia Generalis Plantarum* in 1587. A stalwart among winter vegetables in cool temperate climates, sprouts are extremely hardy and crop heavily, but are rather fiddly to prepare. As with all vegetables, home-grown ones taste far better than those bought from a shop. If you have never eaten sprouts harvested fresh from the garden, try them: they are absolutely delicious.

varieties

Sprouts are divided into early, mid-season and late varieties, harvested from early to mid-autumn, mid-autumn to mid-winter and mid-winter to early spring respectively. 'Earlies' are shorter and faster-growing than the hardier 'lates', which are taller with higher yields. To extend the season, grow one variety from each group if you have space; alternatively, grow mid-season and late types for mid-winter to early spring crops, when other vegetables are scarce.

Although older open-pollinated varieties are very tasty, it is generally accepted that the modern, compact F1 hybrids are a better buy. They produce a heavy crop of uniform 'buttons' all the way up the stem, which remain in good condition for a long period without 'blowing'; plants are also less likely to fall over.

'Braveheart' is vigorous, high yielding and easy to pick, with a nutty flavour. **'Falstaff'** is a vigorous, high-yielding red cultivar with tasty 'buttons'. The red coloration disappears when boiled so steaming is a good method of cooking. **'Noisette'** is a 'gourmet' sprout with a nutty flavour. **'Oliver'** (very early) is a high-yielding variety with large, tasty sprouts and good resistance to powdery mildew. **'Rampart'** (late) has tasty sprouts that last for a long time before 'blowing' but tend to become bitter late in the season. It has good resistance to powdery mildew and some resistance to ringspot. **'Romulus'** grows tall and vigorous, crops well and is very reliable and tasty. **'Rubine'**, a red form, is worth a place in an ornamental border and produces small crops of good-tasting sprouts.

BRUSSELS SPROUT

cultivation

Sprouts need a sheltered, sunny spot; wind rock can be a problem in exposed sites. Soil should be moisture-retentive yet free-draining, with a pH of 6.5.

Propagation

Sow early varieties from late winter to early spring, mid-season varieties from mid- to late spring and late varieties from mid-spring. Sow seeds thinly, 2cm deep in a half tray of moist seed or multi-purpose compost and put them in an unheated greenhouse, cold frame or sheltered spot outdoors to germinate. Transplant seedlings when they are large enough to handle into a larger seed tray, in potting or multi-purpose compost, about 4–5cm apart.

Sowing in modules reduces root disturbance when transplanting. Put 2 seeds in each module and retain the strongest after germination. Sow early varieties from late winter in a propagator at 10–13°C and transplant them into their permanent position after hardening off. Some taller varieties are prone to falling over when grown in modules, but when planted deeply, a long tap root develops.

The tops can be eaten, too

Brussels sprouts can be harvested in autumn and winter

The previous two methods are preferable to sowing in a seedbed, which takes up space that could be used for other crops, and leaves seedlings vulnerable to pests and diseases; I do not recommend it. If necessary, warm the soil with cloches or black polythene. Protect earlier sowings from cold weather; later sowings can be made without shelter. Water before sowing if the soil is dry. Level, firm and rake the seedbed to a fine tilth before sowing seed thinly, 2cm deep in rows 20cm apart, thinning seedlings to 7.5–10cm apart when they are large enough to handle. Transplant into their final position when they are about 10–15cm tall. Gradually harden off those grown under cover before planting them in their final positions.

Growing

Dig in plenty of well-rotted manure or compost several months before planting, particularly on light, poor or heavy soils. The ground should not be freshly manured, as excessive nitrogen causes sprouts to 'blow'.

Plant earlier, smaller varieties about 60cm apart each way, those of moderate size 75cm apart and taller varieties 90cm apart. Wider spacing encourages larger sprouts, improves air circulation and reduces fungal problems, while closer spacing means smaller, compact 'buttons' which will mature at the same time.

Plant with the lowest leaves just above the soil surface. Tug a leaf – if the whole plant moves it has not been planted firmly enough. On light soils, make a drill 7.5–10cm deep, plant sprouts in the bottom and refill it with soil. The extra support makes the plants more stable. Water immediately after transplanting for 3–4 weeks until plants have become established and, if available, mulch with straw to a depth of 5–7.5cm.

Keep the beds weed-free. Watering is not normally needed once plants are established except during drought, when each plant can be given up to 150ml per day to maintain the growth necessary for good cropping. Remove any diseased or yellowing leaves as they appear. Earthing up round the stem base to a depth of 7.5–12.5cm provides support against winter winds, although tall varieties will usually need staking. In exposed gardens, even dwarf varieties need staking.

'Stopping' by removing the growing point is only beneficial for autumn-maturing F1 cultivars being grown for freezing. Plants can be stopped when lowest sprouts reach 1cm diameter, to encourage even development of sprouts on the stem. When left unstopped, sprouts can be picked over a longer period.

Grow sprouts on a 3- or 4-year rotation, preferably following peas and beans (where they benefit from the nitrogen left in the soil). In late summer, feeding with a liquid high-potash fertiliser gives plants a useful boost.

BRUSSELS SPROUT

Plant small lettuces like 'Little Gem' between sprouts for summer and autumn cropping, and winter purslane or land cress for early winter crops.

Lift plants immediately after harvesting, put leaves on the compost heap and shred the stems.

Maintenance

Spring Sow outdoors in seedbeds, thin and keep weed-free.
Summer Transplant outdoors and water during drought.
Autumn Harvest early varieties; stake tall varieties if needed.
Winter Sow early varieties indoors. Harvest later crops.

Protected Cropping

Earlier sowings made outdoors in a seedbed should be protected with cloches or fleece. Continued protection with horticultural fleece provides a physical barrier against pests such as cabbage root fly, flea beetle, aphids and birds.

Harvest sprouts that are tightly closed

Harvesting and Storing

Pick sprouts when those at the base are walnut-sized and tightly closed. Snap them off with a sharp downward tug or cut with a knife, removing 'blown' sprouts and any yellow, diseased leaves. When harvest is over, the tops can be cooked as cabbage. During severe winter weather, lift a few plants to hang in a shed where they can be easily harvested. They will last for several weeks.

Pests and Diseases

Sprouts are very robust yet subject to the usual brassica problems. Downy mildew shows as yellow patches on the leaves, with patches of fluffy mould on the underside in humid conditions. Remove all affected leaves or dust with sulphur. If downy mildew appears on seedlings, improve ventilation and increase spacing.

Powdery mildew is a white powdery deposit over shoots, stems and leaves. In severe cases, plants become yellow and die. It is more of a problem when plants are dry at the roots. Water and mulch, remove diseased leaves and spray with fungicide.

Clubroot, affecting members of the *Brassicaceae* family, is a disease to be avoided at all costs. Roots swell and distort, young plants wilt on hot days but recover overnight, growth is stunted and crops ruined. It is more of a problem on poorly drained, acid soils. Spores remain in the soil for up to 20 years. Never buy brassicas from unknown sources: grow them yourself. Potting plants on into 10–15cm pots allows the roots to become established before they are planted out, which lessens the effects of clubroot. Improve drainage; lime acid soils to create a neutral pH. Earthing up often encourages new roots to form and reduces the effects. Remove all diseased plants, with the whole root system if possible, and destroy them.

Ringspot is worse in cool wet seasons and on well-manured land. It is most evident on older leaves as round brown spots with dark centres. Remove and burn any affected plants and rotate crops.

companion planting

When planted among maturing onions, sprouts benefit from their root residues and the firm soil.

culinary

Steam or boil sprouts briskly for the minimum time required to cook through – they should not turn mushy. Small sprouts can be shredded in salads – 'Rubine' and 'Falstaff' are particularly attractive.

They can be stored for up to 3 days in a polythene bag in the refrigerator. Freeze sprouts only if they are small. Blanch for 3 minutes, cool and drain before drying and packing them into polythene bags.

Stir-fried Sprouts

Serves 4

This can look quite spectacular made with a red variety such as 'Falstaff' or 'Rubine'.

2 tablespoons vegetable oil
2 tablespoons soy sauce
500g Brussels sprouts, trimmed and finely sliced
2 tablespoons hazelnuts, roughly ground
Salt and freshly ground black pepper

Heat the oil in a wok, stir in the soy sauce and, over a high heat, cook the sprouts for 2–3 minutes. Sprinkle over the hazelnuts. Season and serve.

Brassica oleracea (Gongylodes Group). *Brassicaceae*

KOHLRABI

Biennial grown as annual for rounded, swollen roots. Hardy. Value: rich in vitamin C, traces of minerals.

This odd-looking vegetable with a distinctive name has a rounded, swollen stem which, with the leaves removed, looks like a sputnik! Its common name, derived from the German Kohl, meaning cabbage, and rabi, turnip, accurately describes its taste when boiled. Raw, it has a fresh, nutty flavour. Found in northern Europe in the fifteenth century, it may already have existed for centuries, as a similar-sounding vegetable was described by Pliny around AD70. This highly nutritious, tasty vegetable is more drought-resistant than most brassicas, succeeding where swedes and turnips fail. It deserves to be more widely grown and eaten.

varieties

'Blusta' is fast-maturing, has a sweet, nutty flavour and is resistant to bolting. **'Domino'**, an early variety, is good for growing under cover. **'Kongo'** is top quality; fast-growing, very sweet and juicy and high yields. **'Korist'** produces a good-sized crop of mildly flavoured, tender, juicy bulbs. **'Lanro'** is a white-fleshed, green-skinned variety. Juicy and mild-flavoured, it is a good mini vegetable. **'Olivia'** produces good-sized bulbs and is crisp and mild-flavoured. **'Trero'** is sweet, uniform, vigorous and slow to become 'woody'. **'White Vienna'** has a pale green skin and is delicately flavoured.

cultivation

Propagation

As a general rule, green varieties are sown from mid-spring to mid-summer for summer crops and the hardier purple-skinned types from mid-summer to mid-autumn for winter use. Sow successionally for a regular harvest. Early sowings in seed or multi-purpose compost in a propagator at 10–15°C can be made during mid-winter to early spring. Transplant after hardening off in mid-spring when they are no more than 5cm high; if you let them grow taller or sow seed when soil temperatures are below 10°C, they are liable to run to seed. Protect with cloches or horticultural fleece until the plants are established.

Pick when small for the finest flavour

'Purple Vienna'

Water the drills before sowing and sow later crops thinly in drills 1cm deep in rows 30cm apart. Thin seedlings when they are about 2.5cm high and the first true leaves appear, to a final spacing 15–20cm apart. Prompt thinning is vital, as growth is easily checked. Alternatively, plant 3 seeds together in 'stations' 15cm apart and thin to leave the strongest seedling. Kohlrabi can also be grown successfully in modules and then transplanted at their final spacing.

Kohlrabi grown as a mini vegetable is ideal for the small garden. Sow cultivars like **'Rolano'**, **'Logo'**, **'Korist'** and **'Kolibri'** and thin to about 2.5cm apart between the plants and rows. Harvest after 9–10 weeks when about the size of golf balls.

Growing
The ideal situation is a sunny position on light, fertile, humus-rich, well-drained soil. Incorporate organic matter the winter before planting if necessary and lime acid soils to create a pH of 6–7. The ground must be firm before planting, as (in common with other brassicas), kohlrabi does not grow well on loose soil. Lightly fork the area, removing any debris, then gently tread down the surface or firm it with the head of a rake. Finally, rake in a general fertiliser at 90g/sq m and level. Kohlrabi must receive a constant supply of water throughout the season. This is because if growth is checked they can become 'woody'. During drought periods, they need up to 8 litres/sq m of water per week. If growth slows down, liquid feed with a high-nitrogen fertiliser. Keep crops weed-free and mulch with compost to suppress weeds and retain moisture.

Rotate kohlrabi with brassicas.

Maintenance
Spring Prepare the seedbed and sow seed *in situ* under cloches.
Summer Sow regularly for successional cropping. Sow hardier purple varieties later in the season. Weed and water as necessary.
Autumn Sow in mid-autumn and protect with cloches for early winter harvest.
Winter From late winter, sow early crops in modules or trays. Prepare the ground for outdoor sowings.

Protected Cropping
Early and late outdoor crops should be protected with cloches or crop covers.

Harvesting and Storing
Kohlrabi matures rapidly and is ready for harvest 2 months after sowing. Lift when plants are somewhere in size between a golf and a tennis ball. Larger 'bulbs' tend to become woody and unpalatable, but this is less of a problem with newer cultivars.

Harvest as required. In severe weather they can be stored in boxes of sand or sawdust. Remove the outer leaves, keeping the central tuft of leaves to keep them fresh. Some flavour tends to be lost during storage.

Pests and Diseases
As it matures quickly, kohlrabi is untroubled by many of the usual brassica problems, including clubroot.

Birds can cause severe damage, particularly to young plants. You should protect crops with netting, cages, humming wire or with bird scarers.

Flea beetles – 3mm long and black with yellow stripes – nibble holes in leaves of seedlings, checking growth. Dust plants and the surrounding soil thoroughly with derris or insecticide when symptoms appear. They can also be controlled by brushing a yellow sticky trap or piece of wood covered in glue along the tops of the plants: the insects jump out, stick to the glue and can be disposed of.

Cabbage root fly larvae cause stunted growth, wilting and death. Protect the seedlings when transplanting with 12.5cm squares of plastic, cardboard or rubberized carpet underlay, slit from edge to centre and fitted around the stems. This stops adults from laying eggs.

companion planting

Kohlrabi grows well with beet and onions.

container growing

Kohlrabi are ideal for containers, particularly when grown as 'mini vegetables'. Plant in a loam-based compost with a moderate fertiliser content and maintain a regular supply of water. Feed every 2–3 weeks with a general liquid fertiliser.

culinary

There is no need to peel tiny kohlrabi, but peel off the tough outer skin of older globes before cooking. Young ones can be trimmed, scrubbed and boiled whole or sliced for 20–30 minutes, then drained, peeled and served with melted butter, white sauce, or mashed.

Boiled kohlrabi can be made into fritters by frying with egg and breadcrumbs. Add kohlrabi to soups and stews, serve stuffed, cook like celeriac, or eat in a cheese sauce. It complements basil and is excellent steamed.

Kohlrabi globes can also be eaten raw, grated or sliced into salads, and the leaves are good boiled.

Brassica oleracea (Italica Group). *Brassicaceae*

BROCCOLI

Also known as Sprouting Broccoli, Calabrese. Perennial or annual grown for immature flowerheads. Hardy or half hardy. Value: high in beta carotene, vitamin C, folic acid and iron. Moderate levels of calcium.

Said to have originated in the eastern Mediterranean, early forms of broccoli were highly esteemed by the Romans and described by Pliny in the first century AD. It spread from Italy to northern Europe, arriving in England in the eighteenth century. Philip Miller in his *Gardener's Dictionary* of 1724 called it 'sprout cauliflower' or 'Italian asparagus'. Broccoli is an Italian word, derived from the Latin *brachium*, meaning 'arm' or 'branch'. Calabrese, a similar plant with the same botanical origin, grown for its larger immature flowerheads, also takes its name from the Italian – meaning 'from Calabria'. This delicious vegetable was introduced to France by Catherine de Medici in 1560, spreading from there to the rest of Europe. 'Green broccoli' was first mentioned in North American literature in 1806, but was certainly in cultivation long before that. It is said to have been introduced by Italian settlers and is extensively grown around New York and Boston.

'Purple Sprouting Early'

varieties

Old varieties of perennial broccoli are still available; outstanding among them all is **'Nine Star Perennial'**, a multi-headed variety with small white heads. Cropping improves if unused heads are removed before they go to seed.

Sprouting broccoli

This excellent winter vegetable produces a succession of small flowerheads for cropping over a long season from early winter to late spring and is suitable for poor soils and cold areas. 'Purple' varieties are hardier than the 'white', which have a better taste and crop later, but tend to be less productive. Sow **'Summer Purple'** in early spring, **'Rudolph'** from early to late spring and **'Cardinal'** in late spring for a harvest lasting almost twelve months.

'Bordeaux', if sown early, can produce an early flowering harvest from July. **'Purple Sprouting Early'** is easy, prolific and extremely hardy and can be eady for harvesting from late winter. **'Purple Sprouting Late'** is similar, but ready for picking from mid-spring. **'White Sprouting'** is delicious, with shoots like tiny cauliflowers and **'White Sprouting Early'** is the white equivalent of **'Purple Sprouting'**. **'White Sprouting Late'** is ready to harvest from mid-spring.

Calabrese

Also known as American, Italian or green sprouting broccoli, this produces a large central flowerhead surrounded by smaller sideshoots, which develop after the main head has been harvested. Maturing about

3 months after sowing, it crops from summer until the onset of the first frosts.

'Broccoletto' is quick-maturing and sweet, with a single head. **'Fiesta'** is excellent for summer and autumn cropping with large, domed heads of small buds. **'Flash'** produces small to medium buds over a long season. **'Green Sprouting'**, an old Italian variety, matures early. **'Paragon Enhanced'** has hybrid, umbrella like spears and tender, slim, sweet-tasting stems. **'Ramoso'** (**'DeCicco'**), an old Italian variety for spring or autumn cropping, produces heads over a long period. Tasty, tender and freezes well. **'Trixie'** is high yielding with domed heads of small buds.

cultivation

Propagation

Prepare the seedbed for sprouting broccoli by raking the soil to a fine texture. Sow over several weeks from mid- to late spring, planting the earlier varieties first. Sow thinly in drills in a seedbed, 30cm apart and 1cm deep, thinning to 15cm apart before transplanting at their final spacing. Alternatively, sow 2–3 seeds in 'stations' 15cm apart, thinning to leave the strongest seedling.

For early spring crops, sow early maturing cultivars indoors in trays from late summer to early autumn. Transplant seedlings when they are about 13cm tall into an unheated greenhouse or cold frame. Harden off and transplant outdoors from late winter to mid-spring. Alternatively, sow 2 seeds per module and thin to leave the stronger seedling. The final spacing for plants should be about 68–75cm apart in and between the rows. It is worth noting that they take up a lot of space and have a long growing season!

Ready for harvesting

Calabrese can be sown successionally from mid-spring to mid-summer for cropping from early summer to autumn. It does not transplant well and is better sown *in situ*. Sow 2–3 seeds at 'stations', thinning to leave the strongest seedling. Close spacing suppresses side shoots and encourages small terminal spears to form, which are useful for freezing. Wider spacing means higher yields. While they can be as close as 7.5cm apart with 60cm between rows, the optimum spacing is 15cm apart with 30cm between the rows.

Growing

Sprouting broccoli thrives in a warm, sunny position. Soil should be deep, moisture-retentive and free-draining. Nitrogen levels should be moderate; excessive amounts encourage soft, leafy growth. You should avoid shallow or sandy soils and windy sites. Sprouting broccoli tends to be top-heavy, so earth up round the stem to a depth of 7.5–13cm to prevent wind rock, or stake larger varieties. Firm stems loosened by wind or frost.

Keep crops weed-free with regular hoeing or mulch with a 5cm layer of organic matter. Water crops regularly before the onset of dry weather: do not let them dry out. Calabrese needs at least 22 litres/sq m every 2 weeks, though a single thorough watering 2–3 weeks before harvesting is a useful option for those under watering restrictions!

Feed with a general liquid fertiliser, or scatter and water in 15g/sq m of granular fertiliser after the main head has been removed to encourage side shoots to grow.

Maintenance

Spring Sow early broccoli indoors and later crops in seedbeds. Sow calabrese.
Summer Water crops and keep weed-free. Feed calabrese after harvesting the terminal bud.
Autumn Harvest calabrese, remove and dispose of crop debris.
Winter Harvest broccoli.

Steam florets to keep the colour and goodness

Harvesting and Storing

Cut sprouting broccoli when the heads have formed, well before the flowers open, when the stems are 15–20cm long. Regular harvesting is essential, as this encourages side-shoot formation and should ensure a 6–8-week harvest. Never strip the plant completely, or let it flower, as this stops the side shoots forming and makes existing 'spears' woody and tasteless.

Sprouting broccoli can be stored in a polythene bag in the refrigerator. It will keep for about 3 days. To freeze, soak in salted water for 15 minutes, rinse and dry. Blanch for 3–4 minutes, cool and drain. Pack it in containers and then freeze.

Cut the mature central heads of calabrese with a sharp knife, while still firm and the buds tight. This encourages growth of side shoots within about 2–3 weeks. Pick regularly as for sprouting broccoli.

Calabrese can be stored in the refrigerator for up to 5 days, and freezes well.

companion planting

Plant with rosemary, thyme, sage, onions, garlic, beet and chards.

BROCCOLI

culinary

Remove any tough leaves attached to the stalks and wash florets carefully in cold water before cooking. For the best flavour cook immediately after picking in boiling salted water for 10 minutes, or steam by standing spears upright in 5cm of gently boiling water for 15 minutes with the pan covered. Drain carefully and serve hot with white, hollandaise or béarnaise sauce, melted butter, or vinaigrette. An Italian recipe book recommends braising calabrese in white wine or sautéing it in oil and sprinkling with grated Parmesan cheese. In Sicily, it is braised with anchovies, olives and red wine. Broccoli fritters are dipped in batter, then deep-fried.

Stir-fried florets can be blanched for 1 minute and fried with squid and shellfish. On a more mundane (but practical) level, broccoli and calabrese can be used as a fine substitute in cauliflower cheese.

Penne with Broccoli, Mascarpone and Dolcelatte
Serves 4

500g broccoli florets
150g mascarpone cheese
200g dolcelatte cheese
2 tablespoons crème fraîche
1 tablespoon balsamic vinegar
1 tablespoon dry white wine
450g penne
2 tablespoons capers
4 tablespoons black olives
1 tablespoon hazelnuts, crushed
Salt and freshly ground black pepper

Steam the broccoli florets over a pan of boiling water for 2–3 minutes. Run under cold water and set aside. In a heavy pan, gently heat the mascarpone, dolcelatte, crème fraîche, vinegar and wine.

Add the broccoli florets. Cook the pasta until it is just tender and drain well. Pour over the hot sauce, sprinkle with the capers, olives and hazelnuts and toss well. Adjust the seasoning and serve.

Brassica rapa (Rapifera Group). *Brassicaceae*

TURNIP

Biennial grown as an annual for globular, swollen root and young leaves. Hardy. Value: low in calories and carbohydrate; small amounts of vitamins and minerals.

This ancient root crop was known to Theophrastus in 400 BC and many early varieties were given Greek place names. Pliny listed 12 distinct types under *rapa* and *napus* – which became *naep* in Anglo-Saxon, and together with the word 'turn' (meaning 'made round'), gave us the common name. Introduced to Canada in 1541, the turnip was brought to Virginia by the colonists in 1609 and was rapidly adopted by the native Americans. In Britain they have found a role in folklore. In Northern Ireland, turnips were made into lamps for Hallowe'en (31 October), and in the Shetland Islands of Scotland slices were shaped into letters and put into a tub of water for young revellers to retrieve with their mouths; they usually tried to pick the initial of someone they loved.

varieties

Earlies

'Oasis' has virus-resistant white roots and is delicious when raw. **'Purple Top Milan'** produces flattish roots with purple markings and white flesh. Tender when young, early maturing and good for overwintering, it has an excellent flavour. **'Primera'** is succulent and full of flavour. **'Snowball'** is a delicately flavoured, fast-maturing white variety with cut leaves. **'Tokyo Cross'** is an excellent F1 hybrid that matures rapidly, in about 35–40 days, and produces small, tasty white globes. A 'mini vegetable' that is also tasty when larger, it is suitable as a late summer to early autumn crop.

Maincrops

'Golden Ball' (**'Golden Perfection'** or **'Orange Jelly'**) is a small, round yellow variety that should be grown quickly to keep the flesh succulent. Tasty, hardy, excellent for storing. **'Green Globe'** is white-fleshed with round roots; excellent for turnip tops. **'Shogoin'** (Japanese for 'greens' or 'root') is crispy and great for cooking and pickling.

cultivation

Turnips flourish at about 20°C and prefer a sheltered, open site in light, fertile, well-drained but moisture-retentive soil. Summer sowings can be made in moderate shade provided they receive sufficient moisture. Turnips prefer a pH of 5.5–7.0; very acid soil will need liming.

Propagation

Sow early turnips in mid-spring, as soon as the ground is workable, or in late winter or early spring under cloches or fleece.

Prepare the seedbed carefully and sow early turnips thinly in drills about 2–2.5cm deep, in rows 23cm apart. Then thin to a final spacing of 10–12.5cm. They can also be grown in a grid pattern: mark 12.5cm squares in the ground with a cane and sow 3 seeds where the 'stations' cross, thinning after germination to leave the strongest seedling.

Sow maincrop varieties from mid- to late summer in drills 2cm deep and 30cm apart, thinning to 15–23cm apart. It is important to thin seedlings when no more than 2.5cm high, when the first true leaves appear, to ensure the roots develop properly. Firm the soil after thinning; do not thin turnips grown for tops.

When growing for their tops, prepare the seedbed and broadcast seed over a small area or sow thinly in rows 10–15cm apart as soon as soil conditions allow. Sow early cultivars in spring for summer cropping and hardy varieties in late summer or early autumn. Small seedlings of 'maincrops' overwinter and grow rapidly in spring, making them a useful early crop, particularly when covered with cloches or fleece. To ensure a prolonged harvest, make successional sowings of 'earlies' from early spring until early summer.

'Purple Top Milan'

Growing

If the ground has not been manured for the previous crop, double dig in autumn, working plenty of well-rotted organic matter into the soil and allowing it to weather over winter. About a week to 10 days before sowing, remove any weeds or debris and rake general fertiliser into the soil at 60g/sq m. In common with

TURNIP

other brassicas, turnips grow poorly on loose soil, so rake the soil to a fine tilth and firm it with the head of a rake or by carefully treading. If the soil is dry, water thoroughly before sowing and stand on a planting board to avoid compacting the soil. Mark each row with canes or twigs, and label and date the crop.

Keep crops weed-free by hand weeding or careful hoeing. Turnips must have a constant supply of water throughout the growing season, otherwise they tend to run to seed or produce small woody roots, while sudden watering or rain after a period of drought causes them to split. They will need up to 9 litres/sq m per week during dry periods. This improves the size and quality of the crop, but usually reduces the flavour. Turnips should be rotated with other brassicas.

Maintenance

Spring Prepare seedbed, sow early varieties under cloches.

Summer Sow earlies every 2–3 weeks for successional cropping. Keep crops weed-free and water as necessary. From mid- to late summer, sow maincrop varieties.

Autumn Sow 'earlies' under cloches mid- to late autumn. Thin maincrop varieties.

Winter Harvest and store maincrop turnips.

Protected Cropping

Cover ground with cloches, fleece or black polythene for 2–3 weeks in late winter to early spring to warm the ground before sowing early crops. In late summer, protect sowings of early cultivars. Maincrop turnips grown for their tops can be grown under cloches after sowing in autumn and picked during winter.

'Shogoin'

Harvesting and Storing

Harvest early varieties when young and tender. Gather those to be eaten raw when they are the size of a golf ball; any time up to tennis ball-size if they are to be cooked. Hand pull them in the same way as radishes. Maincrop turnips that are lifted in mid-autumn for winter use are much larger, hardier and slower to mature.
To keep the flavour, harvest at maturity as they soon become woody and unpalatable. Turnips can be left in the soil and lifted as required using a garden fork. Keep them in a cool place and use within a few days. In cold wet climates, roots are better lifted to prevent deterioration. Twist off the leaves, remove any soil, put the roots between layers of dry peat substitute, sawdust or sand in a box, then store in a cool shed.

Turnips grown for their tops can be harvested when about 10–15cm high, cutting about 2.5cm above ground level. Keep soil moist and they will resprout several times before finally running to seed.

Pests and Diseases

Flea beetle, 3mm long and black with yellow stripes, nibble holes in leaves of seedlings, checking growth. Large infestations of mealy aphid may kill young plants or cause black 'sooty mould' on leaves. Cabbage root fly larvae feed on roots; transplanted brassicas are particularly vulnerable. Powdery mildew, a white deposit over shoots, stems and leaves, causes stunted growth. In severe cases, leaves yellow and die.

companion planting

Growing with peas and hairy tares deters aphids. Turnips are useful for intercropping between taller crops and for catch-cropping.

container growing

Fast-maturing early varieties grow well in large containers of well-drained, soil-based compost with added organic matter. Water crops well.

medicinal

The liquor from turnips sprinkled with demerara sugar was used in folk medicine to cure colds.

culinary

Both turnips and their leaves (turnip tops) are tasty. Eat early turnips raw in salads or boiled, tossed in butter and chopped parsley. Peel maincrop turnips before cooking. They are good mashed, roasted and in casseroles and soups.

Glazed Turnips

Use small, young turnips. Scrub them and cut into 1cm dice (or use whole). Drop into boiling water for 3 minutes. Drain. Melt a little butter and olive oil in a frying pan, add the turnips, sprinkled with a little sugar, and fry over a high heat, stirring constantly, until browned and caramelised. This is particularly delicious with the 'Snowball' variety.

Turnip Tops

Wash well, removing stringy stalks. Chop roughly into manageable pieces. Steam over boiling water until just tender. Serve warm, tossed in olive oil and lemon juice vinaigrette, sprinkled with a finely chopped garlic clove. Or cook like spinach: put the washed leaves in a pan, add salt, pepper and a small knob of butter and steam for 10 minutes in only the water remaining on the leaves. Drain thoroughly and serve immediately.

Brassica rapa var. *chinensis. Brassicaceae*

PAK CHOI

Also known as Bok Choy. Hardy biennial, grown as an annual. Value: rich in carotenes, calcium, fibre, potassium and folic acid.

***Pak choi*, translated as 'white vegetable' from Cantonese, has been grown in China since the fifth century AD, cultivated in Europe since the eighteenth century and is one of the most familiar oriental vegetables. Relatively few varieties are grown in the west; there are twenty varieties in Hong Kong alone and many more on mainland China and Taiwan. In 1751 a friend of the Swedish Taxonomist Carl Linneaus brought seeds to Europe, at the same time as Jesuit missionaries handed similar races to German scientists working in Russia, where it also spread rapidly.**

varieties

Joy Larkcom, in her definitive work, *Oriental Vegetables*, a 'must-have' book for anyone interested in the ingredients and cultivation of vegetables for eastern cuisine, identifies four types based on their appearance.

Chinese White types
These have thick, light to deep green leaves, usually curving outwards, with wide, white, short and straight leaf blades, often overlapping at the base. They vary in their cold tolerance and tendency to bolt. **'Joi Choi'** is slow to bolt and has good frost resistance.

Soup Spoon types
These are vigorous and versatile, tolerating warm or cool conditions and have narrow 'waists' with thin leaves and leaf stalks; the leaves are shallowly concave, like soup spoons. Most are tall, reaching around 45cm when mature, and have the best flavour. **'Japanese White Celery Mustard'** is particularly delicious.

Green Leaf Stalk types
These varieties produce broad, light green leaf stalks on compact, robust, fast-growing plants.

Canton types
The smallest pak choi come from this group, with dense, white leaf stalks and compact, deep green leaves; they can be harvested as young greens and have a fine flavour. They are more warm-weather tolerant than most and inclined to run to seed in cold weather. **'Canton Dwarf'** is compact with dark green leaves.

cultivation

Propagation
Sow seeds about 1.5cm deep and spaced according to the recommendations on the seed packet, which is dictated by the size of the plant required at harvest. Sow small varieties up to 5cm apart with 18cm between rows; large plants should be up to 45cm apart.

Growing
Most are cool season crops for late summer and autumn, growing best between 15–20°C. Spring-sown seedlings are best grown as 'cut and come again' or for harvesting as young plants. Alternatively, choose cold-tolerant varieties like **'Tai Sai'**. Protect early sowings with crop covers.

Most varieties only tolerate light frost though green-stalk varieties are much tougher. They thrive in rich, fertile, well-drained soil and bolt if the soil dries out though 'Canton' and green-stemmed varieties are the most resistant to drought; keep the soil constantly moist and grow in shade during summer.

Maintenance
Spring Sow seedling crops under cover; harvest early sowings.
Summer Sow seeds in shade and harvest earlier crops.
Autumn Sow seedling crops under cover.
Winter Sow under glass.

Protected Cropping
Grow early- and late-season crops under glass. In winter, seed can be sown in modules at 20°C and temperatures lowered by 50 per cent after germination.

Pak choi

Harvesting and Storing
Pick when the leaves are fresh and crisp, either by removing a few leaves from plants as required or cutting them off at the base, so they can resprout.

Pests and Diseases
They are susceptible to all the standard Brassica problems. Flea beetle is a major problem; grow crops under horticultural fleece or mesh to avoid cabbage root fly, slugs, cutworm, and powdery and downy mildew.

medicinal

Dried leaves have been used to alleviate the effects of dysentery.

culinary

Pak choi has a mild flavour in the way that Swiss chard does. Wash well with several changes of water before use. Pak choi leaves, from 'cut and come again' to mature leaves, stems and flower shoots are used in stir-fries, soups, salads and are an ingredient in many forms of oriental cooking including pickles. Young flower heads from 'bolting' plants can be used as a 'choy sum' substitute.

Tempting, tender and tasty

Brassica rapa var. *japonica*. *Brassicaceae*

MIZUNA & MIBUNA GREENS

Hardy annual. Value: rich in vitamin A and C, folic acid and anti-oxidants.

Mizuna greens are a Chinese vegetable but have long been cultivated in Japan. They are fast growing, productive over a long period of time and can tolerate several degrees of frost without damage; making them an excellent all-round crop for cool conditions. Both 'Mizuna' and 'Mibuna' greens have the characteristics of the parent plant, thought to be a natural hybrid which occurred in times past. There is chromosome evidence from *brassica rapa*, the 'turnip' which has several leafy forms, particularly in Italian cooking where they are known as *cima de rapa*, and *Brassica juncea* or 'Chinese mustard' that is grown as an oil seed in China and Japan.

Mizuna greens add texture and taste to salads

varieties

'Mizuna' greens form a vigorous clump of finely-cut, dark green leaves with white, juicy stems. **'Mibuna'** greens have long been cultivated in Japan; they have slender stems and long strap-like leaves of varying length, which are dark or light green. They are normally referred to as 'early' or 'late' according to the timing of their sowing and harvest.

cultivation

Propagation

Sow outdoors from late spring and late summer to early autumn or under cover, either direct sown or sown in seed trays or a seedbed; transplant 2–3 weeks later. For small plants sow 4cm apart, for medium-sized 22.5cm and large up to 45cm apart.

Growing

Both are tolerant of cool, wet conditions and dislike extreme heat; drought may stunt plants and make the leaves tough. Mibuna and mizuna greens are cool weather plants and ideal for early- and late-season sowings; both need an open, sunny site in moisture-retentive soil with shade if sown in summer. Improve the soil by adding well-rotted organic matter if necessary. Keep plants weed free and watered when necessary to ensure a constant supply of young leaves.

Maintenance

Spring Sow seedlings under cover when the soil warms; sow in trays for transplanting into cloche- or fleece-warmed soil.

Summer Sow in shade, harvest regularly and keep crops well watered.

Autumn Sow under cover for 'cut and come again' seedling crops.

Winter Harvest. Ventilate the greenhouse on hot days.

Protected Cropping

Mizuna and mibuna greens flourish in unheated greenhouses, cloches or cold frames over winter and tolerate low, winter light levels.

Crab salad served on mizuna leaves

culinary

Use Mizuna and mibuna greens in stir-fries and salads. Mizuna can be used in soups, steamed, with fish or poultry, seeds can be dry roasted for use in curries and pickles, sprouted seeds can be used in salads. Use fresh to retain the highest level of nutrients. Wash in several changes of water and dry thoroughly before use. Leaves can be stored for up to five days in the salad compartment of a fridge.

Harvesting and Storing

Mibuna and mizuna greens can be grown as 'cut and come again' crops, juvenile or mature plants. Harvest regularly for a constant supply of tender leaves, up to five cuts can be made per plant, according to the level of maturity and time of year. Older leaves gradually become more fibrous.

Pests and Diseases

Flea beetles and slugs can be a problem, grow under horticultural fleece or mesh.

container growing

They are ideal for growing in containers and growing bags, provided there is adequate space to form a good root system.

Brassica rapa **var.** *komatsuna. Brassicaceae*

KOMATSUNA

Also known as Japanese Mustard Spinach, Turnip Tops. Hardy annual or biennial. Value: rich in calcium, vitamin A, C, folic acid and anti-oxidants.

This oriental vegetable of Japanese origin is another leafy form of the wild turnip; with large, wide, dark green, often glossy leaves it is believed to have been developed from pak choi. Valued for its versatility, it is grown as a food crop in Taiwan and Korea, and as a fodder crop in several Asian countries. It is tough, vigorous, fast growing, impervious to the effects of all but the worst weather and is an ideal winter crop in cool climates.

varieties

'Green Boy' is often listed as a form of pak choi. It is heat tolerant with dark green leaves and thick leaf stems. **'Hybrid Kojisan'** is top quality with bright green leaves. **'Komatsuna'** is an all-season variety, with fleshy, rounded, green stems, and dark green leaves; disease and heat resistant, best in early spring or late summer. Originally found growing near Tokyo. **'Nozawana'** is highly cold- and heat-tolerant with long green leaves. It is popular in Japan for pickling. The variety **'Summer Fest'** is similar. **'Natsu Rakuten – Summer Fest'** hybrid, has thick, dark green leaves and an upright habit. Sow spring to autumn. **'Red Komatsuna'** is ideal as micro- or baby greens; the flat, round leaves are tinted red above and green with red veins below. It is heat-tolerant. Sow spring to autumn. **'Tokyo Early'** has glossy leaves and thin, light green stalks and is slow to bolt. Sow from spring to autumn. **'Toriksan'** is heat-tolerant.

cultivation

Propagation

Sow seeds in wide drills or broadcast, thinning seedlings to leave 2.5cm between 'cut and come again' plants. The spacing of crops growing to maturity depends on the variety; large plants should be up to 45cm apart in and between the rows, those being harvested at other stages of growth can be spaced accordingly and 'thinnings' can be eaten at any stage.

Growing

Komatsuna greens are easy to grow, very prolific and almost the perfect winter-hardy crop, recovering from temperatures down to −14°C. This characteristic ensures that they are less likely to 'bolt' in cold spring weather than other 'oriental' vegetables. They need an open, sunny site and rich, fertile, moisture-retentive soil; incorporate well-rotted organic matter if necessary before planting. Sow 6–12mm deep and keep plants well watered at all times.

'Cut and come again' seedlings can be sown in spring and autumn under cover and in late spring and summer outdoors. Crops being grown to maturity can be sown in late summer and transplanted under cover in autumn or mid-summer, if being grown outdoors, so they are well established before the onset of winter.

Maintenance

Spring Make early sowings under unheated protection; harvest mature crops.
Summer Sow in modules or trays in late summer for transplanting under protection.
Autumn Make late sowings under unheated protection.
Winter Harvest crops.

Protected Cropping

Although very hardy, the quality of the leaves is improved when crops are grown under protection. Protect early and late sowings under cloches, in cold frames or in a cold greenhouse.

Harvesting and Storing

Harvest seedlings from 10cm; cut juvenile plants back to 2cm above ground level and they will regrow. Water plants well to boost regrowth and feed with dilute liquid fertiliser. Cut leaves from mature plants as needed.

Pests and Diseases

Plants are susceptible to the usual Brassica problems like slugs, flea beetle and pigeons. Grow under fleece and protect mature plants with netting.

container growing

Growing bags or containers of organic-based compost can be used for growing 'cut and come again' or semi-mature crops.

culinary

Komatsuna greens have a mellow flavour between spinach and Asian mustard. They can be eaten from 'cut and come again' to maturity; flowering shoots can be eaten too and are sweet and juicy. Steam or lightly boil younger leaves as spinach; older leaves taste stronger and are more like cabbage – they become stronger and hotter as they mature, so are better cooked. Eat young leaves raw in salad, stir-fry, boil, pickle, add to soups or braise.

Broth with Komatsuna and Tofu

Serves 2

2 sheets deep-fried tofu
170g komatsuna, cleaned and trimmed
Pinch of salt

For the Broth:
450ml dashi stock
1 tablespoon sake
1 tablespoon mirin
2 tablespoons light soy sauce
Pinch of salt

Put the tofu in a colander and pour boiling water over it to remove the excess oil. Cut lengthways in half and slice thinly.

Discard any roots of the komatsuna and wash it well. Bring a large pan of water to the boil and add a pinch of salt. Blanch the komatsuna and transfer immediately to a bowl of cold water. Squeeze the komatsuna dry and chop it into 2.5cm lengths.

Put all the ingredients for the simmering broth into a large pan and bring to the boil. Add the tofu and simmer over a medium heat for 2 minutes before adding the chopped komatsuna. Simmer for another 2–3 minutes and serve.

Brassica rapa (Pekinensis Group). *Brassicaceae*

CHINESE CABBAGE

Also known as Chinese Leaves, Celery Cabbage, Pe Tsai, Peking Cabbage. Annual or biennial grown as annual for edible leaves, stems and flowering shoots. Half hardy. Value: moderate levels of folic acid and vitamin C.

Chinese cabbage was first recorded in China around the fifth century AD, and has never been found in the wild. It is thought to have been a spontaneous cross in cultivation between the pak choi and the turnip. Taken to the East Indies and Malaya by Chinese traders and settlers who established communities and maintained their own culture, in the 1400s Chinese cabbage could be found in the Chinese colony in Malacca. By 1751 European missionaries had sent seeds back home, but the vegetable was regarded as little more than a curiosity. Another attempt at introduction was made by a French seedsman in 1845, but the supply became exhausted and the seed was lost. In 1970 the first large-scale commercial crop was produced by the Israelis and distributed in Europe; about the same time it was marketed in the United States as the Napa cabbage, after the valley in California where it was grown. It has become a moderately popular vegetable in the West.

varieties

There are many groups, but three have become popular: the 'tall cylindrical', the 'hearted' or 'barrel-shaped' and the 'loose-headed'.

The cylindrical type has long, upright leaves and forms a compact head, which can be loosely tied to blanch the inner leaves. It is slow-growing, takes about 70 days from sowing to harvest and is most susceptible to bolting. This type is sweet and stores well.

Hearted types have compact, barrel-shaped heads with tightly wrapped leaves and a dense heart. They mature after about 55 days and are generally slow to bolt.

Loose-headed types are lax and open-headed, often with textured leaves. The 'self-blanching' ones have creamy centres and textured leaves. They are also less liable to bolt than headed types. **'Jade Pagoda'** is cylindrical with a firm, crisp head. It takes about 65 days to mature and is cold-tolerant. **'Kasumi'**, a barrel type, has a compact head and is resistant to bolting. **'Monument'** is strong, vigorous, uniform and very tasty. **'Nerva'** is similar, matures quickly, and has dark green leaves with dense heads. **'Ruffles'** is delicious and early-maturing, with lax, pale green heads and a creamy white heart. However, early sowings are liable to bolt. **'Shantung'** has a spreading habit, with tender, light green leaves and a dense heart. **'Tip Top'** is an early variety for spring planting and can be harvested around 70 days after sowing. It is vigorous and produces good-sized heads. **'Wa Wa Sai'** is sweet and very tender. Sow closely for baby leaves.

cultivation

Propagation

A cool-weather crop, Chinese cabbage is more likely to bolt in late spring and summer. Use resistant varieties or sow after mid-summer. The chance of bolting increases if young plants are subjected to low temperatures or dry conditions, or suffer from transplanting check.

Sow the main crop *in situ* from mid- to late summer, 2–3 seeds per 'station', spaced 30–35cm apart in and between the rows. Thin to leave the strongest seedling. Alternatively, sow sparingly in drills and thin.

Otherwise it can be sown in modules or pots, transplanting carefully to avoid root disturbance when there are 4–6 leaves. If the soil is dry, water thoroughly. Broadcast or sow loose-headed types as 'cut and come again' seedlings.

Make first sowings in a cold greenhouse or cold frame in early spring, sow outdoors under cloches or fleece as the weather improves and the soil becomes workable. Summer sowings tend to grow too rapidly and become 'tough', unless it is a cool summer. Make the last sowing under cover in early autumn.

Cut seedlings when they have reached a few centimetres or inches tall, leaving them to resprout. Seeds can also be sprouted, as for alfalfa.

Growing

Chinese cabbage needs a deep, moisture-retentive, free-draining soil with plenty of organic matter. Excessively light, heavy or poor soils should be avoided unless they are improved by incorporating organic matter or grit. Alternatively, grow in raised beds or on ridges. Dig the soil thoroughly before planting; acid soils should be limed, as the ideal pH is 6.5–7.0. Slightly more alkaline soils are advisable where there is a risk of clubroot.

Chinese cabbage prefers an open site, but tolerates some shade in mid-summer. Rotate Chinese cabbage with other brassicas.

Water crops thoroughly throughout the growing season; do not let them dry out. They are shallow-rooted and so need water little and often. Mulching is also advisable. Erratic watering can result in damage to the developing head, encouraging rots. Scatter general fertiliser around the base of transplants or feed with a general liquid fertiliser as necessary to boost growth – this is particularly important on poorer soils. Keep crops weed-free.

Chinese cabbage harvest in Thailand

In late summer, tie up the leaves of hearting varieties with soft twine or raffia. (With self-hearting varieties, this is unnecessary.)

Maintenance

Spring Sow early 'cut and come again' seedling crops under cover. Sow later crops outdoors.
Summer Sow in drills from mid- to late summer. Water a little and often, mulch and keep weed-free. Harvest.
Autumn Sow quick-maturing varieties or 'cut and come again' crops outdoors; sow later crops under cover.
Winter Harvest 'cut and come again' crops grown under cover.

Protected Cropping

Earlier crops can be achieved by sowing bolting-resistant cultivars from late spring to early summer. They need temperatures of 20–25°C for the first 3 weeks after germination to prevent bolting. Harden off, transplant, then protect the plants with cloches or with crop covers.

Late summer-sown crops should be transplanted under cover. Space plants about 13cm apart and grow them as a semi-mature 'cut and come again' crop.

Harvesting and Storing

In autumn, cold-tolerant varieties can stay outside for several weeks provided it is dry. Protect from wet and cold using cloches. Lift developing crops and replant in cold frames, or uproot and lay plants on straw, bracken or similar, covering them with the same material if temperatures fall below freezing point. Ventilate on warm days.

CHINESE CABBAGE

Harvest as 'cut and come again' seedlings, semi-mature or mature plants and for the flowering shoots. Cut seedlings when they are 2.5–5cm tall. Semi-mature or mature plants can be cut with a sharp knife about 2.5cm above the ground and will then resprout; after several harvests they will send up a flower head. Harvest the flowering shoots when they are young, before the flowers open. Harvest mature heads when they are firm.

Chinese cabbage keeps in the fridge for several weeks. Wash thoroughly before storing. Heads can be stored in a cool, frost-free shed or cellar for up to 3 months. When storing, check plants every 3–4 weeks and remove any diseased or damaged leaves immediately.

Pests and Diseases

Chinese cabbage can be affected by any of the usual brassica pests and diseases. You should take precautions in particular against flea beetle, slugs, snails, caterpillars, clubroot and powdery mildew. Crops grow well under horticultural fleece or fine netting.

companion planting

Plant with garlic and dill to discourage caterpillars. Main crops are ideal after peas, early potatoes and broad beans; late crops are good companions for Brussels sprouts. It is good for cropping between slower-growing vegetables.

Chinese cabbages are sometimes grown as sacrificial crops so that slugs, flea beetles and aphids are attracted to them rather than other crops. In the USA they are used among maize, as they attract corn worms.

container growing

The fast-growing varieties give the best results. Use a 25cm pot or container and loam-based compost with added organic matter and a thick basal layer of drainage material. Sow seeds 1cm deep and 2cm apart in shallow pots or modules in mid-spring. Transplant seedlings singly into pots when they have 3–4 leaves.

warning

To minimise the risk of listeria, you should never store Chinese cabbage in plastic bags.

culinary

Cook Chinese cabbage only lightly to retain the flavour and nutrients. Steam, quickly boil, stir-fry – or eat raw. (Seedlings are better cooked.) Leaves blend well in raw salads of lettuce, green pepper, celery, mooli and tomato. They also make a delicious warm salad, stir-fried and mixed with orange. Otherwise use in soups, cook with fish, meat, poultry and use in stuffing.

Outer stalks can be shredded, cooked like celery and tossed in butter. In Korea, China and Japan heads are used to make fermented and salted pickles.

Sweet and Sour Chinese Cabbage

Serves 4

2 tablespoons olive oil
1 onion or several shallots, sliced
2 tablespoons white wine vinegar
2 teaspoons sugar
6 tablespoons chopped tomatoes in their juices
750g Chinese cabbage, finely shredded
Salt and freshly ground black pepper

Heat the oil in a heavy-bottomed pan and cook the onions until soft. Stir in the vinegar, sugar and tomatoes and blend well. Add the Chinese cabbage and seasoning. Cook for 10 minutes with the lid on, stirring occasionally, until the cabbage is tender.

Serve hot.

Capsicum annuum. Solanaceae

PEPPER & CHILLI

Also known as Sweet Pepper, Bell Pepper, Capsicum, Pimento. Annuals and short-lived perennials grown for edible fruits. Half hardy. Value: very rich in vitamin C and beta carotene.

Both hot and mild peppers come from one wild species, which is native to Central and South America. The name capsicum comes from the Latin *capsa*, meaning 'a box'. It is thought that the hot types were the first to be cultivated; seeds have been found in Mexican settlements dating from 7000 BC, and the Aztecs are known to have grown them extensively. They are one of the discoveries made in the New World by Columbus. He thought he had discovered black pepper, which at the time was extremely expensive, and used the name 'pepper' for this new fiery spice. Spanish and Portuguese explorers then distributed the new kinds of peppers around the world.

Sweet peppers were introduced to Spain in 1493 and were known in England by 1548 and Central Europe by 1585. The Spanish use red sweet peppers to make the spice pimentón and for stuffing green olives. Chilli peppers are notorious for their fieriness. Their heat is caused by the alkaloid capsaicin, which is measured in Scoville Units. Mild chillies are around 600 units. Beware the 'Dorset Naga' and 'Bhut Jolokia', both measured at over one million units!

varieties

The larger, bell-shaped, mild-tasting sweet peppers eaten as vegetables are members of the *Capsicum annuum* Grossum Group. Green when immature, different cultivars ripen to yellow, orange, red or 'black'.

'Hungarian Hot Wax'

The smaller, hotter chillies used for flavouring are classed in the *C. a.* Longum Group.

Sweet peppers

'Big Bertha' is one of the largest bell peppers, growing to 18cm long by 10cm wide. Excellent for growing in cooler climates and for stuffing. **'Californian Wonder'** has a mild flavour, is good for stuffing and crops well over a long period. **'Calwonder Wonder Early'** grows well in short seasons and is prolific. **'Gypsy'** is an early cropper with slightly tapered fruits and is also resistant to tobacco mosaic virus. **'Redskin'** is compact and ideal for pots and growbags. **'Sweet Chocolate'** is an unusual chocolate-brown colour and good when frozen whole.

Chilli peppers

'Anaheim' produces tapered, moderately hot fruits over a long period. The moderately hot yellow fruits of **'Hungarian Hot Wax'** ripen to become crimson and are good for salad and stuffing. **'Italian White Wax'** has pointed fruits which pickle well and are mild-tasting when young. **'Large Red Cherry'** is extremely hot, with flattened fruits ripening to cherry red. Good for drying and ideal in curries, pickles and sauces. **'Serrano'** is extremely hot with orange-red fruits. It is prolific and can be dried. **'Tabasco Habanero'** is extremely hot – beware! **'Tam Jalapeno'** has a good, mild flavour and is high yielding. **'Thai Burapa'** is small and pointed; ideal for Asian dishes. **'Thai Denchai'** tapering to 12cm, is great for all Thai dishes. **'Tunisian Baklouti'** has large, tapering pods and is hot.

PEPPER & CHILLI

Chilli peppers drying in India

cultivation

Propagation
Sow seed indoors from late winter to mid-spring in trays, modules or pots of moist seed compost at 21°C. Lower temperature gradually after germination. Transplant into 8–9cm pots when 3 true leaves appear, repotting again into 10–13cm pots when sufficient roots have formed. Move plants into their final position when about 10cm high and the first flowers appear.

Harden off outdoor crops in cool temperate zones and transplant in late spring to early summer when the soil is warm and there is no danger of frost.

Space standard varieties 38–45cm apart; dwarf varieties should be spaced 30cm apart.

Growing
Sweet peppers and chillies flourish outdoors in warmer climates. More successful under cover in cooler zones, they can be grown outside in mild areas or warm microclimates but benefit from protection with cloches or fleece. Chillies tend to be more tolerant of fluctuating temperatures and high or low rainfall and grow in marginally less fertile soil. Blossom drops when night temperatures fall below 15°C. Soil should be moisture-retentive and free-draining on ground manured for the previous crop. Alternatively, dig in plenty of well-rotted organic matter in autumn or winter prior to planting. Rake in a granular general fertiliser at 135g/sq m before planting.

Keep the soil moist and weed-free; mulching is recommended. Feed with a general liquid fertiliser if plants need a boost. Excessive nitrogen can result in flower drop.

If branches are weak and thin, when the plant is about 30cm tall remove the growth tips from the stems to encourage branching. Normally, they branch naturally. Support with a cane if necessary.

Maintenance
Spring Sow seeds under cover. Transplant when the danger of frost is passed or grow in an unheated greenhouse.
Summer Keep crops weed-free; damp down the glasshouse in warm weather. Harvest.
Autumn Cover outdoor crops where necessary. Harvest.
Winter Prepare the ground for outdoor crops.

Protected Cropping
Crops are better grown under glass, polythene, cold frames or cloches in cool temperate climates. In glasshouse borders, prepare the soil as for 'Growing'. Keep it moist but not waterlogged and mist with tepid water to maintain humidity and help fruit set.

Sow in mid-spring for growing in a greenhouse and in late spring for under cloches. Ventilate well during hot weather. If you are growing dwarf varieties on a sunny windowsill, turn pots daily to ensure even growth.

Plants grown under cloches may outgrow their space: rigid cloches can be turned vertically and supported with canes. Alternatively, use polycarbonate sheeting.

Harvesting and Storing
Harvest with scissors or secateurs when sweet peppers or chillies are green, or leave on the plant for 2–3 weeks to ripen and change colour. Picking them when they are green increases the yield.

Peppers may be stored in a cool, humid place for up to 14 days at temperatures of 13–15°C. Towards the end of the season, uproot plants and hang them by the roots in a frost-free shed or greenhouse; fruits will continue to ripen for several months.

The heat of chillies increases with the maturity of the fruit. Fresh chillies keep for up to 3 weeks in the fridge in a paper bag. Or store them in an airtight jar in a dark cupboard. Chillies can either be dried and used whole or ground into powder. Both sweet peppers and chillies freeze successfully.

Pests and Diseases
Red spider mite can be a nuisance under cover. Slugs can damage seedlings, stems, leaves and fruits. Remove any decaying flowers or foliage immediately.

companion planting

Capsicums grow well with basil, okra and tomatoes.

container growing

Grow under cover or outdoors. Plant in 20–25cm pots of loam-based compost with moderate fertiliser levels or in growbags. Dwarf varieties can be grown successfully in pots on windowsills. Keep the compost moist but not waterlogged and mist with tepid water, particularly during flowering to assist fruit set. Water frequently in warm weather, less at other times.

medicinal

Capsaicin increases the blood flow and is used in muscle liniments. It is said to help the body metabolize alcohol, acts as expectorant, and prevents and alleviates bronchitis and emphysema. 10–20 drops of red-hot chilli sauce in a glass of water daily (or hot spicy meals 3 times a week) can keep airways free of congestion, preventing or treating chronic bronchitis and colds. It stimulates endorphins, killing pain and inducing a sense of wellbeing.

warning

Ventilate the kitchen when using chillies. If you have sensitive skin, wear rubber gloves to handle them and always avoid touching your eyes or other sensitive areas after handling.

culinary

Sweet peppers add both colour and taste to the table. They can be eaten raw in salads, roasted or barbecued, fried or stir-fried, stuffed with rice, fish or meat mixtures, and used in countless casseroles and rice dishes.

Chillies are used in chilli con carne, curries and hotpots. Removing the internal ribs and seeds reduces the heat intensity. Paprika is the dried and powdered fruits of sweet peppers. Cayenne pepper is dried and ground powder made from chillies.

Mrs Krause's Pepper Hash

If chilli is too strong, it can cause intestinal burning – cucumber, rice, bread and beans are a good antidote. By coating the tongue, the fat content of yogurt and butter soothes a chilli-burnt mouth. Water only makes things worse.

If you're uninitiated at eating chilli peppers, start with small doses and build up tolerance of heat. Excessively hot peppers can cause jaloproctitis or perianal discomfiture.

Mrs Krause's Pepper Hash
Serves 6

William Woys Weaver, in *Pennsylvania Dutch Country Cooking*, quotes this recipe from Mrs Eugene F. Krause of Bethlehem, who lived in the early part of the twentieth century and was renowned for her peppery hashes.

6 green peppers, deseeded and finely chopped
6 red peppers, deseeded and finely chopped
4 onions, finely chopped
2 small pods hot chilli peppers, deseeded and finely chopped
1½ tablespoons celery seeds
375ml cider vinegar
250g brown sugar
1½ teaspoons sea salt

Combine the peppers, onions, hot chilli peppers and celery seeds in a non-reactive preserving pan. Heat the vinegar in a non-reactive pan and dissolve the sugar and salt in it. Bring to a fast boil, then pour over the pepper mixture. Then cook over a medium heat for 15 minutes, or until the peppers begin to discolour. Pack into hot sterilised preserving jars, seal, and place in a 15-minute water bath. Let the pepper hash mature in the jars for 2 weeks before using.

Jean-Christophe Novelli's Andalouse of Sole
Serves 6

For the piperade:
Carton of sun-dried tomato juice
200g sun-dried tomatoes
3 star anise
Fresh thyme
400ml olive oil
1 large courgette, sliced at an angle
2 medium aubergine, sliced at an angle, then halved
8 large shallots, peeled
8 baby fennel, trimmed
2 red peppers, deseeded and cut in half
2 yellow peppers, deseeded and cut in half
2 green peppers, deseeded and cut in half
10 red cherry tomatoes
1 head of garlic, cloves separated and peeled
Sugar
Freshly ground salt and pepper
Good handful of black olives
20 basil leaves

For the Sole Fillets:
6 large sole fillets
olive oil
1 vanilla pod, split and seeds scraped out

To Serve:
Handful of mixed fresh herbs
Freshly grated Parmesan
200ml truffle oil

To make the piperade, start by heating the tomato juice in a pan with the sun-dried tomatoes, star anise and thyme and cook steadily until it's reduced by one-third.

Heat the olive oil in a pan. Add the vegetables, garlic, a little sugar and the sun-dried tomato juice and season. Cover with a tight-fitting lid and leave over a low heat until the vegetables are just soft to the touch. Remove the vegetables from the heat and add the olives and basil to warm through.

Preheat the grill. Brush the sole fillets with a little olive oil and the vanilla seeds. Season and grill for about 2–3 minutes on each side, skin-side up first. If the fillets are quite thick, they may require 4 minutes on each side.

Carefully arrange the vegetables on the plates. Place a sole fillet on each dish and garnish with fresh herbs, grated Parmesan and a splash of truffle oil.

Chrysanthemum coronarium. Compositae

CHRYSANTHEMUM GREENS

Also known as Shungiku, Chop Suey Greens. Hardy annual. Value: rich in vitamin B, moderate vitamin C and minerals.

There are few Chrysanthemums or their close relatives with culinary value. *Tanacetum balsamita,* a stalwart of the cottage garden, was once classified in the same group and, in times past, its leaves were used to flavour ale while Dioscorides notes that the stalks and leaves of marguerites were 'eaten as other pot herbes are'. It is a mystery how an attractive plant of Mediterranean hillsides, valued as an ornamental for its pretty flowers, spread east to China, Vietnam and Japan, where its foliage became widely eaten as 'greens'. It is not only the leaves that are used; the Japanese dip the flowers in sake and eat them at the beginning of a meal to confer health and long life.

Chrysanthemum in bloom

varieties

There are no cultivars; just three forms are in cultivation. One has pale green, finely-cut, almost feathery leaves, with good low- and high-temperature tolerance and a strong flavour. Another, bushier, high-yielding plant with broader, thicker, shallowly-lobed leaves is tender and mildly flavoured; the third, which is intermediate in appearance between the two, is fast growing, bushy and high-yielding, thriving in both warm and cool climates.

cultivation

Propagation

Sow, in drills or broadcast thinly, lightly covering the seed, every 2–3 weeks during spring and autumn for a constant harvest. Allow 15cm in and between the rows for larger plants. Thinning is not usually required, for young plants, unless sowing is particularly dense, when they should thinned to 4.5cm apart. Plants can also be grown from soft tip cuttings taken from over-wintering plants in spring.

Growing

Chrysanthemum greens are primarily a cool-season crop, tolerating low light levels and withstanding light frosts. They are at their best in spring and autumn as temperatures over 25°C impart bitterness to the leaves. They tolerate most soils but prefer a well-prepared seedbed on rich, fertile, moisture-retentive soil in full sun and a cool, shady position in summer. Plants can be grown as small 'cut and come again' seedlings or leaves harvested from larger individual plants. Keep plants well watered and weed-free by hoeing or hand weeding to prevent competition from annual weeds which suppresses growth.

Maintenance

Spring Sow outdoors once the danger of frost has passed. Harvest winter crops.
Summer Make the final sowings of the season outdoors around 10 weeks before the last frost.
Autumn Sow winter crops under glass; harvest earlier crops.
Winter Harvest earlier sowings.

Protected Cropping

Sow seeds under glass in autumn for cropping over winter in an unheated greenhouse and in early spring for early crops.

Wash thoroughly before eating

culinary

Eat young aromatic shoots and stems raw, in salads, sushi or with pickles; leaves can be cooked or steamed, dipped into tempura batter and deep-fried. Try not to overcook the leaves as they can become bitter. Kikumi, a Japanese pickle, is made from the edible flower petals. Garnish stews with young leaves. Harvest just before use as the leaves wilt quickly.

Harvesting and Storing

As 'cut and come again' crops, the first harvest can be made from 4 weeks after sowing during mid-summer; plants recover rapidly and several cuts can be made from each sowing. Alternatively, cut back the top 20cm of more mature plants leaving them to re-sprout from the base or pick side shoots from individual plants. Shoots can be harvested from flowering plants without any loss of quality and the flowers eaten in salads.

Pests and Diseases

It is pest- and disease-free.

container growing

Can be grown in containers or growing bags throughout the season and respond well to cutting back.

medicinal

Leaves have a calming effect on the stomach and are used as an expectorant.

Cicer arietinum. Papilionaceae

CHICKPEA

Also known as Dhal, Egyptian Pea, Garbanzo, Gram. Annual grown for seed sprouts, seeds, young shoots and leaves. Tender. Value: high in protein, phosphorus, potassium, most B vitamins, iron and dietary fibre.

Chickpeas originated in the northern regions of the fertile crescent. Evidence of their ancient use as a domesticated crop was found at a site in Jericho and dated to around 6500 BC. Seeds excavated in Greece indicate that the chickpea must have been introduced to Europe with the first food crops arriving from the Near East. Today it is cultivated worldwide in sub-tropical or Mediterranean climates as a cool-season crop, needing about 4 to 6 months of moderately warm, dry conditions to flourish. It is the world's third most important pulse after peas and beans, and 80 per cent of the crop is produced in India. It is eaten fresh or dried, made into flour, used as a coffee substitute and grown as a fodder crop. The plant grows about 30cm tall, with compound leaves of up to eight toothed leaflets. Its tiny white- or blue-tinged flowers are followed by a small flat pod containing one or two round seeds, each with a small 'beak' – hence the common name 'chickpea'.

varieties

The following are Indian cultivars: **'Annegeri'** is a semi-spreading, high-yielding variety with yellowish-brown seeds. It is deep-rooting and likes good soil but a coarse tilth is adequate. **'Avrodhi'** has medium-sized brown seeds and is wilt-resistant. **'Bheema'**, a semi-spreading variety, has large, light brown, smooth seeds and is suitable for drought-prone or low-rainfall areas. **'Brown Seeded'** is good for the home gardener, particularly in short, dry seasons. **'Kabuli Black'**, with 2 black seeds per pod, is very hardy, vigorous and fast maturing. It has some tolerance of cold soils and is drought-resistant.

cultivation

Chickpeas need a fertile, well-drained soil in full sun.

Propagation

Seeds can be broadcast or sown in drills during winter in Mediterranean regions or after the rains in subtropical climates. Broadcasting is very simple. Prepare the soil using the 'stale seedbed' method, raking and levelling, allowing the weeds to germinate and hoeing them off before sowing. Then scatter the seed evenly, raking the soil twice – first in one direction, then again at 90 degrees to ensure even coverage.

Chickpeas can also be sown in drills 6–10cm deep with the rows 50cm apart, thinning to 25cm between plants after germination. Alternatively, sow 3–4 seeds in 'stations', 25cm apart, and thin to leave the strongest seedling standing.

Growing

Dig over the area thoroughly before planting, adding organic matter to poor soil. Rake over the area to create a fine tilth and water well before sowing if the seedbed is dry. Alternatively, soak the seeds for an hour. Keep crops weed-free during the early stages; as plants mature, their spreading habit naturally stifles weed growth.

Chickpeas are drought-tolerant, but watering just before flowering and as the peas begin to swell improves productivity. Rotate with other legumes and leave the roots in the ground after harvest to provide nitrogen for the following crop.

CHICKPEA

The flowers are delicate and pretty

Maintenance
Spring Dig over the planting area, adding organic matter where needed.
Summer Keep crops weed-free. Water in prolonged drought, just before flowering and as the peas swell.
Autumn Harvest crops.
Winter Dig over the planting area, adding organic matter where needed.

Protected Cropping
In cooler climates, sow seeds in early spring into small pots of moist seed compost in a glasshouse or on a windowsill. Harden off in late spring and plant outdoors once there is no danger of frost. Growing crops in cloches or polythene tunnels increases the yield.

Harvesting and Storing
Crops are ready after about 4–6 months. Harvest when leaves and pods turn brown; don't leave it too late, or the seeds will be lost when the pods split. Cut the stems at the base and tie them together before drying upside down in a dry, warm place. Collect the dry seeds and store in airtight jars. Peas can also be harvested fresh for cooking, but fresh ones deteriorate rapidly and should be used as soon as possible.

Sprouting Seeds
Always buy untreated chickpeas for sprouting, as seed sold for sowing is often treated with chemical dressings. Soak seeds overnight or for several hours in boiling water, tip into a sieve and rinse. Put several layers of moist paper towel or blotting paper in the base of a jar and cover with a layer of seed. Cut a square from a pair of tights or piece of muslin and cover the top, securing with a rubber band. Place in a bright position, away from direct sunshine, maintaining constant temperatures around 20°C. Rinse the seed 3–4 times a day by filling the jar with water and pouring off again. Harvest after 3–4 days when the 'sprouts' are about 12mm long.

Pests and Diseases
Plants can suffer from root rot. They turn black and finally dry up, leaves fall and the stems desiccate. Ensure the soil is well drained and destroy affected crops immediately. The acidic secretions from the glandular hairs are a good defence against most pests. Gram pod borer caterpillars feed on the crop from seedlings to maturity, damaging seedpods and the immature seeds. Spray with pyrethrum.

container growing

They can be grown in containers, but seed production levels do not make them a worthwhile proposition as a crop plant.

medicinal

The leaves are astringent and used to treat bronchitis. They are also boiled and applied to sprains and dislocated bones; the exudate is used for indigestion, diarrhoea and dysentery. The seeds are a stimulant, tonic and aphrodisiac. In Egypt they are used to gain weight, and to treat headaches, sore throat and coughs. Powdered seed is used as a facepack and also in dandruff treatment.

warning

The whole plant and seed pods are covered with hairs containing skin irritants. You must always wear gloves when harvesting.

culinary

With a protein content of 20 per cent, chickpeas are an important meat substitute and good for children and expectant and nursing mothers. Chickpeas are used fresh or dried. They are ground into 'gram flour' (used in vegan cooking), and the ground meal is mixed with wheat and used for chapatis. Whole chickpeas are fried, roasted (to eat as a snack) and boiled. To make houmous, grind boiled chickpeas into a paste, mix with olive or sesame oil, flavour with lemon and garlic and eat on pitta bread or crackers. Chickpeas are also used to make dahl and are found in spicy side dishes, vegetable curries and soups. The young shoots and leaves are used as a vegetable and cooked like spinach – boiled in soups, added to curries or fried with spices.

Puréed Chickpeas
Serves 4

300g dried chickpeas, soaked overnight
2 tablespoons olive oil
1 onion, sliced finely
3 garlic cloves, crushed
300g tomatoes, peeled, deseeded and chopped
Salt and freshly ground black pepper

Drain the chickpeas, put in a heavy-bottomed saucepan, cover with fresh water and cook until tender. This will take up to 1½ hours, depending on the age of the chickpeas. Drain. Purée through a *mouli légumes*.

Heat the oil in a frying pan, sauté the onion until softened, add the garlic and cook for 30 seconds longer. Add the tomatoes and simmer for 5 minutes before adding the chickpea purée. Season well and serve immediately.

Cichorium endivia. Asteraceae

ENDIVE

Also known as Escarole, Batavian Endive, Grumolo. Annual or biennial grown as annual for blanched hearts and leaves. Value: rich in iron, potassium and beta carotene; moderate vitamin A and B complex.

The origins of this plant are obscure, but it was certainly eaten by the Egyptians long before the birth of Christ and is one of the bitter herbs used at Passover. Mentioned by Ovid, Horace, Pliny and Dioscorides, it was valued by the Greeks and Romans as a cultivated plant. It was introduced to England, Germany, Holland and France around 1548 and was described by several writers.

varieties

There are two types. The upright Batavian, scarole or escarole has large, broad leaves. Curly or fringed frisée is a pretty plant, with a low rosette of delicately serrated leaves. Curled varieties are generally used for summer cropping; the more robust broad-leaved types tolerate cold, are disease-resistant and grow well in winter.

'Broad Leaved Batavian' has tightly packed heads of broad, deep green leaves that become creamy-white when blanched. **'En Cornet de Bordeaux'**, an old variety, is very tasty, extremely hardy and blanches well. **'Green Curled Ruffec'**, a curly type, is tasty blanched and makes a good garnish. Very hardy and cold resistant. **'Green Curled'** (**'Moss Curled'**) produces compact heads of dark green, fringed leaves. **'Lassie'** yields a large amount of deeply cut, well-flavoured leaves. **'Pancalieri'** has very curly, dark green leaves with creamy white hearts. **'Salad King'** is prolific and extremely hardy, with large, dark green, finely cut leaves. **'Sanda'** is vigorous and resistant to tip burn, cold and bolting. **'Scarola Verde'** has a large head of broad, green and white leaves. However it may bolt in hot weather. The pale, green leaves of **'Toujours Blanche'** are deeply cut and finely curled. **'Très Fine Maraichère'** (**'Coquette'**) has finely cut, curled leaves that are mild and delicious, and it grows well in most soils. **'Wallone Frisée Weschelkopf'** (**'Wallone'**) has a large, tightly packed head with finely cut leaves. It is vigorous, hardy and good as a 'cut and come again'.

cultivation

Propagation

Sow thinly, 1cm deep *in situ* or in a seedbed, pots or modules to transplant. Allow 30–38cm between plants and rows.

Endive germinates best at 20–22°C. Sow early crops under cover and shade summer crops.
Sow from early to mid-summer for autumn crops, in late summer for winter crops, using curled or hardy Batavian types. Sow all year round for 'cut and come again' seedlings or semi-mature leaves, making early and late sowings under glass.

Growing

Endive needs an open site, though slimmer crops tolerate a little shade. Soils should be light, moderately rich and free-draining; this is particularly important for winter crops. If necessary, dig in plenty of well-rotted organic matter before planting. Excess nitrogen encourages lush growth and makes plants prone to fungal diseases.

Endive is a cool-season crop, flourishing between 10–20°C, yet it withstands light frosts; hardier cultivars withstand temperatures down to –9°C. Higher temperatures tend to encourage bitterness, though 'curled' types are heat-tolerant. Young plants tend to bolt if temperatures fall below 5°C for long.

Keep crops weed-free; mulch and water thoroughly during dry weather, as dryness at the roots can cause bolting. Use a general liquid fertiliser to boost growth if necessary.

Blanch to reduce bitterness and make leaves more tender. Many newer cultivars have tight heads and some blanching occurs naturally. Damp leaves are likely to rot, so choose a dry period or dry plants under cloches for 2–3 days. Draw the outer leaves together and tie with raffia 2–3 weeks before harvest, placing a tile, piece of cardboard or dinner plate over the centre of the plant, and covering with a cloche to keep off the

Endive is invaluable in winter salads

rain. Alternatively cover the whole plant with a bucket or a flower pot with its drainage holes covered. Blanching takes about 10 days. Blanch a few at a time; they rapidly deteriorate afterwards.

Maintenance
Spring Sow early crops under glass or cloches. Harvest late crops under cover.
Summer Sow curly varieties outdoors in seedbeds for transplanting, or in situ. Keep early-sown crops weed-free and moist. Harvest.
Autumn Sow outdoors and under cover. Harvest.
Winter Sow under cover and harvest.

Protected Cropping
Sow hardy cultivars under cover in trays or modules at 20°C in mid-spring for early summer crops. After germination, maintain a minimum temperature of 4°C for 3 weeks after transplanting – this is to prevent bolting.

For winter and early spring crops, transplant in early autumn from seed trays or modules under cover. Sow 'cut and come again' crops under cover in early spring, and in early autumn.

Endive grows better than lettuce in low light and is also a useful crop for the greenhouse in winter.

Harvesting and Storing
Harvest endive from 7 weeks after sowing, depending on cultivar and season.

'Cut and come again' seedlings may be ready from 5 weeks. With some cultivars, only one or two cuts may be possible before they run to seed. Pick individual leaves as needed or harvest them using a sharp knife about 2cm above the ground, leaving the root to resprout. The whole plant can be lifted in autumn and put in a cool, dark place to blanch.

Leaves do not store well and are better eaten fresh. They last about 3 days in a polythene bag in the salad drawer of a refrigerator.

Pests and Diseases
Protect plants from slugs and control aphids.

Keep winter crops well watered and mulched to prevent tip burn.

Eat endive immediately after harvest

companion planting

Endive is good for intersowing and intercropping.

container growing

Sow directly or transplant into large containers of loam-based compost with added well-rotted organic matter. Keep compost moist and weed-free. Use a general liquid fertiliser to boost growth if necessary.

culinary

Endive is used mainly in salads with – or instead of – lettuce and other greens; the slightly bitter taste and crisp texture gives it more of a 'bite' than the usual lettuce combinations. It suits strongly flavoured dressings. Crisp lardons of bacon or croutons are often included. With mature endive, use the inner leaves for salads; the outer ones can be cooked as greens. Endive can also be braised. Try serving it shredded and dressed with hot crushed garlic, anchovy fillets and a little olive oil and butter.

Warm Red and Yellow Pepper Salad
Serves 6

The slightly bitter taste of curly endive (escarole or frisée) is ideal combined with other lettuces such as lamb's lettuce or watercress in winter months. The sweetness of the peppers in this dish happily complements the endive.

1 large endive, washed and roughly chopped
Big bunch lamb's lettuce, washed
2 large red peppers
3 tablespoons olive oil
3 cloves garlic, crushed
1 tablespoon chopped fresh herbs
Salt and freshly ground black pepper

For the dressing:
3 tablespoons extra virgin olive oil
1 tablespoon white wine vinegar
Salt and freshly ground black pepper

Arrange the lettuces in a large bowl. Core and deseed the peppers and cut them into thin strips. Heat the oil in a heavy pan and sauté the peppers, stirring constantly. Add the garlic after 5 minutes and cook for a further minute.

Add the peppers to the salad. Make the dressing and toss in the herbs and seasoning. Serve the salad alongside plain grilled fish or chicken.

Cichorium intybus. Asteraceae

CHICORY

Also known as Witloof, Belgian Endive, Succory, Sugar Loaf Chicory, Radicchio. Hardy perennial grown as annual for blanched leaves or root. Very hardy. Value: moderate levels of potassium.

A native of Europe through to central Russia and western Asia, chicory has been cultivated for centuries. Pliny tells us that cichorium is a Greek adaptation of the Egyptian name; he also noted its medicinal use as a purgative and the blanching of leaves for salads.

Large-rooted varieties have long been used dried, ground and roasted as a substitute for coffee – particularly popular in England during the Napoleonic wars when a blockade of the French coast cut supplies. It has a distinctive fragrance and is often drunk by those who like the taste of coffee without the caffeine. John Lindley, in the nineteenth century, recorded that roasted chicory was adulterated with a multitude of substances as diverse as marigolds, oak bark, mahogany sawdust and even baked horse liver!

Chicory is a traditional ingredient in Belgium

varieties

Chicory has a distinctive, slightly bitter flavour. Most varieties are hardy and make a good winter crop with colourful, attractive leaves. There are three types. 'Forcing' chicories like 'Witloof' (that is, whiteleaf) produce plump, leafy heads (known as 'chicons') when blanched. 'Red chicory' or 'radicchio' includes older cultivars which respond to the reduced daylength and lower temperatures of autumn by turning from green to red; newer cultivars are naturally red and heart earlier. 'Non-forcing' or 'sugarloaf' types produce large-hearted lettuce-like heads for autumn harvest.

'Brussels Witloof' (**'Witloof de Brussels'**) is one of the most famous forcing types which is also grown for its root. **'Catalogna Frastagliata'** is upright with thin, white ribs. **'Large Rooted Magdeburg'** (**'Magdeburg'**), like **'Brussels Witloof'**, is grown for its root, which is used to make coffee, but young leaves can also be harvested. **'Orchidea Rossa'** is crunchy with good flavour and texture; delicious eaten steamed, roasted or raw. **'Palla rossa Zorzi Precoce'** is a radicchio with a tangy, delicate flavour. It colours better in cool weather. **'Radicchio di Treviso Black Svelta'** has dark red, upright leaves with white midribs. **'Rossa di Treviso'**, another radicchio, has crisp, green leaves that become deep red and veined with white in cooler conditions. It dates back to the sixteenth century and tolerates light frost. **'Rossa di Verona'** is a radicchio with a spreading habit and withstands considerable frost.

cultivation

Propagation

Sow seed of forcing varieties thinly in late spring to early summer in drills 1cm deep and 23–30cm apart. Thin when the first true leaves appear to 20cm apart.

Sow early radicchio under cover in seed trays and modules before hardening off and transplanting

CHICORY

from mid-spring for summer harvest, using early maturing types first. Sow for autumn harvest from early to mid-summer, and in mid- to late summer for transplanting under cover in autumn and cropping over winter. Thin to 23–38cm in and between the rows.

Non-forcing or sugarloaf types can be broadcast or sown in broad drills under cover in late winter. When the soil warms and the weather improves, sow successionally outdoors until late summer. Sow the final crop during early autumn under cover.

Thin to a final spacing of 23–30cm in and between the rows. For a semi-mature 'cut and come again' winter crop, sow seed from mid- to late summer and transplant indoors in autumn.

Growing
Chicories prefer an open, sunny site, but tolerate a little shade. Soil should be fertile and free-draining, with organic matter added for the previous crop: avoid recently manured ground, this causes the roots to fork.

Radicchio tolerates most soils except gravel or very heavy clay. The ideal pH is 5.5–7. Rake soil to a fine texture before sowing. Apply a general balanced fertiliser at 30g/sq m.

Keep weed-free with regular hoeing or mulching. Water thoroughly during dry weather to prevent bolting. Force appropriate varieties in situ if soil is light and winters are mild.

In late autumn to early winter, cut back the leaves to 2.5cm above ground level. Form a ridge of friable soil 15–20cm high over the stumps and cover with straw or leaf mould. After about 8–12 weeks, when the tips are appearing, remove the soil and cut the heads off at about 2.5cm above the neck. Keep the compost moist while the 'chicons' are growing.

Maintenance
Spring Sow early radicchio under cover, transplant from mid-spring. Sow sugarloaf types as 'cut and come again' crops.
Summer Harvest early radicchio and sow maincrop. Sow sugarloaf outdoors for 'cut and come again' and autumn maincrop.
Autumn Harvest radicchio crops sown early to mid-summer; sow late crops for winter. Lift forcing types or force outdoors. Sow sugarloaf indoors.
Winter Force roots indoors successionally until spring. Sow sugarloaf indoors.

Protected Cropping
Non-forcing varieties can be grown under glass, for early or late crops.

In autumn, cover outdoor crops of radicchio with cloches, fleece or similar to extend the growing season.

Forcing
Force 'chicons' indoors if soil is heavy, if winters are severe, or for earlier crops. In mid- to late autumn when the foliage dies down, lift roots carefully, discard any forked or damaged ones and keep those that are at least 3–5cm in diameter at the top. Cut off the remaining leaves to within 1cm of the crown, trim back the side and main roots to about 20–23cm. Pack horizontally in boxes of dry sand, peat substitute or sawdust and store in a cool, frost-free place. For forcing, remove a few roots from storage at a time. Plant about 5 to 6 roots in a 23cm pot of sand or light soil, ensuring that 1cm of the crown is above the surface. Surround the roots with moist peat or compost, leaving the crown exposed above ground.

Radicchio

Water sparingly and cover with a black polythene bag, an empty flower pot with the holes blocked up, or an empty box. Maintain temperatures of 10–15°C. They can also be blanched in a dark cellar or shed or under greenhouse staging (see 'Growing'). The blanched 'chicons' should be ready for cutting within 4 weeks, depending on the temperature. Roots may resprout, producing several smaller shoots which can then be blanched.

Harvesting and Storing
Harvest chicories grown for roots after the first frost. Cut heads of mature sugarloaf varieties with a sharp knife 2.5cm above the soil in late autumn; use immediately or store in a frost-free place.

Allow the plants to resprout as a 'cut and come again' crop under cloches.

Pick radicchio leaves as required, taking care not to over-harvest, as this weakens plants. Alternatively, cut the whole head and leave the roots to resprout.

'Chicons' can be stored in the refrigerator, wrapped in foil or paper to prevent them from becoming bitter. Non-forcing chicory will stay fresh for up to 1 month.

Pests and Diseases
Though seldom troubled by pests and diseases, crops can rot outdoors in cold weather.

companion planting

The blue flowers of chicory are attractive in the ornamental border.

container growing

Chicories can be grown in large containers or growing bags as 'cut and come again'.

medicinal

Chicory is said to be a digestive, diuretic and laxative, reducing inflammation. A liver and gall bladder tonic, it is used for rheumatism, gout and haemorrhoids. Culpeper suggests its use 'for swooning and passions of the heart'.

CHICORY

culinary

All types of chicory make a wonderful winter salad, particularly if you mix the colours of red, green and white. Add tomatoes or a sweet dressing to take away some of the bitter taste. Home-grown chicons stored in the dark tend to be less bitter.

For a delicious light supper dish, pour a robust dressing (made with lemon juice rather than vinegar) over the leaves, add some anchovy fillets, crumbled hard-boiled eggs and top with a handful of kalamata olives.

As an accompaniment to cold meat and game, eat with sliced oranges, onion and chopped walnuts.

With roast meats, braise chicory with butter, lemon juice and cream. Radicchio can also be braised, but loses its colour.

To make a chicory coffee substitute, dry roots immediately after harvest and grind thoroughly.

Chicory with Ham and Cheese Sauce
Serves 4

4 heads chicory
8 slices smoked Bayonne ham
2 tablespoons butter
2 tablespoons plain flour
2 teaspoons Dijon mustard
150ml milk
4 tablespoons grated Gruyère cheese
Single cream
Salt and freshly ground black pepper

Blanch the chicory heads in a pan of salted water for 5 minutes, drain and gently squeeze out as much water as you can. Cut each chicory in half and wrap in a slice of the ham. Then arrange in one layer in an ovenproof dish.

Preheat the oven to 200°C/400°F/gas mark 6. Make the cheese sauce: melt the butter in a heavy pan and stir in the flour. Cook for 2 minutes, then stir in the mustard. Pour in the milk gradually, stirring vigorously as the sauce thickens, then add the cheese and enough cream to accomplish a smooth consistency. Stir over a gentle heat for 5 minutes to let the cheese melt. Season to taste.

Pour the cheese sauce over the chicory wrapped in ham and cook in the oven for 20 minutes until nicely browned. Serve immediately.

Chicory with Ham and Cheese Sauce

Braised Chicory

Braised Chicory
Serves 4

The French, especially in the south-west and in Provence, braise chicory or Belgian endives and serve them with roasted meats such as lamb and beef.

2 tablespoons butter
1 red onion, finely chopped
2 rashers smoked bacon, diced
4 heads chicory, trimmed
150ml chicken stock and white wine combined
Juice of ½ a lemon
Salt and freshly ground black pepper

Liberally coat the sides and bottom of a heavy, lidded casserole with half the butter and heat the remaining butter in a small pan over a medium flame. Gently fry the onion and bacon, and set aside.

Arrange the chicory in the casserole. Add some of the stock and white wine, season with salt and pepper and cover. Allow to sweat over a low heat until just turning colour. Roll them over and cook on the other side.

Add a little more liquid as required. The liquid should evaporate from the chicory by the end, so that they are browned and tender (you may have to remove the lid for a short while). Add the onion and bacon and quickly heat up. Drizzle over the lemon juice, season and serve immediately.

Colocasia esculenta var. *antiquorum*. Araceae

EDDOE

Perennial grown as annual for edible tubers. Tender. Value: rich in starch, magnesium, potassium and vitamin C.

This variety of taro, native of India and Southeast Asia, was first recorded by the Chinese 2,000 years ago. It is now grown throughout the humid tropics. Eddoes flourish in moist soil alongside rivers and streams. The central tuber is surrounded by clusters of smaller tubers which are harvested, making it different from the single-tubered dasheen. The brown, hairy tubers can weigh up to 2.2kg, and when they are sliced reveal flesh which is usually white but can also be yellow, pink or orange. Their taste is similar to a garden potato but with an attractive nutty flavour. Tubers should never be eaten raw as all varieties contain calcium oxalate crystals, a skin irritant.

varieties

'Euchlora' has dark green leaves with violet margins and leaf stems. **'Fontanesii'** produces leaf stems which are dark red-purple or violet. Its leaf blades are dark green with violet veins and margin.

cultivation

Propagation
In the humid tropics, eddoes can be planted any time. In temperate climates, grow under glass or polythene at a minimum temperature of 21°C, and plant in spring. Plant small tubers or a tuber section containing some dormant buds in individual holes 60–75cm apart. Cuttings consisting of the top of a tuber with several leaves and a growth point can be planted directly into the soil and will establish rapidly. Add general fertiliser to the planting hole.

Growing
Eddoes need humus-rich, slightly acid, moisture-retentive soil, in sunshine or partial shade. Cultivate the soil before planting and remove any weeds.

Maintenance
Spring Plant tubers if protected cropping.
Summer Feed every 3–4 weeks with a high-potash fertiliser; additional nitrogen may be needed. During drier periods, irrigate as needed to ensure swelling of the tubers and earth up. Keep weed-free.
Autumn Harvest as required.
Winter Prepare beds for the following year's crop if growing in a greenhouse.

culinary

Tubers can be boiled, baked, roasted, purééd and made into soup. They can also be fried.

Protected Cropping
If you have space in a greenhouse or polythene tunnel, it is worth trying to grow eddoes. Plant presprouted or chitted tubers in spring into growbags or 20–30cm pots containing peat-substitute compost. Maintain heat and high humidity: damp down the glasshouse floor or mist plants with soft tepid water.

Harvesting and Storing
Eddoes take 5–6 months to mature. Harvest when the stems begin to turn yellow and die back. Lift tubers carefully with a garden fork; select some for eating and save others for replanting. If they are undamaged and dried carefully, tubers can be stored for several months.

Pests and Diseases
Eddoes grown outdoors are generally problem-free but when grown under glass they are susceptible to aphids and fungal leaf spots; red spider mite can also be a problem, particularly when humidity is low.

Downy mildew can attack tubers after they have been lifted so it is important to ensure that they are dried well before being stored. Dispose of infected tubers and do not use them for propagation.

warning

When handling and peeling eddoes, be sure to wear gloves or cover the hands with a layer of cooking oil to prevent a nasty rash.

Colocasia esculenta var. *esculenta*. *Araceae*

DASHEEN

Also known as Elephant's Ear, Arvi Leaves, West Indian Kale, Taro, Coco Yam. Herbaceous perennial grown for its edible leaves, shoots and tubers. Tender. Value: tuber rich in starch; leaves high in vitamin A, good source of B2.

In cultivation for around 7,000 years, dasheen is said to have been first grown in India on terraces where rice now flourishes. The common name derives from *de Chine* (from China): the root was imported from Southeast Asia following a competition organised by the Royal Geographical Society to find a cheap food source for the slaves on West Indian sugar plantations.

Dasheen thrives in flooded conditions

varieties

Dasheen have a cylindrical main tuber with fibrous roots and a few side tubers. The upright stems up to 1.8m tall are topped with large, heart-shaped leaves with prominent ribs on the underside. Cultivated types rarely flower and are grouped by the colour of their flesh, ranging from pink to yellow, and leaf stems of green, pinkish purple to almost black.

cultivation

Propagation

Dasheen are propagated from 'tops' with a small section of tuber, small side tubers or 'suckers'. Plant 60cm apart with 100cm between rows, or 60–90cm apart; add general fertiliser to the hole before planting.

Growing

Dasheen tolerate quite heavy, fertile, moisture-retentive soil rich in organic matter with a pH of 5.5–6.5. Dig in compost or well-rotted manure if necessary. As dasheen need plenty of water and tolerate waterlogging, they are ideal for areas by streams and rivers. Where the water table is high, mound or ridge planting is advised. Irrigate heavily during dry weather.

In well-manured soil, a second crop can be planted between the rows 12 weeks before the main crop is harvested.

DASHEEN

Maintenance
Spring Plant presprouted tubers with protection.
Summer Keep crops well fed and watered and weed-free.
Autumn Harvest crops.
Winter Prepare for the following year.

Protected Cropping
Plant pre-sprouted tubers in spring into greenhouse borders. Keep temperatures around 21°C, and maintain high humidity by misting plants with soft tepid water or damping down the greenhouse floor. Do not worry about over watering.

Feed every 3–4 weeks with a high-potash fertiliser; extra nitrogen may be needed if growth slows.

container growing

Dasheen can be grown as a 'novelty' crop in 20–30cm pots of peat-substitute compost. Soak thoroughly after planting and stand the pot in a shallow tray of tepid water throughout the growing season. Treat as for 'Protected Cropping'.

Dasheen roots

culinary

Tubers can be roasted, baked or boiled, served with spicy sauces and in stews. Larger tubers, which tend to be dry and coarse, should be braised and cooked slowly.

Leaves (with midrib removed) can be stuffed, boiled or steamed and eaten with a knob of butter. Avoid particularly large leaves; they are often tough. In the West Indies, Callaloo Soup is made from dasheen leaves, okra, crab meat and coconut milk. Young blanched shoots can be eaten like asparagus.

Palusima
This is a Western Samoan or Polynesian dish. Allow about 200g dasheen per person. Peel and chop roughly, then parboil in plain salted water for 5–10 minutes. Drain, then boil until reduced in coconut milk (enough to come to half the height of the dasheen in the pan) until thickish. Mash. Stuff the mashed dasheen into parboiled leaves and secure with a toothpick. (Alternatively, wrap it in banana leaves, and even spinach or cabbage.) Bake for 15 minutes in a lightly greased dish in a preheated oven at 180°C/350°F/gas mark 4. Serve with any good white fish such as cod.

Pests and Diseases
Taro leaf blight causes circular water-soaked spots on leaves followed by collapse of the plant. Those grown under glass are susceptible to aphids. Red spider mite and downy mildew can also be a problem. Take the necessary precautions.

Harvesting and Storing
Dasheen take 7 to 11 months to mature. Harvest by lifting the main tuber, saving some of the small side tubers for eating and others for replanting. Undamaged tubers can be dried and stored for up to 4 weeks, while washed leaves keep for several days in a refrigerator.

warning

Although selection has, over the years, reduced calcium oxalate levels in the skin of dasheen, it is extremely important to wear gloves or to cover the hands with a layer of cooking oil. This prevents skin irritation when peeling the vegetables.

Always make sure dasheen are cooked thoroughly before eating.

Crambe maritima. Brassicaceae

SEAKALE

Perennial grown for its blanched young shoots. Hardy. Value: an excellent source of vitamin C.

Found on the seashores of northern Europe, the Baltic and the Black Seas, seakale was harvested from the wild and sold in markets long before it came into cultivation. In Victorian times it was seen as an aristocrat of the vegetable garden and widely cultivated by armies of gardeners in the enormous kitchen gardens attached to great houses. Today it is rarely grown, perhaps because the scale on which it was forced for Victorian tables gave it a reputation for being labour-intensive. However, it is easy to grow at home, and quite delicious; so it is high time it experienced a revival!

Two essentials: healthy plants and a blancher

varieties

'Lily White' crops heavily with pale stems and has a good flavour. Unnamed selections of the wild species are also available from nurseries.

cultivation

Propagation

Seakale can be grown from seed, but it is usually propagated from crowns or root cuttings from the side roots, taken in autumn after the leaves have died back. These are called 'thongs'. Buy them from a nursery or select roots that are pencil-thick and 7.5–15cm long.

Make a straight cut across the top of the root and an angled cut at the base (so top and bottom are distinguishable). Store in sand until planting.

Growing

Seakale needs a sunny position on deep, rich, well-drained light soil with a pH of around neutral. The winter before planting, dig in some well-rotted compost; on heavier soils, add horticultural sand or grit, or plant on a raised bed.

SEAKALE

Seakale should crop for about 5–7 years before it needs replacing, so it is vital to prepare the ground thoroughly. Two weeks prior to planting, rake in a general fertiliser. Before planting, in early spring, rub off all the buds to leave the strongest bud and plant thongs or crowns 45cm apart, covered with 2.5–5cm of soil.

Water regularly, feed occasionally, remove flower stems as they appear and keep weed-free. In autumn, cut down the yellowing foliage. From late autumn until midwinter, cover the crowns with a bucket, flower pot (cover the drainage holes) or seakale forcer and surround with manure, straw or leafmould. Stop harvesting in late spring, rake in a general fertiliser and mulch with well-rotted manure or compost. Water well in dry periods. Delaying harvest until the second year lets plants become established.

Maintenance

Spring Force crops outdoors.
Summer Water in dry weather, feed occasionally, keep weed-free.
Autumn Remove the yellowing foliage.
Winter Lift crowns for forcing.

Seakale being forced

Protected Cropping

For an early crop, lift crowns after leaves die back, trim the main root to 15cm and remove the side shoots; plant 10cm apart in boxes or 3 per 23cm pot filled with rich soil from the seakale bed or loam-based compost with a high fertiliser content. Cover with upturned pots, polythene or anything opaque. Keep the compost moist. They should be ready in 5–6 weeks at temperatures of 10–13°C; from 16–21°C they will be ready earlier. Keep them in a cellar, boiler house or under greenhouse benches and maintain a constant supply by lifting crowns regularly.

In frosty weather lift and store crowns in moist sand until required. Dispose of exhausted crowns after use.

Harvesting and Storing

When blanched sprouts are 7.5–20cm long, remove the soil from around the shoots and cut through the stems with a sharp knife, removing a tiny sliver of root. Crops forced outdoors can be harvested in late autumn to early spring. After harvesting, discard the exhausted roots of those forced indoors.

Pests and Diseases

Do not plant in beds infected with clubroot. Flea beetle make round holes in the leaves; cover with mesh or horticultural fleece.

other uses

Seakale is attractive in the flower border. It forms a compact rosette of large, wavy-edged, glaucous leaves, with large bunches of white, honey-scented flowers towering above the foliage in summer.

culinary

Better eaten fresh, seakale lasts for 2–3 days in the salad compartment of the refrigerator.

Wash stems and tie them into bundles using raffia and steam or lightly boil them until tender – overcooking toughens the stems and makes them less palatable.

Victorians served seakale on folded napkins or on toast, drenched in white sauce or melted butter.

Seakale is also delicious served as a starter with béchamel or hollandaise sauce, or covered in lemon-flavoured melted butter.

Young flowering shoots can be eaten once they are lightly boiled.

Seakale Gratinée

Serves 4

750g seakale, washed well

For the béchamel sauce:
50g butter
2 tablespoons flour
1 teaspoon French mustard
150ml milk or crème fraîche
4 tablespoons grated strong Cheddar cheese
Salt and freshly ground black pepper

Rinse the seakale well, tie into bundles and cook until *al dente* in boiling salted water. Drain well. Arrange in an ovenproof dish.

Make the béchamel sauce, flavouring with the Cheddar cheese. Pour the sauce over the seakale and bake in a preheated oven, 200°C/400°F/gas mark 6 for 15 minutes. Serve hot.

Crithmum maritimum. Apiaceae

SAMPHIRE

Also known as Rock Samphire, Sea Fennel, Sea Samphire. Low-growing maritime perennial. Half hardy. Value: good source of iron and vitamin C; moderate iodine.

Rock samphire is found on the shores and cliffs of Europe. Its name comes from 'sampiere', a contraction of the French *herbe de St Pierre* – the fisherman Saint's herb. Collected from the wild for centuries, by the English Tudor period it was widely cultivated in gardens. William Turner wrote: 'Creta marina groweth much in rockes and cliffes beside Dover.' This precarious habitat is mentioned in *King Lear*, where harvesters dangle over the cliffs from a rope. Robert Turner in 1664 described similar dangerous activities on the 200m cliffs of the Isle of Wight, 'yet many adventure it though they buy their sauce with the price of their lives'. Samphire was sent in casks of seawater to London, where wholesalers paid four shillings a bushel, but for the privilege of collecting it and gulls' eggs the Lord of the Manor exacted an annual rent.

varieties

Crithmum maritimum has a woody base, with stems to 60cm and lobed, grey-green leaflets. The white/cream flowers grow in flat clusters and are followed by small oval fruits. The leaves have an aromatic odour which has been likened to the smell of furniture polish!

cultivation

Propagation
Plants are propagated from fresh seed sown in autumn or spring in a sheltered position. Transplant when large enough to handle, or divide plants in spring.

Growing
Its natural habitat is in sand; if you create a satisfactory habitat in a coastal garden, natural colonies may form. Plants often inhabit dry stone walls. They flourish in well-drained, sandy or gritty soil, which is constantly moist and protected from full heat. An open east- or south-facing position is ideal. Mulch with seaweed, or burn seaweed and scatter with the sodium-rich debris. If possible, water with seawater or sea salt solution.

Maintenance
Spring Sow seed in gritty compost or in shallow drills.
Summer Keep soil moist with a saline solution. Harvest.
Autumn Harvest. Protect crops during colder weather.
Winter Mulch with straw or leaves.

Protected Cropping
As a succulent plant, rock samphire needs frost protection. Cover with leaves, straw or cloches.

Harvesting and Storing
Harvest young shoots and leaves by cutting or pulling. Young spring growth is the most tender. Do not harvest excessively from each plant. It is best eaten immediately after harvesting, but will last for up to 2 days in a fridge.

Pests and Diseases
Usually trouble-free.

container growing

Grow in containers of gritty, loam-based compost. Water regularly in summer; in cooler climates, bring plants indoors during the winter.

medicinal

John Evelyn noted 'its excellent vertues and effects against the Spleen. Cleansing the Passages and sharpning appetite.' It was also recommended as a kidney, bladder and general tonic and as a treatment for 'stones'. It is said to be a diuretic, to improve digestion and has been used to encourage weight loss.

culinary

Wash thoroughly in running water before use. Tender young shoots and leaves can be eaten fresh or cooked as a vegetable. They are added raw to salads or dressed with oil and lemon juice as an hors-d'oeuvre.

Pickle young shoots, leaves and stems by filling a jar with samphire cut into 2.5cm lengths, add peppercorns and a little grated horseradish, pour over a boiling mixture of equal parts dry cider and vinegar, and infuse for an hour before sealing. Pickled, it is used as a garnish and as a caper substitute. In Italy it is known as *roscano*.

Cucumis sativus. Cucurbitaceae

CUCUMBER

Climbing or scrambling annual, grown for elongated or round succulent fruits. Tender. Value: moderate potassium and small amounts of beta carotene.

The wild species from central Asia is now rare in nature, yet the world-renowned salad crop has been cultivated for centuries. The first record was in Mesopotamia around 2000 BC in the earliest known vegetable garden, and cucumbers were grown in India a thousand years later. The Romans in the first century AD cultivated them in baskets or raised beds mounted on wheels so they could be moved around 'as the sun moved through the heavens'. When the day cooled, they were moved under frames or into cucumber houses glazed with oiled cloth known as *specularia*. Tiberius found them tasty and was said to have eaten them every day of the year.

Early varieties were quite bitter and were boiled and served with oil, vinegar and honey. They were a common ingredient in soups, stews and as a cooked vegetable until the nineteenth century. Eighteenth-century English recipes include cucumbers stuffed with partly cooked pigeons (with head and feathers left on: the idea was to make the head appear attached to the cucumber); the whole was then cooked in broth and the heads garnished with barberries. Cooks in Georgian England obviously had a rather bizarre sense of humour!

Columbus introduced cucumbers to the New World. They are recorded as being planted in Haiti in 1494 and grown by English settlers in Virginia in 1609. About the same time, French writers Estienne and Liébault warned: 'Beware that your seed be not olde, for if it be 3 years olde, will bring forth radishes.' Obviously their soil was as fertile as their imagination.

varieties

Your seed catalogue will give an idea of the huge number of cultivars to choose from. There are 'greenhouse cucumbers' (though many indoor varieties can grow outdoors in a sheltered position in cooler climates and grow outdoors in warm zones); 'ridge' or outdoor types, which need protection as seedlings but can be grown outdoors in cool temperate climates; pickling cucumbers or gherkins; round varieties; Japanese climbing and bushy types.

Greenhouse or indoor cucumbers

'Crystal Apple' is a small, round, yellow cucumber that is prolific and easy to grow in the unheated greenhouse or outdoors. **'Danimas'** is a vigorous, 'all-female' mini-cucumber, ideal for a slightly heated or cold greenhouse. **'King George'** produces good-quality, long, dark fruit. **'Flamingo'** yields masses of fruit so only requires a few plants. **'Telegraph'** is popular and reliable with smooth skin and good-sized fruits.

Outdoor or ridge cucumbers

'Bianco Lungo di Parigi' produces moderately-sized fruits with creamy-white skin. **'Boothby Blonde'** has been grown for generations by the Boothby family of Maine in USA – tasty, crisp and mild. Harvest at 7.5cm long. Many older varieties have been grown from the seed saved from the previous year's crop. **'Burpless Tasty Green'** is tender, delicious, crisp and the best. **'Bush Champion'** produces bushy, dark green fruit; ideal for patios. **'Long Green Improved'** is robust and highly productive with large, tasty fruits that can also be used for pickling. **'Long White Paris'** is sweet and easy to digest with low acidity. **'Marketmore'** is smooth, tasty and excellent for cooler conditions.

Pickling cucumbers or gherkins

'Athene' is for outdoors or an unheated glasshouse or frame. The sweetness of **'Beit Alpha'** is ideal for salads or pickling. **'Gherkin'** is fast-growing, with masses of small prickly fruits. **'Midget'** is a prolific and compact variety. **'National Pickling'** is short, blunt-ended with small, smooth fruits. It is vigorous and heavy-yielding

Crystal Apple with male and female flowers

and was introduced in 1929 by the National Pickle Packers Association in Britain. **'Vert Petit de Paris'** is a French heritage variety with good flavour and yield. **'West Indian Gherkin'** is a reliable cropper whatever the weather. Use peeled in a salad, or cook unpeeled in a relish or pickle.

cultivation

Propagation
When growing outdoor cucumbers, sow 2 seeds edgeways 1–2cm deep in a 5–7.5cm pot from mid-spring. Place in a propagator or heated greenhouse at 20°C.

Retain the stronger seedling after germination, keep moist with tepid water and tie up small canes. Feed with a general liquid fertiliser to boost growth.

Give plants plenty of light and harden off before transplanting with care to avoid checks, when the danger of frost has passed. Allow 90cm between plants. Protect with cloches or horticultural fleece until established.

Sow successionally to extend the cropping season. Alternatively, pregerminate or sow 2–3 seeds, 1–2cm deep *in situ* under jam jars or cloches from late spring to early summer when the soil temperature is around 20°C. Thin to leave the strongest seedling.

The second method is preferable, as cucumbers do not transplant well.

Growing
'Ridge' cucumbers need rich, fertile, well-drained soil in a sunny, sheltered position. A few weeks before planting, dig out holes or longer ridges at least 45cm wide and 30cm deep. Half fill with well-rotted manure or good garden compost, then return the excavated soil, mounding it to about 15cm above ground level.

Better fruits are obtained when plants are grown against supports, though traditionally they are left to trail over the ground.

Once plants have produced 5–6 leaves, pinch out growing points of the stems, allow 2 laterals to form and pinch out the growing tips again. Ridge cucumbers are insect-pollinated, so do not remove male flowers as you would with greenhouse cultivars.

Grow gherkins like ridge cucumbers, or train them up some netting.

Train Japanese cultivars up trellis, cane tripods, wire or nylon netting. Nip out the growing point when stems reach the top of the support. Water regularly; feed with a high-potash fertiliser every 2 weeks when fruits are forming.

Keep crops weed-free and the soil moist.

Maintenance
Spring Sow under cover from mid-spring.
Summer Plant ridge cucumbers outdoors once frost is past. Maintain high temperatures and humidity indoors. Train and harvest.
Autumn Harvest.
Winter Store in a cool place.

Protected Cropping
Sow under cover or in a propagator from mid-spring at 21–25°C. To avoid erratic germination, seeds can be pre-germinated on moist kitchen towels in a covered plastic container, then placed in an airing cupboard or propagator. After about 3 days, when the seeds have germinated, sow in 7.5cm pots of seed compost. Sow from late spring if no heat can be provided. Fill 8cm pots with seed compost and press the seed down edgeways into each pot about 1cm deep.

Greenhouse cucumbers need a humid atmosphere and temperatures around 20°C. Soils should be rich, moisture-retentive and free-draining, so incorporate well-rotted organic matter if you are planting in the greenhouse border. Even better, grow in 25cm pots of soil-based compost with added organic matter, in growbags, or in untreated straw bales with a bed of compost in the centre.

Insert a cane at the base of the plant and tie it into the roof structure of the greenhouse. Tie in the main stem and remove the growth tip when it reaches the roof, pinching out the laterals 2 leaves beyond each fruit. Greenhouse cucumbers should not be pollinated or they become bitter: remove male flowers. Female flowers are identified by the swelling behind each one. Or choose 'all-female' cultivars.

Mist plants regularly with tepid water and keep the compost moist with tepid water. Feed every 2 weeks with high-potash fertiliser once fruits begin to swell. Outdoor cucumbers can be grown in cold frames or a cold greenhouse in cooler climates. They are better trained above the ground. Ventilate well on warm days and mist regularly.

A cucumber seedling

CUCUMBER

Plenty to enjoy and give away

Harvesting and Storing

Fruits should never be harvested until they are fully ripe, and the sixteenth-century practice of leaving fruit on until they were mottled brown and yellow with a rich flavour may be to your taste!

Outdoor cucumbers crop mid-summer to early autumn: cut with a sharp knife when they are large enough to use. For maximum yields, you should pick fruit regularly. Cucumbers last for several days in the salad drawer of a refrigerator. Cover the cut end with cling film and use as rapidly as possible. Or stand it stalk end down in a tall jug with a little water in the bottom. Like marrows, they can be stored in nets in a cool place.

Pests and Diseases

Red spider mite, aphids, slugs and powdery mildew can be troublesome.

Cucumber mosaic virus shows as mottled, distorted leaves. Burn young infected plants and leaves from older plants. Older plants may recover, though yields will be lower.

culinary

Eat while the stalk end is still firm. Some people remain adamant that peeled cucumbers are best: others think the flavour and appearance of the skin enhance the vegetable.

They are frequently sliced and eaten in salads, and are good in sandwiches with salmon. Mix with yogurt and mint as a side dish to Middle Eastern dishes or curries. Make into a delicious cold soup.

To eat them as a vegetable, peel, seed, dice and stew them in a little water and butter for 20 minutes, until soft. Thicken with cream and serve with mild- or rich-flavoured fish.

Cucumber and Cream Cheese Mousse

Serves 4

½ a cucumber, in chunks
275g cream cheese
1–2 tablespoons mint leaves
2 teaspoons white wine vinegar
2½ teaspoons gelatine
150ml vegetable stock
Salt and freshly ground black pepper
Radicchio leaves and sprigs of fresh mint, to serve

Put the cucumber into a blender with the cream cheese, mint and vinegar and purée until smooth. Dissolve the gelatine in a little stock over a low heat. Leave to cool, then stir in the balance of the stock. Add this to the cream cheese, season and blend. Chill for at least 2 hours before serving, arranged individually with radicchio leaves, garnished with a few fine slices of cucumber and sprigs of mint.

companion planting

Ridge cucumbers thrive in the shade of maize or sunflowers, and grow well with peas and beans, beet or carrots.

Climbing cucumbers flourish, scrambling over sweetcorn and beans.

container growing

Sow indoor types in pots or growbags (see 'Protected Cropping'). Smooth-skinned varieties can be grown in a growbag or container 30cm wide by 20cm deep in a sunny position outdoors. Sow 3 seeds 2.5cm deep in late spring or early summer.
Thin out to leave the strongest seedling; pinch out growing tips when the plant develops 6–7 leaves. Train the side shoots on netting or canes. Keep soil moist. Feed with high-potash fertiliser when fruits form. **'Bush Champion'** is ideal.

medicinal

Cucumbers were used by the Romans against scorpion bites, bad eyesight and to scare away mice.

Wives wishing for children wore cucumbers tied around their waists, and they were carried by midwives and thrown away once the child was born.

Cucurbita maxima, Cucurbita moschata and *Cucurbita pepo. Cucurbitaceae*

PUMPKIN & SQUASH

MARROW, COURGETTE, ZUCCHINI, POTIRON

'On the coast of Coromandel
Where the early pumpkins blow
In the middle of the woods
Lived the Yonghy-Bonghy-Bo...'
Edward Lear (1812–88)

Annuals grown for edible fruits, often colourful, which can be extremely large. Tender. Value: high in beta carotene, moderate amounts of vitamin C and folic acid.

The name 'pumpkin' appeared in the seventeenth century, shortly before Perrault wrote Cinderella, the tale about a poor girl whose fairy godmother turned a pumpkin into a golden coach which took her to the ball. 'Pumpkin' comes from the Greek word for melon – *pepon* or 'cooked by the sun' – while one French name, *potiron*, means 'large mushroom', from the Arabic for morel mushrooms. 'Squash' is an abbreviation of the native North American Indian word *askutasquash*, meaning 'eaten raw or uncooked'.

The squashes originated in the Americas and are believed to have been cultivated for between five and ten thousand years. Wild forms were originally gathered for their seeds and were only later found to have sweet flesh. Many varieties arrived in Europe soon after the discovery of the New World in the sixteenth century. Not only were they eaten, but the seeds were pounded in oatmeal and applied to the face, to bleach freckles and other blemishes.

Estienne and Liébault wrote in 1570: 'To make pompions keep long and not spoiled or rotted, you must sprinkle them with the juice of a houseleek'. In the seventeenth century they were mashed to bulk up bread, or boiled and buttered.

This group contains a wealth of edible and outstandingly ornamental fruits. One of the most beautiful sights in the kitchen garden is that of pumpkins ripening in golden sunshine during the autumn.

PUMPKIN & SQUASH

Provide enough space to spread

varieties

Members of the genus *Cucurbita* are bushy or trailing annuals – sometimes extremely vigorous plants – bearing a wide range of edible and/or ornamental fruit. The fruits from all of these species are often grouped together according to their shape or time of harvest – crookneck, summer squash, winter squash and so on; in practice, categories overlap and some are multi-purpose, being served differently when young and when mature. Additional confusion arises because of local variation in what is grown and what it is called. Here they are classed according to species.

The *Cucurbita maxima* group has large, variable fruits and includes most traditional pumpkins and winter squashes, containing several ornamentals like the banana, buttercup, hubbard and turban types. They tend to have hard skins when mature, and keep well; the yellow flesh needs cooking. They flourish in low humidity from 20–27°C, though some tolerate cooler conditions.

'Atlantic Giant' is not for the faint-hearted as it can grow to 317kg. **'Banana Pink'** is long, broad and curved, with pale pink skin. **'Big Max'** is a massive pumpkin with rough, red-orange skin and bright flesh. It is excellent for pies, exhibitions and as a 'giant' vegetable. **'Buttercup'** is delicious with firm, dense, sweet flesh. The skin is dark green with pale narrow stripes and the flavour is perfect for soups, roasting and pumpkin pie. **'Crown Prince'** is small with tender orange flesh. It's tasty and keeps well. **'Queensland Blue'** is an attractive small variety with blue-grey skin – very tasty. **'Turk's Turban'** (**'Turk's Cap'**) is a wonderful ornamental squash; orange with cream and green markings. The name aptly describes the shape. **'Warted Hubbard'** is a small, round fruit with extraordinary dark green, warty skin and orange-yellow flesh. It keeps well. **'Whangaparoa Crown'** pumpkin is a hard, grey-skinned variety with a pronounced crown and orange flesh. It also stores well.

Cucurbita pepo embraces summer squashes (including courgettes or zucchini and their larger version, the marrow), non-keeping winter pumpkins and ornamental gourds such as custard squash, plus straight and crook-necked types. The fruits are usually soft-skinned, especially when young, and may be served raw when small. One variety, **'Little Gem'**, is slow to mature, taking about 4 months. It doesn't perform well in cooler conditions but is good for storage.

Summer Squash

'Early Golden Summer' Crookneck is an early cropper with bright yellow fruits and excellent flavour. Harvest when 10cm long. **'Early Prolific'** Straight Neck is yellow with finely textured flesh. Pick at 15cm. **'Vegetable Spaghetti'** (**'Spaghetti Squash'**) is pale yellow when mature. Boil or bake fruits whole and then scoop out the flesh inside; it looks like spaghetti. **'White Patty Pan'** (**'White Bush'** Scallop) has an unusual flattened shape with a scalloped edge. It is better harvested and cooked whole when about 7.5cm in diameter. It is bushy and ideal for small gardens. **'Yellow Bush'** Scallop is an old variety with bright yellow skin and coarse, pale yellow flesh that has a distinctive flavour.

Marrow or *Cucurbita pepo*

Even the 'Atlantic Giant' starts small

Marrow

This elongated type of summer squash has long been popular in Britain.

'Long Green Trailing' is a prolific, long-fruited, dark green variety – fine quality and flavour. **'Tiger Cross'** is an early, green bush-type that crops well and yields good-quality fruits. Resistant to cucumber mosaic virus. **'Badger Cross'** is compact, high yielding and crops early. Ideal for containers or small gardens.

Courgette or Zucchini

Varieties of marrow bred for picking small, the following are all bush types and ideal for the smaller garden. **'Ambassador'** is high-yielding with dark green fruit and reliable cropping. **'Defender'** produces high yields of mid-green fruits. Harvest regularly. Resistant to cucumber mosaic virus. **'De Nice à Fruit Rond'**, a round variety, should be picked when the size of a golf ball – delicious flavour. **'Partenon'** fruits without pollenation and is good for early crops or in cool

PUMPKIN & SQUASH

weather. **'Spacemiser'** is a compact and prolific gourmet variety. **'Supremo'** produces very tasty, dark green fruit.

Winter squash

These usually have white or pale yellow flesh, whereas pumpkins have coarse, orange flesh. **'Ebony Acorn'** (**'Table Queen'**) is early-cropping with thin, dark green skin and sweet, pale yellow flesh; delicious baked with honey. It is a semi-bush but can be temperamental. **'Jack be Little'** is a miniature pumpkin with deep ribbed, orange-skinned fruits about 5 x 7.5cm in diameter. You can eat them but they are often more attractive as an autumn decoration. **'Small Sugar'** has rounded orange fruits growing to 18cm diameter. The flesh is tender, yellow and excellent in sweet or savoury dishes. It matures from late autumn.

Cucurbita moschata includes the early butternut, butternut, Kentucky Field and crookneck squashes, harvested in autumn and winter. They are large, rounded and usually have smooth, tough skin. Possibly one of the earliest species in cultivation, they are widely grown and found throughout the tropics. Particularly heat-tolerant.

'Butternut' has pale tan, club-shaped fruits with bright orange flesh. It succeeds in cooler areas and the flavour improves with keeping. **'Early Butternut'** is a curved, narrow fruit with a swollen tip. This bush variety matures rapidly and keeps well. **'Neck'** squash, with straight or curled necks, is tasty, high-yielding and the best for pies. **'Triple Treat'** is a bright orange, round fruit. Its seeds are particularly good eaten raw, fried or roasted. It is easy to carve and is therefore a variety that is often grown for Halloween. **'Waltham Butternut'** has a smooth, pale tan skin, yellow-orange flesh and a nutty taste. It is very good for storing and yields extremely well.

cultivation

Propagation

Sow *in situ* or in pots in cooler climates as seeds do not germinate if the soil is below 13°C.

From mid- to late spring, soak the seed overnight, then sow one seed edgeways, about 2.5cm deep in a 7.5cm pot or module of moist multi-purpose or seed compost and place in a propagator or on a warm windowsill, preferably at 20–25°C. After germination transplant the seedlings when they are large enough to handle into 12.5cm pots, taking care not to damage the roots. Keep compost moist but not waterlogged. Harden off gradually and transplant from late spring to early summer when the danger of frost has passed. Protect with cloches until the plants are established.

Alternatively, sow *in situ* when the soil is warm and workable and there is no danger of frost, from late spring to early summer. Dig out a hole at least 30–45cm square and half fill with well-rotted manure 7–10 days before planting. Sow 2–3 seeds 2.5cm deep in the centre of the mound and cover with a jam jar or cut the base from a plastic bottle and use as a crop cover. After germination, thin to leave the strongest seedling, remove the cover and mark the position with a cane so you know where to water the plant among the mass of stems. Alternatively, prepare the ground as described and transplant seedlings.

Space cultivars according to their vigour. Sow bush varieties on mounds or ridges, 60–90cm apart with 90–120cm between rows. Trailing varieties should be 120–180cm apart with 180–360cm between rows.

Even the flowers are edible

Only female flowers produce fruit

Growing

Pumpkins and squashes need a sunny position in rich, moisture-retentive soil with plenty of well-rotted organic matter and a pH of 5.5–6.8.

It is a good idea to plant through a black polythene mulch laid over the soil with the edges buried to hold it in place. This warms the soil, suppresses weeds, conserves moisture and protects ripening fruit.
A thick layer of straw or horticultural fleece are useful alternatives, or you can lay the ripening fruit on a roof tile, a piece of board or similar to protect it from the soil and prevent rotting.

Hand-pollination is recommended, particularly in cold weather when insect activity is reduced. Female flowers have a small swelling, the embryonic squash, immediately behind the petals, while male flowers have only a thin stalk.

When the weather is dry, remove a mature male flower, fold back or remove the petals and dust pollen on to the stigma of the female flowers. Alternatively, transfer the pollen with a fine paintbrush. Periods of hot weather can reduce the ratio of female to male flowers.

Plants need copious amounts of food and water, particularly when flowers and fruits are forming, up to 11 litres of water per week, but they should never be allowed to become waterlogged.

Feed with a liquid general fertiliser every 2 weeks. Plants grown with a black plastic mulch also need an occasional foliar feed to boost growth.

Pinch out tips of main shoots of trailing varieties when they reach 60cm to encourage branching and trim back those that outgrow their position.

Small trailing types can also be trained over trellis or supports. Where space is limited, push a circle of pegs into the soil and trail the stems around the pegs.

To guarantee large fruits, allow only 2–3 to develop on each plant.

Keep crops weed-free.

Maintenance
Spring Sow seeds under glass or outdoors.
Summer Feed and water copiously; keep crops weed-free. Harvest courgettes and marrows.
Autumn Harvest pumpkins and winter squashes, allow to ripen and protect from frost.
Winter Store winter squashes until midwinter or later.

Protected Cropping
In cooler climates or to advance growth, sow seed indoors and transplant. Protect with cloches or crop covers until they are well established. Use bush varieties for earlier crops.

Harvesting and Storing
Pick courgettes or zucchini and summer squashes when they are about 10cm long and still young and tender. Marrows are harvested when they have reached full size. Push your thumbnail gently into the skin near the stalk; if it goes in easily then the marrow is ready for harvest.

Cut them from the stem, leaving a short stalk on the fruit, and handle with care to avoid bruising. Harvest regularly for continual cropping.

They can be stored in a cool place for about 8 weeks. Courgettes can be kept in a polythene bag in the refrigerator and will stay fresh for about a week.

Plenty of sunshine ensures early ripening

Courgettes are suitable for freezing. Cut into 1cm slices, blanch for 2 minutes, cool, drain and dry. Freeze in polythene bags. Flesh of winter squashes and pumpkins can be cooked, then frozen, without any loss of flavour.

Towards the end of the growing season, remove any foliage that shades the fruits. Harvest pumpkins and winter squashes from late summer to autumn, though they must be brought into storage before the first heavy frosts. On maturity, the foliage rapidly dies, the skin hardens and stem starts to crack.

After harvest, leave them outdoors for about 14 days, as cold weather improves the taste and sugar content, hardening the skin and sealing the stem. Protect from heavy frost with a covering of hessian, straw or similar.

In cooler areas, they can be ripened in a greenhouse or on a sunny windowsill. Pumpkins will last until mid-winter when stored in a frost-free shed.

Store winter squashes at a minimum temperature of 10°C; they can last for up to 6 months. Some Japanese varieties will last even longer.

Pests and Diseases
If fruits show signs of withering, water and feed more often. Aphids, powdery and downy mildew and slugs can be a problem.

Cucumber mosaic virus causes yellow mottling and puckering of the leaves and rotting of the fruit. Destroy infected plants immediately and control aphids, which transmit this virus.

companion planting

Grow courgette and marrow alongside sweetcorn for support and shade, and with legumes, which provide essential nitrogen.

container growing

Courgettes, marrows, bush varieties of other squashes and those that are moderately vigorous can be grown in growbags or containers that are at least 35cm by 30cm deep. Use a loam-based compost with additional well-rotted organic matter. Sow indoors to transplant later or sow directly outdoors. Keep plants well watered and do not allow the compost to dry out. Plants can be grown up strong canes 2m tall. Pinch out growing points when stems reach the top; tie the main stem and side shoots firmly to the supporting canes.

Hand-pollinate for successful cropping.

other uses

Pumpkins are hollowed out and made into Halloween 'Jack o' Lanterns'. Mature marrows can be used for wine making.

medicinal

In Ethiopia, seeds from squashes are used as laxatives and purgatives; they are used worldwide to expel intestinal worms. Eating winter squash and pumpkin is said to reduce the risk of prostate cancer.

warning

Be careful when lifting large squashes: bend from the knees, not the back!

PUMPKIN & SQUASH

culinary

Flowers of all varieties can be used in salads or stuffed with rice or minced meat and fried in batter. Prepare the meat, rice and batter before picking the flowers, as they wilt quickly. They can also be puréed and made into soup. Young shoots are steamed or boiled. Pumpkin flesh is used for pies; their seeds are deep-fried in oil, salted and are known as 'pepitos'. The fruits can be stuffed, steamed, stir-fried, added to curries, made into jam or pickles, and the seeds are edible too.

Pumpkin Kibbeh
Serves 4

A centrepiece of many Middle Eastern meals, these 'balls' should be served warm, rather than hot. This recipe was given to me by Arto der Haroutunian.

200g freshly boiled pumpkin flesh
400g bulgar wheat
150g plain flour
1 onion, finely chopped
Salt and freshly ground black pepper

For the filling:
2 shallots, finely chopped
250g spinach, washed
150g cooked chickpeas, drained
50g walnuts, chopped
50g dried apricots
¼ teaspoon sumac
1 tablespoon lemon juice
Salt and freshly ground black pepper
Oil for frying

In a large bowl, purée the pumpkin, using a fork. Sift the bulgar and flour into the bowl and mix in the onion and seasoning. Leave in a cool place for 10–15 minutes. If the dough is too hard to handle, you may need to add a tablespoon of water and knead well.

To make the filling, sauté the shallots in the oil until they just turn brown. Mix in the spinach and allow to wilt, stirring constantly, for a couple of minutes. Then add the rest of the ingredients, mixing well.

Make the kibbehs with wet hands to prevent the mixture from sticking. Form the bulgar and pumpkin mixture into oval patties – they should be about 8cm long and just big enough to stuff. Create an opening at one end and fill each kibbeh with the spinach and apricot stuffing. Using your fingers, seal up the ends.

Fry the kibbehs in hot oil for a couple of minutes on each side and drain on kitchen paper. There should be enough to make between 20 and 24 kibbehs.

Courgette Omelette
Serves 2

5 eggs
300g courgettes
4 tablespoons olive oil
1 tablespoon fresh (purple) basil leaves
1 tablespoon fresh thyme leaves
Salt and freshly ground black pepper

Whisk the eggs with salt and pepper and set aside. Slice the courgettes into coarse dice. Heat half the oil and sauté the courgettes for a couple of minutes. Then remove from the heat.

In an omelette pan, heat the balance of the oil and pour in the eggs, courgettes and herbs. Stir gently over a low heat while the omelette sets. Turn it on to a plate and slide back into the pan to cook the second side for a minute or so until nicely browned. Serve as a refreshing Sunday supper dish.

Courgette Omelette

Acorn Squash with Balsamic Vinegar

Acorn Squash with Balsamic Vinegar
Allow 100g of acorn squash per person. Cut in half and remove the seeds and fibres. Place in a buttered ovenproof dish and pour over 1 tablespoon balsamic vinegar, 2 tablespoons runny honey and 1 tablespoon lemon juice for each serving. Cook in a preheated oven, 180°C/350°F/gas mark 4 for 40 minutes, turning over halfway.

Custard Marrow with Bacon and Cheese
Serves 2–4

500g custard marrow, roughly diced
2 tablespoons butter
1 small onion, finely sliced
75g smoked back bacon, diced
150ml crème fraîche
4 tablespoons grated mature Cheddar cheese
Salt and freshly ground black pepper

Steam the custard marrow until just tender. Keep warm. Make a sauce by heating the butter and cooking the onion until softened. Add the bacon and continue cooking for 5 minutes, stirring from time to time. Mix in the crème fraîche, the custard marrow and seasoning and pour into a greased ovenproof dish. Top with the cheese and cook in a preheated oven, 225°C/425°F/gas mark 7, for 15 minutes.

Cynara cardunculus. Asteraceae

CARDOON

Also known as Cardon. Perennial grown as annual for 'heart' and blanched leaf midribs. Half hardy. Value: rich in potassium.

This close relative of the globe artichoke is found in the wild through much of the Mediterranean and North Africa. Cultivated versions are valued as a vegetable and in the ornamental garden. When grown as a food crop, the stems are blanched during autumn in a similar manner to celery. Its delights have been enjoyed for centuries; it was grown before the birth of Christ and was esteemed by the Romans, who paid high prices for it in their markets as an ingredient for stews and salads. Cardoons reached England by 1658 and North America by the following century, but never established themselves as a major crop, despite their popularity in Europe. Today they are more likely to be found in the herbaceous border, where the bold angular foliage and tall candelabras of thistle-like purple flowers are outstanding.

The leaves are ready for blanching

varieties

'Gigante di Romagna' is a reliable variety with long stalks. **'Gobbo di Nizza Monferrato'** is a large plant with a strong celery flavour. Use in soups or eat fried, sautéed or raw, dipped in olive oil. **'Plein Blanc Inerma Ameliora'** grows to 120cm tall, with white ribs that are well textured and tasty.

cultivation

Cardoons need a sunny, sheltered site on light, fertile, well-drained soil.

Propagation

Sow *in situ* in mid-spring, planting 3–4 seeds 2.5cm deep in 'stations' 50cm apart with 1.5m between rows. Thin after germination to retain the strongest seedling. In cooler areas or if spring is late and the soil is yet to warm up, sow indoors. Place 3 seeds in 7.5cm pots or modules of moist seed compost in a propagator or glasshouse at 13°C. Thin, leaving the strongest seedling, then harden off before planting outdoors in mid- to late spring, when there is no danger of frost. Water well after planting and protect from scorching sunshine until they have established.

Growing

The autumn or early spring before seed sowing or planting out seedlings, double dig the site, adding well-rotted organic matter. Alternatively, plant in trenches 38–50cm wide and 30cm deep; dig these in late autumn or early spring, incorporating plenty of rotted manure or compost into the base and refilling to 7.5–10cm below the surface. Leave the remaining soil alongside for earthing up. Before planting, rake in general fertiliser at 60g/sq m. In late summer to early

CARDOON

Cardoons make attractive ornamentals

autumn, on a day when the leaves and hearts are dry, begin blanching: pull stems into a large bunch (wear long sleeves and gloves for protection), and tie with raffia or soft string just below the leaves. Wrap cardoons with 'collars' of newspaper, corrugated cardboard, brown wrapping paper or black polythene tied firmly around the stems. Tile drainpipes and plastic guttering are just as effective at excluding the light. Support collars with a stake, particularly on exposed sites. Alternatively, cardoons can be earthed up. Cover stems with dry hay, bracken or straw held firmly at several points with twine and cover with soil, banked at an angle of 45°. The first method is easier, cleaner and quicker.

Cardoons need a regular water supply from early summer through to early autumn, and feeding with liquid general fertiliser every 2 weeks. Keep them weed-free by hand weeding, hoeing or, preferably, mulching with a 5cm layer of organic matter once plants are established.

Maintenance

Spring Sow seeds *in situ* or under glass.
Summer Feed and water regularly. Keep weed-free.
Autumn Harvest crops using a sharp knife. Prepare the planting bed.
Winter Store in a cool, dry place until required.

Protected Cropping

Protect transplants or seedlings under cloches or horticultural fleece if late spring frosts are forecast.

Harvesting and Storing

Blanching takes about 3–4 weeks. When ready to harvest, lift plants with a garden fork, trim off roots and remove outer leaves. Cardoons can remain in the ground until needed, but protect with bracken, straw or other insulating material in moderate frosts. If hard frosts are forecast, lift and store in a cool shed or cellar.

Pests and Diseases

Cardoons are robust and have few problems. They can be affected by powdery mildew, worst when plants are dry at the roots and if nights are cold and days warm and dry. Mice will eat the seeds. An old remedy is to dip the seeds in paraffin; concentrated liquid seaweed is better. Otherwise buy humane traps or a cat!

container growing

Grow in large containers of loam-based compost with added organic matter or in free-draining, moisture-retentive soil. Allow plenty of room for leaves to grow. Water and feed regularly: do not let the compost dry out. Blanch using 'collars'.

culinary

The leaf midribs and thinly sliced hearts are eaten raw in salads, in soups and stews or as an alternative to fennel or celery. They can be boiled in salted water with a squeeze of lemon juice for about 30 minutes until tender. Once cardoons are cut, drop them into water with a squeeze of lemon juice, as the cut surfaces do tend to blacken.

The dried flowers are used as a substitute for rennet in Spain and some parts of South America.

Lamb Tagine with Cardoons

Serves 4

Tagines (or stews) are popular in northern Africa. Cardoons give this traditional Moroccan dish a rich flavour.

750g chunks lamb
3 cloves garlic, crushed
1 teaspoon ground ginger
Pinch saffron
¼ teaspoon turmeric
2–3 tablespoons vegetable oil
2 tablespoons chopped fresh coriander
1 onion, peeled and sliced
750g–1kg cardoons
2 preserved lemons, quartered
4 tablespoons black olives
Juice of 2 lemons
Salt and freshly ground black pepper

Put the lamb, garlic, ginger, saffron, turmeric, oil, chopped coriander and onion in a heavy pan and mix well. Pour over a cup or two of water and bring to the boil.

Skim, if need be, then simmer, covered, for 1 hour, adding more water if necessary, to just cook the lamb.

Then add the cleaned cardoons and enough water to cover them (this is important), and continue cooking for a further 30–40 minutes.

Stir in the preserved lemon quarters and the olives and enough lemon juice, to taste, and ensure the tagine is well mixed. Taste and adjust the seasoning before serving piping hot.

Cynara scolymus. Asteraceae

GLOBE ARTICHOKE

Also known as French Artichoke, Green Artichoke. Tall, upright perennial grown for edible flower buds. Half hardy. Value: 85% water; half carbohydrate indigestible inulin, turning to fructose in storage; moderate iodine and iron content.

Originating in the Mediterranean, globe artichokes were grown by the Greeks and Romans, who regarded them as a delicacy. The common name comes from the Italian *articoclos*, deriving from *cocali*, or pine cone – an apt description of the appearance of the flower bud. Artichokes waned in popularity in the Dark Ages, but were restored to favour when Catherine de Medici introduced them to France in the sixteenth century. From there they spread around the world. Globe artichokes reached the United States in 1806, travelling with French and Spanish settlers.

In Italy, its bitter principal flavours the aperitif Cynar, which is popular as a vermouth and definitely an acquired taste!

Growing to about 1.2–1.5m tall, with a 90cm spread, attractive leaves and large thistle-like flowers, globe artichokes always look wonderful in the flower border and make excellent dual-purpose plants.

 varieties

'Green Globe' has large green heads with thick, fleshy scales and needs winter protection in cooler climates. **'Gros Camus de Bretagne'** is only suitable for warmer climates, but is worth growing for its large, well-flavoured heads. **'Purple Globe'** is hardier than the green form but not as tasty. **'Purple Sicilian'** has small, deep purple-coloured artichokes that are excellent for eating raw when they are very young. However it is not frost-hardy. **'Vert de Laon'** is hardy with an excellent flavour and **'Violetta di Chioggia'**, a purple-headed variety, is excellent in the flower border.

 cultivation

Artichokes need an open, sheltered site on light, fertile, well-drained soil.

Propagation

Artichokes can be grown from seed or divided, but are usually propagated from rooted 'suckers' – shoots arising from the plant's root system. Suckers are bought or removed from established plants in mid-spring. They should be healthy, about 20–23cm long and well

'Purple Globe'

GLOBE ARTICHOKE

rooted, with at least 2 shoots. Clear soil from around the roots of the parent plant and remove them with a sharp knife, cutting close to the main stem between the sucker and parent plant. Alternatively, divide established plants in spring by lifting the roots and easing them apart with 2 garden forks, a spade or an old knife and replanting the sections; these, too, should have at least 2 shoots and a good root system. To keep your stocks vigorous and productive, renew the oldest one-third of your plants every year. This extends the cropping season, too, as mature plants are ready for harvest in late spring to early summer and young plants in late summer.

You can grow from seed and select the best plants, but this is time-consuming, uses valuable space and is not recommended; it is far better to grow proven, named cultivars. If you have the time and inclination, then sow seed in trays of moist seed compost in an unheated glasshouse during late winter or outdoors in early spring. Thin to leave the strongest seedlings and harden off before planting out at their final spacing in late spring. Once flower buds have been produced, retain the best plants for harvest and for future propagation, discarding the rest.

Growing

If necessary, improve the soil by digging in plenty of well-rotted organic matter in spring or autumn before planting. This prevents summer drought and winter waterlogging, conditions that globe artichokes dislike. Before planting, rake in general fertiliser at 60g/sq m.

Plant suckers or divisions 60cm apart with 60–75cm between each row, trimming the leaves back to 12.5cm, which helps to reduce water loss, and shading them from full sun until they are established. Water thoroughly after planting and during periods of dry weather, applying a high-potash liquid fertiliser every 2 weeks when the plants are actively growing.

Keep the beds weed-free and mulch with organic matter in spring. During autumn and winter, if heavy frosts are forecast, protect plants by earthing up with soil, then covering them with a thick layer of straw, bracken or other organic insulation. Remove the covering in spring.

Maintenance

Spring Divide or remove suckers from existing plants and replant.

Summer Keep weed-free and water thoroughly during drought.

Autumn In areas with moderate temperatures, retain leaves and stems as frost protection.

Winter If severe frost is forecast, remove the decayed leaves, earth up, and protect plants with insulating material. Remove the materials in spring before growth begins.

Harvesting and Storing

Each flowering stem normally produces one large artichoke at the tip and several smaller ones below. A few flower heads will be produced in the first year; these are best removed so that the energy goes into establishing the plant, but if you cannot resist the temptation, harvest in late summer. In the second and third years more stems will be produced and are ready for cutting in mid-summer. Harvest when the scales are tightly closed, removing the terminal bud first with 5 or 7.5cm of stem, then the remaining side buds as they grow large enough. Alternatively, remove the lower artichokes for eating when they are about 4cm long.

Once the scales begin to open, globe artichokes become inedible.

Bees are attracted to the flowers

Buds ready to enjoy

Pests and Diseases

Slugs attack young shoots and leaves – the problem is worse in damp conditions. Keep the area free of plant debris, use biological controls, scatter ferric phosphate-based slug pellets around plants or make traps from plastic cartons half-buried in the ground and filled with milk or beer. Lay roof tiles, newspaper, old lettuce leaves or other tempting vegetation on the ground and hand pick regularly from beneath. Or put a barrier of grit around plants, or attract predators such as birds to the garden. Lettuce root aphid can be a problem. Creamy yellow aphids appear on the roots during summer, sucking sap and weakening plants. Water well in dry weather; spray with soft soap.

medicinal

Artichokes are highly nutritious and are especially good for the liver, aiding detoxification and regeneration. They reduce blood sugar and cholesterol levels, stimulating the gall bladder and helping the metabolism of fat. Artichoke is also a diuretic and used to treat hepatitis and jaundice. It was formerly used as a contraceptive and aphrodisiac, but its potency is not recorded!

culinary

Artichokes can be stored for up to a week in a polythene bag in a refrigerator.

The edible parts are the fleshy base of the outer scales, the central 'heart' and the bottom of the artichoke itself. Wash the artichoke thoroughly before use and sprinkle any cut parts with lemon juice to prevent them from turning black. Boil artichokes in a non-metallic pan of salted water with lemon juice for 30–45 minutes, until soft. Check if they are ready by pushing a knife through the heart, or try a basal leaf to check it for tenderness.

Eat artichokes by hand, pulling off the leaves one by one and dipping the base in mayonnaise, hollandaise, lemon sauce, melted butter or plain yogurt before scraping off the fleshy leaf base between your teeth. Pull off the hairy central 'choke', or remove it with a spoon, and then eat the fleshy heart.

Bottoms can be a garnish for roasts, filled with vegetables or sauces. Cook 'Cypriot-style' with oil, red wine and coriander seeds, or toss in oil and lemon dressing as an hors-d'oeuvre. Make a salad of cubed artichoke bottoms and new potatoes (leftovers are suitable) and season well. Toss in mayonnaise and crumble over finely chopped hard-boiled egg and good-quality black olives. Sprinkle with chives and flat-leaved parsley. Whole baby artichokes can be battered and deep-fried or cooked in oil. Eat them cold with vinaigrette.

A seventeenth-century herbalist and apothecary wrote that even the youngest housewife knew how to cook artichokes and serve them with melted butter, seasoned with vinegar and pepper. Florence White, a founder of the English Folk Cookery Association and member of the American Home Economics Association, gives a recipe for artichokes in *Good Things in England* (1929) from the time of Queen Anne:

A Tart of Artichoke Bottoms

'Line a dish with fine pastry. Put in the artichoke bottoms, with a little finely minced onion and some finely minced sweet herbs. Season with salt, pepper and nutmeg. Add some butter in tiny pieces. Cover with pastry and bake in a quick oven. When cooked, put into the tart a little white sauce thickened with yolk of egg and sharpened with tarragon vinegar.'

Artichokes Cypriot-style

Risotto with Artichokes

Risotto with Artichokes

Serves 6

Rose Gray and Ruth Rogers give this recipe in their wonderful *River Café Cookbook*:

8 small globe artichokes, prepared and trimmed (chokes removed if at all prickly)
2 garlic cloves, peeled and finely chopped
3 tablespoons olive oil
1 litre chicken stock
150g butter
1 medium red onion, very finely chopped
300g risotto rice
75ml extra dry white vermouth
175g freshly grated Parmesan
Sea salt and freshly ground black pepper

Cut the artichokes in half and slice as thinly as possible. Fry gently with the garlic in 1 tablespoon of the olive oil for 5 minutes, stirring continuously, then add 120ml water, salt and pepper and simmer until the water has evaporated. Set aside.

Heat the chicken stock and check for seasoning. Melt 90g of the butter in the remaining oil in a large heavy-bottomed saucepan and gently fry the onion until soft, about 15–20 minutes. Add the rice and, off the heat, stir for a minute until the rice becomes totally coated. Return to the heat, add 2 or so ladlefuls of hot stock or just enough to cover the rice, and simmer, stirring, until the rice has absorbed nearly all the liquid. Add more stock as the previous addition is absorbed. After about 15–20 minutes, nearly all the stock will have been absorbed by the rice; each grain will have a creamy coating, but will remain *al dente*.

Add the remaining butter in small pieces, then gently mix in the vermouth, Parmesan and artichokes, being careful not to overstir.

Daucus carota complex *sativus*. *Apiaceae*

CARROT

Swollen-rooted biennial grown as annual for edible orange-red roots. Hardy/half hardy. Value: extremely rich in beta carotene (vitamin A); small amounts of vitamin E.

Though there are white, yellow, purple and violet carrots, most of us are more familiar with orange carrots, which have been known only since the eighteenth century. Domestication is thought to have occurred around the Mediterranean, Iran and the Balkans. The Greeks cultivated them for medicinal uses, valuing them as a stomach tonic. In Roman and early medieval times, carrots were branched, like the roots of wild types; the conical-rooted varieties seem to have originated in Asia Minor around AD1000. Moorish invaders took them to Spain in the twelfth century; they reached North-West Europe by the fourteenth and England in the fifteenth century. Gerard mentions only one yellow variety, purple ones being most popular – even though, when cooked, they turned into a nasty brown colour.

The Elizabethans and early Stuarts used flowers, fruit and leaves as fashion accessories for hats and dresses and carrot tops were highly valued as a substitute for feathers, particularly when they coloured up in autumn. European explorers took the carrot across the Atlantic soon after the discovery of the New World and it was growing on Margarita Island, off the coast of Venezuela, in 1565, arriving in Brazil before the middle of the seventeenth century. The Pilgrim Fathers took it to North America and it was grown by early colonists in Jamestown, Virginia in 1609. Said to make you see well in the dark and make your hair curly, it is now highly valued as a rich source of vitamin A.

CARROT

varieties

There is a huge choice of carrot varieties; many new varieties have been bred for increased sweetness and higher beta carotene content. Carrots are grouped into categories according to their root shape and harvesting time. This is indicated on the packet or in the catalogue.

'Adelaide' is one of the earliest to mature, with cylindrical, stump-shaped roots that are deep orange in colour. **'Amsterdam Forcing 3'** is fast-maturing and produces small, good-quality roots with little core. Ideal for early sowings under unheated cover and good for 'finger' carrots. **'Autumn King 2'** is an excellent 'maincrop' with top-quality roots up to 30cm long. Ideal for eating fresh or winter storage. **'Bangor'** produces large, top-quality cylindrical roots with excellent colour and flavour. It resists greening and cracking and stores well and is one of the best maincrops for the gardener. **'Belgian White'** is a tasty white variety for summer harvesting. **'Carson'** has a rich, crunchy orange core and is one of the best 'Chantenay' hybrids for flavour – a good 'maincrop' for autumn use and winter storage. **'Eskimo'** has excellent cold tolerance, so is good for late cropping or storing in the ground. Rich in vitamin A, it also has good resistance to cavity and leaf spot and is suitable for deep freezing. **'Giant Flakee'** is a reliable old 'maincrop' with huge, juicy roots and exceptional flavour. **'Healthmaster'** grows up to 25cm long and is said to contain up to 35 per cent more beta carotene than any other carrot. The roots are deep orange-red and are resistant to cracking and greening. **'Honeysnack'** is juicy, sweet and delicious. The creamy-yellow roots reach 15cm and are perfect for slicing or cutting into cubes for snacks or salads. The early-maturing Nantes variety **'Jeanette'** has a high level of pest and disease resistance and so is ideal for organic growers. It is well flavoured, vigorous and ideal for salads. **'Kazan' 'Autumn King'** hybrid is one of the best late varieties for colour and flavour and stores well. **'Kingston'** is one of the best for taste, colour and uniformity. Another Nantes type, **'Maestro'**, is vigorous, resists pests (including carrot fly) and is good for organic production. **'Mokum'** is high-yielding, very juicy, crisp and sweet, so ideal for juicing or eating fresh. Grow early under glass or in containers. **'Nantes Frubund'** is the first autumn sowing carrot. It is very cold-resistant and ideal for early spring crops. **'Parmex'**, a round- rooted type, is ideal for shallow or stony soils. **'Paris Market Baron'** is a similar shape. **'Parano'** is a very early bunching type with a smooth skin and delicious cylindrical roots that are ideal for salads. Good carrot fly resistance. **'Samuari'** (**'Red Samurai'**) has sweet, slender, red-skinned roots with pink-tinted flesh that retains its colour after steaming and lends colour to a salad. **'Purple Dragon'** is similar but with an orange core. **'Sugarsnax 54'** and earlier-maturing **'Tendersnax'** both have long, juicy, tender roots that are tantalisingly sweet. Ideal juiced, in salads or lightly cooked. **'Flyaway'**, **'Sytan'** and **'Resistafly'** are reliably carrot fly-resistant and **'Edible leaf carrot'** is grown for the leaves which are good in salads.

'Parmex' – ideal for children

Eat carrots immediately for the finest flavour

cultivation

Propagation

Germination is poor at soil temperatures below 7.5°C: warm the soil before sowing early crops. Another option is available for the small Paris Market varieties – sow 3–4 seeds per module and plant out after hardening off in mid-spring.

Carrot seed is very small and is easier to sow when mixed with sand or 'fluid sown'. Sow sparingly to reduce thinning and associated problems with carrot fly.

Seeds should be in drills 2cm deep in rows 15cm apart; allow 10cm between plants in the rows for early crops and 4–6cm apart for maincrops, depending on the size of roots you require.

Sow all but the **'Berlicum'** and **'Autumn King'** types from mid- to late spring for cropping from late summer to early autumn.

Sow **'Chantenay'**, **'Autumn King'** and **'Berlicum'** types from mid- to late spring for mid- to late autumn crops. Sow **'Autumn King'** and **'Berlicum'** types in late spring for mid- to late winter harvesting.

Growing

Early carrots need an open, sheltered position; maincrops are less fussy. Soils should be deep, light and free-draining, warming early in spring, and with a pH of 6.5–7.5. Carrots are the ideal crop for light sandy soils – you should be able to push your index finger right down into the seedbed.

Let's hope they all germinate

Thin when the soil is damp

Avoid walking on prepared ground to ensure maximum root growth.

Dig in plenty of well-rotted organic matter in the autumn before planting. On heavy or stony soils, grow round or short-rooted varieties, or plant in raised beds or containers.

Rake the seedbed to a fine texture about 3 weeks before sowing and use the 'stale seedbed' method, allowing the weeds to germinate and hoeing off before sowing.

Keep crops weed-free at first by mulching, hand weeding or careful hoeing, to avoid damaging the roots. In later stages of growth, the foliage canopy will suppress weed growth.

Keep the soil moist to avoid root splitting and bolting. Water at a rate of 14–22 litres/sq m every 2–3 weeks, taking particular care with beds surrounded by barriers as a protection against carrot fly, as these create an artificial rain-shadow.

Maintenance

Spring Sow crops successionally from mid-spring when the soil is workable.
Summer Harvest early crops, sow maincrops. Keep well watered.
Autumn Sow under cover for a mid-spring crop. Harvest.
Winter Harvest and store.

Protected Cropping

Sow **'Nantes'** types in an unheated greenhouse, under cloches or fleece in mid-autumn for a mid-spring crop. Sow **'Paris Market'**, **'Nantes'** and **'Amsterdam Forcing'** in mid-spring for early to mid-summer crops.

Harvesting and Storing

Early cultivars are ready to harvest after around 8 weeks, maincrops from 10 weeks. In light soils, roots can be pulled straight from the ground, but on heavier soils they should be eased out with a garden fork. Water the soil beforehand if it is dry.

On good soils, maincrop carrots can be left in the ground until required. Cover with a thick layer of straw, bracken or similar material before the onset of inclement weather to make lifting easier.

Alternatively, lift roots, cut or twist off foliage and store healthy roots in boxes of sand in a cool, dry, frost-free place for up to 5 months. Check regularly and remove any that are damaged.

Freeze finger-sized carrots in polythene bags. Top and tail, wash and blanch for 5 minutes. Cool and rub off the skins.

Carrots will stay fresh for about 2 weeks in a cool room or in a polythene bag in the refrigerator.

Pests and Diseases

Carrot fly is the most serious problem. Attracted by the smell of the juice from the root, their larvae tunnel into the roots, making them inedible. Leaves turn bronze. Sow resistant cultivars, sow sparingly to avoid thinning, thin on a damp overcast day (or water before and after), pinch off the tops of thinnings just above the soil level and dispose of them in the compost heap. Lift and dispose of affected roots immediately. Lift 'earlies' by early autumn and maincrops by mid-autumn. Since these pests do not fly very high, grow carrots under fleece or mesh, or surround with a barrier of fine netting or fleece 60cm high, or grow in raised beds or boxes.

Root and leaf aphids can also be a problem.

companion planting

Intercropping carrots with onions reduces carrot fly attacks; leeks and salsify have also been used with some success.

Mixing with seeds of annual flowers also seems to discourage carrot fly.

Carrots grow well with lettuce, radishes and tomatoes and encourage peas to grow. They dislike anise and dill. If left to flower, carrots attract hoverflies and other beneficial predatory insects to the garden.

container growing

Choose short-rooted or round varieties for containers, window boxes or growbags. Sow in loam-based compost; keep moist throughout the growing season.

medicinal

Reputed to be therapeutic against asthma, general nervousness, dropsy and skin disorders.

Recent research suggests that high intake of beta carotene may slow cancerous growths. Beetroot and carrot juice is reported to prevent diarrhoea.

Carrot juice is tasty and packed with goodness

CARROT

culinary

Young carrots are sweet and can be eaten raw. Older carrots need to be peeled and the hard core discarded. Grated, they can be made into salads They're especially good mixed with raisins and chopped onion and tossed in a French dressing or mayonnaise, thinned with some extra virgin olive oil.

Try freshly plucked baby carrots, rinsed under the tap, topped and tailed, very lightly boiled in as little water as possible, then sprinkled with chopped parsley, a little sugar and freshly ground black pepper, or just a knob of butter. Add a sprinkling of sugar, a tablespoon of butter and a pinch of salt, cover and gently steam until tender. Cook in a cream sauce seasoned with tarragon, nutmeg and dill, or steam with mint leaves.

Cut into sticks as raw snacks, slice for stews and casseroles, pickle, stir-fry, use for jams, wine and carrot cake. Carrots are an essential ingredient in stocks, soups and many sauces.

Balkali Havuçi

Balkali Havuçi
Serves 4

Serve this Armenian-Turkish dish with pilaf and salad; it can be topped with a dollop of yogurt.

500g broad beans, shelled
1 onion, chopped
2 carrots, cut into 2.5cm discs
2 cloves garlic, finely chopped
2 tablespoons chopped fresh dill or mint
1 teaspoon salt
Freshly ground black pepper
Pinch sugar
3 tablespoons extra virgin olive oil

Rinse the beans. Put 300ml water into a large saucepan and bring to the boil; add the onion and beans, bring to the boil, cover and simmer until tender.

Stir in the remaining ingredients and cook until the vegetables are just tender, about 20 minutes more. Serve hot with chunks of good bread as a starter.

Carrot and Raisin Cake
Serves 10–12

This spicy dough mixture can also be used to make cookies; drop heaped tablespoonfuls of dough on to a lightly greased baking sheet 5cm apart and bake for 12–15 minutes until golden. Cool on a rack.

150g plain flour
2 teaspoons baking powder
1 teaspoon ground cinnamon
Pinch mace
Pinch salt
2 heaped tablespoons seedless raisins
100g carrots, grated
Zest of ½ an orange
2 tablespoons orange juice
100g butter
125g brown sugar
2 large eggs

Sift the flour, baking powder, spices and salt into a large bowl and set aside. Mix the raisins, carrots, orange zest and juice and set aside. Cream the butter and sugar thoroughly in a mixer and beat until light. Add the eggs, one at a time, with the blender on slow. Combine the batter with the flour and carrot mixtures and blend well.

Pour into a greased and lined 20cm tin and bake in a preheated oven (180°C/350°F/gas mark 4) for 40–60 minutes. Test with a skewer to ensure the cake is cooked and allow to cool in the tin for 15 minutes before turning out onto a wire rack.

Dioscorea alata. Dioscoreaceae

YAM

Also known as Greater Yam, Asiatic Yam, White Yam, Winged Yam, Water Yam. Twining climber grown for large edible tubers. Tender. Value: rich in carbohydrate and potassium; small amounts of B vitamins.

The 'greater yam' is believed to have originated in east Asia and is widely cultivated as a staple crop throughout the humid tropics. It was said to have reached Madagascar by AD1000 and by the sixteenth century Portuguese and Spanish traders had taken it to West Africa and the New World, often as a food on slave-trading ships. Christopher Columbus knew of the plant as *nyame* and the tubers were regularly used for ships' supplies, because they stored for several months without deteriorating and were easy to handle. There are hundreds of different forms producing tubers with an average weight of 5–10kg, although specimens with massive tubers up to 62kg have been recorded.

varieties

Dioscorea alata has square, four-winged or angled twining stems with pointed, heart-shaped, leaves. Small bulbils are often produced on the stems. The tubers are brown on the outside with white flesh, vary in size and are usually produced singly. There is a great number of cultivars which vary in the colour and shape of the stems, leaves and tubers.

'White Lisbon', one of the most widely grown, is high-yielding, shallow-rooted and tasty and it will store for up to 6 months. **'Belep'**, **'Lupias'**, **'Kinbayo'** and **'Pyramid'** are also high-yielding.

cultivation

Propagation

Yams are usually planted on banks, mounds or ridges at the end of the dry season while they are still dormant, as they need the long rainy season to develop. Small tubers, bulbils or sections with 2–3 buds or 'eyes' taken from the tops of larger tubers are used for propagation. The latter are preferable, as they sprout quickly and produce higher-yielding plants.

They can be 'sprouted' in a shady position before planting 15cm deep and 30–90cm apart on mounds 120cm across or ridges 120cm apart. Spacing depends on the site, soil and the variety.

Growing

Yams need a humid, tropical climate, 150–175cm of rain in a 6–12-month growing season and a site in sun or partial shade. Soils must be rich, moisture-retentive and free-draining, as yams can survive drought but not waterlogging. Dig in plenty of well-rotted organic matter before planting and grow plants up trellising, arbours, poles or netting at least 180cm high, or allow them to grow into surrounding trees.

Keep crops weed-free.

Tropical crops do not grow well below 20°C. Growth increases with temperature and the crucial time for rain is 14–20 weeks after planting, when food reserves are nearly depleted and the shoots are growing rapidly.

YAM

A flowering yam plant

Maintenance

Spring Increase watering as new shoots appear when growing under cover.
Summer Maintain high temperatures and humidity.
Autumn Reduce watering as stems turn yellow.
Winter Keep compost slightly moist.

Protected Cropping

If you are able to provide an environment and cultural conditions similar to those described under 'Growing', yams can be grown indoors. The minimum temperature for active growth is 20°C; damp down greenhouse paths and mist with tepid water during summer, and reduce watering during the resting season as the stems die back, gradually increasing the amount in spring when plants resume active growth.

Harvesting and Storing

Depending on the variety and weather conditions, tubers are ready to harvest from 7–12 months after planting as the leaves and stems die back. Lift them carefully, as damaged tubers cannot be stored.

Tubers are normally dried on a shaded vertical frame or in an open-sided shed before being stored in a dark, cool, airy place where they can last for several months. Do not store yams at temperatures below 10°C.

Pests and Diseases

Leaf spot appears as brown or black spots on stems and leaves. Storage rot can be a serious problem and yam beetles feed on tubers and damage the shoots of newly planted sets. Scale insect also causes problems.

companion planting

Grows well with taro, ginger, maize, okra and cucurbits.

medicinal

Yams have been used as a diuretic and expectorant.

warning

All yams except *Dioscorea esculenta* contain a toxin, dioscorine, which is destroyed by thorough cooking.

culinary

Yam can be peeled and then boiled, mashed, roasted and fried in oil. In West Africa, they are pounded into *fufu* in a similar way to manioc, and are added to a thick soup made of spices, meat, oil, fish and vegetables. Yams are tasty cooked with palm oil, candied, casseroled with orange juice or curried. They can be peeled and boiled in water with a pinch of salt, then brushed with melted butter and grilled or barbecued until brown; serve with more butter.

In China, they are mashed with lotus root, wrapped in lotus leaves and steamed.

They can also be peeled, rubbed with oil and baked at 180°C/350°F/gas mark 4 for 1½ hours, then slit like a baked potato and eaten with seasoned butter or a similar filling.

Yellow Yam Salad

Serves 6

1.5kg yellow yams
1 large onion, sliced
3 tablespoons chopped chives
1 green pepper, deseeded and diced
8 tablespoons mayonnaise
½ teaspoon cayenne pepper

Salt and freshly ground black pepper
4 hard-boiled eggs, chopped
1 tablespoon black olives, stoned and sliced

Clean the yams and cook in boiling salted water until tender but firm. Drain and allow to cool slightly, then cut into cubes. Add the onion, chives and pepper and mix well.

Season the mayonnaise with the cayenne, salt and pepper and stir into the yam mixture, folding in the eggs and olives carefully.

Check the seasoning. Chill for an hour before serving.

Eleocharis dulcis. Cyperaceae

WATER CHESTNUT

Half-hardy aquatic perennial. Value: good source of potassium, B6 and fibre; high in carbohydrates.

***Eleocharis dulcis* has been cultivated for centuries. It has become naturalised on the edge of paddy fields in China and Southeast Asia, is popular in oriental cuisine and is used for making salt in Zimbabwe. Like the tiger nut and lotus root, the water chestnut is notable for remaining crisp even after cooking or canning – this is because its cell walls are cross-linked and strengthened by the presence of aromatic phenolic compounds that also give plants such as cloves their fragrance.**

varieties

Eleocharis dulcis is a tropical or subtropical marginal plant, with dense tufts of thin, hollow, sedge-like leaves that grow to 1m tall.

cultivation

Propagation

Use water chestnuts bought from specialist suppliers or the market; choose ones free of wrinkles and soft spots. Plant in 15cm pots of loam-based compost with the base submerged in warm water in a heated propagator or on a sunny windowsill, at a minimum of 26°C, from mid-spring. Transfer them into their final position as temperatures rise. Alternatively, plant them by hand directly into their final position, 30–40cm apart, once the danger of frost has passed.

Growing

Water chestnut needs an open, sunny site and to be grown in water. Use a paddling pool, an old bath or similar, or create a temporary pond by digging a hole or trench in the ground and lining it with polythene. The 'pond' should be deep enough to allow a 20cm layer of loam or garden soil (with a low organic matter content) to be spread over the base with 10cm of water above. If a paddling pool is being used, sweep the standing area before inflating and put a layer of old carpet, corrugated cardboard or similar under the base to prevent it from being punctured by stones.

Plant water chestnut when the soil and water has warmed; a small pond or swimming pool heater boosts water temperatures and is useful in cooler summers. It needs plenty of warmth to grow, at least six months in temperatures from 30–35°C is ideal, though it can crop in cooler conditions.

Maintenance

Spring Plant corms in a pot and put in a warm place.
Summer Keep water levels topped up in the container.
Autumn Harvest corms.
Winter Store propagation material in damp conditions.

Protected Cropping

In cooler climates, growing under glass or in polythene tunnels encourages heavier cropping. Starting crops off under protection, even if they are then grown outdoors, extends the growing season.

Harvesting and Storing

Tubers form from late summer and are ready to harvest when the stems turn yellow and die down in autumn. Allow the water to evaporate, then drain any remaining water; it is easier to harvest tubers when soil is moist, rather than waterlogged. Store tubers for propagation the following year in a damp, frost-free place.

Pests and Diseases

There are no problems with pests and diseases.

medicinal

Chinese herbalists use water chestnuts for sweetening breath. The juice of the tuber is said to be bactericidal.

culinary

The onion-shaped tubers of the water chestnut are sweet, nutty and crisp. Remove the papery tunic before cooking them, or eating them raw or lightly cooked. A familiar ingredient in Chinese cooking, they can be seasoned or combined with other vegetables and are eaten in chop suey. Soaked lightly in salted water and threaded on to thin bamboo stalks, they are sold by street vendors in China. Dried tubers are ground into flour that is used for thickening, coating vegetables or giving a crispy coating to deep-fried food. They store for up to two weeks in a polythene bag in the fridge, but are better eaten fresh.

Stir-fried Prawns with Vegetables

Serves 4

250g raw tiger prawns, peeled
1 teaspoon salt
1 tablespoon egg white
2 tablespoons cornflour paste
600ml vegetable oil
50g mangetout
1 carrot, peeled and finely sliced
75g water chestnuts from a tin, drained and sliced
75g straw mushrooms from a tin, drained and halved lengthways
2 spring onions, cut into 1cm pieces
2 tablespoons peeled and finely sliced fresh ginger
½ teaspoon sugar
1 tablespoon light soy sauce
2 teaspoons rice wine
A few drops sesame oil

Cut the prawns in half lengthways and remove the black vein. Mix the prawns with the salt, egg white and cornflour paste.

Heat the oil in a wok, then deep-fry the prawns for 1 minute until they turn pink. Remove them with a slotted spoon on to some kitchen paper to drain.

Pour off the oil, leaving roughly 1 tablespoon in the wok. Add the mangetout and carrot and stir-fry for about 1 minute. Then add the rest of the vegetables, the ginger, sugar, soy sauce and rice wine and stir well. Stir-fry for 1 more minute.

Drizzle over the sesame oil and serve piping hot with noodles or boiled rice.

Eruca vesicaria subsp. *sativa*. *Brassicaceae*

ROCKET

Also known as Rocket Salad, Roquette, Italian Cress, Rucola, Arugula. Annual grown for tender edible leaves. Hardy. Value: high in potassium and vitamin C.

Rocket has been cultivated since Roman times and is native to the Mediterranean and Eastern Asia, though it grows in many areas. *Eruca* means 'downy-stemmed'; *vesicaria*, 'bladder-like', describes the slender seed pods. Introduced to North America by Italian settlers, the spicy leaves were particularly popular in Elizabethan England.

varieties

Eruca vesicaria subsp. *sativa*, an erect plant growing to 1m, has hairy stems, broadly toothed leaves and cross-shaped, creamy flowers. **'Apollo'** and **'Runway'** are fast growing forms and **'Sky Rocket'** combines the flavour of wild rocket with the fast growth of salad rocket.

cultivation

Propagation

Sow seed successionally every 2–3 weeks, from mid-spring to early summer, in drills 12mm deep and 30cm apart. Thin when large enough to handle, until 15cm apart. Rocket germinates at fairly low temperatures. In warmer climates sow in winter or early spring.

Growing

Rocket needs rich, moisture-retentive soil in partial shade. It may need extra shading in hot weather, otherwise it produces less palatable leaves. Keep weed-free and water regularly.

Maintenance

Spring Sow seeds from mid-spring; thin when large enough to handle.
Summer Continue sowing until mid-summer. Water and feed as necessary. Harvest regularly to encourage tender growth and keep from bolting.
Autumn Harvest, prepare the ground for the following year's sowing and sow seeds for protected crops.
Winter Harvest winter crops.

Protected Cropping

Although rocket is a hardy plant, protect against severe frosts with cloches. Sow autumn/winter crops from late summer in a cool greenhouse, cold frame, or under cloches.

Harvesting

Plants are ready to harvest after 6–8 weeks. Pick frequently to encourage a regular supply of good-quality leaves and to prevent plants from running to seed in hot weather. Discard any damaged leaves. Either pull leaves as required or treat as a 'cut and come again' crop, cutting the plant 2.5cm above the ground.

Pests and Diseases

Rocket is usually trouble-free, but flea beetle can damage seedlings. Protect with horticultural fleece or mesh.

culinary

The increasingly popular leaves are delicious in salads – younger leaves are milder. They can be lightly boiled or steamed, added to sauces, stir-fried, sautéed in olive oil and tossed with pasta. The flowers are edible and can decorate salads.

Penne with Merguez and Rocket
Serves 4

400g penne
2 tablespoons olive oil
300g merguez sausages, cut into bite-size pieces
1 red onion, thinly sliced
2 tablespoons dry white wine
12 cherry tomatoes, halved
85g rocket leaves, washed and shredded
Salt and freshly ground black pepper
50g freshly grated Parmesan

Bring a pan of salted water to the boil and cook the penne. Meanwhile, heat the oil in a large frying pan, add the merguez and onion and fry for 2–3 minutes. Add the wine and simmer for 10 minutes. Then add the tomatoes. When the pasta is cooked, drain well and toss in the sauce with the rocket. Mix well, season and serve at once with Parmesan.

container growing

Rocket is not the perfect plant for pots, but it can be grown in a soil-based compost with low fertiliser levels with added peat-substitute compost. Sow seed *in situ*. Water thoroughly.

medicinal

Young leaves are said to be a good tonic and are used in cough medicine. Dioscorides described rocket as 'a digestive and good for ye belly'.

Foeniculum vulgare var. *azoricum. Apiaceae*

FLORENCE FENNEL

Also known as Sweet Fennel, Finocchio. Biennial or perennial grown as annual for pungent swollen leaf bases, leaves and seeds. Half hardy. Value: good source of potassium; small amounts of beta carotene.

This outstanding vegetable with a strong aniseed flavour, swollen leafbases (known as 'bulbs') the texture of tender celery, and delicate feathery leaves has been cultivated for centuries as an ornamental vegetable. Its close relative, wild fennel, which lacks the swollen base, is used as a herb. When Portuguese explorers first landed on Madeira in 1418, they found the air fragrant with the aroma of wild fennel, so the city of Funchal was named after *funcho*, the Portuguese name for the plant. Florence fennel was popular with the Greeks and Romans, whose soldiers ate it to maintain good health – while the ladies used it to ward off obesity. In medieval times, seeds were eaten during Lent to alleviate hunger, and dieters still chew raw stalks to suppress their appetite. The first records of its cultivation in England date from the early eighteenth century, when the third Earl of Peterborough cultivated and ate it as a dessert.

In 1824 Thomas Jefferson received seeds from the American consul in Livorno and sowed them in his garden in Virginia. He enthused: 'Fennel is beyond every other vegetable, delicious... perfectly white. No vegetable equals it in flavour.' However, it has become a weed in those countries where conditions are particularly favourable.

varieties

'Amigo' is early cropping and vigorous with medium, white bulbs. **'Heracles'** is fast-maturing and produces top-quality bulbs. **'Perfection'**, a French variety, has medium-sized bulbs and a delicate aniseed flavour. Resistant to bolting, it is ideal for early sowing. **'Pronto'** is an early variety with round, white bulbs. **'Sirio'**, from Italy, is compact with large, sweet white bulbs and matures rapidly. **'Sweet Florence'** is moderately sized and should be sown from mid-spring to late summer. **'Victoria'** produces dense, top-quality, greenish-white bulbs. **'Zefa Fino'** is particularly vigorous and ornamental with uniform bulbs of excellent quality.

cultivation

Propagation

Fennel thrives in a warm climate and is inclined to bolt prematurely if growth is checked by cold, drought or transplanting. For early sowings, choose cultivars that are bolt-resistant.

Grow successionally from late spring to late summer for summer and autumn crops. Sow thinly in drills 1.5cm deep and 45–50cm apart, thinning when seedlings are large enough to handle to a final spacing of 23–30cm apart.

Where possible, fennel is better sown *in situ* to reduce problems with bolting. Modules are preferred for plants started under cover. Growing in trays is fine if transplants are treated carefully enough.

Growing

Fennel flourishes in temperate to sub-tropical climates, though mature plants can withstand light frosts. The largest bulbs are formed during warm, sunny summers. Plants should be grown as rapidly as possible, so incorporate a slow-release general fertiliser into the soil before planting at 30–60g/sq m.

Plants need a sunny, warm, sheltered position and well-drained, moisture-retentive, slightly alkaline soil. A light, sandy soil with well-rotted organic matter dug in the winter before planting is ideal. Stony soils and heavy clays should be avoided. Never allow the soil to

FLORENCE FENNEL

A fine filigree of fennel foliage

dry out; mulch in spring and hand weed around bulbs to avoid damage.

When the stem bases start to swell, earth up to half their height to blanch and sweeten the bulbs. Or tie cardboard 'collars' round the base.

Maintenance

Spring Sow early crops under cover. Warm the soil with cloches before sowing early crops outdoors.
Summer Keep the soil constantly moist and weed-free. Harvest mature bulbs.
Autumn Cover later crops in order to prolong the growing season.
Winter Transplant later sowings for an early winter crop.

Protected Cropping

Cover early outdoor crops with cloches. Sow from mid-spring, in modules or trays of seed compost at 16°C. Pot on seedlings grown in trays when very small, with a maximum of 4 leaves, into 7.5cm peat pots. Transplant in the pots a month after hardening off.

Harden off and plant out those grown in modules at a similar size. Late sowings can be made for transplanting under cover in early winter. Plants do not always produce 'bulbs', but the leaves and stems can still be used in cooking.

Harvesting and Storing

Harvest about 15 weeks after sowing or 2–4 weeks after earthing up, when the bulbs are plump, about 5–7.5cm across and slightly larger than a tennis ball. Cut the bulb with a sharp knife, just above the ground, and the stump should resprout, producing small sprigs of ferny foliage. Bulbs do not store and should be eaten when fresh.

Leaves can be harvested throughout summer and used fresh in salads, or deep-frozen.

Pests and Diseases

Slugs can damage young plants. Lack of water, fluctuating temperatures and transplanting can cause bolting.

companion planting

Allow a few plants to flower; they are extremely attractive to a large number of beneficial insects which prey on garden pests.

Fennel has a detrimental effect on beans, kohlrabi and tomatoes.

container growing

Grow single plants in a container at least 25cm wide by 30cm deep containing loam-based compost with added sharp sand. Apply a general liquid fertiliser monthly from late spring to late summer.

medicinal

An infusion aids wind, colic, urinary disorders and constipation. Recent research indicates that fennel may reduce the effects of alcohol.

Use in an eye bath or as a compress to reduce inflammation. Chew to sweeten breath or infuse as a mouthwash or gargle for gum disease and sore throats, to alleviate hunger and ease indigestion.

warning

Do not take excessive doses of the oil; nor should it be given to pregnant women.

culinary

Use bulbs in the same way as celery, removing green stalks and outer leaves. For salads, slice inner leaf stalks and chill before serving. (To ensure that bulbs are crisp, slice and place in a bowl of water and ice cubes in the fridge for an hour.)

Fennel can be parboiled with leeks and is suitable for egg and fish dishes. Steam, grill or boil and serve with cheese sauce or butter. Infuse fresh leaves in oil or vinegar, add to a bouquet garni or snip as a garnish over soups or salads. Gives characteristic flavour to *finocchiona*, an Italian salami, and the French liqueur, *fenouillette*.

Fennel Sautéed with Peas and Red Peppers

Serves 4

2 tablespoons olive oil
2 sweet red peppers, deseeded and cut in thin strips
500g fennel, trimmed and finely sliced
2 cloves garlic, crushed
250g peas, shelled
Salt and freshly ground black pepper

In a wok or heavy-based frying pan heat the oil, add the peppers and fennel and cook for 10–15 minutes over a moderate heat until crunchy, stirring occasionally. Add the garlic and peas and continue cooking for 2 minutes. Season and serve.

Glycine max (syn. *Glycine soja*). *Leguminosae*

SOYA BEAN

Also known as Soy Bean. Annual grown for seed sprouts and seeds. Half hardy. Value: rich in potassium, protein, fibre, vitamins E, B and iron. Seed sprouts are rich in vitamin C.

One of the most nutritious of all vegetables, this Asian native is thought to have been the plant that Chinese emperor Shen Nung used to introduce people to the art of cultivation. It is mentioned in his *Materia Medica* from around 2900 BC. Soya beans were first known in Europe through Engelbert Kaempfer, physician to the governor of the Dutch East India company on an island off Japan in 1690–92. The Japanese guarded their culture, but, by bribing the guards and picking plants along the route, Kaempfer was able to get his botanical specimens. Benjamin Franklin sent seeds back from France to North America in the late eighteenth century. One of the first Americans to be interested in soya beans was Henry Ford, who saw their potential for manufactured goods and is said to have eaten soya beans at every meal, had a suit made from 'soy fabric' and sponsored a sixteen-course soya bean dinner at the 1934 'Century of Progress' show in Chicago.

varieties

Glycine max is a herb, usually with trilobed leaves and white to pale violet flowers. The pods, containing 2–4 seeds, are mainly on the lower parts of the stem. **'Black Jet'** is early-maturing and one of the best. The seeds have a good flavour and it is ideal for a short growing season. **'Beer Friend'** is a cool-climate variety with seventy days to maturity. It has good flavour and productivity and is a perfect accompaniment to a pint of beer. **'Friskeby'**, a Swedish variety, is ideal for cool climates and **'Prize'** is widely used for sprouting. **'Sayamusume'** produces robust plants, large, moderately sweet seed and consistent crops. **'Ustie'**, bred for the British and similar climates, is self-pollinating and 100 per cent GM-free.

cultivation

Propagation

Sow when danger of frost has passed and the soil has warmed. Sow 2–3 seeds 2.5cm deep in heavy soil and 3cm deep in lighter soils in 'stations' 7.5–10cm apart with 45–60cm between rows. After germination, thin, leaving the strongest seedling. Alternatively, sow thinly and thin to the final spacing when large enough.

Sprouting Seed

Seeds can also be sprouted. Use untreated seed and remove any that are damaged or mouldy. Soak overnight in cold water. The following morning rinse them thoroughly. Put a layer of moist kitchen roll over the base of a flat-bottomed bowl, tray or 'seed sprouter'. Spread over a layer of soya beans 1cm deep and cover with cling film. Exclude light by putting the bowl in a dark cupboard or wrapping it in newspaper or tinfoil. Temperatures should be 20–25°C. Check that the absorbent layer stays damp, rinsing morning and night. They should be ready to harvest in 4–10 days, when shoots are 2.5–5cm long. Remove sprouts from the shell, and rinse well before eating.

Growing

Soya beans flourish in an open site with rich, free-draining soil and a pH of 5.7–6.2, although there are now cultivars to suit most soils. Where necessary, lime soils and dig in well-rotted organic matter before planting. They are not frost hardy; they prefer 20–25°C;

Just a little taster

exceeding 38°C may retard growth.

Keep plants well watered during drought and early stages of growth. Remove weeds regularly, or suppress by mulching in spring.

Maintenance

Spring Sow seeds under cover or outdoors when there is no danger of frost.
Summer Keep crops weed-free. Water during drought.
Autumn Harvest before the pods are completely ripe.
Winter Hang in bunches in a cool shed and collect seed when the pods open.

Protected Cropping

In cooler climates, sow seeds in pots, modules or trays of seed compost, planting out when the soil is warm. Harden off before transplanting and protect under cloches until established. In cooler climates, sow crops in a heated glasshouse.

Harvesting and Storing

Soya needs a hot summer and fine autumn for the seeds to ripen. Harvesting should be carefully timed so that the seeds are ripe but the pods have not yet split. If conditions are unfavourable, pull up plants when the pods turn yellow and hang up to ripen in a dry place. Do not harvest when plants are wet, as the seeds are easily bruised. Harvest for green beans as soon as pods are plump and the seeds are almost full size.

Pests and Diseases

Fungal diseases can be a problem if plants are harvested when wet and pods are bruised or broken.

container growing

This is only worth growing in containers as a 'novelty' crop. Use 20–25cm pots and loam-based compost with moderate fertiliser levels, adding well-rotted organic matter. Keep well watered and weed-free.

other uses

Soya oil is used in a range of products, including ice cream, margarine, soaps and paint; milk substitute is made from the crushed beans, and fermented they make soy sauce and tofu, and also Worcestershire sauce. Also firefighting foam and meat substitute. A valuable plant indeed!

medicinal

Said to control blood sugar levels, lower cholesterol, regulate the bowels and relieve constipation.

warning

Soya beans should be cooked before drying; they can cause stomach upset.

culinary

Used mainly when green or sprouted for salads or stir-fries. Remove shell by plunging into boiling salted water for 5 minutes, then allow to cool and squeeze out the seeds. Cook for 15 minutes, then sauté with butter. Juvenile pods can be cooked and eaten whole.

Soak dried beans before eating. A shortcut is to cover with water in a kettle, boil for 2 minutes, allow to stand for one hour, then cook until tender.

Soya Bean and Walnut Croquettes

Serves 4

100g soya beans, soaked and well rinsed
1 onion, finely chopped
1 clove garlic, crushed
25g butter
100g walnuts, whizzed in the liquidiser
½ teaspoon dried thyme
50g wholewheat breadcrumbs
1 tablespoon tomato purée
2 tablespoons chopped fresh parsley
½ teaspoon mace
1 egg
Salt and freshly ground black pepper
Wholewheat flour
1 egg, beaten
Dried breadcrumbs
Vegetable oil, for frying

Cook the beans until very tender, drain, then mash with a fork, enough to break them up. Fry the onion and garlic in the butter for 10 minutes, then remove from the heat and stir in the beans. Add the walnuts and thyme, together with the breadcrumbs, tomato purée, parsley, mace and egg. (You may need to add more liquid; otherwise use fewer breadcrumbs.) Mix well and season to taste.

Using your hands, shape into small croquettes, then roll in the flour, dip into the egg and roll in the crumbs. Fry in hot oil until crisp and drain on kitchen paper. Serve hot. Particularly good with a spicy tomato sauce.

Helianthus tuberosus. Asteraceae

JERUSALEM ARTICHOKE

Also known as Girasole, Sunchoke. Tall perennial grown as annual for edible tubers. Hardy. Value: high in carbohydrate but mostly inulin, turning to fructose in storage; moderate vitamin B1, B5, low in calories.

This vegetable is not from Jerusalem, nor is it any relative of the globe artichoke. 'Jerusalem' is said to be a corruption of the Italian *girasole* or 'sunflower' – a close relative; the nutty-flavoured tubers were thought to taste similar to globe artichokes, hence the adoption of that name. Frost-hardy, these tall, upright perennials are native to North America, where they grow in damp places. The tubers contain a carbohydrate which causes flatulence; in 1621 John Goodyear wrote that 'they stirre and cause a filthie loathsome wind within the bodie'. In the 1920s they were a commercial source of fructose and were expected to replace beet and cane as a source of sugar.

varieties

'Boston Red' has large, knobbly tubers with rose-red skin. **'Dwarf Sunray'** is a short-stemmed, crisp, tender variety that does not need peeling. It flowers freely and is good for the ornamental border. **'French Mammoth White'** is round, white-skinned and frost-hardy. **'Fuseau'**, a traditional French variety, has long, smooth, white tubers. The plants are compact, reaching 1.5–1.8m. **'Gerrard'**, purple-skinned and round, is smooth and easy to peel. **'Golden Nugget'** has tubers that are good for slicing. **'Originals'** produces traditional rounded tubers and also makes an excellent windbreak or temporary screen in the garden. **'Stampede'** is a quick-maturing variety with large tubers and is also very cold hardy. The tubers of **'Sun Choke'** have a fresh nutty flavour and are excellent raw in salads, cooked or creamed.

cultivation

Jerusalem artichokes prefer a sunny position, but will grow in shade. They tolerate most soils, though tubers are small on poor ground; the best are grown on sandy, moisture-retentive soil.

They will grow in heavy clay provided it is not extremely acid or subject to winter waterlogging. The fibrous root system makes them useful for breaking up uncultivated ground. The tall stems make excellent temporary screens or windbreaks in sheltered areas, but need staking in more exposed sites.

Propagation

Tubers bought from the greengrocer's can be used for planting. Choose tubers the size of hen's eggs, and plant from early to late spring, when the soil becomes workable, 10–15cm deep, 30cm apart and 90cm between rows; cover the tubers carefully. During harvest, save a few tubers to replant or leave some in the soil for the following year.

Growing

Incorporate organic matter in autumn or early winter before planting. Earth up the base of stems to improve stability when plants are 30cm high. Water during dry weather. Remove flower buds as they appear. Shorten stems to 1.5–1.8m in late summer to stop them from being blown over; on windy sites they

JERUSALEM ARTICHOKE

Jerusalem artichoke makes a wonderful windbreak

may also need staking. On poor soil, feed with liquid general fertiliser every 2–3 weeks.

Maintenance

Spring Plant tubers.
Summer Keep weed-free, water and stake if necessary. Remove flower buds.
Autumn Cut back stems and begin harvesting.
Winter Save tubers for next year's crop.

Harvesting and Storing

In autumn, as the foliage turns yellow, cut back stems to within 7.5–15cm of the ground. Use the cut stems as a mulch to protect the soil from frost, making lifting easier; alternatively, cover with straw.

Lift tubers from late autumn to mid-winter. They keep better in the ground, but in cold climates or on heavy ground, lift in early winter and store for up to 5 months in a cool cellar in moist peat substitute or sand.

Pests and Diseases

Slugs tend to hollow out tubers. Set traps, aluminium sulphate pellets, or pick off manually.

Sclerotinia rot causes stem bases to become covered with fluffy white mould. Lift and burn diseased plants; water healthy plants with fungicide. Cutworms eat stems at ground level; damaged plants wilt. Keep crops weed-free, cultivate well or scatter an appropriate insecticide in the soil before planting.

medicinal

The carbohydrate inulin is difficult to digest; tubers are low calorie and suitable for diabetics.

warning

Jerusalem artichokes can become invasive. After harvesting, lift even the smallest tuber from the ground.

culinary

Jerusalem artichokes are versatile vegetables when the weather is cold: they can be kept in the ground until you want to cook them and then dug up root by root. Fresh tubers have a better flavour, but they become more digestible if stored; they will keep in a polythene bag in the salad drawer of the fridge for anything up to 2 weeks.

There is no need to peel them painstakingly unless you want a very smooth, creamy-white purée – the vitamins are, after all, just below the skin. (If you do want them peeled, steaming or boiling knobbly varieties makes the job easier.) To serve as a vegetable, scrub tubers immediately after lifting, boil for 20–25 minutes in their skins in water with a teaspoon of vinegar; peel before serving if desired. The addition of a little grated nutmeg always does wonders in bringing out the unusual flavour.

To make rissoles, form boiled, mashed artichokes into flat cakes and deep-fry. Jerusalem artichokes can also be fried, baked, roasted or stewed – or eaten raw. Artichokes gratinéed in a sauce made with good, strong Cheddar make an excellent accompaniment to plain meat dishes such as baked ham or roast lamb. Parboil the artichokes and drain when they are just tender. Roughly slice and layer into a dish. Pour over a béchamel sauce flavoured with French mustard and well-matured cheese and bake in a hot oven (200°C/400°F/gas mark 6) until browned.

Artichoke Soup

Serves 4

500g Jerusalem artichokes
2 tablespoons olive oil
2 large onions, sliced
1 large clove garlic, crushed
600ml chicken stock
Strip of orange peel
Salt and freshly ground black pepper
4 tablespoons thick cream

Scrub the artichokes, discard any hard knobs and roughly chop. Heat the oil in a heavy-based pan and add the onions. Cook until translucent, add the garlic, continue cooking for a couple of minutes and add the artichokes. Toss well to coat with oil and pour in the chicken stock. Bring to the boil, add the orange peel and season. Cover and simmer for 15–20 minutes, until cooked. Remove from the heat, discard the peel and blend in a food processor. Return to the pan, adjust the seasoning, stir in the cream and serve.

Hibiscus esculentus (syn. *Abelmoschus esculentus*). *Malvaceae*

OKRA

Also known as Lady's Fingers, Bhindi, Gumbo. Annual grown for its edible pods. Tender. Value: rich in calcium, iron, potassium, vitamin C and fibre.

The okra, a close relative of the ornamental hibiscus, with slender edible pods, has been cultivated for centuries. It is thought to have originated in northern Africa around the upper Nile and Ethiopia, spreading eastwards to Saudi Arabia and to India. One of the earliest records of it – growing in Egypt – describes the plant, its cultivation and uses. It was introduced to the Caribbean and southern North America by slaves who brought the crop from Africa; the name 'gumbo' comes from a Portuguese corruption of the plant's Angolan common name.

varieties

'Cow Horn' has a good flavour and texture and is slow to go fibrous. **'Clemson Spineless'** is a popular, reliable variety with high yields over a long period of dark, fleshy pods. It grows well under cover. **'Dwarf Green Long Pod'** is only about 90cm tall but crops well, producing dark green, spineless pods. **'Emerald'** grows prolifically, with deep green pods on tall plants. It pods early and is slow to go 'seedy'. **'Little Lucy'**, a dwarf variety growing to 60cm, is ideal for containers. Its attractive flowers and 10cm pods are held erect above the plant. **'Mammoth Spineless Long Pod'** is vigorous and high-yielding, with pods that stay tender for a long time. It is excellent fresh or bottled. **'Red Velvet'**, another spectacular red variety, is vigorous, reaching 1.2–1.5m. **'Star of David'**, an Israeli heirloom variety growing to 1.8–2.5m tall, is very tasty and high-yielding. The pods grow to 23cm long, but are better eaten when small; great in soups and stews.

cultivation

A garden plant for the tropics and warm temperate climates, okra can be tried outdoors in cooler climates during hot summers, though success is better guaranteed if it is sown under cover.

Propagation

Soak the seeds in warm water for 24 hours before planting. My friend Robert Fleming, who gardens in Memphis, Tennessee, has been successful with several other methods that reduce germination time from 15 down to 5 days: soak the seed in bleach for 45–60 minutes, rinse, then plant; pour boiling water over seed, soak overnight, then plant; or place 3 seeds in each section of an ice-cube tray, allow to freeze for a few hours, then plant.

When soil temperatures are about 16°C, sow seeds in rows about 60cm apart, leaving the same distance between plants. Alternatively, sow in 'stations' 20–30cm apart, thinning to leave the strongest seedling. Plant seedlings grown in trays or modules into 7.5cm pots when they are large enough to handle and later harden them off ready for transplanting when they are about 10–15cm tall.

Growing

Okra needs a rich, fertile, well-drained soil, so incorporate organic matter several weeks before sowing. Put a stake in place before transplanting and tie in the plant as it grows, pinching out the growing tip on the main stems when plants are around 23–30cm tall to encourage bushy growth. Apply a general liquid fertiliser until plants become established, then change to a liquid high-potash feed every 2 weeks or scatter sulphate of potash around the plant base. In cool temperate conditions, warm the soil for several days before planting outdoors, spacing plants 60cm apart once the danger of frost has passed.

Maintenance

Spring Sow seeds under cover or outdoors in warmer

Okra growing in Southern India

OKRA

climates when the soil is warm enough. Pinch out the growth tips as they appear.

Summer Water, feed and tie in plants to their supporting stakes. Keep crops weed-free. Harvest young pods.

Autumn Protect outdoor crops in cooler climates to extend the cropping season.

Winter Prepare the ground for the following season.

Protected Cropping

In cooler climates, plant okra in heated greenhouses or polythene tunnels from early spring, waiting until mid-spring if heat is not provided. Plant them about 60cm apart in beds or borders. Insert a supporting cane before planting.

Temperatures should be a minimum of 21°C with moderate humidity. Feed plants every 2 weeks with a liquid general fertiliser, changing to a high-potash fertiliser once they are established.

Harvesting and Storing

Harvest with a sharp knife or scissors while seed pods are young, picking regularly for a constant supply of new pods. Handle gently: the soft skin marks easily. Pods keep for up to 10 days wrapped in a polythene bag in the salad drawer of the fridge.

Pests and Diseases

Precautions should be taken against heavy infestations of aphids, which weaken and distort growth, and powdery mildew, which stunts growth and, in severe cases, causes death. This is more of a problem when plants are underwatered. Whitefly can cause yellowing, stickiness and mould formation on the leaves. They form clusters on the leaf undersides and fly into the air when the foliage is disturbed.

companion planting

Okra flourishes when it is grown with melons and cucumbers, as it enjoys the same conditions.

container growing

Okra can be grown in 25cm pots or growbags, in peat-substitute compost, under cover or outdoors. Water regularly and feed every 2 weeks with a high-potash liquid fertiliser during the growing season.

medicinal

The mucilage from okra is effectively used as a demulcent, soothing inflammation. In India, infusions of the pods are used to treat urino-genital problems as well as chest infections.

Okra is also added as an ingredient of artificial blood plasma products.

culinary

Okra is used in soups, stews and curries, can be sautéed or fried and eaten to accompany meat or poultry. To deep-fry, remove the stalks, trim round the 'cone' near the base, simmer for about 10 minutes, drain and dry each pod, then deep-fry until crisp. In the Middle East, pods are soaked in lemon juice and salt, then fried and eaten as a vegetable. In Indian cooking, bhindi is used as a vegetable or as a bhaji to accompany curries.

Okra should not be cooked in iron, brass or copper pans, otherwise it will discolour.

Overcooked, the pods become very slimy. Any 'gluey' texture can be overcome by adding a little lemon juice to the pan and the fine, velvety covering over the pod can easily be removed by scrubbing the pod gently under running water.

Okra can also be eaten raw in salads or used as a 'dip'. Wash pods, carefully trimming off the ends. To reduce stickiness, soak them for about 30 minutes in water with a dash of lemon juice, then drain, rinse and dry.

Okra seed oil is also used in cookery.

Gumbo

Serves 4

This spicy Creole dish is known for its good use of okra.

500g okra
1 tablespoon olive oil
250g cooked ham, cubed
175g onion, finely sliced
175g celery, chopped
1 red pepper, deseeded and chopped
½ tablespoon tomato purée
500g tomatoes, skinned and chopped
1 dried chilli, finely chopped
Salt and freshly ground black pepper

Trim the stalk ends of the okra to expose the seeds, then soak for 30 minutes in acidulated water.
In the meantime, heat the oil and gently cook the ham, onion, celery and pepper until the onion begins to colour. Add the tomato purée, tomatoes and dried chilli, mix in well, and cook over a high heat for a minute or so.

Stir in the drained okra and season well. Cover and cook gently until stewed. Add a little water if the mixture gets dry.

Serve with plain grilled chicken.

Ipomoea aquatica. Convolvulaceae

WATER SPINACH

Also known as Water Convolvulus, Kangkong, Ung Choi, Rau Muong, Entsai Water Spinach. Tropical aquatic/semi-aquatic perennial. Value: high in vitamin A, calcium, phosphorus and potassium.

This close relative of the sweet potato has a long history of cultivation in southeastern China, is believed to have originated in Southern India, grows throughout the tropics and has become a pestilent weed and a threat to waterways and wetlands in places like Florida. Once a staple crop of the poor in Vietnam, it gradually became an important ingredient of Vietnamese cuisine and has potential as a food crop for rabbits grown for meat production in Cambodia. It was first mentioned in *Nanfang Zaomu Zhuang*, a Vietnamese botanical work of the fifth century AD, and is a worthwhile 'novelty' crop for those who like to grow horticultural and culinary oddities.

'Pak Quat' flowers

varieties

There are two types: **'Ching Quat'** is narrow leaved with green stems and white flowers, **'Pak Quat'** has broad, arrow-shaped leaves, white stems and pink flowers. The former is more cold-tolerant and robust; the latter is of better culinary quality.

cultivation

Propagation

In temperate climates they are best grown from seed sown just below the surface of loam-based compost at 22°C; plant several seeds in a 23cm pot, standing in a tray of tepid water from mid-spring. Alternatively, sow directly into their final position from mid-spring or start off in modules, transplanting to the final position when they are at least 10cm high. They can also be propagated from 30cm-long stem cuttings with five or more joints on the stem and the basal cut just below a node, in water or damp sand.

Growing

Water spinach can be grown as an aquatic or in waterlogged compost or rich soil, in a sheltered, sunny site. It stops growing and starts flowering as day length shortens, so is grown as an annual. Transplant seedlings into loam-based compost in an old sink or paddling pool, topped up with water, depending on the volume required. Temperatures should be around 25°C from mid-spring. Growth stops below 10°C and lower temperatures can be damaging to water spinach. Plants grown in soil should be spaced in blocks, with

WATER SPINACH

A farmer in Vietnam collecting water spinach plants

around 15cm between rows. Keep the soil moist, or growth becomes coarse. Add well-rotted compost to aid water retention.

Maintenance

Spring Sow seeds and take cuttings.
Summer Keep compost moist and water levels topped up. Harvest regularly.
Autumn Harvest regularly.
Winter Buy seed for next year.

Protected Cropping

Water spinach is best grown in a polythene tunnel or greenhouse in cooler climates.

Harvesting and Storing

Harvest regularly, removing the tips to encourage tender regrowth or cut back plants to within a few inches of the base to encourage regrowth of young, tender shoots. Towards the end of the season, the whole plant can be lifted for harvesting.

Pests and Diseases

None in cooler climates; they suffer from the same diseases as sweet potato in warmer climates.

container growing

Grow plants in old sinks or paddling pools of water to ensure plenty of soft, tender growth.

medicinal

In Burma, the plant is used as an emetic for opium poisoning; also used as a diuretic and to treat constipation.

culinary

Use immediately after harvesting, as the water-filled stems wilt rapidly. Water spinach is used extensively in Chinese and Malayan cuisine. It is mucilaginous, slightly sweet and takes up the surrounding flavours during cooking. The crispy stems are made into a pickle in the Philippines. The edible young shoots and leaves are eaten raw or cooked and eaten like spinach. Eat water spinach as the Chinese do, with fermented bean curd or shrimp sauce; it can also be fried with garlic, chillies and dried prawns or with cuttlefish and sweet, spicy sauce.

Chicken and Water Spinach with Spicy Peanut Sauce

Serves 4

600g water spinach, stems discarded, washed and dried
2 cooked chicken breasts, shredded
1 tablespoon oil
3 cloves garlic, finely chopped
1 tablespoon Thai red curry paste
125ml thick coconut milk
1 tablespoon fish sauce (nam pla)
1 teaspoon palm sugar, finely shaved, or brown sugar
4 tablespoons dry-roasted peanuts, coarsely ground
Salt

In a large saucepan of boiling water quickly blanch the water spinach for about 2 minutes. Drain and cool it under running water to stop the cooking process, then drain again. Arrange on a large platter and scatter the shredded chicken over it.

Heat the oil in a small pan and stir-fry the garlic for a minute or so until fragrant, then add the red curry paste and continue cooking for another 30 seconds. Pour in the coconut milk and bring up to simmering point. Stir in the fish sauce and the sugar, then add the peanuts, seasoning with salt to taste.

Pour over the water spinach and chicken and serve.

Ipomoea batatas. Convolvulaceae

SWEET POTATO

Also known as Kumara, Louisiana Yam, Yellow Yam. Trailing perennials grown as annuals for starchy tubers and leaves. Tender. Value: excellent source of beta carotene, rich in carbohydrates, moderate potassium and vitamins B and C. Yellow and orange types rich in vitamin A.

The 'sweet potato' is unrelated to the 'Irish' potato but is a relative of the bindweed, in the Morning Glory family. It was cultivated in prehistoric Peru and is now found throughout the tropics; it is also a 'staple' crop in Polynesia. Its arrival there is a mystery; some suggest it was taken there by Polynesians who visited South America, others think it arrived on vines clinging to logs swept out to sea. It was cultivated in Polynesia before 1250 and reached New Zealand by the fourteenth century. Captain Cook and Sir Joseph Banks found the Maoris of the North Island growing it when they landed in 1769. It was grown in Virginia by 1648. Columbus introduced it to Spain and it was widely cultivated by the mid-sixteenth century, predating the 'Irish' potato by nearly half a century. It reached England via the Canary Islands at about the same time and was the 'common potato' in Elizabethan times, some even holding it to be an aphrodisiac.

Closely related to Morning Glory

varieties

There are hundreds of sweet potato varieties worldwide. They are classified under three groups: dry and mealy-fleshed, soft and moist-fleshed, and coarse-fleshed types used as animal feed. White or pale types are floury with a chestnut-caramel flavour, while yellow and orange varieties are sweet and watery.

'Centennial' grows vigorously (to 5m) and prolifically in short seasons. Its high-quality tubers have bright copper-orange skin and deep orange flesh. The deep orange flesh of **'Georgia Jet'** has a superb flavour. It matures early and is very reliable. **'T65'** has creamy-white flesh and is the most reliable in cool climates. **'Beauregard Improved'** is a virus-free clone, smaller than T65, but with orange flesh and a more pronounced flavour. **'Tokatoka Gold'** is large, rounded and smooth-textured, and popular in New Zealand.

cultivation

Propagation

Take cuttings from healthy shoots 20–25cm long. Cut just below a leaf joint, remove basal leaves. Alternatively, pack several healthy tubers in a tray of moist and sharp sand, vermiculite or perlite in a warm greenhouse, or plant into hotbeds. When shoots reach 23–30cm, cut them off 5cm above the soil and take the cuttings as described above. Three potatoes should produce about 24 cuttings, enough for a 10m row.

SWEET POTATO

culinary

A sweet potato contains roughly one and a half times the calories and vitamin C of the 'Irish' potato. Before cooking, wash carefully and peel or cook whole. Parboil and cut into 'chips', grate raw and make into fritters or roast with a joint of meat.

Glazed with butter, brown sugar and orange juice, sweet potatoes accompany Thanksgiving dinner. In Latin America and the Caribbean they are used in spiced puddings, casseroles, soufflés and sweetmeats. The leaves can be steamed.

Tzimmes
Serves 4

This one-pot meal is based on Olga Phklebin's recipe from *Russian Cooking* but adapted to our ingredients.

1–2 sweet potatoes, depending on size
2 large potatoes
Olive oil
1kg stewing steak, cubed
1 large onion, chopped
2 carrots, chopped
750ml vegetable stock
3 tablespoons honey
½ teaspoon ground cinnamon
1 tablespoon plain flour
Chopped fresh parsley, to garnish
Salt and freshly ground black pepper

Peel both sorts of potato and cut roughly. Heat the oil in a heavy pan and sauté the meat well to brown it on all sides. Remove from the pan and keep warm. Brown the onion, adding the carrots and the meat and sufficient stock to cover. Season with salt and pepper and bring to the boil, then simmer for 45 minutes.

Stir in the potatoes, the honey, cinnamon and more seasoning if required. Bring back to the boil and simmer for a further 45 minutes, covered. If the stew is too liquid, allow it to boil fast at this stage. Remove 5 tablespoons of the stock and mix with the flour. Pour this back into the pot, bring to the boil and cook for a further 30 minutes or until the meat is tender and the potatoes are cooked. Sprinkle with parsley and serve with a green salad.

In the humid tropics and subtropics, cuttings are rooted *in situ* at the start of the rainy season. Plant on ridges 15–30cm high and 1–1.5m apart. Just below the ridge top, insert cuttings 23–30cm apart, leaving half of the stem exposed. Or plant small tubers 7.5–10cm deep along the top of the ridge. On sandy, free-draining soil plant cuttings and tubers on level ground.

Growing
Sweet potatoes thrive in a tropical or sub-tropical climate with an annual temperature of 21–26°C. Light frost kills leaves and damages tubers. Ideal annual rainfall is 750–1,200mm, with wet weather in the growing period and dry conditions for the tubers to ripen. Tuber production is fastest and sugar production highest when day lengths exceed 14 hours.

Soils should be moisture-retentive and free-draining with a pH of 5.5–6.5. If necessary, dig in well-rotted organic matter before planting. In cool climates, plant cuttings outdoors when there is no danger of frost. Plant through black polythene after hardening off. Once established, scatter a high-potash granular fertiliser around plants. Occasionally lift vines from the ground to prevent rooting at the leaf joints. Rotate crops.

Maintenance
Spring Take cuttings under cover, keep warm and moist.
Summer Keep cuttings weed-free, water well and feed. Prune as necessary.
Autumn Harvest. Save some tubers for the next crop.
Winter Keep compost slightly moist.

Protected Cropping
Grow under cover in cool-temperate climates at a minimum temperature of 26°C.

Take cuttings as normal from healthy shoots of mature plants, and insert about 4 cuttings round the outside of a 15cm pot filled with cuttings compost. Keep moist with tepid water. Transplant into a greenhouse border once a good root system has formed. Mist regularly. Prune stems longer than 60cm to encourage sideshoots and also from late winter to thin out congested growth.

Harvesting and Storing
In good conditions, tubers ripen in 4–5 months. Lift when slightly immature; otherwise wait until vines begin to yellow. Lift carefully to avoid bruising. Use fresh or store, once dried, in a cool dark place for up to a week.

Pests and Diseases
Leaves can suffer from leaf spot and sooty mould. Black rot appears at the base of the stem and brown rot on the tuber. Check stored tubers regularly. Also susceptible to whitefly or red spider mite under cover.

container growing

Grow in pots or containers 30cm deep and 37.5cm wide using loam-based compost with moderate fertiliser levels. Provide supports.

medicinal

Sweet potatoes and their leaves contain anti-bacterial and fungicidal substances and are used in folk medicine. In Shakespeare's day they were sold in crystallised slices with sea holly ('eringo') as an aphrodisiac. In *The Merry Wives of Windsor*, Falstaff cries: 'Let the sky rain potatoes… hail kissing-comforts and snow eringoes'. The Empress Josephine introduced sweet potatoes to her companions, who were soon serving them to stimulate the passion of their lovers. The results are not recorded!

Lactuca sativa. Asteraceae

LETTUCE

Annual grown for edible leaves. Half hardy to hardy. Value: rich in beta carotene, particularly outer leaves.

The garden lettuce is believed to be a selected form of the bitter-leaved wild species *Lactuca serriola*, which is found throughout Europe, Asia and North Africa. The ancient Egyptians were said to have been the first to cultivate lettuces and there are examples of tomb wall paintings depicting a form of Cos lettuce, which is said to have originated on the Greek island of the same name. They believed it was an aphrodisiac and also used its white sap and leaves in a concoction alongside fresh beef, frankincense and juniper berries as a remedy for stomach ache. The Romans, too, attributed medicinal properties to the lettuce and the Emperor Augustus erected an altar and statue in its honour; they believed that it upheld morals, temperance and chastity. The Romans were said to have introduced it to Britain with their conquering armies and even after many centuries the lettuce is still regarded as the foundation of a good salad.

varieties

The many cultivars are divided into three main categories – cabbage, leaf and Cos types. They can be grown all year round, and many modern cultivars are disease-resistant (see pages 662–665).

Cabbage Lettuce

More tolerant of drought and drier soils than the other kinds, this group includes 'Butterhead' types, with soft buttery-textured leaves which are usually grown in summer, and 'Crisphead' types, which have crisp leaves forming compact hearts.

Butterhead

'Action' has thick, pale green leaves and is resistant to mosaic virus and downy mildew. **'All Year Round'** is tasty and compact, with pale green leaves. It is slow to bolt and hardy. **'Arctic King'** is highly resistant to cold and ideal for autumn or early spring sowing. **'Avon defiance'** is an excellent, high-yielding, dark green lettuce which withstands drought and high temperatures. It is ideal for summer crops, particularly those sown from mid- to late summer, and is resistant to root aphid and downy mildew. **'Buttercrunch'** has compact, crisp, dark green heads and a beautiful 'buttery' heart. It is slow to bolt and heat-resistant. **'Cassandra'** is top quality with pale green leaves and large 'hearts'. Resistant to downy mildew and lettuce mosaic virus, it is good for growing all year round. **'Dolly'** is a large lettuce for mid-summer to mid-autumn cropping. It has good resistance to

Pandero

LETTUCE

'Sangria'

downy mildew and lettuce mosaic virus. **'Kwiek'**, a large-headed variety for winter cropping, is resistant to downy mildew. Sow in late summer for harvesting in midwinter or force as an early spring crop. **'Musette'** has dark green, succulent leaves and is resistant to root aphid, lettuce mosaic virus and downy mildew. **'Sabine'** is resistant to root aphid and downy mildew. The outer leaves of **'Sangria'** are tinged red, the inner a pale green and it has some resistance to mildew and mosaic virus. **'Soraya'** also has some resistance to downy mildew and lettuce mosaic virus. **'Valdor'**, which has firm, dark green hearts, is a very hardy lettuce for overwintering outdoors. **'Yugoslavian Red'** has red tips to the leaves and a bright greeny-yellow heart. It is mild-flavoured and ideal for the ornamental vegetable garden.

Crisphead

'Avoncrisp' has some resistance to mildew and root aphid, but can suffer from 'tip burn'. **'Iceberg'** is ideal for spring or summer sowing and has very crisp tender leaves with large ice-white hearts (the heart has the best-quality leaves). **'Malika'** grows very rapidly and should be harvested soon after it matures. It has some resistance to mildew and bolting. **'Premier Great Lakes'** is large, crisp and matures rapidly. It is heat- and 'tip-burn'-resistant. **'Saladin'** is excellent for summer cropping and is resistant to mildew and bolting. **'Vienna'** has crisp, medium-sized heads of fresh, green leaves and is resistant to leaf aphids and mildew. **'Webb's Wonderful'** is extremely popular, and rightly so. It is a good-quality lettuce which lasts well at maturity.

Leaf lettuce

These are loose leaf varieties which sometimes form an insignificant head. Cut to resprout, or remove single leaves as needed. They stand longer before bolting than other types and can be grown at any time of year; however, growth is slower in winter.

'Cocarde' is a large, bronze, 'oak leaf' type with an excellent flavour. **'Grand Rapids'** has crinkled, pale green leaves and is resistant to tip burn. **'Lollo Blonda'** is similar to **'Lollo Rossa'**, but with fresh pale green leaves and some resistance to lettuce root aphid. **'Lollo Rossa'** adds colour to salads. The leaves are tinged red with serrated, wavy margins and it is good mixed with cos or iceberg lettuces. It is also attractive as an edging plant in the flower border or vegetable plot. **'Oak Leaf'** has several different colour forms, from pale green to brown. It is tasty and ornamental. **'Red Salad Bowl'** has bronze-green to crimson leaves. **'Ruby'** is crinkled and pale green with deep red tints and has good heat resistance. **'Salad Bowl'** was one of the first leaf lettuces with masses of green, deeply lobed leaves which are crisp but tender. Good resistance to bolting.

Cos or Romaine

These flourish in humus-rich, moist soil, take longer to mature than other varieties and are better in cooler weather. Some can overwinter outdoors or be grown as leaf lettuce if closely spaced. They are generally very tasty.

'Bubbles' has heavily textured leaves and is mildew resistant. **'Chartwell'** is one of the tastiest and sweetest with firm hearts. It is downy mildew resistant and slow to bolt. Grows well in hot weather. **'Counter'** is very sweet and crisp and is resistant to tip burn, bremia and bolting. **'Freckles'** is very tasty, with attractive, red mottling over the leaves. It is warm weather-tolerant and slow to bolt. **'Lobjoits Green'**, is a large, good-quality, old variety which is deep green, crisp and tasaty. It can be grown as closely as leaf lettuce but is subject to tip burn. **'Little Gem'** is a compact, quick-maturing, semi-Cos with a firm, sweet heart. It is ideal as a catch crop, for early crops under cover and for small gardens and has remained a popular variety since the late nineteenth century. **'Nymans'** has burgundy leaves that blend towards a bright green heart. It is slow to bolt and has good mildew resistance. **'Pandero'** is a well flavoured, red-leaved baby cos. It is mildew-resistant and ideal as a colourful 'cut and come again'. **'Parris Island'** has upright conical heads with light green outer leaves and a white heart. They have a wonderful flavour. **'Valmaine'** is good for growing as a 'cut and come again' or for close planting and use as leaf lettuce. **'Winter Density'** has dark green heads and crisp, refreshing leaves, and also outstanding cold hardiness.

cultivation

Propagation

Germination is poor, particularly with Butterhead types, if soil temperatures exceed 25°C, with the critical period being a few hours after planting. During hot weather sow in late afternoon or evening, when soil temperatures are lower, water after sowing to reduce soil temperature, shade before and after sowing, germinate in trays or modules in a cool place and transplant, or fluid sow. Lettuces do not transplant well in dry soil and hot weather; where possible, summer sowings are better made *in situ* or in modules.

Sow thinly in drills 1–1.5cm deep; seeds sown too deeply are slow or fail to germinate. Thin to the final spacing when plants are large enough to handle

Sow lettuce early, then transplant

Beauty in a bell jar

(overcrowding checks growth and can cause bolting). In cool weather, thinnings can be transplanted if lifted with care to avoid root damage.

Sow those grown in trays or modules in loam-based potting compost to maintain strong growth and harden off before transplanting when they have 4 to 5 true leaves. Do not transplant when the weather is hot and dry, unless shading can be provided, and do not plant them too deeply.

Sow summer crops from mid-spring to mid-summer. Sow winter-hardy varieties from late summer to early autumn, thinning to the final spacing in spring.
As they do not last long after maturity, maintain a continuous supply of lettuce by sowing successionally, about every 2 weeks, just as the seedlings from the previous sowing appear. Thin when the seedlings are large enough to handle, where possible staggering them in a triangular pattern to make optimum use of the area.

Space small lettuces 20cm apart in and between the rows, butterheads with about 28cm in and between the rows, or 25cm apart in rows 30cm apart. Crispheads should be planted 38cm apart or 30cm apart with the rows 38cm apart. Plant **'Salad Bowl'** or Cos about 35cm apart in and between the rows.

Leaf lettuce can also be grown as 'cut and come again' seedlings, providing 2 or 3 harvests before bolting, while summer crops tend to run to seed more rapidly. Make sure you sow thinly.

Cos varieties like **'Lobjoits Green'** can be grown as leaf lettuce. Sow in rows 13cm apart, thinning to 2.5cm. Sow weekly from late spring to early summer and again in 3 consecutive weeks from late summer. This technique was developed at Horticulture Research International, Wellesbourne, UK.

Growing

Lettuce prefer cool growing conditions, from 10–20°C, and need an open, sunny site on light, rich, moisture-retentive soil, with a pH of around neutral.
They struggle on dry or impoverished soil, so dig in plenty of well-rotted organic matter the autumn before sowing or grow on ground manured for the previous crop. Lightly fork in a base dressing of general fertiliser at 60g/sq m about 10 days before sowing and create a seedbed by raking to a fine tilth.

Hoe and hand weed regularly to remove any weeds. A constant supply of moisture is vital for success. Lettuce need 22 litres/sq m per week in dry weather. Water in the mornings on sunny days, so that the water on the leaves evaporates quickly, reducing the risk of disease. If water is scarce, apply only on the last 7–10 days before harvest.

Boost growth of winter-hardy outdoor crops with a liquid general fertiliser in mid-spring and use the same treatment for slow-growing crops at any time of year. Rotate crops every 2 years to avoid the build-up of pests and diseases.

Maintenance

Spring Sow crops under glass or outdoors under cloches in early spring. Sow later crops outdoors. Harvest early crops.
Summer Sow successionally and harvest overwintered crops. Keep crops weed-free; water as needed. Harvest.
Autumn Sow and protect crops under cloches. Harvest.
Winter Sow crops under cover. Harvest.

Protected Cropping

Sow from late winter to early spring in modules or trays for transplanting under cloches or cold frames from mid- to late spring. Alternatively, they can be sown *in situ* in cold frames or under cloches for growing on or transplanting outdoors in a protected part of the garden and covered with floating cloches.

Protect late-spring and summer transplants under cloches until they become established and protect late-summer sowings under cloches to maintain the quality of autumn crops. Grow hardier varieties under floating cloches or in a glasshouse or a cold frame over winter, ideally with a gentle heat around 7°C.

Germination is better in cool conditions

LETTUCE

Lettuce seedlings in a vegetable garden

Sow from late summer to mid-autumn for transplanting. Earlier sowings can be made outdoors in a seedbed for transplanting; others can be sown in modules or seed trays. Ventilate well to avoid disease problems.

Harvest from late autumn to mid-spring. Cover outdoor winter-hardy crops with glass or floating cloches to ensure a good-quality crop and provide protection during severe weather.

Sow early crops of 'cut and come again' seedlings under cover in late winter or spring, and also from mid-autumn.

Harvesting and Storing

Summer maincrop lettuce are ready to harvest from early summer to mid-autumn, around 12 weeks after sowing. Cos stand for quite a time in cool weather but Butterheads deteriorate within a few days of maturity. Crispheads last for about 10 days before deteriorating. Leaf lettuces can be picked over a long period. Harvest hardy overwintered outdoor crops from late spring to early summer.

Harvest 'cut and come again' seedlings about 4 weeks after sowing and Cos types grown as leaf lettuce about 2.5cm above the ground when they are 7.5–12.5cm high. A second crop can be harvested from 3 to 8 weeks later, depending on the growing conditions.

Cut mature lettuces at the base, just below the lower leaves. Do not squeeze hearting lettuces to check if they are ready for harvesting, as this can damage the leaves: press them gently but firmly with the back of your hand. Pull single leaves from leaf lettuce as required or cut 2.5cm above the base and allow the plant to resprout.

Store lettuce in the refrigerator in the salad drawer or in a polythene bag for up to 6 days. Cos lettuce stores the longest.

Pests and Diseases

Root aphids appear in clusters on the roots and are usually covered in a white powdery wax. The symptoms are stunted growth and yellowing leaves; plants may collapse in hot weather. They are less of a problem in cool, damp conditions. Pick and destroy affected plants or grow resistant varieties.

Aphids can also be a problem.

Leaf tips become brown and dry when affected by tip burn, caused by sudden water loss in warm weather. Water well and shade, if necessary; do not allow lettuces to grow excessively large; grow resistant varieties where possible.

Botrytis or grey mould can be a problem in cool, damp conditions. Do not plant seedlings too deeply, handle transplants carefully to avoid damage, thin early, remove diseased material, improve ventilation and spray with a systemic or copper-based fungicide. Grow resistant varieties.

Bolting or running to seed is caused by check during transplanting, drought, temperatures over 21°C, or long days; otherwise lettuces will only bolt after hearting. Grow resistant varieties or leaf lettuce.
Pigeons and sparrows can damage seedlings. Grow under horticultural fleece, tie thread between bamboo canes, use humming line or other deterrents.

Lettuce mosaic virus is a disease that causes the leaves to become puckered and mottled; veins become transparent and growth is stunted – more of a problem on overwintering crops. Destroy those that are affected, control aphids and grow resistant varieties.

 companion planting

Lettuce grow well with cucumbers, onions, radishes and carrots. Dill and chervil protect them from aphids.

 container growing

Compact varieties are suitable for growing in pots, containers and window boxes. Use a soil-based compost, mix in a slow-release granular fertiliser and water regularly. Sow either *in situ* or in modules for transplanting.

 medicinal

Lettuce is used as a mild sedative and narcotic, and lettuce soup is reported to be effective in treating nervous tension and insomnia. Lettuce sap dissolved in wine is said to make a good painkiller.

Lettuce soothes inflammation – lotions for the treatment of sunburn and rough skin are made from its extracts.

Lettuce can also be used as a poultice on bruises or taken internally for stomach ulcers and for irritable bowel syndrome.

It is also anti-spasmodic and can be used to soothe coughs and bronchial problems; it is reputed to cool the ardour.

culinary

Wash leaves thoroughly before use. Use fresh or wilted in salads, braise with butter and flavour with nutmeg or braise with peas and shallots and serve with butter. Stir-fry with onions and mushrooms, steam and add to chicken soup or make cream of lettuce soup and garnish with hardboiled eggs and a sprinkling of curry powder. Stalks can be sliced and steamed.

Cypriot Salad
Serves 2

50g haloumi cheese, cut into 0.5cm slices
3–4 large lettuce leaves, torn into pieces
4 medium tomatoes
¼ cucumber, sliced
¼ yellow pepper, cored, deseeded and diced
Handful of fresh flat-leaf parsley, roughly chopped
10 black olives, stoned
½ red onion, thinly sliced

For the Dressing:
2 tablespoons extra virgin olive oil
1 tablespoon white wine vinegar
1–2 teaspoons lemon juice
Sea salt and freshly ground black pepper

Toast the cheese under a medium grill on a non-stick baking sheet until golden brown. Turn once and brown the other side. Mix with the other ingredients in a large salad bowl. Shake all the dressing ingredients together in a glass jar until well combined. Pour over the salad, gently toss to mix and serve immediately.

Lettuce-wrapped Minced Prawns
Serves 6–8 as a side dish

30 raw prawns, shelled and deveined
¼ teaspoon salt
2 teaspoons cornflour
2 tablespoons egg white
1 crisp lettuce (Webb's or iceberg)
4 tablespoons peanut or vegetable oil
50g Sichuan zhacai, trimmed and finely chopped
1 teaspoon crushed garlic

Cypriot Salad

LETTUCE

Mince the prawns coarsely and put in a large mixing bowl. Season with salt. Add the cornflour and egg white and stir in the same direction until the egg white is absorbed and the mixture is elastic. Refrigerate, covered, for 1 hour.

Arrange the separated lettuce leaves, preferably cup-shaped, on a plate on the dining table.

Heat a wok until hot. Add 1 tablespoon of oil and the Sichuan *zhacai* (pickled mustard stem, from specialist shops) and cook for a few seconds, stirring. Scoop on to a dish and keep nearby. Wipe the wok clean.

Reheat the wok until the smoke rises. Add the rest of the oil and coat the wok with it. Add the garlic and the prawns, stirring vigorously. Add the *zhacai* and continue cooking until the prawns turn pink and are cooked. Then place on a plate next to the lettuce.

To serve, each person takes a lettuce leaf, spoons some prawn mixture into the centre and wraps the lettuce around it: perfect finger-food.

Papaya Salad
Serves 4

2 cloves garlic, peeled
3–4 small fresh red or green chillies
2 yard-long beans or 20 French beans, chopped into 5cm lengths
175g fresh papaya, peeled, deseeded
1 tomato, cut into wedges
2 tablespoons nam pla (Thai fish sauce)
1 tablespoon granulated sugar
2 tablespoons lime juice
A selection of firm seasonal green vegetables, e.g. iceberg lettuce, cucumber and cabbage, to serve

Pound the garlic in a large mortar, then add the chillies and pound again. Add the long beans, breaking them up slightly. Cut the papaya into fine slivers. Now take a spoon and stir in the papaya. Lightly pound together, then stir in the tomato and lightly pound again.

Add the fish sauce, sugar and lime juice, stirring well, then turn into a serving dish. Serve with fresh raw vegetables – any leaves, such as white cabbage, can be used as a scoop for the spicy mixture.

Braised Lettuce
Serves 4

4–5 lettuces, trimmed
A little oil or butter
50g bacon, roughly diced
1 carrot, roughly diced
1 red onion, sliced
300ml vegetable stock
2 tablespoons chopped fresh thyme and parsley
Salt and freshly ground black pepper

In a saucepan of salted water, cook the lettuces for 5 minutes. Drain and refresh in cold water. Drain again and remove as much liquid as possible: you can blot the leaves lightly with kitchen paper.

Grease an ovenproof dish and sprinkle the base with the bacon, carrot, onion and seasoning. Arrange the lettuces neatly in the dish and pour in the stock. Cover the dish with buttered paper and braise in a preheated oven, 180°C/350°F/gas mark 4, for 30–40 minutes.

Remove the lettuces and other ingredients and keep hot. Then reduce the cooking liquid and pour it over the vegetables. Sprinkle with the fresh herbs and serve very hot.

Papaya Salad

Lactuca sativa var. *augustana*. *Asteraceae*.

CELTUCE

Also known as Stem Lettuce, Asparagus Lettuce, Chinese Lettuce. Annual grown for edible leaves. Value: little nutritive value; low in carbohydrate and calories; a source of potassium.

Introduced from China, where it has been grown for centuries, this 'oriental vegetable' consists of a short-stemmed mutation lettuce. It has been listed in European catalogues since 1885, but the name 'celtuce' was adopted by an American seed company who first offered seeds in 1942. It aptly describes its characteristics: the stems are used like celery; the leaves make a lettuce substitute.

varieties

'Zulu' is a variety for cooler climates, with narrow, dull-textured leaves. Others are sold in seed mixes of broad, dull, glossy or red-leaved varieties.

cultivation

Celtuce needs well-drained, rich, fertile soil, with a pH of 6.5–7.5, around neutral. Celtuce tolerates a range of temperatures from light frosts to over 27°C.

It tends to bolt prematurely in extremely hot conditions, but is still more heat-resistant than lettuce. Grow celtuce as a winter crop in mild areas.

Propagation
For successional cropping, sow every 2 weeks from mid-spring until mid-summer in drills 1cm deep and 30cm apart. When large enough to handle, thin to 9cm.

Germination is poor in temperatures above 27°C. In hot summers, sow in seed trays or modules in a cool, partially shaded position. Modules tend to produce better plants than seed trays; weak seedlings rarely produce good stems. Transplant when 3–4 leaves have been produced, generally after 3–4 weeks.

Mulch after planting to conserve moisture and suppress weeds; the shallow roots are easily damaged by hoeing.

Growing
In poor soils, add copious amounts of organic matter the winter before planting. Celtuce grows well on light soil, but develops a stronger root system and more robust plants on heavier soils.

Water well as leaves develop, to keep them tender. As the stems develop, reduce watering, but take care to keep the supply steady: if the soil becomes too wet or too dry, the stems may crack. Feed with a liquid general fertiliser every 3 weeks.

Maintenance
Spring Sow seed.
Summer Water and remove weeds.
Autumn Transplant seedlings for winter crops.
Winter Grow under cover; harvest mature crops.

Why not give it a try?

CELTUCE

Protected Cropping

Grow in cloches, unheated glasshouses, tunnels or under horticultural fleece to extend the season. Raise early crops by sowing and planting under cover in early spring and late crops by transplanting summer-sown seedlings under cover in autumn.

Harvesting and Storing

Harvest 3–4 months after sowing, when 30cm high and 2.5cm diameter. Cut the stalks, pull up the plant or cut it off at ground level. Trim the leaves from the stem but do not touch the top rosette of leaves, in order to keep the stem fresh.

Celtuce stems can be kept for a few weeks if stored in cool conditions.

Pests and Diseases

Celtuce is susceptible to the same problems as lettuce. Pick off slugs, set traps or use ferric phosphate pellets. Downy mildew is worse in cool, damp conditions: treat with fungicide, and remove infected plants and debris at the end of the season.

companion planting

Plant with chervil and dill to protect from aphids. Interplant between slower-growing crops such as cauliflower, self-blanching celery or Chinese chives.

container growing

Sow in containers or pots in a loam-based compost with moderate levels of added fertiliser.

culinary

Celtuce is excellent raw and cooked. Prepare the stems by peeling off the outer layer. Cut into thin slices for salads, or into larger pieces for cooking. In salads, it can either be eaten raw or cooked and cooled, served with a spicy dressing.

Cook lightly, for 4 minutes at most. Stir-fry with white meat, poultry, fish, other vegetables, or on its own seasoned with garlic, chilli, pepper, soy or oyster sauce. It is also delicious served with a creamed sauce or baked *au gratin*. In China it is used in soups and pickles. Young leaves can be cooked like 'greens'.

Crispy Spring Rolls

Serves 4–6

225g fresh beansprouts, washed and husked
115g bamboo shoots
225g celtuce stems
115g white mushrooms
115g carrots
Vegetable oil, for frying
1 teaspoon sugar
1 tablespoon light soy sauce
1 tablespoon rice wine
20 frozen spring roll skins, defrosted
1 tablespoon plain flour mixed with 1 tablespoon water
salt and freshly ground black pepper

Prepare the beansprouts and roughly chop all the other vegetables to the same size. Heat a little oil in a wok until smoking and stir-fry the vegetables for a minute, then add the sugar, soy sauce, rice wine and seasoning, and cook for a further minute or two. Set aside to cool.

Cut the spring roll skins in half diagonally. Place a teaspoon of the vegetable mixture in the centre of each and roll up neatly, folding in all the corners. Place on a lightly floured plate and brush the upper edge with a little flour paste to seal.

Heat enough oil to deep-fry until steaming and drop in the spring rolls for 3–4 minutes, until crispy, cooking in batches. Drain and serve with chilli sauce.

Lagenaria siceraria. Cucurbitaceae

DOODHI

Also known as Bottle Gourd, Calabash Gourd, White-flowered Gourd, Trumpet Gourd. Vigorous annual climber grown for edible young fruits, shoots and seeds. Tender. Value: little nutritive value; a moderate source of vitamin C; small quantities of B vitamins and protein.

Early evidence for the cultivation of this versatile tropical gourd comes from South America around 7000 BC, though it is thought to have originated in Africa, south of the Sahara, or India. Some sources suggest that it may have dispersed naturally by floating on oceanic currents from one continent to another: experiments have found that seed will germinate after surviving over seven months in seawater. One of the earliest crops cultivated in the tropics, these gourds with their narrow necks have developed in many shapes and sizes, some reaching up to 2m long. The young fruits are edible, but mature shells become extremely hard when dried and have been used to make bottles, kitchen utensils, musical instruments, floats for fishing nets and even gunpowder flasks. In the past *lagenaria* leaves were used as a protective charm when elephant hunting.

varieties

Lagenaria siceraria is a vigorous annual, climbing or scrambling by means of tendrils to more than 10m. The leaves are broad and oval with wavy margins. Its solitary, fragrant, white flowers open in the evenings. Its fruits are pale green to cream or yellow, with a narrow 'neck', and contain white, spongy flesh and flat, creamy-coloured seeds. The names of selected forms (like 'bottle', 'trumpet', 'club' or 'powderhorn' gourd) relate to their use and appearance.

cultivation

They flourish in warm conditions, around 20–30°C and plenty of sunshine, but will grow outdoors in warm temperate climates where humidity levels are moderate to high. Plants need to be trained over supporting structures.

Propagation

Soak seeds overnight in tepid water before sowing on mounds of soil about 30cm apart, containing copious amounts of well-rotted manure. Plant 3 seeds, edgeways, and thin to leave the strongest seedling. Seeds can be sown in a nursery bed and transplanted when they have 2 to 3 leaves. Seeds are usually sown at the start of the rainy season.

Growing

Doodhi needs fertile, well-drained soil, preferably with a pH of 7. Add a granular general fertiliser to the planting hole or scatter it around the germinated seedling.

Train the stems into trees, or over fences, arbours or frames covered with 15cm mesh netting. Erect stakes or trellis on the beds or among the groups of mounds. Alternatively, grow plants in beds 120–180cm square, planting a seedling at each corner and training towards the centre. Pinch out the terminal shoots when they are 3–4cm long, to encourage branching. The best yields are obtained in warm-climate areas with rainfall around 80–120cm per annum, but they grow well in drier regions if they are kept well watered. Feed with a high-nitrogen fertiliser every 3 weeks during the growing season, keep crops weed-free and mulch with a layer of organic matter.

Plants can grow extremely rapidly in hot weather: 60cm in 24 hours has been recorded. Where conditions are suitable they can be grown all year round. To grow outdoors in cooler climates, harden off under cloches or cold frames and plant in a sheltered, sunny position when the danger of frost has passed.

The chance of success is greater in hot summers; they should reach edible size, though they rarely have a long enough season to mature fully.

Maintenance

Spring Sow seeds under glass in warm conditions.
Summer Train vines over trellis or netting. Water regularly; feed as necessary. Damp down the greenhouse to maintain humid conditions and harvest young fruits.
Autumn In cooler climates, harvest mature gourds for preserving before the onset of the first frosts.

Protected Cropping

In cool, temperate climates, sow seeds 6–8 weeks before the anticipated planting out time. Sow 2–3 seeds in a 15cm pot of peat-substitute compost in late winter or early spring at 21–25°C. Germination takes 3–5 days. If the ideal temperatures cannot be achieved and maintained, delay sowing until mid- to late spring. Transplant when seedlings are 10–15cm high. They will grow at temperatures down to 10°C, but flourish at high temperatures and in humid conditions in bright, filtered light.

DOODHI

Damp down the greenhouse floor at least twice a day, depending on outside weather conditions. Dig in plenty of well-rotted manure and horticultural grit or sharp sand into the glasshouse border, or grow in containers.

Keep the compost constantly moist, using tepid water, and feed with a liquid general fertiliser every 2 weeks. Train up a trellis, wires or netting. Flowers will appear from early summer and only 1 or 2 should be allowed to grow to maturity.

Harvesting and Storing
Vines begin to fruit 3–4 months after planting. Harvest immature fruits when a few centimetres long, 70–90 days after sowing. Mature fruits can be harvested and dried slowly as ornaments.

Pests and Diseases
This very robust plant is rarely troubled by disease. Whitefly can occasionally be a problem when plants are grown under cover. In warm, humid climates, anthracnose appears as pinhead-sized, water-soaked lesions on the fruits, combining to form a small black mass. Fruit rot can also be a problem. Spray with a copper-based compound and remove badly affected leaves. Harvest fruits carefully to avoid damage.

container growing

Pot on as plants grow, until they are in 30cm pots, or grow in large containers of loam-based compost with added organic matter and horticultural grit or sharp sand to improve drainage. Keep the compost moist, with tepid water. Growth is better restricted when they are grown in pots and is preferable in the greenhouse border, increasing the chance of successful fruiting.

Stop the shoot tips when the stems are 1.5m long and train the side stems along a wire. (It is worth noting that even with this method, plants still need a considerable amount of space.) Hand-pollination ensures a good crop. (Male and female flowers are on the same plant: females are recognisable by the ovary at the back of the flower, covered in glandular hairs.) Once a flower has formed, allow 2 more leaves to appear, then pinch out the growing tip. If it is a male flower or a fruit is not going to form, cut the stem back to the first leaf. A replacement will be formed.

The ornamental gourd tunnel in Rosemoor Gardens, UK

culinary

Young fruits, which are rich in pectin, are popular in tropical Africa and Asia. They have a mild, somewhat bland taste, are peeled before eating and any large seeds removed. They can be cubed or sliced, sautéed with spices to accompany curries and other Indian dishes and added to stews or curries. Young shoots and leaves can be steamed or lightly boiled. Seeds are used in soups in Africa and are boiled in salt water and eaten as an appetiser in India. The seed oil is used for cooking.

medicinal

The fruit pulp around seeds is emetic and purgative and is sometimes given to horses! Juice from the fruit is used to treat baldness; mixed with lime juice, it is used for pimples; boiled with oil, it is used for rheumatism. Seeds and roots are used to treat dropsy and the seed oil used externally for headaches.

Languas galanga (syn. Alpinia galanga). Zingiberaceae

GREATER GALANGAL

Also known as Greater Galingale, Siamese Ginger. Herbaceous perennial; rhizomatous rootstock used as flavouring. Tender. Value: negligible nutritive value.

Two species, greater and lesser galangal, have been grown for centuries for their pungent, aromatic roots, which are used as spices and medicinally. The earliest records date from around AD550, while Marco Polo noted its cultivation in southern China and Java in the thirteenth century. In the Middle Ages it was known as 'galangale', a name also used for the roots of sweet sedge, whose violet-scented rhizomes are used in perfumery.

The word galangal came from the Chinese, meaning 'a mild ginger from Ko', a region of the Canton province. It is an essential ingredient of many Malaysian and Thai dishes. The spicy rhizomes have a somewhat different use in the Middle East, where they have been used to 'spike' horses!

cultivation

The plant is a herbaceous perennial with pale orange-flushed cream bulbous rhizomes. Stems grow to 2m tall, with dark green lance-shaped leaves up to 50cm long. Flowers are pale green and white with pink markings; the fruit is a round red capsule. There are many different races in cultivation, including types with red and white rhizomes. Lesser galangal is regarded as superior, as it is more pungent and aromatic, but is uncommon in cultivation.

Propagation

Sow fresh seed in pots of peat-substitute compost at 20°C. Keep the compost moist, using tepid water, and transplant seedlings when they are large enough to handle. In tropical and sub-tropical regions they can be sown outdoors in a seedbed (for ground preparation, see 'Growing').

Divide rhizomes in spring, when the young shoots are about 2.5cm long. Remove young, vigorous sections from the perimeter of the clump, using a sharp knife, dust the cuts with fungicide and transplant just below the soil or compost surface. Soak thoroughly with tepid water and maintain high humidity and temperatures.

Growing

Galangal grows outdoors in tropical or sub-tropical climates, thriving in sunshine or partial shade. It needs a rich, free-draining soil, so dig in plenty of well-rotted compost and allow the ground to settle for a few weeks before planting. Allow 90cm between plants. Keep crops weed-free and water during dry periods. Mulch with well-rotted compost when plants have become well established.

Protected Cropping

In temperate zones, plants can be grown in a heated polythene tunnel or greenhouse in bright filtered light, with a minimum temperature of 15°C. Maintain high humidity by misting plants with tepid water and 'damping down' paths. Prepare borders as for 'Growing' and allow 90cm between plants.

Keep the compost constantly moist with tepid water and reduce watering in lower temperatures.

Remove flower stems before they flower and reduce watering. The foliage gradually becomes yellow and

culinary

Before use, scrape or peel off the skin. The raw chopped or minced root is used in Malaysian and Indonesian dishes with bean curd, meat, poultry, fish, curries and sauces. It is also used as a marinade to flavour barbecued chicken. The fruits are a substitute for cardamom, the buds can be pickled and the flowers are eaten raw with vegetables or pickles in parts of Java. In Thai cooking, it is preferred to ginger. In medieval England, a sauce was made from bread crusts, galangal, cinnamon and ginger pulverised and moistened with stock. It was heated with a dash of vinegar and strained over fish or meat.

Galangal Soup
Serves 4

This Vietnamese recipe has many variations. First, make a stock from the following, by simmering for about three hours, topping up as necessary:

1 litre water
1 chicken carcass
4 bulbs lemongrass, bruised
12cm galangal, peeled and sliced
1 onion, roughly sliced
2 red dried chillies
6 kaffir lime leaves
Salt and freshly ground black pepper

Then strain the stock through some fine muslin. Reheat in a heavy saucepan, adding the meat from the chicken carcass, chopped finely, together with *1 tablespoon of fish sauce* and *4 tablespoons of lime juice*.

Simmer for 5 minutes, then add *125g shiitake mushrooms*, left whole. Simmer for a further 3 minutes, then stir in *125ml thick coconut milk*. Do not allow to reboil, and check the seasoning.

Serve hot.

dies back in winter. The compost should remain slightly moist throughout the dormant period. Increase watering when new shoots emerge the following spring. Feed regularly with a liquid general fertiliser every 2 weeks while the plant is actively growing. Mulch plants grown in borders annually in spring with well-rotted manure.

Harvesting and Storing
Rhizomes are ready to harvest after 3–4 years. Lift plants towards the end of the growing season and remove mature rhizomes, retaining younger ones for transplanting. It is a good idea to divide and replant a few each year to ensure a constant supply.

Galangal are better used immediately after harvest but will keep for at least a week in a cool place. Alternatively, they can be frozen whole in a polythene bag and segments removed as required. Keep dried roots in airtight containers in a cool dark place.

Pests and Diseases
Red spider mite can be a problem under glass. Check plants regularly, as small infestations are easily controlled. Speckling, mottling and bronzing of the leaf surface are the usual symptoms; in later stages, fine webbing appears on leaves and stems. They prefer hot, dry conditions. Control by maintaining high humidity or spray with derris or a similar insecticide.

container growing

Plants can be grown in containers under cover in a compost mix of 2 parts each of loam and leaf mould, 1 part horticultural grit or sharp sand and 3 parts medium-grade bark. Repot and divide in spring when the rhizomes outgrow their allotted space.

medicinal

Plants contain cineol, an aromatic, antiseptic substance. The essential oil acts as a decongestant and respiratory germicide and digestive aid. In India it is used as a breath purifier and deodorant, and a paste is made from the rhizomes to treat skin infections. It is also said to be an aphrodisiac. Infusions are taken after childbirth.

other uses

Roots of lesser galangal are used in Russia for flavouring tea and a liqueur called Nastoika. The rhizomes produce a yellow or yellow-green dye.

KAEMPFERIA GALANGA
(Chinese keys)

The rhizomes, common in the markets of Southeast Asia, produce a distinctive cluster of finger-like roots which are pale brown with bright yellow flesh and a pungent aroma. They have a very strong taste and should be used sparingly in green curry paste, sauces, soups and curries.

The rhizomes may be eaten raw when young, or steamed and eaten as a vegetable. Young shoots are cooked as a vegetable, pickled or eaten raw. Chinese keys are used as a carminative, stomachic, expectorant, analgesic, and to treat dandruff and sore throats.

Lens culinaris. Leguminosae

LENTIL

Also known as Split Pea, Masur. Annual herb grown for edible, flattened seeds. Tender. Value: rich in protein, fibre, iron, carbohydrate, zinc and B vitamins.

Presumed to be native to south-west Europe and temperate Asia, lentils are one of the oldest cultivated crop plants. Carbonised seeds found in Neolithic villages in the Middle East have been dated at 7–6000 BC, and it is believed that they were domesticated long before that. By 2200 BC plants appeared in Egyptian tombs; they are referred to in the Bible as the 'mess of pottage' for which Esau traded his birthright (Genesis 25: 30, 34). The English 'lens', describing the glass in optical instruments, comes from their Latin name – its cross-section resembles a lentil seed. Christian Lent has the same origin, as it was traditionally eaten during the fast.

Grown throughout the world, lentils have become naturalised in drier areas of the tropics. Because of their relatively high drought tolerance, they are suitable for semi-arid regions. The quick-maturing plant is rarely more than 45cm tall and has branched stems, forming a small bush.
The white to rose and violet flowers lead to two to three seeded pods.

varieties

Lentils have been selected over many centuries for their size and colour. Today many different races and cultivars exist. Two main races predominate: the larger, round-seeded types usually grown in Europe and North America, and the smaller, flatter-seeded types common in the East.

The choice is yours

Frequently encountered is the split red or Persian lentil, which is extremely tender and quick to cook.

Among round-seeded types grown in Europe, the **'Lentille du Puy'**, a tiny green form, is the tastiest and tenderest. Similar but coarser is the **'Lentille Blonde'** or **'Yellow Lentil'** commonly grown in northern France, while **'German or Brown Lentils'** are coarser still and need lengthy cooking to make them tender. Of varieties grown in North America, **'O'Odham'** has flat grey-brown to tan-coloured seeds and **'Tarahumara Pinks'** from Mexico has mottled seeds and thrives well in semi-arid conditions.

cultivation

Propagation

Prepare the seedbed by removing any debris from the soil surface and raking to a moderate texture. Sow in spring, when the soil is warm, in drills 2.5cm deep, thinning seedlings to 20–30cm apart with 45cm between rows. They can also be broadcast and then thinned after germination to 20–30cm apart.

Growing

Lentils are not frost-hardy but flourish in a range of climatic conditions. They prefer a warm, sunny, sheltered position on light, free-draining, moisture-retentive soil. Sandy soils with added well-rotted organic matter are ideal, though equally good crops are grown on silty soil.

Keep crops weed-free and irrigate if necessary during periods of prolonged drought. Excessive watering can lead to over-production of leaves and poor cropping.

Lentils can be grown as a 'novelty' crop in cool temperate climates, but yields are not high enough to make it worthwhile on a large scale.

Maintenance

Spring Prepare the seedbed and sow when soil is warm.
Summer Keep crops weed-free and irrigate if needed.
Autumn Harvest before the seeds split.
Winter Store seeds or pods in a cool dry place, for use as required.

culinary

As their protein content is about 25 per cent, lentils are an important meat substitute. They also have the lowest fat content of any protein-rich food.

Soak lentils overnight, drain and replace the water. Boil rapidly for 10 minutes, then simmer for 25 minutes until tender. Use in soups, thick broth or grind into flour.

Lentils are commonly used for 'dhal'. Seeds are moistened with water and oil and dried before milling 2–3 times, each time separating the 'chaff' from the meal.

Puy Lentils with Roasted Red Peppers and Goat's Cheese
Serves 4

3 red peppers
175g Puy lentils
1 red onion
1 carrot
Sprig each of fresh parsley, marjoram and thyme
2 tablespoons sun-dried tomatoes, chopped
100g crumbled goat's cheese
1 tablespoon freshly chopped herbs

For the Dressing:
4 tablespoons extra virgin olive oil
1 ½ tablespoons lemon juice
Salt and freshly ground black pepper

Roast the peppers on a baking tray in a preheated oven, 250°C/475°F/gas mark 9, for 30 minutes. Put them into a polythene bag in the fridge. When cooled, deseed and peel the peppers, then cut into strips. Set aside.

Wash the Puy lentils thoroughly and cook them, covered in 350ml water, with the whole onion, carrot and herbs. These lentils cook faster than other types, in about 15 minutes. Drain. Roughly chop the onion and add back into the lentils. Discard the carrot and herbs. While still warm, add the tomatoes and goat's cheese and stir gently. Season.

Make the dressing and toss the lentil mixture in it. Arrange the lentils with the slices of red pepper on individual plates.

Sprinkle with the herbs and serve.

Protected Cropping
In cool temperate climates, sow seeds in spring in trays, pots or modules of seed compost. Keep compost moist and pot on when seedlings are large enough to handle. Harden off before planting outdoors when the soil is warm and workable and there is no danger of frosts. Protect plants under cloches until they are properly established. Seeds can be sprouted.

Harvesting and Storing
Lentils take about 90 days to reach maturity. Harvest as foliage begins to yellow, before the pods split and the seeds are shed. Lift the whole plant and lay on trays or mats in the sunshine to air-dry or put them in an airy shed. When the pods dry and split, remove the seed and store in a cool, dry place.

Pests and Diseases
Lentils suffer from few pests and diseases. Leaf rust can occur; burn plants after harvest and use treated seed.

Companion Planting
Lentils can be grown as a 'green manure'.

container growing

Grow in containers of soil-based compost with moderate fertiliser levels. Water well and keep the plants weed-free. To avoid the need for transplanting, seeds can be sown in a container which can then be moved outdoors into a suitable sunny spot after germination.

other uses

Dried leaves and stems are used as forage crops. Lentils have also been used as a source of commercial starch in the textile and printing industries, the by-products being used as cattle feed. Plants are used fresh, or dried as hay and fodder.

warning

Never eat lentils raw.

A protein-packed field

Lycopersicon esculentum. Solanaceae

TOMATO

Also known as Love Apple. Short-lived perennial grown as annual for fleshy, succulent berry. Half hardy. Value: rich in beta carotene and vitamin C; some vitamin B.

The wild species is believed to have originated in the Andean regions of north and central South America, spreading to Central and North America along with maize during human migrations over 2,000 years ago. The fruits had been cultivated in Mexico for centuries when European explorers found them growing under local names including *tomati, tomatl, tumatle* and *tomatas*. When they were first brought to Europe around 1523, tomatoes were considered to be poisonous, due to their strong odour and bright white, red and yellow berries, and were grown only as ornamentals. Dodoens in his *Historie of Plants* of 1578 records, 'This is a strange plant and not found in this country, except in the gardens of some herborists… and is dangerous to be used.'

In Europe, it was first used for food in Italy. Like many vegetable introductions from the New World, it was considered to be an aphrodisiac. The Italian name *pommi dei mori* was corrupted during translation to the French *pomme d'amour* or 'apple of love', as it was thought to excite the passions. Not all believed it to have this effect. Estienne and Liébault wrote that tomatoes were boiled or fried, but gave rise to wind, choler and 'infinite obstructions' – hardly an inducement to romance!

The use of tomatoes by North American settlers was not recorded until after Independence, but they were regularly used as food by Italian immigrants to New England and French settlers living in New Orleans, who were making ketchup by 1779. Thomas Jefferson was certainly growing them in his garden in 1781 and they were introduced to Philadelphia eight years later.

TOMATO

varieties

Most greenhouse varieties are 'indeterminate', with a main stem that can become several metres or yards long – these are usually grown as 'cordons'. Most of those grown outdoors are bush types which do not need supporting and can be grown under crop covers or cloches. Low-yielding, dwarf varieties are good for pots or window boxes.

As you will see from any seed catalogue, there are many hundreds of tomato varieties. Those for cultivation under cover in a cold greenhouse and outdoor varieties are generally interchangeable. They range in size from the large ribbed, 'beefsteak' types to small 'cherry', 'pear-shaped' and 'currant' tomatoes in a fantastic range of colours including mottled, dark skinned, pink and yellow.

'Ailsa Craig' grows well indoors/outdoors. It is a reliable, tasty, heavy-cropping variety. **'Alicante'**, another indoor/outdoor variety, crops heavily and matures early. It produces smooth, tasty fruit and does well in growbags. The large, black-tinted fruit of **'Black Krim'** have a delicious, smoky flavour. It sets well in heat but is also prone to cracking. However, don't let that deter you. **'Brandywine'** is a delicious old variety and the fruit can become quite sizeable. The skin is rosy pink or tinged slightly purplish-red. It performs best in cooler climates. **'Dombito'** is an excellent variety that produces large, round, sweet 'beefsteak' fruits; perfect for salads or slicing. **'Gardener's Delight'** is extremely popular, producing long trusses of delicious, sweet, 'cherry' tomatoes over a long period. It grows indoors but has better flavour outdoors, is ideal for containers and generally trouble-free. **'Ildi'** is very prolific indeed, producing masses of tiny fruits. **'Jaune Flammée'** is deep orange and shaped like an apricot; prolific, tasty and good for drying. **'Marmande'** is a delicious outdoor variety with deep, red ribbed fruits. It crops early and is resistant to fusarium and verticillium wilt. **'Oaxacan Pink'** has small, flattened, pink fruits. **'Orange Banana'** has orange fruit up to 10cm long, which are excellent for drying. **'Red Alert'** is an early bush variety for the greenhouse or outdoors with small, sweet, oval fruits. **'Roma VF'** is an outdoor bush 'plum' tomato, ideal for paste, ketchup, bottling, soups or juice. It crops heavily and sometimes needs supporting, and has high resistance to fusarium and verticillium wilt. **'San Marzano'**, another Italian tomato, also crops heavily and is good for soups, sauces or garnishing salads. **'Shirley'** is grown commercially and has good-quality, tender, tasty, disease-resistant fruit. Withstanding lower temperatures than most types, it does not suffer from 'greenback', is highly resistant to tobacco mosaic virus, leaf mould and fusarium and is therefore an ideal tomato for organic growers. **'Siberia Tomato'** is a bush variety that crops extremely early, around seven weeks after transplanting and sets fruit at temperatures as low as 5°C. **'Stupice'** crops heavily and extremely early, producing fruit with an excellent balance between sweetness and tartness; delicious! **'Sungold'** is one of the sweetest tomatoes ever produced.

Roma

'Yellow Pearshaped'

'Tasty Evergreen' has light yellow to green skin and the flesh remains green when ripe; succulent, tender and sweet. **'Tigerella'** produces tasty, small orange-red fruits with pale stripes and crops well over a long period. **'Tiny Tim'** is compact, bushy and ideal for pots, window boxes and hanging baskets. The fruits are cherry-sized and tasty. **'Tumbler'** was originally bred for hanging baskets. It has flexible, hanging stems, and the bright red fruits ripen quickly and are sweet to the taste. **'Wapsipinicon Peach'** has fuzzy skin like a peach and was a taste test winner in the USA; delicious and sweet – recommended. **'Yellow Pearshaped'** are well described by their name. The dense clusters of fruit are sweet-tasting and have few seeds and the plants are vigorous and high-yielding. **'Yellow Perfection'** crops early and is prolific, producing tasty, bright yellow fruits – an excellent tomato.

Modern greenhouse varieties such as **'Aromata'**, **'Moravi'** and **'Merlot'** are significantly disease-resistant compared with older sorts.

The 'Black Krim' has a fabulous flavour

There's plenty of light by the window

cultivation

Propagation

A minimum temperature of 16°C is needed for germination, but seedlings can tolerate lower night temperatures if those during the day are above this level.

In cool climates, for growing in heated glasshouses, sow from mid-winter. For growing outdoors or in an unheated greenhouse, sow seed indoors 2cm deep in trays of seed compost, 6–8 weeks before the last frost is due or sow 2–3 seeds in 7.5cm pots or modules, thinning to leave the strongest seedling.

Transplant tray- or module-grown seedlings into 7.5cm pots when 2–3 leaves have formed, keeping the plants in a light, well-ventilated position. Harden off carefully and plant out when there is no danger of frost and air temperatures are at least 7°C with soil temperatures at a minimum of 10°C.

Transplant, with the first true leaves just above the soil level, when the flowers on the first truss appear. Do not worry if your plants have become spindly; planting them deeply stimulates the formation of roots on the buried stems, making the plants more stable. Tomatoes can also be fluid-sown from mid-spring *in situ* or in a cold frame for transplanting after warming the soil. Germinate on moist kitchen towel at 21°C, sowing when the rootlets are a maximum of 5mm long (see advice on fluid sowing, page 646). Sow outdoors in drills and thin to leave the strongest seedling or in 'stations' at their final spacing. Cover the drills with compost and protect them with cloches until the first flowers appear.

When sowing in cold frames, sow seed in rows 12.5–15cm apart, thinning to 10–12.5cm apart in the rows. Transplant carefully when the plants are 15–20cm tall.

Alternatively, buy plants grown individually in pots rather than packed in boxes or trays, using a reliable supplier. Plant 'cordon' types 38–45cm apart, or in double rows with 90cm between each pair of rows. Bush types should be planted 45–60cm apart and dwarf cultivars 25–30cm apart, depending on the variety.

Closer spacing produces earlier crops; wider spacing generally produces slightly higher yields.

Growing

Outdoor tomatoes need a warm, sheltered position, ideally against a sunny wall, in a moisture-retentive, well-drained soil. Add well-rotted organic matter where necessary and lime acid soils to create a pH of 5.5–7. If tomatoes are grown in very rich soil or are fed with too much nitrogen they produce excessive leaf growth at the expense of flowers and fruit. Fruit will not set at night temperatures below 12°C and day temperatures above 33°C. Night temperatures around 25°C may well cause blossom to drop. Tomatoes should be fed with a liquid general fertiliser until established, then with a high-potash fertiliser to encourage flowering and fruiting. Use tepid water to avoid shocking plants.

Keep crops weed-free by hoeing and hand weeding, taking care not to damage the stems, or mulch with a layer of organic matter.

Keep them constantly moist but not waterlogged; erratic watering causes the fruits to split and also encourages blossom end rot, particularly when plants are grown in containers or growbags. Use tepid water. In dry conditions or when the first flowers appear they need about 11 litres water each week. Careful watering and feeding is essential (be restrained!), particularly near harvest time to ensure that the fruits are not excessively

Be prepared to recycle and improvise

watery and have a good flavour. Feed with tomato fertiliser according to the manufacturer's instructions.

Cordons need supporting with canes, strings or a frame. Using a sharp knife or by pinching between the finger and thumb, remove sideshoots when they appear and 'stop' plants by removing the growing tip 2–3 leaves above the top truss when 3–5 trusses have been formed or when the plant has reached the top of the support. The number of trusses on each plant depends on the growing season: in shorter growing seasons, leave fewer trusses. Remove leaves below the lowest truss to encourage air circulation, but not too many.

In seasons when ripening is slow, remove some of the leaves near to the trusses with a sharp knife, exposing fruits to the sunshine.

To keep the fruit clean and stop fruit from rotting, grow bush tomatoes on a mulch of straw, felt or crop covers laid over the soil. Black polythene or even a split bin liner (with the edges anchored by burying them in the soil) absorbs heat, warms the soil, conserves moisture and helps fruit to ripen. Rotate crops annually.

Maintenance

Spring Sow crops under cover and, later, outdoors.
Summer Keep crops watered, fed and weed-free. Shade and ventilate as necessary. Remove any sideshoots. Harvest.
Autumn Ripen later outdoor crops under cloches or indoors.
Winter Harvest crops in midwinter from cuttings taken in late summer.

Protected Cropping

In cooler climates, more reliable harvests can be achieved by growing tomatoes in an unheated greenhouse or polythene tunnels. Rotate crops to avoid the build-up of pests and diseases and sterilise or replace the soil every 2–3 years, or use containers. To assist pollination and fruit set, particularly of plants grown indoors, mist occasionally and tap the trusses once the flowers have formed; do this around midday if possible. (It is not so important for outdoor crops, as wind movement assists pollination.) Shade glasshouses before the heat of summer and ventilate well in warm conditions.

A fascinating innovation is the 'Wall-o-Water', a series of connected plastic tubes filled with water which absorbs heat during the day and warms by radiated heat at night. This is reusable and offers protection down to –8°C, extending the growing season and warming the soil. Adding roughly 1 part bleach to 500 parts water will prevent the formation of algae on the inside of the tubes.

A tomato plant supported by a trellis

For tomatoes at Christmas, take cuttings in mid-summer from sideshoots 13cm long and root them in a container of sharp sand or perlite and water. When roots appear they can be potted into 10cm pots and grown on under cover.

Protect newly planted outdoor tomatoes with cloches or floating crop covers until they become established. Once the first flowers are pushing against the cover, make slits along the centre; about 7–10 days later slit the remainder of the cover and leave as a shelter alongside the plants. 'Indeterminate' varieties can then be tied to canes.

Harvesting and Storing

Harvest fruits as they ripen, about 7–8 weeks after planting for bush types and 10–12 weeks for 'cordon' varieties. Lift and break the stem at the 'joint' just above the fruit. Outdoor crops should be harvested before the first frosts. Towards the end of the season, 'cordon' varieties with unripe fruit can be lifted by the roots and hung upside down in a frost-free shed to ripen.

Alternatively, they can be detached from their support, laid on straw and covered with cloches or put in a drawer or a paper bag with a ripe banana or apple (the ethylene these produce ripens the fruit). Bush and dwarf types can be ripened under cloches.

Tomatoes can be cooked and bottled in airtight jars. To freeze, skin and core when ripe, simmer for 5 minutes, then sift, cool and pack in a rigid container.

Tomatoes stay fresh for about a week in a polythene bag or the salad drawer of a refrigerator. To retain flavour they are better stored at 15°C.

Dry large 'meaty' tomatoes in the sun or the oven. Cut into halves or thirds and put skin side down on a tray and cover with a gauze frame as a protection from insects. Ideal drying conditions are in warm, dry windy weather, but they should be brought indoors at night if dew is likely to form. Humid conditions are not suitable for outdoor drying. Oven-dry just below 65°C until tomatoes are dried but flexible. Stored in airtight containers in a cool place, they can last for up to 9 months. Before use, put the tomatoes in boiling water or a 50:50 mix of boiling water and vinegar and allow them to stand until soft. Drain and marinate for

Growing bags can be productive

several hours in olive oil with added garlic to suit your taste. They last in the marinade for about a month and are excellent with pasta and tomato sauce.

Pests and Diseases

Blossom end rot appears as a hard, dark, flattened patch at the end of the fruit away from the stalk. This indicates a deficiency in calcium, usually caused by erratic watering. It is often a problem with plants grown in growbags. Water and feed regularly, particularly during hot weather. Erratic watering causes fruit to split, which may also happen with sudden growth after overcast weather. Pick and use split fruits immediately.

Greenback – hard, green patches appearing near the stalk, caused by sun scorch and overheating – is more of a problem with plants growing under glass. Sometimes this becomes an internal condition known as whitewall. Shade and ventilate well, water regularly and feed with a high-potash liquid fertiliser.

Curled leaves are caused by extreme temperature fluctuations between day and night. It is often a problem in greenhouses: shade, ventilate and damp down. Close ventilation before temperatures drop. Whitefly, aphids and red spider mite can be a problem, as can potato blight in wet summers: dark blotches with lighter margins appear on the leaves. Spray with a copper fungicide or grow resistant varieties.

'Alicante'

Plants with verticillium wilt droop during the day and usually recover overnight. Lower leaves turn yellow and cut stems have brown markings on the inside. Do not plant when the soil is cold, rotate crops regularly, replace the soil, mist regularly and shade. Mounding moist compost round the stem encourages the formation of a secondary root system. The symptoms and control of fusarium wilt are similar.

Tobacco mosaic and other viruses show as mottling on the leaves, some of which may be misshapen; inside the fruit is browned and pitted and growth is stunted. Dispose of affected plants, wash your hands and sterilise tools by passing them through a flame. Grow the following year's plants in sterilised compost or in growbags.

Tomato moth caterpillars eat the fruit. They are about 4cm long and green or brown with a pale lemon line along the body. They appear from late spring to early summer. Check the underside of leaves and squash the eggs. Check over fruits thoroughly and remove by hand.

The symptoms of magnesium deficiency are yellowing between the leaf veins, older leaves being affected first. Spray, drench or scatter magnesium sulphate around the base. Wherever possible, grow disease-resistant varieties.

container growing

Tomatoes are excellent for containers, pots or growbags, indoors or outside where they can be included in 'edible' displays in window boxes, pots and hanging baskets.

Grow in 23cm pots of loam-based compost with high fertiliser levels. For requirements see 'Growing' and 'Protected Cropping'. Bushy varieties are ideal. Plants in containers dry out rapidly, so careful feeding and watering is essential. Water regularly and remember to label your plants.

Marigolds keep whitefly away

companion planting

Grow with French marigolds to deter whitefly. Tomatoes grow well with basil, parsley, alliums, nasturtiums and asparagus.

medicinal

Tomatoes contain lycopene and are believed to reduce the risk of cancer and appendicitis.

In American herbal medicine, tomatoes have been used to treat dyspepsia, liver and kidney complaints and are also said to cure constipation.

warning

Some doctors believe that tomatoes aggravate arthritis and may be responsible for food allergies. The leaves and stems are poisonous.

TOMATO

culinary

In salads, tomatoes are particularly good with mozzarella, basil and olive oil. They are also good with chives. Some people prefer to peel them first, by immersing in boiled water for 1 minute to loosen the skins. Use in soups, stews, sauces and omelettes.

Tomatoes are wonderful grilled: cut them in half and cover the cut surface with olive oil, pepper and sugar. Grill for 5 minutes. Alternatively, glaze them with wine and brown sugar and grill.

Try a BLT – a cold bacon, lettuce and tomato sandwich. Eat 'currant' and 'cherry' varieties as a snack. Use larger 'beefsteak' varieties as a garnish with steak or hollow out and stuff with shrimps, potato salad, cold salmon, cottage cheese or mashed curried egg. Serve hot or cold.

Green tomatoes can be sliced, dipped in batter and then breadcrumbs, and fried in hot oil, made into chutney or jam or be added to orange marmalade.

Jean-Christophe Novelli's Gazpacho Soup
Serves 6

500g very ripe cherry tomatoes
2 large over-ripe tomatoes
2 tablespoons truffle oil
100ml olive oil
4 teaspoons white wine vinegar
Unrefined caster sugar
Salt and freshly ground black pepper
500g bright red peppers, chopped
6 sprigs fresh coriander

Place the tomatoes, truffle oil and olive oil in a blender and add half the vinegar, 2 teaspoons sugar and a little seasoning. Blitz together, thin with water if necessary, and season to taste. Put through a fine sieve and chill for at least 2 hours.

Rinse out the blender, then whizz the peppers with 50ml water to a purée. Pass through a fine sieve and stir in 20g sugar and the remaining vinegar. Put the sifted purée in a pan and simmer until reduced to a syrup. Allow to cool to room temperature.

Stuffed Tomatoes

To finish, carefully pour the chilled gazpacho soup into a bowl, drizzle some red pepper syrup over the surface and serve garnished with coriander.

Stuffed Tomatoes
Serves 4

In times gone by Catholics all over Europe ate these on fast days, when meat was not allowed.

4 large ripe tomatoes
350g white breadcrumbs (day-old bread)
150ml milk
2 eggs, lightly beaten
2 cloves garlic, finely crushed
2 tablespoons chopped fresh basil
2 tablespoons finely chopped fresh parsley
1 onion, finely chopped
2 tablespoons toasted breadcrumbs
5–6 tablespoons grated Gruyère cheese
Olive oil
Salt and freshly ground black pepper

Remove a slice from the top of each tomato and scoop out the pulp. Season the insides of the tomatoes with salt and pepper, and arrange in a greased baking dish.

Preheat the oven to 180°C/350°F/gas mark 4. To make the stuffing, combine the white breadcrumbs with the milk and eggs in a bowl and add the garlic, herbs and onion. Season with salt and pepper and fill the tomatoes. Sprinkle over the toasted breadcrumbs and Gruyère and drizzle over a little olive oil to prevent burning. Bake until the tomatoes are tender, about 30 minutes.

Manihot esculenta (syn. M. utilissima). Euphorbiaceae

CASSAVA

Also known as Tapioca, Yucca, Manioc. Tall herbaceous perennial grown for edible tubers and leaves. Tender. Value: mainly starch; small amounts of vitamin B, C and protein.

One of the most important food crops in the humid tropics, cassava is believed to have been cultivated since at least 2500 BC. Unknown as a wild plant, it may have originated in equatorial South America in the Andean foothills, the Amazon basin or regions of savannah vegetation. The earliest archaeological records, from coastal Peru, date from 1000 BC. Tubers contain highly toxic cyanide, which is removed by cooking; accounts tell of starving European explorers eating raw manioc and dying at the moment they thought sustenance had been found. The indigenous Indians tipped their arrows and blowpipe darts with its toxic sap; Arawak Indians committed suicide by biting into uncooked tubers rather than be tortured by the Conquistadors.

The Portuguese brought the crop to West Africa, whence it quickly spread, reaching Sri Lanka in 1786, India in 1794 and Java by 1835. An estimated 62 million tonnes of cassava is produced annually, much of it in West Africa, where it is eaten as 'fufu'.

varieties

Manihot esculenta is tall and branched and its stems become woody with age. The leaves are long-stalked, with 5–9 lobes; toxic latex is present in all parts of the plant. The swollen tubers are cylindrical or tapering, forming a cluster just below the soil surface, and weigh 5–10kg.

There are two types. **'White Cassava'** is sweet, soft and used as a source of starch; **'yellow'** varieties are bitter and usually grown as a vegetable. The more primitive bitter varieties contain larger quantities of cyanide, which is washed out by boiling in several changes of water before cooking. In recent selections of 'sweet' varieties, most of the toxin is in the skin, and tubers are edible after simple cooking.

There are well over 100 different forms with local names. **'Nandeeba'**, quick to mature, and **'Macapera'**, used for boiling, are both from Brazil.

cultivation

Propagation
Take cuttings from mature stems 15–30cm long; plant 1.2m apart in rows 1m apart or 'pits' 90–105cm square in a grid pattern. Planting cuttings upright, leaving the top 5cm exposed, gives best results. Plant cuttings at the start of the rainy season.

Growing
Cassava flourishes where the warm rainy season is followed by a dry period. It has good resistance to drought; in a constantly wet climate there is excessive stem growth and tuber formation is poor. Soils should be deep, rich and free-draining. It is often grown on ridges or mounds as it dislikes waterlogging. Dig deeply before planting, adding well-rotted organic matter. Earth up as necessary. Cassava is a heavy feeder and cannot usually be grown for more than 3 years on the same ground.

Maintenance
Spring Dig in organic matter before planting.
Summer Plant cuttings at the start of the rainy season.
Autumn Keep weed-free.
Winter Harvest at maturity.

Typically tropical

CASSAVA

Protected Cropping
In cool temperate zones, plants can be grown as a 'novelty crop' in a hothouse. Prepare the borders and cultivate as for 'Growing'. Water well during the growing season and reduce watering during winter.

Harvesting and Storing
Varieties are harvested from 8 months to 2 years after planting, depending on the locality and variety. Harvest when plants have flowered and the leaves are yellow. Lift the whole plant and carefully remove the tubers. They store for up to 2 years in the ground, but should be used within 4–5 days of lifting.

Pests and Diseases
Whitefly and fungal diseases can be a problem. Bacterial diseases, scale and cassava mosaic are a real problem in Africa. Plants are highly resistant to locusts.

other uses

Cassava is also a source of starch for the manufacture of plywood, textiles, adhesives and paper.

warning

All parts of the plant contain toxic latex. Prepare tubers thoroughly before eating. Inhabitants of Guyana take chillies steeped in rum as an antidote to yucca poisoning, but do not rely on it!

culinary

The fresh root is equivalent in starch to 33 per cent of its weight in rice and 50 per cent in bread, but its nutritional value is unbalanced and high consumption often leads to protein deficiency.

Wash thoroughly and remove the skin and rind with a sharp knife or potato peeler. Boil in several changes of water and allow to dry before cooking.

It can be eaten mashed or boiled as a vegetable or made into dumplings and cakes. Mix with coconut and sugar to make biscuits. The juice from grated cassava is boiled down and flavoured with cinnamon, cloves and brown sugar to make 'cassareep', a powerful antiseptic and essential to the West Indian dish, Pepperpot. Tapioca flour is made from ground chips.

In Africa, the fresh root is washed, peeled, boiled and pounded with a wooden pestle to make 'fufu'. Cook cassava chunks in boiling water for 45–50 minutes, drain, cool, pound into dough and shape into egg-sized balls to add to soups and stews as a traditional African accompaniment.

Roots of 'sweet' forms can be roasted like sweet potato, baked or fried in slices.

Young leaves can be boiled or steamed and eaten with a knob of butter.

Cassava Chips (Singkong)

These thin, crispy wafers are perfect served as a snack or as a garnish. Allow 300g per person to accompany a plain meat course.

Slice the cassava very thinly and leave to dry. Heat some peanut or vegetable oil (do not allow it to smoke) in a wok or a deep pan. Deep-fry 1–3 slices at a time by immediately submerging each below the surface of the oil. Remove quickly from the heat and drain well on layers of paper towels.

Cooked Singkong can be stored in an airtight container for several weeks, but is best served at once.

Medicago sativa. Papilionacae

ALFALFA

Also known as Lucerne, Purple Medick. Grown as annual or short-lived perennial for seed sprouts and young leaf shoots. Hardy. Value: good source of iron and protein.

'Medick' comes from the Latin *Herba medica*, the Median or Persian herb, imported to Greece after Darius found it in the kingdom of the Medes. It was a vital fodder crop of ancient civilizations in the Near East and Mediterranean, and known in Britain by 1757. Today, alfalfa is valued by gardeners as a green manure as well as a nutritious vegetable. Its blooms in the wildflower meadow are rich in nectar, while the leaves are a commercial source of chlorophyll.

varieties

The species *Medicago sativa* is a fast-growing evergreen legume with clover-like leaves; growing ultimately to 1m, it has spikes of violet and blue flowers. It produces quality crops on poor soils, as it is highly effective at fixing nitrogen in the root nodules.

Penetrating up to 6m into the ground, the roots draw up nutrients and aerate the soil. Agricultural varieties are available. If you want to sprout alfalfa seeds, buy untreated seed; that sold for sowing as a crop is usually chemically treated.

cultivation

Alfalfa can be grown as a short-lived perennial, a 'cut and come again' crop or as seedsprouts. It tolerates low rainfall and can be grown in any soil, in temperate to subtropical conditions or at altitude in the tropics.

Propagation

Sow in spring or from late summer to autumn. Thin those grown as perennials to 25cm apart when large

Alfalfa flowering in California

ALFALFA

culinary

Young shoot tips and sprouted seeds can be used raw in salads or cooked lightly.

For a tasty salad, try it with hard-boiled eggs (1 per person), anchovies and capers served on a bed of endive and radicchio leaves. To stir-fry, pour a little oil in a pan, add the alfalfa, stir briskly for 2 minutes; serve immediately.

enough to handle. Sow 'cut and come again' crops, spaced evenly, by broadcasting or in shallow drills 10–12.5cm wide.

Growing
Prepare seedbeds in winter, fork the area, remove debris and stones, rake level. Keep weed-free until established. Cut perennials to within a few inches of the base after flowering and renew every 3–4 years, as old plants become straggly.

Maintenance
Spring Sow seed outdoors.
Summer Harvest young shoots from perennials.
Autumn Sow protected crops.
Winter Prepare seedbed.

Protected Cropping
Sow alfalfa seeds under cloches, horticultural fleece or glass in late summer to autumn to harvest a winter crop.

Harvesting
Harvest 'cut and come again' crops when about 5cm long a few weeks after sowing. Cut back plants regularly to encourage new growth; they provide young growths for many harvests.

Sprouting Seeds
To sprout alfalfa, soak seeds overnight or for several hours, tip seeds into a sieve and rinse. Put several layers of moist paper towel or blotting paper in the base of a jar and cover with a 5mm layer of seed. Cut a square from a pair of tights or piece of muslin to cover the top, securing with a band. Place in a bright position, away from direct sun, maintaining constant temperatures around 20°C. Rinse seed daily by filling the jar with water and pouring off again. Harvest shoots after 3–7 days, when they have nearly filled the jar. Wash and dry sprouts and use as required; do not store more than 2 days.

Pests and Diseases
Rabbits can be a problem.

companion planting

Alfalfa accumulates phosphorus, potassium, iron and magnesium; it keeps grass green longer in drought.

container growing

Sow 'cut and come again' crops in a bright open position in a loam-based compost with low fertiliser levels; water pots regularly.

medicinal

An infusion of young leaves in water is used to increase vitality, appetite and weight. The young shoots, rich in minerals and vitamin B, are highly nutritious and the seeds appear to reduce cholesterol levels.

Healthy eating for all

Momordica charantia. Cucurbitaceae

KARELA

Also known as Bitter Gourd, Balsam Pear, Fu Gua, Momordica, Peria. Annual climber grown for edible fruits and leaves. Tender. Value: fruit a good source of iron, ascorbic acid and vitamin C; leaves and young shoots contain traces of minerals.

This strange-looking fruit with skin the texture of a crocodile has been grown throughout the humid tropics for centuries. Rudyard Kipling's description in Mowgli's 'Song Against People' conveys the plant's vigour, as it climbs to 4m with the aid of tendrils:

I will let loose against you the fleet-footed vines,
I will call in the Jungle to stamp out your lines.
The roofs shall fade before it, the house-beams shall fall;
And the Karela shall cover it all.

The strongly vanilla-scented flowers are followed by the elongated fruit. When ripe, it splits at the tip into three sections, exposing brown or white seeds surrounded by blood-red pulp. Fruits are best eaten young.

Go on, try it!

varieties

Male and female flowers are borne on the same plant. Male flowers are 5–10cm long; females are similar, but with a slender basal bract. The fruit can grow up to 25cm – which ripens to become orange-yellow, though a white variety is grown in India and eastern Asia.

'Baby Doll' has top-quality, small fruit and is easy to grow. **'Bankok Large'** produces dark green fruit of excellent quality and **'Japan Green Spindle'** produces spindle-shaped fruit. **'Taiwan Large'** is high-yielding. **'Taiwan White'** is highly ornamental with white fruit and **'Winter Beauty'** is creamy white in colour.

cultivation

Propagation

In humid tropical climates, sow at the start of the rainy season, placing 2–3 seeds outdoors in 'stations' 90cm apart, in and between the rows. Thin after germination, leaving the strongest seedling. Water plants as needed.

The first flowers appear 30–35 days after sowing. In temperate zones, sow seed in early spring under glass at 20°C in peat-substitute compost. Keep the compost moist with tepid water and repot as necessary when roots become visible through the drainage holes.

Growing

Karela flourish in moderate to high temperatures with sunshine and high humidity. Plant in beds or mounds of rich, moisture-retentive, free-draining soil. Dig in plenty of well-rotted manure or similar organic matter before sowing.

Pinch out the terminal shoots when they are 3–4cm long to encourage branching, then train the stems into trees or over fences, trellis or frames covered with 15cm mesh netting. Keep plants moist and weed-free.

Maintenance

Spring Sow seed under glass in peat-substitute compost.
Summer Keep plants well fed and watered.
Autumn When cropping finishes, add leaves and stems to your compost heap.
Winter Prepare the greenhouse border for the following year's crop.

Protected Cropping

Grow under cover in temperate zones. Plants need hot, humid conditions in bright light. When growing plants in the greenhouse border, prepare the soil as for 'Growing'.

Growth is better restricted in borders and containers. Stop the shoot tips when the main stems are 1.5m long, training the lateral stems along wires or trellis.

Appearances are deceiving

Once a flower has formed, allow 2 more leaves to appear, then pinch out the growing tip. If it is a male flower or a flower does not form, cut the stem back to the first leaf to allow a replacement to form. Flowers should be hand-pollinated.

You should damp down the greenhouse floor regularly during hot weather.

Harvesting and Storing

The first fruits appear about 2 months after sowing, are yellow-green in colour and should be harvested when they are about 2cm long. They can be eaten when longer.

Fruits can be kept in a cool dark place for several days or stored in the salad drawer of a refrigerator for 4 weeks. Karela can also be sliced and dried for use out of season.

Pests and Diseases

Fruit fly is common; spray with contact insecticide or protect the fruits with a piece of paper wrapped round the fruit and tied with string round the stalk. Red spider mite is a common pest under cover. Leaves become mottled and bronzed. Check plants regularly; small infestations are easily controlled – isolate young plants. The mite prefer hot, dry conditions, so keep humidity high or use biological control.

container growing

Plants can be grown indoors in containers containing a rich, well-drained potting mixture of equal parts loam-based compost and well-rotted organic matter with added peat and grit.

Keep the compost moist with tepid water and feed every 2 weeks with a liquid general fertiliser.

medicinal

The fruits are said to be tonic, stomachic and carminative and are a herbal remedy for rheumatism, gout, and diseases of the liver and spleen. In Brazil, the seeds are used as an anthelmintic. Its fruits, leaves and roots are used in India and Puerto Rico for diabetes. In India, leaves are applied to burns and used as a poultice for headaches and the roots used to treat haemorrhoids. In Malaya they are used as a poultice for elephants with sore eyes.

culinary

Remove the seeds from mature fruit, and remove any bitterness by salting. Young fruits do not need to be salted.

Karela are ideal diced in curries, chop suey or pickles, stuffed with meat, shrimps, spices and onions and fried or added to meat and fish dishes. Mature fruits can be parboiled before being added to a dish or cooked like courgettes and eaten as a vegetable.

Young shoots and leaves are cooked like spinach.

Pelecing Peria

Serves 4

Sri Owen gives this recipe in her marvellous book, *Indonesian Food and Cookery*. Some of the ingredients need determination to track down, but it is worth the trouble.

3–4 karela (peria)
Salt
6 cabé rawit (or hot red chillies)
3 candlenuts
2 cloves garlic
1 piece terasi (shrimp paste, available at Thai shops)
1 tablespoon vegetable oil
Juice of 1 lime

Cut the karela lengthwise in half, take out the seeds, then slice like cucumbers. Put the slices into a colander, sprinkle liberally with salt and leave for at least 30 minutes. Wash under cold running water before boiling for 3 minutes with a little salt.

Pound the cabé rawit, candlenuts, garlic and terasi in a mortar until smooth. Heat the oil in a wok or frying pan and fry for about 1 minute. Add the karela and stir-fry for 2 minutes; season with salt and lime juice. Serve hot or cold.

Oxalis tuberosa. Oxalidaceae

OCA

Also known as Iribia, Cuiba, New Zealand Yam. Perennial grown for tubers. Half hardy. Value: about 85% water; some carbohydrates; small amounts of protein.

This is common in the high-altitude Andes from Venezuela to northern Argentina, where it is second only to the potato in popularity. At the northern end of Lake Titicaca, more than 150 steep terraces dating from the Incas are still cultivated. Oca is grown in New Zealand, where it was introduced from Chile in 1869. Today it is rarely found in European or American gardens, though it was once grown as a potato substitute. Tubers form in autumn when day lengths are less than 9 hours.

varieties

Oxalis tuberosa is bushy to 25cm tall, with tri-lobed leaves and orange-yellow flowers. It produces small tubers 5–10cm long which are yellow, white, pink, black or piebald.

cultivation

Propagation
Plant single tubers or slice into several sections, each with an 'eye' or dormant bud, and dust the cut surfaces with fungicide. Plant 12.5cm deep and 30cm apart with 30cm between rows.

In frost-free climates plant in mid-spring; in cooler areas propagate under cover in 12.5–20cm pots of compost before planting out once frosts have passed.

Growing
Oca flourish in deep fertile soils, so incorporate organic matter before planting. Earthing up before planting increases the yield.

Maintenance
Spring Plant tubers.
Summer Water as necessary.
Autumn Tubers can be lifted when needed.
Winter Store tubers in sand.

Protected Cropping
Where early or late frosts are likely, grow under cover to extend the harvest season. Plant in mid-spring in greenhouse borders. Harvest from mid-autumn to early winter.

Extend outdoor cropping by protecting with horticultural fleece or cloches.

Harvesting and Storing
Tubers are formed in autumn and can be stored in sand in a dry frost-free place. Lift just before the first frosts or protect with fleece and harvest later.

Pests and Diseases
Slugs are often a problem.

culinary

Tubers have been selected over the centuries for flavour and reduced levels of calcium oxalate crystals, which otherwise render them inedible. Leave them for a few days to become soft before eating. In South America they are dried in the sun until floury and less acid. If dried for several weeks, they become sweet, tasting similar to dried figs.

The acidity can be removed by boiling in several changes of water. The flavour of tubers even improves once frozen. Oca can be eaten raw, roasted, boiled, candied like sweet potato and added to soups and stews. Use leaves and young shoots in salads or cook them like sorrel.

Oca and Bacon
Clean 500g oca and cut into cubes. Boil in salted water until just tender; drain and combine with 250g smoked bacon which has been diced and fried. Coat with mayonnaise, sprinkle over fresh chives and season. Serve warm.

container growing

Oca grow well in containers, although yields are lower. Tubers should be planted in spring in 30cm pots in loam-based compost with moderate fertiliser levels, with added organic matter. Regular watering is vital, particularly as tubers begin to form. Allow the compost surface to dry out before rewatering. An occasional feed with liquid fertiliser helps to boost growth.

companion planting

Oca grow well with potatoes and can be grown under runner beans, maize or crops of a similar height.

warning

Prepare tubers correctly before eating to remove calcium oxalate crystals.

Pachyrrhizus erosus. Leguminosae

YAM BEAN

Also known as Jicama, Potato Bean, Mexican Water Chestnut, Dou Shou, Sha Kot. Tender, vining perennial. Value: low in starch and calories, high in fibre, a good source of vitamin C and potassium.

Originating in tropical America and naturalised in Florida, the yam bean is very popular in the local cuisine of southern China and the Philippines. In Mexico, the Jicama is one of four foods included in rituals for the 'Festival of the Dead' celebrated on 1 November, alongside sugar, tangerines and peanuts.

Grown as a novelty crop in cooler climates, its attractive flowers also give it considerable value as an ornamental plant.

varieties

Pachyrrhizus erosus is a scrambling perennial with attractive violet and purple flowers. Two major types are grown in Mexico, where it is a popular edible crop; **'Jicama de leche'** with dark skin and spindle-shaped roots, which is rather dry and **'Jicama de agua'** with light skin and a succulent, sweet, watery flavour to the tubers.

cultivation

Propagation

From mid-spring, sow two seeds in a 10cm pot filled with seed compost and place in a propagator or on a warm windowsill at 20°C. Thin after germination, leaving the strongest seedling and transplant regularly as the plants grow, supporting the stems with a stick. Plants grow rapidly: allow as much growing time as possible before they are planted outdoors (if that is what you intend to do); ideally the seeds should be sown around 6 to 8 weeks before the final frosts in spring. Although tubers can be stored at temperatures from 12.5–15°C for replanting, they are very sensitive to chilling. It is much easier to save or buy fresh seeds the following year.

Growing yam bean is day-length sensitive and the edible tubers are not formed until autumn when there are fewer than 9 hours of daylight; unfortunately, this often coincides with the first frosts. Vines are best grown indoors in temperate climates, so they have enough energy to produce sizable tubers at the end of the growing season, before light and cold restrict growth.

If growing plants outdoors, warm the soil with cloches before planting the large ones, covering them with cloches for 2 to 3 weeks until they become acclimatised. Grow them like runner beans, along rows of canes or up 'wigwams', in a sunny position in well-prepared, deep, light, rich, moisture-retentive soil, incorporating plenty of well-rotted organic matter, if necessary. Pinch out growth tips if plants become too large, and remove the flower buds, too; this has the added advantage of encouraging tuber growth.

Maintenance

Spring Sow seeds early in the season.
Summer Keep plants well watered.
Autumn Harvest after the first frosts.
Winter Order seed.

Protected Cropping

Grow in large tubs in the greenhouse or polythene tunnel border, then feed and water well during the growing season. A hot, humid environment encourages heavy cropping, so they are best grown under protection in temperate climates.

Harvesting and Storing

Cropping is most productive in temperate zones during hot summers. Harvest the tubers as late in the season as possible, once the top growth has been frosted. Harvest seed pods while immature.

Pests and Diseases

Bean seed weevil can be a problem; the seeds show evidence of tiny holes but usually germinate if sown. Alternatively, dry the seed, in silica gel, then freeze them for 48 hours.

culinary

Like water chestnut, yam bean is very refreshing and thirst quenching when peeled and eaten raw. Tubers can be fried, braised, boiled, eaten in casseroles, used as a bamboo shoot, water chestnut or mooli substitute, added to custards, puddings and fruit salads or eaten as a snack. In California it is commonly sliced and added to salads, in Mexico it is marinated with lime then served topped with chilli powder; sometimes sour orange is used as a lime substitute. Crunchiness is retained after cooking. The best-quality roots are about 2kg in weight; older tubers become 'woody' growing up to 2m long and weighing 20kg, at this stage they become fibrous and inedible. Only the young pods are edible, boil them thoroughly before eating and use as French beans.

container growing

Plants can be grown in raised beds or very large containers, if preferred.

warning

Mature seed pods contain rotenone: do not eat.

Pastinaca sativa. Apiaceae

PARSNIP

Biennial grown as annual for edible root. Hardy. Value: some carbohydrate, moderate vitamin E, smaller amounts of vitamins C and B.

This ancient vegetable is thought to have originated around the eastern Mediterranean. Exactly when it was introduced into cultivation is uncertain, as references to parsnips and carrots seem interchangeable in Greek and Roman literature: Pliny used the word *pastinaca* in the first century AD when referring to both. Tiberius Caesar was said to have imported parsnips from Germany, where they flourished along the Rhine – though it is possible that the Celts brought them back from their forays to the east long before that. In the Middle Ages, the roots were valued medicinally for treating problems as diverse as toothache, swollen testicles and stomach ache. In sixteenth-century Europe parsnips were used as animal fodder, and the country name of 'madneps' or 'madde neaps' reflects the fear that delirium and madness would be brought about by eating the roots.

Introduced to North America by early settlers, they were grown in Virginia by 1609 and were soon accepted by the American Indians, who readily took up parsnip growing. They were used as a sweetener until the development of sugar beet in the nineteenth century; the juices were evaporated and the brown residue used as honey. Parsnip wine was considered by some to be equal in quality to Malmsey and parsnip beer was often drunk in Ireland. In Italy pigs bred for the best-quality Parma ham are fed on parsnips.

Avonresister

varieties

Roots are 'bulbous' (stocky, with rounded shoulders), 'wedge' types (broad and long-rooted) or 'bayonet' (similar, but long and narrow in shape).

'Alba' has small, thin, wedge- and bayonet-shaped roots and good canker resistance; top quality. **'All American'** has wedge-shaped roots and is sweet-tasting. **'Arrow'** should be sown in succession from mid-spring to mid-summer for young roots and harvested from mid- to late summer for small roots. **'Avonresister'** is small with bulbous roots, is sweet-tasting, performs well on poorer soil and grows rapidly. It also has excellent resistance to canker and bruising. **'Bugi Bijeli'** is a Yugoslavian variety with a large, sweet root. **'Cobham Improved Marrow'** is wedge-shaped, medium in size and well-flavoured, and resistant to canker. **'Exhibition Long'** is extra-long with an excellent flavour. **'Gladiator'** produces large, vigorous, well-shaped and fine-flavoured roots with good canker resistance. **'Javelin'** is wedge- or bayonet-shaped, high-yielding and canker-resistant; 'fanging' rarely occurs. **'Student'** has long slender roots and is very tasty. It originated around 1810 from a wild parsnip found in the grounds of the Royal Agricultural College, Cirencester, England. **'Tender and True'** is an old variety that is very tasty, tender and sweet. It has very little core. **'White Gem'** has wedge-shaped to bulbous smooth roots, with delicious flesh and good canker resistance. It is ideal for heavier soils.

PARSNIP

cultivation

Propagation

Parsnip seed must always be sown fresh, as it rapidly loses the ability to germinate. Seeds are also renowned for erratic, slow germination in the cold, wet conditions which often prevail early in the year when they are traditionally sown.

To avoid poor germination, sow later, from mid- to late spring, depending on the weather and soil conditions, when the ground is workable and temperatures are over 7°C.

Warm the soil with cloches or horticultural fleece a few weeks before sowing. Rake in a granular general fertiliser at 60g/sq m 1–2 weeks before sowing. Rake the seedbed to a fine texture before sowing. Sow *in situ* on a still day so the light, papery seeds are not blown away.

In dry conditions, water drills before sowing 2–3 seeds at the recommended final spacing in 'stations', thinning to leave the strongest seedling after germination. Sow radishes between the stations to act as a marker crop indicating where slower-germinating parsnip seeds are sown.

Parsnip – an essential ingredient in winter

Alternatively, parsnip seeds can be sown thinly in drills 5mm to 1cm deep. Sow shorter-rooted varieties in rows 15–20cm apart, thinning them to 5–10cm, and larger types in rows 30cm apart, thinning to 13–20cm.

In stony soil, make a hole up to 90 x 15cm with a crowbar, fill with finely sieved soil or compost, sow 2–3 seeds in the centre of the hole and thin to leave the strongest seedling. Grow short-rooted varieties in shallow soils.

Growing

Parsnips thrive in an open or lightly shaded site on light, free-draining, stone-free soil which was manured the previous year.

Traditionally, parsnips are not grown on freshly manured soil as this causes 'fanging', or forking of the roots; however, recent research has not supported this. Dig the plot and add plenty of well-rotted organic matter in autumn or early winter the previous year. Deep digging is particularly important when growing long-rooted varieties.

The ideal pH is 6.5–7.0; lime where necessary, as roots grown in acid soil are prone to canker. Rotate with other roots.

Keep crops weed-free, by hoeing or hand weeding carefully to avoid damaging the roots, or by mulching with well-rotted organic matter. Do not let the soil dry out, as erratic watering causes roots to split. Water at 16–22 litres/sq m every 2 weeks during dry weather when the roots are swelling.

Maintenance

Spring Rake the seedbed to a fine tilth, apply general fertiliser, sow seed in modules, 'fluid sow' (see page 646), or sow *in situ* when the soil is warm.
Summer Keep crops weed-free by mulching, hand weeding or careful hoeing.
Autumn Harvest crops.
Winter Cover crops with straw or bracken before the onset of inclement weather. Dig the soil for the following year's crop.

Protected Cropping

Possible germination problems can be avoided by pre-germinating and fluid-sowing seed, or by sowing in modules under cover and transplanting before the tap root starts to develop.

Parsnip flowers attract beneficial insects

Harvesting and Storing

Most parsnips are a long-term crop and occupy the ground for around 8 months – a factor worth bearing in mind if your garden is small. Roots are extremely hardy and can remain in the ground until required.

Harvest from mid-autumn onwards, covering plants with straw, bracken or hessian for ease of lifting in frosty weather. Make sure you lift them carefully with a fork to avoid root damage.

Lift all your roots by winter and store them in boxes of moist sand, peat substitute or wood shavings in a cool shed.

Parsnips have a better flavour when they have been exposed for a few weeks to temperatures around freezing point. (This changes stored starch to sugar, increasing sweetness and improving the flavour.) Stored in a polythene bag in the refrigerator, they remain fresh for up to 2 weeks.

Wash, trim and peel roots, then cube and blanch in boiling water for 5 minutes before freezing them in polythene bags.

Pests and Diseases
Parsnip canker is a black, purple or orange-brown rot, often starting in the crown, which can be a problem during drought, when the crown is damaged or the soil is too rich. There is no chemical control. Sow crops later, improve drainage, keep the pH around neutral, rotate crops and sow canker-resistant varieties. Carrot fly can also be a problem.

companion planting

Sow rapidly germinating plants such as lettuces between rows. Parsnips grow well alongside peas and lettuce, provided they are not in the shade.

Plant next to carrots and leave a few to flower the following year, as they attract beneficial insects.

container growing

Shorter-rooted varieties can be grown in large containers of loam-based compost. Longer types are grown in large barrels, making a deep hole in the compost, as described above for stony soil. Make sure that containers are well drained.

medicinal

In Roman times, parsnip seeds and roots were regarded as an aphrodisiac.

Seeds only have a short life

culinary

In seventeenth-century England, there are records of parsnip bread and 'sweet and delicate parsnip cakes'. They were often eaten with salt fish and were a staple during Lent.

Scrub or peel parsnips, and use boiled, baked, mashed or roasted with beef, pork or chicken. They combine particularly well with carrots.

Parsnips can be lightly cooked and eaten cold. Parboil and fry like chips or slice into rings, dip in batter and eat as fritters. Grate into salads, add chopped and peeled to casseroles or soups. Or parboil, drain, then stew in butter and garnish with parsley. They are also good parboiled, then grilled with a sprinkling of Parmesan. Try steaming them whole, slicing them lengthwise and pan-glazing with butter, brown sugar and nutmeg, or garnishing them with chopped walnuts and a dash of sweet sherry.

Purée of Parsnips
Boil some parsnips and mix them with an equal quantity of mashed potatoes, plenty of salt and freshly ground black pepper, a little grated orange zest, a splash of thick cream and enough butter to make a smooth dish. Sprinkle with chopped flat-leaf parsley and serve piping hot. (Puréed parsnips are also wonderful combined with carrots and seasoned with nutmeg.)

Curried Parsnip Soup
Serves 4–6

1kg parsnips
1 large onion, sliced
1 tablespoon butter
2 cloves garlic, crushed
1 good teaspoon curry powder
1 x 400g can chopped tomatoes
1 litre vegetable stock
1 bay leaf
Sprig each of fresh thyme and parsley
4–6 teaspoons yogurt
Salt and freshly ground black pepper
Chopped fresh parsley, to garnish

Clean the parsnips and peel, if old. Chop them roughly. Add the onion to a large soup pot with the butter and garlic and sauté over a medium heat until lightly browned (this helps to give the flavour of Indian cuisine). Stir in the curry powder and continue to cook for 1 minute, stirring constantly.

Add the parsnips and stir, coating them well in the curry and onion mixture, and add the tomatoes, stock and herbs. Stir thoroughly. Season, bring to the boil and simmer, covered, for about 15–20 minutes, until the parsnips are tender.

Remove the herbs. Then liquidise the soup, adjust the seasoning and garnish with yogurt and parsley. This is delicious served with crusty bread.

Petroselinum crispum var. *tuberosum*. Apiaceae

HAMBURG PARSLEY

Also known as Parsley Root, Turnip-rooted Parsley. Swollen-rooted biennial, usually grown as annual for roots and leaves. Hardy. Value: root contains starch and sugar; leaves high in beta carotene, vitamin C and iron.

Hamburg parsley is an excellent dual-purpose vegetable. The tapering white root looks and tastes similar to parsnip and can be eaten cooked or raw. The finely cut, flat, dark green leaves resemble parsley; they are coarser in texture, but contribute a parsley flavour and can be used as garnish. As a bonus, it grows in partial shade, is very hardy and is an ideal winter vegetable.

Popular in Central Europe and in Germany, it is one of several vegetables and herbs known as *Suppengrun,* or 'soup greens', which are added to the water when beef or poultry is boiled and later used for making sauce or soup. Introduced from Holland to England in the eighteenth century, this versatile vegetable enjoyed only a relatively brief period of popularity, but should certainly be more widely grown by Britain's gardeners.

Hamburg parsley roots

cultivation

Hamburg parsley needs a long growing season to develop good-sized roots. Grow it in moisture-retentive soil in an open or semi-shaded position.

Propagation

As seeds are sown early in the year, it may be necessary to cover the bed with cloches or polythene before sowing to warm up the soil. Sow from early to mid-spring in drills 1cm deep and 25–30cm apart, thinning seedlings to a final spacing of 20cm.

They can also be sown in 'stations', planting 3–4 seeds in clusters 23cm apart. Thin to leave the strongest seedling when they are large enough to handle.

Seeds can also be sown in modules, but should be planted in their final position before the tap root begins to form.

To extend the harvesting season, make a second sowing in mid-summer, which will provide crops early the following year.

Growing

Dig in plenty of well-rotted organic matter the autumn before planting or plant in an area where the soil has been manured for the previous crop.

Rake the soil finely before sowing. Germination is often slow and it is important that the seedbed is kept free of weeds, particularly in the early stages.

Use a hoe with care as the plants become established to avoid damaging the roots. Mulching established plants is a sensible option as this stifles weed growth and conserves moisture. A 2.5cm layer of well-rotted compost will perform the task perfectly. Root splitting can occur if plants are watered after the soil has dried out, so water thoroughly and regularly during dry periods to maintain steady growth.

Maintenance

Spring Sow seeds outdoors, warming the soil if necessary. Seeds can also be sown in modules.
Summer Keep crops weed-free, water during drought and make a second sowing in mid-summer for early crops the following year.
Autumn Harvest as required.
Winter Cover crops with bracken or similar protection to make lifting easier in severe weather.

The tops are also tasty

Harvesting and Storing

This is such a hardy vegetable that it can remain in the ground until required any time from early autumn to mid-spring. When frosts are forecast, cover with straw, bracken or a similar material to make lifting easier. After removing the foliage, store in a box of sand or peat substitute in a cool shed or garage; some of the flavour is lost when roots are stored in this way, so they are better left in the ground, if possible.

Pests and Diseases

This tough vegetable is generally trouble-free, but it can suffer from parsnip canker. This appears as dark patches on the root. Control is impossible, but the following measures will help. Dig up and dispose of affected plants immediately, ensure that the soil is well drained, rotate crops and maintain soil at a pH around neutral.

companion planting

When 'station sowing', fill the spaces by planting 'Tom Thumb' or a similar lettuce variety. Sowing radishes between 'stations' is also useful, because their rapid germination indicates the position of the Hamburg parsley and prevents you accidentally removing any newly germinated seedlings when weeding.

container growing

Hamburg parsley can be grown in containers if they are deep enough not to dry out rapidly and provide sufficient space for the roots to form. Regular watering is vital and plants should also be kept out of scorching, bright sunshine.

warning

Hamburg parsley should not be eaten in large amounts by expectant mothers or those with kidney problems.

medicinal

The leaves are a good source of vitamins A and C and contain similar properties to traditional garnishing parsley. They reduce inflammation, are used for urological conditions such as cystitis and kidney stones, and help indigestion, arthritis and rheumatism. After childbirth the leaves encourage lactation; the roots and seeds promote uterine contractions. The leaves are also an excellent breath freshener – powerful enough to counter the effects of garlic!

culinary

Hamburg parsley has a flavour reminiscent of celeriac, and is frequently used in Eastern European cooking.

Prepare for cooking by removing the leaves and fine roots, then gently scrubbing to remove the soil: don't peel or scrape. Try Hamburg parsley sliced, cubed and cooked like parsnips (sprinkle cut areas with lemon juice to prevent discoloration). It is delicious roasted, sautéed or fried like chips, as well as boiled, steamed or added to soups and stews. It can also be grated raw in winter salads.

Dried roots can be used as a flavouring. Dry them on a shallow baking tray in an oven heated to 80°C; allow to cool before storing in an airtight jar, in a dark place. Wash the leaves and use as flavouring or garnish.

Croatian Hamburg Parsley Soup

Serves 8

500g Hamburg parsley
1 turnip, peeled
1 large onion, peeled
1 good-sized leek, cleaned
50g butter
2½ litres vegetable stock
Salt and freshly ground black pepper
4 tablespoons sour cream
2 tablespoons each fresh chives and dill

Chop all the vegetables coarsely. Heat the butter in a heavy-bottomed pan and sauté the vegetables until softened. Stir to prevent burning. Pour over the stock, season with salt and pepper and simmer until the vegetables are tender.

Put through a *mouli légumes* or liquidise, and return to the pan. Stir in the sour cream and the herbs, allow to heat through again and adjust the seasoning. Do not allow to boil. Serve piping hot, with croutons.

Phaseolus aureus. Papilionaceae

MUNG BEAN

Also known as Green Gram, Golden Gram. Annual grown for its seed sprouts and seeds. Tender. Value: sprouts: rich in vitamins, iron, iodine, potassium, calcium; seeds: rich in minerals, protein and vitamins.

This major Indian pulse crop was introduced to Indonesia and southern China many centuries ago and is grown throughout the tropics and sub-tropics, being sown towards the end of the wet season to ripen during the dry season. The leaves and stems are used to make hay and silage and the seeds are used for cattle feed. Mung beans came to prominence in the West with an increased interest in oriental cuisine: bean sprouts are an easy-to-grow major crop. Among other uses, their flour is a soap substitute and they serve as a replacement for soya beans in the manufacture of ketchup.

varieties

Phaseolus aureus is a sparsely leaved annual to about 90cm with yellow flowers and small, slender pods containing up to 15 olive, brown or mottled seeds.

Shoots are so easy to grow

cultivation

Propagation

Rake in a general fertiliser at 60g/sq m about 10 days before sowing. Seeds can be broadcast or sown in drills on a well-prepared seedbed, thinning to 20–30cm apart with 40–50cm between the rows.

Growing

Mung beans flourish on rich, deep, well-drained soils and dislike clay. Add organic matter if necessary and lime to create a pH of 5.5–7.0. Water well throughout the growing season. Ideal growing conditions are between 30–36°C and with moderate rainfall.

Sprouting Seeds

Use untreated seeds for sprouting, as those treated with fungicide are poisonous. Remove any that are discoloured, mouldy or damaged. Soak them overnight in cold water and the following morning rinse in a colander. Place a layer of moist kitchen roll or damp flannel over the base of a flat-bottomed tray or 'seed sprouter', then spread a layer of mung beans about 1cm deep over the surface.

Cover the container with polythene and put the bowl in a dark cupboard, or wrap it to exclude light.

Temperatures of 13–21°C should be maintained for rapid growth: an airing cupboard is ideal. Rinse the sprouts daily, morning and night.

Buy a 'seed sprouter' or make one from a large yogurt pot or ice cream tub with holes punched in the bottom and use the lid as a drip tray. This makes for easier daily rinsing of the seedlings.

MUNG BEAN

culinary

Bean sprouts can be eaten raw, in salads, with a suitable dressing, or lightly cooked for about 2 minutes only in slightly salted water. (Be warned: overcooked shoots lose their taste.)

To stir-fry, use only a little oil, stirring briskly for about 2 minutes. In Oriental cooking, bean sprouts combine well with other vegetables, eggs, red meat, chicken and fish and can be used to stuff savoury pancakes, egg rolls and tortillas.

Stir-fried Mung Beans
Serves 2

1 bag mung beans
Groundnut oil, for frying
2 cloves garlic, crushed
250ml chicken stock
1 tablespoon light soy sauce
1 heaped teaspoon cornflour
2–3 spring onions, roughly chopped

Rinse the mung beans well. Heat the oil and the garlic in a wok, then add the mung beans. Fry for 1 minute, stirring constantly.

Cover with chicken stock and cook until tender and the liquid has evaporated. Then stir in the soy sauce and the cornflour. Cook for 2 minutes more.

Garnish with the spring onions and serve with rice.

Maintenance

Outdoors Broadcast or sow seed in drills; water when needed, as the soil dries out. Harvest before the pods split.

Sprouting Keep seeds moist. Harvest regularly.

Harvesting and Storing

Harvest shoots from 3 days onwards, when they are about 2.5–5cm long. If the seed coats remain attached to the sprouts, soak them in water, then 'top and tail', removing the seed and shoot tip. Store in the fridge. When growing for seed, gather before they split, to prevent the seeds being lost.

Pests and Diseases

Powdery mildew is a problem when plants are dry at the roots. Mulch and water regularly, remove diseased leaves immediately, spray with bicarbonate of soda, improve air circulation by thinning crops and destroy plant debris at the end of the season.

medicinal

The seeds are said to have a cooling and astringent effect on fever and an infusion is used as a diuretic when treating beriberi. In Malaya it is prescribed for vertigo.

warning

Do not let sprouted seeds become waterlogged, as they rapidly become mouldy.

Somebody buy some, please!

Phaseolus coccineus. Papilionaceae

RUNNER BEAN

Also known as Scarlet Runner. Perennial climber grown as annual in temperate climates for edible pods and seeds. Half hardy. Value: moderate levels of iron, vitamin C and beta carotene.

A native of Mexico, the runner bean has been known as a food crop for more than 2,200 years. In the late sixteenth century, Gerard's *Herball* mentions it as an ornamental introduced by the plant collector John Tradescant the Elder: 'Ladies did not…disdain to put the flowers in their nosegays and garlands', and in the garden it was grown around gazebos and arbours. Vilmorin Andrieux commented in 1885: 'In small gardens they are often trained over wire or woodwork, so as to form summer houses or coverings for walks.' Philip Miller, keeper of Chelsea Physic Garden, is credited with being the first gardener to cook them.

It was the fashion in seventeenth-century England to experiment with soaking seeds. Mr Gifford, minister of Montacute in Somerset, noted in his diary: 'May 10th 1679, I steep'd runner beans in sack five days, then I put them in sallet-oyle five days, then in brandy four days and about noon set them in an hot bed against a south wall casting all liquor wherein they had been infused negligently about the holes, within three hours space, eight of the nine came up, and were a foot high with all their leaves, and on the morrow a foot more in height… and in a week were podded and full ripe.' You could try this for yourself – or perhaps you would prefer to stick to today's more conventional methods!

varieties

Older cultivars were rather stringy unless eaten young; newer varieties are 'stringless'. Besides the traditional tall-growing climbers, there is also a choice of dwarf varieties that do not need supporting and are ideal for smaller gardens, early crops under cloches and in exposed sites.

Non-stringless types

'Enorma' is ornamental, with red flowers. It has long pods and is very tasty. **'Liberty'** has good-quality pods to 45–50cm long. **'Painted Lady'**, a variety grown since the nineteenth century, has delicate red and white flowers and long pods.

Stringless types

'Aintree' has red flowers that are followed by good -quality green pods. Tolerant of high temperatures. **'Celebration'** has pink flowers that are followed by plenty of straight, well-flavoured, good-quality pods. **'Desiree'** has white flowers and seeds and is high-yielding and tasty. **'Kelvedon Marvel'** is tasty, matures early and crops heavily. **'Lady Di'** grows long, slim stringless pods with outstanding taste and tenderness. **'Red Rum'** is very early, high-yielding, tasty and resistant to halo blight. **'Scarlet Emperor'** is a traditional 'all rounder', tolerating a wide range of conditions. **'Snowy'** is white flowered and 'sets' even in hot conditions where night temperatures are above 15°C; pods to 30cm. **'White Lady'** is less likely to be attacked by birds. Its white flowers are followed by huge crops of straight fleshy pods (late). **'Wisley Magic'** is one of the heaviest yielding red-flowered varieties, producing masses of pods with a traditional flavour.

'White Lady'

'Scarlet Emperor'

Dwarf varieties

'Hammonds Dwarf Scarlet' is ideal for the small garden. The pods are easy to harvest but the tops may need to be pinched out. **'Hestia'** produces bicoloured flowers and long, slim pods. Ideal for pots or containers.

cultivation

Propagation

Runner beans are not frost-tolerant and need a minimum soil temperature of 10°C to germinate. Sow from late spring to early summer 5–7.5cm deep, 15cm apart in double rows 60cm apart. To shelter pollinating insects they are better grown in blocks or several short rows, rather than a single long one. Allow 1–1.75m between rows for ease of harvesting, depending on the cultivar.

Climbing types can be encouraged to bush by pinching out the main stems when they are around 25cm; the sideshoots can be pinched out at the second leaf joint for a bushy plant needing little or no staking.

Seed sown outdoors crops in about 14–16 weeks, depending on climatic and cultural conditions.

Growing

Runner beans are not frost-hardy and are less successful in cooler areas unless you have a suitable microclimate. They flourish from 14–29°C, needing a warm, sheltered position to minimise wind damage and encourage pollinating insects.

Soil should be deep, fertile and moisture-retentive. Dig a trench at least 30cm deep and 60cm wide in late autumn to early winter before planting, adding plenty of well-rotted organic matter to the backfill. Before sowing, rake in 60–90g/sq m of granular general fertiliser.

Water well when flowering

'Painted Lady'

Keep crops weed-free during early stages of growth. Mulch after germination or after transplanting. Watering is essential for good bud set: the traditional method of spraying flowers has little effect. As the first buds are forming and again as the first flowers are fully open plants need 5–11 litres/sq m. Crops should be rotated.

Climbers can be up to 3m or more tall and a good sturdy support should be in place before sowing or transplanting. Traditionally, a 'wigwam' of canes, or a longer row of crossed canes, were used; these may need supporting strings at the end of each row, like tent guy ropes. Do improvise: I once saw a wonderful structure like a V-shaped frame, designed so the beans would hang down and make picking easier. Use canes, strong wooden stakes, steel tubes, or make frameworks of netting. There should be one cane or length of strong twine for each plant.

Maintenance

Spring Sow crops under cover or outdoors in late spring. Erect supports.
Summer Keep crops weed-free and well watered. Harvest regularly.
Autumn Continue harvesting until the first frosts.
Winter Prepare the soil for the following year's crop.

Protected Cropping

For earlier crops and in cooler climates, sow in boxes or tall pots of seed or multi-purpose compost under cover from mid-spring, harden off and plant out from late spring. Protect indoor and outdoor crops with cloches or horticultural fleece until they are established.

Harvesting and Storing

Harvest from mid-summer to mid-autumn. Picking is essential to ensure regular cropping, high yields and to avoid any 'stringiness'. Runner beans freeze well.

Pests and Diseases

Slugs, black bean aphid and red spider mite may prove troublesome.

Grey mould (*botrytis*) and halo blight may cause problems in wet or humid weather.

Mice may eat seeds: dip in concentrated liquid seaweed.

Root rots may kill plants in wet or poorly drained soils. Rotate crops to avoid this.

companion planting

Runner beans are compatible with all plants except the *Allium* family.

Grow with maize to protect the latter from corn army worms. Nitrogen-fixing bacteria in the roots improve soil fertility. After harvesting, cut off tops, leave roots in the soil or add to compost heaps. They thrive with brassicas; Brussels sprout transplants are sheltered and grow on once beans die back.

Late in the season, their shade can benefit celery and salad crops if enough water is available.

container growing

Runner beans can be grown in containers at least 20cm wide by 25cm deep, depending on the vigour of the cultivar.

Sow seeds 5cm deep and 10–13cm apart indoors in late spring, moving the container outdoors into a sunny site. Water frequently in warm weather, less often at other times. Feed with liquid general fertiliser if plants need a boost.

Stake tall varieties, pinching out the growing points when plants reach the top of their supports. Otherwise use dwarf varieties.

culinary

Wash, top and tail, pull off stringy edges, slice diagonally and boil for 5–7 minutes. Drain and serve with a knob of butter. Alternatively, cook whole and slice after cooking.

Runner Bean Chutney

Makes 1.5kg

Clare Walker and Gill Coleman give this recipe in *The Home Gardener's Cookbook*:

1kg runner beans
500g onions
1 level tablespoon salt
1 level tablespoon mustard seeds
75g sultanas
1½ level teaspoons ground ginger
1 level teaspoon turmeric
6 dried red chillies, left whole
600ml spiced vinegar
500g demerara sugar

Wipe, top, tail and string the beans and cut them into small slices. Place in a preserving pan with the peeled and finely chopped onions. Add 300ml water, the salt and mustard seeds and simmer gently for 20–25 minutes until the beans are just tender.

Then add the sultanas, ginger, turmeric, chillies and spiced vinegar, bring back to the boil and simmer for a further 30 minutes, or until the mixture is fairly thick.

Stir in the sugar, allow to dissolve, then boil steadily for about 20 minutes until the chutney is thick. Pour into warm, dry jars, cover with thick polythene and seal with a lid, if available. Label and store in a cool, dry place for at least 3 months before using. The chillies can be removed before potting.

Phaseolus vulgaris. Papilionaceae

FRENCH AND DRIED BEANS

Also known as Common, Kidney, Bush, Pole, Snap, String, Green, Wax Bean, Haricot, Baked Bean, Flageolet, Haricot Vert. Annual grown for edible pods and beans. Half hardy. Value: moderate potassium, folic acid and beta carotene. Very rich in protein.

Evidence of the wild form, found in Mexico, Guatemala and parts of the Andes, has been discovered in Peruvian settlements from 8000 BC. Both bush and climbing varieties were introduced to Europe during the Spanish conquest in the early sixteenth century, though the dwarf varieties did not become popular for two more centuries. They were first referred to as 'kidney beans' by the English in 1551, alluding to the shape of their seeds.

Gerard in his *Herball* calls them 'sperage' beans and 'long peason', while Parkinson wrote: 'Kidney beans boiled in water and stewed with butter were esteemed more savory… than the common broad bean and were a dish more oftentimes at rich men's tables than at the poor.' Another writer commented in 1681: 'It is a plant lately brought into use among us and not yet sufficiently known.'

In Europe, 'haricot vert' was used in ships' stores in voyages of exploration during the early 1500s. When European colonists first explored the Americas, they found climbing beans planted with maize, providing starch and protein for indigenous tribes.

'The Prince'

varieties

This group contains considerable variety. Plants are dwarf (ideal for smaller gardens or containers) or climbing, pods are flat, oval or round in cross-section and are green, yellow (waxpods) or purple, or marbled.

The seeds are colourful and often mottled. As well as being grown for the pods, seeds are eaten fresh as 'flageolets', or dried as 'haricots'.

Drying beans

'Cannellino' is a dwarf variety with white seeds. **'Canadian Wonder'**, a Victorian variety, should be picked while young and tender, or dry for the red seeds. **'Czar'** produces long, rough pods; tasty when green or can be dried. **'Saissons'** has flat pods and tasty pale green seeds.

Climbing types

'Kingston Gold' is high-yielding with golden-yellow pods and a first-class flavour. **'Hunter'** is flat-podded, and crops heavily over a long period. **'Kwintus'** produces tender, tasty pods. **'Limka'** is ideal for early or late growing under glass. **'Musica'** can be sown early and is full of flavour. **'Romano'**, an old variety, is tender, tasty and prolific, and excellent for freezing.

Dwarf or bush types

'Annabel' is high-yielding, round and tasty – perfect for the patio. **'Chevrier Vert'**, a classic French flageolet from 1880, is tasty and tender. **'Delinel'** is a prolific and tasty 'filet'-type with stringless pods to 15cm long. **'Golddeleaf'** (very early) is prolific with pale yellow pods. **'Purple Queen'** is delicious, producing heavy yields of glossy purple pods. **'Purple Teepee'** is high-yielding, ornamental and easy to spot when harvesting, as pods are held high above the foliage. **'Royalty'** crops heavily with dark purple pods that turn green when cooked and are delicately flavoured.

'Safari' produces slim, tasty beans but low yields. The dark green pods of **'Sprite'** freeze well. **'The Prince'**, an old favourite, has magnificent taste and is also good for freezing.

cultivation

Propagation

French beans dislike cold wet soils and are inclined to rot; for successful germination, do not sow until the soil is a minimum of 10°C.

For early sowings or in cold weather, warm the soil 3–4 weeks beforehand with cloches or black polythene.

Sow successionally from mid-spring to early summer in staggered drills 4–5cm deep with 22cm between the rows and plants for optimum yields. Climbing varieties should be in double rows 15cm apart with 60cm between rows.

Growing

French beans flourish in a sheltered, sunny site on a light, free-draining, fertile soil where organic matter was added for the previous crop. Alternatively, dig in plenty of well-rotted organic matter in late autumn or winter before planting.

'Purple Queen'

The plants need a pH of 6.5–7.0; lime acid soils if necessary. Rake the soil to a medium tilth about 10–14 days before sowing, incorporating a balanced granular fertiliser at 30–60g/sq m.

Keep crops weed-free or mulch when the soil is moist. Earth up round the base of the stems for added support and push twigs under mature bush varieties to keep pods off the soil, or support plants with pea sticks. Support climbing varieties in the same way as runner beans.

Keep well watered during drought; plants are particularly sensitive to water stress when the flowers start to open and as pods swell. Apply 13–18 litres/sq in per week.

Maintenance

Spring Sow early crops under cover or outdoors in late spring. Support climbers. Transplant when 5–7.5cm tall.
Summer Keep crops weed-free, mulch, water in drought. Harvest regularly.
Autumn Continue harvesting until the first frosts. Protect later crops with cloches or fleece.
Winter Prepare the soil for the following year's crop.

Protected Cropping

For earlier crops and in cooler climates, sow in boxes, modules or pots of seed compost under cover from mid-spring. Warm the soil before transplanting. Harden off and plant out from late spring, depending on the weather and soil conditions. Protect crops sown indoors or *in situ* with cloches, polytunnels or with horticultural fleece until established.

For very early harvest, sow in pots in a heated greenhouse at 15°C from late winter. Sow 4 seeds near the edge of a 23–25cm pot containing loam-based compost with moderate fertiliser levels.

Seeds can also be pre-germinated on moist kitchen towel in an airing cupboard or similar. Keep moist with tepid water, ventilate in warm weather and harvest in late spring.

Harvesting and Storing

Plants are self-pollinating, so expect a good harvest. Pick when pods are about 10cm long, when they snap easily, before the seeds are visible. Pick regularly for maximum yields. Cut them with a pair of scissors or hold the stems as you pull the pods to avoid uprooting the plant.

French beans are great when lightly boiled or steamed

For dried or haricot beans, leave pods until they mature, sever the plant at the base and dry indoors. When pods begin to split, shell the beans and dry on paper for several days. Store in an airtight container.

French beans freeze well. Wash and trim young pods, blanch for 3 minutes, freeze in polythene bags or rigid containers. You should use within 12 months. They keep in a polythene bag in the refrigerator for up to a week and last about 4 days in a cool kitchen.

Pests and Diseases

Slugs, black bean aphid and red spider mite can be troublesome. Grey mould (*Botrytis*) and halo blight may be problems in wet or humid weather. Mice may eat seeds. Root rots can kill plants in wet or poorly drained soils. Rotate crops.

companion planting

French beans do well with celery, maize, cucurbits, sweetcorn and melons. Intercrop with brassicas.

container growing

French beans can be grown in containers that are at least 20cm wide by 25cm deep. Sow seeds 4–5cm deep indoors in mid-spring, moving the container outdoors into a sheltered, sunny site. Feed with liquid general fertiliser if necessary. Stake climbing varieties; pinch out growing points when plants reach the top of supports.

medicinal

One cup of beans per day is said to lower cholesterol by about 12 per cent.

FRENCH BEAN

culinary

Fresh French beans have such a delicate flavour that they hardly need more than boiling in water and serving as an accompaniment to meat and other dishes. Wash, top and tail pods and cook whole in boiling salted water for 5–7 minutes, preferably within an hour of harvesting. Cut large flat-podded types into 2.5cm slices. Alternatively, steam them, serve cold in salads or try them stir-fried with other vegetables.

For haricot beans, place fresh beans in cold water, bring to the boil, remove from the heat and allow to stand for an hour. Drain and serve as a hot vegetable, or in vinaigrette as a salad.

Mussel Soup with French Beans
Serves 4

1.5kg mussels, cleaned and debearded
300ml dry white wine
3 tablespoons olive oil
1 onion, finely chopped
200g French beans, finely chopped
Salt and freshly ground black pepper
3 plum tomatoes, peeled, seeded and diced
Bunch fresh coriander, leaves picked and chopped

Start by discarding any mussels that won't close when they are lightly tapped. Heat a large pan over a high heat. Tip in the mussels and cover for 10–15 seconds, then pour over the wine and 300ml of water. Cover and cook, shaking the pan every now and then, for 5 minutes or until the mussels have opened. Discard any that stay closed.

Tip the mussels into a colander set over a large bowl. Strain the cooking liquid and reserve. When cool enough to handle, remove the mussels from their shells and reserve, discarding the shells.

Heat one tablespoon of the oil in a pan and add the onion. Cook for a few minutes until softened, then pour in the reserved liquid and bring to the boil. Season to taste. Add the beans, mussels and tomatoes and just warm through. Ladle into bowls, scatter over the coriander and drizzle over the remaining olive oil. Serve immediately.

French Bean, Roquefort and Walnut Salad
Serves 4

300g French beans, topped and tailed
250g Roquefort cheese
125g walnuts
1 small radicchio plus 2 little gem lettuces
3 tablespoons extra virgin olive oil
1 tablespoon balsamic vinegar
1 clove garlic, crushed
Salt and freshly ground black pepper

Wash the beans and steam them over a pan of boiling water until just crunchy. Keep warm. Crumble the Roquefort and lightly crush the walnuts. Wash the lettuces thoroughly and shake dry. Make a dressing with the remaining ingredients.

Arrange the lettuces in a large bowl, top with the beans, walnuts and cheese, and pour over the dressing. Toss and serve while the beans are still warm.

French Bean, Roquefort and Walnut Salad

Phyllostachys spp. and others. *Graminae/Poaceae*

BAMBOO

Woody-stemmed, evergreen perennial. Hardy to tropical. Value: high in fibre, Vitamin C, E and B6 and several other elements, including phosphorus, potassium and zinc.

From creeping, grass-like species to giant fast-growing varieties climbing towards the heavens, bamboos are not only edible and ornamental but one of the most useful plants on earth. These fascinating, long-lived members of the grass family have influenced history and have many practical uses: from construction and scaffolding to the manufacture of fishing rods, paper and medicine. Bamboo has also inspired works of art, and writings about bamboo in Chinese literature date back many centuries. Silkworm eggs were first smuggled out of the country in a bamboo cane and the Chinese have drilled to 1000m for oil using bamboo canes since at least 200 BC. Bamboos rarely flower, but when they do, every specimen of that particular clone does so simultaneously, wherever it is in the world. Bamboo shoots play a prominent role in the diets of many countries – generations of pandas can't be wrong!

Harvest bamboo before this stage

varieties

There are clump-forming and spreading species of bamboo, so choose garden specimens carefully – those with spreading roots are often vigorous and invasive.

Phyllostachys dulcis has masses of green stems that grow rapidly to around 8 x 4m. It is highly productive in hot summers – produces sweet shoots, is one of the best for eating and thrives in sun or part shade. ***P. edulis*** can grow to a maximum of 20 x 5m but is usually smaller. Clump forming and slow to establish, with green canes. ***Phyllostachys edulis* 'Heterocycla'** grows to 20 x 5m; not for the faint hearted or those with small gardens. Happy in sun or part shade, ***P. aurea*** grows to 5 x 3m, the young shoots are bright green, becoming pale creamy-yellow and matt yellow in sun. Very graceful, yet tough and vigorous in warm climates. ***P. aureosulcata*** grows to 9 x 3m. The green canes with golden grooves often zigzag prominently at the base – very hardy and thrives in sun or part shade. ***P. nigra***, the legendary black-caned bamboo, grows to 12 x 5m, though usually much smaller. Young canes are green and their transition to black is faster in sunshine; ideal in smaller gardens or containers. ***P. nigra* f. *henonis*** – the elegant canes of this clump-forming specimen reach 4 x 4m or more. It has dark green, glossy leaves and brown-yellow canes when mature and is excellent as a specimen plant. ***P. nuda*** grows to 5 x 2.5m and is tough and vigorous with dark green stems. ***P. rubromarginata*** is compact and upright with slender, pale olive-yellow canes. Grows to 6 x 1.5m and is ideal for the smaller garden. ***P. viridiglaucescens*** grows to 4m or more, in clumps up to 3m wide. It is leafy and graceful with tall, glossy, green canes.

BAMBOO

Pleioblastus hindsii forms clumps to 2m and sometimes dense thickets of olive-green canes up to 4m tall. Good in sun and dense shade, for maritime conditions and as a windbreak in sun or shade. ***Pseudosasa japonica*** is adaptable, hardy and forms clumps to 3m and occasionally dense thickets of olive-green canes to 5m tall, which arch towards the tip. Good in shade, as a hedge or windbreak and easy to grow.

cultivation

Propagation
Lift and divide clumps or rhizomes in spring, using a spade, and transplant in well-prepared ground.

Growing
Bamboos prefer rich, moist soils but are extremely tolerant, provided conditions are not waterlogged. Clear the ground of perennial weeds, add plenty of well-rotted organic matter to the soil and soak the root-ball in a bucket of water before planting slightly lower than the level of the plant in the pot, then water in well. Mulch well after planting, using well-rotted organic matter, then do so annually, each spring, after feeding with general fertiliser. Keep the soil moist, particularly during the first 2–3 years, until the plants are established. Bamboos in containers need regular watering during the growing season; do not let the roots dry out in winter, particularly during mild periods.

Bamboo can be grown in a container

Although regular harvesting thins the clump, remove about a third or more of the old and weak stems at the base in spring to encourage regrowth.

Maintenance
Spring Cut out stems that were damaged over winter. Harvest young shoots.
Summer Water if necessary during drought.
Autumn Clear the ground ready for planting next spring.
Winter Order new varieties from plant catalogues.

Protected Cropping
Plants grown in containers can be protected under cover during winter and early spring, advancing shoot production for an earlier harvest.

container growing

Grow in a 50:50 loam-based compost and peat substitute, or multi-purpose compost; mix with added grit for stability and a good layer of drainage material in the base. In cold winters, wrap the pot with hessian or bubble wrap and put the pot in a sheltered position. In spring, feed with slow-release fertiliser and repot into a container one size larger, if necessary. Bamboos dislike being transplanted into pots which are considerably larger than the existing root ball.

Harvesting and Storing
Harvest shoots that are 13mm or more in diameter – plants may take more than 5 years to produce canes of this size – removing a few from each clump so plant growth is not affected; up to one third of the new shoots can be removed without damaging the plants. New shoots grow rapidly, so check clumps daily and harvest from early to mid-spring, cutting shoots off at ground level with a sharp knife or clearing the soil and cutting at the point where they join the roots of spreading varieties. Shoots should be no more than 30cm high; after this stage they become fibrous and inedible. Developing shoots that are blanched by covering with a box, bucket or layer of mulch are tenderer when eaten.

Pests and Diseases
In temperate climates, bamboos are almost disease-free and are resistant to honey fungus. Rabbits, squirrels and deer eat new shoots. Oriental bamboo spider mite can be a problem; spray to increase humidity or with an environmentally friendly insecticide.

culinary

Most bamboos are edible but some are tastier than others. The sweetest are *Phyllostachys dulcis* and *Phyllostachys edulis*. Use the shoots immediately after harvest, in stir-fries and salads, or store them in a bowl of water; once they dry out, the taste deteriorates rapidly. Slice them lengthwise, removing the leafy outer sheath before cooking. Some varieties have a bitter flavour (called *egumi* in Japan) which can be removed by boiling in water for half an hour, either before or after the sheath has been removed. If there is still a hint of bitterness, cook the shoots again in fresh water. If you cannot find fresh bamboo shoots, you can buy them in jars or tins instead.

Fried Bamboo Shoot
Serves 3–4

1 tender bamboo shoot, about 30cm long
2 teaspoons bicarbonate of soda
1 teaspoon fenugreek seeds
4 dried red chillies
Salt
150ml vegetable oil

Rinse the bamboo shoot and chop it into small cubes. Add the bicarbonate of soda and mix well.

Fry the fenugreek seeds and chillies in a dry pan until they release their flavour and add a little salt. Pour in the oil, add the cubed bamboo shoot and continue cooking on a low heat. When the bamboo shoot dries up, add a couple of tablespoons of water and stir well. Serve piping hot.

Pisum sativum. Papilionaceae

PEA

Also known as Garden Pea, English Pea. Climbing or scrambling annual grown for seeds, pods and shoot tips. Half hardy. Value: good source of protein, carbohydrates, fibre, iron and vitamin C.

Like many legumes, peas are an ancient food crop. The earliest records are of smooth-skinned types, found in Mediterranean and European excavations dating from 7000 BC. The Greeks and Romans cultivated and ate peas in abundance and it was the Romans who were said to have introduced them to Britain. In classical Greece they were known as *pison*, which was translated in English to 'peason'; by the reign of Charles I they became 'pease' and this was shortened to 'pea' in the eighteenth century. During the reign of Elizabeth I types seen as 'fit dainties for ladies, they come so far, and cost so dear' were imported from Holland.

In England 'pease pudding', made from dried peas, butter and eggs, was traditionally eaten with pork and boiled bacon. It was obviously quite versatile, hence the nursery rhyme beginning, 'Pease pudding hot, pease pudding cold, pease pudding in the pot nine days old'. Peas were eaten dried or ground until the sixteenth century, when Italian gardeners developed tender varieties for cooking and eating when fresh. It took until the following century before this was accepted by fashionable England.

varieties

Peas are usually listed according to the timing of the crop – early, second early (or early maincrop) and maincrop types – but some descriptions refer to the pea itself, or to the pod.

Earlier varieties are lower-growing than later types, which are taller and consequently higher-yielding. Smooth-seeded types are hardy and are used for early and late crops. Wrinkle-seeded varieties are less hardy and generally sweeter.

'Petit pois' are small and well flavoured. 'Semi-leafless' peas have more tendrils than leaves, becoming intertwined and self-supporting as they grow.

'Sugar peas' or 'mangetout' – varieties of *Pisum sativum* var. *macrocarpon* – are grown for their edible immature pods; of these, the 'Sugar Snap' type are particularly succulent and sweet. Some varieties can be allowed to mature and the peas eaten.

Early Peas
'Early Onward' is a heavy cropper with large blunt pods and wrinkled seeds. **'Feltham First'** is an excellent early round-seeded variety with large, well-filled pods. **'Kelvedon Wonder'** (early or main crop) produces pods packed with delicious peas. Dwarf.

Maincrop peas
'Alderman', an old variety, produces heavy crops to 1.5m with exquisite flavour. **'Cavalier'** produces huge crops and is highly resistant to mildew. Easy to harvest, with wrinkled seeds and fusarium resistance. **'Darfon'** is a high-yielding 'petit pois' type, its pods packed with small peas. It resists downy mildew and fusarium. **'Dorian'** produces up to ten tasty peas per pod and harvests over a long period. **'Hurst Green Shaft'** is sweet-tasting and heavy-cropping, maturing over a long period. Wrinkle-seeded and downy mildew and fusarium-resistant.

Sugar peas
'Oregon Sugar Pod' is sweet and tasty; harvest as

'Early Onward'

Leaves can be harvested for their shoot tips

the peas form. Fusarium-resistant. **'Reuzensuiker'** is a compact plant, needing little support. Its pods are wide and fleshy, and very sweet. **'Sugar Snap'** produces succulent, sweet edible pods or can be grown on for peas. Very sweet. Fusarium-resistant.

cultivation

Propagation

Germination is erratic and poor on cold soils; do not sow outdoors when soil temperatures are below 10°C.

Sow earlies or second earlies successionally every 14–28 days from mid-spring to mid-summer. Avoid excessively hot times; these can affect germination.

According to the growing conditions, earlies mature after about 12 weeks, second earlies (or early maincrop) take 1–2 weeks longer, and maincrops take another 1–2 weeks longer again. As an alternative option, sow groups of earlies, second earlies and maincrops in mid- to late spring; the length of time taken for each type to reach maturity will give you a harvest over several weeks.

In mid-summer, with at least 12 weeks before the first frosts are expected, sow an early cultivar for harvesting in autumn, and where winters are mild, sow earlies in mid- to late autumn for overwintering. Cloche protection may be necessary later in the season.

Peas can be sown in single V-shaped rows 2.5–5cm deep and 5cm apart, double rows 23cm apart or broad or flat drills 25cm wide. The distance between the rows or pairs of rows should equal the ultimate plant height.

Peas can also be sown in strips 3 rows wide with the seeds 11cm apart in and between the rows and with 45cm between the strips, or in blocks 90–120cm wide with seeds 5–7.5cm apart, or in compost in guttering that is then slid into place.

Yields are higher when plants are supported. Use pea sticks made from brushwood or netting. Place supports down one side of a single row, on either side or down the centre of wide drills and around the outside of blocks of plants.

Growing

Peas are a cool-season crop, flourishing at 13–18°C, so crops will be higher in cooler summer temperatures. They do not tolerate drought, excessive temperatures or waterlogged soil. Peas should be grown in an open, sheltered position on moisture-retentive, deep, free-draining soil with a pH of 5.5–7.0.

Incorporate plenty of organic matter in the autumn or winter prior to sowing or plant where the ground was manured for the previous year's crop.

Keep crops weed-free by hoeing, hand weeding or mulching (which also keeps the roots cool and moist). Earth up overwintering and early crops to provide extra support.

Unless there are drought conditions, established plants do not need watering until the flowers appear, then, for a good harvest, they will need 22 litres/sq m each week until the harvest is complete.

Maintenance

Spring Sow early maincrops.
Summer Sow maincrops early in the season. Harvest, keep weed-free and water. Sow earlies for an autumn harvest.
Autumn Sow early overwintering crops under cover. Prepare the ground for the following year.
Winter Sow early crops under cover.

Protected Cropping

Warm the soil before sowing treated seed in spring. Sow the seed of dwarf cultivars under cover in early spring. Remove the covers when the plants need supporting.

Early spring and late autumn sowings can be made under cover, as flowers and pods cannot withstand frosty nights.

Harvesting and Storing

Harvest early types from late spring to early summer and maincrops from mid-summer to early autumn. Pick regularly to ensure a high yield when the pods are swollen. Harvest those grown for their pods when the peas are just forming. If peas are to be dried, leave them on the plant as long as possible, lifting just before the seeds are shed; hang in a cool airy place or spread the pods out on trays to dry until they split and the peas can be harvested. Store in airtight containers.

Freeze young peas of any variety. Shell and blanch for 1 minute. Allow to drain, cool and freeze in polythene bags or containers. Use within 12 months.

A first-class pea support

Peas in a polythene bag in the refrigerator stay fresh for up to 3 days.

Pests and Diseases

Birds and mice can be a problem: net crops.

Pea moths are common, their larvae eating the peas. Protect with crop covers at flower bud stage. Autumn, early and mid-summer sowings often avoid problems. Use bifenthrin or organic spray or cover with fleece.

Pea thrips attack developing pods, making pods distorted and silvery; peas do not develop. Spray with pyrethrum.

Mice may also be troublesome, eating seeds, particularly with over-wintered crops; trap them.

Powdery mildew, downy mildew and fusarium wilt can be a problem: sow resistant varieties.

companion planting

Peas grow well with other legumes, root crops, potatoes, cucurbits and sweetcorn.

container growing

Sow early dwarf varieties successionally through the season in large containers of loam-based compost with moderate fertiliser levels. Provide support, keep weed-free and well watered.

medicinal

Peas are said to reduce fertility, prevent appendicitis, lower blood cholesterol and control blood sugar levels.

culinary

Garden peas are eaten fresh or dried. When small and tender, they can be eaten raw in salads. Peas are traditionally boiled or steamed with a sprig of mint. Eat with butter, salt and pepper or herbs. Serve in a cream sauce with pearl onions, with celery, orange, carrots, wine or lemon sauce. Mangetout should be boiled for 3 minutes (or steamed), tossed in butter and served. Young pea shoot tips can be cooked and eaten.

Jean-Christophe Novelli's Cappuccino Soup

Serves 4

25g butter
1 small onion, diced
60g smoked pancetta, diced
450g fresh or frozen peas
800ml light chicken stock
1 clove garlic
1 sprig fresh mint
A pinch of caster sugar
4 tablespoons whole milk
Cep powder
Salt and freshly ground black pepper

Heat the butter in large pan, add the onion and sweat gently until softened. Add the pancetta and quickly fry until crisp. Add the peas and stock, bring to the boil and cook until the peas are soft to the touch. Pour into a blender, add the garlic, mint, sugar and some seasoning and blitz until smooth.

Warm up the milk and froth up using cappuccino frother or a hand blender. Pour the soup into individual soup terrines or coffee cups. Top with frothed milk and finish with crispy pancetta and a dusting of cep powder to resemble a large cappuccino.

Pasta and Mangetout Salad

Serves 4

500g pasta – penne or fusilli
4 tablespoons olive oil
300g mangetout, topped and tailed
1 small onion, finely sliced
2 cloves garlic, crushed
250g tuna, drained and flaked
2 tablespoons thick cream
2 tablespoons chopped fresh flat-leaf parsley
Salt and freshly ground black pepper

Cook the pasta in boiling salted water for 10 minutes, or until *al dente*, and drain. Drizzle over 1 tablespoon of the oil and toss well. Allow to cool.

Meanwhile, cook the mangetout in a steamer for 2–3 minutes; they should remain crunchy. In a separate pan, heat the oil and sauté the onion for a couple of minutes, then add the garlic and continue cooking for 1 minute. Remove from the heat and allow to cool.

Put the pasta into a large serving bowl and mix in all the ingredients. Taste for seasoning and serve.

Portulaca oleracea subsp. *sativa*. *Portulacaceae*

PURSLANE

Also known as Summer Purslane. Annual grown for succulent shoot tips, stems and leaves. Half hardy. Value: rich in beta carotene, folic acid, vitamin C; contains useful amounts of essential fatty acids.

Purslane has been grown for centuries in China, India and Egypt, and is now widespread in the warm temperate and tropical regions of the world. It was once believed to protect against evil spirits and 'blastings by lightning or planets and burning of gunpowder'. Its name in Malawi translates as 'buttocks of the wife of a chief', referring to the plant's succulent, rounded leaves and juicy stems! The cultivated form has an erect habit and larger leaves than the wild species.

varieties

Portulaca oleracea subsp. *sativa* is a vigorous, upright annual growing to 45cm tall with thick, succulent stems, spoon-shaped leaves and bright yellow flowers. *P. o.* var. *aurea* is a yellow-leaved, less hardy form. It is more succulent, but has less flavour. Attractive in salads and as an ornamental.

cultivation

Propagation

Sow in seed trays indoors in late spring and transplant seedlings into modules when large enough to handle. Harden off and plant when there is no danger of frost, 15cm apart.

In frost-free climates or for later crops, sow directly, thinning to 15cm apart. Sow in late spring for a summer crop and in late summer for autumn cropping.

Growing

Easily cultivated, purslane thrives in a sunny, warm, sheltered site on light, well-drained soil. Add organic matter and sand to improve drainage if needed. Remove flowers as they appear.

Maintenance

Spring Sow protected crops, or *in situ* once the danger of frost is passed.
Summer Keep plants weed-free and water as necessary. Harvest regularly.
Autumn Cut back mature plants to allow regrowth.
Winter In late winter sow early crops under glass.

Protected Cropping

To extend the season, sow in early to mid-spring and early to mid-autumn under cover. Make earlier and late summer sowings under cover as a 'cut and come again' crop.

Harvesting

Pick young shoot tips, stems and leaves when about 3–5cm long. 'Cut and come again' crops are ready to harvest after about 5 weeks. Regular picking encourages young growth. As older plants deteriorate towards the end of the growing season, cut them back to within 5cm of the ground, water them well and they should resprout.

Pests and Diseases

Purslane is prone to slug damage, particularly when young. Damping off can be a problem if sown at low temperatures or in cold soil.

The *aurea* variety

culinary

Wash thoroughly; growing close to the ground, leaves can be gritty. It can be lightly cooked, although the taste is not memorable. Older leaves can be pickled.

Purslane Salad

Serves 4

Use young buds and stems as well as the leaves.

4 ripe nectarines or peaches
Hazelnut oil
Handful purslane
15 hazelnuts, toasted
½ teaspoon coriander seeds, crushed

Slice the nectarines or peaches and arrange on a plate brushed with hazelnut oil. Add the purslane leaves. Trickle over a little more oil, sprinkle with the chopped nuts and season with crushed coriander.

container growing

Plant seedlings or sow seed in pots or containers when there is no danger of frost, using a soil-based compost with a low fertiliser content. Continue to water regularly.

medicinal

A traditional remedy for dry coughs, swollen gums and, infused in water, for blood disorders. Research indicates that its high levels of fatty acids may help to prevent heart attacks and stimulate the body's immune system.

warning

Expectant mothers and those with digestive disorders should not eat purslane in large quantities.

Raphanus sativus. Brassicaceae

RADISH

Annual or biennial grown for edible swollen roots, seed pods and leaves. Hardy. Value: low in calories, moderate vitamin C, small amounts of iron and protein.

Thought to be native to Asia, yet domesticated in the Mediterranean, this reliable little salad vegetable has been in cultivation for centuries. Depicted in the pyramid of Cheops, it was cultivated by the Egyptians in 2780 BC and Herodotus noted that labourers working on the pyramids received 'radishes, onions and garlic' as their rations. By 500 BC it was grown in China, reaching Japan 200 years later. Pliny records that 'models of turnips, beetroots and radishes were dedicated to Apollo in the temple at Delphi, turnips made of lead, beets of silver and radishes of gold,' while Horace wrote of 'lettuces and radishes such as excite the languid stomach.' John Evelyn, too, wrote: 'Radishes are eaten alone with salt only, as conveying their pepper in them.' The fiery flavour is due to the presence of mustard oil. Although radishes are usually red, there are also black-, purple-, yellow- and green-skinned types.

A bunch of organic radishes, including 'Long White Icicle'

varieties

The fast-growing salad types are ready to harvest in about 4 weeks. Larger, slow-growing types, often with long cylindrical roots, include large overwintering varieties and the oriental varieties known as 'mooli' or 'daikon'.

Salad and overwintering radishes

'Cherry Belle' is round, with crisp, white, mild flesh. Tolerant of poorer soils, it is slow to go woody and keeps well. **'China Rose'** is a well-flavoured winter variety with bright pink roots and white flesh. **'French Breakfast 3'** is long, mild, sweet and tender. Harvest at maturity or it becomes hot and woody. **'Hong Vit'** is mild with pink stems. **'Long Black Spanish'**, a winter variety, has wonderful dark skin and is extremely hot. **'Mantang Long'** has green and white skin, magenta flesh and a sweet, nutty flavour. **'München Bier'** is grown for its tasty green pods, which are eaten raw or stir-fried. **'Saisai'** is an oriental leaf radish that came top in a taste test; tender, juicy and succulent – use as young leaves in salads and stir-fries. **'Scarlet Globe'** is justifiably popular for its mild flavour and good quality; can be sown early under cover. **'Short Top Forcing'** is a bright red variety, excellent for winter sowing under cover.

Mooli or daikon

'Mooli' and **'daikon'** are general terms for a group of long, white radishes (*Raphanus sativus* var. *longipinnatus*) that need cool temperatures and short day lengths to flourish.

'April Cross' is crisp, juicy and mild. **'Long White Icicle'** is tender with a pungent, almost nutty taste. **'Minowasa Summer'**, a Japanese variety, is long and mild. **'Tsukushi'** is a hardy, spring cross, white, with firm roots. Sow from late June for winter harvest.

cultivation

Propagation

Radishes are one of the easiest and quickest vegetables to grow. Sow successionally every 2 weeks from when the soil becomes workable, in early to mid-spring, to early autumn. Sow thinly in drills 1cm deep and 15cm apart, thinning to 2.5cm apart about 10 days after they appear. Alternatively, broadcast seed and thin to 2.5cm apart. Radishes dislike being overcrowded.

Sow overwintering radishes in summer, 2cm deep in rows 23–30cm apart; depending on the cultivar, thin to 15–23cm. Sow mooli/daikon from mid- to late summer.

Small radishes can be grown as 'cut and come again' seedlings. Harvest when seedlings are 5–7.5cm tall; if you allow them to grow to 20–22.5cm, the leaves can be cooked like spinach.

Growing

As radishes are a cool-weather crop, grow earlier and later crops in an open site, but plant summer crops in light shade, surrounded by taller plants. They flourish in a light, moisture-retentive, free-draining soil which was manured for the previous crop, with a pH of 6.5–7.5. Dig the ground thoroughly before preparing the seedbed and remove any stones, particularly when growing longer-rooted varieties.

Rake a slow-release granular fertiliser into the seedbed at 30g/sq m before sowing the first crops and before planting winter varieties.

Rapid growth is essential for tasty, tender roots, so supply plenty of water and do not let the seedbed dry out. Overwatering encourages the production of leaves, rather than roots, and erratic watering causes roots to become woody or split. During drought, water weekly at 11 litres/sq m. Hoe and hand-weed regularly.

Maintenance

Spring Sow crops outdoors when soil conditions allow.
Summer Sow successionally; keep crops well watered and weed-free. Sow winter crops.
Autumn Grow later crops under cover.
Winter Harvest overwintering crops. Sow early crops under cover.

Protected Cropping

Grow early and late crops of summer varieties under cloches to extend the cropping season. Ventilate and water thoroughly.

Grow summer crops under floating cloches to protect them.

container growing

Radishes are easily grown in containers. These should be 15–30cm wide by 20cm deep. Use a loam-based compost or free-draining, moisture-retentive garden soil. Water well and liquid-feed with general fertiliser if necessary. They can also be sown in growbags, thinning until they are 2.5–5cm apart.

Harvesting and Storing

Pull immediately they mature, after 8–10 weeks, otherwise they will become woody or run to seed.

Thin the rows regularly for the best results

Later crops can be stored. Twist off leaves and store in boxes of dry sand or sawdust in a frost-free place. Overwintering types can be left in the ground and lifted as needed. Protect with straw or bracken. For ease of lifting, these can be allowed to grow over 35cm long without being coarse.

Kept in a polythene bag in the refrigerator, radishes stay fresh for about a week.

Pests and Diseases

Flea beetle, cabbage fly and slugs can be a problem.

companion planting

Radishes grow well with chervil, peas and lettuce and thrive with nasturtium and with mustard.

Because of their rapid growth, radishes make an excellent 'indicator crop'. Sown in the same row as slow-germinating crops like parsnips or parsley, they mark where the maincrop has been sown, make weeding easier, and can be harvested without disturbing the developing plants.

medicinal

Radishes can be eaten to relieve indigestion and flatulence, as well as being taken as a tonic herb and an expectorant.

RADISH

culinary

Radishes are usually eaten raw, whole, grated or sliced into salads. Alternatively, wash and trim the root, remove the leaves and all but the bottom 2.5cm of stalk to use as a 'handle', and enjoy it with rough bread, creamy butter, salt, cheese and a pint of good ale! (In Germany, you'll find them on the bar instead of peanuts.) They can also be sliced and used instead of onions in hamburgers.

Seedling leaves are eaten raw and older leaves are cooked like spinach.

Long white summer radishes and winter radishes can be eaten raw but are usually cooked and added to casseroles, stews or curries.

Winter varieties are also pickled. Peel, then cube or slice in slightly salted water for about 10 minutes. Drain thoroughly and serve tossed in butter. Thin slices of winter varieties or larger summer types can be stir-fried.

For extra crispness, put them into a bowl of water with a few ice cubes for a couple of hours.

Slice radishes and tangerines, mandarins or oranges into small pieces, sprinkle lightly with salt, chopped fennel leaves and lemon juice. This interesting recipe is an acquired taste!

To make a 'radish rose', remove the stalk and make a number of vertical cuts from the stalk almost to the root. Place in iced water for 30 minutes; the 'petals' will open.

Immature green seed pods can be eaten raw, cooked or pickled. Pick when crisp and green. Scrape or peel and 'top and tail' mooli or daikon before use. In India they are eaten cooked or raw. In Japan shredded mooli is a traditional accompaniment to *sashimi* (raw fish) and is eaten with *sushi*.

As with summer radishes, a thin slice makes an excellent and crunchy mustard substitute with a roast beef sandwich.

Radish and Scallop Soup
Serves 4

2 bunches radishes, topped and tailed
10 scallops, trimmed
2 spring onions, chopped
2 tablespoons butter
600ml fish stock and milk combined
2 bay leaves
Pinch cayenne pepper
1 tablespoon chopped fresh parsley
4 tablespoons double cream

First wash the radishes thoroughly and cut into small dice. Then cut the scallops in half and set to one side.

In a soup pot, cook the onions in the butter until soft, then add the scallops. Cook them for 30 seconds on each side over a gentle heat then add the radishes.

Pour over the stock and milk mixture. Add the bay leaves, cayenne and parsley and cook for 15 minutes over a gentle heat, just simmering.

Remove the soup from the heat. Remove the scallops from the soup, using a slotted spoon; cut them into slivers and return them to the pan. Lastly, stir in the parsley and the double cream and serve immediately.

Rheum x hybridum. Polygonaceae

RHUBARB

Also known as Pieplant. Large perennial herb grown for pink edible stems. Half hardy. Value: very low in calories, contains small amounts of vitamins.

The earliest records of rhubarb date from China in 2700 BC and there are references to its cultivation in Europe in the early 1700s. It was originally grown for its medicinal use as a powerful purgative; the annual value of imports to England for this purpose was once estimated to be £200,000. It was first mentioned as a food plant in 1778 by the French, for making tarts and pies. Rhubarb was introduced to Maine around 1790; from there it spread to New England and Massachusetts. Forcing and blanching were discovered by chance at the Chelsea Physic Garden in 1817 after crowns were covered in debris when a ditch was cleared!

varieties

A greater range of cultivars is to be found in specialist nurseries. **'Early Champagne'** (**'Early Red'**) produces long, delicious scarlet stalks that are excellent for winemaking and early forcing. **'Fulton's Strawberry Surprise'** is vigorous and bright red and won joint first prize in an RHS taste test. **'Merton Foremost'** came second, **'Grandad's Favorite'** (vigorous, thick, sweet stems) was third and **'Stein's Champagne'** was fourth; bright red, vigorous stems – very tasty. **'Glaskin's Perpetual'** is vigorous, tasty and crops over a long period. **'The Sutton'**, introduced by Suttons Seeds in 1893, is tasty and good for forcing. **'Timperley Early'** is very early, vigorous and bred for forcing. **'Valentine'** is hardy and vigorous with tender, rose-coloured stalks. Perfect for pies and jams, it has a wonderful flavour. **'Victoria'**, a reliable old variety, is variable in size. Harvest from late spring.

cultivation

Propagation

Rhubarb can be grown from seed, but the results are invariably poor. It is better to lift and divide mature crowns of known varieties or buy virus-free plants from a reputable nursery.

Plants should be divided every 2 or 3 years; if plants are any older, take divisions from the outer margins. Lift dormant crowns in winter after the leaves have died back and divide with a spade or knife. Each 'set' should be about 10cm across, with plenty of fibrous roots and at least one bud.

Plants tend to establish more rapidly if transplanted into 25cm pots of multi-purpose compost for 3 months prior to planting out. Transplant, 75–90cm apart, from late autumn (the best time) to early spring, with the bud tip covered by 2.5cm of soil. Plant cultivars which have large buds with the buds slightly above ground to prevent rotting. Plants remain productive for several seasons; their decline is marked by the production of masses of thin stalks.

Sow in drills about 2.5cm deep and 30cm apart, thinning to 15cm. Plant out the strongest in autumn or the following spring. Eating quality will be variable.

Growing

Rhubarb flourishes in an open, sunny position in deep, fertile, well-drained soil with a pH of 5.0–6.0. It is ideal for cool temperate climates. It is very hungry, with deep roots, so ensure that the soil contains well-rotted manure or compost. On very heavy soils, plant on ridges or raised beds.

Mulch plants every winter with a good thick layer of well-rotted compost or manure. Do not allow them to flower unless you wish to save the seed, as this affects cropping the following year. Keep weed-free and watered, removing dead leaves instantly. In early spring, scatter a balanced general fertiliser around the crowns.

'Timperley Early'

RHUBARB

Maintenance
Spring Force early crops under a bucket or similar.
Summer Harvest stems.
Autumn When stems die back, remove all plant debris.
Winter Mulch with well-rotted compost or manure.

Protected Cropping
For early crops lift a few crowns in late autumn, leave them above ground and let them be frosted, then bring indoors into a cool place for forcing. Put in a large container packed with soil, or plant under greenhouse staging. Exclude light with an upturned box or bucket 38–45cm high to allow for stem growth. They can also be forced in bin liners. Keep compost moist. Dispose of exhausted forced crowns after harvest.

Alternatively, from late winter, cover *in situ* with a 15cm layer of straw or leaves or with an upturned bin, bucket or blanching pot covered with straw or strawy manure. Harvest in early to mid-spring. Do not harvest from a crown for at least 2 years after forcing.

Harvesting and Storing
Do not harvest until 12–18 months after planting, taking only a few 'sticks' in the second season and more in later years. Cropping can last from early spring to mid-summer. To harvest, hold the stems near the base and twist off. Avoid breaking the stems, as it can cause fungal problems. Do not over-pick; it can weaken the plant.

To freeze, chop the stems into sections and place on an open tray, then freeze for 1 hour before packing into polythene bags. This prevents the sections from sticking together. They can be stored for up to a year.

Pests and Diseases
Honey fungus may appear as white streaks in dead crown tissue; brown toadstools appear round the base. Dig out and burn diseased roots. Crown rot damages terminal buds and makes stems spindly. Dig out and burn badly infected plants. Do not replant in the area. Virus disease has no cure. Dig up and burn.

companion planting

Rhubarb is reported to control red spider mite. A traditional remedy suggests putting rhubarb in planting holes to control clubroot. An infusion of leaves is effective as an aphicide and to check blackspot on roses.

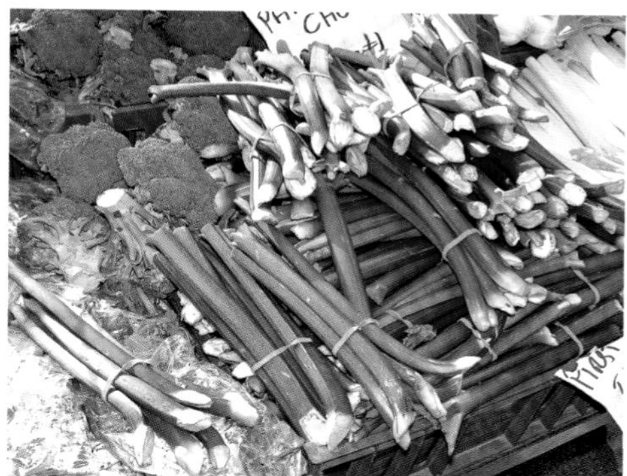

Main course or dessert?

medicinal

Rhubarb is an astringent, stomachic and potent laxative. Dioscorides recommended it for chest, stomach and liver complaints, and ringworm. By the sixteenth century, in western Europe, it was taken as an infusion with parsley as a cure for venereal disease.

warning

Do not eat the leaves, which are extremely poisonous!

culinary

Forced rhubarb is tender and needs less sugar. Cook stems slowly with sugar. Very little or no water is required; do not overcook them. Do not use aluminium pans.

Rhubarb can be stewed for fruit pies, bottling or preserving, for fools, mousses and rhubarb crumble, and is delicious served with duck. The flavour can be improved by adding orange juice, marmalade or cinnamon. It can also be puréed with apple. Claudia Roden's *Middle Eastern Food* demonstrates that it is unexpectedly wonderful stewed with beef or lamb in Persian *khoresh*.

Preserved Rhubarb

3.5kg rhubarb
3.5kg preserving sugar
Zest and juice of 2 lemons
50g blanched almonds

Cut the rhubarb into 2.5cm lengths and cook gently in a preserving pan until the juices start to run. Add the sugar, lemon juice and zest and the almonds. Stir until the sugar dissolves, then boil until a good colour and thickened. Pot up into sterilised jars and seal.

Rhubarb Sorbet
Serves 4

400g rhubarb
150g caster sugar
Juice of ½ a lemon

Cut the rhubarb into 2.5cm lengths and put into a heavy-bottomed pan. Add 50ml water. Warm gently until the juices run, then stir in the sugar and lemon juice and simmer, covered, until tender. Freeze, whisking several times as it freezes, to break up the ice crystals. If you use a sorbetière, churn until smooth. Remove from the freezer 15 minutes before serving and leave in the fridge.

Rorippa nasturtium-aquaticum. Brassicaceae

WATERCRESS

Also known as Summer Watercress. Usually aquatic perennial grown for pungent, edible leaves and stems. Hardy. Value: excellent source of beta carotene, vitamins C and E, calcium, iron and iodine.

This highly nutritious aquatic herb, a native of Europe, North Africa and Asia, has been cultivated as a salad plant since Roman times and is grown throughout the world's temperate zones. It has become a weed in North America and New Zealand. Pliny records the Latin derivation of its original generic name as *Nasus tortus*, meaning 'writhing nose' – referring to its spicy taste and pungent odour; *officinale* is often applied to plants with medicinal uses. Watercress was listed as an aphrodisiac in Dioscorides's *Materia Medica* of AD77.

It was mentioned in early Irish poetry around the twelfth century – 'Well of Traigh Dha Bhan, Lovely is your pure-topped cress,' and, 'Watercress, little green-topped one, on the brink of the blackbirds' well...' Early references to the shamrock are believed to be to watercress. Evidence to support this comes from Ireland's County Meath and Shamrock Well, the watercress of which was still remembered in the 1940s as 'the finest in the district'. Watercress was also known in Ireland as 'St Patrick's Cabbage'. The first records of commercial cultivation are from Germany, around 1750, France, between 1800 and 1811, and near Gravesend in England, around 1808.

cultivation

Found in and alongside fast-flowing rivers and streams, watercress has fleshy, glossy leaves on long stalks with 5–10 leaflets. Its long stems creep or float on the surface and root easily. Small whitish-green flowers appear in flat-topped clusters from mid-spring to early autumn.

The best watercress is grown in pure, fast-flowing chalk or limestone streams with slightly alkaline water. This avoids the risk of contamination from pollution, which can cause stomach upsets.

Propagation

Propagated watercress is from shop-bought material but the cutway **'Aqua'** can be grown from seed. Cuttings 10cm long take about a week to root in a glass of water.

If you live by a fast-flowing stream, plant rooted cuttings 15cm apart in the banks. Firm well to prevent them from being dislodged.

To grow watercress in the garden, dig a trench 60cm wide and 30cm deep, and put a 15cm layer of well-rotted farmyard manure or compost into the base. (Do not use sheep manure, as it can carry dangerous liver fluke.) Mix in a little ground limestone if your soil is not alkaline, then cover with 7.5–10cm of soil. Plant cuttings 15cm apart in mid-spring. Alternatively, in spring, mark out an area and dig in well-rotted organic matter and ground limestone, firm and soak with water, then scatter seed thinly on the surface. Water daily.

Seeds can also be sown indoors from mid- to late spring, in a propagator or trays of peat-substitute seed compost on a window sill. Cover the seeds with 3mm of compost, keep it constantly moist at around 10–15°C.

Transplant 3–4 seedlings into a 7.5cm pot when large enough to handle, then plant out 10–15cm apart from mid-spring onwards. When plants deteriorate, replace them with fresh cuttings.

Growing

Plants grown in the garden need a bright, sheltered position away from direct sunshine; never allow them to dry out, or plants run to seed. As a cool-season crop,

WATERCRESS

The leaves turn coarse after flowering

watercress grows most actively in spring and autumn, and during the winter in warmer climates. Occasional feeding with a dilute high-nitrogen liquid fertiliser or liquid seaweed may be needed. Do not grow in stagnant or still water.

Maintenance

Spring Take cuttings or sow seed mid- to late spring.
Summer Do not let compost-grown plants dry out. Keep weed-free. Harvest as needed. Remove flower heads as they appear.
Autumn Continue to harvest.
Winter Protect with cloches for continuous growth.

Protected Cropping

Cover plants with cloches, fleece or polythene tunnels before the first frosts. Make a watercress bed in an unheated greenhouse over winter, or grow in pots.

Harvesting and Storing

Younger leaves near the stem tips have the best flavour. Harvest lightly during the first season and annually towards autumn if plants are to be overwintered, cutting regularly for a constant supply of bushy shoots. Use leaves fresh or store in the salad compartment of a refrigerator for 2–3 days.

Pests and Diseases

Watercress is rarely troubled, but caterpillars of cabbage white butterfly may cause problems. Cover crops with netting.

container growing

Grow in large pots or containers of moist peat-substitute compost, with a layer of gravel in the base, spacing 10–15cm apart. Keep compost moist by standing the pot in a bowl of water which is replaced daily. Grow plants on a bright windowsill, but away from direct sunshine.

medicinal

Watercress has been valued for its medicinal qualities since antiquity. It has been eaten to cure rheumatism, and used as a diuretic and as an expectorant for catarrh, colds and bronchitis; it is a stimulant, a digestive and a tonic to promote appetite, counteract anaemia and to lower blood-sugar levels in diabetes. Externally it can be used as a hair tonic. Rubbed on the skin it is said to remove rashes. Culpeper recommended the bruised leaves or juice for clearing spots and freckles and a poultice was said to heal glandular tumours and lymphatic swellings.

Traditionally it was taken as a spring tonic. In the past, in isolated parts of the British Isles where the diet was predominantly shellfish and salt meat, it was often grown to prevent scurvy, and was so mentioned by Philip Miller in his *Gardener's Dictionary* of 1731.

warning

Gathering from streams is not recommended if sheep are grazing nearby, as there is a risk of liver fluke. Fluke can be destroyed by thorough cooking.

culinary

As with its relative, the radish, the hot, spicy taste of watercress comes from mustard oil. Remove discoloured leaves and wash thoroughly, shaking off excess water. Eat in salads and stir-fries, or liquidise to make chilled soup. It makes a perfect garnish for sandwiches. Chop finely and add to butter, mashed potatoes, dumplings or a white sauce. Also sauté in butter for 10 minutes and serve as a vegetable.

Among her many tastebud-tingling recipes, Jane Grigson suggests cutting orange segments into quarters and mixing with watercress, olive oil vinaigrette and black pepper; add walnuts or black olives and eat as a salad with ham, duck or veal.

Salmon with Watercress Sauce

Serves 6

6 salmon fillets, trimmed
4 tablespoons butter
4 tablespoons finely chopped shallots
2 large bunches watercress, plus extra to garnish
150ml double cream
Salt and freshly ground black pepper

Steam the salmon fillets, covered, on a steamer rack over boiling water until cooked; this should take 10 minutes or so.

Meanwhile, prepare the sauce by melting the butter in a heavy frying pan and sautéing the shallots until softened. Add the watercress and, constantly stirring, allow the watercress to wilt for about 2 minutes; it should retain its bright green colour. Stir in the cream and seasoning and bring to the boil. Remove from the heat and blend in a liquidiser until smooth. Then reheat gently.

Arrange the salmon fillets in the centre of individual plates and spoon sauce over each, garnishing with a little fresh watercress.

Scorzonera hispanica. *Asteraceae*

SCORZONERA

Also known as False Salsify, Spanish Salsify. Grown as perennial for shoots, flower buds and flowers; annual for cylindrical tapering roots. Hardy. Value: contains indigestible carbohydrate inulin, which, when converted to fructose in storage, increases calorific content (27 calories per 100g); small amounts of vitamins and minerals.

Scorzonera is very similar to salsify, though scorzonera is perennial, not biennial, its skin is darker, the roots narrower and the flavour is not as strong. The name scorzonera may have come from the French '*scorzon*' or serpent, as the root was used in Spain to cure snake bites. Another interpretation suggests it comes from the Italian '*scorza*', bark, and '*nera*', black, describing the roots. Native to central and southern Europe through to Russia and Siberia, scorzonera was known by the Greeks and Romans, who took little interest in its cultivation; it arrived in England by 1560 and in North America by 1806. It is widely grown in Europe as an excellent winter vegetable. The leaves have been used as food for silkworms.

varieties

'Duplex' produces long, tasty roots. **'Flandria Scorzonera'** has long roots, growing to 30cm, with strongly flavoured flesh. **'Habil'** is long-rooted with a delicious flavour. **'Lange Jan'** (**'Long John'**) has long, tapering, dark brown roots. **'Long Black'** is similar, but with black roots. **'Russian Giant'** lives up to its name, with long roots and a subtle, delicate flavour.

cultivation

Propagation

Sow fresh seed *in situ* from mid- to late spring, in drills 1–2cm deep with rows 15cm apart. Alternatively, sow 2 or 3 seeds in 'stations' 15cm apart, thinning to leave the strongest seedling when large enough to handle. Or sow in late summer for use early the following autumn.

Growing

It flourishes in a sunny position on a deep, light, well-drained soil which should have been manured for a previous crop. Do not grow on freshly manured or stony ground, as this causes 'forking'. A pH of 6.0–7.5 is ideal, so lime the soil if necessary.

On heavy or stony soils, fill a narrow trench about 30cm deep with finely sieved soil or free-draining compost, so

The roots are very tasty when cooked

SCORZONERA

Flowering scorzonera

that the roots grow straight. Dig soil deeply and rake in 60–90g/sq m general balanced fertiliser 10 days before sowing.

Remove weeds around plants by hand, as roots bleed easily when damaged by a hoe. Mulching once the roots have established helps to smother weeds, conserves moisture and reduces the risk of bolting during dry weather. Water at a rate of 16–22 litres/sq m per week.

Roots can be left in the ground to produce 'chards' (edible shoots) the following spring. In autumn cut off old leaves, leaving 1–2.5cm above the soil. Earth up the roots to a depth of about 15cm so the developing shoots are blanched during the winter. In late spring, scrape away the soil and harvest when the shoots are 12–15cm long. They can also be blanched by covering to a similar depth with straw or leaves in early spring. Roots too small to harvest in the first year can be left to mature the following year.

Maintenance

Spring Sow thinly, from mid- to late spring. Thin when large enough to handle.
Summer Keep crops weed-free and water thoroughly as needed to keep soil moist.
Autumn Leave roots in the ground and lift carefully with a fork as needed.
Winter Continue harvesting. Prepare the ground for the next crop.

Protected Cropping

Scorzonera is hardy, but protection with straw or cloches before the onset of severe weather makes lifting much easier.

Harvesting and Storing

Plants need at least 4 months to reach maturity and are ready to harvest from mid-autumn to mid-spring. In a good year the roots may grow to 40cm, but are more usually about 20cm long.

Roots can either be left in the ground until needed or lifted – with care, as they are easily damaged. Clean and store in boxes of sand or sawdust in a cool place. They last up to 1 week in a fridge. Mature plants flower in spring or summer of the second year. The edible buds can be harvested with about 8cm of stem.

If, while lifting, you see that your crop has many forked roots, the remaining plants can be kept for their young shoots and buds.

Pests and Diseases

Scorzonera suffers few problems but may develop 'white blister', which looks like glistening paint splashes. Affected plants become distorted. Spray with Bordeaux mixture or destroy.

companion planting

Scorzonera repels carrot root fly and the flowers attract beneficial insects.

container growing

In shallow or stony soils, plants can be grown in deep containers of loam-based compost. Water regularly.

medicinal

The name derived from the French reflects its reputation as an antidote to snake venom, '...and especially to cure the bitings of vipers (of which there may be very many in Spaine and other hot countries),' wrote Gerard in his *Herball* of 1636. This is not proven.

culinary

Roots can be baked, puréed, dipped in batter, sautéed and made into croquettes and fritters, deep-fried or served *au gratin* with cheese and breadcrumbs.

Boiling allows you to appreciate the flavour fully. They discolour when cut, so drop into water with a dash of lemon juice, then boil for 25 minutes in salted water with a tablespoon of flour added. Peel after boiling as you would a hard-boiled egg. Toss with melted butter and chopped parsley. Young 'chards' can be served raw in salads. The flower stalks and buds are cooked and eaten like asparagus.

Sautéed Scorzonera

Allow 150–175g per person of cleaned scorzonera; it should not be peeled. Chop roughly and cook in a heavy frying pan in a little extra virgin olive oil until *al dente*, turning to cook every side. Drain on kitchen paper and sprinkle with lemon juice mixed with 1 crushed garlic clove and 1 tablespoon finely chopped flat-leaf parsley.

Sechium edule. Cucurbitaceae

CHAYOTE

Also known as Choko, Chow Chow, Christophine. Vigorous, scrambling, tuberous rooted perennial, grown for edible fruit and seed. Tender. Value: 90% water; low in calories; some vitamin C.

In good conditions, this climber spreads to 15m and produces huge tubers. It originated in central America; 'chayote' comes from the Aztec *chayotl*, while in the West Indies it is called 'christophine' after Columbus, who reputedly introduced it to the islands. The pear-shaped fruits contain a single nutty-flavoured seed, much prized by cooks.

varieties

'Ivory White' is a small, pale-skinned variety.

cultivation

Chayote needs rich, fertile, well-drained soil.

Propagation

Propagate cultivars from soft tip cuttings in spring at 18°C. Alternatively, plant the whole fruit laid on its side, at a slight angle, with the narrow end protruding from the soil.

Growing

Grow on mounds 30–40cm high and 90 x 90cm apart; cover a shovel full of well-rotted manure with 15cm of soil. Lightly mulch.

Alternatively, grow on beds 3m square; dig in organic matter, plant seeds in the corners and grow vines towards the centre. Or, plant in 90cm wide ridges. Train the stems into trees, over trellising, fences, or 15cm mesh netting. Water regularly in dry weather; optimum growth is during the wet season. A day length of just over 12 hours is required for flowering.

In the humid tropics, it grows better in moderate temperatures at altitude.

Germinating chayote fruit

CHAYOTE

Fruits are even better when suspended above ground

Maintenance

Spring Sow seed or take cuttings.
Summer Feed and water.
Autumn Harvest.
Winter Store fruit for next year's crop.

Protected Cropping

In cool temperate regions, grow under glass in bright light with moderate temperatures and humidity. In warmer areas, start off indoors and plant when the danger of frost has passed. Grow in the greenhouse border or in containers.

Harvesting and Storing

In tropical climates, plants last for several years, fruiting from 3 to 4 months after sowing, all year round. Harvest by cutting the stalk above the fruit with a knife. Fruit reaches its maximum size 25–30 days after fruit set. It will keep up to 3 months in a cool place.

Pests and Diseases

Root knot nematode causes wilting; powdery mildew can appear on leaves and stems; and red spider mite affects plants grown under glass.

container growing

Propagate in spring from seed or cuttings. Pot on into loam-based compost with a high fertiliser content, adding well-rotted manure and grit to improve drainage. Water thoroughly, feed fortnightly with general liquid fertiliser and, once established, with a high-potash fertiliser. Train the growth on to a trellis.

medicinal

Chayote is said to be good for stomach ulcers. It contains some trace elements.

culinary

This versatile vegetable can be made into soups, boiled, candied, puréed (spiced with chilli powder) or added to stews, curries and chutneys. Its seeds can be cooked in butter, the young leaves cooked like spinach, and the tuber eaten when young. Its flesh stays firm after cooking. It makes a good substitute for avocado in a salad and is ideal for those on a diet.

For a stuffing, try a well-flavoured bolognese sauce; add boiled chayote flesh and stuff back into halved chayote shells. Sprinkle with Cheddar and bake in an oven preheated to 180°C/350°F/gas mark 4, for 30 minutes.

Chayote in Red Wine

Serves 6

Jane Grigson gives a recipe for this pudding.

150g sugar
300ml water
6 pear-sized chayotes, peeled and left whole
150ml red wine
5cm cinnamon stick
4 cloves
Lemon juice
Whipped cream and icing sugar

Use a pan that will hold the chayote in a single layer. Put the sugar and water on to dissolve and simmer for 2 minutes. Carefully add the chayote, then the wine and spices.

Cover and simmer until tender. Remove the chayote to a bowl and arrange upright, like pears. Reduce the liquid until syrupy and add a little lemon juice to bring out the distinctive flavour. Strain the juice over the chayote and serve with whipped cream lightly sweetened with icing sugar.

Solanum melongena. Solanaceae

AUBERGINE

Also known as Brinjal, Garden Egg, Eggplant, Guinea Squash, Pea Aubergine. Short-lived perennial grown as annual for fruits. Tender. Value: small amounts of most vitamins and minerals; very low in calories, with 3% carbohydrates and 1% protein.

This glossy-skinned fruit was known to sixteenth-century Spaniards as the 'apple of love'. In contrast, many botanists of the time called it *mala insana* or 'mad apple', because of its alleged effects. The Chinese first cultivated aubergines in the fifth century BC and they have been grown in India for centuries, yet they were unknown to the Greeks and Romans. Moorish invaders introduced them to Spain and the Spaniards later took them to the New World. 'Aubergine' is a corruption of the Arabic name *al-badingan*.

varieties

Fruits vary in shape from large, purple-skinned types to small, rounded white fruits 5cm in diameter. Most modern F1 hybrids are bushy and grow about 1m tall. Unlike older varieties, they are almost spineless.

'Black Beauty' produces high yields of good-quality dark purple fruits over a long period. **'Calliope'** is a dense, spineless plant. The white-skinned, purple striped fruits reach 10cm at maturity. **'Mohican'** and **'Baby Rosanna'** grow golf-ball-sized fruit and are ideal for containers in the greenhouse or on the patio. **'Diamond'**, from Ukraine, has dark purple fruit and flourishes in short summers and cool climates. The white and lavender-striped fruit of **'Listada de Gandida'** are very tasty. **'Long Purple'** produces good yields of dark purple fruits about 15cm long. **'Moneymaker'** is a superb early variety with very tasty fruits. Tolerant of lower temperatures, it can be grown indoors as well as outside. **'Ova'** produces masses of small, white-skinned fruits. **'Pintung Long'** is one of the best – its ornamental, dark lavender fruits grow to 30cm long and are tender and full of flavour. **'Rossa Bianca'**, a classic gourmet variety, produces white, violet-tinted fruit with a mild, creamy taste. **'Rosita'** bears gorgeous glossy, lavender pink fruit that taste sweet. **'Striped Toga'** is unusual for its orange, green-striped fruit that are only 7.5cm long at maturity. **'Thai Green'**, with long, slender fruit, is renowned for its drought tolerance. **'Thai Green Pea'** is a tall plant covered with masses of tiny green fruit with a strong aubergine flavour. Ideal for stir-fries, curries and soups. The fruit of **'Thai Yellow Egg'** are a bright golden yellow and egg-sized; perfect for Thai dishes. **'Thai Long Green'** grows highly ornamental slim, light lime-green fruit to 30cm long: a gourmet variety; tender and absolutely delicious. **'Violette di Firenze'** needs warmth to ripen fully. Its unusual yet very attractive dark mauve fruits make this an ideal plant for growing in a 'potager'.

'Listada de Gandida'

cultivation

Aubergines need long, hot summers, and are the ideal crop for warm climates. In cooler regions, they will grow outside, but better harvests come from those protected indoors. Constant temperatures between 25–30°C with moderate to high humidity are needed for optimum flower and fruit production. Below 20°C growth is often stalled.

Propagation

Temperatures of 15–21°C are needed for good germination. Sow seed in early spring, in trays, pots or modules of moist seed compost in a propagator or warm glasshouse or on a windowsill. Soaking seed in warm water for 24 hours before sowing helps germination. When 3 leaves appear, pot on plants grown in seed trays into 5–7.5cm pots, repotting as required until they are ready to plant outdoors or under cover. If you are growing aubergines outdoors, sow seeds 10–12 weeks before the last frosts are expected.

Growing

Aubergines flourish in a sunny, sheltered position on fertile, well-drained soil. Before planting, fork in a slow-release general fertiliser at 30–60g/sq m, improve the soil with the addition of organic matter and, in cooler climates, warm the soil before planting and leave the cloches in place until the plants are established for 2–3 weeks, allowing them to 'harden off' before removing the cloches.

Space plants 50–60cm apart. 'Pinch out' the growth tip when plants are 40cm tall, or 23–30cm for smaller varieties. Stake the main stem or support branches with string, if necessary, as fruit begins to mature. Mulch outdoor plants to conserve moisture and suppress weeds.

Feed plants with a liquid general fertiliser until they are established, then water with a high-potash liquid feed

AUBERGINE

Beautiful in bloom and fruit

every 10 days, once the first fruits are formed. When the flowers begin to open, a light spray with tepid water helps pollination; for large, high-quality fruits, allow only 4–5 to form on each plant, after which any new side shoots should be removed.

Maintenance
Spring Sow seeds under glass.
Summer Once frosts are over, transplant outdoors. Retain 4–6 fruits per plant, harvesting as they mature.
Autumn Protect outdoor crops from early frosts.
Winter Order seed for the following growing season.

Protected Cropping
When growing plants indoors, mist them regularly with tepid water or 'damp down' the paths on hot days. Keep the compost moist throughout the growing season, but take care to avoid waterlogging.

Harvesting and Storing
Harvest when the skin is shiny: overripe fruits have dull skin and are horribly bitter. Using a knife, cut the fruit stalks close to the stem. They will keep for 2 weeks in a cool, humid place or in a refrigerator.

Pests and Diseases
Aubergines are susceptible to the typical problems of crops grown under glass.

Check plants for aphids, whitefly and red spider mite. Powdery mildew can stunt growth, and in severe cases leaves become yellow and die. Verticillium wilt turns lower leaves yellow; plants wilt, but recover overnight.

companion planting

Aubergines flourish alongside the herbs thyme and tarragon, and peas.

container growing

Aubergines can be grown in 20–30cm pots or in growbags. Keep temperatures around 15–18°C; water regularly, keep the compost moist and feed with a half-strength high-potash fertiliser every other watering.

medicinal

In Indian herbal medicine, white varieties are used to treat diabetes and as a carminative. The Sanskrit *vatin-ganah* means 'anti-wind vegetable'. *Kama Sutra* prescribes it in a concoction for 'enlarging the male organ for a period of 1 month'. Neither claim is proven!

warning

Always remove the fruit's bitter principle, as it irritates the mucous membranes.

culinary

The large 'berries' contain a bitter principle in the flesh: slice large varieties, sprinkle with salt and leave for 30 minutes before rinsing. Small and newer varieties do not need this treatment. Rub cut surfaces with lemon juice to prevent discoloration.

Aubergines can be made into soups, puréed, stewed, stuffed, fried and pickled. Slices can be dipped in batter to make fritters, or drizzled with olive oil and grilled or roasted. In the Middle East the skin is burned off over a naked flame, giving the flesh a smoky flavour.

In Provence, the vegetable stew *ratatouille* is made from aubergines, garlic, peppers, courgettes, onions and coriander seeds, all cooked in olive oil. The Greek *moussaka* contains minced meat and aubergines. In the Caribbean, small white varieties are stewed in coconut milk and sweet spices. Oriental aubergines have a sweetness that does not suit European cooking; use them in stir-fries.

Sautéed Aubergines with Mozzarella
Serves 4

This dish goes well with plain meats.

4 small, long, thin aubergines
1 clove garlic, crushed
1 tablespoon chopped parsley
Salt and freshly ground black pepper
50g toasted breadcrumbs
3 tablespoons olive oil
100g buffalo mozzarella, cut into 5mm slices

Cut the aubergines in half lengthwise. Score the flesh deeply, but do not cut the skin. Arrange in a shallow pan, skin side down. Mix the garlic, parsley, salt and pepper, the breadcrumbs and half the olive oil and press this into the scored aubergines. Drizzle over the rest of the oil and place in a preheated oven (180°C/350°F/gas mark 4) until tender, about 20 minutes. Raise the heat to 200°C/400°F/gas mark 6 and, as the oven warms, arrange the mozzarella on top of the aubergines. Return to the oven for 5 minutes. Serve this dish immediately the mozzarella melts.

Solanum tuberosum. Solanaceae

POTATO

Also known as Common Potato, Irish Potato. Perennial, grown as annual for edible starchy tubers. Half hardy. Value: rich in carbohydrates, magnesium, potassium; moderate amounts of vitamins B and C.

'Then a sentimental passion of a vegetable fashion must excite your languid spleen. An attachment à la Plato for a bashful young potato or a not too French French bean.'
Sir William Gilbert (1836–1911)

The world's fourth most important food crop after wheat, maize and rice, the potato is a nutritious starchy staple grown throughout temperate zones. Hundreds of varieties have been developed since 5000 BC, when potatoes were first cultivated in Chile and Peru. The name derives from *batatas*, the Carib Indian name for the sweet potato, or from *papa* or *patata*, as it was called by South American Indians. The Spaniards introduced potatoes to Europe in the sixteenth century and Sir John Hawkins is reputed to have brought them to England in 1563. Extensive cultivation did not start until Sir Francis Drake brought more back in 1586, after battling with the Spaniards in the Caribbean. Sir Walter Raleigh introduced them to Ireland and later presented some to Elizabeth I. Her cook is said to have discarded the tubers and cooked the leaves, which did not help its popularity!

In England and Germany, potatoes were considered a curiosity; in France they were believed to cause leprosy and fever. However, in 1773 the French scientist Antoine Parmentier wrote a thesis extolling the potato's virtues as a famine food after eating them while a prisoner of war in Prussia. He established soup kitchens to feed the malnourished; potato soup is now known as *Potage Parmentier*, and there is also *Omelette Parmentier*. He created 'French fries', which were served at a dinner honouring Benjamin Franklin, who was unimpressed; it was Thomas Jefferson who introduced French fries to America at a White House dinner.

Parmentier presented a bouquet of potato flowers to Louis XVI, who is said to have worn one in his buttonhole. Marie Antoinette wore them in her hair, which made it highly fashionable. By the early nineteenth century, the potato had become a staple in France.

Ireland's climate and plentiful rain produced large crops. Potatoes were propagated from small tubers which were passed from one household to another, so the whole crop came from a few original plants. These were susceptible to potato blight, and devastating crop failure in the 1840s caused the death of more than 1.5 million people. Almost a million others emigrated to North America. Without the potato famine, John Kennedy and Ronald Reagan may never have been presidents of the United States.

During the American Civil War, potatoes were sent to the prisons and front lines. By eating the potatoes in their skins, soldiers received adequate supplies of vitamin C. The common name 'spud' came from a tool which was once used to weed the potato patch.

varieties

Potatoes are classified as 'first early', 'second early' (or 'mid-crop') and 'maincrop' varieties. Early and second early varieties grow rapidly, taking up less space for a shorter time than maincrops, so are better for small gardens. Yields are usually lower. They are also unaffected by some of the diseases afflicting maincrops. Second earlies are planted about a month after earlies. Maincrop are for immediate consumption or winter storage.

Potatoes come in a huge range of shapes, sizes, colours and textures. The skin may be red, yellow, purple or white and the flesh pale cream or yellow,

Do not eat the leaves!

POTATO

'Maxine'

mottled or blue. Their texture may be waxy or floury and shapes, variously, knobbly, round and oval. It is a wonderful time to grow potatoes. Hundreds of heritage varieties are available in catalogues; many from smaller, specialised producers and breeders are aiming to create more blight-resistant varieties. Grow some new varieties each year alongside your favourites for taste and performance. Use them or lose them!

'Arran Pilot' is an old favourite and a reliable cropper with tasty, firm, waxy flesh. **'Belle de Fontenay'** is a very old and rare French early variety that is excellent for salads. The yellow tubers are small and kidney-shaped with a waxy texture and fine flavour. **'Cara'** a late maincrop with pinkish-red tubers, is good for baking and very disease-resistant. **'Charlotte'** is a delicious second early salad variety with waxy flesh and a superb flavour hot and cold. **'Desirée'** is a popular maincrop with pink skin and pale yellow flesh that is good for chips and baking: crops well on most soils, but prefers medium to heavy; susceptible to mosaic and common scab. **'Golden Wonder'**, a late maincrop with floury, yellow flesh, is ideal for crisps and good for baking but usually disintegrates when boiled. It grows well in moist, humid climates, is resistant to scab but susceptible to slug damage and drought. **'Harlequin'**, a cross between **'Charlotte'** and **'Pink Fir Apple'**, with pale yellow flesh, is an early maincrop salad potato with a delicious flavour – a taste test winner, both hot and cold. **'Highland Burgundy Red'** is an early maincrop with burgundy flesh and a mild taste. **'Maris Bard'** is a very early first early, with white skin and white waxy flesh. It is high-yielding and good under cover. Excellent quality and good virus resistance. **'Maris Piper'** is a prolific second early with waxy flesh when cooked. **'Mayan**

'Maris Piper'

Gold' is an early maincrop, *'phureja'*-type with different origins to traditional potatoes. Yellow-skinned with oval tubers, golden flesh and a nutty flavour and creamy but dry texture, it is good for chips, roasting and pie toppings but not for roasting. **'King Edward'**, another famous high-yielding, good-quality maincrop, is good for baking but susceptible to blight, wart disease and drought. **'Lady Christl'** is an excellent good-textured first early with good disease resistance. **'Maxine'** is a maincrop that produces high yields of large, round, red-skinned tubers. The white waxy flesh remains firm when cooked; excellent for French fries. **'Mimi'**, a first early, produces tiny tubers that are superb in salads. Ideal for containers. First early **'Pentland Javelin'** produces high yields of oval, white-skinned tubers with white waxy flesh; resistant to common scab and golden eelworm. **'Pink Fir Apple'** is a wonderful old late maincrop. The unusual elongated tubers are pink-skinned with pale yellow flesh. Remaining firm when cold, they are good in salads and make good chips. **'Ratte'**, a French classic early maincrop with a nutty flavour, is good for steaming. **'Red Duke of York'**, a first early with delicious yellow flesh, is a good all-rounder with a long harvesting season. **'Sarpo Axona'** is similar to **'Sarpo Mira'** but with more regular-shaped tubers and creamier flesh. **'Sarpo Mira'** is heavy cropping with floury tubers that are good 'all rounders' in the kitchen. It is also tolerant of a range of soils, not affected by slugs and 100per cent blight resistant. (Both late maincrops.) **'Sharpes Express'**, one of the best first early potatoes, has a wonderful flavour and is good for chips and new potatoes. **'Wilja'** is a high-yielding second early with long white tubers and pale yellow waxy flesh. Good for salads and excellent for cooking. Resistant to blight, it also has some resistance to scab and blackleg.

cultivation

Propagation

Potatoes are normally grown from small tubers known as 'seed potatoes'. Buy 'certified' virus-free stock from a reputable supplier to be certain of obtaining a good-quality crop.

First earlies are 'sprouted' (or 'chitted') about 6 weeks before planting to extend the growing season, which is particularly useful for early cultivars and in cooler climates. It is worth chitting second earlies and maincrops if they are to be planted late. Put a single layer of potatoes in a shallow tray or egg box with the 'rose end' (where most of the 'eyes' or dormant buds are) upwards, then put the tray in a light, cool, frost-free place to encourage growth. At 7°C it takes about 6 weeks for shoots around 2.5cm long to form. At this stage, they are ready for planting. They can be planted when the shoots are longer, but need handling with care, as they are easily broken off. For a smaller crop but larger potatoes, leave 3 shoots per tuber on earlies; otherwise leave all of the shoots for a higher yield.

Plant earlies from mid-spring and maincrops from late spring, when there is no longer any danger of hard frosts and soil temperatures are 7°C.

Make trenches or individual holes 7–15cm deep, depending on the size of the tuber. Plant them upright with the shoots at the top and cover with at least 2.5cm of soil. Take care not to damage the shoots. Earlies can be planted a little deeper to give more protection from cooler weather. Ideally the rows should face north-south, so that both sides receive sunshine.

Plant earlies 30–38cm apart with 38–50cm between the rows and maincrops 38cm apart with 75cm between the rows. Alternatively, plant earlies 35cm apart and second earlies and maincrops 30–38cm apart in and between the rows. Spacing can be varied according to the size of the tubers and also according to their subsequent cropping potential.

While the traditional method of propagation is very common, seed is also available, which is easy to handle and produces healthy crops. Seeds are sown indoors before potting on, hardening off and transplanting. 'Plantlets' produced by tissue culture are also available; these are healthy, virus-free and vigorous. These, too, will need growing on before hardening off and then transplanting.

Cutting large potatoes in half is not recommended; nor is the method I once saw being used – hollowing the tubers to leave a thin layer of flesh, then planting the skins only!

Growing

Potatoes flourish in an open, sunny, frost-free site on deep, rich, fertile, well-drained soil. They grow better in cool seasons, when temperatures are between 16–18°C, and are tolerant of most soils.

Lighter soils are better for growing earlies. Add organic matter to sandy soils. Alkaline soils or heavy liming encourages scab. Grow resistant varieties and lime acid soils gradually to create a pH of 5–6, or cultivate in raised beds or containers. If necessary, dig in plenty of well-rotted organic matter in autumn, then rake to a rough tilth 10 days before planting, adding a general granular fertiliser at 90–120g/sq m.

Potatoes are an excellent crop for new or neglected gardens; while the crop may not be large, the root system breaks up the soil and improves its structure.

'Mimi'

POTATO

Keep crops weed-free until they are established, when the dense canopy of foliage suppresses weeds.

Potatoes are 'earthed up'; this prevents exposure to light, which makes them green and inedible, and also disturbs germinating weeds. When the plants are about 20–23cm tall, use a rake or spade to draw loose soil carefully around the stems to a depth of 10–15cm. Alternatively, begin earthing up in stages when the plants are about 10cm tall, adding soil every 2–3 weeks.

To avoid having to earth up, plant small tubers about 12.5cm deep and 23–25cm apart on level ground. Wide spacing means tubers are not forced to the surface, yet still produce a moderate crop.

Potatoes need at least 500mm of rainfall over the growing season for a good crop. During dry weather water earlies every 12–14 days at 16–22 litres/ sq in. Except when there are drought conditions, do not water maincrops until the tubers are the size of marbles (check their development by scraping back the soil below a plant). At this point, a single, thorough soaking with at least 22 litres/sq m encourages the tubers to swell, increases the yield and makes them less prone to scab.

Fresh means full of flavour

Potatoes need a constant supply of water; erratic watering causes malformed, hollow or split potatoes. An organic liquid feed or nitrogenous top dressing helps plants to become established during the early stages of growth.

Early potatoes can be grown under black polythene, which also makes earthing up unnecessary. Prepare the soil, lay a sheet of black polythene over the area, anchor the edges by covering with soil and plant your potatoes through crosses cut in the polythene. Alternatively, plant the potatoes, cover them with plastic and when the foliage appears, make a cut in the polythene and pull through.

Where early frosts and windy conditions do not occur and the ground is excessively weedy or eelworm is a problem, potatoes can be grown in a bed of compost and straw. Clear the ground and cover the soil with a good layer of well-rotted manure or compost. Space the potatoes as required and cover them with a 5–7.5cm layer of straw or hay, adding more as the potatoes grow, to a maximum of 15cm. At this point, spread a 3–4in layer of lawn mowings over the area to exclude light from the developing tubers.

Rotate earlies every 3 and maincrops every 5 years.

Maintenance

Spring Chit potatoes before planting.
Summer Keep crops weed-free and water as necessary. Harvest and use, or store.
Autumn Plant winter crops under cover. Prepare the soil for the following year.
Winter Harvest winter crops.

Protected Cropping

To advance early crops and protect them from frost, cover early potatoes with cloches or floating crop covers, anchored by burying the edges under the soil. When the shoots appear, cut holes in the plastic and pull the foliage through. After 3–4 weeks cut the cover, leaving it in place to allow the potatoes to become acclimatised.

Protect the foliage and stems ('haulm') from heavy frosts by covering them at night with a layer of straw, bracken or newspaper, or a thin layer of soil. Light frosts do not generally cause problems.

For winter crops of new potatoes, in warm areas, plant earlies in mid-summer and cover with cloches in autumn. Alternatively, grow them in the borders of a frost-free greenhouse at 7–10°C. Provide warmth if necessary.

 companion planting

Growing horseradish in large sunken pots near to potatoes controls some diseases. Plant with sweetcorn, cabbage, beans and marigolds. Grow with aubergines, which are a greater attraction to colorado beetle. Protect against scab by putting grass clippings and comfrey leaves in the planting hole or trench.

Harvesting and Storing

Earlies are ready to harvest from early to mid-summer, second earlies from late summer to early autumn and maincrops mature from early to mid-autumn. Harvest earlies when they are about the size of a hen's egg: their readiness is often indicated by the flowers opening. Remove some soil from the side of a ridge and check them for size before lifting the root. Insert a flat-tined fork into the base of the ridge and lift the whole plant, bringing the new potatoes to the surface.

Harvest those grown under black polythene by folding back the sheeting: the crop of potatoes will be lying on the surface. Collect as required. Scrape away compost from those grown in containers to check their size before harvesting them.

Leave healthy maincrop potatoes in the soil for as long as possible, but beware of slugs! In early autumn cut back the haulm to about 5cm above the soil, or wait until the foliage dies down naturally; leave the tubers in the ground for 2 weeks for the skins to harden before lifting.

Cut back the haulm and work along the side of each ridge, lifting the potatoes with a fork. Harvest on a dry sunny day when the soil is moderately moist, leave in the sun for a few hours to dry, then brush off the soil. If the weather is poor, dry them under cloches or on trays indoors. Store healthy tubers in paper or hessian sacks or boxes in a dark, cool, frost-free place. They should keep until spring. Check tubers weekly, removing those that are damaged or diseased.

When harvesting, ensure that all potatoes, however small, are removed from the soil, to prevent future

POTATO

problems occurring with pests and diseases. New potatoes do not store, but can be frozen: blanch whole in boiling water for 3 minutes; cool, drain, pack into rigid containers and freeze. Chips and French fries can also be frozen.

Pests and Diseases

Common potato scab shows as raised, corky scabs on the surface of the tuber. It does not usually affect the whole potato and can be removed by peeling. It is common in hot, dry summers, on light, free-draining and alkaline soils. Water well in dry conditions, add plenty of organic matter to the soil before planting, avoid excessive liming, or rotating after the soil has been limed for brassicas. Do not put infected potatoes or peelings on the compost heap. Grow resistant cultivars.

Potato blight appears on the leaves as brown patches, often with paler margins. The infection can spread to the stems and through the tuber, making it inedible. The disease spreads rapidly in warm, humid conditions and maincrops are more susceptible. Grow resistant cultivars and avoid overhead watering; before the problem appears, apply a systemic fungicide every 2 weeks from midsummer. Alternatively, use copper sulphate sprays such as 'Bordeaux mixture'. Earthing up creates a protective barrier, slowing the infection of tubers. Lift early potatoes as soon as possible and in late summer, cut back the haulms of infected plants just above the ground and burn the infected material or put it in the dustbin. Leave tubers in the ground for 2 weeks before lifting.

'Maxine' leaves showing the first signs of blight

With potato cyst eelworm ('golden nematode' or 'pale eelworm') growth is checked and yields can be severely reduced; badly affected plants turn yellow and die. Lift and burn infected plants, rotate crops, and do not grow potatoes or tomatoes in the soil for at least 6 years. Try to grow resistant varieties.

Wireworms are about 2.5cm long and golden brown in colour. They tunnel into the tubers, making them inedible. They are fairly easy to control. Cultivate the soil thoroughly over winter to expose to the weather and birds, keep crops weed-free, and lift maincrops as early as possible.

Blackleg shows when the upper leaves roll and wilt and the stem becomes black and rotten at the base. The tubers may be rotten. It is more severe in wet seasons; do not plant on waterlogged land. Remove affected plants immediately; burn or put in the dustbin. Do not store damaged tubers.

Slugs tunnel into the tubers; damage is more severe the longer they remain in the ground. Use biological control.

container growing

Potatoes can be grown in any container if it is at least 30cm wide and deep and has drainage holes. It is possible to buy a potato barrel for this purpose – some models have sliding panels for ease of harvesting – but an old dustbin, flower pot or similar is just as good. Potatoes can also be grown in black bin liners. Wider containers can, of course, hold more potatoes. Place a 10–12.5cm layer of compost or good garden soil in a container, stand 2–3 chitted potatoes on the surface and cover with a layer of compost about 10–15cm deep. When the shoots are about 15cm tall, cover with another 10–15cm layer of compost, leaving the tips showing. Continue earthing up until the shoots are 5–7.5cm below the rim of the container.

Winter crops of first earlies can also be grown in containers. Plant in mid-summer for harvesting at Christmas. Cover the haulms in frosty weather or grow under glass.

Be sure to keep crops well fed and watered.

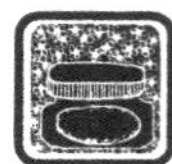

other uses

Potatoes are made into flour and turned into bread. They can be boiled and the starch turned to glucose, which can be fermented to produce strong alcohol, like the poteen made in Ireland. In the past, they have also been a source of starch powder for whitening wigs. The juice of mature potatoes is particularly excellent for cleaning silks, cotton, wool and even furniture.

medicinal

Potatoes are said to be good for rheumatism. One traditional cure for sciatica and lumbago is to carry a potato in your pocket.

The juice from a raw potato or the water in which potatoes have been boiled is said to relieve gout, rheumatism, lumbago, sprains and bruises.

Uncooked peeled and pounded potatoes are said to make a soothing plaster to scalds or burns when applied cold.

Potatoes contain little fat and provide more potassium *pro rata* than bananas, while the average potato contains as many kilojoules as most apples or a glass of orange juice and can be eaten by those on a diet.

warning

Green potatoes contain the toxic alkaloid solanine, which can cause vomiting and stomach upsets. Do not eat green potatoes. Be aware that the tomato-like fruits and leaves of the plant are also poisonous.

culinary

There are more than 500 ways of serving potatoes. They can be boiled, steamed, baked, roasted, mashed, sautéed, fried and cooked *au gratin*. In the past, potatoes have been preserved and candied or mashed with butter, sherry, egg yolks, nutmeg and a little sugar and baked until brown. Rub the skin from new potatoes under running water and boil for 12 minutes with a sprig of mint. Drain and toss in butter, then eat hot or cold in salads. Scrape old potatoes; peeling removes much of the vitamin C just below the surface.

Cut potatoes into 1cm strips, wash in ice-cold water, drain and dry thoroughly. Fry until golden in a shallow frying basket in a pan one-third full of oil at 125°C.

The difference between French fries and chips is their size and the method of cooking. French fries are very thinly sliced and deep-fried in oil. Chips are sliced thickly and were originally cooked in animal fat. Nowadays, the only difference is the size. Steaming or microwaving generally retains the flesh colour in unusual potatoes.

Mashed Potatoes with Olive Oil
Serves 4

For those on a diet, this at least has fewer calories than the traditional dish!

I kg potatoes, peeled
8–10 cloves garlic, unpeeled
8 tablespoons olive oil, preferably extra virgin
Salt and freshly ground black pepper

Put the potatoes and garlic cloves in a pan, cover with water and add salt. Bring to the boil and cook until tender, about 20 minutes, and then drain, reserving the potato cooking liquid. Boil this separately in the pan to reduce to roughly 250ml. Meanwhile peel the garlic cloves.

Mash the potatoes and garlic. Return the pan to a gentle heat and beat in the olive oil and enough potato liquid to give the potatoes the right texture. Season with salt and pepper.

To make Irish champ, add 115–175g of cooked and roughly chopped cabbage or spring greens to the above.

Rösti
Serves 4

750g waxy potatoes
1 small onion, finely chopped
50g smoked bacon, diced
6–8 tablespoons goosefat or olive oil
Salt and freshly ground black pepper

Clean the potatoes and boil them in lightly salted water until just tender. Drain, and when cool enough to handle, peel. Grate coarsely.

In a heavy pan, gently sauté the onion and bacon in half the fat. Fold this into the potatoes with the seasoning. Form the potatoes into one big 'cake'. In the same pan, heat the balance of the fat and gently slide the potatoes into the pan. Cook over a high heat until the first side is browned. Place a plate over the pan and turn the potatoes on to the plate; then gently slide the 'cake' back into the pan to allow the second side to cook. Serve piping hot. This classic Swiss dish can be eaten with Emmenthal melted on top (put the browned potatoes, covered with cheese slices, under the grill for a few minutes).

Make smaller individual rösti if you prefer

Potato Salad with Goat's Cheese Dressing

Potato Salad with Goat's Cheese Dressing
Serves 4

The South Americans have excellent ways of cooking potatoes, including this rich salad which makes a good summer first course.

750g potatoes, fairly small
½ a red pepper, deseeded and diced
1 small onion, peeled and finely chopped
2 hard-boiled eggs, peeled and diced
Handful black olives, stoned
2 tablespoons chopped fresh flat-leaf parsley

For the dressing:
175g well-flavoured goat's cheese
2 cloves garlic, peeled
2 tablespoons yogurt
2 tablespoons lemon juice
1 jalapeño pepper, deseeded
¼ teaspoon turmeric
½ teaspoon ground cumin

Cook the potatoes in their skins in boiling water until tender; drain and peel.

Cut into generous dice and put into a serving dish. While the potatoes cook, mix the dressing ingredients in a blender, puréeing until smooth. Combine the potatoes with the red pepper and onion. Add the dressing and toss. Decorate the top with the diced eggs, olives and parsley.

Spinacia oleracea. Chenopodiaceae

SPINACH

Fast-growing annual, grown for highly nutritious leaves. Hardy. Value: high in iron, beta carotene and folic acid; rich in vitamins A and C.

Spinach is thought to be native to South-West Asia. Unknown to the Greeks and Romans, it was first cultivated by the Persians, was grown in China by the seventh century AD and reached Europe around 1100, after its introduction to Spain by the Moors. The prickly-seeded form was known in thirteenth-century Germany and by the following century it was commonly grown by European monastery gardens. Smooth-seeded spinach was first described in 1522. Its first mention in England comes in William Turner's *New Herball*, in the sixteenth century: 'An herbe lately found and not long in use.' The name comes from Old French *espinache*, derived from its Arabic and Persian name.

Children were encouraged to eat spinach because of its effect on Popeye the Sailor. The idea that it contains exceptional levels of iron originated from Dr E. von Wolf in 1870. His figures went unchecked until 1937, when it was discovered that the iron content was one-tenth that claimed by him as a result of a misplaced decimal point!

Harvest when young for 'cut and come again' crops

varieties

Traditionally there are two main groups. Prickly seeded varieties, which have lobed leaves, are regarded as hardier than the round, smooth-seeded types, which cope better with higher temperatures and are used for summer cropping. Modern cultivars are generally more adaptable. **'Atlanta'** has thick, deep green leaves for spring sowing. It does not overwinter well. **'Bloomsdale'** (**'Bloomsdale Long Standing'**) is a Savoy-leaved spinach with fleshy, tasty leaves. It is cold-tolerant, slow to bolt and crops over a long period. **'Dominant'** is a good all-rounder for spring or autumn sowing: thick-leaved, prolific and resistant to bolting. **'Giant Winter'**, a robust late autumn and winter variety, withstands some frost and is highly disease-resistant. **'Medania'** is for summer cropping. It withstands hot, dry weather and is resistant to downy mildew. **'Spartacus'** is dark green, vigorous, yet slow to mature. Suitable for 'cut and come again' crops. **'Space'** has smooth, dark green leaves and is resistant to downy mildew; ideal for any season. **'Tetona'**, has bright green, rounded leaves, is slow to bolt and ideal for early and late sowing or 'baby' leaves; mildew resistant.

cultivation

Spinach grows best in cool conditions, at temperatures around 16–18°C. It thrives in a bright position, but is better in light shade and moist ground if your garden is hot and sunny.

Propagation

Soak seed overnight to speed up germination. My friend Robert Fleming wraps seed in a wet paper towel and places it in a sealed container in the fridge for 5 days before planting. Germination usually takes 5 days. As a cool-season plant, it will not germinate above 30°C. Sow summer crops every 2–3 weeks from early to late spring for a summer harvest.

Sow in late summer, for cropping in winter until early summer. Sow *in situ* 2.5cm apart and 1–2cm deep in rows 30cm apart. When large enough to handle, thin to 15–25cm. It also works as a 'cut and come again' crop.

In colder areas, it is better to sow in seed trays or modules for transplanting.

Sow thinly in broad drills the width of a hoe or in rows 10cm apart. Leave unthinned unless growth is slow, when they can be thinned to 5cm apart. Plants can be thinned to 10–12.5cm apart for larger leaves, though they are more inclined to bolt.

Growing
Spinach needs a rich, fertile, moisture-retentive soil, so, where necessary, dig in well-rotted organic matter before planting. On poor soils leaves are stunted, bitter and prone to bolt. The ideal pH is 6.5–7.5; lime acid soils.

Never allow the soil to dry out. Water regularly before the onset of dry weather and grow bolting-resistant cultivars. Water every 2 weeks with a high-nitrogen liquid fertiliser.

Maintenance
Spring Sow early crops successionally.
Summer Keep the soil moist and weed-free. Sow winter crops from late summer.
Autumn Protect late crops under cloches; sow 'cut and come again' crops under glass.
Winter Protect winter crops under cloches or fleece or sow in cold frames.

Protected Cropping
Hardy varieties for winter and early spring cropping are better protected in cold frames, cloches or under horticultural fleece.

Harvest spinach regularly

Harvesting and Storing
Harvest from 5 weeks after sowing, cutting the outer leaves first. To harvest as a 'cut and come again' crop, remove the heads 2.5cm above the ground and allow them to resprout. Pick off the leaves carefully; the stems and roots are easily damaged. Alternatively, pull up whole plants and re-sow. Use the leaves fresh, or freeze. They can be washed and stored for up to 2 days in the refrigerator.

Pests and Diseases
Aphids and slugs can be troublesome, especially among seedlings. Downy mildew appears as yellow patches with fluffy mould on the underside of the leaf. Bolting caused by drying out or hot weather is most common cause of failure.

container growing

Spinach is ideal for containers 20–45cm wide and 30cm deep. Water regularly in dry weather. Avoid scorching sunshine.

companion planting

Good with beans, peas, sweetcorn and strawberries.

medicinal

Spinach is said to be good for anaemia, problems of the heart and kidney and in low vitality and general debility. Iron is in a soluble form, so any water left after cooking can be left to cool before drinking.

Research has shown that those who eat spinach daily are less likely to develop lung cancer. Of all the vegetable juices, spinach juice is said to be the most potent for the prevention of cancer cell formation.

other uses

The water drained from spinach after cooking is said to make good matchpaper. It was used to make touchpaper for eighteenth- and nineteenth-century fireworks, as paper soaked in it smouldered well.

culinary

Wash leaves thoroughly. Eat immediately after harvesting. Use young leaves raw in salads, lightly cooked or steamed. Place spinach in a pan without adding water, cover and stand on moderate heat so the leaves do not scorch.

After 5 minutes, stir and cook over high heat for 1 minute. Drain in a colander and cut with a knife or fish slice.

Spinach can also be cooked with olive oil, cheese, cream, yogurt, ham, bacon and anchovies, besides being seasoned with nutmeg, pepper and sugar. All will bring out the flavour.

Scrambled Eggs with Raw and Cooked Spinach
Serves 6

1kg young spinach
2 tablespoons olive oil
Pinch grated nutmeg
2 tablespoons butter
12 eggs, lightly beaten with 1 tablespoon thick cream
1 tablespoon chopped fresh chives
Salt and freshly ground black pepper

Wash the spinach and remove any tough stalks. Heat the oil in a heavy pan, then wilt most of the spinach, keeping back a couple of handfuls for decoration. Season with nutmeg, salt and pepper. Set aside and keep warm.

In the same pan, melt the butter and allow to just brown. Over a low flame, pour in the eggs and season. Stir constantly until just set (overcooked scrambled eggs are bad news!). Mound the eggs in the centre of individual plates and sprinkle with chopped chives. Arrange the cooked spinach around the outside, then make a further ring from the raw spinach leaves. Serve hot.

Stachys affinis. Labiatae

CHINESE ARTICHOKE

Also known as Japanese Artichoke, Crosne. Dwarf herbaceous perennial grown as annual for edible tubers. Hardy. Value: very high in potassium.

From the same family as mint, lavender and many other herbs, Chinese artichokes are not grown for their aromatic foliage but for the small, ridged tubers at the tips of creeping underground stems. A native of Japan and China, it has rough, oval leaves and white to pale pink flowers that appear in small spikes. It was introduced from Peking to France by a physician in the late nineteenth century. It has never been hugely popular, probably because a large area is needed for decent quantities, but it is worth trying, if only as a 'novelty' crop.

cultivation

Chinese artichokes flourish in an open, sunny site on light, fertile, yet moisture-retentive soil.

Propagation

For early crops, sprout the tubers in shallow trays or pots of potting compost in a moderately warm room, then plant out when soil is warm. Increase your stock by planting tubers individually in small pots of soil-based compost with added organic matter, transplanting as required. Place in 20–23cm pots for the rest of the growing season. Feed and water regularly.

Growing

In spring, plant the tubers vertically 4–7.5cm deep; on light soils up to 15cm deep. Plant large tubers only,

Odd-looking, but edible

Lift the plants to harvest the tubers

30cm apart each way or 20cm apart in rows 40cm apart. When they are 30–60cm high, earth up round the stems to about 7.5cm. Keep weed-free at first; as the plants mature, the leaf canopy naturally suppresses weed growth. Feed with a liquid general fertiliser every 2 weeks and remove flowering spikes to concentrate energy into fattening the tubers. Trim back the foliage. Rampant growth make Chinese artichokes an ideal low-maintenance crop for a spare corner of the garden.

Maintenance

Spring Plant tubers when the soil has warmed.
Summer Keep crops weed-free, water and feed.
Autumn Harvest after the first frosts, or about 5 months.
Winter Protect crops with hessian, straw or sacking to enable harvesting to continue.

culinary

The tubers have a delicate, nutty flavour. Only about 5cm long, they are rather fiddly to cook. Boiling is simplest, but they can also be fried, stir-fried, roasted, added to soup or eaten raw in salads.

Creamed Chinese Artichokes
Serves 4

Jane Grigson, in her *Vegetable Book*, quotes this recipe from *La Cuisine de Madame Saint-Ange*.

500g Chinese artichokes
1 tablespoon butter
½ teaspoon lemon juice
300ml double cream
Salt and freshly ground white pepper
Pinch nutmeg
Double cream, to serve

Boil the artichokes for about 5 minutes, drain, add the butter and lemon juice and cook gently for 5–7 minutes. Bring the cream to the boil, then add it to the artichokes, stirring in well. Season with nutmeg and salt and pepper. Cover and leave for 15 minutes over a moderate heat, until the cream has reduced by a quarter. Just before serving, stir in 3 spoonfuls of double cream and quickly remove from the heat.

Harvesting and Storing

Harvest as required from mid-autumn to early spring, after the foliage has died back or when plants have been in the ground for 5–7 months. Lift them just before use, as they quickly shrivel. When forking through the ground during the final harvest, you should remove even the smallest tubers, or they will rapidly become weeds.

Pests and Diseases

Chinese artichokes are usually pest-free, but you should take precautions against lettuce root aphid, which can cause wilting.

Tetragonia tetragonoides (syn. T. expansa). Aizoaceae

NEW ZEALAND SPINACH

Creeping perennial, usually grown as annual for edible leaves. Half hardy. Value: high in iron, beta carotene and folic acid.

This plant was unknown in Europe until 1771, when it was introduced to the Royal Botanic Gardens at Kew by the great botanist Joseph Banks after his voyage on board the *Endeavour* with Captain Cook. By 1820 it was widely grown in the kitchen gardens of Britain and France, but it did not appear in New Zealand literature until the 1940s, and there is no evidence of its use by the Maoris. It is unrelated to the true spinach, but its leaves are used in a similar manner – hence the borrowed name – though their flavour is milder than that of true spinach. A low-growing perennial with soft, fleshy, roughly triangular, blunt-tipped leaves, about 5cm long, it makes an attractive plant for edging and ground cover. It flourishes in hot dry conditions, without running to seed.

cultivation

Propagation

The seed coat is extremely hard and germination is slow unless the seeds are soaked in water for 24 hours before sowing. Sow seeds individually, 2.5cm deep in 7.5cm pots indoors from mid- to late spring; repot as necessary. Plants are very cold-sensitive and so should be hardened off carefully before planting out once the danger of frost has passed.

Alternatively, sow seed *in situ* when there is no danger of frost, 1–2cm deep in drills about 60cm apart, thinning to a similar distance between plants. Seeds can also be sown, 2 or 3 per 'station', 60cm apart, thinning to leave the strongest. Pinch out tips of young plants to encourage bushy growth.

Plants usually self-seed and appear in the same area the following year.

Growing

New Zealand spinach flourishes in an open site on light, fertile soil with low levels of nitrogen and grows well in dry, poor conditions. It tolerates high temperatures, but is very susceptible to frost.

Keep seedlings weed-free with regular hoeing. Once plants have been properly established, their dense foliage will act as a ground cover and suppress weeds. Water regularly during dry spells to encourage vigorous growth. It does not bolt in hot weather, but should be watered well to encourage growth. On poor soils, occasional feeding with a half-strength solution of general liquid fertiliser is beneficial, but take care not to overfeed.

Maintenance

Spring Sow seed under cover for transplanting once frosts are over.
Summer Keep crops weed-free, water and harvest as required.
Autumn Protect plants to extend the growing season.
Winter Prepare the ground for planting in spring.

Protected Cropping

Protecting with cloches extends the harvesting season in cooler areas before frosts turn plants into a soggy mess.

Harvesting and Storing

The first young leaves and shoots are ready to harvest from about 6 weeks after sowing. Pick regularly to maintain a constant supply of young growth and to extend the harvesting season, which should last up to 4 months. Young leaves should be used immediately after cutting, can be stored for up to 2 days in a polythene bag in the fridge, or may be frozen. Wash, blanch for 2 minutes, allow to cool and drain before packing into polythene bags and placing in the freezer.

Pests and Diseases

New Zealand spinach is very robust but downy mildew may be a problem.

container growing

Crops can be grown in large containers of soil-based compost. Add grit to improve drainage. Water plants well throughout the growing season.

companion planting

It is sometimes grown as a 'green manure' to improve the soil structure.

culinary

The young leaves and shoots are steamed, lightly boiled and eaten with a knob of butter, or eaten raw in salads. They are better eaten fresh, as the flavour deteriorates rapidly with age.

Tetragonolobus purpureus. Papilionaceae

ASPARAGUS PEA

Also known as Winged Pea. Sprawling annual winged bean, ornamental but grown for edible pods. Half hardy. Value: small amounts of protein, carbohydrate, fibre and iron.

A wildflower of open fields and wasteland in Mediterranean countries, this plant may have arrived in Britain with imported grain. It has since become naturalised. Gerard records its cultivation before 1569 for its dark crimson, pea-like flowers; only later were its edible merits discovered, but it has never gained widespread popularity. The 1807 edition of Miller's *Gardener's Dictionary* notes: 'Formerly cultivated as an esculent plant, for the green pods…it is now chiefly cultivated in flower gardens for ornament.'

Asparagus pea in bloom

cultivation

Propagation

Sow seed outdoors from mid- to late spring, once the soil warms and frosts are over. Drills should be 2cm deep and 38cm apart; thin to 20–30cm apart once they are large enough to handle. In cooler areas, sow under cloches or in the glasshouse in trays or modules of moist seed compost.

Harden off and transplant outdoors when plants are 5cm tall, spacing them about 20–30cm apart.

Growing

Asparagus peas flourish in an open, sheltered, sunny site on light, rich, free-draining soil. Keep weed-free and water as required. As they are lax in habit, growing up pea sticks or a similar support saves space and makes harvesting easier. Rotate them with other legumes.

Maintenance

Spring Sow seed in mid- to late spring.
Summer Harvest regularly.
Autumn Harvest until early autumn.
Winter Prepare ground for the following year's crop.

Protected Cropping

In cooler areas, sow under glass for transplanting later.

Harvesting and Storing

Harvest from mid-summer to early autumn, when pods are 2.5cm long. Pick regularly to prolong the harvest. Longer, older pods are stringy and inedible. Pods that do not bend easily are unsuitable for eating. Harvesting is easier in the evening, when the leaves close and the small green pods are more visible.

Pests and Diseases

Protect crops from birds using humming line or similar bird scarers. They can also suffer from downy and powdery mildew.

culinary

The subtle flavour is similar to asparagus. Wash the pods, top and tail if necessary, then steam for up to 5 minutes (check them after about 3 minutes, as they need only light cooking). Drain well before eating. Mix with butter and cream or serve on toast. Alternatively, cook them in butter until they are tender, drain and serve.

Tragopogon porrifolius. *Asteraceae*

SALSIFY

Also known as Oyster Plant, Vegetable Oyster, Goat's Beard. Grown as biennial for shoots, flower buds and flowers; annual for cylindrical tapering edible roots. Hardy. Value: contains inulin, turning to fructose in storage; small amounts of vitamins and minerals.

It is difficult to know whether to grow this plant for its delicious roots, or to grow it as a fragile ornamental by letting it flower (the flowers are also edible). Its Latin and common names are apt; *tragos* means goat and *pogon*, beard, describes the tuft of silky hairs on the developing seedheads; *porrifolius* means 'with leaves like a leek'. 'Salsify' comes from the old Latin name *solsequium*, from the way the flowers follow the course of the sun.

Salsify is a relatively recent crop. In the thirteenth century it was harvested from the wild in Germany and France. It was not cultivated until the early sixteenth century – in Italian gardens. The English plant collector John Tradescant the Younger recorded it in 1656. It was taken to North America in the late nineteenth century.

varieties

The species *Tragopogon porrifolius* has a thin, creamy-white, tapering taproot up to 38cm long. It has narrow leaves and long stems topped by beautiful dull purple flowers. These are followed by thistle-like seedheads. **'Geante Noire de Russie'** is an extremely long, black-skinned variety with white roots and a superb flavour. **'Giant'** is consistently long-rooted with a delicate taste of oysters. **'Sandwich Island'** is a selected type with strongly flavoured roots and smooth skin.

Salsify roots aren't always easy to peel

cultivation

Propagation

Always use fresh seed; its viability declines rapidly. Sow as early as possible, from early to mid-spring. Germination can be somewhat erratic, so it is better to sow 2–3 seeds in 'stations' 1–2cm deep and 10cm apart with 30cm between the rows. After germination, when the seedlings are large enough to handle, thin to leave the strongest seedling. Alternatively, sow the seeds in drills and thin to the above spacing.

Growing

Salsify thrives in an open situation on light, well-drained, stone-free soil which was manured for the previous crop (roots tend to fork when they are grown on freshly manured soil). On stony or heavy soils, dig a trench about 30cm deep and 20cm wide, filling it with finely sieved sandy soil or compost and sharp sand to give the roots room to grow.

Lighten heavier soils by thoroughly digging over the area the autumn before planting, adding well-rotted organic matter and sharp sand. A pH of 6.0–7.5 is ideal; lime acid soils. Before sowing, rake the seedbed to a fine tilth and incorporate a slow-release general fertiliser at 60–90g/sq m.

Regular watering ensures good-quality roots and, importantly, stops bolting in dry conditions. It also prevents roots from splitting, which occurs after sudden rainfall or watering after drought. Salsify should be kept weed-free, particularly during the early stages of growth. Hand weed, as the roots are easily damaged. Alternatively, use an onion hoe or mulch moist soil with a layer of organic matter to stifle weed growth and conserve moisture. Roots can be left in the ground to produce edible shoots or 'chards' the following spring. In autumn, cut back the old leaves to within 1–2.5cm of the soil and earth up to about 12.5–15cm. In mid-spring the following year, the shoots are ready to harvest when they appear. On heavier soils, 'chards' can be blanched with a covering of bracken, straw or dry leaves at least 12.5cm deep when new growth appears in spring. An upturned bucket or flower pots with the drainage holes covered work just as well.

Maintenance

Spring Rake over the seedbed, incorporating general fertiliser. Sow seeds in 'stations'.
Summer Keep the crop weed-free, mulch and water during dry periods.
Autumn Lift crops as required. Cut back the tops and cover those being grown for 'chards'.
Winter Dig over soil manured for the previous crop. Lift and store crops grown in colder areas.

Harvesting and Storing

Harvest from mid-autumn. Lift as required; protect roots with a layer of straw, bracken or a similar material before severe frosts or snow. They can be lifted in autumn and stored in boxes of peat substitute, sand or sawdust, in a cool, frost-free garage, shed or cellar. Lift carefully with a garden fork; the roots 'bleed' and snap easily.

If you begin lifting and find the crop misshapen, leave the roots buried in the ground and harvest the shoots and buds the following year. Next time, grow in sieved soil.

Harvest 'chards' in early to mid-spring: scrape away the soil and cut the blanched shoots when they are 12.5–15cm long. Or lift when 15cm tall, without earthing up, though they are not as tender.

Pests and Diseases

Salsify is usually trouble-free but may suffer from white blister, which looks like glistening paint splashes and can distort plants. Spray with Bordeaux mixture and destroy any plants that are badly affected. Aster yellows cause deformed new growth and the leaf veins or whole leaf become yellow. Control the aster leaf hopper, which spreads the disease.

container growing

Salsify can be grown in containers, but it is impractical to produce them in large quantities. However, if you are desperate, fill a large wooden box, tea chest or plastic drum with sandy, free-draining soil or a loam-based compost with added sharp sand. Ensure there are sufficient drainage holes. Keep crops weed-free and water regularly.

companion planting

Salsify grows well with mustard and, planted near carrots, discourages carrot fly.

medicinal

Salsify is said to have an antibilious effect and to calm fevers. In folk medicine, it is used to treat gall bladder problems and jaundice.

Salsify in Batter

culinary

Salsify has a delicate oyster-like flavour and can be cooked in a variety of ways (there is bound to be a recipe that suits you). Peel after boiling until tender, 'skimming' under cold running water as you would a hard-boiled egg.

Cut roots discolour, so drop them into water with a dash of lemon juice, then boil for 25 minutes in salted water with lemon juice and a tablespoon of flour. Drain and skin, toss with melted butter and chopped parsley and indulge.

Try them with a light mornay sauce, deep-fried or served *au gratin* with cheese and breadcrumbs. Perhaps sautéing them in butter and eating with brown sugar is more to your liking?

Roots can also be baked, puréed, creamed for soup or grated raw in salads. They stay fresh in a refrigerator for about a week.

'Chards' can be served raw in salads or lightly cooked. Young flowering shoots can be eaten pickled, raw or cooked and eaten cold like asparagus, with oil and lemon juice. The subtle flavour makes them an excellent hors d'oeuvre.

Pick flower buds just before they open, with about 7.5cm of stem attached, lightly simmer and eat when they are cool.

Salsify in Batter

Serves 4

750g salsify, topped, tailed and cleaned
Vegetable oil, for frying
Lemon wedges, to serve

For the Batter:
100g plain flour
1 teaspoon baking powder
1 egg
1 tablespoon olive oil
¼ teaspoon harissa
½ teaspoon ground cumin
¼ teaspoon dried thyme
Salt and freshly ground black pepper

Cut salsify roots in half and boil in salted water for 30 minutes. Rinse under cold water. Peel and cut into 7.5cm pieces.

Make the batter: whiz all the ingredients with 150ml of water in a food processor for 30 seconds.

Heat the oil in a large frying pan. Dip the salsify in the batter and fry in oil until crisp and golden, turning. Keep warm while cooking the remainder. Serve hot with lemon wedges.

Salsify in Ham with Cheese Sauce

Serves 4

8 salsify, peeled and trimmed
8 slices cooked ham
2 tablespoons butter
2 tablespoons flour
1 teaspoon Dijon mustard
300ml milk
4 tablespoons grated Gruyère cheese
Pinch nutmeg
Salt and freshly ground black pepper

Boil the salsify for 5 minutes and drain. Cut into roughly 10cm lengths. When cool enough to handle, wrap each in a slice of ham and arrange in a greased baking dish.

Melt the butter in a heavy pan and stir in the flour to make a roux. Cook for a minute or two and stir in the mustard. Slowly add the milk, then the Gruyère, and cook until the cheese is melted.

Season with nutmeg, salt and pepper and pour over the salsify. Bake in a preheated oven, 200°C/400°F/gas mark 6, for 20 minutes, until browned.

Valerianella locusta. Valerianaceae

LAMB'S LETTUCE

Also known as Corn Salad, Mache, Lamb's Tongues. Low-growing, extremely hardy annual grown for edible leaves. Value: good source of beta carotene, vitamin C and folic acid. Very few calories.

In spite of its delicate appearance, lamb's lettuce is an extremely hardy plant, particularly valued as a nutritious winter salad crop. Its attractive bright green, rounded leaves have a slightly nutty taste. Depending on the authority, it was named lamb's lettuce because sheep are partial to it, or because it appears during the lambing season. Another name, corn salad, comes from its regular appearance as a cornfield weed.

Gerard wrote: 'We know the Lamb's Lettuce as loblollie; and it serves in winter as a salad herb among others none of the worst.' He also noted that 'foreigners using it in England led to its cultivation in our gardens'. It has been popular for centuries in France, where it is known as *salade de prêtre* (as it is often eaten in Lent), *doucette* ('little soft one', referring to the velvety leaves), and *bourcette*, describing their shape. In England it declined in popularity in the 1700s and a nineteenth-century commentator noted: '… it is indeed a weed, and can be of no real use where lettuces are to be had.' Before the appearance of winter lettuce varieties, lamb's lettuce was the main winter salad; it was at one time classified in the same genus.

varieties

There are two forms, the 'large' or 'broad-leaved' and the darker, more compact 'green' type, which is popular in western Europe, but less productive.

'Cavallo' has deep green leaves and crops heavily. **'Grote Noordhollandse'** is very hardy. **'Large Leaved'** is tender and prolific. **'Valeriana d'Orlanda'** is an Italian variety with larger leaves than **'Verte de Cambrai'**, which is a traditional French type with small leaves and good flavour. **'Verte d'Etampes'** has unusual, attractive savoyed leaves. **'Vit'** is very vigorous with a mild flavour.

cultivation

Propagation

Lamb's lettuce can be grown as single plants or 'cut and come again' seedlings. Sow seeds successionally from mid- to late spring for summer crops in drills 10–15cm apart and 1cm deep, thinning when seedlings have 3–4 seed leaves to about 10cm apart. Sow winter crops successionally from mid- to late summer. Seeds can also be sown in broad drills, broadcast, or in seed trays or modules for transplanting. Leave a few plants to bolt, and transplant seedlings into rows.

Growing

Lamb's lettuce flourishes in a sheltered position in full sun or light shade and needs a deep, fertile soil for rapid and continuous growth. It tolerates most soils, provided they are not waterlogged. Create a firm seedbed and rake in a slow-release general fertiliser at 30–60g/sq m before sowing in spring and summer. Overwintered crops growing on the same site should not need it.

Keep crops weed-free, particularly during the early stages, and water thoroughly to encourage soft growth.

Maintenance

Spring Sow successionally.
Summer Continue sowing and harvesting; keep crops weed-free and water well.

Harvest before the plant starts flowering

LAMB'S LETTUCE

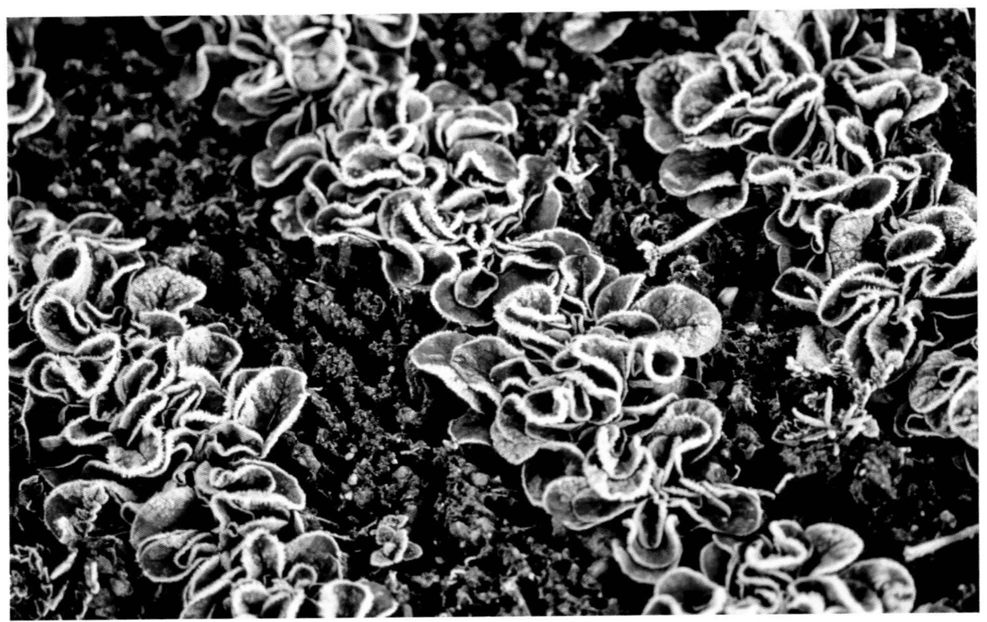
Productive even in winter

Autumn Sow crops under cover. Continue harvesting.
Winter When growth is slower, do not overpick.

Protected Cropping
Although it is extremely hardy, growing lamb's lettuce under cloches, polythene tunnels or horticultural fleece, or in an unheated glasshouse, encourages better quality and higher productivity. Make the first sowings in early autumn.

Harvesting and Storing
Harvest seedlings, pick leaves as required, or lift whole plants when they are mature – about 3 months after planting.

Do not weaken by removing too many leaves at one picking. This is particularly important with outdoor winter crops.

Leaves can be blanched for a few days before picking by covering with a box or pot to remove any bitterness. Young flowers can be eaten in salads.

Pests and Diseases
Some varieties are susceptible to mildew. Protect seedlings from birds and slugs.

companion planting

As plants take up very little space, lamb's lettuce is ideal sown between taller crops.

container growing

Can be grown in 25cm pots of loam-based compost with a low fertiliser content. Sow seeds thinly 1cm deep every 3 weeks from early spring to early summer and again from early to mid-autumn, thinning to leave the strongest seedlings at a final spacing of 10cm apart.

Keep the compost moist when plants are growing vigorously; reduce watering in cooler conditions.

medicinal

With its high beta carotene, vitamin C and folic acid content, it is regarded by many as a winter and early spring tonic, and is useful when other nutritional vegetables are scarce.

culinary

Thinnings can be used in salads. Wash the leaves thoroughly to remove soil. In seventeenth-century Europe, lamb's lettuce was often served with cold boiled beetroot or celery. The leaves are used as a substitute for lettuce and combine well with it, also complementing fried bacon or ham and beetroot. Leaves can also be cooked like spinach.

Lamb's Lettuce and Prawn Salad
Serves 4

A bag of lamb's lettuce, or a mixture of winter salad leaves
16–20 large fresh prawns, peeled
Toasted sesame seeds

For the vinaigrette:
Lemon juice
Balsamic vinegar
Extra virgin olive oil
Mustard
Salt and freshly ground black pepper
Chopped fresh herbs, such as sage, parsley and chives

Wash the salad leaves and arrange on individual plates. Season the prawns with salt and pepper and coat generously with sesame seeds. In a heavy-bottomed frying pan, gently cook them, turning after 3 minutes, when golden, to cook the other side. Arrange on the salad. Make the vinaigrette by mixing the ingredients to taste, sprinkle over and serve.

Vicia faba. Papilionaceae

BROAD BEAN

Also known as Fava Bean, Horse Bean, Windsor. Annual grown for seeds, young leaf shoots and whole young pods. Value: low protein, good source of fibre, potassium, vitamins E and C.

They are thought to have originated around the Mediterranean, and the oldest remains of domesticated broad beans have been dated to 6800–6500 BC. By the Iron Age, they had spread throughout Europe. The Greeks dedicated them to Apollo and thought that overindulgence dulled the senses; Dioscorides wrote that they are 'flatulent, hard of digestion, causing troublesome dreams'.

They may be the origin of the term 'bean feast', being a major part of the annual meal given by employers for their staff to make merry. In the vernacular this has been changed to 'beano'. Britain has many folk sayings concerning the sowing date. Huntingdonshire country wisdom states: 'On St Valentine's Day, beans should be in clay', and it was generally accepted that four seeds were sown – 'One for rook, one for crow, one to rot, one to grow.' Flowers were considered to be an aphrodisiac – 'there ent no lustier scent than a beanfield in bloom' – while the poet John Clare wrote: 'My love is as sweet as a bean field in blossom, beanfields misted wi' dew.'

Hardy and prolific, they are the national dish of Egypt. An Arab saying goes, 'Beans have satisfied even the pharaohs.' How about you?

varieties

Broad beans are classified as 'dwarf', with small, early-maturing pods that are excellent under cloches or in containers, 'longpods', which are also hardy, and 'windsors', with broad pods that usually mature later and have a better flavour. There are green- and white-seeded forms among the longpods and windsors.

'Aquadulce' is a reliable hardy longpod for autumn and spring sowing. **'Aquadulce Claudia'** is similar, with medium to long pods and white seeds, but less susceptible to blackfly; ideal for freezing. **'Bunyards' Exhibition'** is a reliable heavy cropper with a sweet subtle flavour and delicate texture. Good for freezing. **'Imperial Green Longpod'** produces pods around 38cm long with up to 9 large green beans. **'Masterpiece Green Longpod'** is a good-quality, green-seeded bean that freezes well. **'Jubilee Hyson'** crops early with long pods of pale green beans; outstanding quality and flavour. **'Red Epicure'** has beautiful deep chestnut-crimson seeds. Some of the colour is lost in cooking, but the flavour is superb. **'Sweet Lorane'** is a prolific, delicious small-seeded bean that is also cold-hardy. **'The Sutton'**, a prolific dwarf variety, grows to around 30cm high, is ideal for small gardens and under cloches – excellent flavour. **'Witkiem Manita'** is perfect for spring sowing, with sweet, succulent beans.

cultivation

Propagation

As they germinate well in cool conditions, broad beans can be sown from mid- to late autumn and overwintered outdoors for a mid-spring crop. Seedlings should be 2.5cm high when colder weather arrives. 'Aquadulce' cultivars are particularly suitable. When the

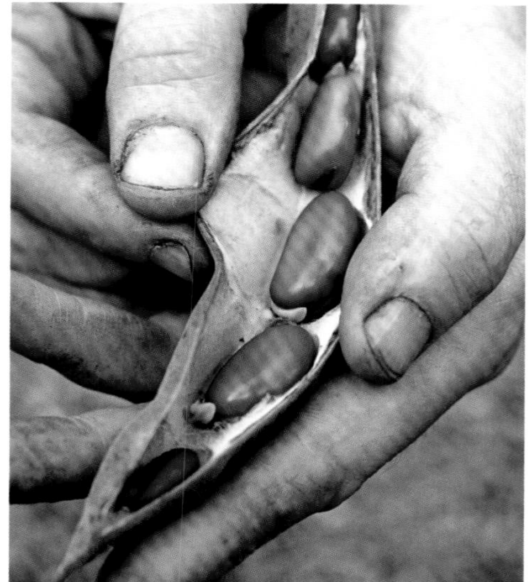

Pods filled with flavour

soil is workable, sow every 3 weeks from late winter to late spring for successional cropping.

Plant seeds 5cm deep in rows or make individual holes with 30cm in and between the rows. Or grow them in staggered double rows, with 23cm in and between the rows or in blocks 90cm square with 20–30cm between plants. Tall plants should be 30–45cm apart. Sow extra seeds at the end of rows, for later use as transplants.

Growing
Broad beans flourish in an open, sunny site but overwintering crops need more shelter. Soils should be deeply dug, moisture-retentive and well-drained, having been manured for the previous year's crop.

Alternatively, add well-rotted organic matter in late autumn before sowing. They need a pH of 6–7, so lime where necessary. Rotate with other legumes. Rake a granular general fertiliser into the seedbed 1 week before sowing. Mulch, hoe or hand weed regularly, especially at first.

Watering should be unnecessary except in drought, but for good-quality crops, plants need 22 litres/sq m per week from flower formation to harvest end.

Support taller plants with stakes. Planting several staggered rows allows plants to support one another. Dwarf types can be supported with brushwood.

Maintenance
Spring Sow early crops under glass, followed by later sowings until late spring.
Summer Harvest from late spring to late summer. Keep crops weed-free and water after flower formation.
Autumn Sow outdoor crops for overwintering.
Winter Protect outdoor crops in severe weather.

Protected Cropping
Where plants cannot be overwintered outdoors, sow under cover in pots or boxes from midwinter. Harden off and plant out in mid-spring. Remove cloches from protected early crops when the beans reach the glass.

Harvesting and Storing
Pick when the beans begin to show, before the pods are too large (and tough). Pull with a sharp downward twist or cut with scissors. Pods harvested at 5–7.5cm long can be treated as snap peas and eaten whole,

but picking at this stage reduces yields. Broad beans freeze well, especially green-seeded varieties. Wash, blanch for 3 minutes, and freeze in polythene bags or rigid containers. Use within 12 months. Keep them in a polythene bag in the refrigerator for 1 week.

Pests and Diseases

Black bean aphid is the most common problem. Control by pinching off the top 7cm of stem when the first beans start to form. This also encourages an earlier harvest.

Mice and Jays steal the seeds. Trap and tolerate.

companion planting

Plant with summer savory to discourage black bean aphid and among gooseberries to discourage gooseberry sawfly. The flowers are very attractive to bees. Broad beans are a good 'nurse' crop for developing maize and sweetcorn. Interplant with brassicas, which benefit from the nitrogen-fixing roots. Dig in plant debris as a 'green manure' to increase soil fertility. Seedling beans protect early potato shoots from wind and frost.

container growing

Grow dwarf cultivars in containers of low-fertiliser loam-based compost at least 20–45cm wide by 25cm deep. Sow seeds 5cm deep and 10–13cm apart. Water regularly in warm weather, less at other times.

culinary

Steam or lightly boil beans to eat with ham, pork and chicken. Eat with sautéed onions, mushrooms and bacon; dress with tomato sauce, warm sour cream or lemon butter and dill.

Broad Beans with Summer Savory

Serves 4

A dish which reflects the pure tastes of summer.

500g young broad beans, shelled weight
2 sprigs plus 2 tablespoons summer savory
2 tablespoons butter
Salt and freshly ground black pepper

Put the beans into a pan of boiling, salted water with the sprigs of summer savory and cook until just done, about 3–5 minutes. Drain and put the pan back on a gentle heat. Add the butter and toss well as it melts. Season with more summer savory and salt and pepper.

Bissara

Serves 4

An Algerian dish with subtle flavours, this is based on one from the late, great foodie, Arto der Haroturian.

675g young broad beans, shelled weight
1 small green chilli, chopped
1 teaspoon paprika
1 teaspoon cumin seeds, crushed in a pestle and mortar
2 cloves garlic, peeled
8 tablespoons extra virgin olive oil
Juice of ½ a lemon
Salt and freshly ground black pepper
A little olive oil, lemon juice and paprika, to garnish

Boil the beans in lightly salted water until just tender. Drain and place in a food processor, together with the chilli, paprika, cumin and garlic. Purée by slowly drizzling in the oil and lemon juice. Season and pour into a bowl.

Garnish with a little olive oil, lemon juice and paprika. Serve bissara with warm bread.

Wasabia japonica. Cruciferae

WASABI

Also known as Japanese Horseradish, Shan Kui. Hardy herbaceous perennial. Value: high in protein, potassium and moderate amounts of calcium and vitamin C.

This Japanese native, found in boggy ground by mountain streams, is highly prized for its horseradish-flavoured roots, which are traditionally ground into a fine powder using a shark skin grater or *oroshi* before being turned into a paste. It has been cultivated in Japan since the tenth century AD; the highest prices are paid for natural water-grown *sawa* or semi-aquatic wasabi, the price and quality of field-grown or *oka* is correspondingly lower. At the time of writing, dried, field-grown wasabi rhizomes, cultivated on the Pacific coast of the USA, are retailing at $130 for 100g. As demand outstrips supply beware of inferior products made of horseradish or Chinese mustard and food colouring – or grow your own.

A true vegetable delicacy

varieties

Wasabia japonica grows up to 40cm tall with glossy, almost heart-shaped leaves the size of small dinner plates, with coarsely-toothed margins and spikes of small white flowers in late spring.

cultivation

Propagation

Repropagate plants every two years, as younger roots have a better flavour. Sow seeds in early spring in loam-based compost; transplant into individual pots when the 'true' leaves are large enough to handle.

Grow them in the greenhouse for the first year, standing in trays of water so the compost stays constantly moist. Plant out in late spring or early summer of the second year after acclimatising plants to outdoor conditions.

Alternatively, divide healthy plants in early spring as they start into growth, grow them on in an unheated greenhouse or cold-frame and plant out in late spring the following year. Small offsets can also be detached from well-established plants in spring.

Growing

The plants will grow in boggy ground of any pH, or in fresh, pollution-free, running water, in sunshine or part shade, particularly in warmer climates. Cultivation techniques are similar to watercress (see page 178) and roots should be protected with bracken, straw or fleece over winter. The optimum water temperature is 10–13°C with air temperatures of 8–20°C, though plants still grow in less than perfect conditions.

Maintenance

Spring Sow seeds, divide or remove 'offsets'.
Summer Ensure plants remain moist. Harvest plants; retain divisions for replanting.
Autumn Remove leaves as they die down.
Winter Cover plants to protect from frost.

Harvesting and Storing

Lift rhizomes as needed and use immediately, for the finest flavour.

Worth their weight in gold

container growing

Grow in large pots of loam-based compost kept constantly moist by standing in a bowl of water which is refreshed daily.

medicinal

Wasabi is used to stimulate digestion, has anti-bacterial qualities, reduces mucus, and controls asthma and other congestive disorders. Isothiocyantates help detoxify the liver and gut; it is also a powerful antioxidant and has anti-cancer properties.

culinary

Use freshly dug rhizomes, as the flavour deteriorates rapidly once the root has been cut; it is similar but superior to horseradish and mid-green in colour. Rhizomes harvested at 15–24 months old are considered to be the best. It is finely grated and turned into a paste and is the ideal complement to raw fish in *sashimi* or *sushi* dishes, added to soups or mixed with soy sauce to make dips. The leaves, flowers and leaf stems are also cooked and eaten. In Japan, the leaves and leaf stalks, flowers and rhizomes are soaked in salt water, mixed with sake lees and used to make a pickle called *wasabe-zuke*. Wasabi can also be bought in powdered form, in small tins. This recipe comes from *The Japanese Kitchen* by Kimiko Barber.

Wasabi and Avocado Dip

Serves 4

2 tablespoons wasabi powder
1 ripe avocado
2 tablespoons good mayonnaise
1 teaspoon light soy sauce
Salt (optional)

Mix the wasabi powder with 2 tablespoons water. Peel the avocado, remove the stone and roughly mash with a fork. Place the wasabi paste, mashed avocado mayonnaise and soy sauce into a food processor and whizz until smooth and creamy. Have a taste and season with salt if necessary.

Zea mays. Graminae

SWEETCORN

Also known as Maize, Corn on the Cob, Indian Corn, Baby Maize. Annual grown for kernels of edible seeds. Half hardy. Value: high in carbohydrates and fibre, moderate protein and B vitamin content.

Wild plants of maize have never been found, but it is believed that the crop was first cultivated in Mexico around 7000 BC. Primitive types, smaller than an ear of wheat, were found in caves at Tehuacan in southern Mexico dating from around 3500 BC, yet they were bred by Mayan and Inca farmers almost to the size of modern varieties. Early American civilisations were based on maize, and life for the Aztecs revolved round the *milpa* or cornfield. Multicoloured types with blue, scarlet, brown and almost black seeds predominated in South America. It became a staple crop in North America after AD800. The first types introduced to Europe in the sixteenth century from Central America were valued for both the cobs and the yellow meal. Corn flourished in Spain, France, Italy, the Balkans and Portugal.

Sweetcorn is a sweet form of maize, a starchy crop used for grain and fodder. The yellow-golden varieties are commonly cultivated for the kitchen, but those with multicoloured seeds are extremely ornamental. Sweetcorn and maize are the third most important cereal in the world, after wheat and rice, and there are more than 500 different by-products. The seeds of corn also provide us with popcorn, with its exploding grains; starch extracted after milling has been used as laundry starch; the inner husks are used for making cigarette papers and the pith has been used for making explosives and packaging material. Best of all is delicious fresh sweetcorn eaten from the cob, dripping with butter; take the advice of Edward Bunyard: 'The principle of the lathe is adopted in eating them.'

varieties

'Normal' varieties were the first selections and have the best seed vigour. 'Supersweets' have very high sugar levels, a longer storage life than 'normal' varieties but poorer germination. 'Sugar Enhanced' are not as sweet as 'Supersweet' but are better for storing and 'Extra Tender Sweets' are sweeter and more tender. If you aim to grow 'supersweet' varieties they should be at least 8m from ordinary varieties to prevent cross pollination. Alternatively, grow sugar enhanced types – if cross pollination occurs, the cobs will be more like traditional varieties.

Supersweet varieties
'Earlibird' has good vigour and is a top-quality variety for cob size and flavour. **'Prelude'** is vigorous and high yielding, with cobs up to 25cm long. **'Ovation'** is a mid-season variety with large tasty cobs up to 20cm long. **'Northern Extra Sweet'** is reliable in British summers and good in cooler climates – one of the earliest to mature and with top-quality kernels. Do not let it overripen.

Extra Tender Sweet varieties
'Swift', voted the best variety in independent trials, produces large cobs with exceptionally sweet, tender and succulent kernels. Does not need to be isolated from other varieties. **'Lark'** has moderate vigour, very sweet cobs and performs well in colder soils.

Sugar Enhanced varieties
'Minipop' is a baby corn with sweet cobs. Plant in rows rather than blocks and harvest before they swell. Eat lightly steamed or with dips. **'Tuxedo'**, a tall variety, performs well under drier conditions and is exceptionally

SWEETCORN

drought resistant. **'Sweet Nugget'** grows to a moderate length, golden yellow cobs that taste very sweet. Ideal in cooler conditions.

Others

'Honey Bantam' grows exceptionally early, 17cm long, with pretty yellow and white kernels that are very sweet indeed. **'Kelvedon Glory'** is an old variety with long, light golden cobs that are filled with flavour. **'Red Strawberry'** has tiny red kernels on compact cobs (only 5cm long). Put them in the microwave and they explode into fluffy popcorn.

cultivation

Propagation

Seeds do not germinate when soil temperatures are below 10°C. Better germination and earlier crops are achieved by sowing from mid-spring under cover in pots or modules. Plant 2–3 seeds 2.5cm deep in pots of moist seed compost from mid- to late spring, at 13°C. After germination, thin to leave the strongest seedling. Sweetcorn is sensitive to root damage and disturbance, so thin with care, holding the strongest seedling while removing weaker ones. Harden off and plant out once there is no longer any danger of frost.

'Minipop'

Plants should be 35cm apart in and between the rows or spaced 25–30cm apart in rows 60cm apart.

To avoid erratic germination, seeds can be pregerminated. Put a few sheets of tissue paper or similar material in the bottom of a seedtray, moisten with water, place seeds evenly over the surface and cover with another layer of moist tissue. Put the tray in a loosely knotted polythene bag in an airing cupboard or warm room. When the seeds swell and tiny rootlets have formed, sow a single seed 2.5cm deep in a 9cm pot of seed compost or sow singly outdoors. Sow a few in pots to replace those that do not germinate outdoors.

Alternatively, sow *in situ* from late spring to early summer. When the soil is workable, make a block about 1.2–1.5m, comprising 4 ridges of soil about 30cm high with 35cm between the top of each ridge. 'Station sow' 2–3 seeds in the furrow 35cm apart each way and thin to leave the strongest seedling.

Cut a square of thin polythene or similar slightly larger than the block. Spread it over the furrows, dig a shallow trench round the edge and weigh the margins down with soil to hold the cover in place. As the seedlings emerge, cut crosses in the cover so they can grow through, or remove it and earth up round the stem bases when plants are 45cm high. Alternatively, sow on a level seedbed and protect crops with fleece or cloches, which should be removed as soon as the plants become too large.

'Supersweet' varieties need slightly warmer conditions for germination; soils should be at a minimum of 12°C. Because sweetcorn is wind-pollinated, plants are sown in blocks for effective germination.

The male tassels (flowers) at the top of the plant shed clouds of pollen on to the female tassels or 'silks' below. These are clusters of pale green strands on the ends of the cobs which become sticky before pollination and each strand is attached to a single grain. Each must be pollinated, and sowing in a block ensures effective pollination. On a calm day tap the stems to help pollination. Mini varieties can be sown in rows.

Seeds germinate in warm conditions

Growing

Sweetcorn needs a moisture-retentive, free-draining soil. Dig in plenty of organic matter several weeks before planting, or use ground manured for the previous year's crop. Ideally, the soil should be slightly acid, with a pH of 5.5–7.0. Plants need a long, warm growing season to succeed, so grow fast-maturing, early cultivars in cooler areas. The site should be open, sunny, warm and sheltered from cold winds and frosts.

Rake in a slow-release general fertiliser after transplanting or when preparing the seedbed. Keep compost moist throughout the growing season. Watering is particularly beneficial when they begin to flower, and when the grains begin to swell. Apply 22 litres/sq m per week. Stake plants when growing on exposed sites.

Sweetcorn is shallow-rooted, so take care when hoeing to avoid damaging the roots. It is much better to mulch with a layer of organic matter to suppress weeds and retain moisture. Mulch with black polythene on level ground. Do not remove sideshoots (tillers). Earth up to make plants more stable.

Sweetcorn can easily be part of an ornamental garden

Maintenance
Spring Sow seed under cover or outdoors.
Summer Sow seeds; keep crops weed-free and water. Harvest.
Autumn Harvest.
Winter Prepare the ground for next year's crop.

Protected Cropping
When soil temperatures are below 10°C, sow from mid-spring under cover in pots or modules at 13°C. Harden off and plant out once there is no danger of frost.

Harvesting and Storing
Each plant usually produces one or two cobs. Those sown *in situ* mature later than seed sown indoors. After pollination the silks begin to turn brown, an indication that the cob is maturing. Peel back the leaves and test for ripeness by pushing your thumbnail into a grain: if the liquid runs clear, it is unripe; if it is milky, it is ready to harvest; if it is thick, then it is over-mature and unsuitable for eating. To pick, hold the main stem in one hand and twist off the cob with the other. Sweetcorn is ideal for instant eating. Have a pot of water boiling before you harvest!

Freeze within at least 24 hours of picking, before the sugars in the seeds convert to starch. Supersweet varieties hold their sugar levels longer.

To freeze, blanch cobs for 4–6 minutes, cool and drain, before wrapping singly in foil or cling film and placing in the freezer.

Cobs remain fresh for up to 3 days if they are stored in the refrigerator.

Pests and Diseases
Badgers cause considerable damage and birds can pull up the seedlings and attack the developing cobs. Slugs attack seedlings. Smut appears as large galls on the cobs and stalks in hot dry weather. Cut off and burn the debris immediately after harvesting and do not grow sweetcorn on the site for at least 3 years. Frit fly larvae bore into the growing points of corn seedlings, which then develop twisted and ragged leaves. Growth is stunted and cobs are undersised. Use 'dressed' seed and cultivate the ground thoroughly. Mice enjoy newly planted seeds, especially supersweet types. Trap, or buy a cat.

Enough to go around

Block planting results in good pollination

companion planting

Grow sweetcorn with or after legumes. Runner beans can be allowed to grow over sweetcorn plants. Intercrop corn with sunflowers, allowing cucurbits to trail through them.

The shade they provide is useful to cucumbers, melons, squashes, courgettes, marrows and potatoes. Brussels sprouts, kale, Savoy cabbages, swedes and broccoli can be interplanted. Grow lettuce and other salads, French beans and courgettes between the crops.

medicinal

Corn is said to reduce the risk of certain cancers, heart disease and dental cavities. Corn oil is reported to lower cholesterol levels more successfully than other polyunsaturated oils.

In parts of Mexico, corn is used to treat dysentery. It is also known in American folk medicine as a diuretic and mild stimulant.

SWEETCORN

culinary

To serve whole, strip off outer leaves, leaving 5–7cm of stalk on the ear. Pull off 'silks'. Boil in unsalted water, drain and eat hot with melted butter. Roasted corn is a favourite with most children – it tastes wonderfully succulent compared with the sogginess of tinned corn.

Deep-fry spoonfuls of a mixture of mashed corn, salt, flour, milk and egg for 1–2 minutes or until golden brown to make corn fritters.

Seeds are ground, meal boiled or baked, cob can be roasted or boiled or fermented, maize meal is cooked with water to create a thick mash, or dough.

Tortillas are made by baking in flat cakes until they are crisp. Polenta is made from ground maize.

Dry-milling produces grits, from which most of the bran and germ are separated. Cornflakes are rolled, flavoured seeds. Multicoloured corn adds colour and flavour to sweet dishes and drinks.

Baby corns are particularly decorative and the perfect size for oriental stir-fries, where they combine well with mangetout.

Barbecued Corn on the Cob

This is the way to really taste the flavour of corn. Take 1 corn ear for each person. Boil them in salted water for 7–10 minutes and then drain.

Place each in the centre of a piece of tinfoil. Use 1 tablespoon butter per corn ear and mix in 1 teaspoon fresh chopped herbs, sea salt and freshly ground black pepper; spread over the corn ears. Season well and then seal the foil.

Lay the corns on a medium-hot barbecue or in a charcoal grill (in which case, keep it covered) and roast for 20 minutes, turning the cobs once, halfway through cooking.

Corn Maque Choux

Serves 4

Cajun cooking has become popular in recent years and this dish reveals the true taste of corn.

1 tablespoon butter
1 large onion, roughly chopped
1 green pepper, deseeded and diced
500g corn, cooked and stripped from the cob
2 tomatoes, peeled, deseeded and diced
½ teaspoon Tabasco sauce
Salt and freshly ground black pepper

Heat the butter in a heavy pan and sauté the onion until softened. Add the pepper and continue cooking for 3–4 minutes. Toss in the corn and the remaining ingredients and, over a low heat, simmer for 10 minutes. Adjust the seasoning and serve.

Succotash

Serves 4

The American Indians cooked succotash, and many recipes for it still exist. **'Silver Queen'** is a particularly good variety to use for this dish.

3–4 tablespoons unsalted butter
300g broad beans, cooked
3 or 4 ears roasted corn
250g French beans, topped and tailed
1 medium red onion, finely chopped
300ml vegetable stock
½ red pepper, deseeded and finely diced
1 beef tomato, peeled, deseeded and chopped
Salt and freshly ground black pepper

Melt the butter in a heavy frying pan and sweat the broad beans, corn, French beans and onion over a medium heat for about 3 minutes.

Then add the stock and continue cooking for 5 minutes, before stirring in the remaining ingredients. Taste for seasoning, mix thoroughly and continue cooking for 5 minutes more. Serve hot.

Corn Maque Choux

HERBS
A-Z

Achillea ageratum

ENGLISH MACE

From the family *Asteraceae*.

Native to Switzerland, English mace is now cultivated in northern temperate countries. This culinary herb is little known and under used.

English mace belongs to the *Achillea* genus, named after Achilles, who is said to have discovered the medicinal properties of the genus. There is no direct historical record of English mace itself apart from the fact that it was discovered in Switzerland in 1798.

varieties

Achillea ageratum
English Mace
Hardy perennial. Ht 30–45cm when in flower. Spread 30cm. Clusters of small cream flowers that look very Victorian in summer. Leaves brightish green, narrow and very deeply serrated.

cultivation

Propagation

Cuttings
This is the best method for the propagation of a large number of plants. Take softwood cuttings in late summer; protect from wilting as they will be very soft. Use a seed compost mixed in equal parts with composted fine bark. When well-rooted, harden off and plant out in the garden 30cm apart.

Division
If you require only a few plants it is best to propagate by division. Either divide the plant in early spring – it is one of the first to appear – or in autumn. Replant in the garden in a prepared site. As this is a hardy plant it will not need protection, but if you leave division until the frosts are imminent, winter the divided plants in a cold frame or cold greenhouse.

Pests and Diseases

Mace, in most cases, is free from pests and diseases.

Maintenance

Spring Divide established plants.
Summer Cut back flowers. Take softwood cuttings.
Autumn Divide established plants if needed.
Winter Does not need protection.

Garden Cultivation

This fully hardy plant, which even flourishes on heavy soil, prefers a sunny, well-drained site. It starts the season off as a cluster of low-growing, deeply serrated leaves and then develops long, flowering stems in summer. Cut back after flowering for a fresh supply of leaves and to encourage a second flowering crop. When in flower this plant may need staking in a windy, exposed site.

ENGLISH MACE

Achillea ageratum new spring growth

Harvesting

Cut fresh leaves when you wish. For freezing – the best method of preserving – cut before flowering and freeze in small containers.

Pick the flowers during the summer. Collect in small bunches and hang upside down to dry. Both flowers and leaves dry particularly well.

container growing

For a tall flowering plant this looks most attractive in a terracotta pot. Make sure it has a wide base to allow for its height later in the season. Use a soil-based compost mixed in equal parts with composted fine bark. Water regularly thoughout the growing season and give a liquid feed (according to manufacturer's instructions) in the summer months during flowering. Cut back after flowering to stop the plant from toppling over and encourage new growth. As this plant dies back in winter, allow the compost to become nearly dry, and winter the container in a cold greenhouse or cold frame.

other uses

Flowers in dried flower arrangements.

culinary

The chopped leaves can be used to stuff chicken, flavour soups, stews, and to sprinkle on potato salads, rice and pasta dishes. The leaf has a mild, warm, aromatic flavour and combines well with other herbs.

Chicken with English Mace in Foil

Serves 4

2 tablespoons yogurt
2 tablespoons Dijon mustard
4 chicken breasts
Salt and fresh ground black pepper
Bouquet garni herb oil (or olive oil)
6 tablespoons of chopped English mace leaves
Juice of 1 lemon

Preheat the oven to 190°C/375°F/gas mark 5. Mix the yogurt and mustard together and coat the chicken pieces on all sides. Sprinkle with salt and pepper. Cut 4 pieces of foil and brush with herb or olive oil. Lay the chicken breasts in the foil and scatter a thick layer of English mace on each piece. Sprinkle with lemon juice. Wrap in the foil, folding the ends very tightly so no juices can escape. Lay the packets on a rack in the oven, cook for 30 minutes. Serve with rice and a green salad.

Achillea millefolium

YARROW

Also known as Nosebleed, Millefoil, Thousand Leaf, Woundwort, Carpenter's Weed, Devil's Nettle, Soldier's Woundwort and Noble Yarrow. From the family *Asteraceae.*

Yarrow is found all over the world in waste places, fields, pastures and meadows. It is common throughout Europe, Asia and North America.

This is a very ancient herb. It was used by the Greeks to control haemorrhages, for which it is still prescribed in homeopathy and herbal medicine today. The legend of Achilles refers to this property – it was said that during the battle of Troy, Achilles healed many of his warriors with yarrow leaves. Hence the name Achillea.

It has long been considered a sacred herb. Yarrow stems were used by the Druids to divine seasonal weather. The ancient Chinese text of prophecy, *I Ching*, The Book of Changes, states that 52 straight stalks of dried yarrow, of even length, were spilled instead of the modern way of using three coins.

It was also associated with magic. In Anglo-Saxon times it was said to have a potency against evil, and in France and in Ireland it is one of the Herbs of St John. On St John's Eve, the Irish hang it up in their houses to avert illness.

There is an old superstition, which apparently still lingers in remote parts of Britain and the United States, that if a young girl tickles her nostrils with sprays of yarrow and her nose starts to bleed, it proves her lover's fidelity:

***'Yarrow away,
Yarrow away, bear
a white blow?
If my lover loves
me, my nose will
bleed now.'***

Achillea filipendulina 'Gold Plate'

Achillea millefolium

varieties

Achillea millefolium
Yarrow
Hardy perennial. Ht 30–90cm, spread 60cm and more. Small white flowers with a hint of pink appear in flat clusters from summer to autumn. Its specific name, *millefolium,* means 'a thousand leaf', which is a good way to describe these darkish green, aromatic, feathery leaves.

***Achillea millefolium* 'Fire King'**
Hardy perennial. Ht and spread 60cm. Flat heads of rich, red, small flowers in flat clusters all summer. Masses of feathery dark green leaves. This has an upright habit and is a vigorous grower.

***Achillea filipendulina* 'Gold Plate' AGM**
Hardy perennial. Ht 1.2m, spread 60cm. Large flat heads of small golden flowerheads in summer that dry well for winter decoration. Filigree green foliage.

***Achillea* 'Moonshine' AGM**
Hardy perennial. Ht 60cm, spread 50cm. Flat heads of bright yellow flowers throughout summer. Masses of small feathery grey/green leaves.

cultivation

Propagation

Seed

For reliable results sow the very small seed under cool protection in autumn. Use either a proprietary seeder or the cardboard trick (see page 645) and sow into prepared seed or plug trays. Leave the trays in a cool greenhouse for the winter. Germination is erratic. Harden off and plant out in the garden in spring. Plant 20–30cm apart, remembering that it will spread. As this is an invasive plant, I do not advise sowing direct into the garden.

Division

Yarrow is a prolific grower, producing loads of creeping rootstock in a growing season. To stop an invasion into areas where it is not wanted, divide by digging up a clump and replanting where required in the spring or early autumn.

Pests and Diseases

Yarrow is usually free from both.

Maintenance

Spring Divide established clumps.
Summer Deadhead flowers, and cut back after flowering to prevent self-seeding.
Autumn Sow seeds. Divide established plants.
Winter No need for protection, very hardy plant.

Garden Cultivation

Yarrow is one of nature's survivors. Its creeping rootstock and ability to self-seed ensure its survival in most soils.

It does well in seaside gardens, as it is drought-tolerant. Still, owners of manicured lawns will know it as a nightmare weed that resists all attempts to eradicate it.

Yarrow is the plant doctor of the garden, its roots' secretions activating the disease resistance of nearby plants. It also intensifies the medicinal actions of other herbs and deepens their fragrance and flavour.

Harvesting

Cut the leaves and flowers for drying as the plant comes into flower.

culinary

The young leaves can be used in salads. Here is an interesting salad recipe:

Salad Made with Three Wild Herbs

Equal parts of yarrow, plantain and watercress leaves
A little garlic
½ cucumber
Freshly chopped or dried chives and parsley
1 medium, boiled cold potato
Salad dressing consisting of lemon and cream, or lemon and oil, or lemon and cream and a little apple juice.

Select and clean herbs. Wash carefully and allow to drain. Cut the yarrow and plantain into fine strips. Cube the cucumber and potato into small pieces. Leave the watercress whole and arrange in a bowl. Add herbs and other vegetables and salad dressing and mix well.

container growing

Yarrow itself does not grow well in containers. However, the hybrids, and certainly the shorter varieties, can look stunning. Use a soil-based potting compost mix and feed plants with liquid fertiliser during the flowering season, following the manufacturer's instructions. Cut back after flowering and keep watering to a minimum in winter. No variety is suitable for growing indoors.

medicinal

Yarrow is one of the best-known herbal remedies for fevers. Used as a hot infusion it will induce sweats that cool fevers and expel toxins. In China, yarrow is used fresh as a poultice for healing wounds. It can also be made into a decoction for wounds, chapped skin and rashes, and as a mouthwash for inflamed gums.

warning

Yarrow should always be taken in moderation and never for long periods because it may cause skin irritation. It should not be taken by pregnant women. Large doses produce headaches and vertigo.

other uses

Flowerheads may be dried for winter decoration. Traditionally the leaves of this unassuming herb were used to accelerate the decomposition of garden waste in a compost bin. The leaves were also used as a copper fertiliser.

Infusion of yarrow

Aconitum napellus

MONKSHOOD

Also known as Friar's Cap, Old Woman's Night-cap, Chariots Drawn by Doves, Blue Rocket, Wolf's Bane and Mazbane. From the family *Ranunculaceae*.

Various species of monkshood grow in temperate regions. They can be found on shady banks, in deciduous woodlands and in mountainous districts. They are all poisonous plants.

One theory for the generic name, *Aconitum*, is that the name comes from the Greek *akoniton*, meaning dart. This is because the juice of the plant was used to poison arrow tips and was used as such by the Arabs and ancient Chinese. Its specific name *napellus* means 'little turnip', a reference to the shape of the root. It was the name used by Theophrastus, the Greek botanist (370–285 BC), for a poisonous plant.

This plant has been known throughout history to kill both animals and humans. In the sixteenth century Gerard commented in his *Herball* on its 'fair and good bluey flowers in shape like helmet which are so beautiful that man would think they were of some excellent virtue'. Appearances should not be trusted. 150 years later, Miller in his garden dictionary wrote, 'Monks Hood was in almost all old gardens and not to be put in the way of children less they should prejudice themselves therewith.'

As recently as 1993 a flower seller in the West of England had to be hospitalised after handling monkshood outside pubs in Salisbury and Southampton.

Aconitum napellus flowers

varieties

Aconitum napellus

Monkshood

Hardy perennial. Ht 1.5m, spread 30cm. Tall slender spires of hooded, light blue/indigo/blue flowers in late summer. Leaves mid-green, palm shaped and deeply cut. There is a white flowered version, *A. napellus* 'Albiflorus', which grows in the same way.

***Aconitum napellus* subsp. *napellus Anglicum* Group**

Monkshood

Hardy perennial. Ht 1.5m, spread 30cm. Tall slender spires of hooded, blue/lilac flowers are borne in early/mid-summer. Leaves mid-green, wedge-shaped, and deeply cut. One of a few plants peculiar to the British Isles, liking shade or half-shade along brooks and streams. Grows in only a few areas in southwest Britain. Probably the most dangerous of all British plants.

MONKSHOOD

cultivation

Propagation

Seed

Sow the small seed under protection either in the autumn (which is best) or spring. Use prepared seed or plug trays. Cover with perlite. Germination can be erratic, an all-or-nothing affair. The seeds do not need heat to germinate. In spring, when the seedlings are large enough to handle, plant out into a prepared, shady site 30cm apart. Wash your hands after handling the seedlings; even better wear thin gloves. The plant takes 2–3 years to flower.

Division

Divide established plants throughout the autumn, so long as the soil is workable. Replant in a prepared site in the garden – remember the gloves. You will notice when splitting the plant that the tap root puts out daughter roots with many rootlets. Remove and store in a warm dry place for planting out later.

Pests and Diseases

This plant rarely suffers from pests, and it is usually disease free.

Maintenance

Spring Plant out autumn-grown seedlings.
Summer Cut back after flowering.
Autumn Sow seeds, divide established plants.
Winter No need to protect. Fully hardy.

Garden Cultivation

In spite of the dire warnings this is a most attractive plant, which is hardy and thrives in most good soils.

Position it so that it is not accessible. Plant at the back of borders, or under trees where no animals can eat it or young fingers fiddle with it. It is useful for planting in the shade of trees as long as they are not too dense.

It is important always to teach people which plants are harmful, and which plants are edible. If you remove all poisonous plants from the garden, people will not learn which to respect.

Harvesting

Unless you are a qualified herbalist I do not recommend harvesting.

container growing

Because of its poisonous nature, I cannot wholeheartedly recommend that it be grown in containers. But it does look very attractive in a large container surrounded by heartsease *(Viola tricolor)*. Use a soil-based compost mixed in equal parts with composted fine bark (see page 643), water regularly throughout the summer months. Liquid feed in summer only.

warning

The symptoms of *Aconitum* poisoning are a burning sensation on the tongue, vomiting, abdominal pains and diarrhoea, leading to paralysis and death. Emergency antidotes, which are obtainable from hospitals, are atropine and strophanthin.

medicinal

This plant contains one of the most potent nerve poisons in the plant kingdom and is used in proprietory analgesic medicines to alleviate pain both internally and externally. These drugs can only be prescribed by qualified medical practitioners. Tinctures of monkshood are frequently used in homeopathy.

Under no circumstances should monkshood ever be prepared and used for self-medication.

Aconitum napellus in a meadow

Agastache

ANISE HYSSOP

Also known as Giant Hyssop, Blue Giant Hyssop, Fennel Hyssop, Fragrant Giant Hyssop. From the family *Lamiaceae*.

Anise hyssop is a native of North America. *A. cana*, the Mosquito Plant originates from Mexico and *A. rugosa* is from Korea.

There are few references to the history of this lovely herb. According to Allen Paterson, Past Director of the Royal Botanical Garden in Ontario, it is a close cousin of the bergamots. It is common in North American herb gardens and is certainly worth including in any herb garden for its flowers and scent. The long spikes of purple, blue and pink flowers are big attractions for bees and butterflies.

varieties

***Agastache foeniculum* (Pursh)**
Anise Hyssop
Hardy perennial. Ht 60cm, spread 30cm. Long purple flower spikes in summer. Aniseed-scented mid-green oval leaves.

Agastache cana
Mosquito Plant
Half-hardy perennial. Ht 60cm, spread 30cm. Pink tubular flowers in the summer with aromatic oval mid-green toothed leaves.

Agastache rugosa
Korean Mint
Hardy perennial. Ht 1m, spread 30cm. Lovely mauve/purple flower spikes in summer. Distinctly minty scented mid-green oval pointed leaves.

***Agastache rugosa* 'Golden Jubilee'**
Anise Hyssop Golden Jubilee
Hardy perennial. Ht 50cm, spread 30cm. Lovely mauve/purple flower spikes in summer. Anise-scented golden leaves. Prone to sun scorch, plant in partial shade.

Agastache cana

ANISE HYSSOP

cultivation

Propagation

Seed

The small fine seeds need warmth to germinate: 17°C. Use the cardboard method (see page 645) and artificial heating if sowing in early spring.

Use either prepared seed or plug trays or, if you have only a few seeds, directly into a pot of standard seed compost mixed in equal parts with composted fine bark. Cover with perlite. Germinates in 10–20 days.

One can also sow outside in the autumn when the soil is warm, but the young plants will need protection throughout the winter months.

When the seedlings are large enough to handle, prick out and pot on using a potting compost mixed in equal parts with propagating bark. In mid-spring, when air and soil temperatures have risen, plant out at a distance of 45cm apart.

Cuttings

Take cuttings of soft young shoots in spring, when all the species root well. Use a seed compost mixed in equal parts with composted fine bark. After a full period of weaning cuttings should be strong enough to plant out in the early autumn.

Semi-ripe wood cuttings may be taken in late summer, using the same compost mix. After they have rooted, pot up, and winter in a cold frame or cold greenhouse.

Division

This is a good alternative way to maintain a short-lived perennial. In the second or third year divide the creeping roots in spring, either by the 'forks back-to-back' method, or by digging up the whole plant and dividing it into several segments.

Pests and Diseases

This herb rarely suffers from pests or diseases, although seedlings can be prone to 'dampening off'.

Maintenance

Spring Sow seeds.
Summer Take softwood or semi-ripe cuttings late in the season.
Autumn Tidy up the plants by cutting back the old flower heads and woody growth. Sow seeds. Protect young plants from frost.
Winter Protect half-hardy species (and anise hyssop below –6°C) with either horticultural fleece, bark or straw.

Garden Cultivation

All species like a rich, moist soil and full sun, and will adapt very well to most ordinary soils if planted in a sunny situation. All are short-lived and should be propagated each year to ensure continuity.

Anise hyssop, although hardier than the other species, still needs protection below –6°C. The Mexican half-hardy species need protection below –3°C.

Harvesting

Flowers

Cut for drying just as they begin to open.

Leaves

Cut leaves just before late spring flowering.

Seeds

Heads turn brown as the seed ripens. At the first sign of the seed falling, pick and hang upside down with a paper bag tied over the heads.

container growing

Not suitable for growing indoors. However, anise hyssop and Korean mint both make good patio plants provided the container is at least 25–30cm in diameter. Use a potting compost mixed in equal parts with fine composted bark (see page 643), and a liquid fertiliser feed only once a year after flowering. If you feed the plant beforehand, the flowers will be poor. Keep well watered in summer.

other uses

Anise hyssop and Korean mint both have scented leaves, which make them suitable for potpourris.

culinary

The two varieties most suitable are:

Anise Hyssop

Leaves can be used in salads and to make refreshing tea. Like borage, they can be added to summer fruit cups. Equally they can be chopped and used as a seasoning in pork dishes or in savoury rice. Flowers can be added to fruit salads and cups giving a lovely splash of colour.

Korean Mint

Leaves have a strong peppermint flavour and make a very refreshing tea, said to be good first thing in the morning after a night on the town. They are also good chopped up in salads, and the flowers look very attractive scattered over a pasta salad.

Korean mint tea

Alchemilla

LADY'S MANTLE

From the family *Rosaceae.*

Lady's mantle is a native of the mountains of Europe, Asia and America. It is found not only in damp places but also in dry shady woods.

The Arab *alkemelych* (alchemy) is thought to be the source of the herb's Latin generic name, *Alchemilla*. The crystal dew lying in perfect pearl drops on the leaves has long inspired poets and alchemists, and was reputed to have healing and magical properties, even to preserve a woman's youth provided she collected the dew in May, alone, in full moonlight, naked, and with bare feet as a sign of purity and to ward off any lurking forces.

In the medieval period it was dedicated to the Virgin Mary, hence lady's mantle was considered a woman's protector, and nicknamed 'a woman's best friend'. It was used not only to regulate the menstrual cycle and to ease the effects of menopause, but also to reduce inflammation of the female organs. In the eighteenth century, women applied the leaves to their breasts to make them recover shape after they had been swelled with milk. It is still prescribed by herbalists today.

Alchemilla mollis

varieties

***Alchemilla alpina* L.**
Alpine Lady's Mantle
Known in America as Silvery lady's mantle.
Hardy perennial. Ht 15cm, spread 60cm or more. Tiny, greenish-yellow flowers in summer. Leaves rounded, lobed, pale green and covered in silky hairs. An attractive plant suitable for ground cover, rockeries and dry banks.

Alchemilla conjuncta
Lady's Mantle Conjuncta
Hardy perennial. Ht 30cm, spread 30cm or more. Tiny, greenish-yellow flowers in summer. Leaves star-shaped, bright green on top with lovely silky silver hairs underneath. An attractive plant suitable for ground cover, rockeries and dry banks.

***Alchemilla mollis* AGM**
Lady's Mantle (Garden variety)
Hardy perennial. Ht and spread 50cm. Tiny, greenish-yellow flowers in summer. Large, pale green, rounded leaves with crinkled edges.

Alchemilla xanthochlora (vulgaris)
Lady's Mantle (Wildflower variety)
Also known as Lion's foot, Bear's foot and Nine hooks. Hardy perennial. Ht 15–45cm, spread 50cm. Tiny, bright greenish-yellow flowers in summer. Round, pale green leaves with crinkled edges.

Alchemilla mollis flower

LADY'S MANTLE

cultivation

Propagation

Seed

Why is it that something that self-seeds readily around the garden can be so difficult to raise from seed? Sow its very fine seed in early spring or autumn under protection into prepared seed or plug trays (use the cardboard method on page 645), and cover with perlite. No bottom heat required. Germination can either be sparse or prolific, taking 2–3 weeks. If germinating in the autumn, winter seedlings in the trays and plant out the following spring when the frosts are over, at 45cm apart. Alternatively, sow in spring where you want the plant to flower. Thin the seedlings to 30cm apart.

Division

All established plants can be divided in the spring or autumn. Replant in the garden where desired.

Pests and Diseases

This plant rarely suffers from pests or diseases.

Maintenance

Spring Divide established plants. Sow seeds if necessary.
Summer To prevent self-seeding, cut off flowerheads as they begin to die back.
Autumn Divide established plants if necessary. Sow seed.
Winter No need for protection.

Garden Cultivation

This fully hardy plant grows in all but boggy soils, in sun or partial shade.

This is a most attractive garden plant in borders or as an edging plant, but it can become a bit of a nuisance, seeding everywhere. To prevent this, cut back after flowering and at the same time cut back old growth.

Harvesting

Cut young leaves after the dew has dried for use throughout the summer. Harvest for drying as plant comes into flower.

container growing

All forms of lady's mantle adapt to container growing and look very pretty indeed. Use a soil-based compost, water throughout the summer, but feed with liquid fertiliser (following manufacturer's instructions) only occasionally. In the winter, when the plant dies back, put the container in a cold greenhouse or cold frame, and water only very occasionally. Lady's mantle can be grown in hanging baskets as a centrepiece.

Alchemilla mollis

Leaves laid out for drying

medicinal

Used by herbalists for menstrual disorders. It has been said that if you drink an infusion of green parts of the plant for 10 days each month it will help relieve menopausal discomfort. It can also be used as a mouth rinse after tooth extraction. Traditionally, the alpine species has been considered more effective, although this is not proven.

culinary

Tear young leaves, with their mild bitter taste, into small pieces and toss into salads. Many years ago Marks & Spencer had a yogurt made with lady's mantle leaves! I wish I had tried it.

other uses

Excellent for flower arranging. Leaves can be boiled for green wool dye. They are also used in veterinary medicine for the treatment of diarrhoea.

Allium

WELSH & TREE ONIONS

From the family *Alliaceae*.

These plants are distributed throughout the world. The onion has been in cultivation so long that its country of origin is uncertain, although most agree that it originated in Central Asia. It was probably introduced to Europe by the Romans. The name seems to have been derived from the Latin word *unio*, a large pearl. In the Middle Ages it was believed that a bunch of onions hung outside the door would absorb the infection of the plague, saving the inhabitants. Later came the scientific recognition that its sulphur content acts as a strong disinfectant. The juice of the onion was used to heal gunshot wounds.

Allium fistulosum

Allium cepa Proliferum Group

varieties

There are many, many varieties of onion; the following information concerns the two that have herbal qualities.

Allium fistulosum
Welsh Onion
Also known as Japanese leek.
Evergreen hardy perennial. Ht 60–90cm. Greenish-yellow flowers on second year's growth in early summer. Leaves are green hollow cylinders. This onion is a native of Siberia and extensively grown in China and Japan. The name Welsh comes from *walsch,* meaning foreign.

***Allium cepa* Proliferum Group**
Tree Onion
Also known as Egyptian onion, Lazy man's onions.
Hardy perennial. Ht 90–150cm. Small greenish-white flowers appear in early summer. It grows bulbs underground and then, at the end of flowering, bulbs in the air. Seeing is believing. It originates from Canada. It is very easy to propagate.

cultivation

Propagation

Seed
Welsh onion seed loses its viability within 2 years, so sow fresh in late winter or early spring under protection with a bottom heat of between 15°C and 21°C. Cover with perlite. When the seedlings are large enough, and after a period of hardening off, plant out into a prepared site in the garden at a distance of 25cm apart. The tree onion is not grown from seed.

Division
Each year the Welsh onion multiply in clumps, so it is a good idea to divide them every 3 years in the spring.

Because the tree onion is such a big grower, it is a good idea to split the underground bulbs every 3 years in spring.

Bulbs
The air-growing bulbils of the tree onion have small root systems, each one capable of reproducing another plant. Plant where required in an enriched soil either in the autumn, as the parent plant dies back, or in the spring.

Pests and Diseases
The onion fly is the curse of the onion family, especially in late spring and early summer. The way to try to prevent this is to take care not to damage the roots or leaves when thinning the seedlings and also not to leave the thinnings lying around, as the scent attracts the fly.

Another problem is downy mildew caused by cool, wet autumns; the leaves become velvety and die back. Another disease is white rot, which can be introduced in contaminated soil. If you have either of the above diseases, burn the affected plants and do not plant in the same position again.

Other characteristic diseases are neck rot and bulb rot, both caused by a *Botrytis* fungus that usually occurs as a result of the bulbs being damaged either by digging or hoeing.

Onions are prone to many more diseases but, if you keep the soil fertile and do not make life easy for the onion fly, you will still have a good crop.

Maintenance

Spring Sow the seed, divide 3-year-old clumps of Welsh and tree onions. Plant bulbs of tree onions.
Summer Stake mature tree onions to stop them falling over and depositing the ripe bulbils on the soil.
Autumn Mulch around tree onion plants with well-rotted manure. Use a small amount of manure around Welsh onions.
Winter Neither variety needs protection.

Garden Cultivation

Welsh Onions

These highly adaptable hardy onions will grow in any well-drained fertile soil. Seeds can be sown in spring after the frosts, direct into the ground. Thin to a distance of 25cm apart. Keep well watered throughout the growing season. In the autumn give the area a mulch of well-rotted manure.

Tree Onions

Dig in some well-rotted manure before planting. Plant the bulbs in their clusters in a sunny well-drained position at a distance of 30–45cm apart.

In the first year nothing much will happen (unless you are one of the lucky ones). If the summer is very dry, water well.

In the following year, if you give the plant a good mulch of well-rotted manure in autumn, it grows to 90–150cm and produces masses of small onions.

Harvesting

Welsh onions may be picked at any time from early summer onwards. The leaves do not dry well but can be frozen like those of their cousin, chives. Use scissors and snip them into a plastic bag. They form neat rings; freeze them.

The little tree onions can be picked off the stems and stored; lay them out on a rack in a cool place with good ventilation.

culinary

Welsh onions make a great substitute for spring onions, as they are hardier and earlier. Pull and use in salads or stir-fry dishes. Chop and use instead of chives.

Tree onions provide fresh onion flavour throughout the year. The bulbils can be pickled or chopped raw in salads (fairly strong), or cooked whole in stews and casseroles.

Pissaladière (A French Pizza)

Serves 4–6

4 tablespoons olive oil (not extra virgin)
20 tree onions, finely chopped
1 clove garlic, crushed
1 dessertspoon fresh thyme, chopped
Salt
Freshly ground black pepper
360g once-risen bread dough
250g ripe tomatoes, peeled and sliced
60g canned anchovy fillets, drained and halved lengthways
16 large black olives, halved and pitted

Heat the olive oil in a heavy frying pan, add the onions, cover the pan tightly and fry, gently stirring occasionally for 15 minutes. Add the garlic and the thyme and cook uncovered for 15 minutes, or until the onions are reduced to a clear purée. Season to taste and leave to cool.

Preheat the oven to 200°C/400°F/gas mark 6. Roll the bread dough directly on the baking sheet into a circle 25cm in diameter. Spread the puréed onions evenly over the dough, put the tomato slices on the onions and top with a decorative pattern of anchovy fillets and olives.

Bake for 5 minutes. Reduce the oven temperature to 190°C/375°F/gas mark 5 and continue to bake for 30 minutes or until the bread base is well risen and lightly browned underneath.

Serve hot with a green herb salad.

container growing

Welsh onions can be grown in a large pot using a soil-based compost, but make sure the compost does not dry out. Feed regularly throughout the summer with a liquid fertiliser.

Tree onions grow too tall for containers.

other uses

The onion is believed to help ward off colds in winter and also to induce sleep and cure indigestion. The fresh juice is antibiotic, diuretic, expectorant and antispasmodic, so is useful in the treatment of coughs, colds, bronchitis, laryngitis and gastroenteritis. It is also said to lower the blood pressure and to help restore sexual potency that has been impaired by illness or mental stress.

Allium schoenoprasum

CHIVES

From the family *Alliaceae.*

Chives are the only member of the onion group found wild in Europe, Australia and North America, where they thrive in temperate and warm to hot regions. Although they are one of the most ancient of all herbs, chives were not cultivated in European gardens until the sixteenth century.

Chives were a favourite in China as long ago as 3,000 BC. They were enjoyed for their delicious mild onion flavour and used as an antidote to poison and to stop bleeding. Their culinary virtues were first reported to the West by the explorer and traveller, Marco Polo. During the Middle Ages they were sometimes known as rush-leeks, from the Greek *schoinos* meaning 'rush' and *parson* meaning 'leek'.

Allium schoenoprasum f. *albiflorum*

varieties

Allium schoenoprasum
Chives
Hardy perennial. Ht 30cm. Purple globular flowers all summer. Leaves green and cylindrical. Apart from being a good culinary herb it makes an excellent edging plant.

***Allium schoenoprasum* fine-leaved**
Extra Fine-Leaved Chives
Hardy perennial. Ht 20cm. Purple globular flowers all summer. Very narrow cylindrical leaves, not as coarse as standard chives. Good for culinary usage.

Allium schoenoprasum* f. *albiflorum
White Chives
Hardy perennial. Ht 20cm. White globular flowers all summer. Cylindrical green leaves. A cultivar of ordinary chives and very effective in a silver garden. Good flavour.

***Allium schoenoprasum* 'Forescate'**
Pink Chives
Hardy perennial. Ht 20cm. Pink flowers all summer. Cylindrical green leaves. The pink flowers can look a bit insipid when planted too close to standard chives. Good in flower arrangements.

Allium tuberosum
Garlic Chives, Chinese Chives
Hardy perennial. Ht 40cm. White flowers all summer. Leaf mid-green, flat and solid with a sweet garlic flavour when young. As they get older the leaf becomes tougher and the taste coarser.

CHIVES

cultivation

Propagation

Seed

Easy from seed, but they need a temperature of 19°C to germinate, so if sowing outside, wait until late spring for the soil to be warm enough. I recommend starting this plant in plug trays with bottom heat in early spring. Sow about 10–15 seeds per 3cm cell. Transplant either into pots or the garden when the soil has warmed through.

Division

Every few years in the spring lift clumps (made up of small bulbs) and replant in 6–10-bulb clumps, 15cm apart, adding fresh compost or manure.

Pests and Diseases

Greenfly may be a problem on pot-grown herbs. Wash off gently under the tap or use a liquid horticultural soap. Be diligent, for aphids can hide deep down low amongst the leaves.

Cool wet autumns may produce downy mildew; the leaves will become velvety and die back from the tips. Dig up, split and re-pot affected plants, at the same time cutting back all the growth to prevent it spreading.

Chives can also suffer from rust. As this is a fungus it is essential to cut back diseased growth immediately and burn it. Do not compost. If very bad, remove the plant and burn it all. Do not plant any chives or garlic in that area.

Maintenance

Spring Clear soil around emerging established plants. Feed liquid fertiliser. Sow seeds. Divide established plants.
Summer Remove the flower stem before flowering to increase leaf production.
Autumn Prepare soil for next year's crop. Dig up a small clump, pot, bring inside for forcing.
Winter Cut forced chives and feed regularly.

Garden Cultivation

Chives are fairly tolerant regarding soil and position, but produce the best growth planted 15cm from other plants in a rich moist soil and in a fairly sunny position. If the soil is poor they will turn yellow and then brown at the tips. For an attractive edging, plant at a distance of 10cm apart and allow to flower. Keep newly transplanted plants well watered in the spring, and in the summer make sure that they do not dry out, otherwise the leaves will quickly shrivel. Chives die right back into the ground in winter, but a winter cutting can be forced by digging up a clump in autumn, potting it into a loam-based potting compost mixed in equal parts with composted fine bark, and placing it somewhere warm with good light.

Harvesting

Chives may be cut to within 3cm of the ground 4 times a year to maintain a supply of succulent fresh leaves. Chives do not dry well. Refrigerated leaves in a sealed plastic bag will retain crispness for up to seven days. Freeze chopped leaves in ice cubes for convenience. Cut flowers when they are fully open before the colour fades for use in salads and sauces.

companion planting

Traditionally it is said that chives planted next to apple trees prevent scab, and when planted next to roses can prevent black spot. Hence the saying, 'Chives next to roses creates posies'.

container growing

Chives grow well in pots or on a windowsill and flourish in a window box if partially shaded. Use a soil-based compost mixed in equal parts with composted fine bark. They need an enormous quantity of water and occasional liquid feed to stay green and succulent. Remember too that, being bulbs, chives need some top growth for strengthening and regeneration, so do not cut away all the leaves if you wish to use them next season. Allow to die back in winter if you want to use it the following spring. A good patio plant, easy to grow, but not particularly fragrant.

medicinal

The leaves are mildly antiseptic and when sprinkled onto food they stimulate the appetite and help to promote digestion.

other uses

Chives are said to prevent scab infection on animals.

culinary

Add chives at the end of cooking or the flavour will disappear. They are delicious freshly picked and snipped as a garnish or flavour in omelettes or scrambled eggs, salads and soups. They can be mashed into soft cheeses or sprinkled onto grilled meats. Add liberally to sour cream as a filling for jacket potatoes.

Chive Butter

Use in scrambled eggs, omelettes and cooked vegetables and with grilled lamb or fish or on jacket potatoes.

4 tablespoons chopped chives
100g butter, softened
1 tablespoon lemon juice
Salt and freshly ground black pepper

Cream the chives and butter together until well mixed. Beat in the lemon juice and season. Cover and cool the butter in the refrigerator until ready to use; it will keep for several days.

Ajuga reptans

BUGLE

Also known as Common or Creeping Bugle, Bugle Weed, Babies' Shoes, Baby's Rattle, Blind Man's Hand, Carpenter's Herb, Dead Men's Bellows, Horse and Hounds, Nelson's Bugle, Thunder and Lightning and Middle Comfrey. From the family *Lamiaceae.*

The bugle found in Britain is a native of Europe. It is frequently found in mountainous areas and often grows in damp fields, mixed woodland and meadows. The bugle of North America is a species of *Lycopus* (gypsy weed).

Among the many folk tales associated with bugle is one that its flowers can cause a fire if brought into the house, a belief that has survived in at least one district of Germany.

varieties

Ajuga reptans
Bugle
Hardy evergreen perennial. Ht 30cm, spread up to 1m. Very good spreading plant. Blue flowers from spring to summer. Oval leaves are dark green with purplish tinge. It is this plant that has medicinal properties.

***Ajuga reptans* 'Atropurpurea'**
Bronze Bugle
Hardy evergreen perennial. Ht 15cm, spread 1m. Blue flowers from spring to summer. Deep bronze/purple leaves. Very good for ground cover.

***Ajuga reptans* 'Multicolor'**
Multicoloured Bugle
Hardy evergreen perennial. Ht 12cm, spread 45cm. Small spikes of blue flowers from spring to summer. Dark green leaves marked with cream and pink. Good for ground cover.

cultivation

Propagation

Seeds
Sow the small seed in autumn, or spring as a second choice. Cover only lightly with soil. Germination can be erratic and slow.

Division
This method is easy and the only one suitable for cultivars as bugle produces runners, each one having its own root system. Plant out in autumn or spring. Space 60cm apart, as a single plant spreads rapidly.

Pests and Diseases
Nothing much disturbs this plant!

Maintenance
Spring Clear winter debris around established plants. Dig up runners and replant in other areas. Sow seeds.
Summer Dig up runners to control established plants.
Autumn Sow seed and dig up runners of established plants. Pot on, using a loam-based compost mixed in equal parts with composted fine bark, and winter in a cold frame. Alternatively, replant into a prepared site in the garden.
Winter No protection needed unless it is colder than –20°C.

Garden Cultivation
It will grow vigorously on any soil that retains moisture, in full sun, and it also tolerates quite dense shade. It will even thrive in a damp boggy area near the pond or in a hedgerow or shady woodland area. At close quarters bugle is very appealing and can be used as a decorative ground cover. Guard against leaf scorch on the variegated variety.

Harvesting
For medicinal usage the leaves and flowers are gathered in early summer.

container growing

Bugle makes a good outside container plant, especially the variegated and purple varieties. Use a soil-based compost mixed in equal parts with composted fine bark. Also good in hanging baskets.

culinary

The young shoots of *Ajuga reptans* can be mixed in salads to give you a different taste. Not mine.

medicinal

An infusion of dried leaves in boiling water is thought to lower blood pressure and to stop internal bleeding. Nowadays it is widely used in homeopathy in various preparations against throat irritation, especially in the case of mouth ulcers.

other uses

In some countries it is gathered as cattle fodder.

Ajuga reptans 'Artopurpurea'

Aloe vera

ALOE VERA

From the family *Aloaceae.*

There are between 250 and 350 species of aloe around the world. They are originally native to the arid areas of Southern Africa. In cultivation they need a frost-free environment. Aloe has been valued at least since the fourth century BC when Aristotle requested Alexander the Great to conquer Socotra in the Indian Ocean, where many species grow.

Flower spike of *Aloe vera (Aloe barbadensis)*

varieties

***Aloe vera* (*Aloe barbadensis*) AGM**
Aloe vera
Half-hardy perennial. Grown outside: Ht 60cm, spread 60cm or more. Grown as a house plant: Ht 30cm. Minimum temperature 10°C. Succulent grey/ green pointed foliage, from which eventually grows a flowering stem with bell-shaped yellow or orange flowers.

***Aloe arborescens* 'Frutescens'**
Half-hardy perennial. Grown outside: Ht and spread 2m. Minimum temperature 7°C. Each stem is crowned by rosettes of long, blue/green leaves with toothed edges and cream stripes. Produces spikes of red tubular flowers in late winter and spring.

***Aloe variegata* AGM**
Partridge-breasted Aloe
Half-hardy perennial. A house plant only in temperate climates. Ht 30cm, spread 10cm. Minimum temperature 7°C. Triangular, white-marked, dark green leaves. Spike of pinkish-red flowers in spring.

cultivation

Propagation

Seed
A temperature of 21°C must be maintained during germination. Sow the small seeds in spring onto the surface of a pot or tray, using a standard seed compost mixed in equal parts with sharp horticultural sand, and cover with perlite. Place in a propagator with bottom heat. Germination is erratic – 4 to 24 months.

Division
In summer gently remove offshoots at the base of a mature plant. Leave for a day to dry, then pot into 2 parts compost to 1 part sharp sand mix. Water in and leave in a warm place to establish. Give the parent plant a good liquid feed when returning to its pot.

Pests and Diseases
Overwatering causes it to rot off.

Maintenance
Spring Sow seeds. Give containerised plants a good dust! Spray the leaves with water. Give a good feed of liquid fertiliser.
Summer Remove the basal offshoots of a mature plant to maintain the parent plant. Re-pot mature plants.
Autumn Bring in pots if there is any danger of frost.
Winter Rest all pot-grown plants in a cool room (minimum temperature 5°C); water sparingly.

Garden Cultivation
Aloes enjoy a warm, frost-free position – the full sun to partial shade – and a free-draining soil. Leave 1m minimum between plants.

Aloe vera leaf

Harvesting
Cut leaves throughout the growing season. A plant of more than 2 years old has stronger properties.

container growing

Compost must be gritty and well drained. Don't over water. Maintain a frost-free, light environment.

cosmetic

Aloe vera is used in cosmetic preparations, in hand creams, suntan lotions and shampoos.

medicinal

The gel obtained by breaking the leaves is a remarkable healer. Applied to wounds it forms a clear protective seal and encourages skin regeneration. It can be applied directly to cuts and burns, and is immediately soothing.

warning

It should be emphasised that, apart from external application, aloes are not for home medication. ALWAYS seek medical attention for serious burns.

Aloysia triphylla

LEMON VERBENA

From the family *Verbenaceae*.

Lemon verbena grew originally in Chile.

This Rolls Royce of lemon-scented plants was first imported into Europe in the eighteenth century by the Spanish for its perfume.

varieties

***Aloysia triphylla* AGM (*Lippia citriodora*)**
Lemon Verbena
Half-hardy deciduous perennial. Ht 1–3m, spread up to 2.5m. Tiny white flowers tinged with lilac in early summer. Leaves pale green, lance shaped and very strongly lemon scented.

cultivation

Propagation

Seeds
The seed sets only in warm climates and should be sown in spring into prepared seed or plug trays and covered with perlite; a bottom heat of 15°C helps. Prick out into 9cm pots using a standard seed compost mixed in equal parts with composted fine bark. Keep in pots for the first 2 years before planting specimens in the garden 1m apart.

Cuttings
Take softwood cuttings from the new growth in late spring. The cutting material will wilt quickly so have everything prepared. Take semi-hardwood cuttings in late summer or early autumn. Keep in pots for the first 2 years.

Pests and Diseases
If grown under protection you may have whitefly and red spider mite; either use the relevant predators or spray with a liquid horticultural soap, but not both.

Maintenance
Spring Trim established plants. Take softwood cuttings. In warm climates sow seed.
Summer Trim lightly after flowering to remove the flowers. Take semi-hardwood cuttings.
Autumn Protect from excessive wet and frosts.
Winter Keep plants frost free.

Garden Cultivation
Likes a warm humid climate. The soil should be light, free draining and warm. A sunny wall is ideal. It will need protection against frost and wind, and temperatures below 4°C. If left in the ground, cover the area around the roots with mulching material. In spring give the plant a gentle prune and spray with warm water to help revive it.

New growth can appear very late in spring so never discard a plant until late summer. Once the plant has started re-shooting, cut back all last year's growth to 4cm. Cut off flowers once flowering has finished.

Harvesting
Pick the leaves any time before they start to wither and darken. Leaves dry quickly and easily, keeping their colour and scent. Store in a damp-proof container.

container growing

Choose a container at least 20cm wide and use a soil-based compost mixed in equal parts with composted fine bark. Place the container in a warm, sunny, light and airy spot. Water well throughout the growing season and feed with liquid fertiliser during flowering. Cut back last year's growth to 4cm in the spring only. In winter move the container into a cold greenhouse, and allow the compost to nearly dry out. The plant must be allowed to drop its leaves: this is not a sign that it is dead.

medicinal

A tea last thing at night is refreshing and has mild sedative properties; it can also soothe bronchial and nasal congestion and ease indigestion. However, long-term use may cause stomach irritation.

other uses

The leaves with their strong lemon scent are lovely in potpourris, linen sachets, herb pillows, sofa sacks. The distilled oil made from the leaves is an essential basic ingredient in many perfumes.

culinary

Use fresh leaves to flavour oil and vinegar, drinks, fruit puddings, confectionery, apple jelly, cakes and stuffings. Infuse in finger bowls.

Add a teaspoon of chopped, fresh leaves to home-made ice cream for a delicious dessert.

Althaea officinalis

MARSH-MALLOW

Also known as Mortification Root, Sweet Weed, Wymote, Marsh Malice, Mesh-mellice, Wimote, and Althea. From the family *Malvaceae.*

Marsh-mallow is widely distributed from Western Europe to Siberia, from Australia to North America. It is common to find it in salt marshes and on banks near the sea. The generic name, *Althaea*, comes from the Latin *altheo,* meaning 'I cure'. It may be the *althea* that Hippocrates recommended so highly for healing wounds. The Romans considered it a delicious vegetable, used it in barley soup and in stuffing for suckling pigs. In the Renaissance era the herbalists used marsh-mallow to cure sore throats, stomach trouble and toothache. The soft, sweet marshmallow was originally flavoured with the root of marsh-mallow.

varieties

Althaea officinalis

Marsh-mallow

Hardy perennial. Ht 60–120cm, spread 60cm. Flowers pink or white in late summer/early autumn. Leaves, grey-green in colour, tear-shaped and covered all over with soft hair.

cultivation

Propagation

Seed

Sow in prepared seed or plug trays in the autumn.

Cover lightly with compost and winter outside under glass. Erratic germination takes place in spring. Plant out, 45cm apart, when large enough to handle.

Division

Divide established plants in the spring or autumn, replanting into a prepared site in the garden.

Pests and Diseases

This plant is usually free from pests and diseases.

Maintenance

Spring Divide established plants.
Summer Cut back after flowering for new growth.
Autumn Sow seeds and winter the trays outside.
Winter No need for protection. Fully hardy.

Garden Cultivation

Marsh-mallow is highly attractive to butterflies. A good seaside plant, it likes a site in full sun with a moist or wet, moderately fertile soil. Cut back after flowering to encourage new leaves.

Harvesting

Pick leaves for fresh use as required; they do not preserve well. For use either fresh or dried, dig up the roots of 2-year-old plants in autumn, after the flowers and leaves have died back.

medicinal

Due to its high mucilage content (35 per cent in the root and 10 per cent in the leaf), marsh-mallow soothes or cures inflammation, ulceration of the stomach and small intestine, soreness of throat, and pain from cystitis. An infusion of leaves or flowers serves as a soothing gargle; an infusion of the root can be used for coughs, diarrhoea and insomnia. The pulverised roots may be used as a healing and drawing poultice, which should be applied warm.

Decoction for Dry Hands

Soak 25g of scraped and finely chopped root in 150ml of cold water for 24 hours. Strain well. Add 1 tablespoon of the decoction to 2 tablespoons of ground almonds, 1 teaspoon of milk and 1 teaspoon of cider vinegar. Beat it until well blended. Add a few drops of lavender oil. Put into a small screwtop pot.

culinary

Boil the roots to soften, then peel and quickly fry in butter.

Use the flowers in salads. Leaves are also good in salads and may be added to oil and vinegar, or steamed and served as a vegetable.

Anethum graveolens

DILL

Also known as Dillweed and Dillseed. From the family *Apiaceae*.

A native of southern Europe and western Asia, dill grows wild in the cornfields of Mediterranean countries and also in North and South America. The generic name *Anethum* derives from the Greek *anethon*. 'Dill' is said to come from the Anglo-Saxon *dylle* or the Norse *dilla*, meaning to soothe or lull. Dill was found amongst the names of herbs used by Egyptian doctors 5,000 years ago and the remains of the plant have been found in the ruins of Roman buildings in Britain.

It is mentioned in the Gospel of St Matthew, where it is suggested that herbs were of sufficient value to be used as a tax payment – oh that that were true today! 'Woe unto you, Scribes and Pharisees, hypocrites! For ye pay tithe of mint and dill and cumin, and have omitted the weightier matters of the law.'

During the Middle Ages dill was prized as protection against witchcraft. While magicians used it in their spells, lesser mortals infused it in wine to enhance passion. It was once an important medicinal herb for treating coughs and headaches, as an ingredient of ointments and for calming infants with whooping cough – dill water or gripe water is still called upon today. Early settlers took dill to North America, where it was known as the'Meeting House Seed', because the children were given the seed to chew during long sermons to prevent them feeling hungry.

varieties

Anethum graveolens

Dill

Annual. Ht 60–150cm, spread 30cm. Tiny yellow/green flowers in flattened umbel clusters in summer. Fine aromatic feathery green leaves.

cultivation

Propagation

Seed

Seed can be started in early spring under cover, using pots or module plug trays. Do not use seed trays, as dill does not like being transplanted, and if it gets upset it will bolt and miss out the leaf-producing stage.

The seeds are easy to handle, being a good size. Place four per plug or evenly spaced on the surface of a pot, and cover with perlite. Germination takes 2–4 weeks, depending on the warmth of the surrounding area. As soon as the seedlings are large enough to handle, the air and soil temperatures have started to rise and there is no threat of frost, plant out 28cm apart.

Garden Cultivation

Keep dill plants well away from fennel, otherwise they will cross-pollinate and their individual flavours will become muddled. Dill prefers a well-drained, poor soil in partial shade. Sow mid-spring into shallow drills on a prepared site, where they will be harvested. Protect seedlings from wind. When the plants are large enough to handle, thin out to a distance of about 20cm apart to allow plenty of room for growth. Make several small sowings in succession so that you have a supply of fresh leaves throughout the summer. The seed is viable for 3 years.

The plants are rather fragile and it may be necessary to provide support. Twigs pushed into the ground around the plant and enclosed with string or raffia give better results than attempting to stake each plant.

In very hot summers, make sure that the plants are watered regularly or they will run to seed. There is no need to liquid feed, as this only promotes soft growth and in turn encourages pests and disease.

Pests and Diseases

Watch out for greenfly in crowded conditions. Treat with

DILL

Dill vinegar

a liquid horticultural soap if necessary. Be warned, slugs love dill plants.

Maintenance

Spring Sow the seeds successively for a leaf crop.
Summer Water well after cutting to promote new growth.
Autumn (early) Harvest seeds.
Winter Dig up all remaining plants. Make sure all the seed heads have been removed before you compost the stalks, as the seed is viable for 3 years. If you leave the plants to self-seed they certainly will, and they will live up to their other name of Dillweed.

Harvesting

Pick leaves fresh for eating at any time after the plant has reached maturity. Since it is quick-growing, this can be within 8 weeks of the first sowing. Although leaves can be dried, great care is needed and it is better to concentrate on drying the seed for storage.

Cut the stalks off the flower heads when the seed is beginning to ripen. Put the seed heads upside down in a paper bag and tie the top of the bag. Put in a warm place for a week. The seeds should then separate easily from the husk when rubbed in the palm of the hand. Store in an airtight container and the seeds will keep their flavour very well.

container growing

Dill can be grown in containers, in a sheltered corner in partial shade. However, it will need staking. The art of growing it successfully is to keep cutting the plant for use in the kitchen. That way you will promote new growth and keep the plant reasonably compact. The drawback is that it will be fairly short-lived, so you will have to do successive sowings in different pots to maintain a supply. I do not recommend growing dill indoors -- it will get leggy, soft and prone to disease.

medicinal

Dill is an antispasmodic and calmative. Dill tea or water is a popular remedy for an upset stomach, hiccups or insomnia, for nursing mothers to promote the flow of milk, and as an appetite stimulant. It is a constituent of gripe water and other children's medicines because of its ability to ease flatulence and colic.

other uses

Where a salt-free diet must be followed, the seed, whole or ground, is a valuable replacement. Try chewing the seeds to clear up halitosis and sweeten the breath. Crush and infuse seeds to make a nail-strengthening bath.

culinary

Dill is a culinary herb that improves the appetite and digestion. The difference between dill leaf and dill seed lies in the degree of pungency. There are occasions when the seed is better because of its sharper flavour. It is used as a flavouring for soup, lamb stews and grilled or boiled fish. It can also add spiciness to rice dishes, and be combined with white wine vinegar to make dill vinegar.

Dill leaf can be used generously in many dishes, as it enhances rather than dominates the flavour of many foods.

Before it sets seed, add one flowering head to a jar of pickled gherkins, cucumbers and cauliflowers for a flavour stronger than dill leaves but fresher than seeds. In America these are known as dill pickles.

Gravlax (Salmon marinaded with dill)

This is a traditional Scandinavian dish of great simplicity and great merit. Salmon treated in this way will keep for up to a week in the fridge.

420–800g salmon, middle or tail piece
1 heaped tablespoon sea salt
1 rounded tablespoon caster sugar
1 teaspoon crushed black peppercorns
1 tablespoon brandy (optional)
1 heaped tablespoon fresh dill

Have the salmon cleaned, scaled, bisected lengthways and filleted. Mix remaining ingredients together and put some of the mixture into a flat dish (glass or enamel) large enough to take the salmon. Place one piece of salmon skin side down on the bottom of the dish, spread more of the mixture over the cut side. Add the second piece of salmon, skin up, and pour over the remaining mixture. Cover with foil and place a plate or wooden board larger than the area of the salmon on top. Weigh this down with weights or heavy cans. Put in the refrigerator for 36–72 hours. Turn the fish completely every 12 hours or so and baste (inside surfaces too) with the juices.

To serve, scrape off all the mixture, pat the fish dry and slice thinly and at an angle. Serve with buttered rye bread and a mustard sauce called Gravlaxsas:

4 tablespoons mild, ready-made Dijon mustard
1 teaspoon mustard powder
1 tablespoon caster sugar
2 tablespoons white wine vinegar
6 tablespoons of vegetable oil
3 to 4 tablespoons of chopped dill

Mix the Dijon mustard with the mustard powder, caster sugar and vinegar, then slowly add the oil until you have a sauce the consistency of mayonnaise. Finally, stir in the dill. Alternatively, substitute a mustard and dill mayonnaise.

ANGELICA

Also known as European Angelica, Garden Angelica and Root Of The Holy Ghost. From the family *Apiaceae.*

Angelica in its many forms is a native of Europe, Asia and North America. It is also widely cultivated as a garden plant. Wild angelica is found in moist fields and hedgerows throughout Europe. American angelica is found in similar growing conditions in Canada and north-eastern and northern central states of America.

***Angelica* probably comes from the Greek *angelos*, meaning 'messenger'. There is a legend that an angel revealed to a monk in a dream that the herb was a cure for the plague, and traditionally angelica was considered the most effective safeguard against evil, witchcraft in particular. Certainly it is a plant no self-respecting witch would include in her brew.**

Angelica is an important flavouring agent in liqueurs such as Benedictine, although its unique flavour cannot be detected from the others used. It is also cultivated commercially for medicinal and cosmetic purposes.

varieties

Angelica archangelica
Angelica
Monocarpic (will live until it has successfully flowered and set seed). Ht 1–2.5m, spread 1m in second year. Dramatic second-year flowerheads late spring through summer, greenish-white and very sweetly scented. Bright green leaves, the lower ones large and bi- or tri-pinnate; the higher, smaller and pinnate. Rootstock varies in colour from pale yellowish-beige to reddish-brown.

Angelica atropurpurea
American Angelica
Also known as Bellyache root, High angelica, Masterwort, Purple angelica and Wild angelica. Biennial. Ht 1.2–1.5m. Flowers resemble those of *A. archangelica* – white to greenish-white, late spring through summer. Leaves large and alternately compound. Rootstock purple. The whole plant delivers a powerful odour when fresh.

Angelica gigas
Korean Angelica
Monocarpic. Ht up to 1.2m, spread 1m in second year. Beautiful second or third year spherical deep red umbels with tiny white flowers in late summer. Palmate green foliage.

***Angelica polymorpha* var. *sinensis* (*Anglica sinensis*)**
Chinese Angelica
Also known as Dang Gui, Women's Ginseng. Monocarpic. Ht. up to 1m, spread 70cm. Umbels of small green and white flowers in summer. Palmate green foliage.

cultivation

Propagation

Seed

The fresh seeds are only viable for 3 months from harvest. Sow in early autumn, either where you want it to grow next year, or into plug trays that are placed outside to get the weather. There is no need to protect the seedlings from frost. Plant out in the following spring. If you cannot sow the seed fresh, mix them with horticultural sand and put them in to a plastic bag and then into the fridge. Once the seeds start to germinate in the bag, approximately 2–6 weeks, sow the seeds in groups of three into a prepared site in the garden 1m apart. This plant

Angelica gigas

ANGELICA

cannot be successfully transplanted once the tap root has developed, so choose your site with care.

Pests and Diseases
Remove blackfly easily with liquid horticultural soap.

Maintenance
Spring Clear ground around existing plants. Plant out autumn seedlings. Put old seed in refrigerator.
Summer Cut stems of second-year growth for crystallising. Cut young leaves before flowering to use fresh in salads or to dry for medicinal or culinary uses. If in summer the leaves turn a yellowish-green, it is usually a sign that the plant needs more water.
Autumn The ideal seed-sowing time.
Winter No need for protection.

Garden Cultivation
Angelica dislikes hot humid climates and appreciates a spot in the garden where it can be in shade for some part of every day. But it can be a difficult plant to accommodate in a small garden, as it needs a lot of space. Site at the back of a border, perhaps near a wall where the plant architecture can be shown off. Make sure that the soil is deep and moist. Add well-rotted compost to help retain moisture. Note that angelica dies down completely in winter but green shoots appear quickly in the spring. Angelica forms a big clump of foliage in the first summer and dramatic flowers the second or third, dying after the seed is set. A plant will propagate itself in the same situation if allowed to self-seed.

Harvesting
Harvest leaves – use fresh from spring onwards for culinary and medicinal; for drying, from early summer until flowering. Dry flowers in early summer for flower arrangements. Collect seeds when they begin to ripen. Harvest roots to dry for use medicinally in the second autumn immediately after flowering.

medicinal

It is used to treat indigestion, anaemia, coughs and colds; it has antibacterial and anti-fungal properties. It is said that a tea made from the young leaves is good for reducing tension and nervous headaches and that a decoction made from the roots is soothing for colds and other bronchial conditions. Reputedly crushed leaves freshen the air in a car and may help travel sickness.

culinary

Candied Angelica

Angelica is now best known as a decoration for cakes. The bright emerald, apparently plastic, specimen sold commercially as angelica cannot compare with home-made, pale green candied angelica; this tastes and smells similar to the freshly bruised stem or crushed leaf of the plant.

Angelica stems
Granulated sugar
Caster sugar for dusting

Choose young tender springtime shoots. Cut into 8–10cm lengths. Place in a saucepan with just enough water to cover. Simmer until tender, then strain and peel off the outside skin. Put back into the pan with enough water to cover and bring to the boil, strain immediately and allow to cool.

When cool, weigh the angelica stalks and add an equal weight of granulated sugar. Place the sugar and angelica in a covered dish and leave in a cool place for 2 days.

Put the angelica and the syrup which has formed back in the pan. Bring slowly to the boil and simmer, stirring occasionally, until the angelica becomes clear and has good colour. Strain, discarding all the liquid, then sprinkle as much caster sugar as will cling to the angelica. Allow to dry in a cool oven (100°C/200°F/gas mark 1/4). If not thoroughly dry they become mouldy. Store in an airtight container between sheets of greaseproof paper.

Stewed Rhubarb

If when you cook rhubarb or gooseberries you add young angelica leaves, you will need to add less sugar. Angelica does not sweeten the fruit but its muscatel flavour cuts through the acidity of the rhubarb.

900g rhubarb
225g angelica stems
1 orange juice and rind
150ml water
50g sugar

Put all the ingredients in a pan, bring to the boil, then turn down the heat and simmer for 5 minutes.

container growing

Angelica is definitely not an indoor plant, though if the container is large enough it can be grown as such. Do not over-fertilise and be prepared to stake when in flower. Be wary of the pot toppling over as the plant grows taller.

warning

Large doses first stimulate and then paralyse the central nervous system. The tea is not recommended for those suffering from diabetes.

Anthriscus cerefolium

CHERVIL

From the family *Apiaceae*.

Native to the Middle East, South Russia and the Caucasus, chervil can be cultivated in warm temperate climates. It is now occasionally found growing wild.

Almost certainly brought to Britain by the Romans, chervil is one of the Lenten herbs, thought to have blood-cleansing and restorative properties. It was eaten in quantity in those days, especially on Maundy Thursday.

Gerard, the Elizabethan physician who superintended Lord Burleigh's gardens, wrote in his *Herball* of 1597, 'The leaves of sweet chervil are exceeding good, wholesome and pleasant among other salad herbs, giving the taste of Anise seed unto the rest.'

varieties

Anthriscus cerefolium
Chervil
Hardy annual (some consider it to be a biennial). Ht 30–60cm, spread 30cm. Flowers, tiny and white, grow in clusters from spring to summer. Leaves, light green and fern-like, in late summer may take on a purple tinge. When young it can easily be confused with cow parsley. However, cow parsley is a perennial and eventually grows much taller and stouter, its large leaves lacking the sweet distinctive aroma of chervil.

***Anthriscus cerefolium* 'Crispum'**
Chervil curly leafed
Hardy annual. Grows like the ordinary chervil except that, in my opinion, the leaf has an inferior flavour.

cultivation

Propagation
Seed
The medium-size seed germinates rapidly as the air and soil temperatures rise in the spring provided the seed is fresh (it loses viability after about a year). Young plants are ready for cutting 6–8 weeks after sowing, thereafter continuously providing leaves as long as the flowering stems are removed.

Sow seed in prepared plug module trays if you prefer, and cover with perlite. Pot on to containers with a minimum 12cm diameter. But, as a plant with a long tap root, chervil does not like being transplanted, so keep this to a minimum. It can in fact be sown direct into a 12cm pot, growing it, like mustard and cress, as a 'cut and come again' crop.

Pests and Diseases
Chervil can suffer from greenfly. Wash off gently with a liquid horticultural soap. Do not blast off with a high-pressure hose, as this will damage the soft leaves.

Maintenance
Spring Sow seeds.
Summer A late sowing in this season will provide leaves through winter, as it is very hardy. Protect from midday sun.
Autumn Cloche autumn-sown plants for winter use.
Winter Although chervil is hardy, some cloche protection is needed to ensure leaves in winter.

Garden Cultivation
The soil required is light with a degree of moisture retention. Space plants 23–30cm apart. Semi-shade is best, because the problem with chervil is that it will burst into flower too quickly should the weather become sunny and hot and be of no use as a culinary herb. For this reason some gardeners sow between rows of other garden herbs or vegetables or under deciduous plants to ensure some shade during the summer months.

Harvesting
Harvest leaves for use fresh when the plant is 6–8 weeks old or when 10cm tall, and all the year round if you cover with a cloche in winter. Otherwise, freezing is the best method of preservation, as the dried leaves do not retain their flavour.

CHERVIL

Anthriscus cerefolium – a cut and come again plant

container growing

When grown inside in the kitchen chervil loses colour, gets leggy and goes floppy, so unless you are treating it as a 'cut and come again' plant, plant outside in a large container that retains moisture and is positioned in semi-shade.

Chervil looks good in a window box, but be sure that it gets shade at midday.

medicinal

Leaves eaten raw are rich in Vitamin C, carotene, iron and magnesium. They may be infused to make a tea to stimulate digestion and alleviate circulation disorders, liver complaints and chronic catarrh, and fresh leaves may be applied to aching joints in a warm poultice.

other uses

An infusion of the leaf can be used to cleanse skin, maintain suppleness and discourage wrinkles.

culinary

It is one of the traditional *fines herbes*, indispensable to French cuisine and a fresh green asset in any meal, but many people elsewhere are only now discovering its special delicate parsley-like flavour with a hint of aniseed.

This is a herb especially for winter use because it is easy to obtain fresh leaves and, as every cook knows, French or otherwise, 'Fresh is best'. Use its leaf generously in salads, soups, sauces, vegetables, chicken, white fish and egg dishes. Add freshly chopped towards the end of cooking to avoid flavour loss. In small quantities it enhances the flavour of other herbs. Great with vegetables.

Chervil with broad beans

Apium graveolens

CELERY LEAF

Also known as Wild Celery, Smallage, Ajmud. From the family *Apiaceae.*

Wild celery has been used in both medicine and food for thousands of years. Seeds from this herb were found in the tomb of Tutankhamun (1327BC). Today it can be found growing wild in marshy ground and by rivers in Europe and northern Africa. The whole plant strongly smells of celery – root, leaves and seed – hence the latin *graveolens,* which means 'strong-smelling'. The salad celery, *Apium graveolens* var. *dulce* was originally bred in the seventeenth century from this wild form.

varieties

Apium graveolens
Celery Leaf
Hardy Biennial. Ht 30cm–1m, spread 15–30cm. Umbels of tiny green white flowers early in the second summer followed by ridged grey-brown seeds. The bright, aromatic, mid-green, cut leaves are very similar to its cousin French parsley, with which, when young, it is often confused.

cultivation

Propagation

Seed

Sow the fresh seed in early spring into pots or plug modules using a standard seed compost. Cover with perlite, place under protection at 15°C. Germination takes 2–3 weeks. Alternatively sow seeds in late spring when the air temperature does not drop below 7°C at night into a well-prepared site that has been fed in the previous autumn with well-rotted manure.

Pests and Diseases

Some years this herb has no problems at all then, in other years, it gets attacked by everything. If you see brown tunnel marks on the leaf, this is caused by the celery leaf miner. As soon as you see affected leaves remove them and burn them. If you have a bad infestation cut the plant back to 3cm above the ground and make sure you remove all debris. Like parsley, it can be attacked by carrot fly in early summer. This fly attacks the roots and stem bases. To prevent this, cover the crop with horticultural fleece, and only thin the seedlings in the early morning before the temperature rises. Cover immediately after thinning with horticultural fleece to prevent the fly laying its eggs.

Maintenance

Spring Sow seeds under protection.
Summer Sow seeds into a prepared site. Remove flowers to maintain growth if seed not needed. Liquid feed plants.
Autumn Harvest seeds. Feed the soil for next season's crops.
Winter Pick leaves. Only protect if temperatures fall below –5°C.

CELERY LEAF

Garden Cultivation
Celery leaf is a hungry plant: it likes a good deep soil that does not dry out in summer. Always feed the chosen site well in the previous autumn with well-rotted manure. If you wish to harvest celery leaf all year round have two different sites prepared. For summer supplies, a western or eastern border is ideal because the plant needs moisture and prefers a little shade. For winter supplies a more sheltered spot will be needed in a sunny position, for example against a sunny wall.

Harvesting
Pick the new young shoots throughout the growing season. Harvest the seeds in the late summer once they are fully ripe and start falling from the seed head. If you are saving the seed for sowing the following year be aware that celery is prone to disease and this can be harboured in the seed. So, for propagation, collect from healthy plants only. Dig up the roots of the second-year growth for use medicinally.

companion planting

It is said that this herb helps repel the cabbage white fly from Brassicas.

container growing

Celery leaf is ideal for containers. It can adapt to any deep container as it has a long tap root. Use a soil-based compost. In summer place the container in partial shade to prevent the leaves from becoming tough. Feed and water regularly throughout the growing season.

medicinal

The whole plant is used medicinally: roots, stem, leaf and seeds. It is used to treat osteoarthritis, rheumatoid arthritis and gout. Externally it is used to treat fungal infections. In the East it is used in Ayurvedic medicine to treat asthma, bronchitis and hiccups.

other uses

The essential oil made from the seed is used in perfumery and to make celery salt.

warning

If the plant is infected with the fungus *Sclerotinia sclerotiorum*, contact with the sap can cause dermatitis in sensitive skin. Never take medicinally when pregnant.

Apium graveolens

culinary

The young leaves are delicious in salads or when added to mashed potato. They are also great added to soups or sweated down with onions. In my opinion the mature, tough leaves are inedible. The seeds are also very useful: grind them in a pestle and mortar or add them whole to stews, casseroles and soups. They are also lovely added to dough for flavoursome bread. In India it is an important minor spice.

Celery Seed Bread
Serves 4

450g wholewheat flour, plus extra for dusting
2 teaspoons of sunflower oil plus some extra for the griddle
1 tablespoon of finely chopped young celery leaves
1 pinch of salt
2 teaspoons celery seeds
Water to make the dough

Make a soft dough with the flour, sunflower oil, celery leaves, salt and celery seeds. Add water sparingly as one works the dough. Once the dough is easy to handle, divide into 4 equal-sized balls. Cover a board or surface with a dusting of flour, roll out each ball into a flat disc, adding a further dusting of flour if it sticks. By keeping the dough moving as you roll, you will prevent this happening.

Heat a griddle to hot. Place the bread on the griddle, reduce the heat and cook until the bread is covered in bubbles. Brush the upper surface with oil and turn over. Cook until bubbles appear on the surface. Cook the three remaining breads, serve warm with soups, pickles or yogurts.

Armoracia rusticana

HORSERADISH

From the family *Brassicaceae*.

Native of Europe, naturalised in Britain and North America.

Originally the horseradish was cultivated as a medicinal herb. Now it is considered a flavouring herb. The common name means a coarse or strong radish, the prefix 'horse' often being used in plants to donate a large, strong or coarse plant. In the sixteenth century it was known in England as Redcol or Recole. In this period the plant appears to have been more popular in Scandinavia and Germany, where they developed its potential as a fish sauce. In Britain horseradish has become strongly associated with roast beef.

Armoracia rusticana 'Variegata'

varieties

Armoracia rusticana (Cochlearia armoracia)
Horseradish
Hardy perennial. Ht 60–90cm, spread indefinite! Flowers white in spring (very rare). Leaves large green oblongs. The large root, which is up to 60cm long, 5cm thick and tapering, goes deep into the soil.

***Armoracia rusticana* 'Variegata'**
Hardy perennial. Ht 60–90cm, spread also indefinite. Flowers white in spring (rare in cool climates). Leaves large with green/cream variegation and oblong shape. Large root which goes deep into the soil. Not as good flavour as *A. rusticana*.

cultivation

Propagation

Root cuttings
In early spring cut pieces of root 15cm long. Put them either directly into the ground, at a depth of 5cm, at intervals of 30cm apart, or start them off in individual pots. These can then be planted out when the soil is manageable and frosts have passed.

Division
Divide established clumps in spring. Remember small pieces of root will always grow, so do it cleanly, making sure that you have collected all the little pieces of root. Replant in a well-prepared site.

Pests and Diseases
Cabbage white caterpillars may feed on the leaves during late summer. The leaves may also be affected by some fungus diseases, but this should not be a problem on vigorous plants and leaves should be simply removed and burnt.

Maintenance
Spring Plant cuttings in garden.
Summer Liquid feed with seaweed fertiliser.
Autumn Dig up roots if required when mature enough.
Winter No need for protection, fully hardy.

Garden Cultivation
Think seriously if you want this plant in your garden. It is invasive. Once you have it, you have it. It is itself a most tolerant plant, liking all but the driest of soils. But for a good crop it prefers a light, well-dug, rich, moist soil. Prepare it the autumn before planting with lots of well-rotted manure. It likes a sunny site but will tolerate dappled shade.

If large quantities are required, horseradish should be given a patch of its own where the roots can be lifted and the soil replenished after each harvest. To produce strong, straight roots I found this method in an old gardening book. Make holes 42cm deep with a crow bar, and drop a piece of horseradish 5–8cm long with a crown on the top into the hole. Fill the hole up with good rotted manure. This will produce strong straight roots in 2–3 years, some of which may be ready in the first year.

Harvesting
Pick leaves young to use fresh, or to dry.

If you have a mature patch of horseradish then the root can be dug up any time for use fresh. Otherwise dig up the roots in autumn. Store roots in sand and make sure you leave them in a cool dark place over the winter.

Alternatively, wash, grate or slice and dry. Another method is to immerse the whole washed roots in white wine vinegar.

companion planting

It is said that when grown near potatoes it improves their disease resistance. However, be careful that it does not take over.

HORSERADISH

Horseradish dye

other uses

Chop finely into dog food to dispel worms and improve body tone.

Traditionally, spray made from an infusion of the roots protects apple trees against brown rot. The roots and the leaves produce a yellow dye for natural dyeing. Slice and infuse in a pan of milk to make a lotion to improve skin clarity.

medicinal

Horseradish is a powerful circulatory stimulant with antibiotic properties. As a diuretic it is effective for lung and urinary infections. It can also help with coughs and sinus congestion. It can also be taken internally for gout and rheumatism, containing as it does potassium, calcium, magnesium and phosphorus.

Grate into a poultice and apply externally to chilblains, stiff muscles, sciatica and rheumatic joints, and to stimulate blood flow.

warning

Overuse may blister the skin. Do not use it if your thyroid function is low or if taking thyroxin. Avoid continuous dosage when pregnant or suffering from kidney problems.

culinary

The reason horseradish is used in sauces, vinegars, and as an accompaniment rather than cooked as a vegetable is that the volatile flavouring oil which is released in grating evaporates rapidly and becomes nothing when cooked. Raw it's a different story. The strongest flavour is from root pulled in the autumn. The spring root is comparatively mild. Fresh root contains calcium, sodium, magnesium and vitamin C, and has antibiotic qualities that are useful for preserving food.

It can be used raw and grated in coleslaw, dips, pickled beetroot, cream cheese, mayonnaise and avocado fillings.

The young leaves can be added to salads for a bit of zip.

Make horseradish sauce to accompany roast beef, and smoked oily fish.

Avocado with Horseradish Cream

Fresh horseradish root (approx. 15cm long); preserved horseradish in vinegar can be substituted. If it is, leave out the lemon juice
1 tablespoon butter
3 tablespoons fresh breadcrumbs
1 apple
1 dessertspoon yogurt
1 teaspoon lemon juice
Pinch of salt and sugar
1 teaspoon chopped fresh chervil
½ teaspoon each of fresh chopped tarragon and dill
3–4 tablespoons double cream
2 avocado (ripe) cut in half, stones removed

Peel and grate the horseradish. Melt the butter and add the breadcrumbs. Fry until brown, and add grated horseradish. Remove from heat and grate the apple into the mixture. Add yogurt, lemon juice, salt, sugar and herbs. Put aside to cool. Chill in refrigerator. Just before serving gently fold the cream into the mixture and spoon into the avocado halves. Serve with green salad and brown toast.

Arnica chamissonis Less.

ARNICA

Also known as Mountain Tobacco, Leopards Bane, Mountain Arnica, Wolfsbane and Mountain Daisy. From the family *Asteraceae.*

It is found wild in the mountainous areas of Canada, North America, and in Europe, where it is a protected species. Bees love it.

The name *Arnica* is said to be derived from the word *ptarmikos*, Greek for 'sneezing'. One sniff of arnica can make you sneeze.

The herb was known by Methusalus and was widely used in the sixteenth century in German folk medicine. Largely as a result of exaggerated 18th-century claims by Venetian physicians, it was, for a short time, a popular medicine.

varieties

***Arnica chamissonis* Less.**
Arnica
Hardy perennial. Ht 30–60cm, spread 15cm. Large, single, scented yellow flowers throughout summer. Oval, hairy, light green leaves.

cultivation

Propagation

Seed

Sow the small seed in spring or autumn in either a pot, plug module or seed tray, and cover with perlite. Place trays in a cold frame as heat will inhibit germination. The seed is slow to germinate, even occasionally as long as two years! Once the seedlings are large enough, pot them up and harden them off in a cold frame.

You can get a more reliable germination if you collect the seed yourself and sow no later than early autumn. After potting up, winter the young plants under protection. They will die back in winter. Plant out in the following spring, when the soil has warmed up, 30cm from other plants.

Division

Arnica's root produces creeping rhizomes, which are easy to divide in spring. This is much more reliable than sowing seed.

Pests and Diseases

Caterpillars and slugs sometimes eat the leaves.

Maintenance

Spring Sow seeds. Divide creeping rhizomes.
Summer Deadhead if necessary. Harvest plant for medicinal use.
Autumn Collect seeds and either sow immediately or store in an airtight container for sowing in the spring.
Winter Note the position in the garden because the plants die right back.

Garden Cultivation

Being a mountainous plant, it is happiest in a sandy acid soil, rich in humus, and in a sunny position. Arnica is a highly ornamental plant with a long flowering season. It is ideally suited for large rock gardens, or the front of a border bed.

Arnica chamissonis Less.

Harvesting

Pick flowers for medicinal use in summer, just before they come into full flower. For drying, pick in full flower, with stalks. Collect leaves for drying in summer before flowering. Dig up roots of 2nd/3rd year growth after the plant has fully died back in late autumn/early winter for drying.

medicinal

Arnica is a famous herbal and homeopathic remedy. A tincture of flowers can be used in the treatment of sprains, wounds and bruises, and also to give relief from rheumatic pain and chilblains, if the skin is not broken. Homeopaths say that it is effective against epilepsy and sea sickness, and possibly as a hair growth stimulant. It has also been shown to be effective against salmonella.

warning

Do not take arnica internally except under supervision of a qualified herbalist or homeopath. External use may cause skin rash or irritation. Never apply to broken skin.

other uses

Leaves and roots smoked as herbal tobacco, hence the name Mountain tobacco.

Artemisia abrotanum

SOUTHERNWOOD

Also known as Lad's Love and Old Man. From the family *Asteraceae.*

This lovely aromatic plant is a native of southern Europe. It has been introduced to many countries and is now naturalised widely in temperate zones.

The derivation of the genus name is unclear. One suggestion is that it honours Artemisia, a famous botanist and medical researcher, sister of King Mausolus (353 BC). Another is that it was named after Artemis or Diana, the Goddess of the Hunt and Moon.

In the seventeenth century, Culpeper recommended that the ashes of southernwood be mingled with salad oil as a remedy for baldness.

varieties

***Artemisia abrotanum* AGM**

Southernwood

Deciduous or semi-evergreen hardy perennial. Ht and spread 1m. Tiny insignificant clusters of dull yellow flowers in summer. The abundant olive-green feathery leaves are finely divided and carry a unique scent.

cultivation

Propagation

Seed

It rarely flowers and sets seed, unless it is being grown in a warm climate.

Cuttings

Take softwood cuttings in spring from the lush new growth, or from semi-hardwood cuttings in summer. Use a standard seed compost mixed in equal parts with composted fine bark. Roots well. It can be wintered as a rooted cutting, when it sheds its leaves and is dormant. Keep the cuttings on the dry side, and in early spring slowly start watering. Plant out 60cm apart after the frosts have finished.

Pests and Diseases

It is free from the majority of pests and diseases.

Maintenance

Spring Cut back to maintain shape. Take cuttings.
Summer Take cuttings.
Autumn Trim any flowers off as they develop.
Winter Protect the roots in hard winters with mulch.

Garden Cultivation

Southernwood prefers a light soil containing well-rotted organic material in a sunny position. However tempted you are by its bedraggled appearance in winter (hence its name, Old man) NEVER cut hard back as you will kill it. This growth protects its woody stems from cold winds. Cut the bush hard in spring to keep its shape, but only after the frosts have finished.

Harvesting

Pick leaves during the growing season for use fresh. Pick leaves for drying in mid-summer.

medicinal

It can be used for expelling worms and to treat coughs and bronchial catarrh. A compress helps to treat frost bite, cuts and grazes.

culinary

The leaves can be used in salads. They have a strong flavour, so use sparingly. It also makes a good aromatic vinegar

other uses

The French call it Garde Robe, and use it as a moth repellent. It is a good fly deterrent, too – hang bunches up in the kitchen, or rub it on the skin to deter mosquitoes.

warning

No product containing southernwood should be taken during pregnancy.

Artemisia absinthium

WORMWOOD

Also known as Absinthe and Green Ginger. From the family *Asteraceae.*

A native of Asia and Europe, including Britain, it was introduced into America as a cultivated plant and is now naturalised in many places. Found on waste ground, especially near the sea in warmer regions.

Legend has it that as the serpent slithered out of Eden, wormwood first sprang up in the impressions on the ground left by its tail. Another story tells that in the beginning it was called *Parthenis absinthium*, but Artemis, Greek goddess of chastity, benefited so much from it that she named it after herself – *Artemisia absinthium*. The Latin meaning of *absinthium* is 'to desist from', which says it all.

Although it is one of the most bitter herbs known, it has for centuries been a major ingredient of aperitifs and herb wines. Both absinthe and vermouth get their names from this plant, the latter being an 18th-century French variation of the German *wermut*, itself the origin of the English name wormwood.

Wormwood was hung by the door where it kept away evil spirits and deterred night-time visitations by goblins. It was also made a constituent of ink to stop mice eating old letters.

It was used as a strewing herb to prevent fleas, hence:

> *'White wormwood hath seed, get a handful or twaine,*
> *to save against March, to make flea to refrain.*
> *Where chamber is sweeped and wormwood is strewn,*
> *no flea for his life, dare abide to be knowne.'*

This extract comes from Thomas Tusser's *Five Hundred Pointes of Good Husbandrie*, written in 1573.

Finally, wormwood is believed to be the herb that Shakespeare had in mind when his Oberon lifted the spell from Titania with 'the juice of Dian's bud', Artemis being known to the Romans as Dian or Diana.

Artemisia absinthium

varieties

Artemisia absinthium
Wormwood
Partial-evergreen hardy perennial. Ht 1m, spread 1.2m. Tiny, yellow, insignificant flowerheads are borne in sprays in summer. The abundant leaves are divided, aromatic and grey/green in colour.

***Artemisia absinthium* 'Lambrook Silver' AGM**
Evergreen hardy perennial. Ht 80cm, spread 50cm. Tiny, grey, insignificant flowerheads are borne in long panicles in summer. The abundant leaves are finely divided, aromatic and silver/grey in colour. May need protecting in exposed sites.

Artemisia pontica
Old Warrior
Evergreen hardy perennial. Ht 60cm, spread 30cm. Tiny, silver/grey insignificant flowerheads are borne on tall spikes in summer. The abundant, feathery, small leaves are finely divided, aromatic and silver/grey in colour. This can, in the right conditions, be a vigorous grower, spreading well in excess of 30cm.

***Artemisia* 'Powis Castle' AGM**
Evergreen hardy perennial. Ht 90cm, spread 1.2m. Tiny, greyish-yellow insignificant flowerheads are borne in sprays in summer. The abundant leaves are finely divided, aromatic and silver/grey in colour.

WORMWOOD

cultivation

Propagation

Seed

Of the species mentioned above, only wormwood is successfully grown from seed. It is extremely small and best started off under protection. Sow in spring in a prepared seed or plug tray, using a standard seed compost mixed in equal parts with composted fine bark. Cover the seeds with perlite and propagate with heat, 15–21°C. Plant out when the seedlings are large enough to handle and have had a period of hardening off.

Cuttings

Take softwood cuttings from the lush new growth in early summer; semi-hardwood in late summer. Use a seed compost mixed in equal parts with perlite.

Division

As they are all vigorous growers, division is a good idea at least every 3 to 4 years to keep the plant healthy, to stop it becoming woody and to prevent encroaching. Dig up the plant in spring or autumn, divide the roots and replant in a chosen spot.

Pests and Diseases

Wormwood can suffer from a summer attack of blackfly. If it gets too bad, use a liquid horticultural soap, following manufacturer's instructions.

Maintenance

Spring Sow seeds. Divide established plants. Trim new growth for shape.
Summer Take softwood cuttings.
Autumn Prune back semi-hardwood cuttings to within 15cm of the ground. Divide established plants.
Winter Protect in temperatures below −5°C. Cover with horticultural fleece, straw, bark, anything that can be removed in the following spring.

Garden Cultivation

Artemisias like a light well-drained soil and sunshine, but will adapt well to ordinary soils provided some shelter is given. Planting distance depends on spread.

Wormwood is an overpoweringly flavoured plant and it does impair the flavour of dill and coriander so do not plant nearby.

Harvesting

Pick flowering tops just as they begin to open. Dry. Pick leaves for drying in summer.

container growing

Artemisia absinthium 'Lambrook Silver' and *Artemisia pontica* (old warrior) look very good in terracotta containers. Use a soil-based compost mixed in equal parts with composted fine bark. Only feed in the summer; if you feed too early the leaves will lose their silvery foliage and revert to a more green look. In winter keep watering to the absolute minimum and protect from hard frosts.

medicinal

True to its name, wormwood expels worms, especially round- and thread- worms.

warning

Not to be taken internally without medical supervision. Habitual use causes convulsions, restlessness and vomiting. Overdose causes vertigo, cramps, intoxication and delirium. Pure wormwood oil is a strong poison, although with a proper dosage there is little danger.

other uses

Wormwood can produce a yellow dye.

Antiseptic vinegar

This vinegar is known as the 'Four Thieves' because it is said that thieves used to rub their bodies with it before robbing plague victims.

1 tablespoon wormwood
1 tablespoon lavender
1 tablespoon rosemary
1 tablespoon sage
1.1 litre vinegar

Put the crushed herbs into an earthenware container. Pour in the vinegar. Cover the container and leave it in a warm sunny place for two weeks. Strain into bottles with tight-fitting, non-metal lids. This makes a very refreshing tonic in the bath, or try sprinkling it on work surfaces in the kitchen.

Moth-Repellent

Wormwood or southernwood can be used for keeping moths and other harmful insects away from clothes. The smell is sharp and refreshing and does not cling to your clothes like camphor moth-balls. Mix the following ingredients well and put into small sachets.

2 tablespoons dried wormwood or southernwood
2 tablespoons dried lavender
2 tablespoons dried mint

Artemisia dracunculus

TARRAGON

Also known as Estragon. From the family *Asteraceae.*

A native of southern Europe, tarragon is now found in dry areas of North America, Southern Asia and Siberia.

***Dracunculus* means 'little dragon'. Its naming could have occurred (via the Doctrine of Signatures) as a result of the shape of its roots, or because of its fiery flavour. It was certainly believed to have considerable power to heal bites from snakes, serpents and other venomous creatures. In ancient times the mixed juices of tarragon and fennel made a favourite drink for the Kings of India.**

In the reign of Henry VIII, tarragon made its way into English gardens, and the rhyme, 'There is certain people, and certain herbs, that good digestion disturbs,' could well be associated with tarragon. I love, too, the story that Henry VIII divorced Catherine of Aragon for her reckless use of tarragon.

varieties

Artemisia dracunculus French

French Tarragon

Half-hardy perennial. Ht 90cm, spread 45cm. Tiny, yellow, insignificant flowerheads are borne in sprays in summer but rarely produce ripe seed sets except in warm climates. The leaves are smooth dark green, long and narrow, and have a very strong flavour.

Artemisia dranunculus Russian

Russian Tarragon

Hardy perennial. Ht 1.2m, spread 45cm. Tiny, yellow insignificant flowerheads borne in sprays in summer. The leaves are slightly coarser and green in colour, their shape long and narrow. This plant originates from Siberia, which explains why it is so hardy.

cultivation

Propagation

Seed

Only the Russian and wild varieties produces viable seed. A lot of growers are propagating and selling it to the unsuspecting public as French tarragon. If you really want Russian tarragon, sow the small seed in spring, into prepared seed or plug module trays, using a standard seed compost mixed in equal parts with composted fine bark. No extra heat required. When the young plants are large enough to handle, transfer to the garden, 60cm apart.

Cuttings

Both French and Russian tarragon can be propagated by cuttings. Dig up the underground runners in spring when the frosts are finished; pull them apart, do not cut. You will notice growing nodules: these will reproduce in the coming season. Place a small amount of root – 8–10cm – each with a growing nodule, in a 8cm pot, and cover with compost. Use a seed compost mixed in equal parts with perlite and place in a warm, well-ventilated spot. Keep watering to a minimum. When well rooted, plant out in the garden after hardening off, 60cm apart.

It is possible to take softwood cuttings of the growing tips in summer. Keep the leaves moist, but the

Artemisia dracunculus

TARRAGON

Artemisia dracunculus

compost on the dry side. It works best under a misting unit with a little bottom heat 15°C.

Division

Divide established plants of either variety in the spring.

Pests and Diseases

Recently there has been a spate of rust developing on French tarragon. When buying a plant, look for tell-tale signs – small rust spots on the underneath of a leaf. If you have a plant with rust, dig it up, cut off all foliage carefully, and bin the leaves. Wash the roots free from soil, and pot up into fresh sterile soil. If this fails, place the dormant roots in hot water after washing off all the compost. The temperature of the water should be 44°C; over 46°C will damage the root. Leave the roots in the hot water for 10 minutes then replant in a new place in the garden.

Maintenance

Spring Sow Russian tarragon seeds if you must. Divide established plants. Take root cuttings.
Summer Remove flowers.
Autumn Pot up pieces of French tarragon root as insurance.
Winter Protect French tarragon. As the plant dies back into the ground in winter it is an ideal candidate for either horticultural fleece, straw or a deep mulch.

Garden Cultivation

French tarragon has the superior flavour of the two and is the most tender. It grows best in a warm dry position, and will need protection in winter. It also dislikes humid conditions. The plant should be renewed every 3 years because the flavour deteriorates as the plant matures.

Russian tarragon is fully hardy and will grow in any conditions. There is a myth that it improves the longer it is grown in one place. This is untrue, it gets coarse. It is extremely tolerant of most soil types, but prefers a sunny position, 60cm away from other plants.

Harvesting

Pick sprigs of French tarragon early in the season to make vinegar. Pick leaves for fresh use throughout the growing season. For freezing it is best to pick the leaves in the mid-summer months.

container growing

French tarragon grows well in containers. Use a soil-based compost mixed in equal parts with composted fine bark. As it produces root runners, choose a container to give it room to grow so that it will not become pot bound. At all times make sure the plant is watered, and in the daytime, not at night. It hates having wet roots. Keep feeding to a minimum: overfeeding produces fleshy leaves with a poor flavour, so be mean. In winter, when the plant is dormant, do not water, keep the compost dry and the container in a cool, frost-free environment.

medicinal

No modern medicinal use. Formerly used for toothache. Traditionally, a tea made from the leaves is said to overcome insomnia.

culinary

Without doubt French tarragon is among the Rolls-Royces of the culinary herb collection. Its flavour promotes appetite and complements so many dishes – chicken, veal, fish, stuffed tomatoes, rice dishes and salad dressings, and of course is the main ingredient of sauce béarnaise.

Chicken Salad with Tarragon and Grapes

Serves 4–6

1.3kg cooked chicken
150ml mayonnaise
75ml double cream
1 heaped teaspoon fresh chopped tarragon (1/2 teaspoon dried)
3 spring onions, finely chopped
1 small lettuce
100g green grapes (seedless or de-pipped)
A few sprigs watercress
Salt and freshly ground black pepper

Remove the skin from the chicken and all the chicken from the bones. Slice the meat into longish pieces and place in a bowl.

In another bowl mix the mayonnaise with the cream, the chopped tarragon, and the finely chopped spring onions. Pour this mixture over the chicken and mix carefully together. Arrange the lettuce on a dish and spoon on the chicken mixture. Arrange the grapes and the watercress around it. Serve with jacket potatoes or rice salad.

Atriplex hortensis

ORACH

Atriplex hortensis var. *rubra*

From the family *Chenopodiaceae*.

The garden species of orach, *Atriplex hortensis*, originated in Eastern Europe and is now widely distributed in countries with temperate climates. In the past it was called mountain spinach and grown as a vegetable in its own right.

The red form, *Atriplex hortensis* var. *rubra*, is still eaten frequently in Continental Europe, particularly with game, and was used as a flavouring for breads.

The common orach, *Atriplex patula*, was considered a poor man's pot herb, which is a fact worth remembering when you are pulling out this invasive annual weed.

varieties

Atriplex hortensis
Orach
Hardy annual. Ht 1.5m, spread 30cm. Tiny greenish (boring) flowers in summer. Green triangular leaves.

Atriplex hortensis* var. *rubra
Red Orach
Hardy annual. Ht 1.2m, spread 30cm. Tiny reddish (boring) flowers in summer. Red triangular leaves.

Atriplex patula
Common Orach
Hardy annual. Ht 90cm, spread 30cm. Flowers similar to orach, the leaves more spear-shaped and smaller.

cultivation

Propagation

Seed

For an early supply of leaves, sow under protection in early spring, into prepared plug module trays. Cover with perlite. When the seedlings are large enough and after hardening off, plant out into a prepared site 25cm apart. Alternatively, in late spring, when the soil has started to warm and all threat of frost has passed, sow in rows 60cm apart, thinning to 25cm apart when the seedlings are large enough. As this herb grows rapidly and the leaves can

Atriplex hortensis

become tough, it is advisable to sow a second crop in mid-summer to ensure a good supply of young and tender leaves.

Pests and Diseases
Mostly pest and disease free, but it can suffer from leaf minor damage; bin or burn all affected leaves.

Maintenance
Spring Sow seeds.
Summer Cut flowers before they form.
Autumn Cut seeds off before they are fully ripe to prevent too much self-seeding.
Winter Dig up old plants.

Garden Cultivation
This annual herb produces the largest and most succulent leaves when the soil is rich. So prepare the site with well-rotted manure. For red orach choose a site with partial shade as the leaves can scorch in very hot summers. Water well throughout the growing season. Red orach looks very attractive grown as an instant hedge or garden divide. Remove the flowering tips as soon as they appear. This will help to maintain the shape and, more importantly, prevent the plant from self-seeding. If you wish to save some seed for the following season, collect before it is fully ripe, otherwise you will have hundreds of orach babies all over your garden and next door.

Harvesting
Pick young leaves to use fresh as required. The herb does not dry or freeze particularly well.

container growing

The red-leaved orach looks very attractive in containers; don't let it get too tall. Nip out the growing tip and the plant will bush out, and do not let it flower. Use a good potting compost. Keep the plant in semi-shade in high summer and water well at all times. If watering in high sun be careful not to splash the leaves as they can scorch, especially the red variety.

medicinal

This herb is no longer used medicinally. In the past it was a home remedy for sore throats, gout and jaundice.

culinary

The young leaves can be eaten raw in salads, and the red variety looks most attractive. The old leaves of both species ought to be cooked as they become slightly tough and bitter. It can be used as a substitute for spinach or as a vegetable, served in a white sauce. It is becoming more popular in Europe, where it is used in soups.

Red Orach Soup

450g potatoes
225g young red orach leaves
50g butter
900ml chicken stock
1 clove garlic, crushed
Salt and freshly ground black pepper
4 tablespoons soured cream

Peel the potatoes and cut them into thick slices. Wash the orach and cut up coarsely. Cook the potatoes for 10 minutes in salted water, drain. Melt the butter in a saucepan with the crushed garlic and slowly sweeten; add the red orach leaves and gently simmer for 5–10 minutes until soft (if the leaves are truly young then 5 minutes will be sufficient).

Pour in the stock, add the parboiled potatoes and bring to the boil; simmer for a further 10 minutes. When all is soft, cool slightly then purée in a blender or liquidise.

After blending, return the soup to a clean pan, add salt and pepper to taste and heat slowly (not to boiling). Stir in the sour cream, and serve.

Ballota nigra

BLACK HOREHOUND

Also known as Stinking Horehound, Dunny Nettle, Stinking Roger and Hairy Hound. From the family *Lamiaceae*.

Black horehound comes from a genus of about 25 species mostly native to the Mediterranean region. Some species have a disagreeable smell and only a few are worth growing in the garden. Black horehound is found on roadsides, hedge banks and in waste places throughout most of Europe, Australia and America.

***Ballota nigra*, the black horehound, was originally called *ballote* by the ancient Greeks. It has been suggested that this comes from the Greek word *ballo*, which means 'to reject', 'cast', or 'throw', because cows and other farm animals with their natural instincts reject it. The origin of the common name is more obscure, it could come from the Anglo-Saxon word *har*, which means 'hoar' or 'hairy'.**

varieties

Ballota nigra
Black Horehound
Hardy perennial. Ht 40–100cm, spread 30cm. Purple-pink attractive flowers in summer. The leaves are green and medium-sized, rather like those of the stinging nettle. All parts of the plant are hairy and have a strong, disagreeable smell and taste.

***Ballota pseudodictamnus* AGM**
Half-hardy perennial. Ht 60cm, spread 30cm. White flowers with numerous purple spots in summer. Leaves white and woolly. This plant originated from Crete. The dried calyces look like tiny furry spinning tops; they were used as floating wicks in primitive oil lamps.

cultivation

Propagation

Seed
Sow the seeds direct into the prepared garden in late summer, thinning to 40cm apart.

Division
Divide roots in mid-spring.

Pests and Diseases
Rarely suffers from any pests or diseases.

Maintenance
Spring Dig up established plants and divide; replant where required.
Summer When the plant has finished flowering, cut off the dead heads before the seeds ripen, if you wish to prevent it from seeding itself in the garden.
Autumn Sow seed.
Winter No need to protect.

Garden Cultivation
Black horehound will grow in any soil conditions, though it prefers water-retentive soil – in fact I have seen it growing in hedgerows throughout England. In the garden, place it in a border. The bees love it and the flowers are attractive. Make sure it is far enough back so that you do not brush it by mistake because it does stink.

Harvesting
As this is a herbalist's herb, the leaves should be collected before flowering and dried with care.

container growing

Not recommended, as it is such an unpleasant-smelling plant.

medicinal

Black horehound was used apparently in the treatment of bites from mad dogs. A dressing was prepared from the leaves and laid on the infected part. This was said to have an anti-spasmodic effect. This is not a herb to be self-administered. Professionals use it as a sedative, anti-emetic and to counteract vomiting in pregnancy.

Bellis perennis

DAISY

Also known as Bruisewort, Eye of the Day, Common Daisy, English Daisy and Barinwort. From the family *Asteraceae*.

This humble, attractive flower, beloved by generations of children, was traditionally used medicinally by eminent herbalists such as Gerard who, in the seventeenth century, called it bruisewort and used it as a cure for wounds and disorders of the liver. Currently its medicinal properties as a treatment for bruising are being researched.

varieties

Bellis perennis
Daisy
Hardy perennial. Ht and spread 15–20cm. Numerous small flowers with yellow centres, white petals with a pink blush to the tips. Basal rosette of spoon-shaped, scalloped mid-green leaves.

***Bellis perennis* 'Alba Plena'**
Double-flowered Daisy
Hardy perennial. Ht and spread 15–20cm. Numerous small flowers with yellow centres, double white petals, which often have a pink blush. A basal rosette of spoon-shaped, scalloped mid-green leaves.

cultivation

Propagation

Seed
Sow fresh seed in summer into prepared trays or plug modules, using a standard seed compost. Cover with perlite, no extra heat is needed. Germination takes 10–20 days. Once the seedlings are large enough, pot up using loam-based compost. Alternatively plant out into a prepared site in early autumn. They will flower in their second season.

The double varieties of the species are, in the majority of cases, sterile and can only be grown by division.

Division
Divide the creeping rhizomes in spring or autumn. Once divided, either pot into loam-based compost or replant into a prepared site in the garden.

Pest and Diseases
Rarely suffers in the garden from any pest or disease. Can be prone to powdery mildew in containers, this is usually caused by the compost drying out then being over watered to compensate.

Maintenance
Spring Divide established plants.
Summer Sow fresh seed, deadhead flowers to maintain flowering.
Autumn Divide established plants.
Winter No need for protection, fully hardy.

Garden Cultivation
This is a most tolerant plant; it will survive in most soils. When grown in an ideal situation it requires a loam that does not totally dry out in summer, in full sun to partial shade. When it is planted in the lawn it can be happily mown; this will encourage it to continue flowering.

Harvesting
Pick the leaves when young for using in salads. Pick the flowers in bud or just opening for use in salads and puddings. Dig up the roots for medicinal use in autumn.

container growing

Daisies look very attractive in a container. Use a soil-based compost so that it does not totally dry out on hot summer days.

culinary

New young leaf shoots can be eaten raw or cooked. The flowering buds and flowers, just as they open, can be added to salads; they have a light, honey bitter flavour which combines well with lettuce. They look most attractive in fruit salads.

medicinal

An ointment made from the leaves can be applied to external wounds and bruising. A decoction made from the roots can ease the irritation of eczema. It is also used in homeopathic remedies to treat bruising.

other uses

Traditionally, an insect-repellent spray was made from an infusion of the leaves.

Bellis perennis

Borago officinalis

BORAGE

Also known as Bugloss, Burrage and Common Bugloss. From the family *Boraginaceae.*

Borage is indigenous to Mediterranean countries, but has now been naturalised in Northern Europe and North America. In fact one can find escapees growing happily on wasteland.

The origin of the name is obscure. The French *bourrache* is said to derive from an old word meaning 'rough' or 'hairy', which may describe the leaf, but the herb's beautiful, pure blue flowers are its feature and are supposed to have inspired the painting of the robes of the Madonna, and charmed Louis XIV into ordering the herb to be planted at Versailles.

The herb's Welsh name translates as 'herb of gladness'; and in Arabic it is 'the father of sweat', which we can accept as borage is a diaphoretic. The Celtic word *borrach* means 'courage', however, and in this we have an association more credible by far. The Greeks and Romans regarded borage as both comforting and imparting courage, and this belief so persisted that Gerard was able to quote the tag, *Ego borage gaudia semper ago* in his *Herball*. It was for courage, too, that borage flowers were floated in stirrup cups given to the Crusaders. Clearly, the American Settlers also thought sufficiently highly of borage to take the seed with them on their long adventure. Records of it were found in a seed order of an American in 1631, where it was called burradge.

Borago officinalis 'Alba'

varieties

Borago officinalis
Borage
Hardy annual (very occasionally biennial). Ht 60cm with hollow, bristly branches and spreading stems. The blue or purplish star-shaped flowers grow in loose racemes from early summer to mid-autumn. The leaves are bristly, oval or oblong. At the base they form a rosette; others grow alternately on either side of the stem.

***Borago officinalis* 'Alba'**
White Borage
Hardy annual. Ht 60cm. White star-shaped flowers from late spring through summer. Bristly, oval, oblong leaves. Can be used in the same way as *B. officinalis*.

cultivation

Propagation
Seed
Borage is best grown directly from seed, sown in spring and summer in its final position, as it does not like having its long tap root disturbed. But for an early crop it is as well to start the seeds under protection. In early spring sow singly in small pots. Transplant to final position as soon as possible after hardening off, when the seedling is large enough and all threat of frosts is over.

Pests and Diseases
Blackfly. If you are growing borage as a companion plant this will not worry you, but if it is becoming a nuisance then spray with liquid horticultural soap.

A disease that can be unsightly at the end of the season, is a form of mildew. Dig the plant up and burn.

BORAGE

Maintenance

Spring Sow seeds. They germinate quickly and plants are fully grown in 5–6 weeks.
Summer Sow seeds. Look out for flower heads turning into seeds – collect or destroy if you do not want borage plants all over the garden. Deadhead flowers to prolong flowering season.
Autumn As the plants begin to die back collect up the old plants. Do not compost the flower heads or next year you will have a garden full of unwanted borage.
Winter Borage lasts until the first major frost, and some years it is the last flowering herb in the garden.

Garden Cultivation

Borage prefers a well-drained, light, rather poor soil of chalk or sand, and a sunny position. Sow borage seeds 5cm deep in mid-spring and again in late spring for a continuous supply of young leaves and flowers. Thin seedlings to 60cm apart and from other herbs, as they produce lots of floppy growth.

I have used borage as an exhibit plant at flower shows, and have found that by continuously deadheading the flowers you can maintain a good supply of flowers for longer.

Harvesting

Pick flowers fresh or for freezing or drying when they are just fully opened. Cut the young leaves fresh throughout summer. They do not dry or freeze very successfully. Collect seed before the plant dies back fully. Store in a light-proof container in a cool place.

companion planting

Borage is a good companion plant. The flower is very attractive to bees, helping with pollination, especially for runner beans and strawberries. Borage also attracts blackfly to itself, theoretically leaving the other plants alone. Equally, if planted near tomatoes it is said to control tomato worm.

container growing

It is not suitable for container-growing indoors. However, when planted outside in large containers (like a half barrel), borage can be very effective combined with other tall plants like oxeye daisies, poppies and cornflowers.

culinary

Be brave, try a young leaf. It may be hairy, some would say prickly, but once in the mouth the hairs dissolve and the flavour is of cool cucumber. Great cut up in salads, or with cream cheese, or added to yogurt, or even in an egg mayonnaise sandwich.

They give a refreshing flavour to summer cold drinks. Finally fresh leaves are particularly good to use in a salt-free diet as they are rich in mineral salts. Try them combined with spinach or added to ravioli stuffing. The flowers are exciting tossed in a salad, floated on top of a glass of Pimms No. 1, or crystallised for cake decoration. Also excellent as garnish for savoury or sweet dishes, and on iced soups.

medicinal

In the 1980s borage was found to contain GLA, gamma linoleic acid, an even more valuable medicinal substance than evening primrose oil. But cultivation problems coincided with a dramatic slump in prices when waste blackcurrant pulp, provided a cheaper and richer source of GLA. So hopes for the future of borage as a commercial crop have diminished recently, but it deserves more medicinal research.

Borage tea is said to be good for reducing high temperatures when taken hot. This is because in inducing sweat – it is a diaphoretic – it lowers the fever. This makes it a good remedy for colds and flu, especially when these infect the lungs as it is also good for coughs. Leaves and flowers are rich in potassium and calcium and therefore good blood purifiers and a tonic.

Borage Tea

Small handful of fresh leaves
600ml of boiling water

Simmer for 5 minutes.

Natural Night-cap

3 teaspoons fresh borage leaves
250ml boiling water
1 teaspoon honey
1 slice lemon

Roughly chop the borage leaves, put them into a warmed cup and pour over the boiling water. Cover with a saucer and leave the leaves to infuse for at least 5 minutes. Strain and add the lemon slice and honey. Drink hot just before retiring to bed.

other uses

Dried flowers add colour to potpourris. Children enjoy stringing them together as a necklace. Add to summer flower arrangements. As a novelty burn the whole plant – nitrate of potash will emit sparks and little explosive sounds like fireworks.

Facial Steam for Dry, Sensitive Skin

Place 2 large handfuls of borage leaves in a bowl. Pour over 1.5 litres of boiling water. Stir quickly with a wooden spoon. Using a towel as a tent, place your face about 30cm over the water. Cover your head with towel. Keep your eyes closed and maintain for about 10–15 minutes. Rinse your face with tepid cool water. Use a yarrow infusion dabbed on with cotton wool to close pores.

warning

Prolonged use of borage is not advisable. Fresh leaves may cause contact dermititis.

Brassica juncea

BROWN MUSTARD

Also known as Juncea, Brown Mustard, Indian Mustard, Mustard Greens and Chinese Mustard. From the family *Brassicaceae.*

This herb is mentioned more than once in the Bible; it has been used for thousands of years both medicinally and for culinary purposes.

Historically the Romans used the brown mustard as a pungent salad vegetable and the seeds of black mustard as a spice. The English name 'mustard' is said to derive from the Latin, *mustem ardens*, which translates as 'burning wine', a reference to the heat of mustard and the French practice of mixing unfermented grape juice with the ground seeds.

Brassica juncea 'Rubra'

Brassica juncea 'Red Frills'

varieties

Brassica alba
White Mustard, Yellow Mustard
Hardy annual. Ht 30–60cm. Pale yellow flowers in summer followed by long seed pods with pale yellow seeds. Rough, bristly, oval, deeply lobed green leaves.

Brassica juncea
Brown Mustard
Hardy annual. Ht 40–100cm. Clusters of pale yellow flowers in summer followed by long seed pods with brown seed. Oval, lobed, olive-green leaves with pale green veins with crinkled edges.

***Brassica juncea* 'Rubra'**
Red mustard
Hardy annual. Ht 30cm. Yellow flowers in summer. Oval, red, purple-tinted leaves with indented edges, pungent mustard flavour. Good in salads and sandwiches.

***Brassica juncea* 'Red Frills'**
Mustard Red Frills
Hardy annual. Ht 15cm. Yellow flowers in summer. Attractive dark red, narrow, indented leaves with a mild mustard flavour, great for salads.

Brassica nigra
Black Mustard
Hardy annual. Ht 40–100cm. Pale yellow flowers in summer followed by long seed pods with dark brown seeds. Rough, bristly, oval-lobed green leaves.

cultivation

Propagation

Seed

Sow seeds in the spring for seed and leaf production and the autumn for leaf, into pots or plug modules using a standard seed compost. Cover with perlite and place under protection at 15°C. Germination is in 5–10 days. Plant out after hardening off at a distance of 20cm apart.

Alternatively sow seeds in late spring or early autumn into prepared open ground, when the air temperature does not go below 7°C at night. Germination takes 2–3 weeks. Once the seedlings are large enough, thin to 20cm apart.

Pests and Diseases

All mustards can be attacked by flea beetle, which is a tiny, shiny, dark blue beetle that hops from plant to plant. A tell-tale sign that the plants are being attacked are small round holes on the leaves, which turn pale brown, making the leaves look unattractive. The adult beetles overwinter in leaf litter, emerging in the spring to attack lush seedlings. The other major attack comes in summer when there is a significant migration of adult beetles from oilseed rape fields into gardens; these can come in such numbers that they attack mature plants. The best organic method is to put a horticultural mesh over the plants so the beetles cannot attack the leaves.

Maintenance

Spring Sow seeds.
Summer Pick leaves and flowers.
Autumn Harvest seeds, sow autumn crop.
Winter Prepare ground for next season's crop.

BROWN MUSTARD

Brassica juncea

Garden Cultivation

All brassicas like to feed well, and mustard is no exception. However, do not become too enthusiastic and give them too much nitrogen, because this will make them soft and flabby and prone to attack from pests. Try and prepare your plant plots the autumn before planting. Give the site a good feed of well-rotted manure and compost in the autumn prior to sowing in the spring.

Harvesting

Pick young leaves as required and the flowers just as they open. The seed pods are harvested as they change colour. Dry the pods in a light airy room until totally dried, remove from the pods and store in clean, dry jars which have a tight-fitting top.

companion planting

Mustard is a good companion plant because it germinates quickly so can be sown next to slow germinators, for example parsley, to indicate the row.

There is some evidence that mustard can be grown as a green manure. It is said to be effective in reducing soil-borne root diseases in pea crops.

container growing

Mustard can successfully be grown in a container, using a standard potting compost. Place the container in partial shade. Water and feed regularly from spring until early autumn.

medicinal

Mustard has been and is still used to ease muscular pain, and treat respiratory tract infections. It is a warming stimulant with antibiotic effects. In China the leaves are eaten to ease bladder inflammation. In Korea, the seeds are used in the treatment of abscesses, colds, lumbago, rheumatism and stomach disorders. Current research has found that mustard seeds inhibit the growth of cancerous cells in some animal studies.

other uses

Mustard seeds have been suggested as a possible source of biodiesel in Australia.

warning

Mustard seeds have been known to cause an allergic reaction. Prolonged contact with the skin can cause blistering.

culinary

The mustard leaves have a distinctive, warm peppery taste; the flowers also have a mild mustard flavour. Both are great in salads, stir-fry dishes and they can transform a sandwich into a delight. Try a red mustard and ham sandwich.

Mustard seeds only get their pungency when they are crushed. When the seed is mixed with water and allowed to stand, the strength of flavour increases. If you mix the seed directly with vinegar or salt without previously soaking in water, you kill the flavour and if you boil the seed it will become bitter. The easiest and best way to use mustard seed is to make your own mustard. Use a combination of the black, brown and white seeds. This mustard is very versatile, complementing many dishes and enhancing the flavours of cheese, vegetables, poultry, red meat and fish. The seeds are also great added to salad dressings or sauces, When adding mustards to food, remember that the pungency of mustard is destroyed by heat, so add it to finished sauces or stews. Mustards are particularly good when combined with crème fraîche or cream cheese as dips for vegetables or pretzels.

Dill Mustard

80g black mustard seeds
30g white mustard seeds
600ml water
115g English mustard powder
200ml cider vinegar
50g light brown sugar

1 1/2 teaspoons salt
1 1/2 teaspoons turmeric
50g chopped dill, or any other herb of your choice.

Place the mustard seed in a china or glass bowl, add in 600ml water and soak for 24 hours prior to use. The next day add the mustard powder, vinegar, salt and turmeric. When thoroughly mixed, place the bowl over a saucepan containing water, ensuring the water does not touch the bottom of the bowl. On a very low heat gently cook the mustard seed for 4 hours, stirring occasionally. Check from time to time that the water has not evaporated and never boil the mixture as the mustard will lose its flavour and become bitter. Once cooked allow to cool, then add the chopped dill. Cover the herb mustard and keep in the refrigerator.

Bulbine frutescens

AFRICAN BULBINE

Also known as Burn Jelly Plant, Snake Flower, Cat's Tail, Bulbinella, Ibhucu (Zulu) and Rooiwortel (Afrikaans). From the family *Asphodelaceae.*

This attractive drought-resistant herb is a native of South Africa and can be found growing wild in the desert grasslands of the Northern, Western and Eastern Capes. Indigenous plants play a pivotal role in Africa for traditional healing; this herb has been used by the Zulu for hundreds of years not only to cure rashes, to stop bleeding and as an antidote to poison, but also to treat their sick livestock.

varieties

Bulbine frutescens
African Bulbine
Frost-tender, evergreen perennial. Ht 60cm, spread 1m. Attractive star-shaped yellow or orange single flowers, which grow in a linear cluster around the stem appearing sequentially throughout the summer. The flower has a characteristic hairy stamen. The leaves are succulent, cylindrical, narrow and varying in length. They are glutinous when broken.

Bulbine latifolia (Bulbine natalensis)
Broad-leaved Bulbine
Frost-tender, evergreen perennial. Ht 60cm, spread 1m. Attractive star-shaped yellow flowers which grow on 60cm flowering spikes throughout the summer. Dark green, aloe-like, pointed, succulent triangular leaves which form a basal crown.

cultivation

Propagation

Seed

Sow fresh seed in spring, into prepared seed trays or module plugs, using a seed compost mixed in equal parts with perlite. Place under protection at 20°C. Germination takes 10–20 days; however it is erratic as, like many herbs, it will happily self seed in a warm garden but, in a controlled situation, it sometimes will not perform. During germination, make sure that the compost does not dry out. Once the seedling is well rooted pot up using a seed compost mixed in equal parts with horticultural sand. Grow on until the plant has at least four well-developed leaves before either planting out in warm climates into a prepared site in the garden or potting up in cold climates to grow on as a container plant.

Cuttings

This is the easiest method of propagation in a cold climate. Cuttings can be taken from early spring for plants raised under protection or in late spring for plants grown outside. By looking at the plant you will note that it puts down roots where the clusters of leaves touch the soil. Using a sharp knife, take the cutting, removing some root at the same time. Put the cuttings into a small pot or large module plug, using a seed compost mixed in equal parts with composted

AFRICAN BULBINE

Bulbine frutescens Yellow form

fine bark. Place the container in a warm position. It will root very quickly. Once the plant is established, either pot up using a loam-based compost mixed in equal parts with horticultural sand or, in warm climates, plant out into a prepared site in the garden.

Division
This plant grows rapidly in a container or garden and benefits from being divided. In spring or early summer, before dividing, remove all the flowering stems; this will make it easier to handle and also to see what one is doing. Divide established garden plants by using the two forks back to back method, replanting immediately into a prepared site. Container-raised plants will need to be removed from the pot, then, either using hands or two hand forks, tease the plant apart, repotting into a loam-based potting compost mixed in equal parts with horticultural sand.

Pests and Diseases
This herb is rarely troubled by pests or disease, with the exception of vine weevil if pot grown. There are two methods of getting rid of vine weevil organically. Either repot all the containers, checking each one for vine weevil grubs or, alternatively, watering the containers with a natural predator called *Steinernema Krauseii* in late summer, early autumn, when the night temperature does not fall below 10°C and before the vine weevil grubs have grown large enough to cause serious damage. This predator is a nematode, a microscopic worm, which infects the weevil grubs with a fatal bacterial disease. Prior to using this nematode, make sure the compost in the containers is moist all the way down by watering well a few hours before you intend to add the biological control.

Maintenance
Spring In early spring remove flowering spikes, sow fresh seeds and divide established clumps.
Summer Remove flowering spikes that have finished flowering, take cuttings.
Autumn Remove flowering spikes.
Winter Do not over water, but do not allow the compost to become totally dry. Protect when temperatures drop below 0°C.

Garden Cultivation
This herb is a good drought-loving plant; it can be planted out in cold climates in the early summer and then lifted before the frosts. Plant in a well-drained fertile soil that was fed with well-rotted manure in the previous spring. Position in full sun or semi-shaded and protected from cold winds.

Harvesting
The leaves can be picked for use throughout the year.

container growing

This useful attractive herb, which can flower all year round, is ideal for growing in a warm conservatory as it requires a good light. Pot up using a loam-based potting compost mixed in equal parts with horticultural sand. Water regularly, and feed monthly, from spring until autumn, with a liquid fertiliser following the manufacturer's instructions.

medicinal

The leaves are filled with a clear gel similar in appearance and consistency to *Aloe vera*. This gel can be used directly on minor burns, wounds, cuts, abrasions, stings and rashes. It can also be used to treat eczema, cracked lips and herpes. An infusion or tincture made from the roots of *Bulbine latifolia* is taken to quell sickness and diarrhoea, and it is also used to treat urinary complaints and rheumatism.

Currently the medicinal properties of this herb are under research, and the leaf gel is being used to aid the healing of post-operative scars.

other uses

A red dye can be obtained from the roots of *Bulbine latifolia*.

warning

Do not take internally when pregnant.

Bulbine frutescens Yellow form

Buxus

BOX

Also known as Boxwood and Bushtree. From the family *Buxaceae.*

Box is a native plant of Europe, Western Asia and North Africa. It has been cultivated widely throughout the world and is found in North America along the Atlantic coast, especially as an ornamental and hedging plant.

The common name, box, comes from the Latin *buxus*, which is in turn derived from the Greek *puxus*, meaning 'a small box'. At one time, boxwoods were widespread in Europe but the demand for the wood, which is twice as hard as oak, led to extensive felling. Its timber is close-grained and heavy, so heavy in fact that it is unable to float on water. The wood does not warp and is therefore ideal for boxes, engraving plates, carvings, and musical and navigational instruments.

It is not used medicinally now but the essential oil from box was used for the treatment of epilepsy, syphilis and piles, and also as an alternative to quinine in the treatment of malaria. A perfume was once made from its bark and a mixture of the leaves with sawdust has been used as an auburn hair dye.

When it rains, box gives off a musky smell evocative of old gardens, delicious to most people, but not Queen Anne who so hated the smell that she had the box parterres in St James's Park (planted for her predecessors, William and Mary, in the last years of the seventeenth century) torn out.

varieties

***Buxus balearica* AGM**
Balearic Box
Half-hardy evergreen. Ht 2m, spread 1.5m. Suitable for hedging in mild areas. Has broadly oval, bright green leaves. Planting distance for a hedge 30–40cm.

Buxus microphylla
Small-leaved Box
Hardy evergreen. Ht 1m, spread 1.5m. Forms a dense mass of round/oblong, dark green, glossy leaves. Attractive cultivar 'Green Pillow', which is good for formal shaping. Planting distance for a hedge 15–23cm.

***Buxus sempervirens* AGM**
Common Box
Hardy evergreen. Ht and spread 5m. Leaves glossy green and oblong. Good for hedges, screening, and topiary. Planting distance for a hedge 37–45cm.

***Buxus sempervirens* 'Elegantissima' AGM**
Variegated Box
Hardy evergreen. Ht and spread 1m. Good variegated gold/green leaves. Susceptible to scorch in hard winter. Trim regularly to maintain variegation. Very attractive as a centre hedging in a formal garden or as specimen plants in their own right. Planting distance for a hedge 30–40cm.

***Buxus sempervirens* 'Latifolia Maculata' AGM**
Golden Box
Hardy evergreen. Ht and spread 1m. The new growth is very golden, and as it matures it becomes mid-green. Planting distance for a hedge 30–40cm.

***Buxus sempervirens* 'Suffruticosa' AGM**
Dwarf Box
Hardy evergreen. Ht and spread 45cm. Evergreen dark shrub that forms tight, dense mass. Slow grower. This is the archetypal edging in a formal herb garden for patterns and parterres of the knots. It is trimmed to about 15cm in height when used for hedging. Planting distance for hedge 15–23cm.

Buxus sempervirens 'Latifolia Maculata'

Buxus sempervirens 'Suffruticosa'

Buxus wallichiana
Himalayan Box
Hardy evergreen. Ht and spread 2m. Slow growing. Produces long, narrow, glossy, bright green leaves. Planting distance for a hedge 30–40cm.

cultivation

Propagation
Cuttings
Box is cultivated from cuttings taken in spring from the new growth. Use a standard seed compost mixed in equal parts with composted fine bark. Keep the cuttings moist but NOT wet, and in a cool place, ideally in shade. They will take 3–4 months to root. If you have a propagator then they will take 6–8 weeks at 21°C. Use a spray to mist the plant regularly. When rooted, pot up the young plants using a soil-based compost or plant out, as per the distances mentioned under varieties.

Alternatively, take semi-ripe cuttings in summer, using the same compost mix as above. Rooting time is approximately 2 months longer than for the softwood cuttings, but only 3–4 weeks longer with heat.

Pests and Diseases
Box can be attacked by the green box sucker nymph, which is noticeable by a white deposit and curled leaves. Cut back the plant, then spray with horticultural soap. Box blight, *Cylindrocladium buxicola* is a fungal disease that can cause serious defoliation. There is no reliable organic remedy. However, I know of one gardener who cut his box plants back, then fed weekly with an organic fertiliser for one year and it recovered.

Maintenance
Spring Take stem cuttings. Trim fast growers such as *Buxus sempervirens* 'Latifolia Maculata'.
Summer Take semi-ripe cuttings. Trim hedges to promote new growth.
Autumn Trim if needed but not hard.
Winter Does not need protection.

Garden Cultivation
Known for its longevity, it is not uncommon for it to live 600 years. Box flourishes on limestone or chalk; it prefers an alkaline soil. But it is very tolerant of any soil provided it is not waterlogged. Box is also fairly tolerant of position, surviving sun or semi-shade.

Box hedges are a frame for a garden, outlining, protecting and enhancing what they enclose. Because one will be planting the box plants closer together than when planting individual plants it is important first to feed the soil well by adding plenty of well-rotted manure or garden compost. Width of bed for a boundary hedge should be 60cm; for small internal hedges 30cm wide. Planting distances vary according to variety (see specific varieties).

If using common box or golden box, the more vigorous growers, trim in spring and prune at the end of summer or, in mild climates, in early autumn. A late cut may produce soft growth that could be damaged by frost, which would then look ugly in winter. The slower varieties need a cut only in the summer. In general the right shape for a healthy, dense hedge is broad at the base, tapering slightly towards the top, rounded or ridged but not too flat, to prevent damage from heavy snow.

container growing

Box, especially the slow-growing varieties, lends itself to topiary and looks superb in containers. *Buxus sempervirens* 'Elegantissima' looks very attractive in a terracotta pot. They are very easy to maintain. Use a soil-based compost. Feed with a liquid fertiliser in spring; water sparingly in winter.

other uses

Boxwood is a favourite timber with cabinet makers, wood engravers and turners because it has a non-fibrous structure.

A mahogany box with boxwood detailing

medicinal

It is advisable not to self-administer this plant; it should be used with caution. Box is used extensively in homoeopathic medicines; a tincture prepared from fresh leaves is prescribed for fever, rheumatism and urinary tract infections.

warning

Animals have died from eating the leaves. All parts of the plant, especially the leaves and seeds, are poisonous. It is dangerous to take internally and should never be collected or self-administered. Symptoms of poisoning are vomiting, abdominal pain and bloody diarrhoea.

Calamintha

CALAMINT

From the family *Lamiaceae*.

***Calamintha* originated in Europe. It is now well established throughout temperate countries, but sadly it is still not a common plant. Calamint has been cultivated since the seventeenth century. Herbal records show that it used to be prescribed for women.**

Calamintha nepeta

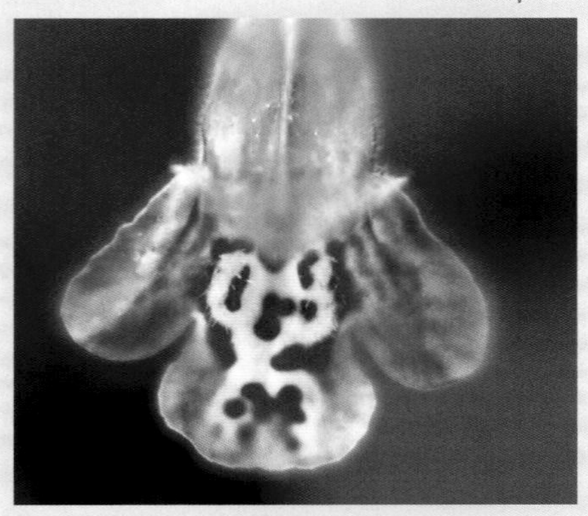

varieties

Calamintha grandiflora
Calamint
Hardy perennial. Ht 37cm, spread 30cm. Square stems arise from creeping rootstock. Dense whorls of lilac pink flowers appear mid-summer to early autumn above mint-scented, toothed, oval green leaves.

***Calamintha grandiflora* 'Variegata'**
Hardy perennial. As *C. grandiflora* but with cream variegated leaves.

Calamintha nepeta
Lesser Calamint
Perennial. Ht 30–60cm, spread 30cm. Small purple/white flowers from summer to early autumn. Stems and leaves pale grey and covered in fine downy hairs. Its wonderful aromatic scent attracts butterflies and bees.

Calamintha nepeta

Calamintha sylvatica* subsp. *ascendens
Mountain Balm, Mountain Mint
Hardy perennial. Ht 30cm, spread 20cm. Pale purple flowers in dense whorls from late summer to early autumn. Leaves mid-green, oval, finely toothed and mint scented. A tisane can be made from the leaves.

cultivation

Propagation

Seed
Sow calamint's fine seeds in spring or autumn, either in their eventual flowering position or in trays, using a standard seed compost mixed in equal parts with composted fine bark and covered lightly with perlite. If autumn sowing in trays leave them outside to overwinter, covered with a sheet of glass. As germination can be tricky, autumn sowing is sometimes more successful because subjecting the seeds to all weathers – thereby giving the hot and cold treatment – can trigger the process (stratification, see page 644). When the seedlings are large enough to handle, prick out and plant up into pots, using a potting compost mixed in equal parts with propagating bark. Alternatively plant them directly into the chosen site in late spring after hardening off.

Cuttings
Take cuttings of young shoots in spring. This is an especially good method for the variegated *grandiflora*. They take easily, but keep in the shade until fully rooted and do not allow to dry out. Plant out in their final position when fully hardened off.

Division
Once the plants are established they can be divided in the spring or autumn, either by lifting the whole plant or by the double fork method. Replant immediately either into a prepared site or into pots using a potting compost mixed in equal parts with propagating bark. If this method is chosen in the autumn, keep in a cold frame all winter.

Pests and Diseases
Rarely suffers from pests or diseases.

Maintenance
Spring Sow seeds. Take cuttings from new growth.
Summer Cut back after first flowering and keep the plant tidy. Give a feed of liquid fertiliser, which can promote a second flowering.
Autumn Sow seeds. Cut back new growth after second flowering.
Winter Protect new growth in frosts below –4°C. Use horticultural fleece, bracken, straw or pine needles.

Garden Cultivation
These plants are indigenous to the limestone uplands and like a sunny position in well-drained soil that is low in nutrients. The leaves of *Calamintha grandiflora* 'Variegata' scorch easily and need some shade.

Harvesting
Pick leaves either side of flowering for use fresh or dried.

container growing

Unsuitable for growing indoors, but can look good in containers outside. Use a potting compost mixed in equal parts with propagating bark and a container with a diameter no less than 12cm. *Calamintha grandiflora* 'Variegata' looks particularly striking in a terracotta pot.

culinary

The young minty leaves of the lesser calamint can be added to salads and used to make a refreshing tea.

medicinal

Infuse dried leaves as a tea for colic, and an invigorating tonic. Use fresh leaves in a poultice for bruises.

Calomeria amaranthoides

INCENSE PLANT

From the family *Asteraceae*.

This fascinating, highly aromatic herb is a native of Australia. Historically this plant came to England via the famous plant collector Sir Joseph Banks who gave some seed to Lady Hume, hence its former name *Humea elegans*, which was only changed in 1993 to *Calomeria amaranthoides*. Sadly it is now rarely grown in parks or large private gardens because of its tendency to cause skin irritation and because of its high pollen count, which can cause asthma attacks.

varieties

Calomeria amaranthoides (Humea elegans)
Incense Plant
Tender biennial (sometimes annual). Ht up to 1.8m, spread 90cm. Tiny, delicate, coral flower bracts, very numerous on large branches. Large, oblong mid-green leaves.

warning

Leaves can cause irritation and the same kind of burns as rue (see page 387). The scent can cause breathing difficulties and when in flower, it has a high pollen count and can trigger asthma attacks.

cultivation

Propagation
Seed
Being a biennial, this is grown from the small seed, which is viable for only a short time. Collect from the plants in the summer, when ripe, and sow immediately into prepared seed or plug module trays using a standard seed compost mixed in equal parts with composted fine bark. Leave the seeds uncovered. Overwinter in a cold frame and cover the seed tray with glass or polythene. Germination is lengthy and very erratic. Pot on seedlings as soon as they appear, taking care not to injure the roots. Grow young plants in a cool, frost-free environment, and keep the roots almost dry through winter. In spring gradually encourage growth by watering and potting on.

Pests and Diseases
As a container-grown plant, it suffers from greenfly and red spider mite. Keep an eye out for these and use a horticultural liquid soap as soon as they appear.

Maintenance
Spring Prick out first year's plants. Pot up second year's.
Summer Feed and water regularly. Collect seeds off second year's plants and sow immediately.
Autumn Protect first year's plants.
Winter Protect plants from frost. Keep watering to a minimum.

Garden Cultivation
Do not plant outside until the night temperature no longer falls below 4°C. Plant in an area protected from the wind; even here, a stake is recommended. It prefers a light soil and a sunny position. All in all, it makes a better indoor plant, where the marvellous scent can be enjoyed.

Harvesting
Collect flowers for drying in summer.

container growing

The incense plant is very ornamental and is the ultimate pot plant, growing to over 1.5m. It is, however, rarely seen because it needs a good deal of attention and protection. Use a good peat-free potting compost mixed in equal parts with composted fine bark. Regularly pot up one size of pot at a time and regularly liquid feed throughout its short life until a pot size of 30cm in diameter is reached. Place in full sun and water regularly throughout the growing season.

other uses

Use in potpourris.

Calendula officinalis

MARIGOLD

Also known as Souci, Marybud, Bulls Eye, Garden Marigold, Holligold, Pot Marigold and Common Marigold. From the family *Asteraceae*.

Native of the Mediterranean and Iran. Distributed throughout the world as a garden plant.

This sunny little flower – the 'merrybuds' of Shakespeare – was first used in Indian and Arabic cultures, before being 'discovered' by the ancient Egyptians and Greeks. The Egyptians valued the marigold as a rejuvenating herb, and the Greeks garnished and flavoured food with its golden petals. The botanical name comes from the Latin *calendae*, meaning 'the first day of the month'.

In India wreaths of marigold were used to crown the gods and goddesses. In medieval times they were considered an emblem of love and used as chief ingredient in a complicated spell that promised young maidens knowledge of whom they would marry. To dream of them was a sign of all good things; simply to look at them was thought to drive away evil humours.

In the American Civil War, marigold leaves were used by the doctors on the battlefield to treat open wounds.

Calendula officinalis

varieties

Calendula officinalis
Marigold
Hardy annual. Ht and spread 60cm. Daisy-like, single or double flowers, yellow or orange: from spring to autumn. Light green, aromatic, lance-shaped leaves.

***Calendula officinalis* Fiesta Gitana Group AGM**
Hardy annual. Ht and spread 30cm. Daisy-like, double flowers, yellow and/or deep orange: from spring to autumn. Light green, lance-shaped leaves.

cultivation

Propagation

Seeds
Sow in autumn under protection directly into prepared pots or singly into plug module trays, in a standard seed compost mixed in equal parts with composted fine bark, covering lightly with compost. Plant out in the spring after any frost, 30–45cm apart. Alternatively in spring sow direct onto a prepared site in the garden.

Pests and Diseases

Slugs love the leaves of young marigolds. Keep night-time vigil with a torch and a bucket, or lay beer

MARIGOLD

Calendula officinalis Fiesta Gitana Group

traps. In the latter part of the season, plants can become infested with blackfly. Treat in the early stages by brushing them off and cutting away the affected areas, or later on by spraying with a horticultural soap. Very late in the season the leaves sometimes become covered with a powdery mildew, which should be removed and burnt.

Maintenance

Spring Sow seeds in the garden.
Summer Deadhead to promote more flowering.
Autumn Sow seeds under protection for early spring flowering.
Winter Protect young plants.

Garden Cultivation

Marigold is a very tolerant plant, growing in any soil that is not waterlogged, but prefers, and looks best in, a sunny position. The flowers are sensitive to variations of temperature and dampness. Open flowers forecast a fine day. Encourage continuous flowering by deadheading. It self-seeds abundantly but seems never to become a nuisance.

Harvesting

Pick flowers just as they open during summer, both for fresh use and for drying. Dry at a low temperature. Pick leaves young for fresh use; they are not much good preserved.

Dried flowers make a colourful oil

container growing

Marigolds look very cheerful in containers and combine well with other plants. Well suited to window boxes, but not so in hanging baskets, where they will become stretched and leggy.

Use a standard potting compost mixed in equal parts with composted fine bark. Pinch out the growing tips to stop the plant from becoming too tall and leggy. Deadhead flowers to encourage more blooms.

medicinal

Marigold flowers contain antiseptic, anti-fungal and anti-bacterial properties that promote healing. Make a compress or poultice of the flowers for burns, scalds, or stings. Also useful in the treatment of varicose veins, chilblains and impetigo. A cold infusion may be used as an eyewash for conjunctivitis, and can be a help in the treatment of thrush. The sap from the stem has a reputation for removing warts, corns and calluses.

other uses

There are many skin and cosmetic preparations that contain marigold. Infuse the flowers and use as a skin lotion to reduce large pores, nourish and clear the skin, and clear up spots and pimples. The petals make a pale yellow dye (see page 675).

culinary

Flower petals make a very good culinary dye. They have been used for butter and cheese, and as a poor man's saffron to colour rice. They are also lovely in salads and omelettes, and make an interesting cup of tea. Young leaves can be added to salads.

Sweet Marigold Buns

Makes 18

100g softened butter
100g caster sugar, plus extra to sprinkle
2 eggs, size 1 or 2
100g self-raising flour
1 teaspoon baking powder
2 tablespoons fresh marigold petals

Put the butter, sugar, eggs, sifted flour and baking powder into bowl, and mix together until smooth and glossy. Fold in 1½ tablespoons of marigold petals. Turn the mixture into greased bun tins or individual paper cake cases. Sprinkle a few petals onto each bun with a little sugar. Bake in an oven 160°C/325°F/gas mark 3 for approximately 25–30 minutes.

Capparis spinosa

CAPER

Also known as Caper Bush, Kapparis, Tápara, Umabusane. From the family *Capparaceae.*

This trailing evergreen shrub is now native in the Mediterranean but probably originated in the Middle East. It can be seen in the most unlikely places, from the ruins of ancient walls to the rubble alongside a newly built hotel. The Greek name *kapparis* is from the Persian *kabar*, hence 'caper'. The first recorded use of the caper bush was for medicinal purposes in 2000 BC by the Sumerians. Since then it has been used not only medicinally, but also as a useful condiment in the kitchen.

varieties

Capparis spinosa

Caper

Tender evergreen shrub. Ht up to 1m, spread up to 1.5m. Masses of green buds (it is these that are pickled) are followed by very pretty solitary white, four-petalled flowers with long pink/purple stamens from early summer until autumn. Leaves are oval, mid-green with a hint of brown, with two spines at the base of the leaf. The leaves grow on long stems which have been known to reach over 1.5m in length.

Capparis spinosa* var. *inermis

Spineless caper

Tender evergreen shrub. Ht and spread up to 1.5m. Masses of edible green buds followed by solitary white four-petalled flowers with long pink/purple stamens from early summer until autumn. Leaves are oval, mid-green with a hint of brown.

cultivation

Propagation

Seed

Caper seeds are miniscule, and once germinated they take a long time to grow into transplantable seedlings. Fresh caper seeds germinate readily – but only in low percentages. Once the seeds dry they become dormant and are notably more difficult to germinate, so patience is required. Start by immersing the dried seeds in hot water, 40°C, and leave to soak for 1 day. Carefully remove the seeds. Place them on some damp white kitchen towel, which makes them easier to see. Then put them into a sealed container and keep in the refrigerator for 2–3 months. After refrigeration, soak the seeds again in warm water overnight and then sow into prepared modules, plug trays, filled with a seed compost mixed in equal parts with perlite. Sow the seeds on the surface of the compost and cover with perlite. Keep warm at a minimum of 10°C. Plants raised from seed will not flower until their fourth or fifth year.

Cuttings

The best and most reliable method is to raise plants

Capparis spinosa var. *inermis*

from cuttings. Take cuttings from the new spring growth. Put them into a prepared modules or plug trays filled with seed compost mixed in equal parts with composted fine bark for extra drainage. Put the tray on a heated propagator at 18°C, making sure the cuttings do not dry out. Once well rooted pot up using a loam-based compost mixed in equal parts with propagating bark.

Pests and Diseases
This herb is rarely attacked by pests. Young cuttings and seedlings can keel over if the watering is too much or too little.

Maintenance
Spring Take cuttings.
Summer Harvest flowering buds.
Autumn Sow fresh seed.
Winter Protect from excessive wet in winter.

Garden cultivation
A simple rule of thumb is that the caper bush can be planted where the olive tree grows. It will thrive in lean, well-drained soil in a hot, sunny location with little or no water. It hates damp, cold, wet winters so if you live in northern Europe you will need to grow it in a container and place the container in a sheltered position for the winter. As an ornamental plant caper bushes can be an attractive loose groundcover, a specimen small shrub can be used as an espalier, which presents the flower buds well for picking. The caper bush is salt-tolerant and will flourish along shores. As flowers are borne on first-year branches, cut back plants back annually in the autumn. It tolerates the cold down to −7°C, however the growing tips can be damaged even in slight frost. The damaged growing tips can be cut off in the spring.

Harvesting
The flower buds are picked early in the morning for pickling and salting. The roots are dug up in autumn, the bark is then stripped from the roots and dried prior to use.

container growing

If the plant is pruned well in the autumn it will look wonderful growing in a container. For ease and self preservation, I advise growing the spineless variety, *Capparis spinosa* var. *inermis*. Pot up using a loam compost mixed in equal parts with horticultural grit.

culinary

Throughout the Mediterranean you can buy fresh capers from the vegetable markets and stores from early summer for a few months. They are bright green tightly closed flower buds, the smallest having the best flavour. However, when eaten fresh they do not taste particularly good; the flavour comes only after they have been pickled. This is due to the development of an organic acid called capric acid, which is an important flavouring in the kitchen. When you buy capers fresh they will still have the stems attached, so remove them, place the capers on a plate or tray, cover with sea salt and each time you pass the plate, give it a shake. After 2 days place them in a colander and rinse well under running water. Pack the capers into jars, add a few fennel seeds to each jar and a few immature flowers. Fill the jar up with white wine vinegar and seal with a non-metallic top. Leave for a month before use.

medicinal

The parts used medicinally are the bark from the roots and the flower buds. The bark is used to treat diarrhoea and rheumatism. In South Africa the roots are reputedly used to treat insanity, snake bites, chest pains, jaundice and malaria. The buds are used to treat coughs.

warning

Capparis spinosa has incredibly sharp spikes: wear gloves when handling this plant.

The attractive stamens of *Capparis spinosa*

Carum carvi

CARAWAY

From the family *Apiaceae.*

Caraway is a native of Southern Europe, Asia and India and thrives in all but the most humid warm regions. It is commercially and horticulturally cultivated on a wide scale, especially in Germany and Holland.

Both the common and species names stem directly from the ancient Arabic word for the seed, *karawya*, which was used in medicines and as a flavouring by the ancient Egyptians. In fact fossilised caraway seeds have been discovered at Mesolithic sites, so this herb has been used for at least 5,000 years. It has also been found in the remains of Stone Age meals, Egyptian tombs and ancient caravan stops along the Silk Road.

Caraway probably did not come into use in Europe until the thirteenth century, but it made a lasting impact. In the sixteenth century when Shakespeare, in *Henry IV*, gave Falstaff a pippin apple and a dish of caraways, his audience could relate to the dish, for caraway had become a traditional finish to an Elizabethan feast. Its popularity was further enhanced 250 or so years later when Queen Victoria married Prince Albert, who made it clear that he shared his countrymen's particular predilection for the seed in an era celebrated in England by the caraway seed cake.

No herb as ancient goes without magical properties of course, and caraway was reputed to ward off witches and also to prevent lovers from straying, a propensity with a wide application – it kept a man's doves, pigeons and poultry steadfast too!

varieties

Carum carvi

Caraway

Hardy biennial. Ht in first year 20cm, second year 60cm; spread 30cm. Flowers white/pinkish in tiny umbellate clusters in early summer. Leaves feathery, light green, similar to carrot. Pale, thick, tapering root comparable to parsnip but smaller. This plant is not particularly decorative.

cultivation

Propagation

Seed

Easily grown; best sown outdoors in early autumn when the seed is fresh. Preferred situation full sun or a little shade, in any reasonable, well-drained soil. For an acceptable flavour it must have full sun. If growing caraway as a root crop, sow in rows and treat the plants like vegetables. Thin to 20cm apart and keep weed free. These plants will be ready for a seed harvest the following summer; and the roots will be ready in their second autumn. Caraway perpetuates itself by self-sowing and can, with a little control, maintain the cycle.

If you want to sow in spring, do it either direct in the garden into shallow drills after the soil has warmed, or into prepared plug module trays to minimise harmful disturbance to its tap root when potting up. Cover with perlite. Pot up when seedlings are large enough to handle and transplant in the early autumn.

Pests and Diseases

Caraway occasionally suffers from carrot root fly. The grubs of these pests tunnel into the roots. The only organic way to get rid of them is to pull up the plants and bin them. To prevent the fly laying the eggs in the first place cover the crop with horticultural mesh.

Caraway seeds

Maintenance

Spring Weed well around autumn-sown young plants. Sow seeds.
Summer Pick flowers and leaves.
Autumn Cut seed heads. Dig up 2nd-year plants. Sow seeds.
Winter Does not need much protection unless it gets very cold.

Garden Cultivation

Prepare the garden seedbed well. The soil should be fertile, free draining and free of weeds, not least because it is all too easy to mistake a young caraway plant for a weed in its early growing stage. Thin plants when well established to a distance of 20cm.

Harvesting

Harvest the seeds in summer by cutting the seed heads just before the first seeds fall. Hang them with a paper bag tied over the seed head or over a tray in an airy place. It was once common practice to scald the freshly collected seed to rid it of insects and then dry it in the sun before storing. This is not necessary. Simply store in an airtight container.

Gather fresh leaves when young for use in salads. They are not really worth drying.

Dig up roots in the second autumn as a food crop.

container growing

Caraway really is not suitable for growing in pots.

other uses

Pigeon fanciers claim that tame pigeons will never stray if there is baked caraway dough in their coot.

An infusion of caraway seeds

medicinal

The fresh leaves, roots and seeds have digestive properties. Chew seeds raw or infuse them to sharpen appetites before a meal, as well as to aid digestion, sweeten the breath, and relieve flatulence.

An infusion can be made from 3 teaspoons of crushed seeds with 1/2 cup of water.

culinary

When you see caraway mentioned in a recipe it is usually the seed that is required. Caraway seed cake was one of the staples of the Victorian tea table. Nowadays caraway is more widely used in cooking, and in savouries as well as sweet dishes. The strong and distinctive flavour is also considered a spice. It is frequently added to sauerkraut, and the German liqueur, Kummel, contains its oil along with cumin.

Sprinkle over rich meats, goose, Hungarian beef stew – as an aid to digestion. Add to cabbage water to reduce cooking smells. Add to apple pies, biscuits, baked apples and cheese.

Serve in a mixed dish of seeds at the end of an Indian meal to both sweeten the breath and aid the digestion.

Caraway root can be cooked as a vegetable, and its young leaves chopped into salads and soups.

Caraway and Cheese Potatoes

Serves 4

4 large potatoes
100g grated Gruyère cheese
2 teaspoons caraway seeds

Scrub but do not peel the potatoes. Cut them in half lengthwise. Wrap in a boat of greaseproof foil and sprinkle each half with some of the grated cheese and a little caraway. Preheat the oven to 180°C/350°F/gas mark 4 and cook for 35–45 minutes, or until the potatoes are soft.

Carlina acaulis

CARLINE

Also known as Carline Thistle, Stemless Thistle, Dwarf Thistle. From the family *Asteraceae.*

This native herb of the Mediterranean is steeped in history. It is said that Charlemagne had a vision that the plant would ward off the plague, and in medieval times it was considered a good antidote to poisons. Today, in France, it is still known as *baromètre* because the flower closes at the approach of rain.

varieties

Carlina acaulis
Carline
Short-lived hardy perennial. Ht 5–10cm, spread up to 15cm. Stemless, large solitary, creamy white flower, which is surrounded by creamy pointed flower bracts in late summer of the second season. Oblong, dark green pinnate prickly leaves.

Carlina vulgaris
Carline Thistle
Biennial. Ht 20–60cm, spread up to 15cm. Groups of two to five yellow to purplish-brown flowers each surrounded by cream, linear, stiff-pointed bracts in the summer of the second season. Leathery dark green, narrow oblong, prickly leaves. Native to Britain and northwest Europe. Traditionally used as a purgative and in magical incantations.

cultivation

Propagation

Seed
Sow fresh seed in autumn into pots or plug modules using a standard loam-based seed compost. Cover lightly with compost, place in a cold frame. Germination takes 4–6 months. Plant out in spring once the seedlings are large enough to handle.

Division
Wearing gloves for protection from the prickles, divide plants in the second spring prior to flowering, using two small hand forks back to back. Replant in a prepared sunny site in the garden.

Pests and Diseases
Rarely suffers from pests and diseases.

Maintenance
Spring Divide first-year plants.
Summer Pick flowers for drying.
Autumn Sow seeds.
Winter No need for protection, hardy to –20°C .

Garden Cultivation
This drought-tolerant, very hardy plant needs to be planted in full sun in a low-nutrient, free-draining soil. In the right conditions it will self-seed. It is now endangered in many Mediterranean countries; do not take from the wild.

Harvesting
Dig up the root of the second-year plant in early autumn for making decoctions or drying. Pick the flower buds before they open for use in the kitchen. Pick the flowers when just opening for drying. Wear gloves to protect your hands.

Carlina vulgaris are popular in dried flower arrangements (right)

container growing

This herb looks most attractive when flowering. Use a loam-based potting compost mixed in equal parts with horticultural grit. Place the container in full sun.

medicinal

A salve made from the root can be used to treat several skin complaints including acne. Today the roots are used in veterinary medicine to stimulate the appetite of cattle.

other uses

The plant is popular in dried flower arrangements as the dried heads keep their appearance indefinitely.

culinary

It may seem unbelievable when you look at this plant that the flower bud is quite edible. It is eaten in the same manner as the globe artichoke *(Cynara scolymus)*. In Southern Italy there is a recipe for stuffing the small bud with cheese and eggs before it is fried. Before cooking the prickles are brushed off with a stiff brush.

warning

If the root is taken in large doses it is emetic and a purgative. Wear gloves when handling this plant.

Catha edulis

KHAT

Also known as Bushman's Tea, Qat, Chat and Miraa, From the family *Celastraceae*.

This herb is indigenous to East Africa. Historically the ancient Egyptians considered the khat plant a 'divine food', which was capable of releasing humanity's divinity. The generic name *Catha* is derived from the Arabic name for this plant, *Khat*, and the specific name *edulis* is a Greek word meaning 'edible'.

varieties

Catha edulis

Khat

Evergreen tropical and subtropical shrub or small tree. Ht up to 10m. Clusters of minute creamy-white to greenish flowers in spring. These are followed by three-lobed capsules, which split in late summer to release one to three narrowly winged seeds. Dark green shiny leaves with a dull underside.

cultivation

Propagation

Seed

If you are able to obtain fresh seed, which is rarely available outside of Africa, sow in spring. Place the tray under protection at 20°C; germination takes 4–6 weeks. Once the seedling is established, pot up using a potting compost mixed in equal parts with horticultural sand. Grow on in a container for two years before planting out into a prepared site in the garden or, in cold climates, potting on as a container plant.

Cuttings

In early summer take cuttings from the non-flowering shoots and insert into small pots or plug module trays using a seed compost mixed in equal parts with propagating bark. Once well rooted, pot up using a soil-based compost mixed in equal parts with horticultural sand. Grow on for 2 years before planting out into a prepared site. In cold climates grow on as a container plant.

Pests and Diseases

Rarely suffers from pests or diseases in warm climates. When grown as a container plant it can be prone to scale and red spider mite. To treat, either use the relevant predators or horticultural soap following the manufacturers' instructions. Do not use both.

Maintenance

Spring In warm climates sow seed.
Summer Take cuttings from new growth.
Autumn Cut back after flowering. Protect from excessive wet and cold weather.
Winter Protect from frost.

Garden Cultivation

In hot climates this herb can make an attractive garden plant, giving height in the border and all-year-round interest. Plant in full sun in a well-drained soil.

Harvesting

In early summer pick the new shoots as required for using fresh.

Bunches of khat shoots

container growing

Khat can be grown successfully in a container using soil-based compost mixed in equal parts with horticultural sand. Water and feed with a liquid fertiliser regularly throughout the growing season. Protect from frost and cold rains from autumn onwards and reduce watering; do not allow the compost to totally dry out. In spring reintroduce watering once the night temperature does not drop below 5°C. Place the container outside in a warm, sunny, sheltered position once the night temperature is above 10°C.

medicinal

In Africa this herb is used to treat chest complaints, asthma and coughs. It is taken in old age to stimulate and improve mental functions and communication skills. However, as with many beneficial medicinal herbs this can, and has been abused; see the warning.

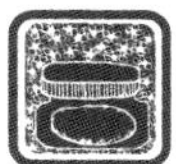

other uses

The bark is used as an insect repellent.

warning

This herb has been banned in many countries. In 1980 the World Health Organization classified khat as a drug of abuse that can produce mild to moderate psychic dependence. Overuse can cause aggression and develop personality problems. It should never be taken when pregnant or breastfeeding.

Cedronella canariensis

BALM OF GILEAD

Also known as Canary Balm. From the family *Lamiaceae.*

Although this herb originates from Madeira and the Canary Islands, as indicated by its species name, balm of Gilead is now established in many temperate regions of the world. Many plants have been called balm of Gilead, the common link being that they all have a musky, eucalyptus, camphor-like scent.

The Queen of Sheba gave Solomon a balm of Gilead, which was *Commiphora opobalsamum*, an aromatic desert shrub found in the Holy Land. Today this plant is rare and protected, its export prohibited.

The balm of Gilead mentioned in the Bible ('Is there no balm in Gilead; is there no physician there?') was initially held to be *Commiphora meccanensis*, which was an aromatic shrub. However, some now say it was oleo-resin obtained from *Balsamodendron opobalsamum*, a plant now thought to be extinct. Whatever is the case, the medicinal balm of Gilead is *Populus balsamifera*. This is balsam poplar, a tree found growing in several temperate countries, which smells heavenly in early summer, while the herb now known as balm of Gilead is *Cedronella canariensis*. This is said to have a similar scent to the Biblical shrubs, perhaps the reason for its popular name.

BALM OF GILEAD

Cedronella canariensis

varieties

Cedronella canariensis* syn. *cedronella triphylla
Balm of Gilead
Half-hardy perennial, partial evergreen. Ht 1m, spread 60cm. Leaves with strong eucalyptus scent, three lobes and toothed edges, borne on square stems. Pink or pale mauve, two-lipped flowers throughout summer. Black seed heads.

cultivation

Propagation
Seed
The fairly small seeds should be sown in spring directly on the surface of a prepared pot, plug or seed tray. Use a standard seed compost mixed in equal parts with composted fine bark. Cover with a layer of perlite. It is a temperamental germinator, so bottom heat of 20°C can be an asset. If using heat, remember not to let the compost dry out, and only water with a fine spray when needed. The seedlings will appear anytime between 2 and 6 weeks. When 2 leaves have formed, prick out the seedlings and plant them in position 1m apart.

Cuttings
More reliable than seed. They take readily either in early summer before flowering from new growth or in early autumn from the semi-ripe wood. Use a standard seed compost mixed in equal parts with composted fine bark.

Pests and Diseases
Since it is aromatic, aphids and other pests usually leave it alone, but the seedlings are prone to damping off.

Maintenance
Spring Sow seeds under protection. In a warm garden a mature plant can self-seed; rub the leaves of any self-seedlings to see if it is balm of Gilead or a young nettle (but don't get stung!). At this stage their aroma is the only characteristic that tells them apart. Plants overwintered in containers should be repotted if root-bound and given a liquid feed.
Summer Cut back after flowering to keep it neat and tidy, and also to encourage new growth from which late cuttings can be taken.
Autumn Take stem cuttings. Collect seed heads.
Winter Protect from frost.

Garden Cultivation
Balm of Gilead grows happily outside in sheltered positions. Plant in a well-drained soil in full sun, preferably against a warm, wind-protecting wall. The plant has an upright habit but spreads at the top, so planting distance from other plants should be approximately 1m.

It is a tender plant that may need protection in cooler climates. If you get frosts lower than −2°C protect the plant for the winter, either by bringing it into a cool greenhouse or conservatory or by covering in an horticultural fleece.

Harvesting
Pick leaves for drying before the flowers open, when they will be at their most aromatic. Either pick flowers when just coming into bloom and dry, or wait until flowering is over and collect the black flower heads (good for winter arrangements). Seeds are ready for extraction when you can hear the flower heads rattle. Store in an airtight container to sow in the spring.

container growing

Balm of Gilead makes an excellent container plant. A 23–25cm pot will be required for a plant to reach maturity. Use a soil-based compost mixed in equal parts with composted fine bark. Liquid feed a mature plant monthly throughout summer.

When grown in a conservatory, the scent of the leaves perfumes the air, especially when the plant is watered or the sun shining on it. Flowers are long lasting and give a good show during the summer. Keep watering to the absolute minimum in the winter months.

medicinal

Crush the leaves in your hand and inhale the aroma to clear your head. Rub the leaves on skin to try and stop being bitten by mosquitoes.

Said to be an aphrodisiac when applied.... no comment.

other uses

Dried leaves combine well in a spicy or woody potpourri with cedarwood chippings, rosewood, pineneedles, small fir cones, cypress oil and pine oil.

Add an infusion of the leaves to bath water for an invigorating bath.

Centella asiatica

GOTU KOLA

Also known as Pennywort, Spadeleaf, Brahmi, Tiger Herb, Tiger Grass. From the family *Apiaceae*.

This herb is indigenous to the subtropical and tropical areas of India, where it is known as Tiger herb. This is because it is said that wounded tigers roll in the leaves, to help themselves heal. It can also be found in marshlands and alongside rivers in Pakistan, Sri Lanka, South Africa, Hawaii and Florida. It has been an important Ayurvedic medicinal herb for thousands of years, and its Ayurvedic name is Brahmi, or 'knowledge'. It was traditionally used to promote wound healing and slow the progress of leprosy, senile decay and loss of memory. It was also reputed to prolong life; for example, a Sinhalese proverb says 'Two leaves a day keep old age away.' In China, Gotu kola is one of the reported 'miracle elixirs of life'. This was attributed to a healer named Li Ching Yun who reputedly lived 256 years by taking a tea brewed from Gotu kola and other herbs. It did not become important in Western medicine until the 1800s.

varieties

Centella asiatica

Gotu Kola

Tender perennial creeping plant. Ht 8cm, spread indefinite. The tiny magenta flowers in summer are surrounded by green bracts, which grow in small umbels near to the soil surface. Bright green kidney-shaped leaves with indented margins.

cultivation

Propagation

Seed

In warm climates this herb will happily self-seed. However, in cool climates it rarely sets seed and, when you propagate from the seeds, the germination is very spasmodic. So it is much easier to raise plants from cuttings.

Cuttings

As this plant tends to root where the stems touch the ground, and it is prolific when growing in damp conditions, it is easy to propagate any time during the growing season from spring until early autumn. Separate the plantlet from the main plant by cutting the stem above the ground, and then gently tease the small roots from the ground. You can then, depending on the size of the root, either pot up into small pots or trim the roots and ease them into a plug tray using a seed compost mixed in equal parts with vermiculite. Once rooted, which takes place very quickly in summer, pot up using a loam-based compost mixed in equal parts with vermiculite, and winter in a frost-free environment.

Division

Divide established plants in summer, either replanting into a prepared site in the garden or repotting using a loam-based compost mixed in equal parts with vermiculite.

Pests and Diseases

When grown as a container plant it can be prone to red spider mite. If this is the case introduce *Phytoseiulus persimilis*, its natural predator, or treat regularly with horticultural soap following the manufacturer's instructions. Do not use both.

Centella asiatica leaves

Centella asiatica flowers

Maintenance

Spring Take cuttings and repot container-raised plants.
Summer Take cuttings. Do not allow the plants to dry out at this time of year.
Autumn Protect young plants and container-grown plants from frost.
Winter Cut off any damaged leaves to prevent the spread of disease.

Garden Cultivation

This plant will only grow successfully outside all year round in tropical or subtropical climates. For those in cooler and cold climates it will tolerate temperatures as low as 10°C; below this it must be grown under protection. Its appearance changes, depending on growing conditions. In shallow water, the plant puts forth floating roots and the leaves rest on top of the water. In dry locations, it puts out numerous small roots and the leaves are small and thin.

Harvesting

Pick the leaves to use fresh from spring until late summer. Pick the leaves for drying in late spring.

container growing

It will adapt happily to being grown in containers, and looks very interesting in a hanging basket where it will cascade. Plant in a loam-based compost mixed in equal parts with vermiculite. Place the container in partial shade, not full sun. Water and feed regularly from early spring until early autumn with a liquid fertiliser following the manufacturer's instructions. This is especially important if you are picking the leaves regularly.

medicinal

This is one of the most important medicinal herbs I grow. It is a rejuvenating, diuretic herb that clears toxins and reduces inflammation. It is being used in the treatment of rheumatism and rheumatoid arthritis. It is commonly used to treat depression, but be warned that it has been reported to also cause depression in well-adjusted individuals.

other uses

As a beauty aid, Gotu kola stimulates the production of collagen and this helps improve the tone of veins near the surface of the skin. It is now being used in face creams which claim to be anti-wrinkle and skin-firming.

warning

Excessive use of this herb taken internally or externally can cause itching, headaches and even unconsciousness. Avoid Gotu kola if you are pregnant or breastfeeding, using tranquillizers or sedatives, or have an overactive thyroid.

culinary

When picking leaves for use in the kitchen, choose the new young tender leaves; the mature ones are dry and tough, especially if grown in dry conditions. Add the young leaves to salads, sandwiches and stir-fry dishes where the dry slightly spicy flavour combines well with fish and vegetables.

Eastern Herb Salad
Serves 4

1 cucumber, peeled, deseeded and cut into 1cm dice
Salt and freshly ground black pepper
2 cloves garlic, crushed
Juice of 1 lemon
3 tablespoons extra virgin olive oil
2 tablespoons young gotu kola leaves, roughly chopped
4 tablespoons purslane leaves, removed from the stalks and lightly chopped
2 tablespoons flat leaf parsley, roughly chopped
2 tablespoons coriander leaves, roughly chopped
2 tablespoons mint, roughly chopped
1 red onion, finely chopped
5 ripe tomatoes, peeled, deseeded and roughly chopped
3 pitta breads, toasted and broken into small pieces or five slices of white bread, toasted then cut into strips

Place the diced cucumber in a colander, sprinkle with salt and leave to drain for 20 minutes. In a large bowl, mix together the garlic, lemon juice and the olive oil to make a dressing. Add the herbs, the diced vegetables, the pieces of bread and toss well to coat with the dressing. Season with salt and pepper, serve immediately. This salad does not keep.

Centaurea cyanus

CORNFLOWER

Also known as Blue Bottle, Bachelors Buttons. From the family *Asteraceae*.

Cornflowers are the national flower of Estonia. Historically it is said that Chiron, an ancient Greek Centaur, taught mankind the healing value of this herb. It was regarded as a good remedy against the poison of scorpions and also as a healing wash for wounds. This pretty traditional European flower is better known as a wild flower, which nearly became extinct in the UK due to the introduction of chemical weed controls in farming.

varieties

Centaurea cyanus
Cornflower
Hardy annual. Ht 30–80cm, spread 15cm. Stunning single and double blue, daisy-like flowers in summer. Lance-shaped, grey-green leaves. The lower leaves are often toothed and covered in fine cotton. There are hybrids with pink, white or purple flowers.

cultivation

Propagation
Seed
In early spring sow seeds into prepared module plug trays using standard seed compost. Place the tray in a warm position; extra bottom heat is not required. Germination takes 14–21 days. Be careful not to over water as the seedlings are prone to damping off. Pot up or plant out when all threat of frost has passed. Alternatively, for the best results, sow in spring where you want the plant to grow and when the air temperature does not go below 10°C at night.

Pests and Diseases
Rarely suffers from pests or diseases.

Maintenance
Spring Sow seeds.
Summer Pick flowers for display and drying.
Autumn Harvest seeds.
Winter Prepare garden for next season's sowing.

Garden Cultivation
Plant in a well-drained soil in a sunny position. If you wish to create a meadow effect, in the autumn prior to sowing, prepare the ground well by clearing all the grass and weeds in a 30cm square area for each plant. Sow 15 seeds per patch, thinning to preferably 2 plants. In autumn allow the plants to die back and self seed naturally.

Harvesting
For use in the kitchen or for drying pick the flowers in summer when they are just half open, before the centre stems are visible. When drying tie into small bunches of about ten flowers per bunch and dry fast, otherwise the colour will fade and the heads will start to crumble.

companion planting

The blue flowers reputedly attract pollinating insects more than other colours, which is beneficial for increasing yield in the vegetable garden.

container growing

They can be grown in large containers using a soil-based potting compost. They look stunning on their own, or mixed with other tall wild flowers like oxeye daisies or poppies. When growing in a container it is worth tying some dark green sewing thread around the established plants to hold them in place and to minimise the damage from high winds or heavy rain.

culinary

The petals of the flowers are edible; they look lovely scattered over green salads or over cold rice dishes.

medicinal

The flower is the part used medicinally; it is considered a tonic and stimulant. The famous French eyewash Eau de Casselunettes was traditionally made from the distilled flowers because of their eye-brightening properties.

other uses

The juice extracted from the petals makes a good blue ink which, when mixed with alum-water, can be used in watercolour paintings. It also makes a very good fabric dye. The dried flowers make up for their lack of scent by looking striking in arrangements or when added to potpourri.

Centaurea cyanus

Centranthus ruber

RED VALERIAN

Also known as American Lilac, Bloody Butcher, Bouncing Bess and Bouncing Betsy. From the family *Valerianaceae.*

A native of central and southern Europe, cultivated widely in temperate climates and has now become widely naturalised. This cheerful plant was a great ornament in Gerard's garden, but he described it in 1597 as 'not common in England'. However, by the early eighteenth century it had become well known.

varieties

Centranthus ruber
Red Valerian
Perennial. Ht 60–90cm, spread 45–60cm. Showy red fragrant flowers in summer. They can also appear in all shades of white and pink. Fleshy, pale green, pointed leaves.

cultivation

Propagation
Seed
Sow the small seeds in early autumn in seed or plug module trays, using a standard seed compost mixed in equal parts with composted fine bark. Cover lightly with compost and leave outside over winter, covered with glass. As soon as you notice it germinating, remove the glass and place in a cold greenhouse. Prick the seedlings out when large enough to handle and pot up using the same mix of compost. Leave the pots outside for the summer, maintain watering until the autumn. No need to feed. Plant out 60cm apart.

Pests and Diseases
This plant does not suffer from pests or diseases.

Maintenance
Spring Dig up self-sown seedlings and replant if you want them.
Summer Deadhead to prevent self-seeding.
Autumn Sow seeds. Plant previous year's seedlings.
Winter A very hardy plant.

Garden Cultivation
Red valerian has naturalised on banks, crumbly walls, rocks and along coastal regions. It is very attractive to butterflies. As an ornamental, it thrives in poor, well-drained, low-fertile soil, and especially on chalk or limestone. It likes a sunny position and self-seeds prolifically.

container growing

Make sure the container is large enough and use a soil-based compost. No need to feed, otherwise you will inhibit its flowers. Position the container in a sunny spot and water regularly.

culinary

Very young leaves are eaten in France and Italy. They are incredibly bitter.

medicinal

A drug is obtained (by herbalists only) from the root, which looks like a huge radish and has a characteristic odour. It is believed to be helpful in cases of hysteria and nervous disorders because of its sedative and anti-spasmodic properties.

warning

Large doses or extended use may produce symptoms of poisoning. Do not take for more than a couple of days at a time, without seeking medicinal advice.

Centranthus ruber

Chamaemelum nobile

CHAMOMILE

From the family *Asteraceae*.

***'I am sorry to say that Peter was not very well during the evening. His mother put him to bed and made some chamomile tea and she gave a dose of it to Peter, one tablespoon full to be taken at bedtime.' – The Tale of Peter Rabbit* by Beatrix Potter**

Chamomile grows wild in Europe, North America, and many other countries. As a garden escapee, it can be found in pasture and other grassy places on sandy soils.

The generic name, *Chamaemelum*, is derived from the Greek *khamaimelon*, meaning 'earth apple' or 'apple on the ground'.

varieties

Chamaemelum nobile
Roman Chamomile
Also known as garden chamomile, ground apple, low chamomile and whig plant. Hardy perennial evergreen. Ht 10cm, spread 45cm. White flowers with yellow centres all summer. Sweet-smelling, finely divided foliage. Ideal for ground cover. Can be used as a lawn, but because it flowers it will need constant cutting.

***Chamaemelum nobile* 'Flore Pleno'**
Double-flowered Chamomile
Hardy perennial evergreen. Ht 8cm, spread 30cm. Double white flowers all summer. Sweet-smelling, finely divided, thick foliage. Good for ground cover, in between paving stones and lawns. More compact habit than Roman chamomile, and combines well with chamomile Treneague.

***Chamaemelum nobile* 'Treneague' syn. *Anthemis nobile* 'Treneague'**
Chamomile Treneague
Also known as lawn chamomile. Hardy perennial evergreen. Ht 6cm, spread 15cm. Non-flowering. Leaves are finely divided and very aromatic. Ideal for ground cover or mow-free lawn. Plant in well-drained soil, free from stones, 10–15cm apart.

Anthemis tinctoria
Dyers Chamomile
Also known as yellow chamomile. Hardy perennial evergreen. Ht and spread 1m. Yellow daisy flowers in the summer. Leaves are mid-green and fern like. Principally a dye plant.

Matricaria recutita
German Chamomile
Also known as scented mayweed, wild chamomile. Hardy annual. Ht 60cm, spread 10cm. Scented white flowers with conical yellow centres from spring to early summer. Finely serrated aromatic foliage. The main use of this chamomile is medicinal.

cultivation

Propagation

Seed
Dyers, Roman and German chamomiles can be grown from seed. In spring, sow onto the surface of a prepared seed or plug tray. Use a standard seed compost mixed in equal parts with composted fine bark. Cover with perlite. Use bottom heat 19°C. Harden off and plant out or pot on.

Cuttings
Double-flowered chamomile and chamomile Treneague can only be propagated by cuttings and division. Take cuttings in the spring and autumn from the offsets or clusters of young shoots. They are easy to grow as they have aerial roots.

Division
All perennial chamomiles planted as specimen plants will benefit from being lifted in the spring of their second or third year and divided.

Pests and Diseases
As all the chamomiles are highly aromatic they are not troubled by pests or diseases.

Chamaemelum nobile 'Flore Pleno'

CHAMOMILE

Maintenance

Spring Collect offshoots, sow seeds. Fill in holes that have appeared in the chamomile lawn. Divide established plants. Give a liquid fertiliser feed to all established plants.
Summer Water well. Do not allow to dry out. In the first season of a lawn, trim the plants to encourage bushing out and spreading. In late summer collect flowers from the dyers chamomile and cut the plant back to 6cm to promote new growth.
Autumn Take cuttings. Divide if they have become too invasive. Cut back to promote new growth. Give the final feed of the season.
Winter Only protect in extreme weather.

Garden Cultivation

All the chamomiles prefer a well-drained soil and a sunny situation, although they will adapt to most growing conditions.

As a lawn plant, chamomile gets more credit than it deserves. Chamomile lawns are infinitely less easy to maintain in good condition than grass lawns. There is no selective herbicide that will preserve chamomile and kill the rest of the weeds. It is a hands-and-knees job with no short-cuts.

Prepare the site well, make sure the soil is light, slightly acid, and free from weeds and stones. Plant young plants in plug form. I use a mix of double-flowered and Treneague chamomile at a distance of 10–15cm apart. Keep all traffic off it for at least 12 weeks, and keep it to the minimum during the whole of the first year.

If all this seems daunting, compromise and plant a chamomile seat. Prepare the soil in the same way and do not sit on the seat for at least 12 weeks. Then sit down, smell the sweet aroma and sip a cool glass of wine. Summer is on hand...

Harvesting

Leaves
Gather in spring and early summer for best results. Use fresh or dry.

Flowers
Pick when fully open, around mid-summer. Use fresh or dry. Dyers chamomile flowers should be harvested in summer for their yellow dye.

companion planting

Chamomile is said to be the plants' physician because when it is planted near ailing plants, it apparently helps them to revive. Roman Chamomile, when planted next to onions, is said to repel flying insects and improve the crop yield.

Traditional Chamomile Infusion

Bring 600ml water to the boil. Add a handful of chamomile leaves and flowers. Cover and let it stand for half a day. This infusion was traditionally used as a spray to prevent dampening off. It was also added to the compost heap as an activator to accelerate decomposition.

container growing

I would not advise growing chamomiles indoors, as they get very leggy, soft and prone to disease. But the flowers can look very cheerful in a sunny window box. Use chamomile 'Flore Pleno', which has a lovely double flower head, or the non-flowering chamomile Treneague as an infill between bulbs, with a standard potting compost mixed in equal parts with composted fine bark.

cosmetic

Chamomile is used as a final rinse for fair hair to make it brighter. Pour 1 litre boiling water over one handful of chamomile flowers and steep for 30 minutes. Strain, cover and allow to cool. It should be poured over your hair several times.

medicinal

German chamomile's highly scented dry flower heads contain up to 1 per cent of an aromatic oil that possesses powerful antiseptic and anti-inflammatory properties. Taken as a tea, it promotes gastric secretions and improves the appetite, while an infusion of the same strength can be used as an internal antiseptic. It may also be used as a douche or gargle for mouth ulcers and as an eye wash.

An oil for skin rashes or allergies can be made by tightly packing flower heads into a preserving jar, covering with olive oil and leaving in the sun for three weeks. If you suffer from overwrought nerves, add five or six drops of chamomile oil to the bath and this will help you relax at night.

Chamomile Tea

Chamomile Tea

1 heaped teaspoon chamomile flowers (dried or fresh)
1 teaspoon honey
Slice of lemon (optional)

Put the chamomile flowers into a warm cup. Pour on boiling water. Cover and leave to infuse for 3–5 minutes. Strain and add the honey and lemon, if required. Can be drunk either hot or cold.

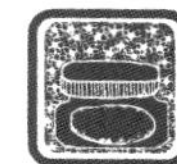

other uses

Dyers chamomile can be used as a dye plant. Depending on the mordant, its colour can vary from bright to olive/brown yellow (see page 675). German and double-flowered chamomile are best for herb pillows and potpourri.

warning

When taken internally, excessive dosage can produce vomiting and vertigo.

Chenopodium bonus-henricus

GOOD KING HENRY

Also known as All Good, Good King Harry, Good Neighbour, Wild Spinach, Lincolnshire Asparagus and Mercury. From the family *Chenopodiaceae*.

Good King Henry comes from a genus *(Chenopodium)* that is distributed all over the world and is found growing in all climates. This species *(C. bonus-henricus)* is native to Europe.

Good King Henry was popular from Neolithic times until the nineteenth century. Its curious name is not taken from the English king, Henry VIII, as might be expected, but from King Henry IV of Navarre, and to distinguish it from the poisonous Bad Henry (*Mercurialis perennis*).

Gerard in the sixteenth century observed that Good King Henry grew in untilled places and among rubbish near common ways, old walls, hedges and fields, and it still does – colonies of the herb can be found on many medieval sites.

varieties

Chenopodiaceae, the goosefoot family, includes 1,500 rather unattractive plants, some of them important edible plants, for example, spinach and beet.

Chenopodium bonus-henricus
Good King Henry
Perennial. Ht 60cm, spread 45cm. Tiny greenish-yellow flowers in early summer. Leaves green and arrow-shaped. Very occasionally a variegated form is found; but the yellow variegation will be difficult to maintain.

Chenopodium album
Fat Hen
Also known as lambs' quarters, white goosefoot, common pigweed, all good and muckweed. Annual. Ht 60cm–1m. Flowers small, greenish-white, summer to mid-autumn. Green lance-shaped leaves. Its seeds have been identified at Neolithic villages in Switzerland and in the stomach of the Iron Age Tollund Man. Rich in fat and albumen, it appears to have been a food supplement for primitive man.

Chenopodium ambrosioides
American Wormseed
Also known as Mexican tea, and in China as fragrant tiger bones. Annual. Ht 60cm–1.25m. Small greenish flowers from late summer to late autumn. Green lance-shaped leaves. This is native to tropical Central America. Introduced through Mexico, it has become naturalised as far north as New England in the USA. It was introduced into Europe in the eighteenth century. Mexican Tea was once included in the United States pharmacopoeia but is now restricted to American folk medicine and mainly used for its essential oil, Chenopodium oil, against roundworm and hookworm.

Warning: *Chenopodium ambrosioides* is poisonous. Use under strict supervision. It causes deafness, vertigo, paralysis, incontinence, sweating, jaundice and death.

GOOD KING HENRY

cultivation

Propagation

Seeds

Sow the fairly small seeds early in spring in prepared seed or plug trays for an early crop. Use a standard seed compost mixed in equal parts with composted fine bark and cover with perlite. No extra heat required. When the seedlings are large enough to handle and after hardening off, plant out in the garden 25cm apart. Can be sown direct in late spring into prepared soil in the garden in late spring in 1cm drills. Allow 45cm between rows. Cover the seeds with 6mm soil. Germination in warm soil, 10–14 days. When large enough to handle, thin to 25cm apart.

Division

Divide established plants in the spring. You will find even small pieces will grow.

Pests and Diseases

Does not usually suffer from these.

Maintenance

Spring Lift and divide established plants. If you wish to grow as an asparagus, blanch the shoots from early spring onwards. As they emerge, earth up with soil. Divide and re-pot container-grown plants.

Summer Give a liquid feed if a second crop of leaves is required.

Autumn Cut back dying foliage and give the plant a mulch of compost.

Winter No need for protection.

Garden Cultivation

Good King Henry will tolerate any soil, but if planted in a soil rich in humus, dug deep and well drained in a sunny position, the quality and quantity of the crop will be much improved. Sow directly.

Keep well watered in dry months. In autumn cover beds with a thin layer of manure. Beds should be renewed every 3 to 4 years.

Harvesting

Allow plants one year to develop before harvesting. From mid-spring the young shoots can provide an asparagus substitute crop. They should be cut when they are about 15cm long. Harvest the flowering spikes as they begin to open. Later in the season gather the larger leaves as a spinach substitute as required. Freeze only when used as an ingredient in a cooked dish.

container growing

Can be grown outside in a large container, in a soil-based potting compost mixed in equal parts with composted fine bark. Needs to be kept well watered throughout the summer and fed once a week to maintain a supply of leaves. Divide each spring and re-pot the divisions in fresh compost.

medicinal

The seeds have a gentle laxative effect, making them suitable relief for a slightly constipated condition. A poultice (or ointment) cleanses and heals skin sores.

other uses

Good King Henry is a cough remedy for sheep. The whole plant is used to fatten poultry. Seed is used commercially in the manufacture of shagreen, an artificially granulated untanned leather, often dyed green. The whole plant of fat hen can be used as a red or golden dye (above).

warning

Sufferers of kidney complaints or rheumatism should avoid medicinal preparations containing extracts from this plant.

culinary

The leaves of good King Henry and fat hen are rich in iron, calcium and vitamins B1 and C, and are particularly recommended for anaemic subjects.

Like all low-growing leaves, good King Henry must be washed with great care; the slightest suspicion of grit in the finished dish will ruin the meal. Use 2 or 3 changes of water.

Eat young leaves raw in salads. Alternatively, cook in casseroles, stuffings, soups and purées and savoury pies. The leaves are more nutritious than those of spinach or cabbage. Steam flower spikes and toss in butter before serving, like broccoli. To blanch the shoots, dip in hot water, rinse immediately in cold water. Alternatively, cut shoots 15cm long. Steam or boil very quickly. Peel if necessary. Serve hot with melted butter, or cold with a vinaigrette.

The seed of fat hen can be ground into flour and used to make into a gruel.

Convallaria majalis

LILY OF THE VALLEY

Also known as Our Lady's Tears, Fairy's Bells, May Lily, Ladder to Heaven and May Bells. From the family *Convallariaceae*.

Lily of the valley is a native of Europe, North America and Canada and has been introduced throughout the world in moist cool climates.

According to European folk tales, lily of the valley either originated from the Virgin Mary's tears, shed at the foot of the Cross, or from those shed by Mary Magdalen when she found Christ's tomb.

From the Middle Ages onwards the flowers formed the traditional part of a bride's bouquet and are associated with modesty and purity.

In the sixteenth century they were used medicinally and called Convall Lily. The Elizabethan physician Gerard has this amazing recipe: 'Put the flowers of May lilies into a glass and set it in a hill of ants, firmly closed for 1 month. After which you will find a liquor that when applied appeaseth the paine and grief of gout.'

varieties

***Convallaria majalis* AGM**
Lily of the Valley
Hardy Perennial. Ht 15cm, spread indefinite. White, bell shaped, scented flowers, late spring to early summer. Mid to dark green oval-shaped leaves.

***Convallaria majalis* 'Flore Pleno'**
Double Flowered Lily of the Valley
Hardy Perennial. Ht 23–30cm, spread indefinite. Creamy white, double, bell-shaped, scented flowers that are larger than the species, late spring to early summer. Mid to dark green oval-shaped leaves.

***Convallaria majalis* 'Fortin's Giant'**
Fortin's Giant Lily of the Valley
Hardy Perennial. Ht 45cm, spread indefinite. White, larger than the species, bell-shaped, scented flowers, late spring to early summer. Mid- to dark green- wide oval-shaped leaves. A robust growing form of the Lily of the valley. Widely thought to be a triploid form.

Convallaria majalis* var. *rosea
Pink Lily of the Valley
Hardy Perennial. Ht 15cm, spread indefinite. Pale pink, bell-shaped, scented flowers, late spring to early summer. Mid to dark green oval-shaped leaves.

***Convallaria majalis* 'Vic Pawlowski's Gold'**
Vic Pawlowski's Gold Lily of the Valley
Hardy Perennial. Ht 15cm, spread indefinite. White, bell-shaped, scented flowers, late spring to early summer. Mid to dark green oval-shaped leaves with attractive thin golden strips running the length of the leaves. This has a more consistent and better variegation than *Convallaria majalis* 'Albostriata'.

LILY OF THE VALLEY

cultivation

Propagation

Seeds

Ripe seeds are seldom formed, and the scarlet berries are highly poisonous, so better to propagate by division.

Division

The plant produces crowns on creeping rhizomes. Divide in the autumn after the plant has finished flowering and the leaves have died back.

Pests and Diseases

Lily of the valley is free from most pests and diseases.

Maintenance

Spring In very early spring, bring pots into the house for forcing.
Summer Do nothing!
Autumn When the plant has died back fully dig up the rhizomes for splitting. Pot up crowns for forcing.
Winter No need for protection.

Garden Cultivation

Unlike its name, it should be grown not in an open valley but in partial shade. Ideal for growing under trees or in woodlands or in the shade of a fence provided there is not too much competition from other plants.

To get the best flowers, prepare the site well. The soil should be deeply cultivated with plenty of well-rotted manure, compost or leaf mould. Plant in autumn, 15cm apart, before the frosts make the soil too hard. Place the crowns upright in the prepared holes with the tips just below the soil.

Harvesting

Pick the flowers when in full bloom for drying so that they can be added to potpourri.

container growing

This plant can be happily grown in pots as long as it is kept in the shade and watered regularly. Use a loam-based compost mixed in equal parts with propagating bark. Feed with liquid fertiliser only during flowering. In winter let the plant die down, and keep it in a cool place outside.

Convallaria majalis

medicinal

This plant, like the foxglove, is used in the treatment of heart disease. It contains cardiac glycosides which increase the strength of the heartbeat while slowing and regularising its rate, without putting extra demand on the coronary blood supply.

culinary

None – all parts of the plant are poisonous.

warning

Lily of the valley should only be used as prescribed by a qualified practitioner and it is restricted. All parts of the plant are poisonous.

other uses

The flowers of the Lily of the valley are often used in bridal arrangements not only for their wonderful scent but also because in the language of flowers they signify a 'return to happiness'. I did not know this fact when I married over 30 years ago where I wore a wreath of Lily of the valley flowers in my hair and carried them as my bouquet.

An essential oil is obtained from the flowers and used in the perfumery industry. The leaves can be used to make a natural dye. The young leaves in spring will give a green dye and the mature autumn leaves will give a yellowish brown dye.

Lily of the Valley soap and cream

Coriandrum sativum

CORIANDER

Also known as Chinese Parsley, Yuen Sai, Pak Chee, Fragrant Green, Dhania (seed), Dhania Pattar and Dhania Sabz (leaves). From the family *Apiaceae.*

A native of southern Europe and the Middle East, coriander was a popular herb in England up until Tudor times. Early European settlers in America included seed among the beloved items they took to the New World, as did Spaniards into Mexico.

Coriander has been cultivated for over 3,000 years. Seeds have been found in tombs from the 21st Egyptian Dynasty (1085–945 BC). The herb is mentioned in the Old Testament – 'when the children of Israel were returning to their homeland from slavery in Egypt, they ate manna in the wilderness and the manna was as coriander seeds' – and it is still one of the traditional bitter herbs to be eaten at the Passover when the Jewish people remember that great journey.

Coriander was brought to Northern Europe by the Romans who, combining it with cumin and vinegar, rubbed it into meat as a preservative. The Chinese once believed it bestowed immortality and in the Middle Ages it was put in love potions as an aphrodisiac. Its name is said to be derived from *koris,* Greek for 'bedbug', since the plant smells strongly of the insect.

varieties

Coriandrum sativum
Coriander
Tender annual. Ht 60cm. White flowers in the summer. The first and lower leaves are broad and scalloped, with a strong, strange scent. The upper leaves are finely cut and have a different and yet more pungent smell. The whole plant is edible. This variety is good for leaf production.

***Coriandrum sativum* 'Leisure'**
Tender annual. Ht 60cm. Much as *C. sativum*; whitish flowers in summer; also suitable for leaf production.

***Coriandrum sativum* 'Sabor'**
Tender annual. Ht 70cm. Flowers white with a slight pink tinge in summer. This variety is best for seed production.

cultivation

Propagation
Seed
Coriander is grown from seed. Thinly sow its large seed directly into the soil in shallow drills. Lightly cover with fine soil or compost, and water in. Look for results after a period of between 5 and 10 days. Seed sowing may be carried out as often as required between early spring (under glass), and late autumn. When large enough to handle, thin out the seedlings to leave room for growth.

Sowing into seed trays is not recommended because coriander plants do not transplant well once the tap root is established. If they get upset they bolt straight into flower, missing out the leaf production stage altogether.

If a harvest of fresh leaves is required, space the plants 5cm apart; if of seed, 23cm apart.

Pests and Diseases
Being a highly aromatic plant, coriander is usually free from pests. In exceptional circumstances it is attacked by greenfly. If so, do not be tempted to pressure hose the pests off, as this will destroy the leaves. Either wash off gently under the tap, and shake the plant carefully to remove excess water on the leaves, or use a liquid horticultural soap.

CORIANDER

Coriandrum sativum

Maintenance
Spring Sow seeds.
Summer Sow seeds, cut leaves.
Autumn Cut seed heads. Sow autumn crop in mild climates. Dig up old plants.
Winter Once the seed heads have been collected, the plant should be pulled up.

Garden Cultivation
Coriander grows best in a light, well-drained soil, in semi-shade and a dry atmosphere. In fact it is difficult to grow in damp or humid areas and needs a good dry summer at the very least if a reasonable crop is to be obtained. However, do not allow the soil to dry out because this will cause the plant to bolt.

When the plant reaches maturity and the seed sets and begins to ripen, the plant tends to loll about on its weak stem and needs staking. On ripening, the seeds develop a delightful orangey scent, and are used widely as a spice and a condiment. For this reason alone, and because the flavour of home-grown seeds is markedly superior to those raised commercially, coriander deserves a place in the garden. If you live in a mild, frost-free climate, sow in the autumn for an over-winter crop; but make sure the plants are in full sunlight.

Harvesting
Pick young leaves any time. They should be 10cm in height and bright green. Watch seed heads carefully, as they ripen suddenly and will fall without warning. Cut the flower stems as the seed smell starts to become pleasant. Cover bunches of about 6 heads in a paper bag. Tie the top of the bag and hang upside down in a dry, warm, airy place. Leave for 10 days. The seeds should come away from the husk quite easily and be stored in an airtight container. Coriander seeds keep their flavour well.

container growing

Coriander can be grown in containers inside with diligence, but the plants produced will be weak and straggly. It is ideal for growing in large, deep pots outside, in partial shade on a patio or by a kitchen door. Fill the container with a standard potting compost mixed in equal parts with composted fine bark, water in well, then sow the seeds thinly and cover lightly wiith compost. Once the leaves are large enough, start cropping. To maintain a supply prepare another large container and sow 20 days later. Water your containers regularly in the morning, not at night, as like many other herbs, coriander does not like having wet feet. Do not allow the compost to dry out because this will cause the plant to bolt to seed.

medicinal

Coriander is good for the digestive system, reducing flatulence, stimulating the appetite and aiding the secretion of gastric juices. It is also used to prevent gripe caused by other medication such as senna or rhubarb. Bruised seed can be applied externally as a poultice to relieve painful joints and rheumatism.

culinary

The leaves and ripe seeds have two distinct flavours. The seeds are warmly aromatic, the leaves have an earthy pungency.

Coriander seeds are used in garam masala (a mixture of spices) and in curries. Use ground seed in tomato chutney, ratatouille, frankfurters, curries, apple pies, cakes, biscuits and marmalade. Add whole seeds to soups, sauces and vegetable dishes.

Add fresh lower leaves to curries, stews, salads, sauces and as a garnish. Delicious in salads, vegetables and poultry dishes. A bunch of coriander leaves with a vinaigrette dressing goes particularly well with hard-boiled eggs.

Mushrooms and Coriander

Serves 2

500g button mushrooms
2 tablespoons cooking oil
300ml dry white wine
2 teaspoons coriander seeds
1 clove garlic
2 tablespoons tomato purée
Salt and pepper
Coriander leaf for garnish

Wipe mushrooms and slice in half. Put the oil, wine, coriander seeds and garlic in a large saucepan. Bring to the boil and cover and simmer for 5 minutes. Add the mushrooms and tomato purée. Cook for 5 minutes, by which time the vegetables should be tender. Remove the mushrooms and put in a serving dish. Boil the liquid again for 5 minutes and reduce it by half. Pour over the mushrooms. When cool, sprinkle with some chopped coriander leaf.

Crithmum maritimum

SEA FENNEL

Also known as Samphire, Crest Marine, Kritamo. From the family *Apiaceae*.

This native of the Atlantic, Mediterranean and Black Sea is mentioned in Shakespeare's *King Lear*, where it is referred to as that 'dreadful trade' of the samphire gatherer. Rock samphire was at one time cultivated in English gardens for its seed pods and sold in London, where it was called crest marine.

varieties

Crithmum maritimum
Sea Fennel
Hardy perennial. Ht 30cm, spread 20cm. Flat umbels of tiny white-green flowers in summer. Aromatic, succulent sea-green, triangular leaves with long rounded linear lance-shaped segments which grow in small groups along the branch.

cultivation

Propagation

Seed
In autumn sow fresh seeds into prepared small pots or module plug trays, using a seed compost mixed in equal parts with horticultural sand, under protection at 10°C. Germination takes 2–3 weeks; if there is no germination within this period, place the container in a refrigerator for 4 weeks, then return to 10°C. Germination should occur within 4–6 weeks. Once the seedlings are large enough, pot up using a loam-based compost mixed in equal parts with horticultural grit. Winter seedlings in a frost-free environment. Plant out in the following spring.

Division
Divide established plants in spring. In the garden, dig up the whole plant, then divide. Once divided, replant in to a well-prepared site. Container-raised plants can be divided gently so as not to destroy the crown, and repotted into a loam-based compost mixed in equal parts with horticultural grit.

Pests and Diseases
These plants, being aromatic, are not prone to pest damage. However, in cold wet winters, when grown in soil that does not drain well, they can be prone to rot.

Maintenance
Spring Divide established plants. Pick young leaves.
Summer Pick leaves until it starts to flower.
Autumn Sow fresh seeds.
Winter Protect from hard frosts below –3°C.

Garden Cultivation
As this is a seaside plant that grows literally in the crevices between rocks, it is essential to prepare the site well, making sure it is well drained, adding extra grit if necessary. Plant in a sunny position and protect from cold winds. In winter cover the crown with straw, not mulch or compost, which would cause it to rot.

Harvesting
In early summer pick the leaves either to use fresh or to pickle.

container growing

Sea fennel will grow happily in containers. Use a soil-based compost mixed in equal parts with grit. Place in a sheltered, sunny position.

medicinal

This herb is very high in vitamin C; it also has digestive and purgative properties. It is under research for treating obesity and is in a number of herbal products.

culinary

The leaves can be eaten fresh or cooked. Prior to cooking, remove any leaves that have begun to turn slimy and any hard parts of the stalk. The leaves have an aromatic salty flavour which combines well in salads or cooked in butter. They can also be used to make sauces and aromatic pickles.

other uses

The seeds produce a fragrant oil which is widely used in modern perfumery and medicine. A number of leading manufactures of face creams purport to having included this herb for hydrating the skin.

Curcuma longa

TURMERIC

Also known as Indian Saffron, Haldi, Haridra. From the family *Zingiberaceae.*

Of all the herbs I grow, this is one of the most traditional and versatile. The exact origin of turmeric is not known. It most probably came from western India where records show it has been used for at least 2,500 years. The yellow and yellow-orange colours obtained from the roots are sacred and auspicious. It is important in Hindu and Buddhist ceremonies, being associated with fertility and prosperity, and brings good luck if applied to a bride's face and body as part of the purification ritual before a wedding. The roots may be given as a present on special occasions, such as a visit to a pregnant woman. The use of turmeric is prohibited in a house of mourning.

varieties

Curcuma longa

Turmeric

Tropical, subtropical, herbaceous perennial. Ht 1m, spread indefinite. The yellow/white flowers, with pink tinges to the tips of the petals, appear surrounded by pale green bracts in spring on a single stem. The flowers are sterile and do not produce viable seed. Aromatic long, up to 60cm, mid green, oval leaves. The root is a large rhizome.

cultivation

Propagation

Cuttings

A word of warning before you start taking root cuttings: wear gloves and an apron because, once the root is cut, it produces a yellow sap which will stain your fingers and can permanently stain cloth.

Unless you live in the tropics, the best source for fresh turmeric root is from Asian and Caribbean shops. However, be aware that the quality can be variable. Often, after it has been air-freighted, the cold temperatures have killed the growth, alternatively it may have been treated with chemicals to inhibit sprouting. Choose a fresh, plump, juicy-looking root which has a tooth bud growing on one side. Choose a shallow container not much larger than the root; fill with a seed compost mixed in equal parts with horticultural grit. Place the root in the container with the tooth bud facing up, and cover the roots with compost making sure the tooth bud is peeping through the compost. Place the pot in a plastic bag, seal, and place in a warm place or in a propagator at 20°C. The shoots should emerge in 3–4 weeks but may take longer depending on the warmth. Once the shoots emerge, remove the plastic bag. Keep the container in a warm place, minimum temperature 18°C, but not in direct sunlight until fully established. Only pot up one size of pot, using a potting compost mixed in equal parts with horticultural grit when the plant looks as if it is bursting out of the container; it likes being pot bound.

Division

If you live in the tropics, or have a plant raised in a container, it can be divided in the spring. In the garden

TURMERIC

Curcuma longa flower with its green bracts

use two forks back to back to divide the rhizome. Once divided replant into a prepared site in the garden. Pot-raised plants should be lifted from the container and excess compost removed so you can see the rhizome. Choose a section of rhizome with a growing bud; slice the root using a sharp knife. Pot up the cutting into a small container which just fits the cutting using a seed compost mixed in equal parts with horticultural grit.

Pests and Diseases
Red spider mite can be an occasional problem on older plants when grown under glass; regular misting and keeping the leaves well-washed will reduce this. If it gets out of hand use a horticultural soap spray following the manufacturer's instructions.

Maintenance
Spring Divide rhizomes.
Summer Pick leaves as required.
Autumn Cut back on watering of container plants.
Winter Grow on at a minimum of 18°C.

Garden Cultivation
This herb can only be grown outside in the tropics. It requires a minimum temperature of 18°C at night. It is suitable for growing in a conservatory, heated greenhouse or a well-lit east- or west-facing window, not in direct midday sun, as this will cause leaf scorch.

Harvesting
Outside the tropics, this only produces a small amount of fresh rhizome, so is most usefully used medicinally or as a cosmetic. Harvest the rhizome in late summer/ early autumn. Only turmeric rhizome, when cured commercially, has the aroma and colour necessary for cooking. The leaves can be used as a flavouring; pick as required throughout the growing season.

container growing

Turmeric is ideal for growing in a container in cool and cold climates. However, you will not be able to harvest much useful root from your plant, although you will be able to use the leaves for flavouring. Pot up using a loam-based potting compost mixed in equal parts with horticultural grit. Be mean with the container size; do not over pot, as this can cause the rhizome to rot in cool climates. In summer, place the plant in partial shade, water and liquid fertilise regularly. In dry weather plants will benefit from a daily light misting with rainwater. In the autumn cut back on the watering, keeping the compost fairly dry. Keep the plant frost free at a minimum of 18°C.

medicinal

Turmeric has been used medicinally for thousands of years; it is an important Ayurvedic herb used to treat inflammation, coughs and gastric disorders. It is also a very good first-aid remedy in the home; a paste can be used as a quick household antiseptic, for cuts, grazes and minor burns. It is also used as a decoction to calm the stomach and can be applied externally to remove hair and alleviate itching.

other uses

Turmeric has been used as dye for centuries, it is used to colour medicine, confectionery, paints, varnishes, silk and cotton.

warning

If the roots are cut, the yellow sap will stain fingers or cloth indelibly.

culinary

Turmeric is an essential ingredient in Indian cuisine. It is used in virtually every Indian meat, vegetable and lentil dish with the exception of greens, because when cooked with green vegetables it turns them grey and bitter. Turmeric has been known as poor-man's saffron as it offers a less expensive alternative yellow colouring. The flavour of the cured turmeric can vary, and this is dependent on how it is used. When added to oil before the main ingredient, the flavour is pungent, when added after the main ingredient it is more subtle.

The leaves have a warm, rich, sweet aroma and can be used fresh to wrap fish or sweets before steaming.

Cymbopogon citratus

LEMON GRASS

Also known as Fever Grass, Bhustrina and Takrai. From the family *Poaceae*.

This important culinary and medicinal herb, which can be found throughout the tropics, is indigenous to Southeast Asia where it is used extensively to produce an essential oil. It is also a snake repellent and a versatile garden plant. There are records showing that the Persians were using it as a tea in the first century BC. I have been lucky enough to see it growing and flowering naturally in the Caribbean where they primarily use it for reducing fevers.

varieties

Cymbopogon citratus
Lemon Grass
Half-hardy perennial, evergreen in warm climates. Ht 1.5m, spread 1m. Lax panicles of awnless spikelets appear throughout the summer. However, it rarely flowers in cold climates or in cultivation. Lemon-scented linear, grey/green leaves up to 90cm in length. Robust cream/beige cane-like stems.

Cymbopogon nardus
Citronella Grass
Half-hardy perennial, evergreen in warm climates. Ht 1.8m and spread 1m. Lax panicles of awnless spikelets appear throughout the summer. However it rarely flowers in cold climates. Lemon-scented drooping, flat, blue/grey/green leaves up to 60cm in length. This species is cultivated for its medicinal and insect-repellent properties. It is also grown around buildings in Africa, as a snake repellent. The oil is used to perfume soaps.

Cymbopogon martini* var. *motia
Palmarosa, Rosha, Indian Geranium
Half-hardy perennial, evergreen in warm climates. Ht 1.5m and spread 1m. Lax panicles of awnless spikelets appear throughout the summer. However, it rarely flowers in cold climates. Rose-scented linear grey/green leaves up to 50cm in length. The oil made from this species is used to perfume cosmetics and soap. It is also used in Ayurvedic medicine to treat fevers and infectious diseases.

cultivation

Propagation

Seed
In spring, sow seeds into prepared seed or module plug trays and place under protection at 20°C (68°F). Germination takes 15–25 days. Once the seedlings are large enough, pot up using a loam-based compost and grow on until well established. Either plant out in the garden in warm climates or, in cool climates, grow on as a container plant.

Cuttings
Take cuttings, in spring, from a plant that is more than a year old, which has an established crown. Gently remove the swollen lower stems from the crown. Remove any grass from the stem and cut the stem back to 10cm. Place in a prepared module plug tray or a very small pot using a seed compost; do not be tempted to over pot the cutting as this will cause it to rot. Place under protection or in a warm position away from cold draughts. Once rooted, pot up into a small pot using a loam-based compost.

Division
In the garden use two forks back to back and gently tease the plant apart, replanting immediately into a prepared site. Divide container plants either with your fingers, or two small forks, teasing the crown apart, repotting into a pot which fits snugly around the roots. This plant is happiest when pot bound. Use a loam-based potting compost.

Pests and Diseases

Outside the tropics this herb can be prone to rot and mildew. To prevent this, in winter, keep container-grown plants nearly dry and in a well-ventilated, frost-free room.

LEMON GRASS

Lemon grass stems

Maintenance
Spring Sow seeds. Divide or take root cuttings of established plants. Feed container-raised plants regularly.
Summer Do not allow the plants to dry out. Maintain feeding until late summer.
Autumn In cool climates, to prevent disease, cut back the grass leaving the stems.
Winter Protect from frost. Keep watering to the minimum.

Garden Cultivation
Lemon grass can be grown outside where the night temperature does not fall below 8°C (48°F). Plant in any soil, including a heavy soil, as long as the summers are hot and wet and the winters are warm and dry. In low light levels the plant can become dormant. In spring, prune back all the old growth and thick stalks to 10cm.

Harvesting
The fresh leaves and lower stems can be cut throughout the summer to use fresh or to dry. The stems can be stored whole in the refrigerator in a plastic bag for up to two weeks. Alternatively the stems and the leaves can be frozen for use within five months.

container growing

An excellent container plant. Use a loam-based compost which should not be allowed to dry out in summer. In winter, bring the plant into a frost-free environment of 5°C minimum. When the light levels and night temperatures drop the plant will go 'dormant', the grass gradually turns brown and the outside leaves shrivel. Reduce the watering to a minimum and cut back the grass to 10cm above the stems. In early spring, as the day lengthens and the temperatures rise, you will notice new grass starting to grow. Cut off all dead growth. Repot if necessary and liquid feed weekly.

medicinal

A tea made from fresh leaves is very refreshing as well as being a stomach and gut relaxant. It is also a good antidepressant and helps lift the spirits if one is in a bad mood. The essential oil is antiseptic, antibacterial, antifungal and deodorising.

other uses

Valued for its exotic citrus fragrance, it is commercially used in soaps, perfumes and as an ingredient in sachets. Also used as an insect repellent.

warning

Do not take the essential oil internally without supervision.

culinary

The fresh leaves and stalks have been traditionally used in Thai, Vietnamese and Caribbean cooking. The lemon flavour complements curries, seafood, garlic and chillies.

Vegetable and Lemon Grass Soup
Serves 4

2 tablespoons light olive oil
1 large onion, finely chopped
2 cloves garlic, finely sliced
4 carrots, scrubbed then finely sliced or, if in a hurry, grated.
150g button mushrooms, sliced
4–6 stems of lemon grass, cut into 4cm lengths and bruised
1 teaspoon ginger, finely grated
1 litre chicken stock or vegetable stock
1 tablespoon very finely chopped lemon grass leaves
Salt to taste

Heat the oil in a heavy-bottomed saucepan, add the onions, garlic and carrots, stir-fry for 3 minutes, stirring all the time. Add the mushrooms and stir-fry for a further minute. Pour in the water, stir and add the lemon grass stems, ginger and salt to taste. Bring to the boil, then reduce the heat, cover and simmer for about 4–7 minutes until all the vegetables are cooked but still crunchy.

With a slotted spoon, remove the lemon grass stalks, add the very finely chopped lemon grass leaves, stir and serve.

Dianthus

PINKS

Dianthus gratianopolitanus

Also known as Clove Pink and Gillyflower. From the family *Caryophyllaceae*.

Dianthus comes from the words *dios*, meaning divine, and *anthos*, meaning flower. Both the Romans and Greeks gave pinks a place of honour and made coronets and garlands from the flowers. In the seventeenth century it was recognised that the flowers could be crystallised, and the petals started being used in soups, sauces, cordials and wine.

varieties

***Dianthus deltoides* AGM**
Maiden Pink
Evergreen hardy perennial. Ht 15cm, spread 30cm. Small cerise, pink or white flowers are borne singly all summer. Small, narrow, lance-shaped, dark green leaves.

***Dianthus gratianopolitanus* AGM**
Cheddar Pink
Evergreen hardy perennial. Ht 15cm, spread 30cm. Very fragrant pink/magenta flat flowers are borne singly all summer. Small, narrow, lance-shaped, grey/green leaves. This variety is protected in the UK.

Dianthus plumarius
Pinks
Evergreen hardy perennial. Ht 15cm, spread 30cm. Very fragrant flowers, white with dark crimson centres, are borne singularly all summer. Loose mats of narrow, lance-shaped, grey/green leaves. This species is related to the Cheddar pink and the origin of the garden pink.

cultivation

Propagation

Seed
Sow the small seeds when fresh in autumn into prepared seed or plug module trays using a standard seed compost, and cover with perlite. Winter seedlings in a cold frame prior to planting out in the following spring 30cm apart. Pinks raised from seed can be variable in height, colour and habit.

Cuttings
Take stem cuttings in spring and heel cuttings in early autumn. Place the cuttings into prepared plug module trays or a small pot using a seed compost mixed in equal parts with fine composted bark. Do not over water the cuttings while they are rooting as this will cause them to rot.

Division
Established plants can be divided in early autumn after flowering.

Layering
Established plants can be layered in late summer and then lifted in the following spring.

Pests and Diseases

During propagation young plants can be prone to rot; this is usually caused by fungus due to the compost being too wet. Infected plants must be removed as soon as this is spotted as it can spread to other plants very quickly. Aphids can be a problem; spray with horticultural soap.

Maintenance

Spring Lift layers. Take stem cuttings.
Summer Deadhead flowers.
Autumn Take heel cuttings. Divide established plants. Sow seed. Layer established plants.
Winter No need for protection.

Garden Cultivation

Pinks prefer to be planted in a well-drained soil, which does not become waterlogged in winter, and in full sun.

Harvesting

Pick the flowers just as they open for either using the petals fresh or for crystallising.

container growing

Pinks grow well in containers. Use a soil-based compost mixed with 25 per cent composted fine bark and 25 per cent fine horticultural grit.

culinary

Before eating the petals, they must be removed individually from the flower. You will notice that each petal has a white heel (see below); this must also be removed as it is very bitter. The prepared petals can be added to salads, sandwiches and fruit pies.
They can be used to flavour jams, sugars and syrups. The crystallised petals can be used to decorate cakes and puddings.

medicinal

An excellent nerve tonic can be made from the petals, either as a cordial or infused in white wine.

other uses

Add dried petals to potpourris, scented sachets and cosmetic products.

Dianthus flowers, showing the bitter-tasting white heel

Digitalis purpurea

FOXGLOVE

Also know as Fairy fingers, Fairy gloves and Deadmen's bells. From the family *Scrophulariaceae*.

Foxgloves are a common wild flower in temperate climates throughout the world, seeding freely in woods and hedgerows. The principal common name is probably derived from the Anglo-Saxon *foxglue* or *foxmusic* after the shape of a musical instrument. In 1542 Fuchs called it Digitalis after the finger-like shape of its flowers, but he considered it a violent medicine. It was not until the late eighteenth century, after William Withering used foxglove tea in Shropshire as an aid for dropsy, that its reputation as a medicinal herb grew.

Digitalis purpurea f. *albiflora*

varieties

Digitalis purpurea

Foxglove, wild, common

Short-lived perennial, grown as a biennial. Ht 1–1.5m, spread 60cm. Flowers all shades of pink, purple and red in second summer. Rough, mid- to dark green leaves.

Digitalis purpurea* f. *albiflora

White Foxglove

Short-lived perennial, grown as a biennial. Ht 1–1.5m, spread 60cm. Tubular white flowers all summer in the second season. Rough, mid- to dark green leaves.

cultivation

Propagation

Seed

For the best germination, sow the very fine seeds in autumn, using the cardboard method (see page 645), either directly onto the prepared ground, or into pots or plug module trays, which have been filled with a standard seed compost. Sow on the surface and do not cover the seeds with compost. Place the container in a cold frame, cold greenhouse or under the eaves of the house. Winter the seedlings in the containers, only protecting if temperatures fall below −10°C. In the following spring either prick out or plant in the garden 45cm apart. They will flower in the second season.

Pests and Diseases

Foxgloves rarely suffer from pests or diseases.

Maintenance

Spring Thin seedlings from autumn sowing.
Summer If self seeding is not required, remove flowering spikes after flowering.
Autumn Sow fresh seeds. Dig up self-sown seedlings for potting on or replanting.
Winter No need for protection.

Garden Cultivation

This is one of the most poisonous plants in the flora, so choose its position with care. Foxgloves will grow in most conditions, even dry exposed sites, but do best in semi-shade and a moist but well-drained acid soil enriched with leaf mould. Water well in dry weather and remove the centre spike after flowering to increase the size of the flowers on the side shoots.

Harvesting

Do not harvest unless you are a trained herbalist or pharmacist.

container growing

Foxgloves are not ideally suited for growing in containers, mainly because of their flowering spike, which can be damaged in high winds. If you do wish to grow them in a container use a soil-based compost and water regularly throughout the summer months.

medicinal

Foxgloves are grown commercially for the production of a drug, the discovery of which is a classic example of a productive marriage between folklore and scientific curiosity. Foxgloves contain glycosides which are extracted from the second-year leaves to make the heart drug digitalis. For more than 200 years digitalis has provided the main drug for treating heart failure. It is also a powerful diuretic.

warning

The whole plant is poisonous – seeds, leaves and roots. Even touching the plant can cause rashes, headaches and nausea. DO NOT USE without medical direction.

ECHINACEA

Also known as Coneflower, Purple Coneflower and Black Sampson. From the family *Asteraceae*.

Echinacea is a herb that has been used by the Native American Indians for hundreds of years for everything from snake bites and wounds to respiratory infections. Its generic name *Echinacea* comes from the Greek *echinos,* 'hedgehog', referring to the central golden cone, which becomes more pointed and prickly as the flower matures. It is only in the past decade that modern research has confirmed its medicinal properties; however there is still some argument over its validity. Despite this, it remains in much demand. This has had a major impact on the wild species, which has now become endangered due to over collection.

Echinacea pallida

Echinacea purpurea

varieties

Echinacea angustifolia

Narrow-leafed Echinacea, Black Sampson

Hardy herbaceous perennial. Ht 60cm, spread 30cm. Single flowers with long, thin purple or, rarely, white petals and a spiky central cone, borne throughout the summer until early autumn. Mid-green linear leaves. In its natural habitat this echinacea has become an endangered species. The American Indians regarded this herb as a cure-all.

Echinacea pallida

Echinacea, Coneflower

Hardy herbaceous perennial. Ht 80cm, spread 45cm. Single flowers with long mauve/pink, narrow, drooping petals and a spiky central cone, throughout the summer until early autumn. The leaves are oval, narrow, dark green and veined.

Echinacea purpurea

Echinacea, Purple Coneflower

Hardy herbaceous perennial. Ht 1.2m, spread 45cm. Single honey-scented flowers with long mauve/pink, narrow, drooping petals and a spiky central cone, throughout the summer until early autumn. The leaves are oval, narrow, dark green and deeply veined. This species is cultivated for its medicinal properties.

cultivation

Propagation

Seed

Sow seeds in early spring into prepared seed trays, plug modules or a small container using a seed compost mixed in equal parts with perlite. Cover the seeds with perlite. Place in a warm position or in a propagator at 18°C. If no germination has occurred after 28 days, place the container outside for a further 21 days, then place back under cover, out of direct sunlight. Germination should then occur within the next 20 days. Once the seedlings are large enough, pot into a loam-based compost mixed in equal parts with composted fine bark. Once rooted, harden off, plant out into a prepared site in the garden 30cm apart.

Division

Divide established plants in the late autumn when all the foliage has died back and the plant is dormant. Use the two forks back to back method. Replant immediately either into a prepared site in the garden, or alternatively, pot up using a loam-based potting compost mixed in equal parts with composted fine bark. Container-raised plants can also be divided at this time of year.

Pests and Diseases

Echinacea, in general, is not prone to pests and diseases. In spring, young plants can be attacked by slugs and snails, so it is worth doing a couple of night patrols with a torch to remove any that you may find in the crown of the plant. In a damp, warm late summer it can suffer from powdery mildew; if this happens cut off any affected parts, bin do not add to the compost.

Maintenance

Spring Sow seeds. Feed established plants lightly with well-rotted manure.
Summer Cut back stems as the blooms fade to encourage further flower production.
Autumn Divide established plants.
Winter No need for protection from the cold, only from excessive wet.

Echinacea purpurea

Garden Cultivation

Echinacea grows wild on the fertile plans of North America, so, to keep it thriving in the garden, plant in a fertile loam soil which is free draining and in plenty of sun. It will adapt to most soils with the exception of excessive wet conditions and cold wet clay soils, which can cause the roots to rot. After flowering, cut back the plants, leaving 8cm of growth above ground, collect the seeds and keep the flower heads for drying. Lightly mulch established plants with well-rotted manure in the spring. Spring growth and young echinacea plants are a snail delicacy, so it is worth checking around the plants daily in spring.

Harvesting

Pick the flowers and leaves during flowering before the cone is fully formed and the petals have started to fall back. When the petals have died back pick the seeds heads, dry well. If you wish to use the plant medicinally, dig up the roots of 4-year-old plants of *Echinacea angustifolia*, or *purpurea*, or *pallida* in autumn for drying and for making fresh tinctures.

container growing

Echinacea adapts happily to being grown in containers. Use a loam-based potting compost mixed in equal parts with composted fine bark. Divide pot-bound plants in the autumn, or pot up one size of pot. Place in full sun for the growing season. Feed regularly with a liquid fertiliser following the manufacturer's instructions. In winter if you live in a damp, wet cold climate, place the container under the eves of the house, or by a wall, to shelter it from the rain.

medicinal

Echinacea has the ability to raise the body's resistance to infections by stimulating the immune system. It is reputedly very effective in preventing colds and flu or reducing their severity. A tincture made from the root is used to treat severe infections, and a decoction made from the root can be used as a gargle to treat sore throats and other throat infections . A decoction, or the juice extracted from the flowers, can be used externally to treat minor wounds, burns, boils and skin infections, including chilblains.

other uses

The cone part of the flower head dries very well and looks most attractive in floral arrangements.

warning

If you are allergic to plants in the Asteraceae family, for example chrysanthemums, marigolds or daisies, then you could be allergic to echinacea.

People who are suffering from progressive systemic auto-immune disorders should not take this herb without full consultation.

Dried roots and flowers

Echium vulgare

VIPER'S BUGLOSS

Also known as Bugles, Wild Borage, Snake Flower, Blue Devil, Blueweed, Viper's Grass and Snakeflower. From the family *Boraginaceae*.

This plant originates from the Mediterranean region and is now widespread throughout the northern hemisphere, being found on light porous stones on semi-dry grassland, moorlands, and waste ground. It is regarded as a weed in some parts of America. To many American farmers this will seem an understatement; they consider it a plague.

The common name, viper's bugloss, developed from the medieval Doctrine of Signatures, which ordained that a plant's use should be inferred from its appearance. It was noticed that the brown stem looked rather like a snake skin and that the seed is shaped like a viper's head. So, in their wisdom, they prescribed it for viper bites, which for once proved right; it had some success in the treatment of the spotted viper's bite.

varieties

Echium vulgare
Viper's Bugloss
Hardy biennial. Ht 60–120cm. Bright blue/pink flowers in the second year. The leaves are mid-green and bristly.

cultivation

Propagation
Seed
Viper's bugloss is easily grown from seed. Start it off in a controlled way in spring by sowing the small seed into a prepared seed or plug module tray. Cover the seed with perlite. When the seedlings are large enough to handle, and after a period of hardening off, plant out into a prepared site in the garden at about 45cm apart.

Pests and Diseases
It rarely suffers.

Maintenance
Spring First year, sow seeds; second year, clear round plants.
Summer Second year, pick off flowers as they die so that they cannot set seed.
Autumn First year, leave well alone. Second year, dig up plants and bin. Do not compost unless you want thousands of viper's bugloss plants all over your garden.
Winter No need to protect first-year plant.

Garden Cultivation
This colourful plant is beautifully marked. Sow the seed in spring directly into the garden. It will grow in any soil and is great for growing on dry soils and sea cliffs. With its long tap root, the plant will survive any drought but cannot easily be transplanted except when very young. The disadvantage is that it self-seeds and is therefore invasive.

Harvesting
Gather flowers in summer for fresh use.

container growing

Because it is a rampant self-seeder, it is quite a good idea to grow it in containers. For the first year it bears only green prickly leaves and is very boring. However, the show put on in the second year is full compensation. Use a soil-based compost; no need to feed. Over-feeding will prohibit the flowering. Very tolerant of drought; nevertheless do water it regularly. Dies back in winter of first year – leave the container somewhere cool and water occasionally.

culinary

The young leaves are similar to borage, but they have lots more spikes. It is said you can eat them when young, but I have fought shy of this. The flowers look very attractive in salads. They can also be crystallised.

medicinal

The fresh flowering tips can be chopped up for making poultices for treating whitlows and boils. Infuse lower leaves to produce a sweating in fevers or to relieve headaches.

Echium vulgare

Elettaria cardamomum

CARDAMOM

Also known as Ela, Ilaichi. From the family *Zingiberaceae*.

Cardamom is indigenous to Southern India where it grows abundantly under the forest canopy. Until the nineteenth century, the world's supply of cardamom came mainly from the wild, in an area known as the Cardamom Hills in Western Ghat, India. The fruits have been traded in India for at least 1,000 years. It was known as the Queen of Spices, with black pepper being the King. Cardamom is the third most expensive spice after saffron and vanilla. It is traded internationally in the form of whole fruits, and to a lesser extent as seeds. Early Arabs enjoyed cardamom seeds in their coffee, a practice which continues today, and which explains why cardamoms are mentioned so often in Sir Richard Burton's translation of *The Arabian Nights*. In India it is common practice to offer cardamoms at the end of a meal, as a digestive and for freshening the breath.

varieties

Elettaria cardamomum
Cardamom
Tropical, subtropical, evergreen perennial. Ht 3m, spread indefinite. In summer the flowers grow on a single stem. They are creamy white, with deep purple lines over the lower lip. They are followed by pale green to fawn fruits, each having three chambers, containing several small aromatic seeds which start white and ripen to black. Aromatic, up to 60cm long, mid-green, lance-shaped leaves. The root is a thick branching rhizome.

cultivation

Propagation

Seed
Outside the tropics it is very difficult to get fresh seed, which remains viable for only 7–10 days once harvested; therefore it is far easier to propagate by division. If you can get fresh seed, sow in autumn into prepared plug modules, using a seed compost. Place in a warm place or propagator at 24°C. Germination takes 14–21 days. Plants raised from seed take 3–5 years to flower.

Division
Divide established plants in late spring, early summer. Replant into a prepared site in a tropical garden or tropical greenhouse. Alternatively, if dividing a pot plant, replant using a loam-based potting compost mixed equally with composted fine bark. Plants in the tropics that have been divided take 3 years to flower; however, it rarely flowers outside the tropics, even when raised in a tropical house, due to the light levels.

Pests and Diseases

Red spider mite can be an occasional problem in older plants; regular misting with soft rain water and keeping the leaves well-washed will reduce this. If it gets out of hand use a horticultural soap spray following the manufacturer's instructions.

If plants are too cold their leaves turn brown. If the leaves develop brown tips at any time (even if the plant is kept warm) it is a sign of over watering. If the leaf develops creamy patches, this can be caused by too much sun and is a form of scorch.

Maintenance

Spring Divide established plants or pot plants.
Summer Spray leaves with rain water.
Autumn Sow fresh seeds.
Winter Protect from cold weather.

Cardamom fruits and the smaller seeds contained within them

CARDAMOM

Garden Cultivation
When grown outdoors in the tropics, it is grown in the shade of the trees, in a deep fertile soil. It needs a minimum annual rainfall of 150cm, and a short dry season. It takes 3 years before the plant will flower and set seed.

In the northern hemisphere it is best grown as a pot plant indoors or in a heated greenhouse, even in hot summers as it needs a minimum temperature of 18°C.

Harvesting
In tropical climates the fruit is harvested by hand from the third year onwards. In cool climates pick fresh leaves for use as required.

container growing

Cardamom can only be grown indoors in climates outside the tropics and subtropics. It will not flower, although you can use the leaves for flavouring. Grow your plant in loam-based potting compost mixed in equal parts with composted fine bark. Cardamoms can be fussy: they do not like draughts, sudden changes of temperature or direct sunlight. Grow them in a warm, steamy, shady place, like a warm bathroom and mist the plant daily with rainwater. Alternatively, stand the pot on a big saucer of pebbles which are kept moist, to encourage a humid atmosphere around the plant. In winter, keep the plant warm at a minimum of 20°C and cut back on the watering, but do not allow the plant to dry out. Regularly liquid feed during the growing season with a general-purpose foliage fertiliser following the manufacturer's instructions.

medicinal

The medicinal properties of cardamom are found in the seeds, which have pain-relieving, anti-inflammatory, and antispasmodic properties that are often used to treat urine retention and stomach disorders. In Ayurvedic medicine it is often used to improve the flavour and quality of medicine and as an expectorant.

Seeds extracted from the pod, chewed after a meal, freshen the breath and aid digestion.

other uses

Cardamom oil made from the seeds is used in cosmetics, soaps, lotions and perfumes.

culinary

In India and throughout South Asia the cardamom fruit is used not only in savoury dishes, where it is an essential ingredient in garam masala, but also in sweets where it is often combined with rose water and thickened milk. The bright lime green pods are the best culinary variety and the true cardamom. There are brown pods that are confusingly also called cardamom, but these come from a plant called *Amomum subulatum*, and they are used in savoury dishes, especially rice dishes.

The leaves don't smell the same as the seeds; they have a warm sweet aromatic scent and can be used to wrap around fish, rice or vegetables to add flavour during cooking. The long stalks are useful to tie the leaves together to make a neat parcel. When used with fish they keep it beautifully moist.

Garam Masala

There are many versions of Garam masala, some are spicy, some are hot; here is a fragrant version that is good with chicken and vegetables.

2 teaspoons cardamom seeds (removed from pods)
1 teaspoon cumin seeds
1 teaspoon whole black peppercorns
2 x 5cm cinnamon sticks
½ teaspoon whole cloves
¼ nutmeg, grated.

Heat a small frying pan and add each spice individually with the exception of the nutmeg. As each one starts to smell fragrant, remove it from the pan onto a plate and allow to cool . Put all the cooled spices into a pestle and mortar or an electirc blender and grind to a fine powder. Add the finely grated nutmeg. Store in a glass jar with an airtight lid in a dark cupboard. Use within 4 weeks.

Equisetum arvense

HORSETAIL

Also known as Mare's Tail, Shave Grass, Bottle Brush and Pewter Wort. From the family *Equisetaceae*.

The Horsetail is a plant left over from prehistoric times. By the evidence of fossil remains, it has survived almost unchanged since the coal seams were laid. The Romans always used it to clean their pots and pans, not just to make them clean but also, thanks to the silica, to make them non-stick. The plant was used in the Middle Ages as an abrasive by cabinet makers and to clean pewter, brass and copper, and for scouring wooden containers and milk pans.

varieties

Equisetum arvense
Horsetail
Hardy perennial. Ht 45cm. The plant does not flower. It grows on thin creeping rhizomes producing 20cm long grey/brown fertile shoots with 4–6 sheaths in spring. The shoots die off and the spores are spread just like those of ferns.

cultivation

Propagation
I am not sure that this is necessary because it is so invasive, but if you do require a supply of horsetail it may be of merit.

Cuttings
Take root cuttings in summer and place into prepared seed or plug module trays using a standard seed compost mixed in equal parts with fine composted bark. Plant into a prepared site the following spring.

Pests and Diseases
This herb is pest and disease free.

Maintenance
Spring Remove any fertile shoots to prevent the plant spreading.
Summer Cut back plants as they begin to die back to prevent the spores spreading.
Autumn Cut down to the ground.
Winter No need for protection; very hardy.

Garden Cultivation
If horsetail is to be introduced into the garden at all, and to be honest I do not recommend it, it is best to confine it to a strong container.

Harvesting
Pick the green/brown shoots that look like minute Christmas trees in summer for drying.

Equisetum arvense fertile shoots

container growing

This is the only sane way to grow this herb. Choose a large tough container; dustbins are ideal. Fill it with a soil-based potting compost. If you wish to sink it into the garden make sure that the rim is at least 15cm above the surface so that the rhizomes cannot penetrate or creep over the top. Be sure to cut back in summer to prevent spread by the spores. No need to feed and it requires little watering. It can look attractive!

culinary

It has been eaten as a substitute for asparagus, but I do not recommend it unless you are stuck on a desert island and there is no other food available.

medicinal

This plant is a storehouse of minerals and vitamins, so herbalists recommend it in cases of amnesia and general debility. It also enriches the blood, hardens fingernails and revitalises lifeless hair. Its astringent properties help to strengthen the walls of the veins, tightening up varicose veins and help guard against fatty deposits in the arteries.

other uses

The dried stems can be used to scour metal and polish pewter and fine woodwork. The whole plant yields a yellow ochre dye.

warning

If you wish to take horsetail medicinally do not self-administer; consult a herbalist.

Eriocephalus africanus

SOUTH AFRICAN WILD ROSEMARY

Also known as Wild Rosemary, Snowbush, Kapokbos. From the family *Asteraceae*.

This amazing drought-loving plant, which is well known in the Cape of South Africa, has a special place in my Herb Farm because, while the majority of my herbs are becoming dormant in winter, it goes into full flower, which lifts everyone's spirits during the short grey days of the winter months. These flowers are followed by the most attractive seedheads, which are covered in white tufts, hence its Afrikaans name *Kapokbos*, which is derived from *Kapok*, meaning 'snow'. I had been growing this herb for a number of years before realizing its full potential as a medicinal herb; for example it has been used for many hundreds of years by the Khoi people of southwestern Africa as a diuretic.

varieties

Eriocephalus africanus

South African Wild Rosemary

Half-hardy evergreen shrub. Ht and spread 1m. Clusters of small white flowers with magenta centres from early to late winter, which are followed by seeds covered in masses of tiny white hairs that make them look fluffy. Small, needle-like, silver-haired, oval, slightly succulent, aromatic leaves which grow in tufts along the branch.

cultivation

Propagation

Seed

In spring sow fresh seeds into prepared plug modules, or small containers using a seed compost mixed in equal parts with perlite. Place in either a warm place or a propagator at 20°C, germination takes 10–15 days. Once large enough to handle, pot up using a loam-based compost mixed with 25 per cent horticultural sand. In cool climates, winter young plants in a frost-free environment prior to planting out in the following spring.

Cuttings

In late spring take softwood cuttings from the growing tips and insert into prepared plug modules, using a seed compost mixed in equal parts with perlite. In cool climates grow under protection for the first year prior to planting out in the following spring.

Pests and Diseases

Very rarely suffers from pests or diseases. In cold climates excessive water in winter can cause the plant to rot.

Maintenance

Spring Prune after flowering in late spring to encourage new growth. Sow seeds. Take cuttings.

Summer Feed container-grown plants regularly.

Autumn In cold climates protect from heavy autumn rains.

Winter Protect from excessive wet in winter and when temperatures fall consistently below 3°C during the day.

Garden Cultivation

This amazingly drought-tolerant herb will adapt to most

SOUTH AFRICAN WILD ROSEMARY

Eriocephalus africanus seedhead

soils with the exception of heavy cold clay and marshy types of soil. The ideal situations are full sun and a well-drained soil. It makes an ideal coastal plant as it likes the sea spray and wind. In Mediterranean and other warm climates it can be grown as a hedge, or clipped into ball shapes. In situations where temperatures fall below 3°C at night it is advisable to grow this herb in a container.

Harvesting
Pick the leaves to use fresh or to dry after flowering from spring until early autumn. The twiggy branches are also used in some medicinal decoctions; pick these as and when required. Harvest the seeds when they start to drop.

companion planting

Because this herb flowers in winter it is a most beneficial late nectar plant and therefore attracts insects to the garden, which increases pollination of late-flowering plants.

container growing

This wild rosemary makes a spectacular container plant as it cascades beautifully over the pot; another plus is that it does not mind a bit of neglect. Use a loam based compost mixed with horticultural sand, 75 per cent loam, 25 per cent sand. Repot every spring and give the plant a good hair cut after flowering. Water regularly throughout the summer and feed monthly with a general-purpose liquid fertiliser from spring until first flowering in early autumn.

medicinal

Wild rosemary has traditionally been used as a medicine for many ailments like coughs and colds, flatulence and colic, and as a diuretic and a diaphoretic. It is said to have similar qualities to rosemary *(Rosmarinus officinalis)*. An infusion of the leaves can be used in the bath to help one relax, and also in a foot bath to stimulate the start of the menstrual period and to relieve swollen legs. An infusion of the leaves and twigs can be used to control dandruff and to stimulate hair growth.

other uses

When dried it can be added to sachets and potpourris.

warning

Do not take medicinally during pregnancy.

culinary

The leaves can be used in a very similar way to rosemary *(Rosmarinus officinalis)*, especially with lamb dishes and vegetable stews as the flavour is fairly similar.

Special Lamb Stew
Serves 4–6

100g dried chickpeas, soaked overnight, or use tinned
4 tablespoons light olive oil
1 large onion, finely chopped
500g lean lamb, cubed
2 lemons, juice only
2 bay leaves
3 sprigs South African wild rosemary, roughly 10cm long
3 leeks, cleaned and chopped
250g spinach
200g French flat-leaf parsley
1 fresh lime, sliced
Salt and black pepper

Strain the soaked chickpeas, cook in fresh unsalted water for 15–20 minutes, drain, rinse under cold fresh water and set aside. Alternatively, open can, drain, rinse under cold fresh water and set aside.

Heat 2 tablespoons of olive oil in a large, heavy pan, slowly sauté the onion, add the lamb and brown on all sides, season with salt and pepper, pour over the lemon juice, add enough boiled hot water to just cover the meat. Add the bay leaves and South African wild rosemary. Cover and simmer for 30 minutes. Heat the remaining oil, sauté the leeks, spinach and parsley until just cooked. Add this to the meat together with the chickpeas and sliced lime. Check that the liquid just covers the lamb, adding extra if required. Simmer all the ingredients for a further hour, checking from time to time that nothing is sticking to the bottom of the pan. Just before serving remove the remaining twigs of South African wild rosemary and find the limes, which can be used to decorate the serving plate. Serve with rice and either a green salad or green beans.

Ferula assa-foetida

ASAFOETIDA

Also known as Devil's Dung, Food of the Gods and Hing. From the family *Apiaceae*.

This herb, indigenous to the Middle East, has for thousands of years been renowned for its gum resin which is extracted from its tuberous roots. Asafoetida gets its name from the Persian *aza*, 'resin', and the Latin *foetidus*, 'stinking', which is highlighted by one of its common names, Devil's dung. Ironically it is also called Food of the Gods, because minute quantities of the sulphur-smelling resin can enhance the flavour of many foods.

varieties

Ferula assa-foetida
Asafoetida
Perennial. Ht 2m, spread 1.5m. Flowers from the 4th year in early spring, with flat umbels of tiny yellow flowers, followed by small brown seeds. The flowering spikes can reach 4m. Large finely divided green/grey sulphuric, garlic-scented leaves. This plant often dies after flowering.

cultivation

Propagation
Seed
Sow fresh seeds in late summer, into prepared plug modules or small pots using a seed compost mixed in equal parts with vermiculite. Once the seedlings are large enough pot up using a loam-based compost mixed in equal parts with horticultural sand.

Pests and Diseases
Rarely suffers from pests or diseases. Container-raised plants can be attacked by aphids, however. Treat with a horticultural soap spray following the manufacturer's instructions.

Maintenance
Spring Stake flowering spikes when grown on an exposed site.
Summer Sow seeds; harvest root from 4th year on.
Autumn Mulch established plants with well-rotted manure.
Winter Protect from excessive wet.

Garden Cultivation
Plant in a warm, well-drained soil and a sunny position. Because of the height when in flower and because it dislikes being moved, position the plant with care. Do give it some protection from the prevailing wind, bearing in mind that it flowers in early spring.

Harvesting
Harvest the seed and resin in the summer from 4-year-old plants. Cut off the stems, and make successive slices through the roots. A milky smelly liquid will exude from the cuts. When dry it forms a resin which turns from creamy, greyish-white to reddish-brown as it is exposed to the air. One root, after successive slicing, can yield up to 1kg of resin.

container growing

Only pot up one size of pot at a time; this plant does not like being over potted. Use a loam-based compost mixed in equal parts with horticultural sand. Place the container in full sun and liquid feed regularly throughout the growing season following the manufacturer's instructions.

medicinal

In Ayurvedic and Eastern herbal medicine asafoetida gum resin is used to treat bloating, wind, indigestion and constipation. It also helps lower blood pressure and thins the blood. Owing to its foul taste and smell it is usually taken in pill form.

other uses

A mixture of garlic and asafoetida apparently makes the ultimate insect repellent. I think it would also make the ultimate human repellent. In Afghanistan it is said that asafoetida when rubbed over boots keeps snakes away.

warning

Do not administer to very young children or babies.

culinary

Asafoetida is used throughout Southern India in the preparation of beans, peas and lentils, which are collectively known as *dal*. The best way to use asafoetida in the kitchen is to buy it already prepared in an airtight container. Add a minute pinch of resin to hot oil before adding the other ingredients; this calms the aroma and balances the other ingredients.

Filipendula

MEADOWSWEET

Also known as Bridewort, Meadow Queen, Meadow-wort, and Queen of the Meadow. From the family *Rosaceae*.

Meadowsweet can be found growing wild in profusion near streams and rivers, in damp meadows, fens and marshlands, or wet woodlands to 1,000m altitude. It is a native of Europe and Asia that has been successfully introduced into, and is naturalised in, North America.

varieties

Filipendula ulmaria
Meadowsweet
Hardy perennial. Ht 60–120cm, spread 60cm. Clusters of creamy-white flowers in mid-summer. Green leaf made up of up to 5 pairs of big leaflets separated by pairs of smaller leaflets.

***Filipendula ulmaria* 'Aurea'**
Golden Meadowsweet
Hardy perennial. Ht and spread 30cm. Clusters of creamy-white flowers in mid-summer. Bright golden yellow, divided leaves in spring that turn a lime colour in summer. Susceptible to sun scorch.

***Filipendula ulmaria* 'Variegata'**
Variegated Meadowsweet
Hardy perennial. Ht 45cm and spread 30cm. Clusters of creamy-white flowers in mid-summer. Divided leaf, dramatically variegated green and yellow in spring. Fades a bit as the season progresses.

Filipendula vulgaris
Dropwort
Hardy perennial. Ht 60–90cm, spread 45cm. Summertime clusters of white flowers (larger than meadowsweet). Fern-like green leaves.

cultivation

Propagation

Seed
Sow in prepared seed or plug module trays in the autumn. Use a standard seed compost mixed in equal parts with composted fine bark. Cover lightly with compost (not perlite) and winter outside under glass. Check from time to time that the compost has not become dry as this will inhibit germination. Stratification is helpful but not essential (see page 644). Germination should take place in spring. When the seedlings are large enough to handle, plant out, 30cm apart, into a prepared site.

Division
The golden and variegated forms are best propagated by division in the autumn. Dig up established plant and tease the plantlets apart; they separate easily. Replant in a prepared site, 30cm apart, or pot up decorative varieties in a loam-based potting compost.

Pests and Diseases
Meadowsweet can be prone to powdery mildew; cut bark and remove infected leaves.

Maintenance
Spring Remove winter debris around new growth.
Summer Cut back after flowering.
Autumn Divide established plants, sow seed for wintering outside.
Winter No need for protection.

Garden Cultivation
Meadowsweet adapts well to the garden, but does prefer sun/semi-shade and a moisture-retentive soil. If your soil is free-draining, mix in plenty of well-rotted manure and/or leaf mould, and plant in semi-shade.

Harvesting
Gather young leaves for fresh or dry use before flowers appear. Pick flowers just as they open, use fresh or dry.

container growing

Golden and variegated meadowsweet look attractive in containers, but use a loam-based compost to make sure moisture is retained. Position in partial shade to inhibit drying out and sun scorch. The plant dies back in winter so leave it outside where the natural weathers can reach it. If you live in an extremely cold area, protect the container from damage by placing in a site protected from continuous frost, but not warm. Liquid feed only twice during flowering.

other uses

A black dye can be obtained from the roots by using a copper mordant (see also p675). Use dried leaves and flowers in potpourris.

medicinal

The whole plant is a traditional remedy for an acidic stomach. The fresh root is used in homeopathic preparations and is effective on its own in the treatment of diarrhoea. The flowers, when made into a tea, are a comfort to flu victims.

culinary

A charming local vet gave me meadowsweet vinegar to try. Much to my amazement it was lovely, and combined well with oil to make a different salad dressing, great when used with a flower salad. The flowers make a very good wine, and add flavour to meads and beers. Add the flowers to stewed fruit and jams to impart a subtle almond flavour.

Galega officinalis

GOAT'S RUE

Also known as French Lilac, Italian Fitch and Professor-weed. From the family *Papilionaceae*.

This ancient herb, indigenous to central and southern Europe and western Asia, has been used for hundreds of years to treat plagues and infections. Historically it was also recommend for snake bites. The name *Galega* comes from the Greek *gala,* meaning 'milk' because of its reputation for increasing lactation. The common name goat's rue originates from the leaves, which smell unpleasant when crushed.

varieties

Galega officinalis

Goat's Rue

Herbaceous perennial. Ht 1–1.5m, spread up to 1m. Attractive clusters of white or mauve flowers in summer, followed by long seed pods. Green, compound, divided, lance-shaped leaflets.

cultivation

Propagation

Seed

Sow the easy-to-handle seeds in the spring into prepared seed or plug module trays using a seed compost, and cover with perlite. Germination takes 10–20 days without extra heat. Once the seedlings are strong enough, either pot on using a loam-based potting compost or alternatively in mid-spring, when the air and soil temperature has risen, plant out at a distance of 75cm apart.

Division

This is a good method of propagation for this herb as it prevents the plant from becoming too large and it encourages it to put on new growth. In the second or third year divide the root ball either by using two forks back to back or by digging up the whole plant and dividing. Once divided replant in a well-prepared site.

Pests and Diseases

Rarely suffers from pests. However, it is prone to powdery mildew, especially when planted against a wall or fence.

Leaves of *Galega officinalis*

Maintenance

Spring Sow seeds, divide established plants.
Summer Cut back if growing too straggly; this will promote a second flowering.
Autumn Collect seeds for drying.
Winter No need for protection.

Garden Cultivation

This fully hardy herb will grow in most soils. It prefers a deep soil that does not dry out in summer and allows the roots to become well established. This is important as they act as an anchor to stop it from being blown over in the summer when it is in full flower. If the plant becomes invasive or outgrows its position, cut it back hard; this will keep it under control and encourage flowering at a lower height, which can look most effective.

Harvesting

All the aerial parts of the plant are harvested in summer just before flowering, then dried for medicinal use.

container growing

I do grow this herb in a container for exhibiting at the flower shows, using a loam-based potting compost. However, for home display, I would not recommend it as it needs repotting at least 3 times in the growing season. Also, it grows very tall, requiring a large container, which makes it awkward to handle.

medicinal

Used medicinally to reduce the blood sugar levels and as a useful diuretic. It is also used to increase lactation in nursing mothers.

other uses

The leaves and stem are used as an animal food supplement to increase milk yield.

warning

Only to be used under professional supervision when treating diabetes.

Galium odoratum

SWEET WOODRUFF

Also known as New Mowed Hay, Rice Flower, Ladies In The Hay, Kiss Me Wuick, Master Of The Wood, Woodward and Woodrowell. From the family *Rubiaceae*.

This is a native of Europe and has been introduced and cultivated in North America and Australia. It grows deep in the woods and in hedgerows.

Records date back to the fourteenth century, when woodruff was used as a strewing herb, as bed-stuffing and to perfume linen.

On May Day in Germany, it is added to Rhine wine to make a delicious drink called 'Maibowle'.

varieties

Galium odoratum* syn. *Asperula odorata
Sweet Woodruff
Hardy perennial. Ht 15cm, spread 30cm or more. White, star-shaped flowers from spring to early summer. The green leaves are neat and grow in a complete circle around the stem. The whole plant is aromatic.

cultivation

Propagation

Seed
To ensure viability, only use fresh seed. Sow in early autumn into prepared seed or plug module trays, and cover with compost. Water in well. Seeds require a period of stratification (see page 644). Once the seedlings are large enough, either pot or plant out as soon as the young plants have been hardened off. Plant 10cm apart.

Root Cuttings
The rootstock is very brittle and every little piece will grow. The best time to take cuttings is after flowering in the early summer. Lay small pieces of the root, 2–4cm long, evenly spaced, on the compost in a seed tray. Cover with a thin layer of compost, and water. Leave in a warm place, and the woodruff will begin to sprout again. When large enough to handle, split up and plant out.

Pests and Diseases
This plant rarely suffers from pests and diseases.

Maintenance
Spring Take root cuttings after flowering.
Summer Dig up before the flowers have set, to check spreading.
Autumn The plant dies back completely in autumn. Sow seeds.
Winter Fully hardy plant.

Garden Cultivation
Ideal for difficult places or underplanting in borders, it loves growing in the dry shade of trees right up to the trunk. Its rich green leaves make a dense and very decorative ground cover, its underground runners spreading rapidly in the right situation. It prefers a rich alkaline soil with some moisture during the spring.

Harvesting
The true aroma (which is like new-mown hay) comes to the fore when it is dried. Dry flowers and leaves together in early summer.

container growing

Make sure the container is large enough, otherwise it will become root-bound very quickly. Use a soil-based compost mixed in equal parts with composted fine bark. Only feed with liquid fertiliser when the plant is flowering. Position the container in semi-shade and do not over water.

culinary

Add the flowers to salads. The main ingredients for a modern-day May Wine would be a bottle of hock, a glass of sherry, sugar, and strawberries, with a few sprigs of woodruff thrown in an hour before serving.

medicinal

A tea made from the leaves is said to relieve stomach pain, act as a diuretic, and be beneficial for those prone to gall stones.

Sweet woodruff tea

warning

Consumption of large quantities can produce symptoms of poisoning, including dizziness and vomiting.

Ginkgo biloba

GINKGO

Also known as Maidenhair Tree and Bai Guo. From the family *Ginkgoaceae*.

This beneficial herb is a living fossil, thought to be one of the oldest trees on the planet, dating back to when dinosaurs lived. Chinese monks are credited with keeping the tree in existence, as a sacred herb. The name *Ginkgo* is derived from the Japanese word *ginkyo*, meaning 'silver apricot', referring to the fruit, and *biloba* translates as 'two-lobed', which refers to the split in the middle of the fan-shaped leaf blades.

varieties

***Ginkgo biloba* AGM**
Ginkgo
Hardy deciduous tree. Ht 40m, spread 20m. It is dioecious, meaning that it bears male catkins and female flowers on different trees in early summer. The female flowers are followed by small fruits. Fan-shaped green leaves are sometimes whole, but often have a single, central, vertical slit.

cultivation

Propagation

Seed
In autumn, prepare the seeds by removing the pith; wear gloves, as the pith can cause dermatitis, then wash in a mild detergent. Sow immediately, singly into small pots using a loam-based seed compost mixed in equal parts with horticultural grit. Cover with grit and place in a cold frame. Germination takes 4 months or longer. Grow on in a container for a minimum of four years before planting into the garden. You will only know the sex of your seed-raised plant when it flowers, which takes approximately 20 years.

Cuttings
Take cuttings from new growth in summer or semi-ripe growth in early autumn. Place in a seed compost mixed in equal parts with horticultural grit. Keep frost-free until rooted. Grow as a container plant in exactly the same way as a seed-raised plant.

Pests and Diseases
Rarely suffers from pests or diseases.

Maintenance
Spring Repot or top dress container-raised plants.
Summer Take cuttings from new growth.
Autumn Sow fresh seeds. Take cuttings.
Winter Fully hardy; no extra protection needed.

Garden Cultivation
Plant in full sun to partial shade in any fertile soil with the exception of heavy, cold, wet clay. To produce fruit (in warm climates only) a male and female tree need to be planted near each other.

Harvesting
In autumn pick the leaves for drying as they turn colour from green to yellow. Also harvest the ripe small fruits from the female tree, and extract the kernel (nut) which is dried for medicinal use or used fresh in the kitchen.

container growing

Once the young plant is over two years old, use a loam-based potting compost mixed in equal parts with horticultural sand. Water and liquid feed regularly throughout the growing season.

medicinal

Ginkgo is one of the bestselling Western herbal medicines and is taken to improve the memory. The dried leaves are used to improve circulation, ease tinnitus and in the treatment of asthma. Currently the leaves are being researched as a treatment for Alzheimer's disease.

warning

If taken in excess, it can cause a toxic reaction. Wear gloves when preparing fresh nuts. Children can be allergic to cooked nuts.

culinary

There is current debate regarding the toxicity of the fresh nut, as to whether it can be eaten raw or whether it needs to be cooked. I would err on the side of caution as most recipes say cooked, where it is used in soups, stir-fries and roasted. These nuts are available from Eastern supermarkets, where they are often called 'White Nuts'.

Glcyrrhiza glabra

LIQUORICE

Also known as Licorice, Sweet Licorice and Sweetwood. From the family *Papilionaceae*.

This plant, which is a native of the Mediterranean region, is commercially grown throughout the temperate zones of the world and extensively cultivated in Russia, Iran, Spain and India. It has been used medicinally for 3,000 years and was recorded on Assyrian tablets and Egyptian papyri. The Latin name *Glycyrrhiza* comes from *glykys*, meaning 'sweet', and *rhiza*, meaning 'root'.

It was first introduced to England by Dominican friars in the sixteenth century and became an important crop. The whole of the huge cobbled courtyard of Pontefract Castle was covered by top soil simply to grow liquorice. It is sad that Pontefract cakes are made from imported liquorice today.

varieties

Glycyrrhiza glabra

Liquorice

Hardy perennial. Ht 1.2m, spread 1m. Pea-like, purple/blue and white flowers borne in short spikes on erect stems in late summer. Large greenish leaves divided into oval leaflets.

cultivation

Propagation

Seed

The seedlings often damp off. In cooler climates the seed tends not to be viable. Root division is a much easier method.

Division

Divide when the plant is dormant, making sure the root has one or more buds. Place into pots half filled with a loam-based potting compost. Cover with compost. Water well and leave in a warm place until shoots appear. Harden off, then plant out in early spring or autumn. If the latter, winter in a cold greenhouse or cold frame.

Maintenance

Spring Divide established plants.
Summer Do nothing.
Autumn Divide established plants if necessary.
Winter In very cold winters protect first-year plants.

Garden Cultivation

Liquorice needs a rich, deep, well-cultivated soil.

Plant pieces of the root, each with a bud, directly into a prepared site 15cm deep and 1m apart in early spring or in autumn during the dormant season, if the ground is workable and not frosty. Liquorice does best in long, hot summers, but will need extra watering if your soil is very free draining.

Harvesting

Harvest roots for drying in early winter from established 3- or 4-year-old plants.

container growing

Never displays as well as in the garden. Use a soil-based compost. Feed throughout the growing season and water until it dies back.

culinary

Liquorice is used as a flavouring in the making of Guinness and other beers.

medicinal

The juice from the roots provides commercial liquorice. It is used either to mask the unpleasant flavour of other medicines or to provide its own soothing action on troublesome coughs. The dried root, stripped of its bitter bark, is recommended as a remedy for colds, sore throats and bronchial catarrh.

Liquorice is a gentle laxative and lowers stomach acid levels, so relieving heartburn. It has a remarkable power to heal stomach ulcers because it spreads a protective gel over the stomach wall and in addition it eases spasms of the large intestine. It also increases the flow of bile and lowers blood cholesterol levels.

warning

Large doses of liquorice cause side effects, notably headaches, high blood pressure and water retention.

Liquorice sticks

Helichrysum italicum

CURRY PLANT

From the family *Asteraceae*.

This plant is from southern Europe and has adapted well to damper, cooler climates. It is the sweet curry scent of its leaves that has caused its recent rise in popularity.

varieties

Helichrysum italicum* AGM syn. *H. angustifolium
Curry Plant
Hardy evergreen perennial. Ht 60cm, spread 1m. Clusters of tiny mustard-yellow flowers in summer. Narrow, aromatic, silver leaves. Highly scented. Planting distance for hedging 60cm.

***Helichrysum italicum* 'Dartington'**
Dartington Curry Plant
Hardy evergreen perennial. Ht 45cm, spread 60cm. Compact plant with clusters of small yellow flowers in summer. Grey/green, highly scented, narrow leaves (half the size of *H. italicum*). Its compact upright habit makes this a good plant for hedges and edging in the garden. Planting distance for hedging 30cm.

***Helichrysum italicum* 'Korma'**
Korma Curry Plant
Hardy evergreen perennial. Ht 60cm, spread 1m. Broad clusters of small bright yellow flowers, produced on upright white shoots. Narrow, aromatic, silver leaves. Planting distance for hedging 60cm.

cultivation

Propagation

Seed
I have not known any *H. italicum* set good seed. For this reason I advise cuttings.

Cuttings
Take softwood cuttings in spring and semi-ripe ones in autumn. Use a seed compost mixed in equal parts with composted fine bark.

Pests and Diseases
Pests give this highly aromatic plant a wide berth, and it is usually free from disease.

Maintenance
Spring Trim established plants after frosts to maintain shape and promote new growth. Take softwood cuttings.

The silver leaves of *Helichrysum italicum* 'Korma'

Summer Trim back after flowering, but not too hard.
Autumn Take semi-ripe wood cuttings.
Winter If the temperature falls below −10°C, protect from frost.

Garden Cultivation
The curry plant makes an attractive addition to the garden and it imparts a strong smell of curry even if untouched. It is one of the most silvery of shrubs and makes a striking visual feature all year round.

Plant in full sun in a well-drained soil. Do not cut the curry plant as hard back as cotton lavender, but it is worth giving it a good hair cut after flowering to stop the larger ones flopping and to keep the shape of the smaller ones.

If it is an exceptionally wet winter, and you do not have a free-draining soil, lift some plants, and keep in a cold greenhouse or cold frame.

Harvest
Pick leaves at any time for using fresh. Or before flowering for drying. Pick the flowers when fully open, in summer, for drying.

container growing

All the curry plants grow happily in large containers (at least 20cm in diameter). Use a soil-based compost mixed in equal parts with composted fine bark. Place in the sun to get the best effect, and do not over water.

culinary

There are not many recipes for the curry plant in cooking, and in truth the leaves smell stronger than they taste, but a small sprig stuffed into the cavity of a roasting chicken makes an interesting variation on tarragon. Add sprigs to vegetables, rice dishes and pickles for a mild curry flavour. Remove before serving.

other uses

The bright yellow button flowers can be used to add colour to potpourris.

Hesperis matronalis

SWEET ROCKET

Also known as Damask Violet and Dame's Violet. From the family *Brassicaceae*.

This sweet-smelling herb is indigenous to Italy. It can now be found growing wild in much of the temperate world as a garden escapee. The old Greek name *Hesperis* was used by Theophrastus, the Greek botanist (370–285 BC). It is derived from *hesperos*, meaning 'evening', which is when the flowers are at their most fragrant.

varieties

Hesperis matronalis

Sweet Rocket

Hardy biennial; very occasionally perennial, sending out new shoots from the rootstock. Ht 60–90cm, spread 25cm. The 4-petalled flowers are all sweetly scented and come in many colours – pink, purple, mauve and white – in the summer of the second year. The leaves are green and lance shaped.

There is a double-flowered form of this plant – *Hesperis matronalis* double-flowered. It can only be propagated by cuttings or division and needs a more sandy loam soil than sweet rocket.

cultivation

Propagation

Seed

Sow the seed in the autumn in prepared seed or plug trays, covering the seeds with perlite. Winter the young plants in a cold greenhouse for planting out in the spring at 45cm apart. Propagated this way it may flower the first season as well as the second. Alternatively, sow seed in the garden in late spring.

Pests and Diseases

This herb is largely free from pests and diseases.

Maintenance

Spring Sow seed outdoors.

Summer In the second year deadhead flowers to prolong flowering.

Autumn Sow seed under protection.

Winter No need to protect.

Garden Cultivation

It likes full sun or light shade and prefers a well-drained fertile soil. The seed can be sown direct into a prepared site in the garden in late spring, thinning to 30cm apart, with a further thinning to 45cm later on if need be.

Harvesting

Pick leaves when young for eating. Pick flowers as they open for using fresh or for drying.

container growing

Sweet rocket is a tall plant. It looks attractive if 3 or 4 one-year-old plants are potted together, positioned to make the most of the scent on a summer evening. Use a standard potting compost mixed in equal parts with composted fine bark and water well in summer months. No need to feed.

culinary

Young leaves are eaten occasionally in salads. Use sparingly because they are very bitter. The flowers look attractive tossed in salads. They can also be used to decorate desserts.

other uses

Add dried flowers to potpourris for pastel colours and sweet scent.

Hesperis matronalis

Helleborus niger

CHRISTMAS ROSE

Also known as Black Hellebore. From the family *Ranunculaceae*.

This ancient herb has had many stories attributed to its beauty and its medicinal properties. It is said to be one of Britain's oldest cultivated plants and it is thought to have been introduced by the Romans when they invaded these isles. There is an old legend, associated with the common name Christmas rose, that it sprouted in the snow from the tears of a young shepherdess who had no gift to give Jesus, the newborn King. An angel, upon hearing her weeping, appeared and brushed away the snow to reveal the most beautiful flower, the Christmas rose.

varieties

***Helleborus niger* AGM**

Christmas Rose

Evergreen perennial. Ht and spread 30cm. Large, nodding cup-shaped pink, purple or white flowers with golden stamens from very early in the New Year until April. Leathery deep green divided basal leaves.

***Helleborus foetidus* AGM**

Stinking Hellebore

Evergreen perennial. Ht and spread 45cm. Clusters of nodding cup-shaped pale green, red-margined flowers from very early in the New Year until April. Deeply divided dark green leaves. This variety is used in homeopathy in the treatment of headaches, psychic disorders, enteritis and spasms.

cultivation

Propagation

Seed

Sow fresh seed in late summer into prepared seed trays, plug modules or small pots using a loam-based seed compost mixed in equal parts with horticultural sand. A word of warning about purchased seed: it must be fresh, old seed is nearly impossible to germinate. Cover with grit, and place the container outside for the winter as the seed needs frost to break its dormancy. Do not worry if the container becomes immersed in snow, as melting snow will aid germination. Germination will be erratic and may take a further season so do not throw away the sown seed after one winter.

Once it has germinated, grow on for one season in its container prior to planting in a well-prepared site in the garden. As this plant hates being transplanted choose your site with care. Alternatively, as anyone who has grown this plant successfully in the garden will know, hellebores will happily self seed. This is by far the easiest way to raise the plants from seed. When each seedling has at least one true leaf, carefully lift it and transplant into a prepared site that has moist, fertile soil in dappled shade. All seed raised plants will take 3–5 years to flower.

Division

Hellebores hate their roots being disturbed, and are slow to re-establish after division. In early autumn,

CHRISTMAS ROSE

Helleborus niger

choose a well-established plant and, with two forks back to back, tease very gently a clump away from the main plant. Carefully firm the soil in around the remaining clump and give the plant a feed with some well-rotted compost. As the original clump has had minimal disturbance it should flower in the early spring. Replant the divided section with the roots down, not spread out, into a well-prepared site that has had some new well-rotted compost dug in, and in partial dappled shade. It is unlikely to flower in the following spring but should do so the year after.

Pests and Diseases

Hellebores suffer from two diseases, black death and black spot. If you notice black streaking between the leaf veins, and seriously distorted stems and leaves, then this is most probably black death virus. Dig up the plants immediately to prevent it spreading and destroy, do not compost. Black spot fungus is identifiable by large, irregular brown or black spots on the leaves and stems which cause the leaf to die. Remove these leaves and burn or bin – again, do not compost.

Hellebores can also suffer from root rot, a fungus that thrives on waterlogged soils. If you notice your plants are not thriving and you know you have very wet soil, then dig up the plants, and add some grit or bark to improve the drainage before replanting.

Maintenance

Spring Transplant self-sown seedlings into permanent positions.
Summer Sow fresh seed.
Autumn Divide established plants.
Winter No need for protection, fully hardy.

Garden Cultivation

The Christmas rose should be planted in a deep, fertile, well-draining but moisture-retaining soil, in partial dappled shade. Under deciduous trees is ideal. In the autumn, it is well worth feeding the plants with well-rotted compost or leaf mould which will encourage it to flower profusely in the New Year.

Harvesting

The root is harvested in autumn, then dried for medicinal use.

companion planting

As this herb flowers so early it is very beneficial for early insects as an early nectar plant; it attracts beneficial predators, which makes it a good companion near the vegetable patch.

container growing

Because the Christmas rose has a very long tap root, choose containers that are deep, a 'long tom' is ideal. Use a soil-based potting compost mixed with 25 per cent sand, and place the container in partial shade. Feed weekly throughout the growing season with a liquid feed following the manufacturer's instructions. Repot up one size of pot, or alternatively divide established plants in autumn once the plant is dormant. Remember if you have divided the plant it is unlikely to flower in the following spring.

medicinal

Helleborus niger is a very poisonous plant that is toxic when taken in all but the smallest doses. So this herb should not be self-administered and should only be taken under professional guidance. This herb contains properties similar to *Digitalis*, and was traditionally used as a heart stimulant and to treat dropsy, nervous disorders and hysteria. The best and safest way to take this herb is homeopathically, where it is used to treat headaches, psychic disorders, enteritis and spasms.

warning

Highly poisonous and toxic plant, not to be self-administered.

Helleborus niger

Humulus lupulus

HOPS

Also known as Hopbind and Hop Vine. From the family *Cannabaceae.*

Native of the Northern temperate zones, cultivated commercially, especially in Northern Europe, North America and Chile.

Roman records from the first century AD describe hops as a popular garden plant and vegetable, the young shoots being sold in markets to be eaten rather like asparagus. Hop gardens did not become widespread in Europe until the ninth century. In Britain the hop was a wild plant and used as a vegetable before it became one of the ingredients of beer. It was not until the sixteenth century that the word hop and the practice of flavouring and preserving beer with the strobiles or female flowers of *Humulus lupulus* were introduced into Britain by Flemish immigrants, and replaced traditional bitter herbs such as alehoof and alecost.

During the reign of Henry VIII, parliament was petitioned against the use of the hop, as it was said that it was a wicked weed that would spoil the taste of the drink, ale, and endanger the people. Needless to say the petition was thrown out. The use of hops revolutionised brewing since it enabled beer to be kept for longer.

Hops have also been used as medicine for at least as long as for brewing. The flowers are famous for their sedative effect and were either drunk as a tea or stuffed in a hop pillow to sleep on.

varieties

Humulus lupulus

Common Hop

Hardy perennial, a herbaceous climber. Ht up to 6m. There are separate female and male plants. The male has yellowish flowers in branched clusters. They are without sepals and have 5 tepals and 5 stamen. The female plant has tiny greenish-yellow, scented flowers, hidden by big scales (see left), which become papery when the fruiting heads are ripe. These are the flowers that are harvested for beer. The mid-green leaves have 3 to 5 lobes with sharply toothed edges. The hollow stems are covered with tiny hooked prickles which enable the plant to cling to shrubs, trees, or anything else. It always entwines clockwise.

***Humulus lupulus* 'Aureus' AGM**

Golden Hop

Hardy perennial, a herbaceous climber. Ht up to 6m. The main difference between this plant and the common hop is that the leaves and flowers are much more golden, which makes it very attractive in the garden and in dried flower arrangements. It has the same properties as the common hop.

cultivation

Propagation

Seed

Beer is made from the unpollinated female flowers. If you grow from seed you will not know the gender for 2 to 3 years, which is the time it takes before good flowers are produced. Obtain seed from specialist seedsmen.

Sow in summer or autumn. The seed is on the medium to large side so sow sparingly; if using plug module trays, 1 per cell. Push the seed in and cover it with the compost. Then cover the tray with a sheet of glass or polythene, and leave somewhere cool to germinate – a cold frame, a cold greenhouse, or a garage. Germination can be very erratic. If the seed is not fresh you may need to give the hot/cold treatment to break dormancy.

Warning: As the seed will be from wild hops these should not be grown in areas of commercial hop growing, because they might contaminate the crop.

Humulus lupulus female flowers

Humulus lupulus male flowers

Cuttings
Softwood cuttings should be taken in spring or early summer from the female plant. Choose young shoots and take the cuttings in the morning as they will lose water very fast and wilt.

Division
In the spring dig up and divide the root stems and suckers of established plants. Replant 1m apart against support.

Pests and Diseases
The most common disease is hop wilt. If this occurs, dig up and burn. Do not plant hops in that area again. Leaf miner can sometimes be a problem. Remove infected leaves immediately. The golden variety sometimes suffers from sun scorch. If this occurs, prune to new growth, and change its position if possible the following season.

Maintenance
Spring Divide roots and separate rooted stems and suckers in spring. Re-pot container-grown plants. Check trellising. Take cuttings.
Summer Sow seed late in the season. Take cuttings.
Autumn Cut back remaining growth into the ground. Give the plants a good feed of manure or compost. Bring containers into a cool place. Sow seed.
Winter No need for protection.

Garden Cultivation
For successful plants the site should be sunny and open, and the soil needs to be rich in humus and dug deeply. It is not generally necessary to tie the plants if good support is at hand. A word of warning: you must dominate the plant. Certainly it will need thinning and encouraging to entwine where you want it to go rather than where it chooses. But remember that it dies back completely in winter. Cut the plant into the ground each autumn and then give it a good feed of manure or compost.

Harvesting
Pick young fresh side shoots in spring. Gather young fresh leaves as required.

Pick male flowers as required. Pick ripe female flowers in early autumn. Dry and use within a few months, otherwise the flavour becomes unpleasant.

container growing

Hops, especially the golden variety, can look very attractive in a large container with something to grow up. Use a soil-based compost mixed in equal parts with composted fine bark, and feed regularly with a liquid fertiliser from late spring to mid-summer. Keep well watered in the summer months and fairly dry in winter. It can be grown indoors in a position with good light such as a conservatory, but it seldom flowers. Provide some form of shade during sunny periods. During the winter months, make sure it has a rest by putting the pot in a cool place, keeping the compost on the dry side. Re-pot each year.

other uses

The leaf can be used to make a brown dye. If you live close to a brewery it is worth chatting them up each autumn for the spent hops, which make either a great mulch or a layer in a compost heap.

culinary

In early spring pick the young side shoots, steam them (or lightly boil), and eat like asparagus. The male flowers can be parboiled, cooled and tossed into salads. The young leaves can be quickly blanched to remove any bitterness and added to soups or salads.

warning

Contact dermatitis can be caused by contact with the flower. Also, hops are not recommended in the treatment of depressive illnesses because of their sedative effect.

medicinal

Hop tea made from the female flower only is recommended for nervous diarrhoea, insomnia and restlessness. It also helps to stimulate appetite, dispel flatulence and relieve intestinal cramps. A cold tea taken an hour before meals is particularly good for digestion. It can be useful combined with fragrant valerian for coughs and nervous spasmodic conditions. Recent research into hops has shown that it contains a certain hormone, which accounts for the beneficial effect of helping mothers improve their milk flow.

To make a hop pillow, sprinkle hops with alcohol and fill a small bag or pillowcase with them (which all in all is bound to knock you out).

Hop pillow

Hyoscyamus niger

HENBANE

Also known as Devil's Eye, Hen Pen, Hen Penny, Hog Bean, Stinking Roger, Symphoniaca, Jusquiamus, Henbell, Belene, Hennyibone, Hennebane, Poisoned Tobacco and Stinking Nightshade. From the family *Solanaceae*.

This native of Europe has become widely distributed worldwide and is found growing on waste ground or roadsides on well-drained sandy or chalk soils.

Two famous deaths are attributed to henbane. Hamlet's father was murdered by a distillation of henbane being poured in his ear, and in 1910 Dr Crippen used hyoscine, which is extracted from the plant, to murder his wife. Every part of the plant is toxic.

Henbane has been considered to have aphrodisiac properties and is the main ingredient in some love potions and witches' brews. It was also placed by the hinges of outer doors to protect against sorcery.

varieties

Hyoscyamus niger

Henbane

Annual/biennial. Ht up to 80cm, spread 30cm. Flowers bloom in summer, are yellow/brown or cream, funnel-shaped, and usually marked with purple veins. Leaves are hairy with large teeth, and the upper leaves have no stalks. The whole plant smells foul.

Hyoscyamus albus

White Henbane

Annual. Ht 30cm, spread 30cm. Its summer flowers are pale yellow marked with violet veins, funnel-shaped. Leaves are identical to *H. niger*.

cultivation

Propagation

Seed

Sow the fairly small seeds on the surface of pots or trays in spring, and cover with perlite. Germination takes 10–15 days. If you want henbane to behave like a biennial, sow in early autumn, keeping the soil moist until germination (which can be erratic, but on average takes about 14–21 days). Winter the young plants in a cold frame or cold greenhouse. Plant out the following spring at a distance of 30cm apart.

Pests and Diseases

This plant in the main is free from pests and diseases.

Maintenance

Spring Sow seed.

Summer Deadhead flowers to maintain plant (wear gloves).

Autumn Sow seed for second-year flowers.

Winter Protect young plants.

Garden Cultivation

Choose the site for planting henbane with care, because it is poisonous. It will tolerate any growing situation but shows a preference for a well-drained sunny site. Sow seeds in late spring. When the seedlings are large enough to handle, thin to 30cm apart. It can look striking in a mixed border.

Harvesting

Collect seed when the head turns brown and begins to open at the end.

container growing

Inadvisable to grow such a poisonous plant this way.

medicinal

This plant was used for a wide range of conditions that required sedation. The alkaloid hyoscine, which is derived from the green tops and leaves, was used as a hypnotic and brain sedative for the seasick, excitable and insane. It was also used externally in analgesic preparations to relieve rheumatism and arthritis. The syrup has a sedative effect in cases of Parkinson's disease.

warning

The whole plant is poisonous. Children have been poisoned by eating the seeds or seed pods. Use preparation and dosage only under strict medical direction.

Hypericum perforatum

ST JOHN'S WORT

Also known as Warrior's wound, Amber, Touch and Heal, Grace of God and Herb of St John. From the family *Clusiaceae*.

This magical herb is found in temperate zones of the world in open situations on semi-dry soils. Whoever treads on St John's Wort after sunset will be swept up on the back of a magic horse that will charge round the heavens until sunrise before depositing its exhausted rider on the ground.

Besides its magical attributes, *Hypericum* has medicinal properties and was universally known as the Grace of God. In England it cured mania, in Russia it gave protection against hydrophobia and the Brazilians knew it as an antidote to snake bite. St John's Wort ('wort', incidentally, is Anglo-Saxon for 'medicinal herb') has been used to raise ghosts and exorcise spirits. When crushed, the leaves release a balsamic odour similar to incense, which was said to be strong enough to drive away evil spirits. The red pigment from the crushed flowers was taken to signify the blood of St John at his beheading, for the herb is in full flower on 24 June, St John's Day.

varieties

Hypericum perforatum

St John's Wort

Hardy perennial. Ht 30–90cm, spread 30cm. Scented yellow flowers with black dots in summer. The small leaves are stalkless and covered with tiny perforations (hence *perforatum*), which are in fact translucent glands. This is the magical species.

cultivation

Propagation

Seed

Sow very small seed in spring into prepared seed or plug module trays, and cover with perlite. Germination is usually in 10–20 days depending on the weather. When the seedlings are large enough to handle and after a period of hardening off, plant out 30cm apart.

Division

Divide established plants in the autumn.

Pests and Diseases

Largely free from pests and diseases.

Maintenance

Spring Sow seeds.
Summer Cut back after flowering to stop self-seeding.
Autumn Divide established clumps.
Winter No need for protection, fully hardy.

Garden Cultivation

Tolerates most soils, in sun or light shade, but it can be invasive in light soils.

Harvesting

Harvest leaves and flowers as required.

container growing

Can be grown in containers, but it is a bit tall so you do need a large clump for it to look effective. Use a loam-based compost. Water in the summer months. Only feed with liquid fertiliser twice during the growing season, otherwise it produces more leaf than flower.

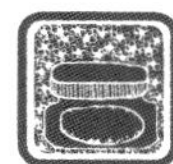

other uses

The flowers release a yellow dye with alum, and a red dye with alcohol.

warning

St John's Wort has sometimes poisoned livestock. Its use also makes the skin sensitive to light.

medicinal

Oil extracted by macerating the flowers in vegetable oil and applied externally eases neuralgia and the pain of sciatica wounds, varicose veins, ulcers and sunburn. Only take internally under supervision.

St. John's Wort oil

Hyssopus officinalis

HYSSOP

'Purge me with Hyssop and I shall be clean.'
(Psalm 51, verse 7)

From the family *Lamiaceae*.

Hyssop is a native of the Mediterranean region, where it grows wild on old walls and dry banks. It is found as a garden escapee elsewhere in Europe and has been cultivated in gardens for about the last 600 years. It was one of the herbs taken to the New World by the colonists to use in tea, in herbal tobacco and as an antiseptic.

Hyssopus officinalis subsp. *aristatus*

There has been much to-ing and fro-ing about whether common hyssop is the one mentioned in the Bible. Some say it was oregano or savory. However, present thinking is that hyssop is flavour of the month, especially since it has been discovered that the mould that produces penicillin grows on its leaf. This may have acted as an antibiotic protection when lepers were bathed in hyssop.

The Persians used distilled hyssop water as a body lotion to give a fine colour to their skin.

Hippocrates recommended hyssop for chest complaints, and today herbalists still prescribe it.

varieties

These are the common hyssops, readily available from nurseries and garden centres.

Hyssopus officinalis
Hyssop, Blue Hyssop
Hardy semi-evergreen perennial. Ht 80cm, spread 90cm. Blue flowers from summer to early autumn. Small, narrow, lance-shaped leaves, aromatic and darkish green.

Hyssopus officinalis* f. *albus
White Hyssop
Hardy semi-evergreen perennial. Ht 80cm, spread 90cm. White flowers from summer to early autumn. Small, narrow, lance-shaped leaves, aromatic, and darkish green in colour.

Hyssopus officinalis* subsp. *aristatus
Rock Hyssop
Hardy, semi-evergreen, perennial. Ht 30cm, spread 60cm. Dark blue flowers from summer to early autumn. Small, narrow, lance-shaped leaves, aromatic and darkish green.

***Hyssopus officinalis* 'Roseus'**
Pink Hyssop
Hardy, semi-evergreen, perennial. Ht 80cm, spread 90cm. Pink flowers from summer to early autumn. Small, narrow, lance-shaped leaves, aromatic and darkish green.

cultivation

Propagation

Seeds
In early spring sow the small seeds in plug module or seed trays under protection, using a standard seed compost mixed in equal parts with composted fine bark. Cover with perlite. If very early in spring, a bottom heat of 15–21°C would be beneficial. When the seedlings are large enough, either pot up or transplant into the garden after a period of hardening off. Plant at a distance of 30cm apart. All varieties can be grown from seed with the exception of rock hyssop, which can only be grown from cuttings. However, if you want a guaranteed pink or white hyssop, cuttings are a more reliable method.

Cuttings
In late spring or early summer, take softwood cuttings from the new lush growth and non-flowering stems.

Pests and Diseases
This genial plant rarely suffers from pests or diseases.

Maintenance
Spring Sow seeds. Trim mature plants. Trim hedges.
Summer Deadhead flowers to maintain supply, trim after flowering to maintain shape. Trim hedges.
Autumn Cut back only in mild areas.
Winter Protect in cold, wet winters and temperatures that fall below –5°C. Use horticultural fleece, straw, bracken, etc.

Garden Cultivation
This attractive plant likes to be planted in conditions similar to rosemary and thyme, a well-drained soil in

HYSSOP

Hyssopus officinalis f. *albus*

a sunny position. The seeds can be sown directly into the ground in very late spring or early summer, when the soil is warm. Thin to 30cm apart if being grown as specimen plants. If for hedging, 18cm. As all parts of the plant are pleasantly aromatic and the flowers very attractive, plant it where it can be seen and brushed against. The flowers are also attractive to bees and butterflies. For these reasons hyssop makes a very good hedge or edging plant. Trim the top shoots to encourage bushy growth. In early spring, trim the plant into a tidy shape with scissors. To keep the plant flowering in summer, remove the dead heads. Cut back to 20cm in autumn in mild areas, or trim back after flowering in cold areas. Keep formal hedges well clipped during the growing season.

Harvesting

Cut young leaves for drying in summer. The flowers should be picked during the summer too, when they are fully opened. The scent is generally improved with drying.

companion planting

It is said that when grown near cabbages it lures away cabbage whiteflies. When planted near vines, it is reputed that the blue flowers of Hyssop attract the pollinating insects, so helping to increase the yield.

container growing

Hyssop is a lovely plant in containers. It is happy in plenty of sunshine and prefers a south-facing wall. It also likes dry conditions and its tough leaves are not affected by the fumes of city centres, making it ideal for window boxes. Equally, it is good on a patio as the scent is lovely on a hot summer's evening. Give it a liquid feed only during the flowering period. Cut back after flowering to maintain shape.

medicinal

An infusion is used mainly for coughs, whooping cough, asthma and bronchitis, and upper respiratory catarrh. It is also used for inflammation of the urinary tract. Externally it can be used for bruises and burns. It was once a country remedy for rheumatism.

warning

Hyssop should not be used in cases of nervous irritability. Strong doses, particularly those of distilled essential oil, can cause muscular spasms. This oil should not be used in aromatherapy for highly strung patients, as it can cause epileptic symptoms. Do not use continuously for extended periods. No form of hyssop should be taken during pregnancy.

culinary

The flowers are delicious tossed in a green salad. In small amounts, leaves aid digestion of fatty foods, but as they are somewhat pungent, use them sparingly. The herb has a slightly bitter, minty taste and is therefore good flavouring in salads or as an addition to game, meats and soups, stews and stuffings. A good idea is to add a teaspoon of chopped leaf to a Yorkshire pudding batter. Hyssop is still used in Gascony as one of the herbs in bouquet garni and for flavouring a concentrated purée of tomatoes preserved for the winter. It is used in continental sausages and also added to American fruit pies, 1/4 teaspoon hyssop being sprinkled over the fruit before the top crust goes on.

When making a sugar syrup for fruit, add a sprig of hyssop as you boil the sugar and water; it adds a pleasant flavour, and the sprig can be removed before adding the fruit. When making cranberry pie, use the leaves as a lining for the dish.

Basque-Style Chicken

Serves 6

4 sweet peppers (2 red, 2 green)
Hyssop olive oil
1.5 kg chicken
5 tablespoons dry white wine
6 onions
4 cloves garlic
4 medium tomatoes, peeled and roughly chopped
1 bouquet garni with a sprig of hyssop
Salt and freshly ground black pepper

De-seed and slice the peppers into thin strips. Gently fry them in a small amount of oil until soft. Remove from pan and put to one side. Joint the chicken and gently fry in the oil, turning all the time. Transfer to a casserole, and season with salt and pepper, moisten with the wine, and leave over a gentle heat to finish cooking. Slice the onions and peel the garlic cloves, and soften without colouring in the olive oil in the frying pan. Then add the tomatoes, peppers and bouquet garni, and season. When reduced almost to a cream, turn into the casserole over the chicken and keep on a low heat until ready to serve, about a further 20–30 minutes.

Inula helenium

ELECAMPANE

Also known as Allecampane, July campane, Elicompane, Dock, Sunflower, Wild sunflower, Yellow starwort, Elfdock, Elfwort, Horse elder, Horse heal and Scabwort. From the family *Asteraceae.*

Elecampane originates from Asia whence, through cultivation, it spread across Western Europe to North America and now grows wild from Nova Scotia to Ontario, North Carolina and Missouri.

Sources for the derivation of the principal common name, Elecampane, and the generic/species name, *Inula helenium*, are not altogether satisfactory, but I have found three possible explanations.

Helen of Troy was believed to be gathering the herb when she was abducted by Paris, hence *helenium*.

Down through the ages the herb was considered as good medicinally for horse or mule as for man; it was even sometimes called horselene. *Inula* could come from *hinnulus*, meaning 'a young mule'.

Finally, the Romans called the herb *Enula Campana* (Inula of the fields) from which Elecampane is a corruption.

According to the Roman writer Pliny, the Emperor Julius Augustus enjoyed elecampane so much he proclaimed, 'Let no day past without eating some of the roots candied to help the digestion and cause mirth'. The Romans also used it as a candied sweetmeat, coloured with cochineal. This idea persisted for centuries, and in the Middle Ages, apothecaries sold the candied root in flat pink sugary cakes, which were sucked to alleviate asthma and indigestion and to sweeten the breath. Tudor herbalists also candied them for the treatment of coughs, catarrhs, bronchitis and chest ailments. Their use continued until the 1920s as a flavouring in sweets.

I have discovered an Anglo-Saxon ritual using elecampane – part medicinal, part magical. Prayers were sung of the *Helenium* and its roots dug up by the medicinal man, who had been careful not to speak to any disreputable creature – man, elf, goblin or fairy – he chanced to meet on the way to the ceremony. Afterwards the elecampane root was laid under the altar for the night and eventually mixed with betony and lichen from a crucifix. The medicine was taken against elf sickness or elf disease.

There is an ancient custom in Scandinavia of putting a bunch of elecampane in the centre of a nosegay of herbs to symbolise the sun and the head of Odin, the greatest of Norse gods.

ELECAMPANE

Inula helenium

varieties

Inula helenium
Elecampane
Hardy perennial. Ht 1.5–2.4m, spread 1m. Bright yellow, ragged, daisy-like flowers in summer. The leaves are large, oval-toothed, slightly downy underneath, and of a mid-green colour. Dies back fully in winter.

Inula hookeri
Hardy perennial. Ht 75cm, spread 45cm. Yellowish-green, ragged, daisy-like flowers slightly scented in summer. Lance-shaped hairy leaves, smaller than *I. magnifica*, and mid-green in colour. Dies back fully in winter.

Inula magnifica
Hardy perennial. Ht 1.8m, spread 1m. Large, ragged, daisy-like flowers. Lots of large, dark green lance-shaped rough leaves. May need staking in an exposed garden. This is often mistaken for *I. helenium*; the leaf colour is the biggest difference, and on average *I. magnifica* grows much larger. Dies back fully in winter.

cultivation

Propagation
Seed
The seed is similar to dandelion; when the plant has germinated you can see the seeds flying all over the garden, which should be all the warning you need... In spring sow on the surface of a pot or plug module tray. Cover with perlite. Germination is 2–4 weeks, depending on the sowing season and seed viability. Prick out and plant 1–1.5m apart when the seedlings are large enough to handle.

Root Division
If the plant grows too big for its position in the garden, divide in the autumn when the plant has died back. As the roots are very strong, choose the point of division carefully. Alternatively, remove the offshoots that grow around the parent plant; each has its own root system, so they can be planted immediately in a prepared site elsewhere in the garden. This can be done in autumn or spring.

Pests and Diseases
It rarely suffers from disease, although if the autumn is excessively wet, as the leaves die back, they may suffer from a form of mildew. Simply cut back and destroy the leaves.

Maintenance
Spring Sow seed. Divide established plants.
Summer Remove flowerheads as soon as flowering finishes.
Autumn Cut back growth to stop self-seeding and to prevent the plant becoming untidy. Remove offshoots for replanting. Divide large plants.
Winter The plant dies back so needs no protection.

Garden Cultivation
Plant in a moist, fertile soil, in full sun, sheltered from the wind (elecampane grows tall and would otherwise need staking). It can look very striking at the back of a border against a stone wall, or in front of a screen of deciduous trees. In a very dry summer it may need watering.

Harvesting
Dig up second- or third-year roots in the autumn; they can be used as a vegetable, or dried for use in medicine. The flowers are good in autumn flower arrange-ments and dry well upside down if you cut them just before the seeds turn brown.

container growing

Elecampane grows too big for most containers and is easily blown over.

culinary

Elecampane has a sharp, bitter flavour. Use dried pieces or cook as a root vegetable.

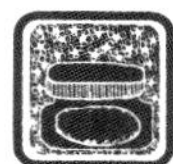

other uses

Your cat may be interested to know that scientific research indicates that elecampane has a sedative effect on mice.

medicinal

The main use is for respiratory complaints, at one time specifically for TB. It is still employed in folk medicine as a favourite constituent of cough remedies, and has always been popular both as a medicine and as a condiment.

In America, elecampane oil is used for respiratory infections, intestinal catarrh, chronic diarrhoea, chronic bronchitis and whooping cough. A decoction of the root has long been used externally for scabies, herpes, acne and other skin diseases, hence its country name Scabwort. Recent research shows that the lactines found in the roots are powerful agents against bacteria and fungi.

Elecampane oil

IRIS

From the family *Iridaceae*.

All those mentioned are native to the northern hemisphere and are cultivated in varying conditions, from dry light soil (orris – *Iris* 'Florentina') to damp boggy soils (blue flag iris – *Iris versicolor*).

The Greek word *iris*, meaning 'rainbow' and the name of the Greek goddess of the rainbow, was appended to describe the plant's variable colours. The iris is one of the oldest cultivated plants – it is depicted on the wall of an Egyptian temple dating from 1500 BC.

In this very large family, three stand out for their beneficial herbal qualities. Orris has a violet-scented root which has been powdered and used in perfumes since the times of the ancient Egyptians and Greeks. The Latin *Iris* 'Florentina' depicts its association with Florence in the early Middle Ages. It is said to be the fleur de lys of French heraldry.

The blue flag iris is the common wetland plant, native to eastern North America and exported from there to Europe. Employed by the Native Americans and early settlers as a remedy for gastric complaints, it was included in the United States Pharmacopeia and is still believed in folk medicine to be a blood purifier of use in eruptive skin conditions. Sometimes the plant is known as liver lily because of its purifying effect.

The root of the yellow flag iris, a native of the British Isles, was powdered and used as an ingredient in Elizabethan snuff. It was taken to America and Australia by the earliest settlers.

Iris pseudacorus

varieties

***Iris* 'Florentina' AGM**
Orris
Also known as Florentine iris and orris root. Hardy perennial. Ht 60cm–1m. Spread indefinite. Large white flowers tinged with pale lavender and with a yellow beard appearing in early to mid-summer. Green, sword-shaped leaves. The rootstock is stout and rhizomatous with a violet scent. Grows well throughout Europe and North America, except in the warm moist climate of Florida and the Gulf Coast.

***Iris germanica* AGM**
Purple Iris
Also known as garden iris, and flag iris. Hardy perennial. Ht 60–90cm, spread indefinite. The fragrant flowers are blue/violet, occasionally white, and form early to mid-summer. Leaves are greyish-green and sword-shaped. The root is thickish rhizome. There are many cultivated varieties. It is grown commercially for the rhizomes and, like *Iris* 'Florentina', is used in perfumery and pharmaceutical preparations.

***Iris pseudacorus* AGM**
Yellow Iris (Yellow Flag)
Perennial. Ht 40cm–150cm, spread indefinite. Flowers are bright yellow with radiating brown veins and very slightly scented. They appear early to mid-summer. The root is a thick rhizome from which many rootlets descend.

***Iris versicolor* AGM**
Blue Flag Iris. Also known as flag lily, fleur de lys, flower du Luce, liver lily, poison flags, snake lily, water flag and wild iris.
Hardy perennial. Ht 30–100cm, spread indefinite. Flowers claret-purple-blue in summer. Large, sword-shaped, green leaves. Root large and rhizomatous.

Iris 'Florentina'

cultivation

Propagation

Seed

All the irises produce large seeds, which take some time to germinate and often benefit from a period of stratification. As the seeds are of a good size, sow directly into an 8cm pot in autumn, using a standard seed compost mixed in equal parts with composted fine bark. Water in well, and cover the pots with cling film (to stop the mice eating the seed). Put outside to get the weather. Check that the compost remains damp. If there is any danger of it drying out, stand the container in water. This is especially important for blue and yellow flag irises.

Division

Divide the rhizome roots in late spring or early autumn. This suits all the varieties. Replant immediately in a prepared site. Leave a decent distance between plants; spread is indefinite.

Pests and Diseases

The only major pest is the iris sawfly, which is attracted to waterside irises. The darkish-grey larvae feed along the leaf-margins, removing large chunks. Pupation takes place in the soil beneath or near the host plants, and the adult sawflies are on the wing during early to mid-summer. Pick off the larvae when seen. This is an annual pest and there is not much one can do to prevent it.

Maintenance

Spring Divide roots of mature plants.
Summer Collect the seeds as soon as ripe.
Autumn Sow seeds and leave outside. Divide roots.
Winter Fully hardy; no need for protection.

Garden Cultivation

Orris and Purple Iris prefer a well-drained, rich soil and a sunny situation. When planting, make sure that part of the rhizomes is exposed.

Yellow and Blue Flag Irises are marsh-loving plants, ideal for those with a pond or ditch or piece of boggy ground. They grow happily in semi-shade but need full sun in order to produce the maximum bloom. In deep shade they will not flower at all but will spread quickly by stout underground rhizomes. A measure of control will be necessary.

Harvesting

The full violet fragrance of orris will not be apparent until the roots are 2 years old. Dig up these rhizomes in autumn and dry immediately.

Gather yellow flag flowers and roots for use as a dye, in early summer and autumn respectively. Dig up blue flag roots in autumn and dry.

container growing

These irises grow on strong rhizomes, so make sure that the container is strong enough, large enough, and shaped so that it will accommodate the plant happily and not blow over. For the bog lovers use peat in the compost mix – 75 per cent peat mixed with 25 per cent composted fine bark, but put lots of gravel and broken crocks in the bottom of the container to make up for loss of weight. For the dry gang, use a soil-based compost. Do not let either compost dry out. They become pot bound very quickly, so split and re-pot every year.

other uses

The violet-scented, powdered root of orris is used as a fresh scent to linen, a base for dry shampoos, a base for tooth powders, in face-packs, as a fixative in potpourris and as a dry shampoo. Flowers of yellow flag make a good yellow dye, while the rhizomes yield a grey or black dye when used with an iron mordant.

medicinal

Orris and yellow flag are rarely used medicinally nowadays. However, herbalists still use the blue flag as a blood purifier acting on the liver and gall bladder to increase the flow of bile, and as an effective cleanser of toxins. It is also said to relieve flatulence and heartburn, belching and nausea, and headaches associated with digestive problems.

culinary

Apparently, if you roast the seed of the yellow flag iris, they make an excellent coffee substitute. Apart from this little gem, I cannot find any culinary uses for irises.

warning

Always wash your hands well after handling this plant as it can cause uncontrollable vomiting and violent diarrhoea. Large doses of the fresh root can cause nausea, vomiting and facial neuralgias.

Isatis tinctoria

WOAD

Also known as Dyers woad, Glastum. From the family *Brassicaceae*.

This ancient dye plant for cloth and wool has been used for thousands of years. It was the only source of blue dye available in Europe until the end of the sixteenth century, when indigo *(Indigofera tinctoria)* became available from India via the spice trade. There is a resurgence in interest in this herb not only because of recent films on Picts, Celts and Romans, but also due to current research into its medicinal properties.

varieties

Isatis tinctoria

Woad

Hardy biennial. Ht in first year 45cm, second year when in flower 130cm, spread 45cm. Clusters of numerous small, bright yellow honey-scented flowers in the second summer, which are followed by pendulous black seeds. Lance-shaped, lightly toothed, blue/green leaves. The tap root in the second season can be 1.5m long.

cultivation

Propagation

Seed

Sow fresh seeds in autumn into prepared plug modules or small pots using a seed compost mixed in equal parts with perlite. Place the container in a cold greenhouse or cold frame. Germination takes 3–4 weeks. In the spring plant out into a prepared site 30cm apart. Alternatively, sow in late spring, when the night time temperature does not fall below 9°C, direct into a well-prepared site. Once the seedlings have emerged, thin to 30cm apart.

Pests and Diseases

Woad seedlings can be attacked by the flea beetle. To prevent this, in spring, cover with a fine horticultural mesh for the first season only.

Maintenance

Spring Protect new seedlings from flea beetle. Mulch second-year plants.

Summer Harvest seeds and roots of second-year plants.

Autumn Sow fresh seeds.

Winter No need for protection; fully hardy.

Garden Cultivation

Plant in a sunny position in a well-drained, well-fed soil. Be aware that this herb has a very long tap root and hates being transplanted once the plant starts to become established. In late summer collect the seeds before they fall as it is renowned for self-seeding.

Harvesting

In late summer pick the leaves, either to dry for medicinal use, or to ferment and then dry for use as a dye. Dig up the roots of the second year's growth in late summer then dry for medicinal use. Harvest the seeds in late summer as they turn dark brown and before they drop.

container growing

Being biennial, this herb looks very boring for the first year and when it goes to flower in the second it becomes very tall, which does not make it an ideal container plant.

medicinal

In Chinese medicine the root of this herb is known as *ban lang gen*, and it is used to treat meningitis and mumps. However be aware that woad is very astringent and poisonous and should only be taken internally under supervision. In 2006 the *New Scientist* reported that Italian biochemists have discovered that woad contains more than 60 times the amount of Glucobrassicin, a type of glucosinolate, in the leaves than broccoli, which might help in the prevention of cancer.

other uses

Traditional blue dye plant.

warning

Astringent and poisonous when taken internally. It is classed as a noxious weed in Australia and the USA.

Woad dye

Jasminum officinale

JASMINE

Also known as Common jasmine, Yasmine, Chambeli, Jessamine and Mogra. From the family *Oleaceae*.

The scent of the jasmine flower wafting over the night air is very evocative. The jasmine flower symbolises femininity. Traditionally Italian girls include jasmine in their bridal head dresses. There is a lovely story that the Duke of Tuscany in the seventeenth century introduced this herb from China, and one of his gardeners gave his betrothed a piece as a token of his love and was promptly dismissed for this misdemeanour. However, his betrothed planted the cutting and their garden was soon festooned with jasmine, which was much sought after by the girls of the village for its aphrodisiac properties.

varieties

***Jasminum officinale* AGM**
Jasmine
Hardy deciduous, occasionally semi-evergreen, climber. Ht up to 10m, spread up to 3m. Small fragrant white flowers all summer, followed by black poisonous berries. Pinnate mid-green leaves with 3 to 9 leaflets

Jasminum grandiflorum
Royal Jasmine, Spanish Jasmine, Jati
Tender evergreen climber. Ht up to 5m. Clusters of highly scented white flowers with a pink tinge all summer. Pinnate dark green leaves with 7 or 9 leaflets.

cultivation

Propagation

Seed
In autumn, extract the seeds from the berries, and sow into prepared pots or plug modules using a standard loam-based seed compost mixed in equal parts with coarse horticultural sand. Cover with sand, and then place outside for the winter. The seeds need frost to break their dormancy. Germination takes 4–6 months, but it can be erratic, so be patient and do not discard the container.

Cuttings
Take cuttings from semi-ripe growth in summer into prepared plug modules or a small pot, using a seed compost mixed in equal parts with propagating bark. Place in a cold frame or cold greenhouse. Once rooted, pot up into a loam-based potting compost, winter in a cold frame, plant out in the following spring.

Pests and Diseases
Rarely suffers from pests or diseases.

Maintenance
Spring Prune established plants.
Summer Take cuttings from semi-ripe growth.
Autumn Sow seeds.
Winter No need for protection, fully hardy.

Garden Cultivation
This sturdy climber, once established, will outgrow a trellis or need continual pruning if grown on a house, therefore is best grown over an arbour, porch, or a large old tree stump in a fertile soil and a sunny position. Prune in spring.

Harvesting
Pick flowers early in the morning, just as they open, for use as a flavouring, for infusions or for making oils.

container growing

As this is a vigorous climber it is not ideally suited to container growing.

medicinal

In Ayurvedic medicine a juice pressed from the leaves is used to remove corns. The leaves are also used as a gargle for mouth ulcers. An infusion made from the flowers is used to cool inflamed and bloodshot eyes. The essential oil made from the flowers is considered an antidepressant and relaxant.

other uses

The essential oil is used in perfumes and in several food flavourings.

warning

Do not take jasmine essential oil internally. The berries are poisonous.

Jasminum officinale berries

Juniperus communis

JUNIPER

Juniperus communis 'Compressa'

From the family *Cupressaceae*.

Juniper is widely distributed throughout the world and grows either as a shrub or a small tree. It is a native of the Mediterranean region, but also grows in the Arctic, from Norway to the Soviet Union, in the North and West Himalayas and in North America. It is found on heaths, moorlands, open coniferous forests and mountain slopes.

This widely distributed plant was first used by the ancient Greek physicians and its use has continued right up to the modern day. It was believed to cure snake bites and protect against infectious diseases like the plague.

The English word 'gin' is derived from an abbreviation of Holland's 'geneva', as the spirit was first called. This in turn stems from the Dutch *jenever*, meaning juniper.

varieties

Juniper is a conifer, a group of trees and shrubs distinguished botanically from others by its production of seeds exposed or uncovered on the scales of the fruit. True to form, it is evergreen and has needle-like leaves.

There are many species and varieties available, *Juniperus communis* being the main herbal variety. On the varieties detailed below, the flowers are all very similar: male flowers are very small catkins; female flowers are small, globose and berry-like, with usually 3–8 fleshy scales. Over a period of 3 years, these turn blue and then finally black as they ripen.

Juniperus communis
Juniper
Hardy evergreen perennial. Ht 30cm–8m, spread 1–4m. The size of the plant is very dependent on where it is growing. Leaves small and bright green, sharply pointed and aromatic.

***Juniperus communis* 'Compressa' AGM**
Juniper Compressa
Hardy evergreen perennial tree. Ht 75cm, spread 15cm. The leaves are small and bluish-green, sharply pointed and aromatic. Very slow growing with an erect habit, ideal for rock gardens or containers.

***Juniperus communis* 'Hibernica' AGM**
Irish Juniper
Hardy evergreen perennial tree. Ht 3–5m, spread 30cm. Leaves small and bluish/silvery-green, sharply pointed and aromatic. Columnar in shape and with a hint of silver in certain lights. Very slow growing.

***Juniperus communis* 'Hornibrookii' AGM**
Juniper Hornibrookii
Hardy evergreen perennial tree. Ht 50cm, spread 2m. Leaves small and darkish green, sharply pointed and aromatic. A big carpeting plant.

***Juniperus communis* 'Prostrata'**
Prostrate Juniper
Hardy evergreen perennial tree. Ht 20–30cm, spread 1–2m. Leaves small and bluish-green, sharply pointed and aromatic. A smaller carpeting plant.

Juniperus communis berries

cultivation

Propagation

Seed

All the species can be propagated by seed. Sow seeds taken from ripe berries in a cold greenhouse, cold frame or cold conservatory in early autumn. As junipers on the whole are extremely slow growing, it is best to grow the seedlings in a controlled environment for 1 or 2 years before planting out in a permanent position in the garden. Start in seed or plug module trays; then, when the seedlings are large enough, pot up into small pots using a soil-based seed compost. This method is the easiest but to be sure of the plant's gender and leaf colour, taking cuttings is more reliable.

Cuttings

It is quite easy to raise juniper from semi-hardwood cuttings taken from fresh current growth in either spring or autumn.

Pests and Diseases

Various rusts attack juniper. If you see swellings on the branches with rusty, gelatinous extrusions, cut the branches out and burn them. Honey fungus attacks many conifers. If this occurs, dig up the plant, making sure you have all the roots, burn it, and plant no further trees in that space.

Maintenance

Spring Plant out 2-year-old plants. Remove any leaders growing incorrectly in late spring/early summer.
Summer Take semi-hardwood cuttings. Harvest berries and store in a dry place.
Autumn Sow seeds. Take cuttings.
Winter Winter young plants in cold frames, or provide added protection.

Garden Cultivation

Juniper likes an exposed, sunny site. It will tolerate an alkaline or neutral soil. Both male and female plants are necessary for berry production. The berries, which only grow on the female bush, can be found in various stages of ripeness on the same plant. Their flavour is stronger when grown in warm climates. To maintain the shape of the juniper, trim with secateurs to ensure that there is no more than one leader, the strongest and straightest. Remember, when trimming, that most conifers will not make new growth when cut back into old wood, or into branches that have turned brown.

Harvesting

Harvest the berries when ripe in late summer. Dry them spread out on a tray, as you would leaves.

container growing

Juniper is slow growing and can look most attractive in pots. Use a soil-based compost, starting off with a suitable-sized pot, only potting up once a year if necessary. If the root ball looks happy, do not disturb it. Do not over water. As the plant is hardy and evergreen, the container will need more protection than the plant during the winter months. Feed during the summer months only with a liquid fertiliser as per the manufacturer's guidelines.

medicinal

Juniper is used in the treatment of cystitis, rheumatism and gout. Steamed inhalations of the berries are an excellent treatment for coughs, colds and catarrh.

warning

Juniper berries should not be taken during pregnancy or by people with kidney problems. Internal use of the volatile oil must only ever be prescribed by professionals.

culinary

Crushed berries are an excellent addition to marinades, sauerkraut and stuffing for guinea fowl and other game birds. Although no longer generally considered as a spice, it is still an important flavouring for certain meats, liqueurs, and especially gin.

Pork Chops marinated with Juniper

Serves 4

Marinade
2 tablespoons olive oil
6 juniper berries, crushed
2 cloves garlic, crushed
Salt and freshly ground black pepper

Mix the oil, juniper berries, garlic and seasoning together in a bowl.

4 pork chops
25g plain flour
275ml dry cider

Lay the pork chops in the base of a shallow dish and cover them with the marinade, turning the chops over once to make sure they are covered. Leave for a minimum of 3 hours, or if possible overnight. Remove the chops from the marinade, and reserve it. Heat a large frying pan and add the reserved marinade. When hot, add the pork chops and cook over a moderate heat for about 20 minutes, turning the chops regularly. When all traces of pink have gone from the meat, remove from the heat, and put the chops on a plate. Return the pan to the heat and stir the flour into the remaining juices. Add the cider and bring to the boil. Return the chops to the sauce in the pan. Heat slowly, and serve with mashed potato and broccoli.

Laurus nobilis

BAY

From the family *Lauraceae*.

Bay is an evergreen tree native to southern Europe, and now found throughout the world.

That this ancient plant was much respected in Roman times is reflected in the root of its family name, Lauraceae, the Latin *laurus*, meaning 'praise', and in its main species name, *Laurus nobilis*, the Latin *nobilis*, meaning 'famous', 'renowned'. A bay wreath became a mark of excellence for poets and athletes, a symbol of wisdom and glory. The word laureate means 'crowned with laurels' (synonym for bay), hence Poet Laureate, of course, and the French *baccalauréat*.

The bay tree was sacred even earlier – to Apollo, Greek god of prophecy, poetry and healing. His priestesses ate bay leaves before expounding his oracles at Delphi. As large doses of bay induce the effect of a narcotic, this may explain their trances. His temple had its roof made entirely of bay leaves, ostensibly to protect against disease, witchcraft and lightning. Apollo's son Asclepius, the Greek god of Medicine, also had bay dedicated to him as it was considered a powerful antiseptic and guard against disease, in particular the plague.

In the seventeenth century, Culpeper wrote that 'neither witch nor devil, thunder nor lightening, will hurt a man in the place where a bay-tree is.' He also wrote that 'the berries are very effectual against the poison of venomous creatures, and the stings of wasps and bees.'

varieties

Laurus azorica* syn. *Laurus canarinsis

Canary Island Bay

Perennial evergreen tree. Ht to 6m. Reddish-brown branches, a colour that sometimes extends to the leaves.

***Laurus nobilis* AGM**

Bay

Also known as sweet bay, sweet laurel, laurel, Indian bay, Grecian laurel. Perennial evergreen tree. Ht up to 8m, spread 3m. Small, pale yellow, waxy flowers in spring. Green oval berries turning black in autumn. The leaves may be added to stock, soups and stews and are among the main ingredients of bouquet garni. *L. nobilis* is the only bay used for culinary purposes.

Laurus nobilis* f. *angustifolia

Willow Leaf Bay

Perennial evergreen tree. Ht up to 7m. Narrow-leafed variety, said to be hardier than *L. nobilis*. This is not strictly true.

***Laurus nobilis* 'Aurea' AGM**

Golden Bay

Perennial evergreen tree. Ht up to 5m. Small, pale yellow, waxy flowers in spring. Green berries turning black in autumn. Golden leaves can look sickly in damp, cooler countries. Needs good protection in winter especially from wind scorch and frosts. Trim in the autumn/spring to maintain the golden leaves.

Umbellularia californica

Californian Laurel

Perenial evergreen tree. Ht up to 18m. Pale yellow flowers in late spring. Very pungent/aromatic leaves. Can cause headaches and nausea when the leaves are crushed. NOT culinary.

cultivation

Propagation

Seed

Bay sets seed in its black berries, but rarely in cooler climates. Sow fresh seed in spring on the surface of either a seed or plug tray or directly into pots. Use a standard seed compost mixed in equal parts with composted fine bark. Keep warm: 21°C. Germination is erratic, may take place within 10–20 days, in 6 months, or even longer. Make sure the compost is not too wet or it will rot the seeds.

Cuttings
Not a plant for the faint hearted. When I started propagating over 30 years ago I thought my bay cuttings were doing really well, but a year later not one had properly struck, and three-quarters of them had turned black and died. A heated propagator is a great help and high humidity is essential. Use either a misting unit or cover the cutting in plastic and maintain the compost or perlite at a steady moisture. Cuttings are taken in late summer/early autumn 10cm long.

Division
If offshoots are sent out by the parent plant, dig them up or they will destroy the shape of the tree. Occasionally roots come with them and these then can be potted up in a standard seed compost mixed in equal parts with composted fine bark. Place a plastic bag over the pot to maintain humidity. Leave somewhere warm and check from time to time to see if new shoots are starting. When they do, remove the bag. Do not plant out for at least a year.

Layering
In spring. A good way of propagating a difficult plant.

Pests and Diseases
Bay is susceptible to sooty mound, caused by the scale insect which sticks both to the undersides of leaves and to the stems, sucking the sap. Get rid of them by hand or use a liquid horticultural soap. Bay sucker distorts the leaf so as soon as you see it, cut back the branch to clean growth.

Maintenance
Spring Sow seeds. Cut back standard and garden bay trees to maintain shape and to promote new growth. Cut back golden bay trees to maintain colour. Check for pests and eradicate at first signs. Give container-grown plants a good liquid feed.
Summer Check that young plants are not drying out too much. In very hot weather, and especially in a city, spray-clean container-grown plants with water. Propagate by taking stem cuttings or layering in late summer.
Autumn Take cuttings of mature plants. Protect container-grown and young garden plants. Cover garden plants in straw or bracken, if in a sufficiently sheltered position, or use horticultural fleece.

Winter In severe winters the leaves will turn brown, but don't despair. Come the spring, it may shoot new growth from the base. To encourage this, cut the plant nearly down to the base.

Garden Cultivation
Bay is shallow rooted and therefore more prone to frost damage. Also, leaves are easily scorched in very cold weather or in strong, cold winds. Protection is thus essential, especially for bay trees under two years old. When planting out, position the plant in full sun, protected from the wind, and in a rich well-drained soil at least 1m away from other plants to start with, allowing more space as the tree matures. Mulch in the spring to retain moisture throughout the summer.

Harvesting
Being evergreen, leaves can be taken all year round. It is fashionable now to preserve bay leaves in vinegars.

culinary

Fresh leaves are stronger in flavour than dried ones. Use in soups, stews and stocks.

Add leaves to poached fish, like salmon.
Put on the coals of a barbecue.
Put fresh leaves in jars of rice to flavour the rice.
Boil in milk to flavour custards and rice pudding.

Bouquet Garni
I quote from my grandmother's cookbook, *Food for Pleasure*, published in 1950: 'A bouquet garni is a bunch of herbs constantly required in cooking.' The essential herbs in a bouquet garni are bay leaf, parsley and thyme.

Laurus nobilis

Berries are cultivated for use in laurel oil and laurel butter. The latter is a vital ingredient of laurin ointment, which is used in veterinary medicine.

container growing

Bay makes a good container plant. Young plants benefit from being kept in a container and indoors for the winter in cooler climates. The kitchen windowsill is ideal. Do not water too much, and let the compost dry out in the winter months.

Large standard bays or pyramids look very effective in half barrels or containers of a similar size. They will need extra protection from frosts and wind, so if the temperature drops below –5°C bring the plants in. To produce a standard bay tree, start with a young containerised plant with a straight growing stem. As it begins to grow, remove the lower side shoots below where you want the ball to begin. Allow the tree to grow to 20cm higher than desired, then clip back the growing tip. Cut the remaining side shoots to about 3 leaves. When the side shoots have grown a further 4/5 leaves, trim again to 2/3 leaves. Repeat until you have a leafy ball shape. Once established, prune with secateurs in late spring and late summer to maintain it.

other uses

Place in flour to deter weevils. Add an infusion to a bath to relieve aching limbs.

medicinal

Infuse the leaves to aid digestion and increase appetite.

Lavandula

LAVENDER

From the family *Lamiaceae*.

Native of the Mediterranean region, Canary Isles and India. Now cultivated in different regions of the world, growing in well-draining soil and warm, sunny climates.

Lavandula stoechas subsp. *stoechas* f. *rosea* 'Kew Red'

Long before the world made deodorants and bath salts, the Romans used lavender in their bath water; the word is derived from the Latin *lava*, 'to wash'. It was the Romans who introduced this plant to Britain and from then on monks cultivated it in their monastic gardens. Little more was recorded until Tudor times when people noted its fragrance and a peculiar power to ease stiff joints and relieve tiredness. It was brought in quantities from herb farms to the London Herb Market at Bucklesbury. 'Who will buy my lavender?' became perhaps the most famous of all London street cries.

It was used as a strewing herb for its insect-repellent properties and for masking household and street smells. It was also carried in nosegays to ward off the plague and pestilence. In France in the seventeenth century, huge fields of lavender were grown for the perfume trade. This has continued to the present day.

varieties

This is another big family of plants that are eminently worth collecting. I include here a few of my favourites.

Lavandula angustifolia* syn. *L. spica*, *L. officinalis
Common/English Lavender
Hardy evergreen perennial. Ht 80cm, spread 1m. Mauve/purple flowers on a long spike in summer. Long, narrow, pale greenish-grey, aromatic leaves. One of the most popular and well known of the lavenders.

***Lavandula angustifolia* 'Alba'**
White Lavender
Hardy evergreen perennial. Ht 70cm, spread 80cm. White flowers on a long spike in summer. Long, narrow, pale greenish-grey, aromatic leaves.

***Lavandula angustifolia* 'Bowles' Early'**
Lavender Bowles
Hardy evergreen perennial. Ht and spread 60cm. Light blue flowers on a medium-size spike in summer. Medium-length, narrow grey-greenish, aromatic leaves.

***Lavandula angustifolia* 'Folgate'**
Lavender Folgate
Hardy evergreen perennial. Ht and spread 45cm. Purple flowers on a medium spike in summer. Leaves as above.

***Lavandula angustifolia* 'Hidcote' AGM**
Lavender Hidcote
Hardy evergreen perennial. Ht and spread 45cm. Dark blue flowers on a medium spike in summer. Fairly short, narrow, aromatic, grey-greenish leaves. One of the most popular lavenders. Often used in hedging, planted at a distance of 30–40cm.

Lavandula x *intermedia* 'Old English'

LAVENDER

Lavandula angustifolia 'Munstead'

***Lavandula angustifolia* 'Loddon Pink' AGM**
Lavender Loddon Pink
Hardy evergreen perennial. Ht and spread 45cm. Pale pink flowers on a medium-length spike in summer. Fairly short, narrow, grey-greenish, aromatic leaves. Good compact habit. 'Loddon Blue' is the same size, same height, with pale blue flowers.

***Lavandula angustifolia* 'Munstead'**
Lavender Munstead
Hardy evergreen perennial. Ht and spread 45cm. Purple/blue flowers on a fairly short spike in summer. Medium length, greenish-grey, narrow, aromatic leaves. This is now a common lavender and used often in hedging, planted at a distance of 30–40cm.

***Lavandula angustifolia* 'Nana Alba' AGM**
Dwarf White Lavender
Hardy evergreen perennial. Ht and spread 30cm. White flowers in summer. Green-grey narrow short leaves. This is the shortest growing lavender and is ideal for hedges.

***Lavandula angustifolia* 'Rosea'**
Lavender Pink/Rosea
Hardy evergreen perennial. Ht and spread 45cm. Pink flowers in summer. Medium length greenish-grey, narrow, aromatic leaves.

***Lavandula angustifolia* 'Twickel Purple'**
Lavender Twickel Purple
Hardy evergreen perennial. Ht and spread 50cm. Pale purple flowers on fairly short spike. Medium length, greenish-grey, narrow, aromatic leaves. Compact.

Lavandula dentata
Fringed Lavender
Sometimes called French lavender. Half-hardy evergreen perennial. Ht and spread 60cm. Pale blue/mauve flowers from summer to early autumn. Highly aromatic, serrated, pale green, narrow leaves. This plant is a native of southern Spain and the Mediterranean region and so needs protecting in cold damp winters. It is ideal to bring inside into a cool room in early autumn as a flowering pot plant.

Lavandula* x *intermedia* Dutch Group syn. *L. vera
Lavender Vera
Hardy evergreen perennial. Ht and spread 45cm. Purple flowers in summer on fairly long spikes. Long greenish-grey, narrow, aromatic leaves.

***Lavandula* x *intermedia* 'Old English'**
Old English Lavender
Hardy evergreen perennial. Ht and spread 60cm. Light lavender-blue flowers on long spikes. Long, narrow, silver/grey/green aromatic leaves.

***Lavandula* x *intermedia* 'Pale Pretender' syn. *L.* 'Grappenhall'**
Lavender Pale Pretender
Hardy evergreen perennial. Ht and spread 1m. Large pale mauve flowers on long spikes in summer. The flowers are much more open than those of other species. Long greenish-grey, narrow, aromatic leaves.

***Lavandula* x *intermedia* 'Seal'**
Lavender Seal
Hardy evergreen perennial. Ht 90cm, spread 60cm. Long flower stems, mid-purple. Long, narrow, silver/grey/green aromatic leaves.

Lavandula angustifolia 'Twickel Purple'

Lavandula pendunculata subsp. *pedunculata*

***Lavandula lanata* AGM**
Woolly Lavender
Hardy evergreen perennial. Ht 50cm, spread 45cm. Deep purple flowers on short spikes. Short, soft, narrow, silver-grey aromatic foliage.

***Lavendula pedunculata* subsp. *pedunculata* AGM**
Lavender Pedunculata
Sometimes known as Papillon. Half-hardy evergreen perennial. Ht and spread 60cm. The attractive purple bracts have an extra mauve centre tuft, which looks like two rabbit ears. The aromatic leaves are long, very narrow and grey. Protect in winter.

***Lavandula stoechas* AGM**
French Lavender
Sometimes called Spanish lavender. Hardy evergreen perennial. Ht 50cm, spread 60cm. Attractive purple bracts in summer. Short, narrow, grey/green, aromatic leaves.

Lavandula stoechas* subsp. *stoechas* f. *leucantha
White French Lavender
As *L. stoechas* except white bracts in summer.

***Lavandula stoechas* subsp. *stoechas* f. *rosea* 'Kew Red'**
Lavender Kew Red
Half-hardy evergreen perennial. Ht and spread 40cm. Cerise crimson flowers and pale pink 'ears'. Short, narrow, grey/green aromatic leaves. Prefers an acidic soil.

Lavandula viridis
Green Lavender, Lemon Lavender
Half-hardy evergreen perennial. Ht and spread 60cm. This unusual plant has green bracts with a cream centre tuft. The leaves are green, narrow, and highly aromatic. Protect in winter.

Lavenders – Small
Grow to 45–50cm:
L. angustifolia 'Folgate', *L. angustifolia* 'Hidcote', *L. angustifolia* 'Loddon Pink', *L. angustifolia* 'Munstead', *L. angustifolia* 'Nana Alba', *L. angustifolia* 'Twickel Purple'

Lavenders – Medium
Grow to 60cm:
L. angustifolia 'Bowles' Early', *L.* x *intermedia* 'Old English'

Lavandula viridis

Lavenders – Large
70cm and above:
L. angustifolia 'Alba', *L.* x *intermedia* 'Seal'.

Half-hardy Lavenders
L. dentata (50cm), *L. lanata* (50cm), *L. stoechas* (50cm), *L. stoechas* subsp. *stoechas* f. *leucantha* (50cm), *L. pedunculata* subsp. *pedunculata* (60cm), *L. viridis* (50cm).

cultivation

Propagation
Seed
Lavender can be grown from seed but it tends not to be true to species, with the exception of *L. stoechas*.

Seed should be sown fresh in the autumn on the surface of a seed or plug tray and covered with perlite. It germinates fairly readily with a bottom heat of 4–10°C. Winter the seedlings in a cold greenhouse or cold conservatory with plenty of ventilation. In the spring, prick out and pot on using a standard seed compost mixed in equal parts with composted fine bark. Let the young plant establish a good-size root ball before planting out in a prepared site in the early summer. For other species you will find cuttings much more reliable.

Cuttings
Take softwood cuttings from non-flowering stems in spring. Root in a standard seed compost mixed in equal parts with composted fine bark. Take semi-hardwood cuttings in summer or early autumn from the strong new growth. Once the cuttings have rooted well, it is better to pot them up and winter the young lavenders in a cold greenhouse or conservatory rather than plant them out in the first winter. In the spring, plant them out in well-drained, fertile soil, at a distance of 45–60cm apart or 30cm apart for an average hedge.

Layering
This is easily done in the autumn. Most hardy lavenders respond well to this form of propagation.

Pests and Diseases
The chief and most damaging pest of lavender is the rosemary beetle. The beetle and its larvae feed on the leaves from autumn until spring. As soon as you see this small shiny beetle, place some newspaper underneath the plant, then tap or shake the branches, which will knock the beetles and larvae on to the paper, making them easy to destroy.

Lavandula angustifolia 'Loddon Pink'

The flowers in wet seasons may be attacked by grey mould (*Botrytis*). This can occur all too readily after a wet winter. Cut back the infected parts as far as possible, again remembering not to cut into the old wood if you want it to shoot again. There is another fungus (*Phoma lavandulae*) that attacks the stems and branches, causing wilting and death of the affected branches. If this occurs dig up the plant immediately and destroy, keeping it well away from any other lavender bushes.

Maintenance
Spring Give a spring hair cut.
Summer Trim after flowering. Take cuttings.
Autumn Sow seed. Cut back in early autumn, never into the old wood. Protect all the half-hardy lavenders. Bring containers inside before frosts start.
Winter Check seedlings for disease. Keep watering to a minimum.

LAVENDER

Lavender herb jelly

Garden Cultivation

Lavender is one of the most popular plants in today's herb garden and is particularly useful in borders, edges, as internal hedges, and on top of dry walls. All the species require an open sunny position and a well-drained, fertile soil. But lavender will adapt to semi-shade as long as the soil conditions are met, otherwise it will die in winter. If you have very cold winter temperatures, it is worth growing lavenders in containers to move inside in winter.

The way to maintain a lavender bush is to trim to shape every year in the spring, remembering not to cut into the old wood as this will not re-shoot. After flowering, trim back to the leaves. In the early autumn trim again, making sure this is well before the first autumn frosts, otherwise the new growth will be too soft and be damaged. By trimming this way, you will keep the bush neat and encourage it to make new growth, so stopping it becoming woody.

If you have inherited a straggly mature plant then give it a good cut back in autumn, followed by a second cut in the spring and then adopt the above routine. If the plant is aged, I would advise that you propagate some of the autumn cuts, so preserving the plant if all else fails.

Harvesting

Gather the flowers just as they open, and dry on open trays or by hanging in small bunches. Pick the leaves any time for use fresh, or before flowering if drying.

container growing

If you have low winter temperatures, lavender cannot be treated as a hardy evergreen. Treated as a container plant, however, it can be protected in winter and enjoyed just as well in the summer. Choose containers to set off the lavender; they all suit terracotta. Use a soil-based compost mixed in equal parts with composted fine bark. The ideal position is sun, but all lavenders will cope with partial shade, though the aroma can be impaired.

Feed regularly through the flowering season with liquid fertiliser, following the manufacturer's instructions. Allow the compost to dry out in winter (not totally, but nearly), and slowly reintroduce watering in spring.

medicinal

Throughout history, lavender has been used medicinally to soothe, sedate and suppress. Nowadays it is the essential oil that is in great demand for its many beneficial effects.

The oil was traditionally inhaled to prevent vertigo and fainting. It is an excellent remedy for burns and stings, and its strong antibacterial action helps to heal cuts. The oil also kills diphtheria and typhoid bacilli as well as streptococcus and pneumococcus.

Add 6 drops of oil to bath water to calm irritable children and help them sleep. Place 1 drop on the temple for a headache relief. Blend for use as a massage oil in aromatherapy for throat infections, skin sores, inflammation, rheumatic aches, anxiety, insomnia and depression. The best oil is made from distillation, and may be bought from many shops.

other uses

Rub fresh flowers onto skin or pin a sprig on clothes to discourage flies. Use flowers in potpourri, herb pillows, and linen sachets, where they make a good moth repellent.

Left: Lavender sachets make good presents and can be used as moth repellents

culinary

Lavender has not been used much in cooking, but as there are many more adventurous cooks around, I am sure it will be used increasingly in the future. Use the flowers to flavour a herb jelly, or a vinegar. Equally the flowers can be crystallised.

Lavender Biscuits

50g caster sugar
100g butter
175g self-raising flour
2 tablespoons fresh chopped lavender leaves
1 teaspoon lavender flowers removed from spike

Cream the sugar and butter together until light. Add the flour and lavender leaves to the butter mixture. Knead well until it forms a dough. Gently roll out on a lightly floured board. Scatter the flowers over the rolled dough and lightly press in with the rolling pin. Cut into small shapes with a cutter. Place biscuits on a greased baking sheet. Bake in a hot oven 450°F/230°C/ gas mark 7 for 10–12 minutes until golden and firm. Remove at once and cool on a wire tray.

Levisticum officinale

LOVAGE

Also known as Love Parsley, Sea Parsley, Lavose, Liveche, Smallage and European Lovage. From the family *Apiaceae*.

This native of the Mediterranean can now be found naturalised throughout the temperate regions of the world, including Australia, North America and Scandinavia.

Lovage was used by the ancient Greeks, who chewed the seed to aid digestion and relieve flatulence. Knowledge of it was handed down to Benedictine monks by the Romans, who prescribed the seeds for the same complaints. In Europe a decoction of lovage was reputedly a good aphrodisiac that no witch worthy of the name could be without. The name is likely to have come from the Latin *ligusticum*, after Liguria in Italy, where the herb grew profusely.

Because lovage leaves have a deodorizing and antiseptic effect on the skin, they were laid in the shoes of travellers in the Middle Ages to revive their weary feet, like latter-day odour eaters.

varieties

Levisticum officinale
Lovage
Hardy perennial. Ht up to 2m, spread 1m or more. Tiny, pale, greenish-yellow flowers in summer clusters. Leaves darkish green, deeply divided, and large toothed.

A close relation, ***Ligusticum scoticum***, shorter with white clusters of flowers, is sometimes called lovage. It can be used in the same culinary way, but lacks its strong flavour and the growth.

cultivation

Propagation

Seed
Sow under protection in spring into prepared plug or seed trays and cover with perlite; a bottom heat of 15°C is helpful. When the seedlings are large enough to handle and after a period of hardening off, transplant into a prepared site in the garden about 60cm apart.

Division
The roots of an established plant can be divided in the spring or autumn. Make sure that each division has some new buds showing.

Pests and Diseases

Leaf miners are sometimes a problem. Watch out for the first tunnels, pick off the affected leaves and destroy them, otherwise broad, dry patches will develop and the leaves will start to wither away. To control heavy infestations, cut the plant down to the ground, burning the affected shoots. Feed the plant and it will shoot with new growth. The young growth is just what one needs for cooking.

Maintenance

Spring Divide established plants. Sow seeds.
Summer Clip established plants to encourage the growth of new shoots.
Autumn Dry seeds. Divide roots.
Winter No need for protection.

Garden Cultivation

Lovage prefers a rich, moist but well-drained soil. Prior to first planting, dig over the ground deeply and manure well. The site can be either in full sun or partial shade. Seeds are best sown in the garden in the autumn. When the seedlings are large enough, thin to 60cm apart. It is important that lovage has a period of dormancy so that it can complete the growth cycle.

Lovage is a tall plant, so position it carefully. It will reach its full size in 3–5 years. To keep the leaves young and producing new shoots, cut around the edges of the clumps.

Harvesting

After the plant has flowered the leaves tend to have more of a bitter taste, so harvest in early summer. I personally believe that lovage does not dry that well and it is best to freeze it (see page 671 and use the parsley technique). Harvest seed heads when the seeds start to turn brown. Pick them on a dry day, tie a paper bag over their heads, and hang upside down in a dry, airy place. Use, like celery seed, for winter soups. Dig the root for drying in the autumn of the second or third season.

container growing

Lovage is fine grown outside in a large container. To keep it looking good, keep it well-clipped. I do not advise letting it run to flower unless you can support it. Remember at flowering stage, even in a pot, it can be in excess of 1.5m tall.

medicinal

Lovage is a remedy for digestive difficulties, gastric catarrh and flatulence. I know of one recipe from the West Country – a teaspoon of lovage seed steeped in a glass of brandy, strained and sweetened with sugar. It is taken to settle an upset stomach! Infuse either seed, leaf or root and take to reduce water retention. Lovage assists in the removal of waste products, acts as a deodoriser, and aids rheumatism. Its deodorising and antiseptic properties enable certain skin problems to respond to a decoction added to bath water. This is made with 45–60g of rootstock in 600ml water. Add to your bath.

warning

As lovage is very good at reducing water retention, people who are pregnant or who have kidney problems should not take this herb medicinally.

Lovage, brandy and sugar settles an upset stomach

Lovage soup

culinary

Lovage is an essential member of any culinary herb collection. The flavour is reminiscent of celery. It adds a meaty flavour to foods and is used in soups, stews and stocks. Also add fresh young leaves to salads, and rub on chicken, and round salad bowls. Crush seeds in bread and pastries, sprinkle on salads, rice and mashed potato. If using the rootstock as a vegetable in casseroles, remove the bitter-tasting skin.

Lovage Soup

Serves 4

25g butter
2 medium onions, finely chopped
500g potatoes, peeled and diced
4 tablespoons finely chopped lovage leaves
850ml chicken or vegetable stock
300ml milk or 1 cup cream
Grated nutmeg
Salt and freshly ground black pepper

Melt the butter in a heavy pan and gently sauté the onions and diced potatoes for 5 minutes until soft. Add the chopped lovage leaves and cook for 1 minute. Pour in the stock, bring to the boil, season with salt and pepper, cover and simmer gently until the potatoes are soft (about 15 minutes). Purée the soup through a sieve or liquidiser and return to a clean pan. Blend in the milk or cream, sprinkle on a pinch of nutmeg and heat through. Do not boil or it will curdle. Adjust seasoning. This soup is delicious hot or cold. Serve garnished with chopped lovage leaves.

Lovage and Carrot

Serves 2

3 carrots, grated
1 apple, grated
2 teaspoons chopped lovage leaves
2 tablespoons mayonnaise
125g plain yogurt
1 teaspoon salt (if needed)
Lettuce leaves
1 onion sliced into rings
Chives

Toss together the grated carrots, apple, lovage, mayonnaise and yogurt. Arrange the lettuce leaves on a serving dish and fill with the lovage mixture. Decorate with a few raw onion rings, chives and tiny lovage leaves.

Lovage as a Vegetable

Treat lovage as you would spinach. Use the young growth of the plant stalks and leaves. Strip the leaves from the stalks, wash, and cut the stalks up into segments. Bring a pan of water to the boil add the lovage, bring the water back to the boil, cover, and simmer for about 5–7 minutes until tender. Strain the water. Make a white sauce using butter, flour, milk, salt, pepper and grated nutmeg. Add the lovage. Serve and wait for the comments!

Lonicera

HONEYSUCKLE

Also known as Woodbine, Beerbind, Bindweed, Evening Pride, Fairy Trumpets, Honeybind, Irish Vine, Trumpet Flowers, Sweet Suckle, and Woodbind. From the family *Caprifoliaceae*.

'Come into the garden, Maud,
I am here at the gate alone;
And the woodbine spices are
wafted abroad,
And the musk of the rose is blown.'
Alfred Lord Tennyson (1809–1892)

Honeysuckle grows all over northern Europe, including Britain, and can also be found wild in North Africa, Western Asia and North America.

Honeysuckle, *Lonicera*, receives its common name from the old habit of sucking the sweet honey-tasting nectar from the flowers. Generically it is said to have been named after the 16th-century German physician, Lonicer.

Honeysuckle was among the plants that averted the evil powers abroad on May Day and took care of milk, the butter and the cows in the Scottish Highlands and elsewhere. Traditionally it was thought that if honeysuckle was brought into the house, a wedding would follow, and that if the flowers were placed in a girl's bedroom, she would have dreams of love.

Honeysuckle's rich fragrance has inspired many poets, including Shakespeare, who called it woodbine after its notorious habit of climbing up trees and hedges and totally binding them up.

'Where oxlips and the nodding violet grows
quite over-canopied with luscious woodbine . . .'
A Midsummer Night's Dream

The plant appeared in John Gerard's sixteenth-century *Herball*; he wrote that 'the flowers steeped in oil and set in the sun are good to anoint the body that is benummed and grown very cold'.

varieties

There are many fragrant climbing varieties of this lovely plant. I have only mentioned those with a direct herbal input.

***Lonicera* x *americana* (Miller)**
American Honeysuckle
Deciduous perennial. Ht up to 7m. Strongly fragrant yellow flowers starting in a pink bud turning yellow and finishing with orangish-pink throughout the summer. The berries are red, and the leaves are green and oval, the upper ones being united and saucer-like.

***Lonicera caprifolium* AGM**
Deciduous perennial. Ht up to 6m. The buds of the fragrant flowers are initially pink on opening; they then change to a pale white/pink/yellow as they age and finally turn deeper yellow. Green oval leaves and red berries, which were once fed to chickens. The Latin species name, *caprifolium*, means goats' leaf, reflecting the belief that honeysuckle leaves were a favourite food of goats. This variety and *Lonicera periclymenum* can be found growing wild in hedgerows.

Lonicera etrusca
Etruscan Honeysuckle
Semi-evergreen perennial. Ht up to 4m. Fragrant, pale, creamy yellow flowers which turn deeper yellow to red in autumn and are followed by red berries. Leaves oval, mid-green, with a bluish underside. This is the least hardy of those mentioned here, and should be grown on a sunny wall, and protected in winter where temperatures fall below −3°C.

Lonicera japonica
Japanese Honeysuckle
Semi-evergreen perennial. Ht up to 10m. Fragrant, pale, creamy white flowers turning yellow as the season progresses, followed by black berries. The leaves are oval and mid-green in colour. In the garden it is apt to build up an enormous tangle of shoots and best allowed to clamber over tree stumps or a low roof or walls. Attempts to train it tidily are a lost cause. Still used in Chinese medicine today.

Lonicera periclymenum
Deciduous perennial. This is the taller grower of the two common European honeysuckles, and reaches a

HONEYSUCKLE

height of 7m. It may live for 50 years. Fragrant yellow flowers appear mid-summer to mid-autumn, followed by red berries. Leaves are oval and dark green with a bluish underside.

cultivation

Propagation

Seed

Sow seed in autumn thinly on the surface of a prepared seed or plug tray. Cover with glass and winter outside. Keep an eye on the compost moisture and only water if necessary. Germination may take a long time, it has been known to take 2 seasons, so be patient. A more reliable alternative method is by cuttings.

Cuttings

Take from non-flowering, semi-hardwood shoots in summer and root in a standard seed compost mixed in equal parts with composted fine bark. Alternatively, take hardwood cuttings in late autumn, leaving them in a cold frame or cold greenhouse for the winter.

Layering

In late spring or autumn honeysuckle is easy to layer. Do not disturb until the following season when it can be severed from its parent.

Pests and Diseases

Grown in too sunny or warm a place, it can become infested with greenfly, blackfly, caterpillars and red spider mites. Use a horticultural soap, and spray the pests according to the manufacturer's instructions. To prevent powdery mildew, avoid dryness around the root.

Lonicera caprifolium

Maintenance

Spring Prune established plants.

Summer Cut back flowering stems after flowering. Take semi-hardwood cuttings.

Autumn Layer established plants. Lightly prune if necessary.

Winter Protect certain species in cold winters.

Garden Cultivation

This extremely tolerant, traditional herb garden plant will flourish vigorously in the most unpromising sites. Honeysuckle leaves are among the first to appear, sometimes mid-winter, the flowers appearing in very early summer and deepening in colour after being pollinated by the insects that feed on their nectar. Good as cover for an unsightly wall or to provide a rich summer evening fragrance in an arbour.

Plant in autumn or spring in any fertile, well-drained soil, in sun or semi-shade. The best situation puts its feet in the shade and its head in the sunshine. A position against a shady wall is ideal, or on the shady side of a support such as a tree stump, pole or pergola. Prune in early spring, if need be. Prune out flowering wood on climbers after flowering.

Harvesting

Pick and dry the flowers for potpourris just as they open. This is the best time for scent although they are at their palest in colour. Pick the flowers for use in salads as required. Again the best flavour is before the nectar has been collected, which is when the flower is at its palest.

container growing

This is not a plant that springs to mind as a good pot plant, certainly not indoors. But with patience, it makes a lovely mop head standard if carefully staked and trained; use an evergreen variety like *Lonicera japonica*. The compost should be a soil-based one. Water and feed regularly throughout the summer and in winter keep in a cold frame or greenhouse and water only occasionally.

warning

The berries are poisonous. Large doses cause vomiting.

Honeysuckle tea

culinary

Add flowers to salads.

medicinal

An infusion of the heavy perfumed flowers can be taken as a substitute for tea. It is also useful for treating coughs, catarrh and asthma. As a lotion it is good for skin infections. Recent research has proved that this plant has an outstanding curative action in cases of colitis.

other uses

Flowers are strongly scented for potpourris, herb pillows and perfumery. An essential oil was once extracted from the plant to make a very sweet perfume but the yield was extremely low.

Honeysuckle flowers

Malva

MALLOW

Also known as High Mallow, Country Mallow, Billy Buttons, Pancake Plant and Cheese Flower. From the family *Malvaceae.*

Native to Europe, Western Asia and North America, it can be found growing in hedgebanks, field edges, and on road sides and wastelands in sunny situations.

The ancient Latin name given to this herb by Pliny was *malacho*, which was probably derived from *malachi*, the Greek word meaning 'to soften', after the mallow's softening and healing properties. Young mallow shoots were eaten as vegetables, and it was still to be found on vegetable lists in Roman times. Used in the Middle Ages for its calming effect as an antidote to aphrodisiacs and love-potions. The shape of its seed rather than its flowers suggested the folk name.

varieties

Malva sylvestris
Common Mallow
Biennial. Ht 45–90cm, spread 60cm. Flower, dark pink or violet form, early summer to autumn. Mid-green leaves, rounded at the base, ivy shaped at stem.

Malva moschata
Musk Mallow
Perennial. Ht 30–80cm, spread 60cm. Rose/pink flowers, late summer to early autumn. Mid-green leaves – kidney-shaped at base, deeply divided at stem – emit musky aroma in warm weather or when pressed.

Malva rotundifolia
Dwarf Mallow
Also known as cheese plant, low mallow and blue mallow. Annual. Ht 15–30cm, a creeper. Purplish-pink, trumpet-shaped flowers from early summer to mid-autumn. Leaves rounded, slightly lobed and greenish. North American native.

cultivation

Propagation

Seed
Sow in prepared seed or plug trays in the autumn. Cover lightly with compost (not perlite). Winter outside, covered with glass. Germination is erratic but should take place in the spring. Plant out seedlings when large enough to handle, 60cm apart.

Cuttings
Take cuttings from firm basal shoots in late spring or summer. When hardened off the following spring, plant out 60cm apart into a prepared site.

Pests and Diseases
Mallows can catch hollyhock rust. There is also a fungus that produces leaf spots and a serious black canker on the stems. If this occurs dig up the plants and destroy them. This is a seed-borne fungus and may be carried into the soil, so change planting site each season.

Maintenance
Spring Take softwood cuttings from young shoots. Sow seeds in the garden.
Summer Trim after flowering. Take cuttings.
Autumn Sow seed under protection.
Winter Hardy enough.

Garden Cultivation
Mallows are very tolerant of site, but prefer a well-drained and fertile soil (if too damp they may well need staking in summer), and a sunny position (though semi-shade will do). Sow where it is to flower from late summer to spring. Press gently into the soil, 60cm apart, and cover with a light compost. Cut back stems after flowering, not only to promote new growth, but also to keep under control and encourage a second flowering. Cut down the stems in autumn.

Harvesting
Harvest young leaves for fresh use as required throughout the spring. For use in potpourris, gather for drying in the summer after first flowering.

container growing

Malva moschata is the variety to grow in a large container. It can look very dramatic and smells lovely on a warm evening. Water well in the growing season, but feed only twice. Maintain as for garden cultivation.

medicinal

Marshmallow *(Althaea officinalis)* is used in preference to the mallows *(Malva* subsp.*)* in herbal medicine. However, a decoction can be used in a compress, or in bath preparations, for skin rashes, boils and ulcers, and in gargles and mouth washes.

culinary

Young tender tips of the common mallow may be used in salads or steamed as a vegetable. Young leaves of the musk mallow can be boiled as a vegetable.

Young leaves of the dwarf mallow can be eaten raw in salads or cooked as a spinach.

Mandragora

MANDRAKE

Mandragora autumnalis

Also known as Devil's Apples, Satan's Apples. From the family *Solanaceae*.

Mandrake is indigenous to the Mediterranean. It is steeped in myth and magic. Historically it is recorded that in 200 BC the Carthaginians created a bait for the Romans who were invading their city, made from wine spiked with mandrake juice. This rendered the Romans insensible and the city was saved. It has been used in witchcraft to make barren women fertile and it is also mentioned in the Bible, Genesis 30, where Rachel, the second wife of Jacob becomes fertile after her sister found mandrake growing in a field.

varieties

☠ *Mandragora officinarum*

Mandrake, Satan's Apple

Hardy herbaceous, perennial. Ht 5cm, spread 30cm. Small bell-shaped, very pale blue flowers in spring, followed by aromatic round, yellow, toxic fruits in late summer. Large, up to 30cm long, oblong rough green leaves.

☠ *Mandragora autumnalis*

Hardy herbaceous, perennial. Height 5cm and spread 30cm. Small bell-shaped pale blue/violet flowers in autumn, followed in the third year by aromatic round, yellow, toxic fruits in late summer. Large, oblong, hairy dark green leaves.

cultivation

Propagation

Seed

Sow the fresh seeds *M. officinarum* in autumn and *M. autumnalis* in spring, into prepared plug modules, using a soil-based seed compost mixed in equal parts with coarse horticultural sand. Cover with coarse sand and place in a cold frame. Germination can take anything from 4 months to 2 years. Once the seedling is large enough to handle, pot up and grow on for two years prior to planting in the garden.

Cuttings

In the early spring (autumn for *M. autumnalis*) when you can see the new shoots appearing, clear around the plant to expose the crown and the top of the roots. Choose a side with new shoot appearing and, using a sharp knife, slice the root including a new shoot. Place the cutting into a small pot filled with a soil-based potting compost mixed in equal parts with coarse horticultural sand. Place the container in a warm position until rooted. Grow on for two years prior to planting into a prepared site.

Pests and Diseases

Slugs can decimate a plant overnight. So check regularly at night and remove any that are found.

Mandrake root

Maintenance

Spring Sow seeds and cuttings of *Mandragora autumnalis.*

Summer Check watering.

Autumn Sow seeds and take cuttings of *Mandragora officinarum.*

Winter Protect from excessive wet.

Garden Cultivation

Plant in a well-drained, deep, fertile soil in partial shade. Because mandrake has a very long tap root, up to 1.2m, it dislikes being transplanted, so choose your site carefully. In summer the plant will become dormant if the soil dries out.

Harvesting

Dig up the roots in autumn or spring when the plant is dormant for medicinal use.

container growing

Mandrake's long tap root makes it not particularly suitable for growing in containers. It is possible, but the plant will be smaller and rarely set fruit. Use a soil-based potting compost that has been mixed in equal parts with coarse horticultural sand.

medicinal

Both the species mentioned have the same medicinal properties. Today, mandrake is rarely used in herbal medicine. However, tinctures made from the root are still widely used in homeopathy as a treatment for asthma and coughs.

warning

The whole plant, including the root, is toxic. Only take under professional supervision. Not to be taken in any form during pregnancy.

Marrubium vulgare

WHITE HOREHOUND

Also known as Horehound and Maribeum. From the family *Lamiaceae*.

Common throughout Europe and America, the plant grows wild everywhere from coastal to mountainous areas.

The botanical name comes from the Hebrew *marrob*, which translates as 'bitter juice'. The common name is derived from the old English *har hune*, meaning 'a downy plant'.

varieties

Marrubium vulgare
White Horehound
Hardy perennial. Ht 45cm, spread 30cm. Small clusters of white flowers from the second year in mid-summer. The leaves are green and wrinkled with an underside of a silver woolly texture. There is also a variegated version.

cultivation

Propagation

Seed
The fairly small seed should be sown in early spring in a seed or plug tray, using a standard seed compost mixed in equal parts with composted fine bark. Germination takes 2–3 weeks. Prick out into pots or transplant to the garden after a good period of hardening off.

Cuttings
Softwood cuttings taken from the new growth in summer usually root within 3–4 weeks. Use a standard seed compost mixed in equal parts with composted fine bark. Winter under protection in a cold frame or cold greenhouse.

Division
Established clumps benefit from division in the spring.

Pests and Diseases

If it is very wet and cold in winter, the plant can rot off.

Maintenance

Spring Divide established clumps. Prune new growth to maintain shape. Sow seed.
Summer Trim after flowering to stop the plant flopping and prevent self-seeding. Take cuttings.
Autumn Divide only if it has dangerously transgressed its limits.
Winter Protect only if season excessively wet.

Garden Cultivation

White horehound grows best in well-drained, dryish soil, biased to alkaline, sunny and protected from high winds. Seed can be sown direct into a prepared garden in late spring, once the soil has started to warm up. Thin the seedlings to 30cm distance apart.

Harvesting

The leaves and flowering tops are gathered in the spring, just as the plants come into flower, when the essential oil is at its richest. Use fresh or dried.

container growing

Horehound can be grown in a large container situated in a sunny position. Use a compost that drains well and do not over water. Only feed after flowering, otherwise it produces lush growth that is too soft.

other uses

Traditionally an infusion of the leaf was used as a spray for cankerworm in trees. Also the infusion was mixed with milk and put in a dish as a fly killer.

medicinal

White horehound is still extensively used in cough medicine, and for calming a nervous heart; its property, marrubiin, in small amounts, normalises an irregular heart beat. The plant has also been used to reduce fevers and treat malaria.

A Traditional Cold Cure

Finely chop 9 small horehound leaves. Mix 1 tablespoon of honey and eat slowly to ease a sore throat or cough. Repeat several times if necessary.

Traditional Cough Sweets

100g fresh white horehound leaves
1/2 teaspoon crushed aniseed
3 crushed cardamom seeds
350g white sugar
350g moist brown sugar

Put the horehound, aniseed and cardamom into 600ml of water and simmer for 20 minutes. Strain through a filter. Over a low heat, dissolve the sugars in the liquid; boil over a medium heat until the syrup hardens when drops are put into cold water. Pour into an oiled tray. Score when partially cooled. Break up and store in wax paper.

Melissa officinalis

LEMON BALM

Also known as Balm, Melissa, Balm Mint, Bee Balm, Blue Balm, Cure All, Dropsy Plant, Garden Balm and Sweet Balm. From the family *Lamiaceae*.

This plant is a native of the Mediterranean region and Central Europe. It is now naturalised in North America and as a garden escapee in Britain.

This ancient herb was dedicated to the goddess Diana, and used medicinally by the Greeks some 2,000 years ago. The generic name, *Melissa*, comes from the Greek word for bee and the Greek belief that if you put sprigs of balm in an empty hive it would attract a swarm; equally, if planted near bees in residence in a hive they would never go away. This belief was still prevalent in medieval times when sugar was highly priced and honey a luxury.

In the Middle Ages lemon balm was used to soothe tension, to dress wounds, as a cure for toothache, mad dog bites, skin eruptions, crooked necks and sickness during pregnancy. It was even said to prevent baldness, and ladies made linen or silk amulets filled with lemon balm as a lucky love charm. It has been acclaimed the world over for promoting long life. Prince Llewellyn of Glamorgan drank melissa tea, so he claimed, every day of the 108 years of his life. Wild claims apart, as a tonic for melancholy it has been praised by herbal writers for centuries and is still used today in aromatherapy to counter depression.

Melissa officinalis 'Aurea'

varieties

Melissa officinalis
Lemon Balm
Hardy perennial. Ht 75cm, spread 45cm or more. Clusters of small, pale yellow/white flowers in summer. The green leaves are oval, toothed, slightly wrinkled and highly aromatic when crushed.

***Melissa officinalis* 'All Gold'**
Golden Lemon Balm
Half-hardy perennial. Ht 60cm, spread 30cm or more. Clusters of small, pale yellow/white flowers in summer. The leaves are all yellow, oval in shape, toothed, slightly wrinkled and aromatic with a lemon scent when crushed. The leaves are prone to scorching in high summer. More tender than the other varieties.

***Melissa officinalis* 'Aurea'**
Variegated Lemon Balm
Hardy perennial. Ht 60cm, spread 30cm or more. Clusters of small, pale yellow/white flowers in summer. The green/gold variegated leaves are oval, toothed, slightly wrinkled and aromatic with a lemon scent when crushed. This variety is as hardy as common lemon balm. The one problem is that in high season it reverts to green. To maintain variegation keep cutting back; this in turn will promote new growth, which should be variegated.

cultivation

Propagation
Seed
Common lemon balm can be grown from seed. The seed is small but manageable, and it is better to start

LEMON BALM

it off under protection. Sow in prepared seed or plug trays in early spring, using a standard seed compost mixed in equal parts with composted fine bark, and cover with perlite. Germination takes between 10 and 14 days. The seeds dislike being wet so, after the initial watering, try not to water again until germination starts. When seedlings are large enough to handle, prick out and plant in the garden, 45cm apart.

Cuttings
The variegated and golden lemon balm can only be propagated by cuttings or division. Take softwood cuttings from the new growth in late spring/early summer. As the cutting material will be very soft, take extra care when collecting it.

Division
The rootstock is easy to divide (autumn or spring). Replant directly into the garden in a prepared site.

Pests and Diseases
The only problem likely to affect lemon balm is a form of fungus similar to mint rust; cut the plant back to the ground and dispose of all the infected leaves, including any that may have accidentally fallen on the ground.

Maintenance
Spring Sow seeds. Divide established plants. Take cuttings.
Summer Take cuttings. Keep trimming established plants. Cut back after flowering to reduce self-seeding.
Autumn Divide established plants, or any that may have encroached on other plant areas.
Winter Protect plants if the temperature falls below −5°C. The plant dies back, leaving but a small presence on the surface of the soil. Protect with a bark or straw mulch or agricultural fleece.

Garden Cultivation
Lemon balm will grow in almost any soil and in any position. It does prefer a fairly rich, moist soil in a sunny position with some midday shade. Keep all plants trimmed around the edges to restrict growth and encourage fresh shoots. In the right soil conditions this can be a very invasive plant. Unlike horseradish, the established roots are not difficult to uproot if things get out of hand.

Harvesting
Pick leaves throughout the summer for fresh use. For drying, pick just before the flowers begin to open when flavour is best; handle gently to avoid bruising. The aroma is rapidly lost, together with much of its therapeutic value, when dried or stored.

container growing

If you live in an area that suffers from very cold winters, the gold form would benefit from being grown in containers. This method suits those with a small garden who do not want a takeover bid from lemon balm. Use a soil-based compost mixed in equal parts with composted fine bark. Only feed with liquid fertiliser in the summer, otherwise the growth will become too lush and soft, and aroma and colour will diminish. Water normally throughout the growing season. Allow the container to become very dry (but not totally) in winter, and keep the pots in a cool, protected environment.

medicinal

Lemon balm tea is said to relieve headaches and tension and to restore the memory. It is also good after meals to ease the digestion, flatulence and colic. Use fresh or frozen leaves in infusions because the volatile oil tends to disappear during the drying process. The isolated oil used in aromatherapy is recommended for nervousness, depression, insomnia and nervous headaches. It also helps eczema sufferers.

other uses

This is a most useful plant to keep bees happy. The flower may look boring to you but it is sheer heaven to them. So plant lemon balm around beehives or orchards to attract pollinating bees.

culinary

Lemon balm is one of those herbs that smells delicious but tastes like school-boiled cabbage water when cooked.

Add fresh leaves to vinegar. Add leaves to wine cups, teas and beers, or use chopped with fish and mushroom dishes. Mix freshly chopped with soft cheeses.

It has frequently been incorporated in proprietary cordials for liqueurs and its popularity in France led to its name 'Tea de France'. It is used as a flavouring for certain cheeses in parts of Switzerland.

Lemon balm with cream cheese

Mentha

MINT

From the family *Lamiaceae*.

The *Mentha* family is a native of Europe that has naturalised in many parts of the world, including North America, Australia and Japan.

Mint has been cultivated for its medicinal properties since ancient times and has been found in Egyptian tombs dating back to 1,000 BC. The Japanese have been growing it to obtain menthol for at least 2,000 years. In the Bible the Pharisees collected tithes in mint, dill and cumin. Charlemagne, who was very keen on herbs, ordered people to grow it. The Romans brought it with them as they marched through Europe and into Britain, from where it found its way to America with the settlers.

Its name was first used in Greek mythology. There are two different stories, the first that the nymph Minthe was being chatted up by Hades, god of the Underworld. His queen Sephony became jealous and turned her into the plant mint. The second that Minthe was a nymph beloved by Pluto, who transformed her into the scented herb after his jealous wife took umbrage.

Mentha longifolia Buddleia Mint Group

Mentha x *piperita* f. *citrata*

varieties

The mint family is large and well known. I have chosen a few to illustrate the diversity of the species.

Mentha aquatica
Water Mint
Hardy perennial. Ht 15–60cm, spread indefinite. Pretty purple/lilac flowers borne all summer. Leaves soft, slightly downy, mid green in colour. The scent can vary from a musty mint to a strong peppermint. This should be planted in water or very wet marshy soil. It can be found growing wild around ponds and streams.

Mentha arvensis* var. *piperascens
Japanese Peppermint
Hardy perennial. Ht 60cm–1m, spread 60cm and more. Loose purplish whorls of flowers in summer. Leaves, downy, oblong, sharply toothed and green-grey; they provide an oil (90 per cent menthol), said to be inferior to the oil produced by *M.* x *piperita*. This species is known as English mint in Japan.

Mentha* x *gracilis* syn. *Mentha* x *gentilis
Ginger Mint
Also known as Scotch mint. Hardy perennial. Ht 45cm, spread 60cm. The stem has whorls of small, 2-lipped, mauve flowers in summer. The leaf is variegated, gold/green with serrated edges. The flavour is a delicate, warm mint that combines well in salads and tomato dishes.

MINT

Mentha longifolia Buddleia Mint Group
Buddleia Mint
Hardy perennial. Ht 80cm, spread indefinite. Long purple/mauve flowers that look very like buddleia (hence its name). Long grey-green leaves with a musty minty scent. Very good plant for garden borders.

Mentha longifolia subsp. _schimperi_
Desert Mint, Eastern Mint
Hardy perennial. Ht 60–80cm, spread indefinite. Long pale mauve flowers in summer. Long narrow grey-green leaves with a highly pepperminted scent and flavour used to make tea. Plant in free-draining soil.

Mentha x piperita
Peppermint, White Peppermint
Also known as mentha d'Angleterre, mentha Anglais, pfefferminze and Englisheminze. Hardy perennial. Ht 30–60cm, spread indefinite. Pale purple flowers in summer. Pointed leaves, darkish green with a reddish tinge, serrated edges. Very peppermint scented. This is the main medicinal herb of the genus. There are two species worth looking out for – *M* x *piperita* black peppermint, with leaves much darker, nearly brown, and *M* x *piperita* 'Black Mitcham' with dark brown, tinged with reddish-brown leaves.

Mentha x piperita f. _citrata_
Eau de Cologne Mint
Also known as orange mint and bergamot mint. Hardy perennial. Ht 60–80cm, spread indefinite. Purple/mauve flowers in summer. Purple-tinged, roundish, dark green leaves. A delicious scent that has been described as lemon, orange, bergamot, lavender, as well as eau de cologne. This plant is a vigorous grower. Use in fruit dishes with discretion. Best use is in the bath.

Mentha spicata var. *crispa*

Mentha x piperita f. _citrata_ 'Chocolate'
Chocolate Peppermint
Hardy perennial. Ht 40–60cm, spread indefinite. Pale purple flowers in summer. Pointed dark green-brown leaves with serrated edges. Very peppermint scented with deep chocolate undertones. Great in puddings.

Mentha x piperita f. _citrata_ 'Lemon'
Lemon Mint
Hardy perennial. Ht 45–60cm, spread indefinite. Purple whorl of flowers in summer. Green serrated leaf, refreshing minty lemon scent. Good as a mint sauce, or with fruit dishes.

Mentha requienii
Corsican Mint
Also known as rock mint. Hardy semi-evergreen perennial. Ground cover, spread indefinite. Tiny purple flowers throughout the summer. Tiny bright green leaves, which, when crushed, smell strongly of peppermint. Suits a rock garden or paved path, grows naturally in cracks of rocks. Needs shade and moist soil.

Mentha spicata
Spearmint
Also known as garden mint and common mint. Hardy perennial. Ht 45–60cm, spread indefinite. Purple/mauve flowers in summer. Green pointed leaves with serrated edges. The most widely grown of all mints. Good for mint sauce, mint jelly, mint julep.

Mentha spicata var. _crispa_
Curly Mint
Hardy perennial. Ht 45–60cm, spread indefinite. Light mauve flowers in spring. When I first saw this mint I thought it had a bad attack of aphids, but it has grown on me! The leaf is bright green and crinkled, its serrated edge slightly frilly. Flavour very similar to spearmint, so good in most culinary dishes.

Mentha longifolia subsp. *schimperi*

Mentha spicata var. _crispa_ 'Moroccan'
Moroccan Mint
Hardy perennial. Ht 45–60cm, spread indefinite. White flowers in summer. Bright green leaves with a texture and excellent mint scent. This is the one I use for all the basic mint uses in the kitchen. A clean mint flavour and scent, lovely when served with yogurt and cucumber.

Mentha suaveolens
Apple Mint
Hardy perennial. Ht 60cm–1m, spread indefinite. Mauve flowers in summer. Roundish hairy leaves. Tall vigorous grower. Gets its name from its scent, which is a combination of mint and apples. More subtle than some mints, so good in cooking.

Mentha suaveolens 'Variegata'
Pineapple Mint
Hardy perennial. Ht 45–60cm, spread indefinite. Seldom produces flowers, all the energy going into producing very pretty cream and green, slightly hairy leaves that look good in the garden. Not a rampant mint. Grows well in hanging baskets.

Mentha suaveolens 'Variegata'

Mentha x *piperita* f. *citrata* 'Chocolate'

Mentha* x *villosa* var. *alopecuroides
Bowles' mint
Bowles' Mint
Hardy perennial. Ht 60cm–1m, spread indefinite. Mauve flowers, round, slightly hairy green leaves, vigorous grower. Sometimes incorrectly called apple mint. Has acquired reputation as the 'connoisseur's culinary mint'. Not sure I agree, but mint tastes do vary.

Pycnanthemum pilosum
Mountain Mint
Hardy perennial. Ht 90cm, spread 60cm. Knot-like white/pink flowers, small and pretty in summer. Leaves long, thin, pointed and grey-green with a good mint scent and flavour. Not a *Mentha*, so therefore not a true mint, and does not spread. Looks very attractive in a border, and is also appealing to butterflies. Any soil will support it provided it is not too rich.

cultivation

Propagation
Seed
The seed on the market is not worthwhile – leaf flavour is inferior and quite often it does not run true to species.

Cuttings
Root cuttings of mint are very easy. Simply dig up a piece of root. Cut it where you can see a little growing node (each piece will produce a plant) and place the cuttings either into a plug or seed tray. Push them into the compost (a standard seed compost mixed in equal parts with composted fine bark). Water in and leave. This can be done any time during the growing season. If taken in spring, in about 2 weeks you should see new shoots emerging through the compost.

Division
Dig up plants every few years and divide, or they will produce root runners all over the place. Each bit of root will grow, so take care.

Corsican mint does not set root runners. Dig up a section in spring and divide by easing the plant apart and replanting.

Pests and Diseases
Mint rust appears as little rusty spots on the leaves. Remove them immediately, otherwise the rust will wash off into the soil and the spores spread to other plants. One sure way to be rid is to burn the affected patch. This effectively sterilises the ground.

Another method, which I found in an old gardening book, is to dig up the roots in winter when the plants are dormant, and clean off the soil under a tap. Heat some water to a temperature of 40–44°C and pour into a bowl. Place the roots in the water for 10 minutes. Remove the runners and wash at once in cold water. Replant in the garden well away from the original site.

Maintenance
Spring Dig up root if cuttings are required. Split established plants if need be.
Summer Give plants a hair cut to promote new growth. Control the spread of unruly plants.
Autumn Dig up roots for forcing. Bring in containers.
Winter Sterilize the roots if rust is evident during the growing season.

Garden Cultivation
Mint is one of those plants that will walk all over the plot if not severely controlled. Also, mint readily hybridises itself, varying according to environmental factors.

If choosing a plant in a nursery or garden centre rub the leaf first to check the scent. Select a planting site in sun or shade but away from other mints. Planted side by side they seem to lose their individual scent and flavour.

To inhibit spread, sink a large bottomless container (bucket or bespoke frame) in a well-drained and fairly rich soil to a depth of at least 30cm, leaving a small ridge above soil level. Plant the mint in the centre.

Harvesting
Pick the leaves for fresh use throughout the growing season. Pick leaves for drying or freezing before the mint flowers.

companion planting

Spearmint or peppermint planted near roses may deter aphids. Buddleia mint will attract hoverflies, which are predators of greenfly.

container growing

Mint is good in containers. Make sure the container is large enough, use a soil-based compost mixed in equal parts with composted fine bark, and do not let the compost dry out. Feed regularly throughout the growing season with a liquid fertiliser. Place the container in semi-shade.

Forcing
One good reason for growing mint in containers is to prolong the season. This is called forcing. In early autumn dig up some root. Fill a container, or wooden box lined with plastic, with compost. Lay the root down its length and cover lightly with compost. Water in and place in a light, warm glasshouse or warm conservatory (even the kitchen windowsill will do). Keep an eye on it, and fresh shoots should sprout within a couple of weeks. This is great for fresh mint sauce out of season.

MINT

Fresh mint tea

other uses

Pick a bunch of eau de cologne mint, tie it up with string, and hang it under the hot water tap when you are drawing a bath. You will scent not only your bath, but the whole house. It is very uplifting (unless you too have a young son, who for some reason thinks it is 'gross').

medicinal

Peppermint is aromatic, calmative, antiseptic, anti-spasmodic, anti-inflammatory, anti-bacterial, anti-parasitic, and is also a stimulant. It can be used in a number of ways for a variety of complaints including gastro-intestinal disorders where anti-spasmodic, anti-flatulent and appetite-promoting stimulation is required. It is particularly useful for nervous headaches, and as a way to increase concentration. Externally, peppermint oil can be used in a massage to relieve muscular pain.

warning

The oil may cause an allergic reaction. Avoid prolonged intake of inhalants from the oil, which must never be used by babies.

culinary

With due respect to their cuisine, the French are always rude about our 'mint sauce with lamb'; they reckon it is barbaric. On the other side of the Channel they use mint less than other countries in cooking. But slowly, even in France, this herb is gaining favour. Mint is good in vinegars and jellies. Peppermint makes a great tea. And there are many uses for mint in cooking with fish, meat, yogurt, fruit, and so on. Here is a recipe for chocoholics like me:

Chocolate Mint Mousse

Serves 2

100g plain dark chocolate
2 eggs, separated
1 teaspoon instant coffee
1 teaspoon fresh chopped mint, either Moroccan, spearmint or curly
Whipped cream for decoration
4 whole mint leaves

Melt the chocolate either in a microwave, or in a double saucepan. When smooth and liquid, remove from heat. Beat egg yolks and add to the chocolate while hot (this will cook the yolks slightly). Add coffee and chopped mint. Leave the mixture to cool for about 15 minutes. Beat the egg whites (not too stiff) and fold them into the cooling chocolate mixture. Spoon into containers. When you are ready to serve put a blob of whipped cream in the middle and garnish with whole leaves.

Mentha pulegium

PENNYROYAL

Also known as European Pennyroyal and Pudding Grass. From the family *Lamiaceae*.

Pulegium is derived from the Latin *pulex*, meaning 'flea' because both the fresh plant and the smoke from the burning leaves were used to exterminate the insect. Long ago, the wise women of the village used pennyroyal to induce abortion. It has since been medicinally used to facilitate menstruation. In the seventeenth century the herbalist John Gerard called it pudding grass. He claimed it would purify corrupt water on sea voyages.

Mentha pulegium 'Upright'

varieties

Mentha pulegium
Pennyroyal
Also known as Cunningham mint and creeping Pennyroyal. Semi-evergreen hardy perennial. Ht 15cm, spread indefinite. Clusters of mauve flowers in summer. Bright green, oval, highly peppermint scented leaves.

***Mentha pulegium* 'Upright'**
Pennyroyal Upright
Semi-evergreen hardy perennial. Ht 30cm, spread indefinite. Clusters of mauve flowers in summer. Bright green, oval, highly peppermint scented leaves.

The American pennyroyal or rock pennyroyal *(Hedeoma pulegioides)* has a similar aroma and usage as the European species.

cultivation

Propagation

Seed
Sow the very fine seed in late spring, using the cardboard technique (see page 645), into prepared plug module trays using a standard seed compost. Cover with perlite. Place under protection at 18°C; germination takes 10–20 days. Leave in the modules to become well established before planting out in early summer 30cm apart.

Cuttings
This plant roots where it touches the ground. In spring dig up a plant, divide it into small sections including some root, and put into prepared plug modules or small pots using a standard seed compost mixed in equal parts with composted fine bark.

Division
Established or invasive plants can be divided in spring and early autumn.

Pests and Diseases
Pennyroyal likes a well-drained soil; when grown in damp conditions it can suffer from mildew. Remove infected plants. If grown in a container cut back on the watering.

Maintenance
Spring Sow seed, divide established plants.
Summer Cut back after flowering.
Autumn Divide established plants. Dig up a plant as insurance against a wet winter and winter in a cold frame or greenhouse.
Winter This mint hates wet winters, it will stand temperatures below –8°C only in a well-drained soil.

Garden Cultivation
Plant in a well-drained fertile soil in a sunny position. This may seem a contradiction, but it does like to be watered well in summer.

Harvesting
Pick leaves as required to use fresh. Pick either side of flowering for freezing. Not worth drying.

container growing

Both species can be happily grown in containers. Use a soil-based compost mixed in equal parts with composted fine bark.

culinary

This mint has a very strong peppermint flavour so use sparingly. It is a good substitute for peppermint in water ice.

medicinal

A hot infusion from the leaves of this plant is an old-fashioned remedy for colds and head colds. The essential oil from this plant is highly toxic and should only be used by professionals, as if incorrectly self-administered it can cause irreversible kidney damage.

other uses

The fresh leaves rubbed onto bare skin is an excellent insect and mosquito repellent. Equally, if you have been bitten by a horsefly or mosquito and rub the fresh leaves onto the bite it alleviates the itching.

warning

Not to be medicinally used in pregnancy or if suffering from kidney disease. May cause contact dermatitis.

Monarda

BERGAMOT

Monarda citriodora

Also known as Oswego Tea, Bee Balm, Blue Balm, High Balm, Low Balm, Mountain Balm and Mountain Mint. From the family *Lamiaceae*.

This beautiful plant with its flamboyant flower is a native of North America and is now grown horticulturally in many countries throughout the world.

The species name *Monarda* honours the Spanish medicinal botanist Dr Nicholas Monardes of Seville who, in 1569, wrote a herbal on the flora of America. The common name, bergamot, is said to have come from the scent of the crushed leaf which resembles the small, bitter Italian Bergamot orange *(Citrus bergamia)*, from which oil is produced that is used in aromatherapy, perfumes and cosmetics.

The wild or purple bergamot *(Monarda fistulosa)* grows around the Oswego river district near Lake Ontario in the United States. The Indians in this region used it for colds and bronchial complaints as it contains the powerful antiseptic, thymol. They also made tea from it, hence Oswego tea, which was drunk in many American households, replacing Indian tea, following the Boston Tea Party of 1773.

varieties

There are too many species and cultivars of bergamot to mention here, so I have included a short selection:

***Monarda* 'Beauty of Cobham' AGM**
Bergamot Beauty of Cobham
Hardy perennial. Ht 75cm, spread 45cm. Attractive dense 2-lipped pale pink flowers throughout summer. Toothed mid-green aromatic leaves.

***Monarda* 'Blaustrumpf'**
Bergamot Blue Stocking
Hardy perennial. Ht 80cm, spread 45cm. Attractive purple flowers throughout summer. Aromatic, green, pointed foliage.

***Monarda* 'Cambridge Scarlet' AGM**
Bergamot Cambridge Scarlet
Hardy perennial. Ht 1m, spread 45cm. Striking rich red flowers all summer. Aromatic mid-green leaves.

Monarda didyma

Monarda citriodora
Lemon bergamot; Wild Bee Balm
Hardy annual. Ht 60cm, spread 45cm. Beautiful mauve surrounded by pale mauve bracts. Intense lemon-scented leaves.

***Monarda* 'Croftway Pink' AGM**
Bergamot Croftway Pink
Hardy perennial. Ht 1m, spread 45cm. Soft pink flowers throughout summer. Aromatic green leaves.

Monarda didyma
Bergamot (Bee Balm Red)
Hardy perennial. Ht 80cm, spread 45cm. Fantastic red flowers throughout summer. Aromatic, mid-green foliage.

***Monarda* 'Schneewittchen'**
Bergamot Snow Maiden
Hardy perennial. Ht 80cm, spread 45cm. Very attractive white flowers throughout summer. Aromatic, mid-green, pointed leaves.

Monarda 'Cambridge Scarlet'

cultivation

Propagation

Seed

Only species will grow true from seed. Cultivars (i.e. named varieties) will not.

Sow the very small seed indoors in the spring on the surface of either seed or plug trays or on individual pots, using a standard seed compost mixed in equal parts with composted fine bark. Cover with perlite. Germination is better with added warmth: 21°C. Thin or transplant the strongest seedlings when large enough to handle. Harden off. Plant in the garden at a distance of 45cm apart.

Cuttings

Take first shoots in early summer, as soon as they are 7.5–10cm long.

Division

Divide in early spring. Either grow on in pots, or replant in the garden, making sure the site is well prepared with well-rotted compost. Planting distance from other plants 45cm.

Pests and Diseases

Bergamot is prone to powdery mildew. At the first sign, remove leaves. If it gets out of hand, cut the plant back to ground level.

Young plants are a *bonne bouche* for slugs!

Maintenance

Spring Sow seeds of species. Divide roots. Dig up 3-year-old plants, divide and replant.
Summer Take cuttings of cultivars and species, when you need them.
Autumn Cut back to the ground, and give a good feed with manure or compost.
Winter All perennial bergamots die right back in winter. In hard winters protect with a mulch.

Garden Cultivation

Bergamot is a highly decorative plant with long-lasting, distinctively fragrant flowers that are very attractive to bees, hence the country name bee balm. All grow well in moist, nutrient-rich soil, preferably in a semi-shady spot; deciduous woodland is ideal. However, they will tolerate full sun provided the soil retains moisture. Like many other perennials, bergamot should be dug up and divided every three years, and the dead centre discarded.

Harvesting

Pick leaves as desired for use fresh in the kitchen. For drying, harvest before the flower opens.

Cut flowers for drying as soon as they are fully opened. They will then dry beautifully and retain their colour.

It is only worth collecting seed if you have species plants situated well apart in the garden. If near one another, cross-pollination will make the seed variable – very jolly provided you don't mind unpredictably mixed colours. Collect the flower heads when they turn brown.

container growing

Bergamot is too tall for a window box, but it can look very attractive growing in a large pot, say 35–45cm across, or tub as long as the soil can be kept moist and the plant be given some afternoon shade.

other uses

Because the dried bergamot flowers keep their fragrance and colour so well, they are an important ingredient in potpourris.

The oil is sometimes used in perfumes, but should not be confused with the similarly smelling bergamot orange.

medicinal

Excellent herb tea to relieve nausea, flatulence, menstrual pain and vomiting.

Aromatherapists have found bergamot oil good for depression, as well as helping the body to fight infections.

culinary

Pick the small flower petals separately and scatter over a green salad at the last moment. Put fresh leaf into China tea for an Earl Grey flavour, and into wine cups and lemonade. The chopped leaves can be added sparingly to salads and stuffings, and can also be used to flavour jams and jellies.

Pork Fillets with Bergamot Sauce

Serves 2

2 large pork fillets
75g butter
2 shallots, very finely chopped
40g flour
4 tablespoons chicken or vegetable stock
4 tablespoons dry white wine
3½ tablespoons chopped bergamot leaves
Salt and freshly ground black pepper
1 tablespoon double cream

Preheat the oven to 200°C/400°F/gas mark 6.

Wash the pork. Pat dry, season and smear with half the butter. Roast in a shallow greased tin for 25 minutes. Allow to rest for 5 minutes before slicing. Arrange slices in a warmed serving dish.

Prepare this sauce while the fillets are in the oven. Sweat the shallots in the remaining butter until soft. Stir in the flour and cook for about a minute, stirring all the time. Whisk in the stock. Simmer until it thickens, stirring occasionally. Then slowly add the wine and 3 tablespoons of the chopped bergamot. Simmer for several minutes then season to taste. Remove from heat, stir in the cream, pour over arranged pork slices and garnish with remaining chopped bergamot. Serve with mashed potato, and fresh green vegetables such as broccoli.

Murraya koenigii

CURRY TREE

Also known as Indian Bay, Nim Leaves, Kahdi Patta and Karapincha. From the family *Rutaceae*.

The curry leaf trees are indigenous to India, and they have naturalised in forests and wasteland throughout the subcontinent, except in the higher parts of the Himalayas. Curry leaves are closely associated with South India where the word 'curry' originates from the Tamil *kari* for spiced sauces. The use of curry leaves as a flavouring for vegetables is described in early Tamil literature dating back to the first century AD. These leaves are absolutely necessary for the authentic flavour of Southern Indian and Sri Lankan cuisine. Please do not confuse this herb with *Helichrysum italicum* (see page 304), which is known in Europe as the curry plant.

varieties

Murraya koenigii
Curry Leaf, Kahdi Patta
Tropical, subtropical, evergreen shrub or small tree. Ht 4.7–6m, spread up to 5m. Clusters of small creamy white sweetly scented, star-shaped flowers from late summer, followed by small black edible fruit, the seed of which is poisonous if digested. Oval, soft, glossy, pinnate aromatic leaves, each leaf divided into 11–21 leaflets. The bark is dark brown, nearly black, and the wood is greyish-white, hard and close grained.

cultivation

Propagation

Seed
Sow fresh ripe seed in autumn into prepared plug module trays using a seed compost mixed in equal parts with perlite. Cover the seeds with perlite, put in a warm place or propagator at 20°C. Germination is erratic, approximately 2–4 months.

Cuttings
Take semi-ripe stem cuttings in late spring or early summer. Place the cuttings into prepared plug module trays using a standard seed compost mixed in equal parts with perlite. If you do not have a covered propagator, either cover the cuttings in white plastic or place in a white plastic bag and put in a warm place. Check the bag regularly and do not allow the compost to dry out; equally do not over water. The cuttings should root in about one month. Once rooted, remove the bag or covering, grow on until fully rooted, then pot up into a small pot, using a soil-based potting compost mixed in equal parts with horticultural grit.

Division
Plants grown outside in the garden often set suckers. In summer, using a spade, these can be removed gently, including some root. Pot up into a pot that fits the size of root using a soil-based potting compost mixed in equal parts with horticultural grit. Replant in the garden after two seasons.

Pests and Diseases
Being an aromatic plant, it is often unaffected by pests. When grown as a pot plant it can be attacked by scale insects. Rub them off by hand, or use a horticultural soap spray following the manufacturer's instructions.

Maintenance
Spring Take cuttings. Increase the watering of pot-raised plants.
Summer Feed container-raised plants weekly. Take cuttings. Remove suckers.
Autumn Sow fresh seeds.
Winter In areas outside the tropics, cut back on watering, protect from frost.

Garden Cultivation
Only in the tropics or subtropics can this herb be grown in the garden. Plant in a fertile, light soil in partial shade. Young plants need to be watered daily for the first 3 years. In dry hot climates protect from the midday sun. This tropical plant will need protection when the night temperatures fall below 13°C.

Harvesting
Pick the fresh leaves for using in cooking and medicine as required. The leaves freeze well.

container growing

In cool climates this herb makes an ideal container

plant. It prefers to be pot bound, so do not over pot. Use a soil-based potting compost mixed in equal parts with horticultural grit. Place the container in partial shade. Water daily and feed weekly throughout the growing season. As the plants grow, keep trimming them regularly to maintain a supply of young leaves for cooking. Water very sparingly during the winter months and do not feed. Reintroduce water and feed as soon as the light levels and temperatures increase in the spring and move the plants to a warm, light place (around 18–20°C).

medicinal

Curry leaves have been used in both Ayurvedic and Hindu medicine for many cures. A paste is made from the leaves to cure eruptions and bites. Fresh leaves eaten raw are reputedly a good cure for dysentery, or drunk as an infusion it is said to stop vomiting. The traditional use of the curry leaf to treat diabetes has attracted a great deal of interest. Recent medicinal research has found special compounds which could be used to make an effective new medicine for treatment of diabetes.

other uses

The twigs from the branches make a good toothbrush. This is a very popular method for cleaning the teeth as it is said that it helps to strengthen the gums and the teeth. The hard wood is used to make agricultural tools.

warning

Seed is poisonous if ingested.

Murraya koenigii

culinary

Curry leaves are usually used fresh but can be found dried or in powder form. In some recipes the leaves are oven-dried, toasted immediately before use or quickly fried in butter or oil; this scented oil is then poured on top of many dishes to add richness and flavour. Equally, the leaves can be dropped into hot oil before adding the main ingredient. Since South Indian cuisine is dominantly vegetarian, curry leaves seldom appear in non-vegetarian food; the main applications are thin lentil or vegetable curries and stuffings for samosas. Because of their soft texture, they are not always removed before serving. Personally, I love the flavour of the leaves and have adapted many dishes to include them. This is a winter family favourite.

Curried Parsnip Soup
Serves 6

8 curry leaves plus 6 leaves for garnish
40g butter
1 tablespoon light olive oil
2 medium onions, chopped
2 cloves garlic, chopped
1 teaspoon ground ginger
1 teaspoon turmeric
700g parsnips, peeled, chopped into small pieces.
1.2 litres vegetable stock

Heat a non-stick frying pan and quickly fry 4 of the curry leaves, remove as soon as you can smell the wonderful aroma. Crush the leaves up. Return the frying pan to the heat, add the butter and oil. Once the butter has melted add the onions and cook gently until they become translucent. Add the garlic and fry for 2 minutes, then add the ginger, dried curry leaves, fresh curry leaves and turmeric, stir then add the parsnips. Toss them so they are covered in the wonderful spices and then transfer all this to a saucepan with the stock. Stir, bring to a very gentle simmer, cook for about 1 hour stirring from time to time. Remove from the heat, then liquidise or food process, return to the saucepan, gently warm, do not boil. Serve in soup bowls garnished with a curry leaf.

Myrrhis odorata

SWEET CICELY

Also known as Anise, Myrrh, Roman Plant, Sweet Bracken, Sweet Fern and Switch. From the family *Apiaceae*.

Sweet cicely was once cultivated as a pot shrub in Europe and is a native of this region and other temperate countries.

The Greeks called sweet cicely *seselis* or *seseli*. It is logical to suppose that 'Cicely' was derived from them, 'sweet' coming from its flavour.

In the sixteenth century John Gerard recommended the boiled roots as a pick-me-up for people who were 'dull'. According to Culpeper, the roots were thought to prevent infection by the plague. In South Wales, sweet cicely is quite often seen growing in graveyards, planted around the headstones to commemorate a loved one.

In the Lake District, sweet bracken (Cicely) was not only used in puddings but also for rubbing upon oak panels to make the wood shine and smell good.

Myrrhis odorata

varieties

Myrrhis odorata

Sweet Cicely

Hardy perennial. Ht 60–90cm, spread 60cm or more. The small white flowers appear in umbels from spring to early summer. The seeds are long, first green, turning black on ripening. The leaves are fern-like, very divided, and smell of aniseed when crushed.

The following plant is called sweet cicely in North America. It is unrelated to the European one, but used in a similar way:

Osmorhiza longistylis

Also known as anise root, sweet anise and sweet chervil. Perennial. Ht 45–90cm. Inconspicuous white flowers appear in loose compound umbels in summer. The leaves are oval to oblong and grow in groups of three. The whole plant has an aniseed odour. Its roots used to be nibbled by children for their anise liquorice flavour.

cultivation

Propagation

Seed

Sow the seed when ripe in early autumn. Use prepared plug or seed trays and, as the seed is so large, sow only one per plug and cover with compost. Then cover the trays with glass and leave outside for the whole winter.

The seed requires several months of cold winter temperatures to germinate. Keep a check on the compost, making sure it does not dry out. When germination starts bring the trays into a cold greenhouse. A spring sowing can be successful provided the seed is first put in a plastic bag mixed with a small amount of damp sharp sand, refrigerated for 4 weeks, and then sown as normal in prepared seed or plug trays. When the seedlings are large enough to handle, which is not long after germination, and after the frosts are over, transplant to a prepared site in the garden, 60cm apart.

Root Cuttings

The tap root may be lifted in spring or autumn, cut into sections each with a bud, and replanted either in prepared plug trays or direct into a prepared site in the garden at a depth of 5cm.

Division

Divide the plant in autumn when the top growth dies down.

Pests and Diseases

Sweet cicely is, in the majority of cases, free from pests and diseases.

Maintenance

Spring Take root cuttings.
Summer Cut back after flowering, to produce new leaves and to stop self-seeding.
Autumn Sow seeds. Divide established plants. Take root cuttings.
Winter No need for protection.

Garden Cultivation

It is one of the first garden herbs to emerge after winter and is almost the last to die down, and is therefore a most useful plant.

If you have a light well-drained poor soil you may find that Sweet cicely spreads all round the garden, and when you try to dig out established plants that the tap root is very long; even a tiny bit remaining will produce another plant. On the soil at my farm, which is heavy clay, it is a lovely, well-behaved plant, however, remaining just where it was planted in a totally controlled fashion.

The situation it likes best is a well-draining soil, rich in humus, and light shade. If the seed is not wanted for propagation or winter flavouring, the whole plant should be cut down immediately after flowering. A new batch of leaves will soon develop.

Sweet cicely is not suitable for growing in humid areas because it needs a good dormant period before winter to produce its root and lush foliage.

Harvesting

Pick young leaves at any time for fresh use. Collect unripe seeds when green; ripe seeds when dark brown. The foliage and seed do not dry or freeze, but the ripe seed stores well in a dry container.

Dig up roots for drying in autumn when the plant has died back.

container growing

As this herb has a very long tap root it does not grow happily in a container. But it can be done. Choose a container that will give the root plenty of room to grow, and use a soil-based compost mixed in equal parts with composted fine bark. Place it in a semi-shady place and keep it well watered throughout the growing season.

medicinal

This herb is now rarely used medicinally. The boiled root is said to be a tonic for both the teenager and the elderly alike.

Sweet cicely wine

culinary

The root can be cooked as a vegetable and served with butter or a white sauce, or allow to cool and chop up for use in salads. Alternatively, it can be eaten raw, or peeled and grated, and served in a French salad dressing. It is difficult to describe the flavour – think of parsnip, add a hint of aniseed. The root makes a very good wine.

Toss unripe seeds, which have a sweet flavour and a nutty texture, into fruit salads. Chop them into ice cream. Use ripe seeds whole for flavouring cooked dishes such as apple pie, otherwise use them crushed.

The leaf flavour is sweet aniseed. Chop finely and stir in salads, dressings and omelettes.
Add to soups, stews and to boiling water when cooking cabbage.

Add to cream for a sweeter, less fatty taste. It is a valuable sweetener, especially for diabetics, but also for the many people who are trying to reduce their sugar intake.

When cooking tart fruit, such as rhubarb, plums, gooseberries, red- or blackcurrants, add 4–6 teaspoons of chopped Sweet cicely leaf. Or, as I do sometimes, mix a handful of large fresh leaves with some lemon balm and add to the boiling water in which the fruit is to be stewed. It gives a delightful flavour and helps to save almost half the sugar needed.

Nepeta

CATMINT

Also known as Catnep, Catnip, Catrup, Catswart and Field Balm. From the family *Lamiaceae.*

The species name may have derived from the Roman town Nepeti, where it was said to grow in profusion.

The Elizabethan herbalist, Gerard, recorded the source of its common name: 'They do call it *herba cataria* and *herba catti* because cats are very much delighted herewith for the smell of it is so pleasant unto them, that they rub themselves upon it and wallow or tumble in it and also feed on the branches and leaves very greedily.'

This herb has long been cultivated both for its medicinal and seasoning properties, and in the hippie era of the late 1960s and 1970s for its mildly hallucinogenic quality when smoked.

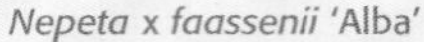

Nepeta x faassenii 'Alba'

Nepeta cataria

varieties

Nepeta cataria, Nepeta* x *faassenii* AGM** and ***Nepeta racemosa are all called catmint, which can be confusing. However, the first is the true herb with the medicinal and culinary properties and, just to be even more confusing, it is known also as dog mint!

Nepeta camphorata
Camphor Catmint
Hardy perennial. Ht and spread 60–75cm. Very different from ordinary catmint and very fragrant. Tiny white blooms all summer. Small, silvery grey, aromatic foliage. Prefers a poor, well-drained, dryish soil, not too rich in nutrients, and full sun. However, it will adapt to most soils except wet and heavy ones.

Nepeta cataria
Dog Mint, Nep-in-a-Hedge
Hardy perennial. Ht 1m, spread 60cm. White to pale pink flowers from early summer to early autumn. Pungent aromatic leaves. This plant is the true herb. In the seventeenth century it was used in the treatment of barren women.

***Nepeta* x *faassenii* 'Alba'**
Hardy perennial. Ht and spread 45cm. Loose spikes of white flowers from early summer to early autumn. Small greyish-green aromatic leaves form a bushy clump.

***Nepeta racemosa* 'Walker's Low'**
Hardy perennial. Ht and spread 80cm. Spikes of lavender blue/purple flowers from late spring to autumn. Small, mildly fragrant, greyish leaves. Marvellous edging plant for tumbling out over raised beds or softening hard edges of stone flags. Combines especially well with old-fashioned roses.

cultivation

Propagation

Seed
Sow its small seed in spring or late summer, either where the plant is going to flower or on to the surface of pots, plug or seed trays. Cover with perlite. Gentle bottom heat can be of assistance. Germination takes 10–20 days, depending on the time of year (faster in late summer). Seed is viable for 5 years. When large enough to handle, thin the seedlings to 30cm. The seed of *N. camphorata* should be sown in autumn to late winter. It will usually flower the following season.

Cuttings
Take softwood cuttings from new growth in late spring through to mid-summer. Do not choose flowering stems.

Division
A good method of propagation, particularly if a plant is becoming invasive. Divide established plants in spring. But beware of cats! The smell of a bruised root is irresistible. Cats have been known to destroy a specimen replanted after division. If there are cats around, protect the newly divided plant.

CATMINT

Nepeta racemosa 'Walker's Low'

Pests and Diseases
These plants are aromatic and not prone to pests. However, in cold wet winters, they tend to rot off.

Maintenance
Spring Sow seeds. Take cuttings. Divide established plants.
Summer Sow seeds until late in the season. Cut back hard after flowering to encourage a second flush. Take cuttings.
Autumn Cut back after flowering to maintain shape and produce new growth. If your winters tend to be wet and cold, pot up and over-winter this herb in a cold frame.
Winter Sow seeds of *Nepeta camphorata*.

Garden Cultivation
The main problem with catmint is the love cats have for it. If you have ever seen a cat spaced-out after feeding (hence catnip) and rolling on it, then you will understand why cat lovers love catmint, and why cat haters who grow it get cross with cat neighbours. The reason why cats are enticed is the smell; it reminds them of the hormonal scent of cats of the opposite sex. Bearing all this in mind, make your choice of planting site carefully. Nepeta make very attractive border or edging subjects. They like a well-drained soil, sun, or light shade. The one thing they dislike is a wet winter; they may well rot off.

Planting distance depends on species, but on average plant 50cm apart. When the main flowering is over, catmint should be cut back hard to encourage a second crop and to keep a neat and compact shape.

Harvesting
Whether you pick to use fresh or to dry, gather leaves and flowering tops when young.

companion planting

Planting *Nepeta cataria* close to vegetables is said to deter flea beetle.

container growing

N. x *faassenii* and all *N. racemosa* look stunning in large terracotta pots. The grey-green of the leaves and the blue-purple of the flowers complement the terracotta, and their sprawling habit in flower completes the picture. Use a soil-based compost mixed in equal parts with composted fine bark. Note that both varieties tend to grow soft and leggy indoors.

Nepeta racemosa 'Walker's Low'

culinary

Use freshly picked young shoots in salads or rub on meat to release their mintish flavour. Catmint was drunk as a tea before China tea was introduced into the West. It makes an interesting cup!

other uses

Dried leaves stuffed into toy mice will keep kittens and cats amused for hours. The scent of catnip is said to repel rats, so put bunches in hen and duck houses to discourage them. The flowers of *Nepeta* x *faassenii*, and *Nepeta racemosa* are suitable for formal displays.

medicinal

Nepeta cataria is now very rarely used for medicinal purposes. In Europe it is sometimes used in a hot infusion to promote sweating. It is said to be excellent for colds and flu. It soothes the nervous system and helps one to sleep. It also helps to calm upset stomachs and counters colic, flatulence and diarrhoea. An infusion can be applied externally to soothe scalp irritations, and the leaves and flowering tops can be mashed for a poultice to be applied to external bruises.

Ocimum basilicum

BASIL

Ocimum basilicum var. *purpurascens*

Also known as Common Basil, St Joseph Wort, and Sweet Basil. From the *family Lamiaceae*.

Basil is native to India, the Middle East and some Pacific Islands. It has been cultivated in the Mediterranean for thousands of years, but the herb only came to Western Europe in the sixteenth century with the spice traders, and to America and Australia with the early European settlers.

This plant is steeped in history and intriguing lore. Its common name is believed to be an abbreviation of *Basilikon phuton*, Greek for 'kingly herb', and it was said to have grown around Christ's tomb after the resurrection. Some Greek Orthodox churches use it to prepare their holy water, and put pots of basil below their altars. However, there is some question as to its sanctity – both Greeks and Romans believed that people should curse as they sow basil to ensure germination. There was even some doubt about whether it was poisonous or not, and in Western Europe it has been thought both to belong to the Devil and to be a remedy against witches. In Elizabethan times sweet basil was used as a snuff for colds and to clear the brain and deal with headaches, and in the seventeenth century Culpeper wrote of basil's uncompromising if unpredictable appeal – 'It either makes enemies or gains lovers but there is no in-between.'

varieties

Ocimum basilicum
Sweet Basil
Annual. Ht 45cm. A strong scent. Green, medium-sized leaves. White flowers. Without doubt the most popular basil. Use sweet basil in pasta sauces and salads, especially with tomato. Combines very well with garlic. Do not let it flower if using for cooking.

***Ocimum basilicum* 'Cinnamon'**
Cinnamon Basil
Annual. Ht 45cm. Leaves olive/brown/green with a hint of purple, highly cinnamon-scented when rubbed. Flowers pale pink. Comes from Mexico and is used in spicy dishes and salad dressings.

***Ocimum basilicum* 'Green Ruffles'**
Green Ruffles Basil
Annual. Ht 30cm. Light green leaves, crinkly and larger than sweet basil. Spicy, aniseed flavour, good in salad dishes and with stir-fry vegetables. But it is not, to my mind, an attractive variety. In fact the first time I grew it I thought its crinkly leaves had a bad attack of greenfly. Grow in pots and protect from frost.

***Ocimum basilicum* 'Horapha'**
Thai Basil, Horapha Basil (Rau Que)
Annual. Ht 42cm. Leaf olive/purplish. Stems red. Flowers with pink bracts. Aniseed in scent and flavour. A special culinary basil from Thailand. Use the leaves as a vegetable in curries and spicy dishes.

***Ocimum basilicum* 'Napolitano'**
Lettuce-leaved Basil
Annual. Ht 45cm. Leaves very large, crinkled, and with a distinctive flavour, especially good for pasta sauce.

Ocimum tenuiflorum

BASIL

Ocimum basilicum 'Cinnamon'

Originates in Naples region of Italy and needs a hot summer in cooler countries to be of any merit.

Ocimum basilicum* var. *purpurascens
Purple Basil, Dark Opal Basil
Annual. Ht 30cm. Strongly scented purple leaves. Pink flowers. Very attractive plant with a perfumed scent and flavour that is especially good with rice dishes. The dark purple variety that was developed in 1962 at the University of Connecticut represents something of a breakthrough in herb cultivation, not least because, almost exclusively, herbs have escaped the attentions of the hybridisers. The variety was awarded the All American Medal by the seedsmen.

***Ocimum basilicum* var. *purpurascens* 'Purple Ruffles'**
Purple Ruffles Basil
Annual. Ht 30cm. Very similar to straight purple basil (above), though the flavour is not as strong and the leaf is larger with a feathery edge. Flowers are pink. It can be grown in pots in a sunny position outside, but frankly it is a pain to grow because it damps off so easily.

Ocimum x citriodorum
Lemon Basil (Kemangie)
Annual. Ht 30cm. Light, bright, yellowish-green leaves, more pointed than other varieties, with a slight serrated edge. Flowers pale, whitish. Lemon basil comes from Indonesia, is tender in cooler climates, and susceptible to damping off. Difficult to maintain but well worth the effort. Both flowers and leaves have a lemon scent and flavour that enhance many dishes.

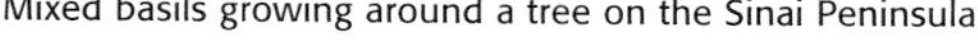

Mixed basils growing around a tree on the Sinai Peninsula

Ocimum basilicum 'Horapha'

Ocimum minimum
Bush Basil
Annual. Ht 30cm. Small green leaves, half the size of sweet basil. Flowers small, scented and whitish. Spread from Chile throughout South America, where, in some countries, it is believed to belong to the Goddess Erzulie and is carried as a powerful protector against robbery and by women to keep a lover's eye from roving. Excellent in pots on the windowsill. Delicious whole in green salads and with ricotta cheese.

***Ocimum minimum* 'Greek'**
Greek Basil (Fine-leaved Miniature)
Annual. Ht 23cm. This basil has the smallest leaves, tiny replicas of the bush basil leaves but, despite their size, they have a good flavour. As its name suggests, it originates from Greece. It is one of the easiest basils to look after – especially good in a pot. Use leaves un-chopped in all salads and in tomato sauces.

Ocimum tenuiflorum* syn. *Ocimum sanctum
Sacred Basil, Kha Prao, Tulsi
Annual. Ht 30cm. A small basil with olive/purple leaves with serrated edges. Stems deep purple. Flowers mauve/pink. The whole plant has a marvellously rich scent. Originally from Thailand, where it is grown around Buddhist temples. Can be used in Thai cooking with stir-fry hot peppers, chicken, pork or beef. The Indian-related form, is considered kingly or holy by the Hindus, sacred to the gods Krishna and Vishnu. It was the herb upon which to swear oaths in courts of law. It was also used throughout the Indian subcontinent as a disinfectant against malaria.

cultivation

Propagation

Seed

All basils can be grown from seed. Sow direct into pots or plug trays in early spring and germinate with warmth. Avoid using seed trays because basil has a long tap root and dislikes being transplanted. Plugs also help minimise damping off, to which all basil plants are prone (see below).

Water well at midday in dry weather even when transplanted into pots or containers: basil hates going to bed wet. This minimises the chances of damping off and will prevent root rot, a hazard when air temperature is still dropping at night.

Plant out seedlings when large enough to handle and the danger of frost has passed. The soil needs to be rich and well drained, and the situation warm and sheltered, preferably with sun at midday. However, prolific growth will only be obtained usually in the greenhouse or in large pots on a sunny patio. I suggest that you plant basil in between tomato plants for the following reasons:

1. Being a good companion plant, it is said to repel flying insects, so may help to keep the tomatoes pest-free.
2. You will remember to use fresh basil with tomatoes.
3. You will remember to water it.
4. The situation will be warm and whenever you pick tomatoes you will tend to pick basil, which will encourage bushy growth and prevent it flowering, which in turn will stop the stems becoming woody and the flavour of its leaves bitter.

Ocimum basilicum

Pests and Diseases

Greenfly and whitefly may be a problem with pot-grown plants. Wash off with liquid horticultural soap. Seedlings are highly susceptible to damping off, a fungal disease encouraged by overcrowding in overly wet conditions in seed trays or pots. It can be prevented by sowing the seed thinly and widely and guarding against an over-humid atmosphere.

Maintenance

Spring Sow seeds in early spring with warmth and watch out for damping off; plant out around the end of the season. Alternatively, sow directly into the ground after any frosts.

Summer Keep pinching out young plants to promote new leaf growth and to prevent flowering. Harvest the leaves.

Autumn Collect seeds of plants allowed to flower. Before first frosts, bring pots into the house and place on the windowsill. Dig up old plants and dig over the area ready for new plantings.

Garden Cultivation

This is only a problem in areas susceptible to frost and where you can't provide for its great need for warmth and nourishment. In such areas plant out after the frosts have finished; choose a well-drained, rich soil in a warm, sunny corner, protected from the wind.

Harvesting

Pick leaves when they are young and always from the top to encourage new growth. If freezing to store, paint both sides of each leaf with olive oil to stop it sticking to the next leaf and to seal in its flavour. If drying, do it as fast as you can. Basil leaves are some of the more difficult to dry successfully and I do not recommend it. The most successful course, post-harvest, is to infuse the leaves in olive oil or vinegar. As well as being very useful in your own kitchen, both the oil and the vinegar make great Christmas presents (see also page 672). Gather flowering tops as they open during the summer and early autumn. Add them fresh to salads, or dry to potpourris.

container growing

Basil is happy on a kitchen windowsill and in pots on the patio, and purple basil makes a good centrepiece in a hanging basket. In Europe basil is placed in pots outside houses to repel flies. Use a standard potting

Ocimum minimum 'Greek'

compost mixed in equal parts with composted fine bark. Water containers before midday but do not over water. If that is not possible, water earlier in the day rather than later and again do not over water.

medicinal

Once prescribed as a sedative against gastric spasms and as an expectorant and laxative, basil is rarely used in herbal medicines today. However, leaves added to food are an aid to digestion and if you put a few drops of basil's essential oil on a sleeve and inhale, it can allay mental fatigue. For those that need a zing it can be used to make a very refreshing bath vinegar, which also acts as an antiseptic.

other uses

Keep it in a pot in the kitchen to act as a fly repellent, or crush a leaf and rub it on your skin, where the juice repels mosquitoes.

Basil oil

culinary

Basil has a unique flavour, so newcomers should use with discretion otherwise it will dominate other flavours. It is one of the few herbs to increase its flavour when cooked. For best results add at the very end of cooking.

Hints and Ideas

1. Tear the leaves, rather than chop. Sprinkle over green salads or sliced tomatoes.

2. Basil combines very well with garlic. Tear into French salad dressing.

3. When cooking pasta or rice, heat some olive oil in a saucepan, remove it from the heat, add some torn purple basil leaves, toss the pasta or rice in the basil and oil, and serve. Use lemon basil to accompany a fish dish – it has a sharp lemon/spicy flavour when cooked.

4. Add to a cold rice or pasta salad.

5. Mix low-fat cream cheese with any of the basils and use in baked potatoes.

6. Basil does not combine well with strong meats such as goat or vension. However, aniseed basil is very good with stir-fried pork.

7. Sprinkle on fried or grilled tomatoes while they are still hot as a garnish.

8. Very good with French bread and can be used instead of herb butter in the traditional hot herb loaf. The tiny leaves of Greek basil are best for this because you can keep them whole.

9. Sprinkle on top of pizzas.

10. Basil makes an interesting stuffing for chicken. Use sweet basil combined with crushed garlic, breadcrumbs, lemon peel, beaten egg and chopped nuts.

Pesto Sauce

One of the best-known recipes for basil, here is a simple version for 4 people.

1 tablespoon pine nuts
4 tablespoons chopped basil leaves
2 cloves garlic, chopped
6 tablespoons sunflower oil or olive oil (not virgin)
75g Parmesan cheese
salt

Blend the pine nuts, basil and chopped garlic until smooth. Add the oil slowly and continue to blend the mixture until you have a thick paste. Season with salt to taste. Stir the sauce into the cooked and drained pasta and sprinkle with Parmesan cheese.

Pesto sauce will keep in a sealed container in the fridge for at least a week. It can also be frozen but it is important, as with all herbal mixtures, to wrap the container with at least two thicknesses of polythene to prevent the aroma escaping.

Oenothera

EVENING PRIMROSE

Also known as Common Evening Primrose, Evening Star, Fever Plant, Field Primrose, King's Cure-all, Night Willowherb, Scabish, Scurvish, Tree Primrose, Primrose, Moths Moonflower and Primrose Tree. From the family *Onagraceae*.

A native of North America, evening primrose was introduced to Europe in 1614 when botanists brought it from Virginia as a botanical curiosity. In North America it is regarded as a weed, but elsewhere as a pretty garden plant.

The generic name, *Oenothera*, comes from the Greek *oinos*, meaning 'wine' and *thera,* 'hunt'. According to ancient herbals the plant was said to dispel the ill effects of wine, but both plant and seed have been used for other reasons – culinary and medicinal – by American Indians for hundreds of years. The Flambeau Ojibwe tribe were the first to realise its medicinal properties. They used to soak the whole plant in warm water to make a poultice to heal bruises and overcome skin problems. Traditionally, too, it was used to treat asthma, and its medicinal potential is under research. Oil of evening primrose is currently attracting considerable attention worldwide as a treatment for nervous disorders, in particular multiple sclerosis. There may well be a time in the very near future when the pharmaceutical industry will require fields of this beautiful plant to be grown on a commercial scale.

The common name comes from the transformation of its bedraggled daytime appearance into a fragrant, phosphorescent, pale yellow beauty with the opening of its flowers in the early evening. All this show is for one night only, however. Towards the end of summer the flowers tend to stay open all day long. (It is called evening star because the petals emit phosphorescent light at night.) Many strains of the plant came to Britain as stowaways in soil used as ballast in cargo ships.

Oenothera fruticosa subsp. *glauca*

varieties

Oenothera biennis
Evening Primrose
Hardy biennial. Ht 90–120cm, spread 90cm. Large evening-scented yellow flowers for most of the summer. Long green oval or lance-shaped leaves. This is the medicinal herb, and the true herb.

***Oenothera fruticosa* subsp. *glauca* AGM**
Perennial. Ht 50cm, spread 45cm. Small clusters of trumpet-shaped golden flowers, buds and flowers, often tinged with red. Lance-shaped mid-green leaves. The flowers are edible.

***Oenothera macrocarpa* AGM**
Hardy perennial. Ht 10cm, spread 40cm or more. Large yellow bell-shaped flowers, sometimes spotted with red, open at sundown throughout the summer. The small to medium green leaves are of a narrow oblong shape.

cultivation

Propagation

Seeds

Sow in early spring on the surface of pots or plug trays, or in late spring direct into a prepared site in the garden. Seed is very fine so be careful not to sow it too thick. Use the cardboard method (see page 645). When the weather has warmed up sufficiently, plant out at a distance of 30cm apart. Often the act of transplanting *O. biennis* will encourage the plant to flower during the first year. This variety is a prolific self-seeder. So once it has been introduced into the garden, it will stay there.

EVENING PRIMROSE

Pests and Diseases
This plant rarely suffers from pests or diseases.

Maintenance
Spring Sow seed.
Summer Deadhead plants to cut down on self-seeding.
Autumn Dig up old roots of second-year growth of the biennials.
Winter No need to protect.

Garden Cultivation
Choose a well-drained soil in a dry, sunny corner for the best results. It is an extremely tolerant plant, happy in most situations, and I have known the seedlings of *O. biennis* appear in a stone wall, so be forewarned.

Harvesting
Use leaves fresh as required. Best before flowering.

Pick the flowers when in bud or when just open. Use fresh. Picked flowers will always close and are no good for flower arrangements.

Collect the seeds as the heads begin to open at the end. Store in jar for sowing in the spring. Dig up roots and use fresh as a vegetable or to dry.

container growing

The lower-growing varieties are very good when grown in window boxes and tubs. Tall varieties need support from other plants or stakes. None are suitable for growing indoors.

culinary

It is a pot herb – roots, stems, leaves, and even flower buds may be eaten. The roots can be boiled – they taste like sweet parsnips, or alternatively pickled and tossed in a salad.

other uses

Leaf and stem can be infused to make an astringent facial steam. Add to hand cream as a softening agent.

Evening Primrose soap

medicinal

Soon this plant will take its place in the hall of herbal fame. It can have startling effects on the treatment of pre-menstrual tension. In 1981 at St Thomas's Hospital, London, 65 women with PMS were treated. 61 per cent experienced complete relief and 23 per cent partial relief. One symptom, breast engorgement, was especially improved – 72 per cent of women reported feeling better. In November 1982, an edition of the prestigious medical journal *The Lancet* published the results of a double-blind crossover study on 99 patients with ectopic eczema, which showed that when high doses of evening primrose oil were taken, about 43 per cent of the patients experienced improvement of their eczema. Studies of the effect of the oil on hyperactive children also indicate that this form of treatment is beneficial.

True to the root of its generic name, the oil does appear to be effective in counteracting alcohol poisoning and preventing hangovers. It can help withdrawal from alcohol, easing depression. It helps dry eyes and brittle nails and, when combined with zinc, the oil may be used to treat acne.

But it is the claim that it benefits sufferers of multiple sclerosis that has brought controversy. It has been recommended for MS sufferers by Professor Field, who directed MS research for the UK Medical Research Council. Claims go further. It is said to be effective in guarding against arterial disease; the effective ingredient, gami-linolelic acid (GLA), is a powerful anti-blood clotter.

It is also said to aid weight-loss; a New York hospital discovered that people more than 10 per cent above their ideal body weight lost weight when taking the oil. It is thought that this occurs because the GLA in evening primrose oil stimulates brown fat tissue.

In perhaps the most remarkable study of all, completed in Glasgow Royal Infirmary in 1987, it helped to relieve the symptoms of 60 per cent of patients suffering from rheumatoid arthritis. Those taking fish oil, in addition to evening primrose oil, fared even better.

The scientific explanation for these extraordinary results is that GLA is a precursor of a hormone-like substance called PGEI, which has a wide range of beneficial effects on the body. Production of this substance in some people may be blocked. GLA has also been found in oil extracted from blackcurrant seed and borage seed, both of which are now a commercial source of this substance.

Oenothera biennis

Origanum

OREGANO & MARJORAM

Also known as Wild Marjoram, Mountain Mint, Winter Marjoram, Winter Sweet, Marjolaine and Origan. From the family *Lamiaceae*.

For the most part these are natives of the Mediterranean region. They have adapted to many countries, however, and a native form can now be found in many regions of the world, even if under different common names. For example, *Origanum vulgare* growing wild in Britain is called wild marjoram (the scent of the leaf is aromatic but not strong, the flowers are pale pink); while in Mediterranean countries wild *Origanum vulgare* is known as oregano (the leaf is green, slightly hairy and very aromatic, the flowers are similar to those found growing wild in Britain).

Oregano is derived from the Greek *oros*, meaning 'mountain', and *ganos*, meaning 'joy' and 'beauty'. It therefore translates literally as 'joy of the mountain'. In Greece it is woven into the crown worn by bridal couples.

According to Greek mythology, the King of Cyprus had a servant called Amarakos, who dropped a jar of perfume and fainted in terror. As his punishment the gods changed him into oregano, after which, if it was found growing on a burial tomb, all was believed well with the dead. Venus was the first to grow the herb in her garden.

Origanum 'Rosenkuppel'

Origanum vulagare 'Compactum'

Aristotle reported that tortoises, after swallowing a snake, would immediately eat oregano to prevent death, which gave rise to the belief that it was an antidote to poison.

The Greeks and Romans used it not only as scent after taking a bath and as a massage oil, but also as a disinfectant and preservative. More than likely they were responsible for the spread of this plant across Europe, where it became known as marjoram. The New Englanders took it to North America, where there arose a further confusion of nomenclature. Until the 1940s, common marjoram was called wild marjoram in America, but is now known as oregano. In certain parts of Mexico and the southern states of America, oregano is, confusingly, the colloquial name for a totally unrelated plant that has a similar flavour.

Sweet marjoram, which originates from North Africa, was introduced into Europe in the sixteenth century and was incorporated in nosegays to ward off the plague and other pestilence.

varieties

Origanum amanum AGM
Hardy perennial. Ht and spread 15–20cm. Open, funnel-shaped, pale pink or white flowers borne above small heart-shaped, aromatic, pale green leaves. Makes a good alpine house plant. Dislikes a damp atmosphere.

Origanum x applii
Winter Marjoram
Half-hardy perennial. Ht 23cm, spread 30cm. Small pink flowers. Very small aromatic leaves which, in the right conditions, are available all year round. Good to grow in a container.

Origanum dictamnus
Ditany of Crete
Hardy perennial. Ht 12–15cm, spread 40cm. Prostrate habit, purplish pink flowers that appear in hop-like clusters in summer. The leaves are white and woolly and grow on arching stems. Pretty little plant, quite unlike the other origanums in appearance. Tea made from the leaves is considered a panacea in Crete.

Origanum 'Kent Beauty'
Hardy perennial. Ht 15–20cm, spread 30cm. Whorls of tubular pale pink flowers with darker bracts appear in summer on short spikes. Round, oval and aromatic leaves on trailing stems, which give the plant its prostrate habit and make it suitable for a wall or ledge. Decorative more than culinary.

Origanum 'Kent Beauty'

Origanum laevigatum AGM
Hardy perennial. Ht 23–30cm, spread 20cm. Summer profusion of tiny, tubular, cerise/pink/mauve flowers, surrounded by red/purple bracts. Aromatic, dark green leaves, which form a mat in winter. Decorative more than culinary.

Origanum majorana
Sweet Marjoram
Also known as Knotted Marjoram or Knot Marjoram. Half-hardy perennial. Grown as an annual in cool climates. Ht and spread 30cm. Tiny white flowers in a knot. Round pale green leaves, highly aromatic. This is the best variety for flavour. Use in culinary recipes that state marjoram. The leaf is also good for drying, retaining a lot of its scent and flavour.

Origanum onites
Pot Marjoram
Hardy perennial. Ht and spread 45cm. Pink/purple flowers in summer. Green aromatic leaves that form a mat in winter. Good grower with a nice flavour. Difficult to obtain the true seed; grows easily from cuttings, however.

Origanum 'Rosenkuppel'
Hardy perennial. Ht 60cm. Clusters of small dark pink flowers in summer. Dark green, oval, slightly hairy leaves. Can be used in cooking, but has an inferior flavour to *Orgianum onites.*

Origanum rotundifolium AGM
Hardy perennial. Ht 23–30cm, spread 30cm. Prostrate habit. The pale pink, pendant, funnel-shaped flowers appear in summer in whorls surrounded by yellow/green bracts. Leaves are small, round, mid-green and aromatic. Decorative more than culinary.

Origanum vulgare
Oregano
Also known as Wild Marjoram
Hardy perennial. Ht and spread 45cm. Clusters of tiny tubular mauve flowers in summer. Dark green, aromatic, slightly hairy leaves, which form a mat in winter. When grown in its native Mediterranean, it has a very pungent flavour, which bears little resemblance to that obtained in the cooler countries. When cultivated in the garden it becomes similar to pot marjoram.

Origanum vulgare 'Aureum Crispum'

Origanum vulgare 'Aureum' AGM
Golden Marjoram
Hardy perennial. Ht and spread 45cm. Clusters of tiny tubular mauve/pink flowers in summer. Golden, aromatic, slightly hairy leaves, which form a mat in winter. The leaves have a warm aromatic flavour when used in cooking; combines well with vegetables.

Origanum vulgare 'Aureum Crispum'
Golden Curly Marjoram
Hardy perennial. Ht and spread 45cm. Clusters of tiny tubular mauve/pink/white flowers in summer. Leaves, small, golden, crinkled, aromatic and slightly hairy, which form a mat in winter. The leaves have a slightly milder savoury flavour (sweeter and spicy) that combines well with vegetable dishes.

Origanum vulgare 'Compactum'
Compact Marjoram
Hardy perennial. Ht 15cm, spread 30cm. Lovely large pink flowers. Smallish green aromatic leaves, which form a mat in winter, have a deliciously warm flavour and combine well with lots of culinary dishes.

Origanum vulgare 'Gold Tip'
Gold-Tipped Marjoram
Also known as Gold Splash
Hardy perennial. Ht and spread 30cm. Small pink flowers in summer. The aromatic leaves are green and yellow variegated. Choose the garden site carefully: shade prevents the variegation. The leaves have a mild savoury flavour.

Origanum vulgare subsp. hirtum 'Greek'
Greek Oregano
Hardy perennial. Ht and spread 45cm. Clusters of tiny tubular white flowers in summer. Grey/green hairy leaves, which are very aromatic and excellent to cook with.

Origanum vulgare 'Nanum'
Dwarf Marjoram
Hardy perennial. Ht 10cm, spread 15cm. White/pink flowers in summer. Tiny green aromatic leaves. It is a lovely, compact, neat little bush, great in containers and at the front of a herb garden. Good in culinary dishes.

Origanum onites

cultivation

Propagation

Seed

The following can be grown from seed: *O. vulgare, O. majorana, O. vulgare* subsp. *hirtum* 'Greek'. The seed is very fine, so sow in spring into prepared seed or plug trays. Use the cardboard trick (see page 645). Leave uncovered and give a bottom heat of 15°C. Germination can be erratic or 100 per cent successful. Watering is critical when the seedlings are young; keep the compost on the dry side. As the seed is so fine, thin before pricking out to allow the plants to grow. When large enough, either pot on, using a standard seed compost mixed in equal parts with composted fine bark, or if the soil is warm enough and you have grown them in plugs, plant into the prepared garden.

Cuttings

Apart from the three species mentioned above, the remainder can only be propagated successfully by cuttings or division. Softwood cuttings can be taken from the new growing tips of all the named varieties in spring. Use a standard seed compost mixed in equal parts with composted fine bark.

Division

A number of varieties form a mat during the winter. These lend themselves to division. In spring, or after flowering, dig up a whole clump and pull sections gently away. Each will come away with its own root system. Replant as wanted.

Pests and Diseases

Apart from occasional frost damage, marjorams and oreganos, being aromatic, are mostly pest free.

Maintenance

Spring Sow seeds. Divide established plants. Take softwood cuttings.
Summer Trim after flowering to stop plants becoming straggly. Divide established plants in late summer.
Autumn Before they die down for winter, cut back the year's growth to within 6cm of the soil.
Winter Protect pot-grown plants and tender varieties.

Garden Cultivation

O. majorana and *O.* x *applii* need a sunny garden site and a well-drained, dry, preferably chalk, soil. Otherwise plant them in containers. All the rest are hardy and adaptable, and will tolerate most soils as long as they are not waterlogged in winter. Plant gold varieties in some shade to prevent the leaves from scorching. For the majority, a good planting distance is 25cm, closer if being used as an edging plant.

Harvesting

Leaves

Pick leaves whenever available for use fresh. They can be dried or frozen, or be used to make oil or vinegar.

Flowers

The flowers can be dried just as they open for dried flower arrangements.

container growing

The oreganos look great in containers. Use a soil-based compost mixed in equal parts with composted fine bark. Make sure that they are not over watered and that the gold and variegated forms get some shade at midday. Cut back after flowering and give them a liquid fertiliser feed.

medicinal

This plant is one of the best antiseptics owing to its high thymol content.

Marjoram tea helps ease bad colds, has a tranquillizing effect on nerves, and helps settle upset stomachs. It also helps to prevent sea sickness. For temporary relief of toothache, chew the leaf or rub a drop of essential oil on the gums. A few drops of essential oil on the pillow will help you sleep.

other uses

Make an infusion and add to the bath water to aid relaxation.

culinary

Marjoram and oregano aid the digestion, and act as an antiseptic and as a preservative.

They are among the main ingredients of bouquet garni, and combine well with pizza, meat and tomato dishes, vegetables and milk-based desserts.

Red Mullet with Tomatoes and Oregano

Serves 4–6

4–6 red mullet, cleaned
3 tablespoons olive oil
1 medium onion, sliced
1 clove garlic, chopped
500g tomatoes, peeled and chopped
1 green or red pepper, seeded and diced
1 teaspoon sugar
1 teaspoon chopped fresh oregano or ½ teaspoon dried oregano
Freshly milled salt and pepper
Oil for baking or shallow frying

Rinse the fish in cold water and drain on kitchen paper. Heat the olive oil in a pan and cook the onion and garlic slowly until golden brown; add the tomatoes, pepper, sugar and oregano, and a little salt and pepper. Bring to the boil, then simmer for 20 minutes until thickened.

Bake or fry the fish. Brush them with oil, place in an oiled ovenproof dish and cook at a moderately hot temperature, 190°C/375°F/gas mark 5 for 7–8 minutes. Serve with the sauce.

Papaver

POPPY

From the family *Papaveraceae*.

The poppy is widely spread across the temperate zones of the world. For thousands of years corn and poppies and civilizations have gone together. The Romans looked on the poppy as sacred to their corn goddess, Ceres, who taught men to sow and reap. The ancient Egyptians used poppy seed in their baking for its aromatic flavour.

The field poppy grew on Flanders Fields after the battles of the First World War and became the symbol of Remembrance Day.

Papaver rhoeas

varieties

***Papaver commutatum* AGM**
Ladybird Poppy
Hardy annual. Ht 30–90cm, spread 45cm. Red flowers in summer, each with black blotch in centre. Leaf oblong and deeply toothed. Native of Asia Minor.

Papaver rhoeas
Field Poppy
Also known as common poppy, corn poppy, blind eyes, blind man, red dolly, red huntsmen, poppet, old woman's petticoat, thunderbolt, and wartflower. Hardy annual. Ht 20–60cm, spread 45cm. Brilliant scarlet flower with black basal blotch from summer to early autumn. The mid-green leaf has 3 lobes and is irregularly toothed.

Papaver somniferum
Opium Poppy
Hardy annual. Ht 30–90cm, spread 45cm. Large pale lilac, white, purple or variegated flowers in summer. The leaf is long with toothed margins and bluish in colour. There is a double-flowered variety, *P. somniferum* var. *paeoniiflorum*.

Meconopsis cambrica
Welsh Poppy
Hardy perennial. Ht 30–60cm. Yellow flowers in summer. Green leaves are divided into many leaflets. It differs from *Papaver* in that seeds are released through slits in the seed heads, not through pepper-pot heads.

cultivation

Propagation
Seed
Sow the very fine seed in autumn onto the surface of prepared seed or plug trays, using a standard seed compost mixed in equal parts with composted fine bark. Cover with glass and leave outside for winter stratification. In spring, when seedlings are large enough, plant out into the garden in groups.

Pests and Diseases
Largely pest and disease free.

Maintenance
Spring Plant out in garden.
Summer Deadhead flowers to prolong flowering and prevent self-seeding.
Autumn Sow seed. Dig up old plants.
Winter No need to protect.

Garden Cultivation
Poppies all prefer a sunny site and a well-drained fertile soil. Sow in the autumn in a prepared site; press seed into the soil but do not cover. Thin to 20–30cm apart. Remove the heads after flowering to prevent self-seeding.

Harvesting
The ripe seeds can be collected from both field and opium poppies, the seed of which is not narcotic. It must be ripe, otherwise it will go mouldy in store.

container growing

Use a standard potting compost mixed in equal parts with composted fine bark. Place in full sun out of the wind, and water well in the summer. Avoid feeding as this will produce lots of soft growth and few flowers.

medicinal

The unripe seed capsules of the opium poppy are used for the extraction of morphine and the manufacture of codeine.

other uses

The oil extracted from the seed of the opium poppy is used not only as a salad oil, and for cooking, but also for burning in lamps, and in the manufacture of varnish, paint and soap.

warning

All parts of the opium poppy, except the ripe seeds, are dangerous and should be used only by trained medical staff.

culinary

Sprinkle the ripe seeds on bread, cakes and biscuits for a pleasant nutty flavour. Add to curry powder for texture, flavour, and as a thickener.

Pelargonium

SCENTED GERANIUMS

From the family *Geraniaceae*.

These form a group of marvellously aromatic herbs which should be used more. Originally native to South Africa, they are now widespread throughout many temperate countries, where they should be grown as tender perennials.

The generic name, *Pelargonium*, is said to be derived from *pelargos*, 'a stork'. With a bit of imagination one can understand how this came about: the seed pods bear a resemblance to a stork's bill.

varieties

There are many different scented geraniums. I am mentioning a few typical of the species that I have a soft spot for. They are very collectable plants.

***Pelargonium* 'Attar of Roses' AGM**
Half-hardy evergreen perennial. Ht 30–60cm, spread 30cm. Small pink flowers in summer. 3-lobed, mid-green leaves that smell of roses.

***Pelargonium* 'Atomic Snowflake'**
Half-hardy evergreen perennial. Ht 30–60cm, spread 30cm. Small pink flowers in summer. Intensely lemon-scented, roundish leaves with silver grey/green variegation.

Pelargonium capitatum
Half-hardy evergreen perennial. Ht 30–60cm, spread 30cm. Small mauve flowers in summer, irregular 3-lobed green leaves, rose scented. This is now mainly used to produce geranium oil for the perfume industry.

***Pelargonium* 'Chocolate Peppermint'**
Half-hardy evergreen perennial. Ht 30–60cm, spread 1m. Small white/pink flowers in summer. Large, rounded, shallowly lobed leaves, velvety green with brown marking and a strong scent of chocolate peppermints! This is a fast grower so pinch out growing tips to keep shape.

***Pelargonium* 'Clorinda'**
Half-hardy evergreen perennial. Ht and spread 1m. Large pink attractive flowers in summer. Large rounded leaves, mid-green and eucalyptus-scented.

Pelargonium crispum
Half-hardy evergreen perennial. Ht and spread 30–60cm. Small pink flowers in summer. Small 3-lobed leaves, green, crispy crinkled and lemon scented. Neat habit.

***Pelargonium crispum* 'Peach Cream'**
Half-hardy evergreen perennial. Ht and spread 30–60cm. Small pink flowers in summer. Small 3-lobed leaves, green with cream and yellow variegation, crispy crinkled and peach-scented.

***Pelargonium crispum* 'Variegatum' AGM**
Half-hardy evergreen perennial. Ht and spread 30–60cm. Small pink flowers in summer. Small 3-lobed leaves, green with cream variegation, crispy crinkled, and lemon-scented.

Pelargonium denticulatum
Half-hardy evergreen perennial. Ht and spread 1m. Small pinky-mauve flowers in summer. Deeply cut palmate leaves, green with a lemon scent.

***Pelargonium denticulatum* 'Filicifolium'**
Half-hardy evergreen perennial. Ht and spread 1m. Small pink flowers in summer. Very finely indented green leaves with a fine brown line running through, slightly sticky and not particularly aromatic, if anything a scent of balsam. Prone to whitefly.

***Pelargonium* Fragrans Group**
Half-hardy evergreen perennial. Ht and spread 30cm. Small white flowers in summer. Greyish green leaves, rounded with shallow lobes, and a strong scent of nutmeg/pine.

***Pelargonium* Fragrans Group 'Fragrans Variegatum'**
Half-hardy evergreen perennial. Ht and spread 30cm. Small white flowers in summer. Greyish green leaves with cream variegation, rounded with shallow lobes and a strong scent of nutmeg/pine.

Pelargonium graveolens
Rose Geranium
Half-hardy evergreen perennial. Ht 60cm–1m. Spread up to 1m. Small pink flowers in summer. Fairly deeply cut green leaves with a rose/ peppermint scent. One of the more hardy of this species, with good growth.

***Pelargonium* 'Lady Plymouth' AGM**
Half-hardy evergreen perennial. Ht and spread 30–60cm. Small pink flowers in summer. Fairly deeply cut greyish green leaves with cream variegation and a rose/peppermint scent.

Pelargonium 'Atomic Snowflake'

Pelargonium 'Chocolate Peppermint'

Pelargonium 'Lemon Fancy'
Half-hardy evergreen perennial. Ht 30–60cm, spread 30–45cm. Smallish pink flowers in summer. Small roundish green leaves with shallow lobes and an intense lemon scent.

Pelargonium 'Lilian Pottinger'
Half-hardy evergreen perennial. Ht 30–60cm, spread 1m. Small whitish flowers in summer. Leaves brightish green, rounded, shallowly lobed with serrated edges. Soft to touch. Mild spicy apple scent.

Pelargonium 'Mabel Grey' AGM
Half-hardy evergreen perennial. Ht 45–60cm, spread 30–45cm. Mauve flowers with deeper veining in summer. If I have a favourite scented geranium, this is it: the leaves are diamond-shaped, roughly textured, mid-green and oily when rubbed and very strongly lemon-scented.

Pelargonium odoratissimum
Half-hardy evergreen perennial. Ht 30–60cm, spread 1m. Small white flowers in summer. Green, rounded, shallowly lobed leaves, fairly bright green in colour and soft to touch, with an apple scent. Trailing habit, looks good in large containers.

Pelargonium 'Prince of Orange'
Half-hardy evergreen perennial. Ht and spread 30–60cm. Pretty pink/white flowers in summer. Green, slightly crinkled, slightly lobed leaves, with a refreshing orange scent. Prone to rust.

Pelargonium quercifolium
Oak-Leafed Pelargonium
Half-hardy evergreen perennial. Ht and spread up to 1m. Pretty pink/purple flowers in summer. Leaves oak-shaped, dark green with brown variegation, and slightly sticky. A different, spicy scent.

Pelargonium 'Rober's Lemon Rose'
Half-hardy evergreen perennial. Ht and spread up to 1m. Pink flowers in summer. Leaves greyish-green – oddly shaped, lobed and cut – with a rose scent. A fast grower, so pinch out the growing tips regularly to maintain shape.

Pelargonium 'Royal Oak' AGM
Half-hardy evergreen perennial. Ht 38cm, spread 30cm. Small pink/purple flowers in summer. Oak-shaped, dark green leaves with brown variegation, slightly sticky with spicy scent. Very similar to *P. quercifolium*, but with a more compact habit.

***Pelargonium tomentosum* AGM**
Half-hardy evergreen perennial. Ht 30–60cm, spread 1m. Small white flowers in summer. Large rounded leaves, shallow lobed, velvet grey-green in colour with a strong peppermint scent. Fast grower, so pinch out growing tips regularly to maintain shape. Protect from full sun.

Mixed *Pelargonium*

cultivation

Propagation

Seed
Although I have known these plants to be grown from seed, I do not recommend this method. Cuttings are much more reliable for the majority. However, if you want to have a go, sow in spring in a standard seed compost mixed in equal parts with composted fine bark at a temperature no lower than 15°C.

Cuttings
All scented geraniums can be propagated by softwood cuttings which generally take very easily in the summer. Take a cutting about 10–15cm long and strip the leaves from the lower part with a sharp knife. At all costs do not tear the leaves off as this will cause a hole in the stem and the cutting will be susceptible to disease, such as blackleg. This is a major caveat for varieties such as *Pelargonium crispum* 'Variegatum'. Use a sharp knife and slice the leaf off, insert the cutting into a tray containing equal parts seed compost and composted bark. Water in and put the tray away from direct sunlight. Keep an eye on the compost, making sure it does not thoroughly dry out, but only water if absolutely necessary. The cuttings should root in two to three weeks. Pot up into separate pots containing a standard potting compost mixed in equal parts with composted fine bark. Place in a cool greenhouse or cool conservatory for the winter, keeping the compost dry and watering only very occasionally. In the spring re-pot into larger pots and water sparingly. When they start to produce flower buds give them a liquid feed. In early summer pinch out the top growing points to encourage bushy growth.

Pests and Diseases
Unfortunately pelargoniums suffer from a few diseases.

1. Cuttings can be destroyed by blackleg disease. The cutting turns black and falls over. The main cause of this is too much water. So keep the cuttings as dry as possible after the initial watering.

2. Grey mould (*Botrytis*) is also caused by the plants being too wet and the air too moist. Remove damaged leaves carefully so as not to spread the disease, and burn. Allow the plants to dry out, and increase ventilation and spacing between plants.

3. Leaf gall appears as a mass of small proliferated shoots at the base of a cutting or plant. Destroy the plant, otherwise it could affect other plants.

4. Geraniums, like mint and comfrey, are prone to rust, especially on *P.* 'Prince of Orange'. Destroy the affected plants, or it will spread to others.

SCENTED GERANIUMS

5. Whitefly. Be vigilant. If you catch it early enough, you will be able to control it by spraying with a liquid horticultural soap. Follow manufacturer's instructions.

Maintenance

Spring Sow seed. Trim, slowly introduce watering, and start feeding. Re-pot if necessary.
Summer Feed regularly. Trim to maintain shape. Take cuttings.
Autumn Trim back plants. Bring in for the winter to protect from frost.
Winter Allow the plants to rest. Keep watering to a minimum.

Garden Cultivation

Scented pelargoniums are so varied that they can look very effective grown in groups in the garden. Plant out as soon there is no danger of frost. Choose a warm site with well-drained soil. A good method is to sink the re-potted, over-wintered geraniums into the soil. This makes sure the initial compost is correct, and makes it easier to dig up the pot and bring inside before the first frost.

Harvesting

Pick leaves during the growing season, for fresh use or for drying. Collect seeds before the seed pod ripens and ripen in paper bags. If allowed to ripen on the plant, the pods will burst, scattering the seeds everywhere.

container growing

Scented pelargoniums make marvellous pot plants. They grow well, look good, and smell lovely. Pot up as in 'Propagation', using a standard potting compost mixed in equal parts with composted fine bark. Place so that you can rub the leaves as you walk past.

other uses

In aromatherapy, geranium oil is relaxing but use it in small quantities. Dilute 2 drops in 2 teaspoons of soy oil for a good massage, or to relieve pre-menstrual tension, dermatitis, eczema, herpes or dry skin.

warning

None of the crispums should be used in cooking as it is believed that they can upset the stomach.

culinary

Before artificial food flavourings were produced, the Victorians used scented pelargonium leaves in the bottom of cake tins to flavour their sponges. Why not follow suit? When you grease and line the bottom of a 20cm sandwich tin, arrange approximately 20 leaves of either *P.* 'Lemon Fancy', *P.* 'Mabel Grey', or *P. graveolens*. Fill the tin with a sponge mix of your choice and cook as normal. Remove the leaves with the lining paper when the cake has cooled. Scented pelargonium leaves add distinctive flavour to many dishes although, like bay leaves, they are hardly ever eaten, being removed after the cooking process. The main varieties used are *P. graveolens*, *P. odoratissimum*, *P.* 'Lemon Fancy' and *P.* 'Attar of Roses'.

Geranium Leaf Sorbet

12 scented Pelargonium graveolens *leaves*
75g caster sugar
300ml water
Juice of 1 large lemon
1 egg white, beaten
4 leaves for decoration

Wash the leaves and shake them dry. Put the sugar and water in a saucepan and boil until the sugar has dissolved, stirring occasionally. Remove the pan from the heat. Put the 12 leaves in the pan with the sugar and water, cover and leave for 20 minutes. Taste. If you want a stronger flavour bring the liquid to the boil again add some fresh leaves and leave for a further 10 minutes. When you have the right flavour, strain the syrup into a rigid container, add the lemon juice and leave to cool. Place in the freezer until semi-frozen (approximately 45 minutes) – it must be firm, not mushy – and fold in the beaten egg white. Put back into freezer for a further 45 minutes. Scoop into individual glass bowls, and decorate each with a geranium leaf.

Rose Geranium Punch

1.2 litres of apple juice
250g sugar
6 leaves of Pelargonium graveolens
6 drops of green vegetable colouring (optional)
4 limes

Boil the apple juice, sugar and geranium leaves for 5 minutes. Strain the liquid. Cool and add colouring if required. Thinly slice and crush limes, add to the liquid. Pour onto ice in glasses and garnish with geranium leaves.

Rose Geranium Butter

Butter pounded with the leaves makes a delicious filling for cakes and sweet biscuits. It can also be spread on bread and topped with apple jelly.

Perilla frutescens

SHISO

Also known as Beefsteak Plant, Chinese Basil, Wild Sesame, Rattle Snake Weed, Egoma and Zi su. From the family *Lamiaceae*.

The common name 'beefsteak plant' is in reference to the large purple shiso leaves looking like a slice of raw beef. This name originated in the US, where it has become a naturalised wild plant in many southern and eastern states. Shiso has been used in Chinese medicine for thousands of years to treat morning sickness. Until recently, however, the culinary uses of this Far Eastern herb were relatively unknown in Western countries. Even in the early 1980s purple shiso was being used in the UK only as a spectacular spot bedding plant rather than in the kitchen. It has now become a chef's designer herb, often served in baby leaf form, which intensifies its unique flavours.

Perilla frutescens var. *purpurascens*

varieties

Perilla frutescens* var. *frutescens
Zi su, Egoma
Hardy annual. Height up to 1.5m, spread 60cm. Mauve flowers in summer. Large aromatic, green with a hint of brown, leaves that have a deep purple underside. This is an important culinary and medicinal herb from Korea.

***Perilla frutescens* var. *crispa* AGM**
Green Shiso, Aojiso and Perilla
Hardy annual. Height up to 1.2m, spread 60cm. Pink flowers during summer. Aromatic, anise-flavoured, deeply cut bright green leaf that has crinkled edges.

Perilla frutescens* var. *purpurascens
Purple Shiso, Beefsteak Plant
Hardy annual. Height up to 1.2m, spread 60cm. Pink flowers in summer. Deeply cut, aromatic, dark purple leaf with crinkled bronzed edges. There can be considerable variation in seed-raised plants; the flowers can be red and the leaves smoother.

cultivation

Propagation

Seed

In spring sow seeds into prepared seed trays, module plugs or small pots using a seed compost mixed in equal parts with perlite, and place under protection at 20°C. Germination takes 1–2 weeks. Do not over water once germinated as seedlings are prone to damping off, especially when the nights are cold. Once large enough to handle, pot up using a potting compost mixed in equal parts with composted fine bark. Grow on until seedlings are large enough to plant out into a prepared site 30cm apart. Alternatively sow into prepared open ground in late spring, when the air temperature does not go below 8°C at night. Germination takes 14–20 days.

Pests and Diseases

Beware of caterpillars, especially in late summer. Hand pick off as soon as spotted.

Maintenance

Spring Sow seeds.
Summer Keep pinching out the growing tips to

SHISO

Perilla frutescens var. *frutescens*

maintain shape and to produce a bushy plant.
Autumn Harvest leaves, then seeds.
Winter Clean the seed for next year's sowing.

Garden Cultivation
Plant in a fertile well-drained soil on a site that has been prepared with well-rotted compost or leaf mould in the previous autumn. Plant in the late spring, in sun or partial shade. In arid climates, regular irrigation will be necessary.

Harvesting
Pick leaves to use fresh as required throughout the growing season. Pick flowering tops in late summer. Harvest the seeds in late autumn.

container growing

This herb looks lovely growing in containers, especially the purple-leaved varieties. Plant in a potting compost mixed in equal parts with composted fine bark. Place the container in a sunny position. Water regularly and liquid feed weekly through the summer, following the manufacturer's instructions.

medicinal

Shiso has been used for centuries in Oriental medicine. It is a pungent, aromatic, warming herb. An infusion of the plant is useful in the treatment of asthma, colds, coughs and lung afflictions, constipation, food poisoning and allergic reactions, especially from seafood. An infusion made from the stems of this herb is a traditional Chinese remedy to alleviate morning sickness; however this should only be taken under the guidance of a Chinese doctor or fully trained herbalist.

other uses

When walking or hiking in the country, rub the leaves directly on your skin and clothes and on to your dog to repel ticks. Before using, rub a bit on the back of your hand to see if you are allergic to it.

warning

Do not take medicinally during pregnancy. Can cause contact dermatitis. Do not plant where horses or cattle can eat this plant as it can cause respiratory failure.

culinary

This herb is one of the few aromatic plants to have established itself in Japanese cuisine, although it is a relatively newcomer to the European kitchen. It is sometimes confused with basil *(Ocimum)*, however the flavour is very different being, in my opinion, a mixture of cumin, mint and nutmeg with a hint of plums for the purple variety and anise for the green variety. Purple shiso is used as a dye for pickling fruit and vegetables, as a side dish with rice in the form of a dried powder, as an ingredient in cake mixes and as flavouring in beverages. The flower heads are used as a condiment with sushi. Green leaf shiso, the variety most commonly seen in Japanese markets, is used as a vegetable. The leaves are used as a wrapping for rice cake, in salads and tempura. The seeds from this variety are used as a condiment and with pickles. The essential oil extracted from the leaf and flowering parts contains a substance which is used in confectionery that is many times sweeter than sugar.

Shiso, Mooli and Kiwi Fruit Salad

1 mooli (This is a Japanese white radish which has a crunchy texture and a mild peppery flavour and can be found in large supermarkets, or in Asian or Caribbean shops.)

Lemon juice
3 tablespoons light olive oil
1 tablespoon white wine vinegar or rice vinegar
3 kiwi fruits, peeled and sliced
A good handful (approx 24 tips) of young purple shiso new growing tips, 6 tips put aside and the remainder finely chopped
Salt and freshly ground black pepper

Peel and slice the mooli thinly, sprinkling the slices with lemon juice to prevent discoloration. Make the dressing with the oil and vinegar, season to taste. Arrange the sliced kiwi fruit and the mooli on a plate. Scatter over the chopped shiso, then drizzle the dressing over the salad, and decorate with the remaining purple shiso tips.

Persicaria odorata

VIETNAMESE CORIANDER

Also known as Rau Ram, Laksa Plant and Vietnamese Coriander. From the family *Polygonaceae*.

Vietnamese Coriander is indigenous throughout the tropics and subtropics of South and Eastern Asia, where it is used in the kitchen and as a herbal remedy. It is a member of the knotweed family of plants that are notoriously invasive and more often than not considered as weeds. It should not be confused with a native wild flower found in Northwest Europe called *Persicaria bistorta*, common bistort, which, has pink flowers and leaves that smell of vegetables when crushed, as opposed to the warm fragrant spicy scent of the tropical Vietnamese coriander leaves.

varieties

Persicaria odorata
Vietnamese Coriander
Tropical evergreen perennial. Ht. 45cm, spread indefinite. Attractive small creamy white flowers in summer until late autumn, rarely produced under cultivation and in cold climates. Highly aromatic leaves when crushed, narrow pointed, with a brown/maroon V-shaped marking near the base.

Persicaria bistorta
Common Bistort
Hardy perennial. Ht. 20–100cm, spread indefinite. Dense clusters of small pink flowers in summer until early autumn. Hairless triangular mid-green leaves.

Medicinally the root of this species is used to treat staunch blood flower and contract tissues. It is one of the most astringent of all medicinal plants.

cultivation

Propagation

Cuttings

Cutting can be taken from spring until late summer. Take the cutting just below the stem joint, place in prepared module plug trays or a small pot filled with a standard seed compost. Put in a sheltered warm position, bottom heat is not required; it will root within 10 days in mid-spring. Alternatively, as this plant will root anywhere that the stem touches the ground, it is possible to take a stem cutting with some roots attached. Grow on in exactly the same way as taking a standard cutting. Once the cuttings are fully rooted in warm climates plant out into the garden into a prepared site; in cool and cold climates pot up and grow on as a container plant.

Division

Garden and container-grown plants will need dividing to keep the plant healthy and productive and also to stop it either invading the garden or outgrowing the pot. With both situations you can be as vigorous as you like, either replanting divisions into a prepared site in the garden or repotting into a loam-based potting compost.

Pests and Diseases

This highly aromatic plant rarely suffers from any pests or diseases.

Persicaria odorata flower

VIETNAMESE CORIANDER

Maintenance
Spring Reintroduce regular watering. Take cuttings.
Summer Cut back to produce new growth.
Autumn Protect from frosts.
Winter In cool and cold climates cut back on watering.

Garden Cultivation
As this herb is a tropical plant it will need protection when temperatures fall below 7°C at night. It can be grown outside in summer, planted in a rich fertile soil, in partial shade, but it will need lifting well before the winter before the first hint of frost. Be warned in warm climates: if you introduce this plant to your garden it will be more invasive than mint.

Harvesting
Pick the leaves to use fresh as required throughout the growing season.

container growing

Vietnamese coriander grows very happily in containers; use a loam-based potting compost. Place the container in a warm greenhouse, conservatory or on a windowsill. Protect from the midday sun as this will and can scorch the leaves, which will then taste bitter. To maintain good-quality leaves, either divide the plant annually or pot up a size of pot. Water and liquid feed regularly from spring until autumn. Outside the tropics it is wise to reduce the watering and stop feeding in the winter months.

medicinal

Throughout the Far East, Vietnamese coriander is used to treat indigestion, flatulence and stomach aches. It is also reputedly eaten by Buddhist monks to suppress sexual urges.

other uses

Currently in Australia there is interesting research into the essential oil of Vietnamese coriander, called Kesom oil, for use in food flavouring.

culinary

In some countries this herb goes under the name Vietnamese mint, which is a misnomer in all aspects as it is certainly not a member of the mint family (Lamiaceae), nor does it in any way have a mint flavour. The first time I ate the leaves of this herb I was totally taken by surprise. To begin with the taste is mild with a hint of lime and spice, then as the flavour developed it became hot and peppery. This herb is never cooked, it is used as a fresh leaf condiment, always added at the end of cooking. In Asia it is usually eaten raw as a salad or herb accompaniment. It combines well with meat, vegetables and fruit. It is important in Vietnamese cooking being used copiously with noodle soups (pho) where large heaps of the leaves are dipped into the soup using chopsticks. Personally I like it scattered over stir-fried vegetables, which then gives the dish an extra zing.

Stir-fried Vegetables
Serves 4

10g dried Chinese mushrooms, soaked in warm water for 20 minutes.
4 tablespoons sesame seed oil
100g white cabbage, washed and finely sliced
100g carrots, peeled and sliced into thin strips
100g cucumber, cut into thin strips
100g bamboo shoots.
4 tablespoons chicken or vegetable stock
2 tablespoons soy sauce
pinch of sugar
Salt and freshly ground black pepper to taste
10 Vietnamese coriander leaves, finely chopped.

Drain the Chinese mushrooms and cut into small pieces. Heat the oil in a large frying pan or wok, add the cabbage, stir-fry for 2 minutes. Add the mushrooms, carrots, cucumber and bamboo shoots, stir-fry for a further 2 minutes. Add the stock, soy sauce, and pinch of sugar, salt and pepper to taste. Stir-fry for 2 minutes to heat through, add the chopped Vietnamese coriander leaves and serve with rice or noodles.

Petroselinum

PARSLEY

Also known as Common Parsley, Garden Parsley and Rock Parsley. From the family *Apiaceae.*

Best-known of all garnishing herbs in the West. Native to central and southern Europe, in particular the Mediterranean region, now widely cultivated in several varieties throughout the world.

The Greeks had mixed feelings about this herb. It was associated with Archemorus, the Herald of Death, so they decorated their tombs with it. Hercules was said to have chosen parsley for his garlands, so they would weave it into crowns for victors at the Isthmian Games. But they did not eat it themselves, preferring to feed it to their horses. However, the Romans consumed parsley in quantity and made garlands for banquet guests to discourage intoxication and to counter strong odours.

It was believed that only a witch or a pregnant woman could grow it, and that a fine harvest was ensured only if the seeds were planted on Good Friday. It was also said that if parsley was transplanted, then misfortune would descend upon the household.

varieties

Petroselinum crispum
Parsley
Hardy biennial. Ht 30–40cm. Small creamy white flowers in flat umbels in summer. The leaf is brightish green and has curly toothed edges and a mild taste. It is mainly used as a garnish.

***Petroselinum crispum* French**
French Parsley
Also known as Broad-Leafed Parsley.
Hardy biennial. Ht 45–60cm. Small creamy white flowers in flat umbels in summer. Flat dark green leaves with a stronger flavour than *P. crispum*. This is the one I recommend for culinary use.

Petroselinum crispum* var. *tuberosum
Hamburg Parsley
Also known as Turnip-Rooted Parsley. Perennial, grown as an annual. Root length up to 15cm. Leaf, green and very similar to French parsley. This variety, probably first developed in Holland, was introduced into England in the early eighteenth century, but it was only popular for 100 years. The plant is still frequently found in vegetable markets in France and Germany, where it is sold as a root vegetable.

Warning: In the wild there is a plant called fool's parsley *(Aethusa cynapium)*, which both looks and smells to the novice like French parsley. Do not be tempted to eat it as it is extremely poisonous.

cultivation

Propagation

Seed

In cool climates, to ensure a succession of plants, sow seedlings under cover only in plug trays or pots. Avoid seed trays because it hates being transferred. Cover with perlite. If you have a heated propagator, a temperature of 18°C will speed up germination. It takes 4–6 weeks without bottom heat and 2–3 weeks with. When the seedlings are large enough and the air and soil temperature have started to rise (about mid-spring), plant out 15cm apart in a prepared garden bed. Alternatively, in late spring sow into a prepared site in the garden, in drills 30–45cm apart and about 3cm deep. Germination is very slow. Keep

Parsley seed tea

the soil moist at all times, otherwise the seed will not germinate. As soon as the seedlings are large enough, thin to 8cm and then 15cm apart.

Pests and Diseases

Slugs love young parsley plants. There is a fungus that may attack the leaves. It produces first brown then white spots. Where this occurs the whole stock should be destroyed. Get some fresh seed.

Maintenance

Spring Sow seed.
Summer Sow seed. Cut flower heads as they appear on second-year plants.
Autumn Protect plants for winter crop.
Winter Protect plants for winter picking.

Garden Cultivation

Parsley is a hungry plant, it likes a good deep soil, not too light and not acid. Always feed the chosen site well in the previous autumn with well-rotted manure.

If you wish to harvest parsley all year round, prepare two different sites. For summer supplies, a western or eastern border is ideal because the plant needs moisture and prefers a little shade. For winter supplies, a more sheltered spot will be needed in a sunny place.

The seeds should be sown thinly, in drills. If at any time the leaves turn a bit yellow, cut back to encourage new growth and feed with a liquid fertiliser. At the first sign of flower heads appearing, remove them if you wish to continue harvesting the leaves. Remember to water well during hot weather. In the second year parsley runs to seed very quickly. Dig it up as soon as the following year's crop is ready for picking, and remove it from the garden.

Hamburg or turnip parsley differs only in the respect that it is a root not a leaf crop. When the seedlings are large enough, thin to 20cm apart. Water well all summer. The root tends to grow more at this time of year, and unlike a lot of root crops the largest roots taste the best. Lift in late autumn, early winter. They are frost resistant.

Harvesting

Pick leaves during first year for fresh use or for freezing (by far the best method of preserving parsley).

Dig up roots of Hamburg parsley in the autumn of the first year and store in peat or sand.

container growing

Parsley is an ideal herb for containers, it even likes living inside on the kitchen windowsill, as long as it is watered, fed, and cut. Use a standard potting compost mixed in equal parts with composted fine bark. Curly parsley can look very ornamental as an edging to a large pot of nasturtiums. It can also be grown in hanging baskets, (keep well watered), window boxes (give it some shade in high summer), and containers. That brings me to the parsley pot, the one with six holes around the side. Do not use it. As I have already said, parsley likes moisture, and these containers dry out too fast as the holes in the side are small and make it very difficult to water; also, parsley has too big a tap root to be happy.

medicinal

All parsleys are a rich source of vitamins, including vitamin C. They are also high in iron and other minerals and contain the antiseptic chlorophyll. It is a strong diuretic suitable for treating urinary infections as well as fluid retention. It also increases mothers' milk and tones the uterine muscle. Parsley is a well-known breath freshener, being the traditional antidote for the pungent smell of garlic. Chew raw, to promote a healthy skin. Use in poultices as an antiseptic dressing for sprains, wounds and insect bites.

other uses

A tea made from crushed seeds kills head lice vermin. Pour it over the head after washing and rinsing, wrap your head in a towel for 30 minutes and then allow to dry naturally. Equally, the seeds or leaves steeped in water can be used as a hair rinse.

warning

Avoid medicinal use during pregnancy. There is an oil produced from parsley, but it should be used only under medical supervision.

culinary

Parsley is a widely used culinary herb, valued for its taste as well as its rich nutritional content. Cooking with parsley enhances the flavour of other foods and herbs. In bland food, the best flavour is obtained by adding it just before the end of cooking.

As so many recipes include parsley, here are some basic herb mixtures.

Fines Herbes

You will see this mentioned in a number of recipes and it is a classic for omelettes.

1 sprig parsley, chopped
1 sprig chervil, chopped
Some chives cut with scissors
1–2 leaves French tarragon, chopped

Combine all the herbs and add to egg dishes.

Fish Bouquet Garni

2 sprigs parsley
1 sprig French tarragon
1 sprig fennel (small)
2 leaves lemon balm

Tie the herbs together in a bundle and add to the cooking liquid.

Boil Hamburg parsley as a root vegetable or grate raw into salads. Use in soup mixes; the flavour resembles both celery and parsley.

Phlomis fruticosa

JERUSALEM SAGE

From the family *Lamiaceae*.

Originates from the Mediter-ranean region but is now cultivated widely as an ornamental garden plant.

The generic name, *Phlomis*, was used by Dioscorides, a Greek physician in the first century whose *Materia Medica* was the standard reference on the medical application of plants for over 1,500 years.

varieties

***Phlomis fruticosa* AGM**
Jerusalem Sage
Hardy evergreen perennial. Ht and spread 1.2m. Whorls of hooded yellow flowers in summer. Grey/green oblongish leaves, slightly wrinkled.

Phlomis italica
Narrow-Leaved Jerusalem Sage
Hardy evergreen perennial. Ht 90cm, spread 75cm. Whorls of lilac/pink flowers in mid-summer, borne at the ends of shoots amid narrow, woolly, grey/green leaves.

Phlomis italica

cultivation

Propagation

Seeds
Sow the medium-sized seed in the autumn into either seed or plug trays and cover with a thin layer of compost. Winter in a cold greenhouse or cold frame. Does not need stratification nor heat, just cool temperature. Germination is erratic. When the seedlings are large enough to handle, prick out into pots using a potting compost mixed in equal parts with composted fine bark. Plant the young plants into the garden when there is no threat of frosts.

Cuttings
Take softwood cuttings in summer from non-flowering shoots; they root easily.

Division
If an established plant has taken over its neighbour's spot, dig up and divide it in the spring; re-plant into a prepared site.

Pests and Diseases
This plant is mostly free from pests and diseases.

Maintenance
Spring Divide established plants if need be.
Summer Cut back after flowering to maintain shape.
Autumn Sow seeds.
Winter Protect outside plants if the winter temperature is persistently below –5°C.

Garden Cultivation
Jerusalem sage is an attractive plant, making a fine mound of grey-furred leaves that cope with all but the most severe winter. A prolific summer flowerer, happy in a dry, well-drained, sunny spot. Cut back each year after flowering (late summer) and you will be able to control and maintain its soft grey dome all year round. Do not trim in the autumn as any frost will damage and in some cases kill the plant.

Harvesting
Pick leaves for drying before plant flowers.

container growing

Jerusalem sage is happy in a large container using a soil-based compost mixed in equal parts with composted fine bark. Be mean on the feeding and watering as it is a drought-loving plant. Trim back especially after flowering to restrict its rampant growth. Protect during the winter in a cool greenhouse or conservatory. Keep watering to the absolute minimum.

other uses

The slightly aromatic leaves are attractive in a potpourri.

culinary

Although not listed amongst culinary herbs, the leaves are pleasantly aromatic. In Greece the leaves are collected from the hillside and, once dried and bundled together with other related species, are hung up for sale. The dried leaves can be used in stews and casseroles.

Phytolacca americana

POKE ROOT

Also known as Red Ink Plant, Virginia Poke Weed, Pigeon Berry, Coccum, Poke, Indian Poke, American Poke and Cancer Root. From the family *Phytolaccaceae.*

This herbaceous plant is a native to the warmer regions of America (especially Florida), Africa and Asia. It has been introduced elsewhere, particularly in the Mediterranean region.

Its generic name is derived from two Greek words: *phyton*, meaning 'plant' and *lac*, meaning 'lake', referring to the purple/blue dye that flows from some of the phytolaccas when crushed.

The herb was introduced to European settlers by the Native Americans, who knew it as pocan or coccum, and used it as an emetic for a number of problems. It acquired a reputation as a remedy for internal cancers and was called cancer root.

varieties

Phytolacca americana* syn. *P. decandra

Poke Root

Hardy perennial. Ht and spread 1.2–1.5m. Shallow, cup-shaped flowers, sometimes pink, flushed white and green, borne in terminal racemes in summer. They are followed by round fleshy blackish-purple berries with poisonous seeds that hang down when ripe. Oval to lance-shaped mid-green leaves, tinged purple in autumn. There is a variegated form with green and white leaves.

Phytolacca polyandra* syn. *P. calvigera

Hardy perennial. Ht and spread 1.2m. Clusters of shallow, cup-shaped, pink flowers in summer, followed by rounded blackish berries with poisonous seeds. Has brilliant crimson stems, oval to lance-shaped, mid-green leaves that turn yellow in summer through autumn. This plant is a native of China.

cultivation

Propagation

Seed

Wearing gloves, sow the seeds fresh in the autumn or spring in prepared seed or plug trays. Cover with perlite. If sown in the autumn, winter the young plants in a cold greenhouse or cold frame. In the spring, after a period of hardening off, plant them out in a prepared site in the garden, 1m apart.

Division

Both species have large root systems that can be divided (wearing gloves) either in autumn or spring.

Pests and Diseases

Largely free from pests and diseases.

Maintenance

Spring Sow seeds. Divide established plants.

Phytolacca americana

Summer Cut off the flowers if you do not want berries.

Autumn Sow seeds. Divide established plants.

Winter Dies back into the ground; no protection needed.

Garden Cultivation

Plant poke root in sun or shade in a moist, fertile soil, sheltered from the wind. This plant can look marvellous in a garden, especially in autumn.

Harvesting

It can be used as a pot herb, the young shoots being picked in the spring. But because it is easy to confuse the identity of species, and toxicity varies among them, only do this if you really know what you are doing. So it is better to err on the side of caution and pick some nice fresh sorrel or red orach instead.

container growing

It is a tall plant, and when in berry is sufficiently heavy to unbalance even a large pot. Keep the poisonous berries out of reach of children. If you choose to try it, use a soil-based compost mixed in equal parts with composted fine bark and water well during the summer months.

medicinal

Herbalists prescribe it for the treatment of chronic rheumatism, arthritis, tonsillitis, swollen glands, mumps and mastitis. An extract from the roots can destroy snails. This discovery is being explored in Africa as a possible means to control the disease bilharzia, which is carried by water snails.

warning

POISONOUS. When handling either seeds, roots or the mature plant, gloves should be worn. It is toxic and dangerous. It should be used only by professionally trained herbalists.

Polemonium caeruleum

JACOB'S LADDER

Polemonium reptans

Also known as Blue Jacket, Charity, Jacob's Walking Stick, Ladder to Heaven, Greek Valerian. From the family *Polemoniaceae*.

This herb is steeped in history. It was known to the ancient Greeks as *Polemonium* and they administered a decoction made from the root mixed with wine in cases of dysentery, toothache and against the bites of poisonous animals. As late as the nineteenth century it was known as Valeranae Graecae or Greek Valerian and was being used in some European pharmacies. It was predominantly used as an anti-syphilitic agent and in the treatment of rabies. To confuse things, the American Shakers called it 'Abscess' and used it for pleurisy and fevers.

varieties

Polemonium caeruleum
Jacob's Ladder
Hardy perennial. Ht and spread 45–60cm. Clusters of attractive cup-shaped lavender-blue flowers in summer. The mid-green leaves are finely divided into small lance shapes.

Polemonium reptans
Also known as false Jacob's ladder or American Greek valerian. Hardy perennial. Ht 20–45cm and spread 30cm. Clusters of attractive cup-shaped blue flowers in summer. The silver/green leaves are finely divided into small lance shapes. Native of Eastern North America. The root of this species is bitter in flavour and is employed as an astringent and as an antidote against snake bites.

cultivation

Propagation
Seed
For flowering early the following spring, sow the fairly small seeds in autumn into a prepared seed or plug module tray using a standard seed compost. Winter the seedlings in a cool/cold, but frost-free greenhouse. Prick out in spring, when the threat of frosts has passed, then plant directly into the garden after hardening off, at a distance of 30cm apart.

Division
Divide established plants in the spring. Dig the whole plant up and ease it in half using two forks back to back. Replant in a prepared site in the garden.

Pests and Diseases
These plants rarely suffer from pests or diseases.

Maintenance
Spring Sow seeds if not previously done, divide established plants if needed.
Summer Deadhead flowers and, after flowering, cut back to prevent self-seeding.
Autumn Sow seeds under protection.
Winter Established plants are hardy and should not need protection.

Garden Cultivation
Jacob's ladder prefers a rich, moisture-retentive soil, with added lime in a sunny position. However it is a most tolerant plant and will adapt to most soils with the exception of those that are very dry.

Harvesting
Cut the flowers just as they open for drying. Dry either in small bunches or individual sprays.

container growing

This herb looks most attractive grown in a container. Use a loam-based potting compost and place the container in semi-shade to help prevent the compost from drying out. Feed regularly only when the plant is flowering using a liquid fertiliser and following the manufacturer's instructions.

medicinal

This herb is rarely used in modern herbal medicine.

other uses

The dried flowers do not smell, but do look attractive in potpourris.

When the roots are combined with olive oil, it makes a black dye.

Polemonium caeruleum

Polygonatum

SOLOMON'S SEAL

Also known as David's Harp, Jacob's Ladder, Lady's Lockets, Lily of the Mountain, Drop Berry, Seal Root or Sealwort. From the family *Convallariaceae*.

This plant's generic name *Polygonatum* is derived from *Poly*, meaning 'many' and *gonu*, meaning 'knee joint', which may refer to the many-jointed rhizome.

King Solomon, wiser than all men, gave his approval to the use of the roots (which are said to resemble cut sections of Hebrew characters) as a poultice for wounds, and to help heal broken limbs.

varieties

☠ *Polygonatum x hybridum* AGM
Solomon's Seal
Hardy perennial. Ht 30–80cm, spread 30cm. White waxy flowers tipped with green hang from arching stems in spring to summer and are followed by blue/black berries. Oval to lance-shaped, mid-green leaves.

☠ *Polygonatum odoratum* AGM syn. *Polygonatum officinale*
Angular Solomon's Seal
Hardy perennial. Ht 60cm, spread 30cm. Produces pairs of fragrant tubular bell-shaped, green-tipped white flowers in spring, which are followed by blue/black berries. Oval to lance-shaped mid-green leaves.

cultivation

Propagation

Seed
In autumn, sow fresh seeds into prepared seed or plug module trays using a standard seed compost mixed in equal parts with fine composted bark. Cover with compost then with glass and leave outside for the winter. Remove the glass as soon as germination starts in early spring. Once the seedlings are large enough to handle, plant out into a well-prepared site, and water regularly throughout the first season.

Division
This is a far more reliable method of propagation. Divide established plants in spring or autumn. Replant into a prepared site in the garden.

Pests and Diseases
If you notice clean-cut holes in the leaves it is likely to be the sawfly caterpillar. This will not irreversibly damage the plant, but it can look devastated. Pick the larvae off or treat with soft horticultural soap, although this is unlikely to completely eradicate the caterpillars.

Maintenance
Spring When the soil is damp, divide established plants.
Summer Make sure the soil does not dry out, and water if necessary.
Autumn Sow fresh seeds, divide established plants.
Winter Protect with a layer of mulch if the winter has a prolonged frost below −10°C.

Garden Cultivation
Plant in a cool shady position in fertile well-drained soil. Dig the soil over before planting with some well-rotted leaf mould, and each winter top dress with extra leaf mould. This plant looks much better planted in large drifts than as single plants.

Harvesting
In the autumn, after the foliage has died back, dig up and dry the roots for medicinal use from a well-established 3-year-old plant.

container growing

Solomon's seal can be successfully grown in a container. Use a soil-based potting compost and top dress in autumn with well-rotted leaf mould. Position the container in semi-shade and water regularly throughout the summer months.

medicinal

The powdered roots and rhizomes make a good poultice for bruises, inflammation and wounds and a good skin wash for rashes and blemishes.

other uses

The plant has for centuries been employed as a cosmetic to clear freckles and as a skin tonic.

warning

All parts of the plant are poisonous and should be taken internally only under the supervision of a qualified medicinal or herbal practitioner. Large doses can be harmful.

Primula veris

COWSLIP

Also known as St Peter's Keys, Palsywort, Coweslop, Fair Bells and Keys of Heaven. From the family *Primulaceae.*

This traditional herb is a native of Northern and Central Europe. In America the plant that is called cowslip is in fact the English marsh marigold *(Caltha palustris)* and is not to be confused with the above. A legend of northern Europe is that St Peter let his keys to Heaven drop when he learned that a duplicate set had been made. Where they fell the cowslip grew, hence the common name Keys of Heaven.

varieties

***Primula veris* AGM**
Cowslip
Hardy perennial. Ht and spread 15–20cm. Tight clusters of fragrant tubular, yellow flowers produced on stout stems in spring. Oval mid-green leaves.

Cowslips are often mistaken for oxlip *(P. elatior)*, which is a hybrid of the cowslip and the primrose *(P. vulgaris)*. Oxlips have large pale yellow flowers in a one-sided cluster.

cultivation

Propagation

Seed
Sow fresh seed in autumn on to the surface of prepared plug module trays or small pots, using a standard seed compost. Cover with perlite or a piece of glass. Place in a cold frame or cold greenhouse. Germination takes 4–6 weeks. If covered with glass, remove immediately you see the seeds break. Winter the young plants in the cold frame or cold greenhouse before planting out in the following spring. If you sow old seed in winter you will need to stratify the seed (the cold treatment; see page 644) to enable the seed to germinate.

Division
All *Primula* divide easily and this is by far the easiest method of propagation. Divide established plants in early autumn either using two hand forks back to back or by digging up a clump, dividing by hand and replanting into a prepared site 15cm apart.

Ensure plants bought come from a cultivated source and are not dug from the wild.

Pests and Diseases

The scourge of all *Primula* plants, when grown in containers, is the vine weevil. To eradicate in pots water with nematodes, following the manufacturer's instructions, in spring and autumn when the soil temperature does not fall below 5°C.

Maintenance

Spring Clear winter debris from around plants.
Summer Deadhead if you do not require seeds.
Autumn Collect fresh seeds and sow. Divide established plants.
Winter No need for protection.

Dried cowslips

Garden Cultivation

Plant cowslips in semi-shade or sun, in a moist but well-drained soil. They look better grown in clumps and drifts rather than singularly.

Harvesting

I am sure I do not need to remind you, please do not dig up wild plants. This is prohibited in many European countries. Pick leaves and flowers as required to use fresh. Dig up the roots of cultivated cowslips for drying in autumn.

container growing

Cowslips adapt happily to being grown in containers outside. Use a loam-based compost. Protect the container from the midday sun.

culinary

Use young leaves and flowers in salads, the leaves can also be added to meat stuffings. The flowers can be used to make wine.

medicinal

A tea made from the flowers is a simple remedy for insomnia and nervous tension. Cowslip syrup was a country remedy for palsy, hence the common name palsywort. The roots have a high saponin content and are used to treat whooping cough and bronchitis.

warning

Primula veris is renowned for causing contact dermatitis.

Primula vulgaris

PRIMROSE

Also known as Early Rose, Easter Rose, First Rose and May Flower. From the family *Primulaceae*.

This herald of spring is a native of Europe. The name Primrose originates from the Latin *prima*, meaning 'first' and *rosa*, meaning 'rose'.

In the Middle Ages, concoctions were made from primroses which were used as a remedy for gout and rheumatism.

Due to the encouragement of sympathetic farming practices, one can now again see this once-endangered wild flower in the hedgerows. However it is still illegal to pick or dig this flower in the wild.

varieties

***Primula vulgaris* AGM**
Primrose
Hardy perennial. Ht and spread 15cm. The fresh yellow, sweetly scented flowers with darker yellow centres are borne singly on hairy stems in early spring. Mid-green, wrinkled, oval leaves.

cultivation

Propagation
Seed
Sow fresh seed, while it is still green and before it becomes dry and turns brown, in late summer on to the surface of prepared plug module trays or small pots, using a standard seed compost. Cover with perlite or a piece of glass. Place in a cold frame or cold greenhouse. Germination takes 4–6 weeks. If covered with glass, remove when you see the seeds break. Winter the young plants in the cold frame or cold greenhouse before planting out in the following spring. If you sow dry brown seeds in winter you will need to stratify the seed (the cold treatment, see page 644). Be patient, it can take 2 years to germinate.

Division
All *Primula* divide easily, and this is by far the easiest method of propagation. Divide established plants in early autumn either using two hand forks back to back or by digging up a clump, dividing by hand and replanting into a prepared site 15cm apart.

Pests and Diseases
The scourge of all *Primula* plants, when grown in containers, is the vine weevil. To eradicate in pots, water with nematodes, following the manufacturer's instructions, in spring and autumn when the soil temperature does not fall below 5°C.

Maintenance
Spring Plant out autumn-germinated plants.
Summer Sow fresh seed.
Autumn Divide established plants.
Winter No need for protection.

Garden Cultivation
Plant primroses in semi-shade, under a deciduous tree or near hedgerows, in a moist but well-drained soil. They look better grown in clumps and drifts rather than singly. When growing primroses in a wild garden, delay cutting the grass until mid-summer, which is after the plants have self-seeded.

Primrose tisane

Harvesting
In spring pick young leaves and flowers to use fresh. In summer collect seed for immediate sowing.

container growing

Primroses adapt happily to being grown in containers outside and are a great pick me up as they flower early in spring. Use a loam-based compost. Position the container in semi-shade.

culinary

The flowers are lovely in green salads, and they can be crystallised to decorate puddings and cakes. The young leaves make an interesting vegetable when steamed and tossed in butter.

medicinal

Medicinally this herb is rarely used today. However, if nothing else is available, a tisane made from the leaves and flowers can be used as a mild sedative.

warning

Primulas are renowned for causing contact dermatitis.

PROSTANTHERA

Also known as Mint Bush. From the family *Lamiaceae*.

These highly attractive aromatic shrubs are natives of Australia.

I have fallen in love with these most generous of flowerers. When I was exhibiting one in flower at the Chelsea Flower Show, some member of the public fell in love with it in equal measure and tried to liberate it from my display!

I can find no historical references other than in the RHS *Dictionary of Gardening*, which states that the generic name, *Prostanthera*, comes from *prostithemi*, 'to append', and *anthera*, meaning 'anther', the pollen-bearing part of the stamen. This therefore alludes to the appendages usually borne by the anthers.

Prostanthera ovalifolia

varieties

***Prostanthera cuneata* AGM**
Evergreen half-hardy perennial. Ht and spread 60–90cm. Very attractive white flowers with purple spots that look rather like little orchids; late spring, early summer. Round, dark green, slightly leathery and shiny, mint-scented leaves. Can withstand a minimum temperature of –2°C.

Prostanthera lasianthos
Victorian Christmas bush.
Evergreen large scrub. Ht 1–6m. Profuse sprays of white pale lilac flowers. Long narrow toothed menthol-scented green leaves.

***Prostanthera ovalifolia* AGM**
Evergreen tender perennial. Reaches a height and spread of 1.2m in its native country. Attractive purple flowers on short leafy racemes throughout spring and summer. Dark green aromatic leaves. Can only withstand a minimum temperature of 5°C.

***Prostanthera rotundifolia* 'Rosea' AGM**
Evergreen half-hardy perennial. A small tree up to 3m tall in its native country; in cooler climates it's a lot smaller. Pretty mauve/purple flowers in spring that last a long time. The dark green leaves (not as dark as *P. cuneata*) are round and mint-scented. Withstands a minimum temperature of 0°C.

cultivation

Propagation
Cuttings
Take cuttings in spring or late summer. Use a standard seed compost mixed in equal parts with composted fine bark. When the cuttings are well rooted, in 8–12 weeks, pot up again using the same mix and keep in containers for the first year.

Pests and Diseases
Over-watering young plants causes root rot – a killer.

Maintenance
Spring Take cuttings.
Summer Cut back after flowering only if necessary.
Autumn Protect from frosts.
Winter Protect from hard frosts and excessive water.

Garden Cultivation
In cool climates with persistent frosts they are better grown in a container. However, if your climate is mild, plant out in the spring in a warm corner, in a lime-free, well-draining soil at a distance of 60–90cm apart. Rain combined with frost is the killer in winter. If you want to make a low hedge out of *Prostanthera cuneata* then plant specimens 45cm apart.

Harvesting
Pick leaves in the summer after flowering for drying and inclusion in potpourris.

container growing

This is a real crowd puller when in flower, and even when not, makes a most attractive aromatic plant. Use a soil-based compost mixed with 25 per cent composted fine bark and 25 per cent peat. Keep young plants on the dry side, but water freely in the growing season.

medicinal

I am sure that a plant such as *P. cuneata* that gives off as much scent, and has obviously so much oil in the leaf, will one day have some use.

Prostanthera lasianthos

Prunella vulgaris

SELF HEAL

Also known as Carpenter's Herb, Sticklewort, Touch and Heal, All Heal, Woundwort, Hercules' Woundwort, Blue Curls, Brownwort and Hock Heal. From the family *Lamiaceae*.

This herb is found growing wild throughout all the temperate regions of the northern hemisphere, including Europe, Asia and North America. It is found on moist, loamy, well-drained soils, in grassland, pastures and open woodland, especially in sunny situations. Now introduced into China and Australia.

In strict 16th-century adherence to the Doctrine of Signatures, whereby it was believed that every plant bore an outward sign of its value to mankind, people noted that the upper lip of the flower was shaped like a hook, and as billhooks and sickles were a main cause of wounds in their agrarian society, they decided that the purpose of the herb was to heal wounds (hence Self Heal). They also saw the shape of the throat in the flower, which was why it was introduced to treat diseases of the throat such as quinsy and diphtheria, a propensity with a precedent in ancient Greece, where physicians used it to cure sore throats and tonsillitis.

varieties

Prunella vulgaris
Self Heal
Hardy perennial. Ht 5–30cm, spread 15–30cm. Clusters of blue/purple flowers all summer. Oval leaves of a bright green.

There is a much rarer white-flowered species, *Prunella laciniata*, which has very deeply cut leaves.

cultivation

Propagation
Seed
Sow the small seeds into prepared seed or plug trays in either spring or autumn and cover with perlite; no extra heat is required. If an autumn sowing, winter the young plants in a cold frame. In spring, when the plants are large enough, plant out at a distance of 15–20cm apart.

Division
This plant grows runners that have their own small root systems and is, therefore, easy to divide. Dig up in the spring or autumn, and split and replant either in the garden or amongst grass.

Garden Cultivation
This plant, which is easy to establish, makes a colourful ground cover with attractive flowers. It is happy in full sun to semi-shade and will grow in most soils, including those that are rather acid, though it does best if the soil is fertile. It can be grown in a lawn, and while the mower keeps its spread and height in check, it will still flower and be much visited by bees and butterflies.

Pests and Diseases
In most cases it is free from pests and diseases.

Maintenance
Spring Sow seed. Divide established plants.
Summer Cut back after flowering to curtail self-seeding.
Autumn Divide established plants. Sow seeds.
Winter No need for protection, fully hardy.

Harvesting
Harvest for medicinal use only. Dry both the leaves and flowers.

container growing

Self heal can be grown in containers using a soil-based compost. However, as it looks a bit insipid on its own, it is better combined with plants like heartsease, poppies and cowslips.

Water well during the growing season, but only feed with liquid fertiliser twice otherwise it will produce too lush a growth.

medicinal

Used in herbal medicines as a gargle for sore throats and inflammation of the mouth. A decoction is used to wash cuts and to soothe burns and bruises.

Prunella vulgaris

Pulmonaria

LUNGWORT

Also known as Jerusalem Cowslip, Adam and Eve, Lady Mary's Tears, Spotted Bugloss, and Spotted Comfrey. From the family *Boraginaceae*.

Lungwort is a native plant of Europe and northern parts of the USA. The markings on the leaves were attributed to the Virgin Mary's milk or her tears; however the generic name, *Pulmonaria*, comes from *pulmo*, meaning 'lung'. The Doctrine of Signatures, which held that each plant must be associated either by appearance, smell or habit, with the disease which it was said to heal, thought that the leaves resembled a diseased lung so used it to treat various lung disorders with some success.

varieties

***Pulmonaria angustifolia* AGM**
Semi-evergreen hardy perennial. Ht 23cm, spread 20–30cm. Pink flowers which turn to bright blue in spring. Lance-shaped mid-green leaves with no markings.

Pulmonaria longifolia
Hardy perennial. Ht 30cm, spread 45cm. Pink flowers which turn purplish-blue in spring. Lance-shaped, slightly hairy, dark green leaves with white spots.

Pulmonaria officinalis
Lungwort
Semi-evergreen hardy perennial. Ht 30cm, spread 60cm. Pink flowers which turn blue in spring. Oval leaves with blotchy white/cream markings on a mid-green slightly hairy surface.

Note The American native Virginian cowslip, *Mertensia virginica,* is known as smooth lungwort. It belongs to the same *Boraginaceae* family, but does not have the same medicinal properties.

cultivation

Propagation
Seed
This plant rarely produces viable seed. However it has been known to self-seed erratically around the garden.

Division
This is by far the easiest way to propagate this herb. Divide established plants either after flowering in late spring or alternatively in early autumn using the two forks back to back technique.

Pests and Diseases
In long, dry, hot summers this herb can be prone to powdery mildew. Cut back and burn infected growth, and then water the plant well.

Maintenance
Spring Dig up and replant seedlings.
Summer Water well, especially if the ground is drying out.
Autumn Divide established plants.
Winter No need for protection; fully hardy.

Garden Cultivation
Lungwort prefers to be planted in semi shade and in a moist but free-draining soil, which has been well fed with leaf mould or well-rotted compost the season prior to planting. The leaf mould is particularly important if you live in a drought area or if your soil is prone to becoming excessively dry in summer.

Harvesting
For medicinal use, pick the leaves after flowering for drying.

container growing

Lungwort can be grown successfully in a container as long as it has enough room for the creeping rhizomes to spread. Use a loam-based potting compost in a frost-hardy container. Winter the plants outside; if brought in they become prone to mildew.

medicinal

Lungwort is a soothing expectorant. The silica contained in the leaves was traditionally used to restore the elasticity of the lungs. Externally the leaves have been used for healing all kinds of wounds.

warning

Like many in the *Boraginaceae* family, lungwort has leaves that can cause contact dermatitis.

Lungwort potpourri

Rheum palmatum

CHINESE RHUBARB

Also known as Da Huang. From the family *Polygonaceae*.

To many people rhubarb may seem a curiously English fruit; it is in fact a vegetable and an important medicinal herb, native to China and Tibet where it has been used for over 2,000 years to treat many digestive problems. It was first mentioned in the Shen Nong *Canon of Herbs* which was written in the Han dynasty (206 BC–AD 23).

varieties

Rheum palmatum
Chinese Rhubarb
Herbaceous perennial. Ht up to 2m in flower, spread 60–75cm. Panicles of deep red flowers in summer followed by 3-winged fruit. Large, up to 90cm, palmate, lobed mid-green leaves. Nearly round green stems are NOT edible.

cultivation

Propagation

Division
The easiest method of propagation is by root division. To start, buy a plant from a reputable nursery. Once you have an established plant, over 3 years old, it can then be divided as necessary. Do this when the plant is dormant, in late autumn. Lift the crown, then with a sharp knife or spade, divide the crown into 10cm sections including as much root as possible and a new bud with each section. Pot up immediately into a soil-based potting compost with the bud tip just below the surface of the compost. Leave the container in a cold frame or cold greenhouse until early spring, then plant out into a previously prepared site at a distance of 75cm apart.

Pests and Diseases

Rhubarb can suffer from crown rot, which causes the rotting of the stalks and leaves. If you see infected plants, dig up and burn and do not replant in that position again.

Maintenance

Spring Plant out divided plants.
Summer Remove flower heads once they form seeds to prevent self seeding.
Autumn Divide established plants. Feed plants with well-rotted manure.
Winter No need for protection, fully hardy.

Garden Cultivation

Plant in a sunny position, in a deep fertile soil that has been fed with well-rotted manure in the previous autumn. If you are growing this herb for use medicinally it will need to stay in position for at least 6 years, so choose your site with care.

Harvesting

In the autumn lift 6-year-old rhizomes for medicinal use. Either dry or create tinctures.

container growing

Rhubarb is not ideally suited for growing in containers because of its long root system.

culinary

Chinese rhubarb is not a culinary herb.

medicinal

The rhizomes are used to improve digestion, for liver and gall bladder complaints and as a laxative. It can be made into a mouthwash for mouth ulcers but it tastes revolting! In homeopathy it is used to treat teething children.

other uses

Traditionally an infusion of the leaves was used as an effective spray to control aphids and to check black spots on roses. As this infusion is poisonous when used in the kitchen garden, the gardeners would not harvest the treated vegetables for a minimum of 7 days.

warning

The leaves are very toxic; on no account eat them. Do not take as a medicine during pregnancy or when breastfeeding.

Ground Chinese rhubarb root

Ricinus communis

CASTOR OIL PLANT

Also known as Palma-Christi, Eranda and Rendi. From the family *Euphorbiaceae*.

The castor oil plant is one of the first medicinal plants known to man. Its seeds have been found in Egyptian tombs and it is mentioned in papyrus scrolls written in 1500 BC. The Greek historian Herodotus noted its use as a lamp oil, a purpose for which it is still employed in the temples of Southern India. The lampblack produced by the combination of castor oil with wicks dipped in herbal preparations, has provided the universal eye cosmetic of India known as kohl. In the eighteenth century this herb became popular in Europe as a laxative.

varieties

Ricinus communis
Castor oil plant
Half-hardy evergreen shrub, grown as an annual in cool climates. Ht up to 4m. The separate male and female flowers are borne together in dense upright branching heads. The male flowers have yellow stamens and the green petal-less female flowers have beard-like red stigmas, which are followed by fruits each containing 3 seeds. The seeds are poisonous. The large palmate leaves when young are often red, turning green as they mature.

cultivation

Propagation
Seed
Sow the large, fresh seeds, in early spring individually into prepared pots or module plug trays using a standard seed compost mixed in equal parts with perlite. As the seeds are large, push them into the compost and cover with perlite. Place under protection at 21°C; germination takes 14–21 days. Once large enough to handle, pot up using a standard loam-based compost. Plant into a prepared site in the garden once the night time temperature does not drop below 10°C. When grown in cool climates it will die with the first frosts. Wearing gloves, clear the plant from the ground, dig the site over and feed with well-rotted manure.

Pests and Diseases
This herb rarely suffers from disease and is not affected by pests; because the leaves have insecticidal properties, it actually repels insects.

Maintenance
Spring Sow seeds.
Summer Stake flowering heads in exposed sites.
Autumn Harvest seeds.
Winter When grown as an annual, feed soil where the plant was grown.

Garden Cultivation
This frost-tender fast-growing evergreen shrub is often grown just for its attractive foliage in the garden. Plant in a sheltered position in full sun, in a rich fertile soil. It may need support in exposed areas.

Harvesting
The seed capsules can be gathered throughout the year when nearly ripe and then put out in the sun to dry and mature before the oil is extracted.

companion planting

It is reputedly a good mole repellent.

medicinal

The oil extracted from the seeds is well known for its strong laxative properties. It is also used as a massage oil in India where it is used on tender joints.

other uses

The oil extracted from the seeds is used in numerous ways: in the manufacture of high-grade lubricants, as a fabric coating, as a leather preservative and in the manufacturing of inks, dyes and fibres.

warning

Do not eat the seeds: they are extremely poisonous. Do not take castor oil during pregnancy. Wear gloves when handling this plant, because it can cause an allergic reaction.

ROSEMARY

Rosmarinus officinalis 'Majorca Pink'

From the family *Lamiaceae.*

Rosemary is a shrub that originated in the Mediterranean area and is now widely cultivated throughout the temperate regions. The ancient Latin name means sea-dew. This may come from its habit of growing close to the sea and the dew-like appearance of its blossom at a distance. It is steeped in myth, magic and folk medicinal use. One of my favourite stories about rosemary comes from Spain. It relates that originally the blue flowers were white. When the Holy family fled into Egypt, the Virgin Mary had to hide from some soldiers, so she spread her cloak over a rosemary bush and knelt behind it. When the soldiers had gone by she stood up and removed her cloak and the blossoms turned blue in her honour. Also connected to the Christian faith is the story that rosemary will grow for 33 years, the length of Christ's life, and then die.

In Elizabethan days, the wedding couple wore or carried a sprig of rosemary as a sign of fidelity. Also bunches of rosemary were tied with coloured ribbon tipped with gold and given to guests at weddings to symbolise love and faithfulness.

Rosemary was burnt in sick chambers to freshen and purify the air. Branches were strewn in courts of law as a protection from gaol fever. During the plague people used to wear it in neck pouches to sniff as they travelled, and in Victorian times it was carried in the hollow handles of walking sticks for the same reasons.

varieties

Rosmarinus officinalis

Rosemary

Evergreen hardy perennial. Ht and spread 1m. Pale blue flowers in early spring to early summer and then sometimes in early autumn. Needle-shaped dark green leaves are highly aromatic.

Rosmarinus officinalis* var. *albiflorus

Rosemary White

Evergreen hardy perennial. Ht and spread 80cm. White flowers in early spring to early summer and then sometimes in early autumn. Needle-shaped dark green leaves are highly aromatic.

***Rosmarinus officinalis* var. *angustissimus* 'Benenden Blue' AGM**

Rosemary Benenden Blue

Evergreen hardy perennial. Ht and spread 80cm. Dark blue flowers in early spring to early summer and then sometimes again in early autumn. Leaves are fine needles and fairly dense on the stem, good aroma.

Rosmarinus officinalis 'Blue Lagoon'

Rosmarinus officinalis 'Aureus'
Rosemary Golden
Evergreen hardy perennial. Ht 80cm, spread 60cm. It hardly ever flowers, but if it does they are pale blue. The thin needle leaves are green splashed with gold. If you did not know better you would think the plant was suffering from a virus. It still looks very attractive.

Rosmarinus officinalis 'Blue Lagoon'
Rosemary Blue Lagoon
Evergreen hardy perennial. Ht 40cm, spread 80cm. Striking small dark blue flowers in spring. Short dark green needle-shaped aromatic leaves. Arching semi-prostate habit.

Rosmarinus officinalis 'Boule'
Rosemary Boule
Evergreen hardy perennial. Ht 30cm, spread 80cm. Small mid-blue flowers in spring. Short, dark green needle-shaped aromatic leaves with a white underside. Prostate hanging habit. Ideal for walls and containers.

Rosmarinus officinalis 'Fota Blue'
Rosemary Fota Blue
Evergreen hardy perennial. Ht and spread 80cm. Very attractive dark blue flowers in early spring to early summer and then sometimes again in early autumn. Very well-spaced, narrow, needle-like dark green leaves; the plant has a fairly prostrate habit.

Rosmarinus officinalis 'Sissinghurst Blue'

Rosmarinus officinalis 'Majorca Pink'
Rosemary Majorcan Pink
Evergreen half-hardy perennial. Ht and spread 80cm. Pink flowers in early spring to early summer and then sometimes again in early autumn. The needle-shaped dark green leaves are highly aromatic. This is a slightly prostrate form of rosemary.

Rosmarinus officinalis 'Miss Jessopp's Upright' AGM
Rosemary Miss Jessopp's Upright
Evergreen hardy perennial. Ht and spread 2m. Very pale blue flowers in early spring to early summer and then sometimes again in early autumn. This rosemary has a very upright habit, making it ideal for hedges (see page 382). The leaves are dark green needles spaced closely together, making the plant very bushy.

Rosmarinus officinalis 'Primley Blue'
Rosemary Primley Blue
(Not Frimley, which it has been incorrectly called for a few years.)
Evergreen hardy perennial. Ht and spread 80cm. Blue flowers in early spring to early summer and then sometimes again in early autumn. The needle-shaped dark green leaves are highly aromatic. This is a good hardy bushy variety.

Rosmarinus officinalis 'Roseus'

Rosmarinus officinalis Prostratus Group
Rosemary Prostrate
Evergreen hardy perennial. Ht 30cm, spread 1m. Light blue flowers in early spring to early summer and then sometimes again in early autumn. The needle-shaped dark green leaves are highly aromatic. This is a great plant for trailing on a wall or bank.

Rosmarinus officinalis 'Roseus'
Rosemary Pink
Evergreen half-hardy perennial. Ht and spread 80cm. Pink flowers in early spring to early summer and then sometimes again in early autumn. The needle-shaped dark green leaves are highly aromatic.

Rosmarinus officinalis 'Severn Sea' AGM
Rosemary Severn Seas
Evergreen half-hardy perennial. Ht and spread 80cm. Mid-blue flowers in early spring to early summer and sometimes again in early autumn. Highly aromatic needle-shaped dark green leaves. The whole plant has a slightly prostrate habit with arching branches.

Rosmarinus officinalis 'Sissinghurst Blue' AGM
Rosemary Sissinghurst
Evergreen hardy perennial. Ht 1.5m, spread 1m. Light blue flowers in early spring to early summer and then sometimes again in early autumn. The plant has an upright habit and grows very bushy. The needle-shaped dark green leaves are highly aromatic.

cultivation

Propagation

Seed

Rosmarinus officinalis can, with care, be grown from seed. It needs a bottom heat of 27–32°C to be successful. Sow in the spring in prepared seed or plug trays, using a standard seed compost mixed in equal parts with composted fine bark, and cover with perlite. Having got it to germinate, be careful not to over water the seedlings as they are prone to damping off. Harden off the young plant slowly in summer and pot up. Keep it in a pot for the first winter, and plant out the following spring into the required position at a distance of 60–90cm apart.

Cuttings

As cultivated varieties do not come free from seed, this is a more reliable method of propagation as it ensures that you achieve the variety you require.

Softwood: Take these in spring off the new growth. Cut lengths of about 15cm. Use a standard seed compost mixed in equal parts with composted fine bark.

Semi-hardwood: Take these in summer from the non-flowering shoots, using the same compost as for softwood cuttings.

Rosmarinus officinalis 'Fota Blue'

Rosmarinus officinalis Prostratus Group

Layering

Layer established branches in summer.

Pests and Diseases

Rosemary beetle and its larvae feed on the leaves from autumn until spring. Place some newspaper underneath the plant, then tap or shake the branches, which will knock the beetles and larvae on to the paper, making them easy to destroy.

Maintenance

Spring Trim after flowering. Sow seeds of *Rosamarinus officinalis*. Take softwood cuttings.
Summer Feed container plants. Take semi-hardwood cuttings. Layer plants.
Autumn Protect young tender plants.
Winter Put a mulch, or straw or agricultural fleece around all plants.

Garden Cultivation

Rosemary requires a well-drained soil in a sheltered sunny position. It is frost hardy but in cold areas it prefers to grow against a warm, sunny wall. If the plant is young it is worth giving some added protection in winter. If trimming is necessary cut back only when the frosts are over; if possible leave it until after the spring flowering. Sometimes rosemary looks a bit scorched after frosts, in which case it is worth cutting the damaged plants to healthy wood in spring. Straggly old plants may also be cut back hard at the same time. Never cut back plants in the autumn or if there is any chance of frost, as the plant will be damaged or even killed. On average, despite the story about rosemary growing for 33 years, it is best to replace bushes every 5 to 6 years.

Harvesting

As rosemary is evergreen, you can pick fresh leaves all year round as long as you are not greedy. If you need large quantities then harvest in summer and either dry the leaves or make an oil or vinegar.

Hedges

Rosemary certainly makes an effective hedge; it looks pretty in flower, smells marvellous and is evergreen. In fact it has everything going for it if you have the right soil conditions which, more importantly than ever, must be well drained and carry a bias towards lime. The best varieties for hedges are *R.O.* 'Primley Blue' and *R.O.* 'Miss Jessopp's Upright'. Both are upright, hardy and bushy. *R.O.* 'Primley Blue' has a darker blue flower and I think is slightly prettier. Planting distance 45cm apart. Again, if you need eventually to trim the hedge, do it after the spring flowering.

companion planting

If planted near carrots it is reputed to repel carrot fly. It is also said to be generally beneficial to sage.

container growing

Rosemary does well in pots and this is the preferred way to grow it in cold districts. The prostrate and less hardy varieties look very attractive and benefit from the extra protection offered by a container. Use a soil-based compost mixed in equal parts with composted fine bark, and make sure the compost is well drained. Do not over water, and feed only after flowering.

Rosmarinus officinalis 'Boule'

ROSEMARY

Rosemary infusion

other uses

Put rosemary twigs on the barbecue; they give off a delicious aroma. If you have a wood-burning stove, a few twigs thrown onto it makes the whole house smell lovely.

Rosemary is used in many herbal shampoos and the plant has a long reputation as a hair tonic. Use an infusion in the final rinse of a hair wash, especially if you have dark hair, as it will make it shine.

medicinal

Like many other essential oils, rosemary oil has anti-bacterial and anti-fungal properties, and it helps poor circulation if rubbed into the affected joints. The oil may be used externally as an insect repellent. It also makes an excellent remedy for headaches if applied directly to the head.

Rosemary tea makes a good mouthwash for halitosis and is also a good antiseptic gargle. Drunk in small amounts it reduces flatulence and stimulates the smooth muscle of the digestive tract and gall bladder and increases the flow of bile. Cover a teaspoon of chopped leaves with a cup of boiling water; cover and stand for 5 minutes. An antiseptic solution of rosemary can be added to the bath to promote heathy skin. Boil a handful in 500ml of water for 10 minutes.

warning

The oil should not be used internally. Also, extremely large doses of the leaf are toxic, possibly causing abortion, convulsions and, very rarely, death.

culinary

This is one of the most useful of culinary herbs, combining well with meat, especially lamb, casseroles, tomato sauces, baked fish, rice, salads, egg dishes, apples, summer wine cups, cordials, vinegars and oils.

Vegetarian Goulash
Serves 4

2 tablespoons rosemary olive oil
2 medium onions, sliced
1 dessertspoon wholemeal flour
1 tablespoon paprika
275ml hot water mixed with 1 teaspoon tomato purée
400g tin Italian tomatoes
2 sprigs rosemary, 10cm long
225g cauliflower sprigs
225g new carrots, washed and cut into chunks
250g new potatoes, washed and cut into halves
1/2 green pepper, de-seeded and chopped
150ml soured cream or Greek yogurt
Salt and freshly milled black pepper

Heat the rosemary oil in a flameproof casserole, fry the onions until soft, then stir in the flour and three-quarters of the paprika. Cook for 2 minutes. Stir in the water, tomatoes and sprigs of rosemary. Bring to the boil stirring all the time. Add all the vegetables and the seasonings. Cover and bake in the preheated oven (190°C/375°F/gas mark 5) for 30–40 minutes. Remove from the oven, carefully take out the rosemary sprigs and stir in the soured cream or yogurt, plus the remaining paprika. Serve with fresh pasta and/or garlic bread.

Rumex

SORREL

Rumex acetosa seed

Also known as Bread and Cheese, Sour Leaves, Tom Thumbs, Thousand Fingers and Sour Sauce. From the family *Polygonaceae*.

Sorrel is a native plant of Europe, Asia and North America. It has naturalised in many countries throughout the world on rich, damp, loamy, acid soils. The generic name, *Rumex*, comes from the Latin *rumo*, 'I suck'. Apparently, Roman soldiers sucked the leaves to relieve thirst, and their doctors used them as a diuretic.

The name sorrel comes from the old French word *surelle*, meaning 'sour'. The Tudors considered the herb to be one of the best English vegetables; Henry VIII held it in great esteem. In Lapland, sorrel juice has been used instead of rennet to curdle milk.

varieties

Rumex acetosa

Sorrel

Also known as broad-leafed sorrel, common sorrel, garden sorrel, meadow sorrel, and confusingly (see below), French sorrel. Hardy perennial. Ht 60–120cm, spread 30cm. The flowers are small, dull and inconspicuous; colour greenish, turning reddish-brown as the fruit ripens. The mid-green leaves are lance-shaped with 2 basal lobes pointing backwards.

Rumex acetosella

Sheep's Sorrel

Hardy perennial. Ht 15–30cm, spread indefinite (can be very invasive). The flowers are small, dull and inconspicuous; colour greenish, turning brown as the fruit ripens. The mid-green leaves are shaped like a barbed spear. It grows wild on heaths and in grassy places, but is rarely found on chalky soil.

Rumex scutatus

Buckler-Leaf Sorrel

Also known as French sorrel. Hardy perennial. Ht 15–45cm, spread 60cm. The flowers are small, dull and inconspicuous; colour greenish, turning brown as the fruit ripens. The mid-green leaves are shaped like squat shields.

Rumex scutatus 'Silver Shld' is a variegated form of *Rumex scutatus*.

cultivation

Propagation

Seed

For an early crop start off under protection in early spring. Sow into prepared seed or plug trays, using a standard seed compost mixed in equal parts with composted fine bark, and covering the seeds with perlite. Germination is fairly quick, 10–20 days without extra heat. When the seedlings are large enough and the soil has started to warm up, plant out 30cm apart.

Rumex scutatus

Division

Sorrel is easy to divide and it is a good idea to divide broad-leaf sorrel every other year to keep the leaves succulent. Autumn is the best time to do this, replanting in a prepared site.

Pests and Diseases

Wood pigeons, slugs and occasionally leaf miners attack sorrel, but should cause no problems with established plants. Remove the affected leaves, and put out traps for the slugs.

Maintenance

Spring Sow seed, under protection, in early spring and outdoors from mid-spring.

Summer Cut off flowers to maintain leaf production and prevent self-seeding. In a hot summer, water regularly to keep the leaves succulent.

Autumn Divide established plants.

Winter Fully hardy.

Garden Cultivation

This perennial herb likes a rich acid soil which retains moisture in full sun to partial shade. Sow the seeds in late spring into a prepared site. When germinated, they seedlings out to a distance of 7cm and finally to a distance of 30cm apart. Can be grown under cloches to provide leaf throughout the year. The plant tends to run to seed quickly so, to keep the leaves fresh and succulent, remove flowerheads as they appear.

In warm summers or warm climates, sorrel leaves tend to become bitter as the season progresses. A mulch will keep the soil cooler and, once the season cools down, the flavour will improve. Grow buckler-leaf sorrel with its smaller leaf, as it is less susceptible.

If sorrel is causing a problem in your garden simply add a few applications of lime to eradicate it.

Harvesting

Pick young leaves throughout the growing season for fresh use and for freezing. Sorrel does not dry well.

container growing

The buckler variety makes a good low-growing pot plant. Use a soil-based compost mixed in equal parts with composted fine bark, and make sure the container has room for the plant to spread. It is a very useful culinary herb, so for those with a small garden or who live on a chalk soil this makes an ideal container plant. Remember to keep cutting off flowers to keep leaves tender. Water well in the growing season, and feed with liquid fertiliser, especially if you are picking a lot.

medicinal

Sorrel is considered to have blood cleansing and blood improving qualities in a similar way to spinach, which improves the haemoglobin content of the blood. It also contains vitamin C. A leaf may be used in a poultice to treat certain skin complaints, including acne.

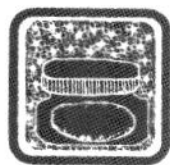

other uses

Sorrel is a good dye plant; with an alum mordant it makes a yellow or green dye (see also pages 675). Use juice of the leaf to remove rust, mould and ink stains from linen, wicker and silver.

warning

Care has to be taken that sorrel is not used in too great a quantity or too frequently. Its oxalic acid content may be damaging to health if taken in excess. Very large doses are poisonous, causing severe kidney damage.

The herb should not be used medicinally by those predisposed to rheumatism, arthritis, gout, kidney stones or gastric hyperacidity. The leaf may cause dermatitis.

Rumex scutatus 'Silver Shield'

Sorrel and lettuce soup

culinary

This is an excellent herb with which to experiment. Use sparingly in soups, omelettes, fish sauces and with poultry and pork. It is useful for tenderizing meat. Wrap it around steaks or add pounded leaf to a marinade.

Eat leaves raw in salads, especially the buckler-leaf sorrel, but reduce the vinegar or lemon in any accompanying dressing to compensate for the increased acidity. Cook like spinach, changing the cooking water once to reduce acidity.

A Green Sauce

Wash a handful each of sorrel and lettuce leaves and a handful of watercress. Cook in a little water with a whole peeled onion until tender. Remove onion and discard. Allow the (mushy) leaves to cool then add 1 tablespoon (15ml) of olive oil, 1 tablespoon (15ml) of wine vinegar, pepper and salt. Stir until creamy. Serve with fish or cold poultry.

Sorrel and Lettuce Soup

Serves 4

100g sorrel
100g lettuce
50g French parsley
50g butter
100g potatoes, peeled and sliced
600ml chicken stock
4 tablespoons thin cream

Wash the sorrel, lettuce and French parsley, pat dry and roughly chop. Heat the butter in a heavy pan and add the sorrel, lettuce and parsley. Stew very gently for about 5 minutes, and then add the potato. Mix all together, pour over the heated stock and simmer covered for 25 minutes. Put in a liquidiser, or, if you are a purist, through a coarse food mill. Return to the pan and heat gently (do not boil). Swirl in some cream just before serving.

Ruta graveolens

RUE

Also known as Herb of Grace and Herbygrass. From the family *Rutaceae*.

Rue is a native of Southern Europe, especially the Mediterranean region, and is found growing in poor, free-draining soil. It has established itself in North America and Australia in similar conditions. It has also adapted to cooler climates and is now naturalised in Northern Europe. Rue was known as herb of grace, perhaps because it was regarded as a protector against the Devil, witchcraft and magic.

It was also used as an antidote against every kind of poison from toadstools to snake bites. The Romans brought it across northern Europe to Britain, where it did not gain favour until the Middle Ages, when it was one of the herbs carried in nosegays by the rich as protection from evil and the plague. Also, like rosemary, it was placed near the judge before prisoners were brought out, as protection from the pestilence-ridden gaols and gaol fever.

It was famous for preserving eyesight and was said to promote second sight, perhaps acting on the third eye. Both Leonardo da Vinci and Michelangelo are supposed to have said that their inner vision had been enhanced by this herb.

Ruta graveolens

varieties

Ruta graveolens
Rue
Hardy evergreen perennial. Ht and spread 60cm.
Yellow waxy flowers with 4 or 5 petals in summer.
Small rounded lobed leaves of a greeny blue colour.

***Ruta graveolens* 'Jackman's Blue'**
Rue Jackman's Blue
Hardy evergreen perennial. Ht and spread 60cm.
Yellow waxy flowers with 4 or 5 petals in summer.
Small rounded lobed leaves of a distinctive blue colour.

***Ruta graveolens* 'Variegata'**
Variegated Rue
Hardy evergreen perennial. Ht and spread 60cm.
Yellow waxy flowers with 4 or 5 petals in summer.
Small rounded lobed leaves with a most distinctive cream/white variegation, which is particularly marked in spring, fading in the summer unless the plant is kept well clipped. I have known people mistake the variegation for flowers and try to smell them, which shows how attractive this plant is. Smelling it at close quarters is not, however, a good idea as this plant, like other rues, can cause the skin to blister.

Ruta graveolens 'Jackman's Blue'

cultivation

Propagation

Seed

In spring sow the fine seed using the cardboard trick (see page 645) in prepared plug or seed trays. Use a standard seed compost mixed in equal parts with composted fine bark, and cover with perlite. You may find that a bottom heat of 20°C is helpful. Germination can be an all or nothing affair, depending on the source of the seed. Young seedlings are prone to damping off, so watch the watering, and just keep the compost damp, not wet.

Unlike many variegated plants, the variegated rue will be variegated from seed. When the seedlings are large enough, plant out into a prepared site in the garden at a distance of 45cm.

Cuttings

Take cuttings of new shoots in spring or early summer. 'Jackman's Blue' can only be propagated from cuttings. Use a standard seed compost mixed in equal parts with composted fine bark for the cuttings; again, do not over water.

Pests and Diseases

Rue is prone to whitefly followed by black sooty mould. Treat the whitefly with a liquid horticultural soap as soon as the pest appears, following manufacturer's instructions. This should then also control the sooty mould.

Ruta graveolens 'Variegata'

Maintenance

Spring Cut back plants to regain shape. Sow seed. Take softwood cuttings.
Summer Cut back after flowering to maintain shape.
Autumn The variegated rue is slightly more tender than the other two varieties, so protect when frosts go below −5°C.
Winter Rue is hardy and requires protection only in extreme conditions.

Garden Cultivation

All the rues prefer a sunny site with a well-drained poor soil. They are best positioned away from paths or at the back of beds where people won't brush against them accidentally, especially children, whose skin is more sensitive than adults. In the spring, and after flowering in the summer (not autumn), cut back all the plants to maintain shape, and the variegated form to maintain variegations.

Harvesting

Pick leaves for use fresh. No need to preserve.

container growing

Rue can be grown in containers; use a soil-based compost mixed in equal parts with composted fine bark. Position the container carefully so that one does not accidentally brush the leaves. Although it is drought tolerant, in containers it prefers to be watered regularly in summer. Allow to dry in winter, watering only once a month. Feed plants in the spring with liquid fertiliser following the manufacturer's instructions.

warning

Handling the plant can cause allergic reactions or phytol-photodermatitis. If you have ever seen a rue burn, it really is quite serious, so do heed this warning.

To minimise this risk do not take cuttings off the plants either when they are wet after rain or when in full sun, as this is when the plant is at its most dangerous. Wait until the plant has dried out or the sun has gone in; alternatively wear gloves.

Must only be used by medical personnel and not at all by pregnant women, as it is abortive. Large doses are toxic, sometimes precipitating mental confusion, and the oil is capable of causing death.

culinary

I seriously cannot believe that people enjoy eating this herb; it tastes incredibly bitter. It can be added finely chopped with discretion to egg, fish or cheese dishes.

medicinal

This ancient medicinal herb is used in the treatment of strained eyes, and headaches caused by eye strain. It is also useful for nervous headaches and heart palpitations, for treating high blood pressure and helping to harden the bones and teeth. The anti-spasmodic action of its oil and the alkaloids explains its use in the treatment of nervous digestion and colic. The tea also expels worms.

Salvia

SAGE

From the family *Lamiaceae*.

***'How can a man grow old who has sage in his garden?'* Ancient Proverb**

This large family of over 750 species is widely distributed throughout the world. It consists of annuals, biennials and perennials, herbs, sub-shrubs and shrubs of various habits. It is an important horticultural group. I have concentrated on the medicinal, culinary and a special aromatic species.

Salvia lavandulifolia

The name *Salvia* is derived from the Latin *salveo*, meaning 'I save' or 'I heal', because some species have been highly regarded medicinally.

The Greeks used it to heal ulcers, consumption and snake bites. The Romans considered it a sacred herb to be gathered with ceremony. A special knife was used, not made of iron because sage reacts with iron salts. The sage gatherer had to wear clean clothes, have clean feet and make a sacrifice of food before the ceremony could begin. Sage was held to be good for the brain, the senses and memory. It also made a good gargle and mouthwash and was used as a toothpaste.

There are many stories about why the Chinese valued it so highly, and in the seventeenth century Dutch merchants found that the Chinese would trade three chests of China tea for one of sage leaves.

varieties

I have only chosen a very few species to illustrate; they are the main ones used in cooking and medicine – with two exceptions, with which I begin.

***Salvia argentea* AGM**
Silver Sage
Hardy biennial. Ht up to 120cm in second season. Pale pinkish-white flowers in second season. Splendid large silvery grey leaves covered in long silvery white hairs. Protect from excessive rainwater.

Salvia elegans* 'Scarlet Pineapple' syn. *S. rutilans
Pineapple Sage
Half-hardy perennial. Ht 90cm, spread 60cm. Striking red flowers, mid- to late summer. The leaves are green with a slight red tinge to the edges and have a glorious pineapple scent. This sage is subtropical and must be protected from frost. In temperate climates it is basically a house plant and if kept on a sunny windowsill can be used throughout the year. It can only be grown from cuttings. This is an odd sage to cook with; it does not taste as well as it smells. It is fairly good with apricots as a stuffing for pork, otherwise my culinary experiments with it have not met with great success.

Salvia lavandulifolia
Narrowed-Leaved Sage
Also known as Spanish Sage. Hardy evergreen perennial. Ht and spread 45cm. Attractive blue flowers in summer. The leaves are green and textured, small, thin, and oval in shape and highly aromatic. This is an excellent sage to cook with, very pungent. It also makes a good tea. Can only be grown from cuttings.

Salvia officinalis
Sage
Also known as common sage, garden sage, broad-leaved sage, and sawge. Hardy evergreen perennial. Ht and spread 60cm. Mauve/blue flowers in summer. The leaves are green and textured, thin and oval in shape and highly aromatic. This is the best-known sage for culinary use. Can be easily grown from seed. There is also a white-flowering sage, *Salvia officinalis* 'Albiflora', which is quite rare.

Salvia officinalis broad-leaved

Salvia officinalis 'Purpurascens'

***Salvia officinalis* broad-leaved**

Broad-Leaved Sage

Hardy evergreen perennial. Ht and spread 60cm. Very rarely flowers in cool climates; if it does, flowers are blue/mauve in colour. The leaves are green and textured, larger than the ordinary sage, with an oval shape and highly aromatic. Good for cooking. Can only be grown from cuttings.

***Salvia officinalis* 'Purpurascens' AGM**

Purple/Red Sage

Hardy evergreen perennial. Ht and spread 70cm. Mauve/blue flowers in summer. The leaves are purple and textured, a thin oval shape and aromatic. If you clip it in the spring, it develops new leaves and looks really good but flowers only a small amount. If you do not clip it and allow it to flower it goes woody. If you then cut it back it does not produce new growth until the spring, so can look a bit bare. So what to do? There is also a variegated form of this purple sage *Salvia officinalis* 'Purpurascens Variegata'. Both of these can only be grown from cuttings.

***Salvia officinalis* 'Tricolor'**

Tricolor Sage

Half-hardy evergreen perennial. Ht and spread 40cm. Attractive blue flowers in summer. The leaves are green with pink, white and purple variegation, with a texture. They are small, thin, and oval in shape and highly aromatic. It has a mild flavour, so can be used in cooking. Can only be grown from cuttings.

Salvia viridis* var. *comata

Painted Sage, Red-topped Sage

Hardy annual. Ht 45cm. Striking red, blue, purple-pink bracts and tiny bi-coloured flowers in summer. Green, oval rough-textured leaves.

Salvia officinalis 'Tricolor'

cultivation

Propagation

Seed

Common and painted sage can be grown successfully in the spring from seed sown into prepared seed or plug trays using a standard seed compost mixed in equal parts with composted fine bark and covered with perlite. The seeds are a good size. If starting off under protection in early spring, warmth is of benefit – temperatures of 15–21°C. Germination takes 2–3 weeks. Pot up or plant out when the frosts are over at a distance of 45–60cm apart.

Cuttings

This is a good method for all variegated species and the ones that do not set seed in cooler climates. Use a standard seed compost mixed in equal parts with composted fine bark. Take softwood cuttings in late spring or early summer from the strong new growth. All forms take easily from cuttings; rooting is about 4 weeks in summer.

Salvia viridis var. *comata*

Layering

If you have a well-established sage, or if it is becoming woody, layer established branches in spring or autumn.

Pests and Diseases

Sage grown in the garden does not suffer over much from pests and disease. Sage grown in containers, especially pineapple sage, is prone to red spider mite and leaf hopper. When you see these pests, treat with a liquid horticultural soap as per the instructions.

Maintenance

Spring Sow seeds. Trim if needed, and then take softwood cuttings.
Summer Trim back after flowering.
Autumn Protect all half-hardy sages, and first-year plants.
Winter Protect plants if they are needed for fresh leaves.

Salvia viridis var. *comata*

Salvia argentea

Garden Cultivation

Sage, although predominately a Mediterranean plant, is sufficiently hardy to withstand any ordinary winter without protection, as long as the soil is well drained and not acid, and the site is as warm and dry as possible. The flavour of the leaf can vary according to how rich, damp, etc, the soil is. If wishing to sow seed outside, wait until there is no threat of frost and sow direct into prepared ground, spacing the seeds 23cm apart. After germination thin to 45cm apart. For the first winter cover the young plants with horticultural fleece or a mulch.

To keep the plants bushy prune in the spring to encourage young shoots for strong flavour, and also after flowering in late summer. Mature plants can be pruned hard in the spring after some cuttings have been taken as insurance. Never prune in the autumn as this can kill the plant. As sage is prone to becoming woody, replace the plant every 4–5 years.

Harvesting

Since sage is an evergreen plant, the leaves can be used fresh any time of the year. In Mediterranean-type climates, including the southern states of America, the leaves can be harvested during the winter months. In cold climates this is also possible if you cover a chosen bush with horticultural fleece as this will keep the leaves in better condition. They dry well, but care should be taken to keep their green colour. Because this herb is frequently seen in its dried condition people assume it is easy to dry. But beware, although other herbs may lose some of their aroma or qualities if badly dried or handled, sage seems to pick up a musty scent and a flavour really horrible to taste – better to grow it in your garden to use fresh.

companion planting

Sage planted with cabbages is said to repel cabbage white butterflies. Planted next to vines it is reputed to be beneficial.

container growing

All sages grow happily in containers. Pineapple sage is an obvious one as it is tender, but a better reason is that if it is at hand one will rub the leaves and smell that marvellous pineapple scent. Use a soil-based compost mixed in equal parts with composted fine bark for all varieties, feed after flowering, and do not over water.

other uses

The dried leaves, especially those of pineapple sage, are good added to potpourris.

Sage tea

medicinal

For centuries, sage has been esteemed for its healing powers. It is a first-rate remedy as a hot infusion for colds. Sage tea combined with a little cider vinegar makes a gargle which is excellent for sore throats, laryngitis and tonsillitis. It is also beneficial for infected gums and mouth ulcers. The essential oil is obtained by steamed distillation of the fresh or partially dried flower stems and leaves. It is used in herbal medicine but more widely in toilet waters, perfumes and soap, and to flavour wine, vermouth and liqueurs.

warning

Extended or excessive use of sage can cause symptoms of poisoning. Although the herb seems safe and common, if you drink the tea for more than a week or two at a time, its strong antiseptic properties can cause potentially toxic effects.

Salvia officinalis 'Purpurascens'

culinary

This powerful healing plant is also a strong culinary herb, although it has been misused and misjudged in the culinary world. Used with discretion it adds a lovely flavour, aids digestion of fatty food, and being an antiseptic it kills off any bugs in the meat as it cooks. It has long been used with sausages because of its preservative qualities. It also makes a delicious herb jelly, or oil or vinegar. But I like using small amounts fresh. The original form of the following recipe comes from a vegetarian friend of mine. I fell in love with it and have subsequently adapted it to include some other herbs.

Hazelnut and Mushroom Roast
Serves 4

A little sage oil
Long grain brown rice (measured to the 150ml mark on a glass measuring jug)
275ml boiling water
1 teaspoon salt
1 large onion, peeled and chopped
110g mushrooms, wiped and chopped
2 medium carrots, pared and roughly grated
1/2 teaspoon coriander seed
1 tablespoon soy sauce
110g wholemeal breadcrumbs
175g ground hazelnuts
1 teaspoon chopped sage leaves
1 teaspoon chopped lovage leaves
Sunflower seeds for decoration
A 900g loaf tin, lined with greaseproof paper

Preheat the oven (180°C/ 350°F/gas mark 4).

Heat 1 dessertspoon of sage oil in a small saucepan, toss the rice in it to give it a coating of oil, add boiling water straight from the kettle and the teaspoon of salt. Stir, and let the rice cook slowly for roughly 40 minutes or until the liquid is absorbed.

While the rice is cooking, heat 1 tablespoon of sage oil in a medium sized frying-pan, add the onions, mushrooms, carrots, the coriander seed and soy sauce. Mix them together and let them cook for about 10 minutes.

Combine the cooked brown rice, breadcrumbs, hazelnuts, sage and lovage; mix with the vegetables and place the complete mixture in the prepared loaf tin. Scatter the sunflower seeds on top and bake in the oven for 45 minutes. Leave to cool slightly in the tin. Slice and serve with a home-made tomato sauce and a green salad.

Sambucus

ELDER

Also known as Boun-tree, Boon-tree, Dogtree, Judas Tree, Scores, Score Tree, God's Stinking Tree, Black Elder, Blackberried, European Elder, Ellhorne and German Elder. From the family *Caprifoliaceae*.

Elder grows worldwide throughout temperate climates. Its common name is probably derived from the Anglo Saxon *Ellaern* or *Aeld*, which mean 'fire' or 'kindle', because the hollow stems were once used for getting fires going. The generic name, *Sambucus*, dates from ancient Greek times and may originally have referred to *sambuke*, a kind of harp made of elderwood. Pipes were made from its branches too, possibly the original Pan pipes. People thought that if you put it on the fire you would see the Devil. They believed it unlucky to make cradle rockers out of it, that the spirit of the tree might harm the child. Again, farmers were unwilling to use an elder switch to drive cattle and one folktale had it that elder would only grow where blood had been shed. Planting it outside the back door was a sure way of protecting against evil, black magic, and keeping witches out of the house, which would never be struck by lightning. It was thought that Christ's cross was made of elderwood.

Sambucus nigra

varieties

Sambucus canadensis
American Elder
Also known as black elder, common elder, Rob elder, sweet elder. Deciduous hardy perennial. Ht 1.5–3.6m. Numerous small white flowers in flat cymes throughout summer. Berries are dark purple in early autumn; its leaves are long, sharply toothed and bright green.

Warning: All parts of the fresh *S. canadensis* can poison. Children have even been poisoned by chewing or sucking the bark. Once cooked, however, flowers and berries are safe. Some Native American tribes use a tea made from the root-bark for headaches, mucous congestion, and to promote labour in childbirth.

***Sambucus canadensis* 'Aurea'**
Deciduous hardy perennial. Ht and spread 4m. Creamy white flowers in summer, red fruits in early autumn. Large golden yellow leaves.

Sambucus ebulus
Dwarf Elder
Also known as blood Elder, danewort, wild elder, walewort. Deciduous hardy perennial. Ht 60–120cm, spread 1m. White flowers with pink tips in summer.

spread 1m. White flowers with pink tips in summer.
Black berries in early autumn. Its green leaves are oblong, lance-shaped, and toothed around the edges. Dwarf elder grows in small clusters in Europe and in Eastern and Central States of America.

Warning: All parts of *S. ebulus* are slightly poisonous and children should be warned not to eat the bitter berries. It has a much stronger action than its close relative, common *S. nigra*. Large doses cause vertigo, vomiting and diarrhoea, the latter, denoted colloquially as 'the Danes', being the origin of Danewort. Nowadays *S. ebulus* is rarely used and should be taken internally only under strict medical supervision.

Sambucus nigra
Common Elder
Also known as European elder, black elder, bore tree. Deciduous hardy perennial. Ht 6–7m, spread 4.5m. Spreading branches bear flat heads of small, star-shaped, creamy-white flowers in late spring and early summer, followed in early autumn by drooping purplish-black juicy berries. The leaves of *Sambucus nigra* are purgative and should not be taken internally; decoctions have an insecticidal effect.

***Sambucus nigra* 'Aurea' AGM**
Golden Elder
Deciduous shrub. Ht and spread 6m. Flattened heads of fragrant, star-shaped, creamy-white flowers from early to midsummer. Black fruits in early autumn. Golden yellow, oval, sharply toothed leaves usually in groups of five.

Sambucus racemosa
Red Elder
Deciduous hardy perennial. Ht and spread 3–4m. Brown bark and pale brown pith. Flowers arranged in dense terminal panicles of yellowish cream. *Racemosa* refers to the flower clusters. The fruits are also distinct in being red in drooping clusters. It rarely fruits freely.

Red berried elder is native to central and southern Europe. It has naturalised in Scotland, northern USA and Canada. The fully ripe fruits are used medicinally. Bitter tasting, they may be used fresh or dried, and are high in vitamin C, essential oil, sugar and pectins. Fruits are a laxative and the leaves are a diuretic. This is the most edible and tasty of the elders.

Warning: The seeds inside the berries of *S. racemosa* are poisonous before being cooked.

Elderflower sorbet

cultivation

Propagation

Seed

Sow ripe berries 2cm deep in a pot outdoors using a soil-based seed compost. Plant seedlings in semi-shade in the garden when large enough to handle.

Cuttings

Take semi-hardwood cuttings in summer from the new growth. Use a seed compost mixed in equal parts with composted fine bark, and winter these cuttings in a cold frame or cold greenhouse. When rooted, either pot on or plant out into a prepared site 30cm apart.

Take hardwood cuttings of bare shoots in autumn and replant in the garden 30cm apart. The following autumn lift and replant.

Pests and Diseases

Rarely suffers from pests or diseases apart from black-fly, which can be treated with horticultural soap.

Maintenance

Spring Prune back golden and variegated elders.
Summer Take semi-ripe cuttings.
Autumn Take hardwood cuttings. Prune back hard.
Winter Established plants do not need protection.

Garden Cultivation

Elder tolerates most soils and *S. nigra* is very good for chalky sites. They all prefer a sunny position. Elder grows very rapidly indeed and self-sows freely to produce new shoots 120cm long in one season. It is short-lived. It is important to dominate elder otherwise it will dominate your garden. Cut back in late autumn, unless it is gold or variegated, when it should be pruned in early spring before growth begins.

Harvesting

Handle flower heads carefully to prevent bruising, spread out to dry with heads down on a fine net without touching one another. Pick the fruits in autumn, as they ripen, when they become shiny and violet.

container growing

Golden varieties of elder can look good in containers, as long as the containers are large enough and positioned to give the plants some shade, to stop the leaves scorching. Use a soil-based compost. Keep well watered, feed with a liquid fertiliser.

other uses

Elderflower water whitens and softens the skin, removes freckles. The fruits make a lavender or violet dye when combined with alum. The flowers and berries of *Sambucus nigra* are used in industry for cosmetics. The wood from the adult plant is highly prized by craftsmen.

medicinal

Elderflowers reduce bronchial and upper respiratory catarrh and are used in the treatment of hay fever. Externally a cold infusion of the flowers may be used as an eye wash for conjunctivitis and as a compress for chilblains. A gargle made from elderflower infusion or elderflower vinegar alleviates tonsillitis and sore throats. Elderflowers have a mild laxative action and in Europe have a reputation for treating rheumatism and gout. The berries are a mild laxative and sweat-inducing. 'Elderberry Rob' is traditionally made by simmering the berries and thickening with sugar as a winter cordial for coughs and colds.

Elderberry Conserve (for neuralgia and migraine)

500g elderberries
500g sugar

Boil the elderberries with the least quantity of water to produce a pulp. Pass through a sieve and simmer the juice gently to remove most of the water. Add the sugar and stir constantly until the consistency of a conserve is produced. Pour into a suitable container. Take two tablespoons as required.

culinary

Only common elder is used for culinary purposes, and its berries should not be eaten raw, nor fresh juice used. Be sure to cook very slightly first.

Elderflower Cordial

Pick flowers on a dry sunny day, as the yeast is mainly in the pollen.

4.5 litres water
700g sugar
Juice and thinly peeled rind of 1 lemon
30ml cider or wine vinegar
12 elderflower heads

Bring the water to the boil and pour into a sterilised container. Add the sugar, stirring until dissolved. When cool add the lemon juice and the rind, vinegar and elderflowers. Cover with several layers of muslin and leave for 24 hours. Filter through muslin into strong, sterilised glass bottles. This drink is ready after 2 weeks. Serve chilled.

warning

Use parts of plants from this genus with care; many are poisonous. See warnings under relevant species.

Sambucus nigra

Sanguisorba minor

SALAD BURNET

Also known as Drumsticks, Old Man's Pepper and Poor Man's Pepper. From the family *Rosaceae.*

This herb is a native of Europe and Asia. It has been introduced and naturalised in many places elsewhere in the world, especially Britain and the United States. Popular for both its medicinal and culinary properties, it was taken to New England in the Pilgrim Fathers' plant collection and called pimpernel. It is found in dry, free-draining soil in grassland and on the edges of woodland. The name *Sanguisorba* comes from *sanguis*, meaning 'blood', and *sorbere*, meaning 'to soak up'. It is an ancient herb, which has been grown in Britain since the sixteenth century. Traditionally it was used to staunch wounds. In Tudor times salad burnet was planted along borders of garden paths so the scent would rise up when trodden on.

varieties

Sanguisorba minor

Salad Burnet

Evergreen hardy perennial. Ht 20–60cm, spread 30cm. Produces small spikes of dark crimson flowers in summer. Its soft mid-green leaves are divided into oval leaflets.

Sanguisorba officinalis

Great Burnet

Also known as drumsticks, maidens hairs, red knobs, and redheads. Perennial. Ht up to 1.2m, spread 60cm. Produces small spikes of dark crimson flowers in summer. Its mid-green leaves are divided into oval leaflets. This wild plant is becoming increasingly rare due to modern farming practices.

Sanguisorba minor flower heads

cultivation

Propagation

Seed

Sow the small flattish seed in spring or autumn into prepared seed or plug trays and cover the seeds with perlite; no need for extra heat. If sown in the autumn, winter the seedlings under protection and plant out in spring to a prepared site, 30cm apart. If spring sown allow to harden off and plant out in the same way. When used as an edging plant it needs to be planted at 20cm intervals.

Division

It divides very easily. Dig up an established plant in the early autumn, cut back any excessive leaves, divide the plant and replant in a prepared site in the garden.

Pests and Diseases

This herb is, in the main, free from pests and diseases.

Maintenance

Spring Sow seeds.
Summer Keep cutting to stop it flowering, if being used for culinary purposes.
Autumn Sow seeds if necessary. Divide well-established plants.
Winter No protection needed, fully hardy.

Garden Cultivation

This is a most attractive, soft-leaf evergreen and is very useful in both kitchen and garden. That it is evergreen is a particular plus for the herb garden, where it looks most effective as an edging plant. It also looks good in a wild flower garden, where it grows as happily as in its original grassland habitat.

With no special requirements, it prefers chalky soil, but it will tolerate any well-drained soil in sun or light shade. It is deep rooting and very drought resistant.

The art with this plant is to keep cutting, which stops it flowering and encourages lots of new growth.

Harvesting

Pick young tender leaves when required. Not necessary to dry leaves (which in any case do not dry well), as fresh leaves can be harvested all year round.

container growing

Salad burnet will grow in containers, and will provide an excellent source of soft evergreen leaves throughout winter for those with no garden. Use a soil-based compost. Water regularly, but not too frequently; feed with liquid fertiliser in the spring only. Do not over-feed otherwise the leaf will soften and lose its cool cucumber flavour, becoming more like a spinach. For regular use the plant should not be allowed to flower. Cut back constantly to about 15cm to ensure a continuing supply of tender new leaves.

medicinal

Chewing the leaf assists digestion. An infusion of the whole plant is used for treating haemorrhoids and diarrhoea.

Infusion of salad burnet

culinary

The leaves of salad burnet have a nutty flavour and a slight taste of cucumber. The young leaves are refreshing in salads and can be used generously – they certainly enhance winter salads. Tender young leaves can also be added to soups, cold drinks, cream cheeses, or used (like parsley) as a garnish or to flavour casseroles – add at the beginning of cooking. The leaves also make an interesting herbal vinegar.

Salad burnet combines with other herbs, especially rosemary and tarragon. Serve in a sauce with white fish.

Salad Burnet Herb Butter

This butter is lovely with grilled fish, either cooked under the grill or on the barbecue, and adds a cucumber flavour.

1½ tablespoons chopped salad burnet
1 tablespoon chopped garden mint (spearmint)
75g butter
Salt and freshly ground black pepper
Lemon juice

Mix the chopped herb leaves together. Melt the butter in a saucepan, add the herbs and simmer on a very low heat for 10 minutes. Season the sauce to taste with salt and pepper, and a squeeze (no more) of lemon. Pour over grilled fish (plaice or sole).

other uses

Because of its high tannin content, the root of great burnet can be used in the tanning of leather.

warning

Great burnet should never be taken in large doses.

Santolina

COTTON LAVENDER

Also known as Santolina and French Lavender. From the family *Asteraceae*.

Cotton lavender is a native of Southern France and the Northern Mediterranean area, where it grows wild on calcareous ground. It is widely cultivated, adapting to the full spectrum of European and Australian climates and to warm-to-hot regions of North America, surviving even an Eastern Canadian winter on well-drained soil.

The Greeks knew cotton lavender as *abrotonon* and the Romans as *habrotanum,* both names referring to the tree-like shape of the flying branches. It was used medicinally for many centuries by the Arabs. It was valued in medieval England as an insect and moth repellent and vermifuge.

The plant was probably brought into Britain in the sixteenth century by French Huguenot gardeners, who were skilled in creating the knot garden so popular among the Elizabethans. Cotton lavender was used largely in low clipped hedges, and as edging for the geometrical beds.

Santolina chamaecyparissus 'Small-Ness'

varieties

Despite its common name, this is not a member of the *Lavandula* family, but is a member of the daisy family.

***Santolina chamaecyparissus* AGM**
Cotton Lavender
Hardy evergreen shrub. Ht 75cm, spread 1m. Yellow button flowers from mid-summer to early autumn, silver coral-like aromatic foliage.

***Santolina pinnata* subsp. *neapolitana* 'Edward Bowles'**
Cotton Lavender Edward Bowles
Hardy evergreen shrub. Ht 75cm, spread 1m. Cream button flowers in summer. Feathery, deep-cut, grey/green foliage.

***Santolina chamaecyparissus* 'Lemon Queen'**
Cotton Lavender Lemon Queen
As 'Edward Bowles', but feathery, deep-cut grey foliage.

***Santolina chamaecyparissus* 'Small-Ness'**
Hardy evergreen shrub. Ht and spread 45cm. Aromatic, deeply cut blue/green foliage. Very neat, tight habit.

***Santolina pinnata* subsp. *neapolitana* AGM**
Cotton Lavender Neopolitana
As 'Edward Bowles'.

***Santolina rosmarinifolia* subsp. *rosmarinifolia* 'Primrose Gem' AGM**
Cotton Lavender Primrose Gem
Hardy evergreen shrub. Ht 60cm, spread 1m. Pale yellow button flowers in summer. Finely cut green leaves.

Santolina rosmarinifolia subsp. *rosmarinifiolia* 'Primrose Gem'

Santolina rosmarinifolia* subsp. *rosmarinifolia
Cotton Lavender, Holy Flax, Virens
As 'Primrose Gem'. Bright yellow button flowers in summer. Finely cut, bright green leaves.

cultivation

Propagation

Seed
Seed is not worth the effort as germination is poor.

Cuttings
Take 5–8cm soft stem cuttings in spring before flowering, or take semi-ripe stem cuttings from mid-summer to autumn. Use a standard seed compost mixed in equal parts with composted fine bark. They root easily without the use of any rooting compound.

Pests and Diseases
Compost or soil that is too rich will attract aphids.

Maintenance
Spring Cut straggly old plants hard back. Take cuttings from new growth.
Summer I cannot stress enough that after flowering the plants should be cut back or the bushes will open up and lose their attractive shape.
Autumn Take semi-ripe cuttings, protect them from frost in a cold frame or greenhouse.
Winter Protect in only the severest of winters.

Santolina pinnata subsp. *neapolitana*

Garden Cultivation
This elegant aromatic evergreen is ideal for the herb garden as a hedging or specimen plant in its own right. Plant in full sun, preferably in sandy soil. If the soil is too rich the growth will become soft and start to lose colour. This is particularly noticeable with the silver varieties.

Planting distance for an individual plant 45–60cm, for hedging 30–38cm. Hedges need regular clipping to shape in spring and summer. Do not cut back in the autumn in frosty climates, as this can easily kill the plants. If temperatures drop below −15°C protect with horticultural fleece or a layer of straw, spruce or bracken.

Harvesting
Pick leaves and dry any time before flowering. Pick small bunches of flower stems for drying, in late summer. They can be dried easily by hanging the bunches upside down in a dry, airy place.

container growing

Cotton lavender cannot be grown indoors, however as a patio plant, a single plant clipped to shape in a large terracotta pot can look very striking. Use a soil-based compost mixed in equal parts with composted fine bark. Place pot in full sun. Do not over-feed with liquid fertiliser or growth will be too soft.

culinary

Cotton lavender *(S. chamaecyparissus)* makes an interesting addition to shortbread biscuits instead of rosemary. Interesting being the operative word.

medicinal

Although not used much nowadays, it can be applied to surface wounds, hastening the healing process by encouraging scar formation. Finely ground leaves ease the pain of insect stings and bites.

Santolina pinnata subsp. *neapolitana* 'Edward Bowles'

other uses

Lay in drawers, under carpets, and in closets to deter moths and other insects, or make a herbal moth bag.

Herbal Moth Bag

A handful of wormwood
A handful of spearmint
A handful of cotton lavender
A handful of rosemary
1 tablespoon of crushed coriander

Dry and crumble the ingredients, mix together and put in a muslin or cotton bag.

Herbal moth bag, using cotton lavender

Saponaria officinalis

SOAPWORT

Also known as Bouncing Bet, Bruisewort, Farewell Summer, Fuller's Herb, Joe Run By The Street, Hedge Pink, Dog's Clove, Old Maid's Pink and Soaproot. From the family *Caryophyllaceae*.

Soapwort, widespread on poor soils in Europe, Asia and Northern America, was used by medieval Arab physicians for various skin complaints. Fullers used soapwort for soaping cloth before it went on the stamps at the mill, and sheep were washed with a mixture of the leaves, roots and water before being shorn.

varieties

Saponaria officinalis
Soapwort
Hardy perennial. Ht 30–90cm, spread 60cm or more. Compact cluster of small pretty pink or white flowers in summer to early autumn. The leaf is smooth, oval, pointed and mid-green in colour.

***Saponaria officinalis* 'Rubra Plena'**
Double-Flowered Soapwort
Hardy perennial. Ht 90cm, spread 30cm. Clusters of red, ragged, double flowers in summer. The leaves are mid-green and oval in shape.

***Saponaria ocymoides* AGM**
Tumbling Ted
Hardy perennial. Ht 2.5–8cm, spread 40cm or more. Profusion of tiny, flat, pale pink/crimson flowers in summer. Sprawling mats of hairy oval leaves.

cultivation

Propagation

Seed
Only soapwort and tumbling Ted can be grown from seed. Sow in autumn into prepared seed or plug trays and cover with compost. Place glass over container and leave out over winter. Usually germinates in spring, but can be erratic. When large enough, plant 60cm apart.

Cuttings
Softwood cuttings of the non-flowering shoots can be taken from late spring to early summer.

Division
The creeping rootstock is easy to divide in the autumn.

Garden Cultivation
Plant it in a sunny spot, in a well-drained poor soil; in rich soil it can become very invasive. Do not plant near ponds because the creeping rhizomes excrete a poison.

Pests and Diseases
Soapwort is largely free from pests and disease.

Maintenance
Spring Take cuttings.
Summer Cut back after flowering to encourage a second flowering and to prevent self-seeding.
Autumn Divide established plants. Sow seed.
Winter Fully hardy.

Harvesting
Pick the leaves when required. Dig up the roots in the autumn and dry for medicinal use.

container growing

S. ocymoïdes is the best variety for container growing. Use a soil-based compost. Water well during the growing season, but only feed twice. In winter keep in a cold greenhouse with minimum watering.

medicinal

It has been used not only for treating skin conditions such as eczema, cold sores, boils and acne, but also for gout and rheumatism. It is probably effective due to the anti-inflammatory properties of its saponins.

warning

This herb should only be prescribed by a qualified herbalist because of the high saponin content, which makes it mildly poisonous.

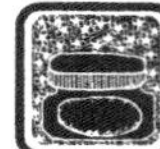

other uses

The gentle power of the saponins in soapwort makes the following shampoo ideal for upholstery and delicate fibres.

Soapwort Shampoo

15g dried soapwort root or two large handfuls of whole fresh stems
750ml water

Crush the root with a rolling pin or roughly chop the fresh stems. If using dried soapwort, prepare by soaking first overnight. Put the soapwort into an enamel pan with water and bring to the boil, cover and simmer for 20 minutes, stirring occasionally. Allow to stand until cool and strain through a fine sieve.

Soapwort Shampoo

Scutellaria

SKULLCAP

Scutellaria lateriflora

Also known as Helmet Flower, Mad Dog Weed, Blue Skullcap and Blue Pimpernel. From the family *Lamiaceae*.

The various varieties of skullcap are natives of different countries. They are found in America, Britain, India, and one grows in the rain forests of the Amazon.

The name *Scutellaria* is derived from *scutella,* meaning 'a small shield', which is exactly how the seed looks.

The American Indians used *Scutellaria lateriflora* as a treatment for rabies. In Europe it was used for epilepsy.

varieties

Scutellaria galericulata
Skullcap
Hardy perennial. Ht 15–50cm, spread 30cm and more. Small purple/blue flowers with a longer spreading lower lip in summer. Leaves bright green and lance-shaped with shallow round teeth. This plant is a native of Europe.

Scutellaria minor
Lesser Skullcap
As *S. galericulata* except ht 20–30cm, spread 30cm and more. Small purple/pink flowers. Leaves lance-shaped with four rounded teeth. Found on wet land.

Scutellaria lateriflora
Virginian Skullcap
As *S. galericulata* except ht 30–60cm, spread 30cm and more. Leaves oval and lance-shaped with shallow, round teeth. Native of America.

cultivation

Propagation

Seed
Sow the small seeds in autumn into prepared seed or plug trays and cover the seeds with compost. Leave the tray outside under glass. If germination is rapid, winter the young seedlings in a cold greenhouse. If there is no germination within 10–20 days, leave well alone. The seed may need a period of stratification. In the spring, when the plants are large enough, plant out into a prepared site in the garden 30cm apart.

Scutellaria galericulata

Root Cuttings
Take cuttings from the rhizomous root. In spring dig up an established clump carefully, for any little bits of root left behind will form another plant. Ensure each cutting has a growing node; place in a seed tray and cover with compost. Put into a cold greenhouse to root.

Division
Established plants can be divided in the spring.

Pests and Diseases
Skullcap is normally free from pests and disease.

Maintenance
Spring Divide established plants. Take root cuttings.
Summer Cut back to restrain.
Autumn Sow seeds.
Winter: No need for protection, fully hardy.

Garden Cultivation
Tolerates most soils but prefers a well-drained, moisture-retentive soil in sun or semi-shade. Make sure this plant gets adequate water.

Harvesting
Dry flowers and leaves for medicinal use only.

container growing

It can be grown in containers but ensure its large root system has room to spread. Use a soil-based compost. Feed only rarely with liquid fertiliser or it will produce too lush a growth and inhibit flowering. Leave outside in winter in a sheltered spot, allowing the plant to die back.

medicinal

S. lateriflora is the best medicinal species; the two European species are a little less strong. It is used in the treatment of anxiety, nervousness, depression, insomnia and headaches. The whole plant is effective as a soothing antispasmodic tonic and a remedy for hysteria. Its bitter taste also strengthens and stimulates the digestion.

warning

Should only be dispensed by a trained herbalist.

Satureja

SAVORY

From the family *Lamiaceae*.

Savory is a native of southern Europe and North Africa, especially around the Mediterranean. It grows in well-drained soils and has adapted worldwide to similar climatic conditions. Savory has been employed in food flavouring for over 2,000 years. Romans added it to sauces and vinegars, which they used liberally as flavouring. The Ancient Egyptians, on the other hand, used it in love potions. The Romans also included it in their wagon train to northern Europe, where it became an invaluable disinfectant strewing herb. It was used to relieve tired eyes, for ringing in the ears, indigestion, wasp and bee stings, and for other shocks to the system.

Satureja spicigera

varieties

***Satureja coerulea* AGM**
Purple-Flowered Savory
Semi-evergreen hardy perennial. Ht 30cm, spread 20cm. Small purple flowers in summer. The leaves are darkish green, linear and very aromatic.

Satureja hortensis
Summer Savory
Also known as bean herb. Half-hardy annual. Ht 20–30cm, spread 15cm. Small white/mauve flowers in summer. Aromatic leaves, oblong, pointed, and green. A favourite on the Continent and in America, where it is known as the bean herb. It has become widely used in bean dishes as it helps prevent flatulence.

Satureja montana
Winter Savory
Also known as mountain savory. Semi-evergreen hardy perennial. Ht 30cm, spread 20cm. Small white/pink flowers in summer. The leaves are dark green, linear and very aromatic.

Satureja spicigera
Creeping Savory
Perennial. Ht 8cm, spread 30cm . Masses of small white flowers in summer. The leaves are lime greenish and linear. This is a most attractive plant and is often mistaken for thyme or even heather.

cultivation

Propagation

Seed
Only summer and winter savory can be grown from seed, which is tiny, so it is best to sow into prepared seed trays under protection in the early spring, using the cardboard method (see page 645). The seeds should not be covered as they need light to germinate. Germination takes about 10–15 days – no need to use bottom heat. When the seedlings are large enough, and after a period of hardening off (making quite sure that the frosts have finished), they can be planted out into a prepared site in the garden, 15cm apart.

Cuttings
Creeping, purple-flowered and winter savory can all be grown from softwood cuttings in spring, using a standard seed compost mixed in equal parts with composted fine bark. When these have rooted they should be planted out – 30cm apart for creeping savory, 15cm apart for the others.

Division
Creeping savory can be divided, as each section has its own root system similar to creeping thymes. Dig up an established plant in the spring after the frosts have finished and divide into as many segments as you require. Minimum size is only dependent on each having a root system and how long you are prepared to wait for new plants to become established. Replant in a prepared site.

Pests and Diseases
Being an aromatic plant, savory is, in the main, free from pests and diseases.

Maintenance
Spring Sow seed. Take softwood cuttings. Divide established plants.
Summer Keep picking and do not allow summer savory to flower, if you want to maintain its flavour.
Autumn Protect from prolonged frosts.
Winter Protect.

Garden Cultivation
All the savories featured here like full sun and a poor, well-drained soil. Plant summer savory in the garden in a warm, sheltered spot and keep picking the leaves to

Satureja montana

stop it getting leggy. Do not feed with liquid fertiliser, otherwise the plant will keel over.

S. montana can make a good edging plant and is very pretty in the summer, although it can look a bit sparse in the winter months. Again, trim it from time to time to maintain shape and promote new growth. Creeping savory does not like cold wet winters, or for that matter clay soil, so on this nursery I grow it in a pot (see below). If, however, you wish to grow it in your garden, plant it in a sunny rockery or a well-drained, sheltered corner.

Harvesting

For fresh use, pick leaves as required. For drying, pick those of summer savory before it flowers. They dry easily.

container growing

All savories can be grown in containers, and if your garden suffers from prolonged cold wet winters it may be the only way you can grow this delightful plant successfully. Use a soil-based compost mixed in equal parts with composted fine bark. Pick the plants continuously to maintain shape, especially the summer savory, which can get straggly. If you are picking the plants a lot they may benefit from a feed of liquid fertiliser, but keep this to a minimum as they get over eager when fed. Summer savory, being an annual, dies in winter; creeping savory dies back; the winter savory is a partial evergreen. So, the latter two will need protection in winter. Place them in a cool greenhouse or conservatory. If the container cannot be moved, wrap it up in paper or horticultural fleece. Keep watering to the absolute minimum.

culinary

The two savories that are used in cooking are winter and summer savory. The other varieties are edible but their flavour is inferior. Summer and winter savory combine well with vegetables, pulses and rich meats. These herbs both stimulate the appetite and aid digestion. The flavour is hot and peppery, and so the leaves should be added sparingly in salads.

Summer savory can replace both salt and pepper and is a great help to those on a salt-free diet. It is a pungent herb and until one is familiar with its strength it should be used carefully. Summer savory also makes a good vinegar and oil. The oil is used commercially as a flavouring, as is the leaf, which is an important constituent of salami.

The flavour of winter savory is both coarser and stronger; its advantage is that it provides fresh leaves into early winter.

medicinal

Summer savory is the plant credited with medicinal virtues and is said to alleviate the pain of bee stings if rubbed on the affected spot. Infuse as a tea to stimulate appetite and to ease indigestion and flatulence. It is also considered a stimulant and was once in demand as an aphrodisiac. Winter savory is also used medicinally but is inferior.

Beans with Garlic and Savory

Serves 3–4

200g dried haricot beans
1 Spanish onion
1 carrot, scrubbed and roughly sliced
1 stick celery
3 tablespoons olive oil
1 tablespoon white wine vinegar
1 clove garlic, crushed
2 tablespoons chopped summer savory
2 tablespoons chopped French parsley

Soak the beans in cold water overnight or for at least 3–4 hours. Drain them and put them in a saucepan with plenty of water. Bring to the boil slowly. Add half the peeled onion, the carrot and celery, and cook until tender. As soon as the beans are soft, drain and discard the vegetables. Mix the oil, vinegar and crushed garlic. While the beans are still hot, stir in the remaining half onion (thinly sliced), the chopped herbs, and pour over the oil and vinegar dressing. Serve soon after cooling. Do not chill.

Sempervivum

HOUSELEEK

Also known as Bullocks Eye, Hen and Chickens, Jupiter's Eye, Jupiter's Beard, Live For Ever, Thunder Plant, Aaron's Rod, Healing Leaf, Mallow Rock and Welcome-husband-though-never-so-late. From the family *Crassulaceae*.

Originally from the mountainous areas of central and southern Europe, now found growing in many different areas of the world, including North America.

The generic name *Sempervivum* comes from the Latin *semper vivo* meaning 'to live for ever'. The species name, *tectorum*, means 'of the roofs', there being records dating back 2,000 years of houseleeks growing on the tiles of houses. The plant was said to have been given to man by Zeus or Jupiter to protect houses from lightning and fire. Because of this the Romans planted courtyards with urns of houseleek, and Charlemagne ordered a plant to be grown on every roof. This belief continued throughout history and in medieval times the houseleek was thought to protect thatched roofs from fire from the sky and witchcraft. In the Middle Ages the plant was often called Erewort and employed against deafness. When the settlers packed their bags for America they took houseleek with them.

Sempervivum 'Commander Hay'

Sempervivum tectorum with offsets

varieties

Fourty years ago, this genus of hardy succulents had 25 species. Now, due to re-classification, it has over 500 different varieties. As far as I am aware only houseleek has medicinal properties.

***Sempervivum tectorum* AGM**
Houseleek
Hardy evergreen perennial. Ht 10–15cm (when in flower) otherwise it is 5cm, spread 20cm. Flowers are star-shaped and pink in summer. The leaves, grey/green in colour, are oval, pointed and succulent.

Some other *Sempervivum* worth collecting:

***Sempervivum arachnoideum* AGM**
Cobweb Houseleek
Hardy evergreen perennial. Ht 10–12cm, when in flower, otherwise it is 5cm, spread 10cm. Flowers are star-shaped and pink in summer. The leaves, which are grey/green in colour, are oval, pointed and succulent. The tips of the leaves are covered in a web of white hairs.

***Sempervivum* 'Commander Hay' AGM**
Hardy evergreen perennial. Ht 10–15cm, when in flower, otherwise it is 5cm, spread 20cm. Flowers star-shaped, pink in summer. The leaves are deep maroon in colour, oval, pointed and succulent.

HOUSELEEK

Sempervivum montanum
Hardy evergreen perennial. Ht 8–15cm, when in flower, otherwise 5cm, spread 10cm. Flowers star-shaped and deep red in summer. Leaves grey/green in colour, oval, pointed and succulent.

cultivation

Propagation
Seed
Most species hybridise readily, so seed cannot be depended upon to reproduce the species true to type. When you buy seed it often says, 'mixture of several species and varieties' on the packet. It can be good fun to sow these as long as you do not mind what you get; it is even more fun trying to name them as they develop.

The seed is very small, so start off in a seed or plug tray in spring. Sow on the surface. Do not cover except with a sheet of glass. No need for bottom heat. Use a standard seed compost mixed in equal parts with composted fine bark.

Offsets
All the houseleeks produce offsets that cluster around the base of the parent plant. In spring gently remove them and you will notice each has its own root system. Either put straight into a pot, using a standard seed compost mixed in equal parts with composted fine bark, or plant where required. Plant 23cm apart.

Pests and Diseases
Vine weevil, this scourge of the garden, is very destructive to a number of plants and one of these is houseleek. You will know they have been when you see rosettes lying on their sides with no roots. See page 660 for methods of destroying the pests.

Maintenance
Spring Sow seeds. Pot up or re-plant offsets.
Summer Collect seeds if they are required from flowering plants.
Autumn Remove offsets if the plant is becoming too invasive, and pot up for following season's display.
Winter No need for protection.

Garden Cultivation
Basically the soil should be well-drained and thin, as houseleeks prefer very little to no soil. They will grow anywhere, on weathered rocks and screes and of course rock gardens. Another good place to plant them is between paving stones, or in between other creeping plants like thymes. They can take many years to flower, and when they do they die, but by then there will be many offsets to follow.

Harvesting
Pick leaves to use fresh as required. There is no good way of preserving them.

Sempervivum mixed in flower

Juice from a houseleek soothes burns and insect bites

container growing

If the Romans could do it, so can we. Houseleeks do look good in containers and shallow stone troughs. The compost must be poor and very well drained. Use a soil-based compost mixed with 50 per cent horticultural grit and 25 per cent composted fine bark. No need to feed, and do not over water.

medicinal

The leaves are an astringent and when broken in half can be applied to burns, insect bites and other skin problems. Press the juice from the leaf onto the infected part. My son, when he goes on hikes or is building dens, always has some in his pockets for when he gets stung by nettles as houseleeks are more soothing than dock when applied.

To soften skin around corns, bind one leaf for a few hours, soak foot in water in attempt to remove corn. Repeat as necessary. Infuse as a tea for septic throats, bronchitis and mouth ailments. It is also said that chewing a few leaves can ease toothache.

culinary

The leaves can be added to salad dishes. I think it would be polite to say that it is an acquired taste.

Solidago

GOLDENROD

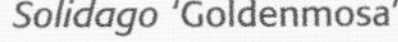
Solidago 'Goldenmosa'

Also known as Woundwort, Aaron's Rod, Cast The Spear and Farewell Summer. From the family *Asteraceae*.

This plant is widely distributed throughout Europe including the British Isles, and North America. It is common from the plains to the hills, but especially where the ground is rich in silica.

Its generic name, *Solidago*, is derived from the Latin word *solido,* which means 'to join' or 'make whole', a reference to the healing properties attributed to goldenrod.

The plant, originally called Heathen Wound Herb in Britain, was first imported from the Middle East, where it was used by the Saracens, and it was some time before it was cultivated here. In Tudor times it was available in London but at a price, its expense due to the fact that it was still available only as an import. Gerard wrote, 'For in my remembrance, I have known the dry herb which comes from beyond the sea, sold in Bucklesbury in London for half a crown an ounce,' and went on to say that when it was found growing wild in Hampstead wood, no one would pay half a crown for 100cwt of it, a fact which the herbalist felt bore out the old English proverb, 'Far fetch and dear, bought is best for ladies.'

From Culpeper, around the same time, we know that goldenrod was used to fasten loose teeth and as a remedy for kidney stones (which it still is).

Solidago 'Goldenmosa'

varieties

Solidago odora
Sweet Goldenrod
Also known as aniseed-scented goldenrod, blue mountain tea, common goldenrod and woundweed. Perennial. Ht 60cm–1.2m, spread 60cm. Golden-yellow flowers on a single stem from mid-summer to autumn. The green leaf is linear and lance-shaped.

Solidago nemoralis
Grey Goldenrod
Also known as dyer's weed, field goldenrod and yellow goldenrod. Perennial. Ht 60cm–1m, spread 60cm. Yellow flowers on large terminals on one side of the panicle. Leaves greyish-green or olive-green.

Solidago virgaurea
Goldenrod
Also known as European goldenrod. Perennial Ht 30–60cm, spread 60cm. Small yellow flowers from summer to autumn. The green leaves are lance-shaped.

***Solidago* 'Goldenmosa' AGM**
Golden Mimosa
Perennial. Ht 1m, spread 60cm. Sprays of mimosa-like yellow flowers from summer to autumn. Lance-shaped green leaves. Attractive border plant. Has no herbal use.

GOLDENROD

cultivation

Propagation

Seeds

Sow in plug or seed trays in spring. As seed is fine, sow on the surface of a standard seed compost mixed in equal parts with composted fine bark, and cover with perlite. Germination within 14–21 days without bottom heat. Prick out, harden off, and plant out into prepared site in the garden at a distance of 45cm. Remember, the plant will spread.

Division

Divide established plants in spring or autumn. Dig up the plant, split into required size, half, third, etc., and replant in a prepared site in the garden.

Pests and Diseases

This plant rarely suffers from pests or diseases.

Maintenance

Spring Sow seeds.
Summer Enjoy the flowers. If you have rich soil, the plants may become very tall and need support in exposed sites.
Autumn Divide mature plants.
Winter No need for protection.

Garden Cultivation

It is an attractive plant and has been taken into cultivation as a useful late-flowering ornamental. It is ideal for the herbaceous border, as it spreads rapidly to form clumps.

In late summer, sprays of bright yellow flowers crowd its branching stems amongst sharply pointed hoary leaves. When planting in the garden, it prefers open conditions and soils that are not too rich and are well drained. It tolerates sites in sun, semi-shade and shade, and, being a wild plant, it can be naturalised in poor grassland.

Sow seed thinly in spring or autumn in the chosen flowering position, having prepared the site, and cover lightly with soil. When the seedlings are large enough, thin to 30cm distance apart. (The plant will spread and you may have to do a second thinning.) If sown in the autumn, the young plants may, in very cold temperatures, need added protection. Use a mulch that they can grow through the following spring, or which can be removed.

Solidago virgaurea

Harvesting

Collect the flowering tops and leaves in summer. Dry for medicinal use.

container growing

Goldenrod can be grown in containers, but being a tall plant, it looks much more attractive in a garden border. Use a soil-based compost mixed in equal parts with composted fine bark, and in the summer only give it liquid fertiliser and water regularly.

In winter, as the plant dies back, move the container into a cool and airy place that is completely protected from frost, but not warm. Keep the compost on the dry side.

medicinal

Goldenrod is used in cases of urinary and kidney infections and stones, and catarrh. It also helps to ease backache caused by renal conditions because of its cleansing, eliminative action. It is used to treat arthritis.. A cold compress is helpful on fresh wounds because of its anti-inflammatory properties. Sweet goldenrod is used as an astringent and as a calmative. The tea made from the dried leaves and flowers is an aromatic beverage and can be used to improve the taste of other medicinal preparations. Native Americans applied lotions made from goldenrod flowers to bee stings.

Goldenrod in potpourri

Stachys officinalis

BETONY

Also known as Lousewort, Purple Betony, Wood Betony, Bishop's Wort and Devil's Plaything. From the family *Lamiaceae*.

This attractive plant, native to Europe, is still found growing wild in Britain.

Betony certainly merits inclusion in the herb garden, but is thought by some to be one of the plant world's frauds. There are so many conflicting stories, all of which are well worth hearing. I leave it to you to decide what is fact or fiction.

The ancient Egyptians were the first to attribute magical properties to betony. In England, by the tenth century, the Anglo-Saxons had it as their most important magical plant, claiming it as effective against the Elf sickness. In the eleventh century it was mentioned in the *Lacnunga* as a beneficial medicinal plant against the Devilish affliction of the body. Later, Gerard wrote in his *Herball*, 'Betony is good for them that be subject to the falling sickness,' and went on to describe its many virtues, one of them being as 'a remedy against the biting of mad dogs and venomous serpents'.

In the eighteenth century it was still used in the cure of diverse afflictions, including headaches and drawing out splinters, as well as used in herbal tobacco and snuff. Today, betony retains an important place in folk medicine, though its true value is seriously questioned. We owe the name to the Romans, who called the herb first *Bettonica* and then *Betonica*.

varieties

Stachys officinalis
Betony
Hardy perennial. Ht 60cm, spread 25cm. Dense spikes of pink or purple flowers from late spring through summer. Square hairy stems bear aromatic, slightly hairy, round, lobed leaves.

***Stachys officinalis* 'Alba'**
White Betony
Hardy perennial. Ht 60cm, spread 25cm. White flowers from late spring through summer.

cultivation

Propagation

Seed
Grows readily from seed, which it produces in abundance. Sow in late summer or spring in planting position and cover very lightly with soil. Alternatively, sow seeds in trays and prick out seedlings into small pots when large enough to handle.

Division
Divide roots of established plants in spring or autumn, replant at a distance of 30cm from other plants. Alternatively pot up using a standard seed compost mixed in equal parts with composted fine bark.

Pests and Diseases
Apart from the occasional caterpillar, which can be picked off, this plant is pest and disease free.

Maintenance
Spring Sow seeds. Divide established plants.
Summer Plant out spring seedlings.
Autumn Cut back flowering stems, save seeds, divide established plants.
Winter No protection needed.

Garden Cultivation
A very accommodating plant, it will tolerate most soils, but prefers some humus. Flourishes in sun or shade, and in fact it will put up with all but the deepest of shade. A wild plant, but it has for centuries been grown in cottage gardens.

BETONY

Dried leaves, flowers and root of betony

In the wild flower garden it is a very colourful participant and establishes well either in a mixed bed or in grassland. It is also an excellent plant for the woodland garden.

Harvesting

Collect leaves for drying before flowering in late spring and early summer. Use leaves fresh either side of flowering. Pick flowers for drying and for use in potpourris just as they start to open. Collect through flowering season to use fresh.

Collect and save seed in early autumn. Store in a dry, dark container.

container growing

Betony grows to great effect in half a beer barrel and combines well with other wild flowers such as poppies, oxeye daisy, chamomile. Use a soil-based compost mixed in equal parts with composted fine bark. I do not advise it for growing indoors or in small containers.

other uses

The fresh plant provides a yellow dye. A hair rinse, good for highlighting greying hair, can be made from an infusion of the leaves.

warning

Care must be taken if it is taken internally because in any form the root can cause vomiting and violent diarrhoea.

Betony leaves

medicinal

Today opinions differ as to its value. Some authorities consider it is only an astringent while others believe it is a sedative. It is however now chiefly employed in herbal smoking mixtures and herbal snuffs. As an infusional powder, it is used to treat diarrhoea, cystitis, asthma and neuralgia. Betony tea is invigorating, particularly if prepared in a mixture with other herbs. In France it is recommended for liver and gall bladder complaints.

Stevia rebaudiana

STEVIA

Also known as Sweet Leaf, Yerba Dulce, Honeyleaf and Caa'-ehe. From the family *Asteraceae*.

This fascinating and controversial subtropical and tropical herb is indigidous to South and Central America where it has been used for hundreds of years. In the sixteenth century the invading Spaniards noticed that the Guarani Indians of Paraguay were using this herb not only to sweeten their drinks, but also in herbal remedies and for making snacks. The Indians had many local names for stevia, one of which was Caa'-ehe; this and all the other local names have some form of reference to the sweetness of the leaf. In 1931 two French chemists, Bridel & Lavieille, extracted 'stevioside' from the leaf, a compound which is 300 times sweeter than sucrose, which is now being used in many countries throughout the world.

varieties

Stevia rebaudiana

Stevia

Subtropical evergreen perennial, grown as an annual in cold climates. Ht and spread 45cm. Clusters of small white flowers in late summer, early autumn. Mid-green, oval leaves with serrated edges, which have an intense sugary taste, with a liquorice after taste, when eaten.

cultivation

Propagation

Seed

Sow fresh seeds in spring into prepared module plug trays or small pots using a standard seed compost mixed in equal parts with perlite. Cover the seeds with perlite. Place under protection at 18°C; germination is erratic. As soon as the first few seeds emerge, remove from the heat, keep the container warm and the remaining seeds may germinate over the next few months. Once the seedlings are large enough, pot up using a soil-based potting compost mixed in equal parts with composted fine bark.

Cuttings

This is by far the most reliable method of propagation for this herb. Take semi-ripe cuttings from non-flowering shoots in early summer and place into prepared module plug trays using a seed compost mixed in equal parts with propagating bark. Place in a warm environment (bottom heat is not necessary), and it should root within 2 weeks. Once rooted, pot up using the same methods as for seed-raised plants.

Pests and Diseases

This herb, when grown in the tropics, rarely suffers from pests and diseases. In cold climates it is difficult to winter as it is prone to mildew and rot, especially if over watered. If it is grown under protection in spring and summer, make sure that the leaves do not become scorched and watch out for greenfly, especially on new growth. If the infestation becomes invasive, spray with a horticultural soap following the manufacturer's instructions.

Maintenance

Spring Sow seeds.
Summer Take cuttings.
Autumn Bring plants inside to protect from frost.
Winter Cut back on the watering of container plants in cool and cold climates.

Garden Cultivation

Outside the tropics this herb can only be grown as an annual in the garden. Plant in full sun – it will not tolerate shade – in a light fertile soil that does not dry out in summer. If you know your soil is prone to drying, after planting add a good deep mulch, preferably leaf mould or well-rotted compost.

Harvesting

Pick the leaves to use fresh as required from spring until late summer. Pick the leaves for drying in early summer.

container growing

Stevia is ideally suited for growing in a container. Plant in a soil-based potting compost mixed in equal parts with composted fine bark. Place the container in a warm sunny position. Water and feed regularly throughout the growing season. In autumn, in cold and cool climates, bring the container into a warm greenhouse, conservatory or well-lit windowsill. Cut back the plant to 10cm and reduce the watering to minimal, but not totally dry. Re-introduce watering and pot up one size of pot in the spring. Place the container outside once all frosts have passed.

medicinal

For today's society this herb has phenomenal potential in the treatment of obesity, high blood pressure and for those, such as diabetics, who require a natural sweetener and for those on a low carbohydrate diet. Due to lack of clinical trials it is not yet recognised, however there is currently lots of different scientific research being carried out worldwide and I would not be surprised to find it being included in many herbal preparations in the near future.

warning

This herb is banned for sale as a food or food ingredient in the UK and EU. In the USA it can only be sold as a dietary supplement and not as a sweetener.

Stevia rebaudiana under cultivation

culinary

I have read many arguments regarding the safety of stevia in food production, and they have left me puzzled and frustrated. The Japanese, who have banned all artificial sweeteners, have been using stevia as a commercial sweetener for the past 30 years without a single case of documented stevia toxicity or adverse reaction, yet it has been banned by the USA and European agencies, although the USA have agreed that it can be used as a food supplement in the home.

The leaves can be used fresh or dried; they have a very very sweet, slightly liquorice flavour. The flavour and intensity of the sugar content can vary due to growing position, sun strength, watering and age of plant. To test how strong the fresh leaves are, place a minute amount on the end of your tongue, this will immedately indicate strength. Alternatively, take one leaf, pour over 300ml of boiled water, let it stand for 10 minutes, then taste the water with a teaspoon. If not sweet enough for your taste or recipe add another leaf, reheat the water to just below boiling, let it stand for a further 10 minutes then taste again. Repeat this until you get the flavour you want. Always start with a few leaves because you can add but you cannot take away once you have created a sweet water. Once you have the flavour you require you can then use this sweet water in drinks, with fruit, water ices and other puddings.

Sutherlandia frutescens

CANCER BUSH

Also known as Kankerbos, Kankerbossie, Unwele, Phetola Mukakana and Lerumo-lamadi. From the family *Papilionaceae*.

The first time I saw this plant flower I knew that it was special, however I had no idea how special until I heard the BBC world news in November 2001 where it was proclaimed as a beneficial herb in the treatment of AIDS. This native of South Africa is regarded as the most profound and multi-purpose of their native medicinal plants. It has been used for centuries as a medicine; the Zulu sangomas, or traditional healers, know it as *unwele*, 'the great medicine', which they used to ward off the effects of the devastating 1918 influenza pandemic which claimed 20 million lives worldwide. While the Afrikaaners call it the *kankerbossie* or cancer bush, because of its properties in treating people suffering with internal cancers and wasting.

varieties

Sutherlandia frutescens

Cancer Bush

Evergreen tender shrub. Ht up to 1.2m, spread 1m. Bright scarlet flowers, from early to mid-summer, which grow in terminal clusters and are followed by inflated seed pods that are pale green, ripening to beige, tinged with red, in which there are small flat black seeds. Each grey, silver slightly hairy leaf is divided into 13–21 leaflets.

cultivation

Propagation

Seed

The cancer bush self-seeds readily around the garden. However, as with many herbs that do this, it does need encouragement when grown under controlled conditions in a greenhouse or cold frame. Sow either in autumn using fresh seed, or in the following spring with dried seeds. Prepare the seeds in exactly the same way that one would prepare the seeds of sweet peas, as they are from the same botanical family. Prior to sowing, scarify the seeds lightly by rubbing them gently on sandpaper and then soak them overnight in hand-hot water. Finally sow the prepared seed into plug modules, or small pots, using a standard seed compost mixed in equal parts with perlite. Cover the seeds with just perlite as this helps prevent damping off. Place under protection at 15°C; germination takes 2 to 3 weeks. When sowing in the autumn, winter the seedlings in a frost-free environment. Once the seedlings are large enough, pot up using a soil-based compost mixed in equal parts with horticultural sand. Harden off prior to planting in the garden or before growing on in a container. Plants raised from seed will take 3 years before they flower.

Cuttings

Take cuttings from non-flowering shoots from April until June. Put into prepared plug modules or very small pots, using a standard seed compost mixed in equal parts with perlite. Place the cuttings in a warm position or on a propagator at 15°C; rooting takes 2–3 weeks. Once roots have established, pot up using a soil-based potting compost mixed in equal parts with horticultural sand.

CANCER BUSH

Sutherlandia frutescens

Pests and Diseases

The plant is quite pest resistant in the garden. When container grown, however, it can be prone to red spider mite *(Tetranychus urticae)*. The best way to control this is to use a biological predator *Phytoseiulus persimilis,* which feeds on the eggs and the active stages of red spider mite. This predator does not work well in cold weather, therefore use from late spring until early autumn, well before the first frosts.

Container plants in the winter can suffer from powdery mildew, usually due to poor air circulation and the root ball being too dry. It also can suffer from downy mildew, which again is especially a problem in cold damp autumns where the air circulation is stagnant and the plant's leaves have not been picked up. To prevent both of these mildews, check the plants regularly, and lift the container off the floor onto bricks, to stop osmosis from the damp floor. Check the watering regularly to make sure that the root ball is just damp but not over-dry or sodden and, if the leaves do drop, make sure that they are removed and not left to rot on the surface of the pot.

Maintenance

Spring Sow seeds.
Summer Harvest leaves. Take cuttings.
Autumn Harvest leaves and seeds. Bring in containers.
Winter Protect from frosts. Check air circulation.

Garden Cultivation

This stunning plant makes a very good focal point in the garden. In the northern hemisphere and areas where the night temperature falls constantly below 10°C, it can only be grown as an annual, or as a container plant that is sunk into the border for the growing season and lifted in the autumn before the first frosts. With either method, plant in full sun in any good fertile, free-draining soil. The Cancer bush, once established, is fast growing; it is not particularly long lived, approximately five years, so it is worth taking cuttings each summer as insurance.

Harvesting

The leaves are harvested in summer and then dried for medicinal use.

companion planting

The cancer bush is a member of the legume (pea) family. Plants of this family fix atmospheric nitrogen via their roots in the soil, which then becomes a benefit for other plants.

container growing

This herb grows happily in containers; use a soil-based potting compost mixed in equal parts with horticultural sand. Water regularly and feed weekly with a liquid fertiliser throughout the growing season. In winter, in cool and cold climates, cut back on the watering and place in a frost-free environment that has good air circulation.

medicinal

The cancer bush is a traditional Cape remedy that is taken as a bitter tonic and used for numerous conditions including chickenpox, piles, backache and rheumatism.

In 2001 the Medical Research Council of South Africa performed clinical trials to assess the immune-boosting properties of the cancer bush to support the anecdotal evidence that this plant can improve the quality of life of thousands of people both with HIV and full-blown AIDS. A multi-disciplinary team headed by Dr Nigel Gericke, a botanist, medical doctor and indigenous plant specialist, found that *Sutherlandia* contained a powerful combination of molecules which have been identified and used in the treatment of patients with cancer, tuberculosis, diabetes, schizophrenia and clinical depression, and as an antiretroviral agent.

warning

This herb should be taken under supervision only.

Seed pods

Symphytum

COMFREY

Also known as Knitbone, Boneset, Bruisewort, Knitback, Church Bells, Abraham, Isaac-and-Jacob (from the variation in flower colour) and Saracen's Root. From the family *Boraginaceae*.

Native to Europe and Asia, it was introduced into America in the seventeenth century, where it has naturalised.

Traditionally known as Saracen's root, common comfrey is believed to have been brought to England by the Crusaders, who had discovered its value as a healing agent – mucilaginous secretions strong enough to act as a bone-setting plaster, and which gave it the nickname Knitbone.

The Crusaders passed it to monks for cultivation in their monastic herb gardens, dedicated to the care of the sick.

Elizabethan physicians and herbalists were never without it. A recipe from that time is for an ointment made from comfrey root boiled in sugar and liquorice, and mixed with coltsfoot, mallow and poppy seed. People also made comfrey tea for colds and bronchitis.

Symphytum ibericum

But times have changed. Once the panacea for all ills, comfrey is now under suspicion medicinally as a carcinogen. In line with its common name 'Bruisewort', research in America has shown that comfrey breaks down the red blood cells. At the same time, the Japanese are investigating how to harness its beneficial qualities; there is a research programme into the high protein and vitamin B content of the herb.

varieties

***Symphytum* 'Hidcote Blue'**
Comfrey Hidcote Blue
Hardy perennial. Ht 50cm, spread 60cm. Pale blue flowers in spring and early summer. Green lance-shaped leaves. Very attractive in a large border.

Symphytum ibericum
Dwarf Comfrey
Hardy perennial. Ht 25cm, spread 1m. Yellow/white flowers in spring. Green lance-shaped leaves. An excellent ground cover plant, having foliage through most winters. This comfrey contains little potassium and no allantoin, the crucial medicinal substance.

Symphytum x *uplandicum*

Symphytum officinale
Comfrey (Wild or Common)
Hardy perennial. Ht and spread 1m. White/purple/pink flowers in summer. This is the best medicinal comfrey and can also be employed as a liquid feed, although the potassium content is only 3.09 per cent compared to 7.09 per cent in 'Bocking 14'. It makes a first-class composting plant, as it helps the rapid breakdown of other compost materials.

Symphytum* x *uplandicum
Russian Comfrey
Hardy perennial. Ht 1m, spread indefinite. Pink/purple flowers in summer. Green lance-shaped leaves. This is a hybrid that occurred naturally in Upland, Sweden. It is a cross between *S. officinale*, the herbalist's comfrey, and *S. asperum*, the blue-flowered, prickly comfrey from Russia. A very attractive form is *S.* x *uplandicum* 'Variegatum', with cream and green leaves.

***Symphytum* x *uplandicum* 'Bocking 4'**
Hardy perennial. Ht 1m, spread indefinite. Flowers near to violet in colour, in spring and early summer. Thick, solid stems. Large green lance-shaped leaves. Not a particularly attractive plant but it contains almost 35 per cent total protein, the same percentage as in soya beans. Comfrey is an important animal feed in some parts of the world, especially in Africa.

***Symphytum* x *uplandicum* 'Bocking 14'**
Hardy perennial. Ht 1m, spread indefinite. Mauve flowers in spring and early summer. Thin stems. Green oval leaves, tapering to a point. This variety has the highest potash content, which makes it the best for producing liquid manure.

cultivation

Propagation

Seed

Not nearly as reliable as root cutting or division. Sow in spring or autumn in either seed or plug trays. Germination slow and erratic.

Root Cuttings

In spring, dig up a piece of root, cut into 2cm sections, and put these small sections into a prepared plug or seed tray.

Division

Use either the double spade method or simply dig up a chunk in the spring and replant it elsewhere.

Pests and Diseases

Sometimes suffers from rust and powdery mildew in late autumn. In both cases cut the plant down and burn the contaminated leaves.

Maintenance

Spring Sow seeds. Divide plants. Take root cuttings.
Summer Cut back leaves for composting, or to use as a mulch around other herbs in the growing season.
Autumn Sow seeds.
Winter None needed.

Garden Cultivation

Fully hardy in the garden, all the comfreys prefer sun or semi-shade and a moist soil, but will tolerate most conditions. The large tap root can cause problems if you want to move the plant. When doing this make sure you dig up all the root because any left behind will reappear later.

Harvesting

Cut leaves with shears from early summer to autumn to provide foliage for making liquid feed. Each plant is able to give four cuts a year if well fed. Cut leaves for drying before flowering. Dig up roots in autumn for drying.

Liquid Manure

Use either *S.* x *uplandicum* or *S.* x *uplandicum* 'Bocking 14' to make a quickly available source of potassium for the organic gardener. One method of extracting it is to put 6kg of freshly cut comfrey into a 90 litre tapped, fibreglass water butt. Do not use metal as rust will add toxic quantities of iron oxide to the liquid manure. Fill up the butt with rain or tap water and cover with a lid to exclude the light. In about 4 weeks a clear liquid can be drawn off from the tap at the bottom. Ideal feed for tomatoes, onions, gooseberries, beans and all potash-hungry crops. It can be used as a foliar feed.

Powdery mildew on comfrey leaf

The disadvantage of this method is that the liquid stinks, because comfrey foliage is about 3.4 per cent protein, and when proteins break down they smell.

An alternative is to bore a hole into the side (just above the bottom) of a plastic dustbin. Stand the container on bricks, so that it is far enough off the ground to allow a dish to be placed under the hole. Pack it solid with cut comfrey, and place something (a heavy lump of concrete) on top to weigh down the leaves. Cover with lid, and in about 3 weeks a black liquid will drip from the hole into a dish.

This concentrate can be stored in a screw-top bottle if you do not want to use it immediately. Dilute it 1 part to 40 parts water, and if you plan to use it as a foliar feed, strain it first.

container growing

Comfrey is not suitable for growing indoors, but it can be grown on a patio as long as the container is large enough. Situate in partial shade and give plenty of water in warm weather.

culinary

Fresh leaves and shoots were eaten as a vegetable or salad and there is no reason to suppose that it is dangerous to do so now, although it may be best to err on the side of caution until suspicions are resolved (see 'Warning' below).

medicinal

Comfrey has received much attention in recent years, both as a valuable healing herb, a source of vitamin B12 and self-proliferate allantoin, and as a potential source of protein. Comfrey is also useful as a poultice for varicose ulcers and a compress for varicose veins, and it alleviates and heals minor burns.

warning

Comfrey is reported to cause serious liver damage if taken in large amounts over a long period of time.

other uses

Boil fresh leaves for golden fabric dye (see also page 675). Comfrey is a good feed for racehorses and helps cure laminitis. For curing septic sores on animals, make a poultice between clean pieces of cotton and tie to the affected places.

Comfrey dye

Tagetes lucida

WINTER TARRAGON

Also known as Mexican Tarragon, Spanish Tarragon and Sweet Mace. From the family *Asteraceae*.

The first time I saw winter tarragon was over 15 years ago in Florida, where I found it on sale in a garden centre. Until then I had never seen an edible *Tagetes* and was amazed by the power and the anise flavour of the leaves, which proved to be great in cooking.

Winter tarragon is indigenous to southern and central Mexico. Historically it is said that the Aztecs used it to flavour *chocólatl*, a coca-based drink and it is still popular today with the Tarahumara Indians of Chihuahua. It is revered by the modern Huichol Indians as a shamanic trance 'tobacco', which is used in their religious rituals.

varieties

Tagetes lucida
Winter Tarragon, Mexican Tarragon
Tender herbaceous perennial. Ht 80cm in warm climates only, spread 45cm. Sweetly scented yellow flowers in late summer. Rarely flowers in cold temperatures. Narrow lance-shaped aromatic mid-green toothed leaves, which have strong aniseed scent and flavour. This herb, in cool climates, dies back in early spring, reappearing in early summer.

Tagetes patula
Wild Mexican marigold, French marigold
Half-hardy annual. Ht 1.2m, spread 45cm. Clusters of single yellow flowers, mid-summer until first frosts. Aromatic, deeply divided, lightly toothed mid-green leaves. This species is said to deter nematodes in the soil, and whitefly from tomatoes. The leaves have been used to flavour food; medicinally it is a diuretic and it improves the digestion. Externally it has been used to treat sore eyes and rheumatism. The flowers have been used to feed poultry and to colour dairy produce and textiles.

cultivation

Propagation

Seed

Sow the seeds in early spring into prepared seed or plug module trays using a standard seed compost mixed in equal parts with perlite. Place under protection at 20°C; germination takes 14–21 days. Once the seedlings are large enough, pot up into a soil-based compost mixed in equal parts with horticultural sand. Alternatively sow in late spring, into prepared open ground, when the air temperature does not fall below 10°C at night; germination takes 2–4 weeks.

Cuttings

Take tip cuttings in early summer, place the cutting into prepared plug module trays using a standard seed compost mixed in equal parts with perlite. Place in a warm environment until rooted, about 2–3 weeks. Once fully rooted, pot up into a soil-based compost mixed in equal parts with horticultural sand.

Pests and Diseases

Rarely suffers from pests. In cold, damp climates it will

die back in late winter; make sure at this time of year that the leaves do not become too damp because this could rot the crown. If you notice mildew, cut off all the remaining growth, raise the container up onto some bricks and put in a well-aired, frost-free place.

Maintenance

Spring Sow seeds. Cut back any winter growth.
Summer Take cuttings.
Autumn Lift garden plants and bring in containers once the night temperature falls below 10°C.
Winter Cut back the watering to minimal.

Garden Cultivation

Winter tarragon can only be grown successfully outside all year round in a tropical or subtropical climate. In cooler climates it must be grown as an annual or lifted in winter. Plant in a well-drained fertile soil in a sunny position.

Harvesting

The leaves can be picked as required throughout the growing season for using fresh. Alternatively pick in mid-summer for drying.

container growing

Winter tarragon is ideal for growing in a container in cool and cold climates. Pot up using a soil-based compost mixed in equal parts with horticultural sand. Once the night temperature falls below 10°C, bring the container into a frost-free environment and cut back the watering to minimal but not totally dry. Repot each spring, after removing any winter growth. Only pot up one size of pot at a time as this herb likes being pot bound.

medicinal

The leaves of this herb are used as a relaxant and to treat diarrhoea, indigestion, malaria and feverishness. It also has the reputation of being a very good hangover cure.

other uses

Burnt leaves make a good insect repellent.

The juice from crushed leaves when applied to ticks makes them easy to remove.

warning

Not to be taken medicinally when pregnant or breast feeding. If taken in excess it can cause hallucinations.

Tagetes patula

culinary

Winter tarragon makes a good substitute for French tarragon (*Artemisia dracunculus*), not only in winter, when French Tarragon has died back into the ground, but also where it is difficult to grow, i.e. in humid climates. The leaves of this tarragon are much stronger and more anise than the French variety so use with care until you get used to it.

Chicken and Winter Tarragon Parcel

Serves 2

2 skinless organic chicken breasts
2 cloves garlic, peeled and sliced
6 x 10cm sprigs winter tarragon, leaves removed and chopped
30g butter
160ml white wine
Extra-wide tin foil

Preheat the oven to 220°C/425°F/gas mark 7. Cut two pieces of tin foil 35cm x 50cm. Place them on top of each other. Then fold in half and seal the two long sides leaving the top open. Place the chicken breasts in the foil bag, with the garlic, chopped tarragon and butter. Check the sides of the bag are well sealed then pour in the white wine and seal the top. Place this on a baking tray and put in the preheated oven for 25 minutes. Once cooked, place the foil bag onto a plate or serving dish before opening to catch all the juices. Serve with baked potatoes and a crisp green salad.

Tanacetum balsamita

ALECOST

Also known as Costmary, Bible Leaf, English Mace and Sweet Mary. From the family *Asteraceae*.

The first syllable in the common name, 'alecost', is derived from the use to which its scented leaves and flowering tops were used in the Middle Ages, namely to clear, preserve and impart an astringent minty flavour to beer. The word 'cost' derives from *kostos*, the Greek for spicy so therefore 'alecost' literally means a spicy herb for ale. With the demise of its use in beer it has now become a rare plant in Europe, however, in North America it can be found as a garden escape growing wild in the Eastern and mid West states.

varieties

Tanacetum balsamita
Alecost
Perennial. Ht. 1m. and spread 45cm.
Clusters of small white yellow-eyed daisy flowers from July–September. Oblong, silver/green mint-scented leaves.

Tanacetum balsamita* subsp. *balsamitoides
Camphor plant
Perennial. Ht. 1m. and spread 45cm.
Clusters of small white yellow-eyed daisy flowers from July–September. Oblong, silver/green camphor-scented leaves.

cultivation

Propagation

Seed
Although possible, as it is difficult to obtain viable seed I recommend raising this plant from division.

Division
Divide established plants in spring or early autumn. Either plant out directly into the prepared garden or pot up into individual pots using a potting compost mixed in equal parts with composted fine bark. If you have a very wet cold soil it is better to winter autumn divided plants in pots, in a cold frame or cold greenhouse.

Pests and Diseases
Both species rarely suffer from pest or diseases.

Maintenance
Spring Divide established plants.
Summer Dead head flowers.
Autumn Cut back the flowers. Divide established plants.
Winter Tidy up dead leaves as these will spread disease if left to rot. Protect pot grown plants from excessive wet.

Garden Cultivation
Both species will adapt to most conditions. However, they prefer a fertile, well-drained soil. Plant at a distance of 60cm (2ft) from other plants and, if possible, in a sunny position because when planted in shade they will not flower.

Harvesting
Pick Alecost leaves for culinary use throughout the growing season. For drying pick the leaves in late spring, before flowering, which is when the leaves have the best sweet scent.

container growing

This herb is not ideal for growing in containers, year round, as it grows tall and rapidly outgrows its pot, so requiring regular repotting throughout the growing season.

culinary

Alecost leaves have a sharp bitter tang which can be overpowering so use sparingly. Add finely chopped leaves to carrots, salads, game, poultry, stuffings and fruit cakes.

medicinal

Traditionally a tea made from the leaves of this herb was used to ease the pain of childbirth. It was also used as a tonic for colds, catarrh, stomach upsets and cramps. Rub a fresh leaf of Alecost on a bee sting or horse fly bite to relieve the pain.

other uses

The leaves and flowers can be used in home brewing to clear, flavour and preserve the beer. Dried leaves make a good addition to potpourri, they can also be added to linen bags where they not only smell good but also act as a good moth repellent. Fresh or dried leaves of Alecost can be added to baths for a refreshing soak.

warning

Do not use Camphor in any culinary dishes.

Tanacetum cinerariifolium

PYRETHRUM

Also known as Dalmatian Pellitory. From the family *Asteraceae*.

This plant is a native of Croatia and can be found growing wild in the hills above the Adriatic sea. It is now cultivated commercially in many parts of the world, including Japan, South Africa and parts of central Europe for its insecticidal properties. Originally the insecticide was known as Dalmatian insect powder because for many hundreds of years the eastern coast of the Adriatic was known as Dalmatia. The insecticide is now known as Pyrethrum and can be found in many proprietary products.

varieties

Tanacetum cinerariifolium* syn. *Chrysanthemum cinerariifolium

Hardy perennial. Ht 30–37cm, spread 20cm. Daisy-like flower, white petals with a yellow centre. Leaves green/grey, finely divided, with white down on the underside.

Tanacetum coccineum* syn. *Chrysanthemum roseum

Hardy perennial. Ht 30–60cm, spread 30cm. Large flowerhead which can be white or red, and sometimes tipped with yellow. Very variable under cultivation. Vivid green leaves. Native of Iran. It was known as Persian Insect Powder, but it is not as strong as *T. cinerariifolium* so fell out of favour. An attractive garden plant.

cultivation

Propagation

Seeds

In spring sow into a prepared seed or plug tray and cover with perlite. Germination is easy and takes 14–21 days. In late spring, when the young plants are large enough and after a period of hardening off, plant out into a prepared site in the garden 15–30cm apart.

Division

Established clumps can be dug up in the spring or autumn and divided.

Pests and Diseases

Mostly free from pests and diseases.

Maintenance

Spring Sow seed. Divide established plants.
Summer Deadhead flowers to prolong season if not collecting seed.
Autumn Divide established plants if necessary.
Winter Fully hardy.

Garden Cultivation

Pyrethrum likes a well-drained soil in a sunny spot; it is drought tolerant and fully hardy in most winters.

Harvesting

The flower heads are collected just as they open and dried gently. When they are quite dry, store away from light. The insecticide is made from the powdered dried flower.

container growing

This herb is very well suited for growing in containers, especially terracotta. Use a soil-based compost mixed in equal parts with composted fine bark. Water well during the summer months, but be mean on the liquid fertiliser, otherwise the leaves will become green and it will stop flowering.

medicinal

This herb is rarely used medicinally. Herbalists have found the roots to be a remedy for certain fevers. Recent research has shown that the flowerheads possess a weak antibiotic.

other uses

This is a useful insecticide because it is non-toxic to mammals and does not accumulate in the environment or in the bodies of animals. It acts by paralysing the nervous system of the insects, and can kill pests living on the skin of man and animals. Sprinkle the dry powder from the flowers to deter all common insects, pests, bed bugs, cockroaches, flies, mosquitoes, aphids and ants. Traditionally the dried flowers were mixed with methylated spirits and water and used as a garden spray. When buying it in a proprietary product, do read the instructions carefully as Pyrethrum is harmful to many beneficial pollinating insects.

warning

When used as an insecticide, it can kill helpful insects and fish, as well as pests. Can cause contact dermititis.

Tanacetum parthenium

FEVERFEW

Also known as Featherfew in America and Febrifuge. From the family *Asteraceae.*

The common name suggests that this herb was used in the treatment of fevers and the old herbalists even call it febrifuge, however, strange as it may seem, the herb was hardly ever employed for this purpose. In the seventeenth century Culpeper advised it use for pains in the head and colds and nowadays it is used in the treatment of migraines.

varieties

Tanacetum parthenium* syn. *Chrysanthemum parthenium
Feverfew
Hardy herbaceous perennial. Height up to 1.2m, spread 45cm. Clusters of small white yellow-eyed daisy flowers from early summer until the first frosts. The leaf is mid green, lobed and divided with lightly serrated edges.

***Tanacetum parthenium* 'Aureum'**
Golden Feverfew
Hardy herbaceous perennial. Height 20–45cm, spread 45cm. Clusters of small white yellow-eyed daisy flowers from early summer until the first frosts. The leaf is golden, lobed and divided with lightly serrated edges. Excellent edging plant.

cultivation

Propagation

Seed
The fine, thin and fairly small seeds tend to stick together especially if they are damp. Mix the seed with a small amount of perlite or dry sand prior to sowing. Sow seeds in autumn or spring, into prepared pots or plug modules using a standard seed compost; do not cover. Place in a cold frame, germination 2–4 weeks. Autumn sowings, winter in the cold frame. Plant out when large enough to handle 30cm (12in) apart.

Cuttings
Take stem cutting in summer, making sure there are not flowers on the cutting material.

Division
Divide established plants in autumn or spring, replanting into a prepared site.

Pests and Diseases
Aphids are the only real problem for feverfew.

Maintenance
Spring Sow seeds.
Summer As flowering finishes, cut back to restore shape and minimise self-seeding. Take stem cuttings.
Autumn Divide established plants. Sow seeds.
Winter No need for protection, frost hardy.

Garden Cultivation
Plant in a well-drained fertile soil in a sunny position. Feverfew is drought tolerant and will adapt to most soils and conditions, however if planted in shade they are unlikely to flower. Golden feverfew can suffer from sun scorch; if this occurs cut back and the new growth will be unaffected.

Feverfew in sachets makes a good moth repellant

Harvesting
Pick the leaves before flowering to use fresh or to dry. Pick the flower heads just as they open for drying.

container growing

Both forms of Feverfew look attractive in containers. Use a soil-based compost mixed in equal parts with composted fine bark.

culinary

The young leaves can be added to salads; however they have a very bitter taste so use sparingly.

medicinal

Feverfew is a renowned remedy for certain types of migraines. Because they are so bitter place 2–3 fresh leaves in between a slice of bread, this will make them more palatable. Do not take for a prolonged period and do not over-eat.

other uses

A decoction made from the leaves is a good basic household disinfectant. The dried leaves, placed in a sachet, makes a good moth repellent.

warning

One side effect associated with taking feverfew is ulceration of the mouth.

Tanacetum vulgare

TANSY

Also known as Bachelor's Buttons, Bitter Buttons, Golden Buttons, Stinking Willy, Hind Heel and Parsley fern. From the family *Asteraceae*.

Tansy is a native to Europe and Asia, and it has managed to become naturalised elsewhere, especially in North America.

The name derives from the Greek *athanasia*, meaning 'immortality'. In ancient times it was used in the preparation of the embalming sheets and rubbed on corpses to save them from earthworms or corpse worm.

varieties

Tanacetum vulgare
Tansy
Hardy perennial. Ht 90cm, spread 30–60cm and more. Yellow button flowers in late summer. The aromatic leaf is deeply indented, toothed and fairly dark green.

Tanacetum vulgare* var. *crispum
Curled Tansy
As *T. vulgare* except ht 60cm, and the aromatic leaf is crinkly, curly and dark green.

***Tanacetum vulgare* 'Isla Gold'**
Tansy Isala Gold
As *T. vulgare* except ht 60cm, and the leaf is golden in colour.

***Tanacetum vulgare* 'Silver Lace'**
Tansy 'Silver Lace'
As *T. vulgare* except ht 60cm, and the leaf starts off white-flecked with green, progressing to full green. If, however, you keep cutting it, some of the variegation can be maintained.

cultivation

Propagation
Seed
Sow the small seed in spring or autumn in a prepared seed or plug tray using a standard seed compost, and cover with perlite. Germination takes 10–21 days. Plant out 45cm apart when the seedlings are large enough to handle. If sown in the autumn, protect over winter.

Division
Divide root runners in either spring or autumn.

Pests and Diseases
Aphids are the only real problem for tansy.

Maintenance
Spring Sow seed. Divide established clumps.
Summer Cut back after flowering to maintain shape and colour.
Autumn Divide established clumps. Sow seeds.
Winter The plant is fully hardy and dies back into the ground for winter.

Garden Cultivation
Tansy needs to be positioned with care as the roots spread widely. The gold and variegated forms are much less invasive and very attractive in a semi-shaded border. They tolerate most conditions provided the soil is not completely wet.

Harvesting
Pick leaves as required. Gather flowers when open.

container growing

Because of its antisocial habit, container growing is recommended. Use a soil-based compost and a large container, water throughout the growing season, and only feed about twice during flowering. In winter keep on the dry side in a cool place.

medicinal

Can be used by trained herbalists to expel roundworm and threadworm. Use tansy tea externally to treat scabies, and as a compress to bring relief to painful rheumatic joints.

warning

Use tansy only under medical supervision. It is a strong emmenagogue, provoking the onset of a period, and should not be used during pregnancy. An overdose of tansy oil or tea can be fatal.

other uses

Rub into the coat of your dog or cat to prevent fleas. Hang leaves indoors to deter flies. Put dried sprigs under carpets. Add to insect-repellent sachets. Sprinkle chopped leaves and flowers to deter ants and mice. It produces a yellow/green woollen dye.

Tanacetum vulgare 'Isla Gold'

Taraxacum

DANDELION

Also known as Pee In The Bed, Lions Teeth, Fairy Clock, Clock, Clock Flower, Clocks and Watches, Farmers Clocks, Old Mans Clock, One Clock, Wetweed, Blowball, Cankerwort, Lionstooth, Priests Crown, Puffball, Swinesnout, White Endive, Wild Endive and Piss-a-beds. From the family *Asteraceae*.

Dandelion is one of nature's great medicines and it really proves that a weed is only a plant out of place! It is in fact one of the most useful of herbs. It has become naturalised throughout the temperate regions of the world and flourishes on nitrogen-rich soils in any situation to a height of 2,000m.

There is no satisfactory explanation why it is called dandelion, dents lioness, tooth of the lion in medieval Latin, and *dent de lion* in French. The lion's tooth may be the tap root, the jagged leaf or the flower parts.

The Arabs promoted its use in the eleventh century. By the 16th it was well established as an official drug. The apothecaries knew it as *Herba taraxacon* or *Herba urinari*, and Culpeper called it piss-a-beds, all referring to its diuretic qualities.

varieties

***Taraxacum officinale* agg.**
Dandelion
Perennial. Ht 15–23cm. Large, brilliant yellow flowers 5cm wide, spring to autumn. The flower heads as they turn to seed form a fluffy ball (dandelion clock). Leaves oblong with a jagged edge.

***Taraxacum kok-saghyz* Rodin**
Russian Dandelion
Perennial. Ht 30cm. Similar to the above. Extensively cultivated during the Second World War: latex was extracted from the roots as a source of rubber.

Taraxacum mongolicum
Chinese Dandelion
Perennial. Ht 25–30cm. Similar to the above. Used to treat infections, particularly mastitis.

cultivation

Propagation

Seed
Grow as an annual to prevent bitterness developing in the plant. Sow seed in spring on the surface of pots or plug trays. Do not use seed trays as the long tap root makes it difficult to prick out. Cover with a fine layer of perlite. Germination will be in 3–6 weeks, depending on seed freshness and air temperature. Plant out when large enough to handle.

Root
In spring and autumn, sections of the root can be cut and put in pots, seed trays or plug trays. Each piece will sprout again, just like comfrey.

Pests and Diseases

Dandelion is rarely attacked by either pest or disease.

Maintenance

Spring Sow seeds for use as an autumn salad herb. Take root cuttings.
Summer Continually pick off the flower buds if you are growing dandelion as a salad crop.
Autumn Put an up-turned flower pot over some of the plants to blanch them for autumn salads. Sow seed for spring salad crop. Take root cuttings.
Winter No protection is needed. For salad crops,

DANDELION

however, if temperatures fall below −10°C, cover with agricultural fleece or 8cm of straw or bracken to keep the leaves sweet.

Garden Cultivation

If the dandelion was a rare plant, it would be thought a highly desirable garden species, for the flowers are most attractive, sweet smelling and a brilliant yellow, and then form the delightful puff balls. Up to that point all is fine. But then the wind disperses the seed all over the garden. And it is very difficult to eradicate when established since every bit of root left behind produces another plant. So, it finds no favour at all with gardeners.

In general, details on how to grow dandelions are superfluous. Most people only want to know how to get rid of them. The easiest time to dig up the plants completely is in the early spring.

Harvesting

Pick leaves as required to use fresh, and flowers for wine as soon as they open fully. Dig up roots in autumn for drying.

Dandelion wine

container growing

Dandelions do look attractive growing in containers, especially in window boxes, if you can stand neighbours' remarks. But in all seriousness the containers will need to be deep to accommodate the long tap root.

medicinal

It is one of the most useful medicinal plants, as all parts are effective and safe to use. It is regarded as one of the best herbal remedies for kidney and liver complaints. The root is a mildly laxative, bitter tonic, valuable in treating dyspepsia and constipation. The leaves are a powerful diuretic. However, unlike conventional diuretics, dandelion does not leach potassium from the body as its rich potassium content replaces what the body loses.

The latex contained in the leaves and stalks is very effective in removing corns and in treating warts and verrucas. Apply the juice from the plant daily to the affected part.

The flowers can be boiled with sugar for coughs, but honey has a greater medicinal value.

other uses

As a herbal fertiliser, dandelion has a good supply of copper. Pick 3 plants completely: leaves, flowers and all. Place in a bucket, pour over 1 litre of boiling water, cover and allow to stand for 30 minutes. Strain through an old pair of tights or something similar. This fertiliser will not store.

A dye, yellow-brown in colour, can be obtained from the root and dandelions are excellent food for domestic rabbits, guinea pigs and gerbils.

There is one thing for which they are useless, however – flower arrangements. As soon as you pick them and put them in water, their flowers close tight.

culinary

Both the leaves and root have long been eaten as a highly nutritious salad. In the last century, cultivated forms with large leaves were developed as an autumn and spring vegetable. The leaves were usually blanched in the same way as endive. Dandelion salad in spring is also considered a blood cleanser, owing to its diuretic and digestive qualities. The leaves are very high in vitamins A, B, C and D, the A content being higher than that of carrots. The flowers make an excellent country wine and dandelion roots provide, when dried, chopped and roasted, the best-known substitute for coffee.

Dandelion and Bacon Salad

Serves 4

225g young dandelion leaves
4 tablespoons olive or walnut oil
1 tablespoon white wine vinegar
sugar
100g streaky bacon, diced
1 clove garlic, crushed
1cm slice white bread, cubed
Salt and freshly ground black pepper
Oil for cooking

Wash and dry the leaves and tear into the salad bowl. Make a vinaigrette using olive oil and vinegar, and season to taste, adding a little sugar if desired. Fry the bacon, crushed garlic and bread in oil until golden brown. Pour the contents of the pan over the leaves and turn the leaves until thoroughly coated. Add the vinaigrette and toss again and serve at once.

Teucrium chamaedrys L.

WALL GERMANDER

Also known as Ground Oak and Wild Germander. From the family *Lamiaceae*.

This attractive plant, native of Europe, is now naturalised in Britain and other countries in the temperate zone. It is found on dry chalky soils.

The Latin *Teucrium* is said to have been named after Teucer, first king of Troy. It is the ancient Greek word for 'ground oak', its leaves resembling those of the oak tree.

In medieval times, it was a popular strewing herb and a remedy for dropsy, jaundice and gout. It was also used in powder form for treating head colds, and as a snuff.

varieties

***Teucrium chamaedrys* L.**
Wall Germander
Evergreen hardy perennial. Ht 45cm, spread 20cm. Pink flowers from mid-summer to early autumn. The leaves are mid-green and shaped like miniature oak leaves with crumpled edges.

Teucrium fruticans
Tree Germander
Evergreen hardy perennial. Ht 1–2m, spread 2–4m. Blue flowers in summer. The leaves are aromatic, grey/green with a white underside.

Teucrium x lucidrys
Hedge Germander
Evergreen hardy perennial. Ht 45cm, spread 20cm. Pink flowers from midsummer to early autumn. The leaves are dark green, small, shiny and oval. When rubbed, they smell pleasantly spicy.

cultivation

Propagation

Seed
Sow the small seeds in spring. Use a prepared seed or plug tray and a standard seed compost mixed in equal parts with composted fine bark. Cover with perlite. Germination can be erratic – from 2–4 weeks. When the seedlings are large enough to handle, plant in a prepared site 20cm apart.

Cuttings
This is a better method of propagating germander. Take softwood cuttings from the new growth in spring, or semi-hardwood in summer. Ensure compost does not dry out or become sodden.

Division
The teucriums produce creeping rootstock in the spring and are easy to divide. Dig up the plants, split them in half, and replant in a chosen site.

Pests and Diseases
Wall germander rarely suffers from pests or diseases.

Maintenance
Spring Sow seeds. Take softwood cuttings. Trim established plants and hedges. Divide established plants.
Summer Trim plants after flowering, take semi-hardwood cuttings.
Autumn Trim hedges.
Winter Protect when temperatures drop below –5°C.

Garden Cultivation
Wall germander needs a well-drained soil (slightly alkaline) and a sunny position. It is hardier than lavender and cotton lavender, and makes an ideal hedging or edging plant. To make a good dense hedge, plant at a distance of 15cm. If you clip the hedge in spring and autumn to maintain its shape, you will never need to cut it hard back.

It can also be planted in rockeries, and in stone walls where it looks most attractive. During the growing season it does not need extra water, even in hot summers, nor does it need extra protection in cold winters.

Harvesting
For drying for medicinal use, pick leaves before flowering, and flowering stems when the flowers are in bud.

container growing

Looks good in containers. Use a soil-based compost mixed in equal parts with composted fine bark. Only feed in the flowering season. Keep on the dry side in winter.

culinary

Used extensively in the flavouring of liqueurs.

medicinal

Its herbal use today is minor. However, there is a revival of interest, and some use it as a remedy for digestive and liver troubles, anaemia and bronchitis.

Teucrium scorodonia

WOOD SAGE

Also known as Gypsy Sage, Mountain Sage, Wild Sage and Garlic Sage. From the family *Lamiaceae*.

This plant is a native of Europe and has become naturalised in Britain and other countries in the temperate zone.

There is not much written about wood sage apart from the fact that, like alecost, it was used in making ale before hops were introduced. However, Gertrude Jekyll recognised its value, and with renewed interest in her gardens comes a revival of interest in wood sage.

varieties

Teucrium scorodonia

Wood Sage

Hardy perennial. Ht 30–60cm, spread 25cm. Pale greenish-white flowers in summer. Soft green heart-shaped leaves, which have a mild smell of crushed garlic.

***Teucrium scorodonia* 'Crispum'**

Curly Wood Sage

Hardy perennial. Ht 35cm, spread 30cm. Pale greenish-white flowers in summer. The leaves are soft, oval and olive green with a reddish tinge to their crinkled edges. Whenever it is on show it causes much comment.

cultivation

Propagation

Wood sage can be propagated by seed, cuttings or division. Curly wood sage can only be propagated by cuttings or division.

Seed

Sow the fairly small seed under protection in autumn or spring in a prepared seed or plug tray. Use a standard seed compost mixed in equal parts with composted fine bark, and cover with perlite. Germination can be erratic, taking from 2–4 weeks. When the seedlings are large enough to handle, pot up and winter under cover in a cold frame. In the spring, after a period of hardening off, plant out in a prepared site in the garden at a distance of 25cm.

Cuttings

Take softwood cuttings from the new growth in spring, or semi-hardwood cuttings in summer.

Division

Both wood sages produce creeping rootstock. In the spring they are easy to divide.

Pests and Diseases

In the majority of cases, it is free from pests and diseases.

Maintenance

Spring Sow seeds. Divide established plants. Take softwood cuttings.

Summer Take semi-hardwood cuttings.

Autumn Sow seeds.

Winter No need for protection, the plants die back for the winter.

Garden Cultivation

It grows well in semi-shaded situations, but also thrives in full sun on sandy and gravelly soils. It will adapt quite happily to clay and heavy soils, but not produce such prolific growth.

Harvesting

Pick young leaves for fresh as required.

container growing

I have grown curly wood sage most successfully in containers. The plain wood sage does not look quite so attractive. Use a soil-based compost mixed in equal parts with composted fine bark, and plant in a large container – its creeping rootstock can too easily become pot-bound. Only feed twice in a growing season, otherwise the leaves become large, soft and floppy. When the plant dies back, put it somewhere cool and keep it almost dry.

medicinal

Wood sage has been used to treat blood disorders, colds and fevers, and as a diuretic and wound herb.

culinary

The leaves of ordinary wood sage have a mild garlic flavour. When young and tender, the leaves can be added to salads for variety. But go steady as they are slightly bitter.

Thymus

THYME

From the family *Lamiaceae*.

This is a genus comprising numerous species that are very diverse in appearance and come from many different parts of the world. They are found as far afield as Greenland and Western Asia, although the majority grow in the Mediterranean region.

This ancient herb was used by the Egyptians in oil form for embalming. The Greeks used it in their baths and as an incense in their temples. The Romans used it to purify their rooms, and most probably its use spread through Europe as their invasion train swept as far as Britain. In the Middle Ages drinking it was part of a ritual to enable one to see fairies, and it was one of many herbs used in nosegays to purify the odours of disease. Owing to its antiseptic properties, judges also used it along with rosemary to prevent gaol fever.

Thymus vulgaris

Mixed thymes

varieties

There are so many species of thyme that I am only going to mention a few of interest. New ones are being discovered each year. They are eminently collectable. Unfortunately their names can be unreliable, a nursery preferring its pet name or one traditional to it, rather than the correct one.

Thymus caespititius
Caespititius Thyme
Evergreen hardy perennial. Ht 10cm, spread 20cm. Pale pink flowers in summer. The leaves narrow, bright green and close together on the stem. Makes an attractive low-growing mound, good between paving stones.

Thymus camphoratus
Camphor Thyme
Evergreen half-hardy perennial. Ht 30cm, spread 20cm. Pink/mauve flowers in summer, large green leaves smelling of camphor. Makes a beautiful compact bush.

***Thymus cilicicus* Boiss. & Bail.**
Sicily Thyme
Evergreen hardy perennial. Ht 5cm, spread 20cm. Pink flowers in summer. The leaves are bright green, narrow and pointed, growing close together on the stem with an odd celery scent. Makes an attractive low-growing mound, and is good between paving stones.

Thymus citriodorus
Lemon Thyme
Evergreen hardy perennial. Ht 30cm, spread 20cm. Pink flowers in summer. Fairly large green leaves with a strong lemon scent. Excellent culinary thyme, combines well with many chicken or fish dishes.

THYME

Thymus citriodorus 'Golden King'
Golden King Thyme
Evergreen hardy perennial. Ht 30cm, spread 20cm. Pink flowers in the summer. Fairly large green leaves variegated with gold, strongly lemon scented. Excellent culinary thyme, combines well with many dishes, like chicken, fish and salad dressing.

Thymus citriodorus 'Silver Queen' AGM
Silver Queen Thyme
Evergreen hardy perennial. Ht 30cm, spread 20cm. Pink flowers in the summer. Fairly large leaves, grey with silver variegation, a strong lemon scent. Excellent culinary thyme, combines well with many dishes, like chicken and salad dressing.

Thymus Coccineus Group AGM
Coccineus Thyme
Also known as creeping red thyme. Evergreen hardy perennial, prostrate form, a creeper. Red flowers in summer. Green small leaves. Decorative, aromatic and good ground cover.

Thymus comosus
Evergreen hardy perennial. Ht 10cm, spread 20cm. Attractive clusters of pink flowers in summer. Small mid-green round leaves. Excellent for growing in gravel.

Thymus doerfleri
Doerfleri Thyme
Evergreen half-hardy perennial. Ht 2cm, spread 20cm. Mauve/pink flowers in summer, grey, hairy, thin leaves, which are mat forming. Decorative thyme, good for rockeries, hates being wet in winter. Originates from the Balkan Peninsula.

Thymus herba-barona

Thymus 'Jekka'

Thymus doerfleri 'Bressingham'
Bressingham Thyme
Evergreen hardy perennial. Ht 2cm, spread 20cm. Mauve/pink flowers in summer. Thin, green, hairy leaves, which are mat forming. Decorative thyme, good for rockeries, hates being wet in winter.

Thymus 'Doone Valley'
Doone Valley Thyme
Evergreen hardy perennial. Ht 8cm, spread 20cm. Purple flowers in summer. Round variegated green and gold leaves with a lemon scent. Very decorative, can be used in cooking if nothing else is available.

Thymus 'Fragrantissimus'
Orange-Scented Thyme
Evergreen hardy perennial. Ht 30cm, spread 20cm. Small pale pink/white flowers in summer. The leaves are small, narrow, greyish green, and smell of spicy orange. Combines well with stir-fry dishes, poultry – especially duck – and even treacle pudding.

Thymus herba-barona
Caraway Thyme
Evergreen hardy perennial. Ht 2cm, spread 20cm. Rose-coloured flowers in summer. Dark green small leaves with a unique caraway scent. Good in culinary dishes especially stir fry and meat. It combines well with beef.

Thymus 'Jekka'
Jekka's Thyme
Evergreen hardy perennial. Ht 10cm, spread 30cm. Clusters of small pink and white flowers in early summer. Prolific mid-green aromatic leaves. Spreading habit. Good culinary flavour.

Thymus polytrichus A. Kern. ex Borbás subsp. _britannicus_
Wild Creeping Thyme
Also known as mother of thyme and creeping thyme. Evergreen hardy perennial. Ht 2cm, spread 20cm. Pale mauve flowers in summer. Small dark green leaves which, although mildly scented, can be used in cooking. Wild thyme has been valued by herbalists for many centuries.

Thymus 'Porlock'
Porlock Thyme
Evergreen hardy perennial. Ht 30cm, spread 20cm. Pink flowers in summer. Fairly large green leaves with a mild but definite thyme flavour and scent. Excellent culinary thyme. Medicinal properties are anti-bacterial and anti-fungal.

Thymus pseudolanuginosus
Woolly Thyme
Evergreen hardy perennial. Ht 2cm, spread 20cm. Pale pink/mauve flowers for most of the summer. Grey hairy mat-forming leaves. Good for rockeries and in stone paths or walls. Dislikes wet winters.

Thymus pulegioides
Broad-Leaved Thyme
Evergreen hardy perennial. Ht 8cm, spread 20cm. Pink/mauve flowers in summer. Large round dark green leaves with a strong thyme flavour. Good for culinary uses, excellent for ground cover and good in hanging baskets.

Thymus pulegioides 'Aureus' AGM
Golden Thyme
Evergreen hardy perennial. Ht 30cm, spread 20cm. Pale pink/lilac flowers. Green leaves that turn gold in summer, good flavour, combining well with vegetarian dishes.

Thymus pulegioides 'Bertram Anderson' AGM
Bertram Anderson Thyme
Evergreen hardy perennial. Ht 10cm, spread 20cm. Pink/mauve flowers in summer. More of a round mound than 'Archer's Gold' and the leaves are slightly rounder with a more even golden look to the leaves. Decorative and culinary, it has a mild thyme flavour.

Thymus 'Redstart'
Redstart Thyme
Evergreen hardy perennial. Ht 5cm, spread 20cm. Bright red flowers in summer. Small dark green round leaves. Decorative, aromatic and good ground cover.

Thymus serpyllum var. _albus_
Creeping White Thyme
Evergreen hardy perennial, prostrate form, a creeper. White flowers in summer. Bright green small leaves. Decorative aromatic and good ground cover.

Thymus serpyllum 'Annie Hall'
Annie Hall Thyme
Evergreen hardy perennial, prostrate form, a creeper. Pale pink flowers in summer. Small green leaves. Decorative, aromatic and good ground cover.

Thymus serpyllum 'Goldstream'
Goldstream Thyme
Evergreen hardy perennial, prostrate form, a creeper. Pink/mauve flowers in summer. Green/gold variegated small leaves. Decorative, aromatic and good ground cover.

Thymus serpyllum 'Lemon Curd'
Lemon Curd Thyme
Evergreen hardy perennial, prostrate form, a creeper. White/pink flowers in summer. Bright green lemon-scented small leaves. Decorative, aromatic and good ground cover. Can be used in cooking if nothing else available.

Thymus serpyllum 'Minimalist'
Minimalist Thyme
Evergreen hardy perennial, prostrate form, a creeper. Pink flowers in summer. Tiny leaves, very compact. Decorative, aromatic and good ground cover. Ideal for growing between pavings and alongside paths.

Thymus serpyllum 'Pink Chintz' AGM

Thymus 'Redstart'

Thymus comosus

Pink Chintz Thyme
Evergreen hardy perennial, prostrate form, a creeper. Pale pink flowers in summer. Grey/green small hairy leaves. Decorative, aromatic, good ground cover. Does not like being wet in winter.

Thymus serpyllum 'Rainbow Falls'
Rainbow Falls Thyme
Evergreen hardy perennial, prostrate form, a creeper. Purple flowers in summer. Variegated green/gold small leaves. Decorative, aromatic and good ground cover.

Thymus serpyllum 'Russetings'
Russetings Thyme
Evergreen hardy perennial, prostrate form, a creeper. Purple/mauve flowers in summer. Small green leaves. Decorative, aromatic and good ground cover.

Thymus serpyllum 'Snowdrift'
Snowdrift Thyme
Evergreen hardy perennial, prostrate form, a creeper. Masses of white flowers in summer. Small green round leaves. Decorative, aromatic and good ground cover.

Thymus vulgaris
Common (Garden) Thyme
Evergreen hardy perennial. Ht 30cm, spread 20cm. Mauve flowers in summer. Thin green aromatic leaves. This is the thyme everyone knows. Use in stews, salads, sauces, etc. Medicinal properties are anti-bacterial and anti-fungal.

Thymus vulgaris 'Silver Posie'
Silver Posie Thyme
Evergreen hardy perennial. Ht 30cm, spread 20cm. Pale pink/lilac flower. The leaves have a very pretty grey/silver variegation with a tinge of pink on the underside. This is a good culinary thyme and looks very attractive in salads.

Thymus zygis
Zygis Thyme
Evergreen half-hardy perennial. Ht 30cm, spread 20cm. White attractive flowers. Small thin grey/green leaves which are aromatic. This is an attractive thyme which is good for rockeries. Originates from Spain and Portugal, therefore does not like cold wet winters.

Upright Thymes
Up to 30cm:
T. caespititius, *T. cilicicus* Boiss. & Bail., *T.* Jekka, *T. pulegioides* 'Bertram Anderson'

30cm and above:
T. camphoratus, *T. citriodorus*, *T.* 'Fragrantissimus', *T.* 'Porlock', *T. vulgaris*

Creeping Thymes
T. Coccineus Group, *T. Doerfleri*, *T.* 'Doone Valley', *T. Herba-barona*, *T. pseudolanuginosus* and all *T. serpyllum*.

cultivation

Propagation
To maintain the true plant, it is better to grow the majority of thymes from softwood cuttings. Only a very few, such as *T. vulgaris* and *T.* 'Fragrantissimus', can be propagated successfully from seed.

Seed
Sow the very fine seed in early spring using the cardboard technique (see page 645) on the surface of prepared trays (seed or plug), using a standard seed compost mixed in equal parts with composted fine bark. Give a bottom heat of 15–21°C. Do not cover. Keep watering to the absolute minimum, as these seedlings are prone to damping off disease. When the young plants are large enough and after a period of hardening off, plant out in the garden in late spring/early summer, 23–38cm apart.

Cuttings
Thymes are easily increased by softwood cuttings from new growth in early spring or summer. The length of the cutting should be 5–8cm. Use a standard seed

compost mixed in equal parts with composted fine bark. Winter the young plants under protection and plant out the following spring.

Division
Creeping thymes put out aerial roots as they spread, which makes them very easy to divide in late spring.

Layering
An ideal method for mature thymes that are getting a bit woody. Use either the strong branch method of layering in early autumn or mound layer in early spring.

Pests and Diseases
Being such an aromatic plant, it does not normally suffer from pests but, if the soil or compost is too rich, thyme may be attacked by aphids. Treat with a liquid horticultural soap. All varieties will rot off if they become too wet in a cold winter.

Maintenance
Spring Sow seeds. Divide established plants. Trim old plants. Layer old plants.
Summer Take cuttings of non-flowering shoots. Trim back after flowering.
Autumn Protect tender thymes. Layer old plants.
Winter Protect containers and only water if absolutely necessary.

Garden Cultivation
Thymes need to be grown in poor soil, in a well-drained bed to give their best flavour. They are drought-loving plants and will need protection from cold winds, hard and wet winters. Sow seed when the soil has warmed and there is no threat of frost. Thin on average to 20cm apart.

It is essential to trim all thymes after flowering; this not only promotes new growth, but also stops the plant from becoming woody and sprawling in the wrong direction.

In very cold areas grow it in the garden as an annual or in containers and then winter with protection.

Harvesting
As thyme is an evergreen it can be picked fresh all year round provided you are not too greedy. For preserving, pick before it is in flower. Either dry the leaves or put them in a vinegar or oil.

container growing

All varieties suit being grown in containers. They like a free-draining poor soil (low in nutrients); if grown in a rich soil they will become soft and the flavour will be impaired. Use a soil-based compost mixed in equal parts with composted fine bark; water sparingly, keeping the container bordering on dry, and in winter definitely dry – only watering if absolutely necessary, when the leaves begin to lose too much colour. Feed only occasionally in the summer months. Put the container in a sunny spot, which will help the aromatic oils come to the leaf surface and impart a better flavour. Trim back after flowering to maintain shape and promote new growth.

medicinal

Thyme has strong antiseptic properties. The tea makes a gargle or mouthwash, and is excellent for sore throats and infected gums. It is also good for hangovers. The essential oil is anti-bacterial and anti-fungal and used in the manufacture of toothpaste, mouthwash, gargles and other toilet articles. It can also be used to kill mosquito larvae. A few drops of the oil added to the bath water helps ease rheumatic pain, and it is often used in liniments and massage oils.

warning

Although a medical dose drawn from the whole plant is safe, any amount of the volatile oil is toxic and should not be used internally except by prescription. Avoid altogether if you are pregnant.

culinary

Thyme is an aid to digestion and helps break down fatty foods. It is one of the main ingredients of bouquet garni; it is good, too, in stocks, marinades and stews; and a sprig or two with half an onion makes a great herb stuffing for chicken.

Poached Trout with Lemon Thyme
Serves 4

4 trout, cleaned and gutted
Salt
6 peppercorns (whole)
4 fresh bay leaves
1 small onion, cut into rings
1 lemon
1 sprig lemon thyme
1 tablespoon chopped lemon thyme leaves
100ml white wine
2 tablespoons fresh snipped garlic chives
75g butter

Place the trout in a large frying pan. Sprinkle with salt and add the peppercorns. Place one bay leaf by each trout. Put the onion rings on top of the trout, cut half the lemon into slices and arrange this over the trout, add the thyme sprig, and sprinkle some of the chopped thyme leaves over the whole lot. Pour in the wine and enough water just to cover the fish. Bring it to the boil on top of the stove and let it simmer uncovered for 6 minutes for fresh trout, 20 minutes for frozen.

Mix the remaining chopped lemon thyme and garlic chives with the butter in a small bowl. Divide this mixture into 4 equal portions. When the trout are cooked lift them out gently, place on plates with a slice of lemon and the herb butter on the top. Serve with new potatoes.

Trapaeolum majus

NASTURTIUM

Also known as Garden Nasturtium, Indian Cress and Large Cress. From the family *Tropaeolaceae*.

Nasturtiums are native to South America, especially Peru and Bolivia, but are now cultivated worldwide.

The generic name, *Tropaeolum*, is derived from the Latin *tropaeum*, meaning 'trophy' or 'sign of victory'. After a battle was finished, a tree-trunk was set up on the battlefield and hung with the captured helmets and shields. It was thought that the round leaves of the nasturtium looked like shields and the flowers like blood-stained helmets.

It was introduced into Spain from Peru in the sixteenth century and reached London shortly afterwards. When first introduced it was known as *Nasturcium indicum* or *Nasturcium peruvinum*, which is how it got its common name, Indian cress. The custom of eating its petals, and using them for tea and salads, comes from the Orient.

varieties

Tropaeolum majus
Nasturtium
Half-hardy annual. Ht and spread 30cm. Red/orange flowers from summer to early autumn. Round, mid-green leaves.

***Tropaeolum majus* Alaska Series**
Nasturtium Alaska (Variegated)
Half-hardy annual. Ht and spread 30cm. Red, orange and yellow flowers from summer to early autumn. Round, variegated (cream and green) leaves.

***Tropaeolum majus* 'Peaches & Cream'**
Nasturtium Peaches & Cream
Half-hardy annual. Ht 20cm, spread 30cm. Cream and orange flowers from summer to early autumn. Round, mid-green leaves.

A few special species of interest:

Tropaeolum peregrinum
Canary Creeper
Tender perennial. Climber: Ht 2m. Small, bright yellow flowers with 2 upper petals that are much larger and fringed from summer until first frost. Grey/green leaves with 5 lobes. In cool areas, best grown as an annual.

Tropaeolum polyphyllum
Hardy perennial. Ht 5–8cm, spread 30cm or more. Fairly small yellow flowers from summer to early autumn. Leaves grey-green on trailing stems. A fast-spreading plant once established. Looks good on banks or hanging down walls.

Tropaeolum speciosum
Flame Creeper
Hardy perennial. Climber: Ht 3m. Scarlet flowers in summer followed by bright blue fruits surrounded by deep red calyxes in autumn. Leaves green with 6 lobes. This very dramatic plant should be grown like honeysuckle with its roots in the shade and head in the sun.

Tropaeolum majus 'Peaches & Cream'

NASTURTIUM

cultivation

Propagation

Seed

The seeds are large and easy to handle. To have plants flowering early in the summer, sow in early spring under protection directly into prepared pots or cell trays using a standard seed compost, and cover lightly with compost. Plugs are ideal, especially if you want to introduce the young plants into a hanging basket; otherwise use small pots to allow more flexibility before planting out. When the seedlings are large enough and there is no threat of frosts, plant out into a prepared site in the garden, or into containers.

Cuttings

Take cuttings of the perennial varieties in the spring from the new soft growth.

Pests and Diseases

Aphids and caterpillars of cabbage white butterfly and its relatives may cause a problem. If the infestation is light, the fly may be brushed off or washed away with soapy water.

Maintenance

Spring Sow seed early under protection, or after frosts in the garden.
Summer Deadhead flowers regularly to prolong the flowering season.
Autumn Dig up dead plants.
Winter Plan next year.

Garden Cultivation

Nasturtiums prefer a well-drained, poor soil in full sun or partial shade. If the soil is too rich, leaf growth will be made at the expense of the flowers. They are frost-tender and will suffer if the temperature falls below 4°C.

As soon as the soil has begun to warm and the frosts are over, nasturtiums can be sown directly into the garden. Sow individually 20cm apart. For a border full of these plants, sow 15cm apart. Claude Monet's garden at Giverny in France has a border of nasturtiums sprawling over a path that looks very effective.

Trapaeolum majus

Harvesting

Pick the flowers for fresh use only; they cannot be dried. Pick the seed pods just before they lose their green colour (for pickling in vinegar). Pick the leaves for fresh use as required. They can be dried, but personally I don't think it is worth it.

companion planting

This herb attracts blackfly away from vegetables such as cabbage and broad beans. It also attracts hover flies, whose larvae attack aphids. Further, it is said to repel whitefly, woolly aphids and ants. Altogether it is a good tonic to any garden.

container growing

This herb is excellent for growing in pots, tubs, window boxes and hanging baskets. Use a standard potting compost mixed in equal parts with composted fine bark. Do not feed, because all you will produce are leaves not flowers, but do keep well watered, especially in hot weather.

medicinal

This herb is rarely used medicinally, although the fresh leaves contain vitamin C and iron as well as an antiseptic substance, which is at its highest before the plant flowers.

warning

Use the herb with caution. Do not eat more than 15g at a time or 30g per day.

culinary

I had a group from the local primary school around the farm to talk about herbs. Just as they were leaving I mentioned that these pretty red flowers were now being sold in supermarkets for eating in salads. When a little boy looked at me in amazement, I suggested he try one. He ate the whole flower without saying a word. One of his friends said, 'What does it taste like?' With a huge smile he asked if he could pick another flower for his friend, who ate it and screamed, 'Pepper pepper...' The seeds, flowers and leaves are all now eaten for their spicy taste. They are used in salads also as an attractive garnish. The pickled flower buds provide a good substitute for capers.

Nasturtium Cream Cheese Dip

100g cream cheese
2 teaspoons tender nasturtium leaves, chopped
3 nasturtium flowers

Blend the cream cheese with the chopped leaves. Put the mixture into a bowl and decorate with the flowers. Eat this mixture as soon as possible because it can become bitter if left standing.

Urtica

NETTLE

Also known as Common Nettle, Stinging Nettle, Devil's Leaf and Devil's Plaything. From the family *Urticaceae*.

This plant is found all over the world. It is widespread on waste-land especially on damp and nutrient-rich soil.

The generic name *Urtica* comes from the Latin *uro*, meaning 'I burn'. The Roman nettle *Urtica pilulifera* originally came to Britain with the invading Roman army. The soldiers used the plants to keep themselves warm. They flogged their legs and arms with nettles to keep their circulation going.

The use of nettles in the making of fabric goes back for thousands of years. Nettle cloth was found in a Danish grave of the later Bronze Age, wrapped around cremated bones. It was certainly made in Scotland as late as the eighteenth century. The Scottish poet, Thomas Campbell, wrote then of sleeping in nettle sheets in Scotland and dining off nettle tablecloths. Records show that it was still being used in the early twentieth century in Tyrol.

In the Middle Ages it was believed that nettles marked the dwelling place of elves and were a protection against sorcery. They were also said to prevent milk from being affected by house trolls or witches.

Settlers in New England in the seventeenth century were surprised to find that this old friend and enemy had crossed the Atlantic with them. It was included in a list of plants that sprang up unaided. Before World War II, vast quantities of nettles were imported to Britain from Germany. During the war there was a drive to collect as much of the home-grown nettle as possible. The dark green dye obtained from the plant was used as camouflage, and chlorophyll was extracted for use in medicines.

varieties

Urtica dioica

Stinging Nettle

Hardy perennial. Ht 1.5m, spread infinite on creeping rootstock. The male and female flowers are on separate plants. The female flowers hang down in clusters, the male flower clusters stick out. The colour for both is a yellowish green. The leaves are green toothed and have bristles. This is the variety that can be eaten when young.

Urtica pilulifera

Roman Nettle

Hardy perennial. Ht 1.5m, spread infinite on creeping rootstock. This looks very similar to the common stinging nettle, but its sting is said to be more virulent.

Urtica urens

Small Nettle

Hardy annual. Ht and spread 30cm. The male and female flowers are in the same cluster and are a greenish white in colour. The green leaves are deeply toothed and have bristles.

Urtica urentissima

Devil's Leaf

This is a native of Timor and the sting said to be so virulent that its effects can last for months and may even cause death. A plant of interest, but not for the garden.

Urtica dioica in flower

NETTLE

cultivation

Propagation

Seed

Nettles can be grown from seed sown in the spring. But I am sure any of your friends with a garden would be happy to give you a root.

Division

Divide established roots early in spring before they put on much leaf growth, and the sting is least strong.

Pests and Diseases

Rarely suffers from pests and disease, well not ones that one would wish to destroy!

Maintenance

Spring Sow seeds, divide established plants.
Summer Cut plants back if they are becoming invasive.
Autumn Cut back the plants hard into the ground.
Winter No need for protection, fully hardy.

Garden Cultivation

Stinging nettles are the scourge of the gardener and the farmer, the pest of children in summer, but are very useful in the garden, attracting butterflies and moths, and making an excellent caterpillar food. They will grow happily in any soil. It is worth having a natural corner in the garden where these and a few other wild flowers can be planted.

Harvesting

Cut young leaves in early spring for use as a vegetable.

medicinal

The nettle has many therapeutic applications but is principally of benefit in all kinds of internal haemorrhages, as a diuretic in jaundice and haemorrhoids, and as a laxative. It is also used in dermatological problems including eczema.

Nettles make a valuable tonic after the long winter months when they provide one of the best sources of minerals. They are an excellent remedy for anaemia. Their vitamin C content makes sure that the iron they contain is properly absorbed.

Nettle hair conditioner

other uses

Whole plants yield a greenish/yellow woollen dye. Traditionally the old gardeners used nettles to make a spray to get rid of aphids, especially blackfly. They soaked the nettles in rain water for a period of time, then used the steeped water to spray the infected plants, it was also said to be a good plant tonic.

Nettles have a long-standing reputation for preventing hair loss and making the hair soft and shiny. They also have a reputation for eliminating dandruff.

Nettle Rinse and Conditioner

Use this as a final rinse after washing your hair, or massage it into your scalp and comb through the hair every other day. Store it in a small bottle in the refrigerator.

1 big handful-size bunch of nettles
500ml water

Wear rubber gloves to cut the nettles. Wash thoroughly and put the bunch into an enamel saucepan with enough cold water to cover. Bring to the boil, cover and simmer for 15 minutes. Strain the liquid into a jug and allow to cool.

warning

Do not eat old plants uncooked, they can produce kidney damage and symptoms of poisoning. The plants must be cooked thoroughly to be safe.

Handle all plants with care; they do sting.

culinary

Nettles are an invaluable food, rich in both vitamins and minerals.

In spring the fresh leaves may be cooked and eaten like spinach, made into a delicious soup, or drunk as a tea.

When cooked, I am pleased to say, nettles lose their sting.

Nettle Soup

Serves 4

250g young nettle leaves
50g oil or butter
1 small onion, chopped
250g cooked potatoes, peeled and diced
1 teaspoon each (mix, fresh, chopped) sweet marjoram, sage, lemon thyme
1 dessertspoon fresh chopped lovage
900ml milk
2 tablespoons cream and French parsley, chopped (optional)

Pick only the fresh young nettle leaves, and wear gloves to remove from stalks and wash them. Heat the oil in a saucepan, add the chopped onion, and slowly sweat until clear. Then add the nettles and stew gently for about a further 10 minutes. Add the chopped potatoes, all the herbs and the milk and simmer for a further 10 minutes. Allow to cool then put all the ingredients into a liquidiser and blend. Return to a saucepan over gentle heat. Add a swirl of cream to each bowl and sprinkle some chopped French parsley over the top. Serve with French bread.

Valeriana officinalis

VALERIAN

Also known as All Heal, Set All, Garden Heliotrope, Cut Finger and Phu. From the family *Valerianaceae*.

This herb is indigenous to Europe and West Asia and is now naturalised in North America. It can be found in grasslands, ditches, damp meadows and close to streams. The name may come from the Latin *valere*, 'to be healthy', an allusion to its powerful medicinal qualities. The root is the medicinal part of this herb which, when dug up, literally stinks, hence another of its common names 'phu'. But to cats the scent of this herb is sheer elixir, even more than catmint *(Nepeta cateria)*, and it is said the Pied Piper of Hamelin carried the root to entice the cats to eat the rats of Hamelin. A tincture of valerian was employed in the First and Second World Wars to treat shell-shock and nervous stress.

varieties

Valeriana officinalis
Valerian
Hardy herbaceous perennial. Ht up to 1.5m, spread 1m. Clusters of small, sweetly scented white flowers that are often tinged with pink in summer. Leaves are mid-green, deeply divided and toothed around the edges. Short conical rootstock which, when broken or cut, exudes a strong, rather unpleasant aroma.

Other varieties of special interest and well worth looking out for are:-

Valeriana jatamansii* syn. *Valeriana wallichii
Spikenard, Nard, Indian Valerian, Tagara
Hardy herbaceous perennial. Ht 10–25cm and spread 45cm. Clusters of small white flowers in summer. Mid-green, toothed, heart-shaped leaves. In India the roots are used medicinally to treat hysteria, hypochondria and as a tranquiliser, and an essential oil is extracted from the sweet-smelling roots for use in perfumery. I was given this plant by Roy Lancaster, the well-known plantsman, who collected it in 1978 on one of his expeditions to the Himalayas. It has happily adapted to our climate. He grows his plant, from which he gave me a cutting, in the garden and I am growing it in a container using a soil-based compost mixed with 25 per cent horticultural grit. This herb is often confused with, and medicinally substituted for, *Nardostachys grandiflora* (see below).

Nardostachys grandiflora
Nard, Spikenard, Musk Root
Hardy herbaceous perennial. Ht and spread 25–30cm. Clusters of small pink to pale purple flowers in summer. Narrow, basal, lance-shaped, mid-green leaves. The roots of this plant have a sweet musk scent. This herb is a member of the Valerianaceae family, and often confused with *Valeriana jatamansii*. It is indigenous to the alpine Himalayas and has been used for thousands of years as a nervine tonic. It is mentioned in the Song of Solomon and was the substance used to anoint the feet of Jesus at the Last Supper.

VALERIAN

Valeriana jatamansii

cultivation

Propagation

Seed

In spring, sow the fairly small seeds into prepared seed or module plug trays, using a standard seed compost. Press the seeds into the soil, cover with perlite. Do not cover with soil as this will delay germination. Place in a cold frame or cold greenhouse. Germination takes 3–4 weeks. Plant out when large enough to handle, 60cm apart.

Division

Divided established plants in the autumn, or early spring, using two forks back to back, replanting into a well-prepared site.

Pests and Diseases

Valerians rarely suffer from pests and diseases.

Maintenance

Spring Sow seeds. Divide established plants.
Summer Cut back after flowering to prevent self-seeding.
Autumn Divide established plants if required.
Winter A very hardy plant, does not need protection.

Garden Cultivation

The roots of this herb like to be kept cool and damp in summer, so plant in any soil that does not dry out in high summer, in sun or partial shade. Choose the position in your garden with care and remember, if you have to move the plant, that the scent of the broken roots will attract all the neighbourhood cats.

Harvesting

Dig up the roots of a second- or third-year plant. Wash and remove the pale fibrous root. Cut the root into manageable slices for drying.

companion planting

It is said that valerian makes a good companion plant when planted near vegetables; the roots release phosphorus and stimulate earthworm activity.

container growing

Valerian can be grown in a container, but make sure that it is large enough to accommodate the root system. Use a soil-based potting compost. Position the container in partial shade so that the compost does not dry out. Water regularly throughout the growing season, especially in summer. Divide the plants each autumn; this will prevent the roots rotting in the pots in wet winters.

medicinal

This herb has been used for thousands of years as a sedative and relaxant. The dried roots are prepared into tablets, powder, capsules or tinctures which are then used as a safe, non-addictive relaxant that reduces nervous tension and anxiety and promotes restful sleep.

This is an extremely useful herb for treating anxious or restless pets. I have had a number of cats and dogs and one dog in particular, Hampton, hated going in the car. After treating him with a few drops of valerian tincture, diluted in some water, prior to the journey he became much less frightened and slept peacefully throughout the journey, which made the family holiday even more enjoyable. I have also used it with the cats, especially if one had been injured in a fight, before cleaning their wounds or taking them to the vet for treatment. Cats are notoriously fussy about their food, but adore valerian, which they find very soothing. Make a decoction by crushing one teaspoon (5ml), of dried root, which is then added to 1 litre of cold water. Leave for 24 hours. Strain the decoction into a clean, sterilised bottle. This will keep for 48 hours in a refrigerator. For a medium-sized, stressed or anxious cat, add 3 drops of the decoction to a small amount of water, which they will drink with relish.

other uses

An infusion of the root sprayed onto the soil is said to attract earthworms. The root has been used as a bait in rat traps and to catch wild cats. Add the mineral-rich leaves to the compost heap.

warning

Do not take for an extended period. Do not take during pregnancy.

Verbena officinalis

VERVAIN

Also known as Holy Herb, Simpler's Joy, Pigeon's Grass, Burvine, Wizard's Herb, Herba Sacra, Holy Plant, European Vervain, Enchanter's Plant and Herba The Cross. From the family *Verbenaceae.*

This herb is a native of Mediterranean regions. It has now become established elsewhere within temperate zones and, for that matter, wherever the Romans marched.

It is a herb of myth, magic and medicine. The Egyptians believed that it originated from the tears of Isis. The Greek priests wore amulets made of it, as did the Romans, who also used it to purify their altars after sacrifice. The Druids used it for purification and for making magic potions.

Superstition tells that when you pick vervain, you should bless the plant. This originates from a legend that it grew on the hill at Calgary, and was used to staunch the flow of Christ's blood at the Crucifixion.

In the Middle Ages it was an ingredient in a holy salve, a powerful protector against demons and disease: 'Vervain and Dill hinders witches from their will.'

varieties

Verbena officinalis

Vervain

Hardy perennial. Ht 60–90cm, spread 30cm or more. Small pale lilac flowers in summer. Leaves green, hairy and often deeply divided into lobes with curved teeth. This plant is not to be muddled with lemon verbena *(Aloysia triphylla)*.

cultivation

Propagation

Seed

Sow the small seeds in early spring in a prepared seed or plug tray using a standard seed compost. Cover with perlite. No need for extra heat. When the seedlings are large enough, and after hardening off, plant out in a prepared site, 30cm apart.

Division

An established plant can be divided either in the spring or autumn. It splits easily with lots of roots.

Pests and Diseases

If soil is too rich or high in nitrates, aphids can attack.

Maintenance

Spring Sow seeds. Divide established plants.
Summer Cut back after flowering to stop it self-seeding.
Autumn Divide established plants.
Winter No need for protection; fully hardy.

Verbena officinalis

Garden Cultivation

Vervain can be sown direct into the garden in the spring in a well-drained soil and a sunny position. It is better to sow or plant in clumps because the flower is so small that otherwise it will not show to advantage. But beware its capacity to self-seed.

Harvesting

Pick leaves as required. Cut whole plant when in bloom. Dry leaves or whole plant if required.

container growing

Vervain does nothing for containers, and containers do nothing for vervain.

medicinal

Vervain has been used traditionally to strengthen the nervous system, dispel depression and counter nervous exhaustion. It is also said to be effective in treating migraines and headaches of the nervous and bilious kind. Chinese herbalists use a decoction to treat sup-pressed menstruation, and for liver problems and urinary tract infections.

culinary

In certain parts of France, a tea is made from the leaves. Use with caution.

warning

Avoid during pregnancy.

Viola tricolor

HEARTSEASE

Also known as Wild Pansy, Field Pansy, Love Lies Bleeding, Love In Idleness, Herb Trinity, Jack Behind The Garden Gate, Kiss Me Behind The Garden Gate, Kiss Me Love, Kiss Me Love At The Garden Gate, Kiss Me Quick, Monkey's Face, Three Faces Under A Hood, Two Faces In A Hood and Trinity Violet. From the family *Violaceae*.

Heartsease is a wild flower in Europe and North America, growing on wasteland and in fields and hedgerows.

In the Middle Ages, due to the influence of Christianity and because of its tricolour flowers – white, yellow and purple – heartsease was called Trinitaria or Trinitatis Herba, the herb of the Blessed Trinity.

In the traditional language of flowers, the purple form meant memories, the white loving thoughts, and the yellow, souvenirs.

Viola tricolor

varieties

Viola arvensis
Field Pansy
Hardy perennial. Ht 5–10cm. The flowers are predominantly white or creamy, and appear in early summer. The green leaves are oval with shallow, blunt teeth.

Viola lutea
Mountain Pansy
Hardy perennial. Ht 8–20cm. Single-coloured flowers in summer vary from yellow to blue and violet. The leaves are green and oval near the base of the stem, narrower further up.

Viola tricolor
Heartsease
Hardy perennial, often grown as an annual. Ht 15–30cm. Flowers from spring to autumn. Green and deeply lobed leaves.

cultivation

Propagation
Seed
Sow seeds under protection in the autumn, either into prepared seed, plug trays or pots using a standard seed compost. Do not cover the seeds. No bottom heat required. Winter the seedlings in a cold frame or cold greenhouse. In the spring harden off and plant out at a distance of 15cm apart.

Maintenance
Spring Sow seed.
Summer Deadhead flowers to maintain flowering.
Autumn Sow seed for early spring flowers.
Winter No need to protect.

Garden Cultivation
Heartsease will grow in any soil, in partial shade or sun. Sow the seeds from spring to early autumn where they are to flower. Press into the soil but do not cover.

Harvesting
Pick the flowers fully open from spring right through to late autumn. Use fresh or for drying. The plant has the most fascinating seed capsules, each capsule splitting into 3. The best time to collect seeds is midday when the maximum number of capsules will have opened.

container growing

Heartsease look very jolly in any kind of container. Pick off the dead flowers as this appears to keep the plant flowering for longer.

medicinal

An infusion of the flowers has long been prescribed for a broken heart. Less romantically, it is also a cure for bed-wetting. An ointment made from it is good for eczema and acne and also for curing milk rust and cradle cap. Herbalists use it to treat gout, rheumatoid arthritis and respiratory disorders. An infusion of heartsease leaves added to bath water has proved beneficial to suffers of rheumatic disease.

warning

In large doses, it may cause vomiting.

other uses

Cleansing the skin and shampooing thinning hair.

culinary

Add flowers to salads and to decorate sweet dishes.

Viola

VIOLET

From the family *Violaceae*.

There are records of sweet violets growing during the first century AD in Persia, Syria and Turkey. It is a native not only of these areas but also of North Africa and Europe. Violets have been introduced elsewhere and are now cultivated in several countries for their perfume.

This charming herb has been much loved for over 2,000 years and there are many stories associated with it. In a Greek legend, Zeus fell in love with a beautiful maiden called Io. He turned her into a cow to protect her from his jealous wife Juno. The earth grew violets for Io's food, and the flower was named after her.

The violet was also the flower of Aphrodite, the goddess of love, and of her son, Priapus, the god of gardens. The ultimate mark of the reverence in which the Greeks held sweet violet is that they made it the symbol of Athens.

For centuries perfumes have been made from the flowers of sweet violet mixed with the violet-scented roots of orris, and the last half of the nineteenth century saw intense interest in it – acres were cultivated to grow it as a market garden plant. Its main use was as a cut flower. No lady of quality would venture out without wearing a bunch of violets. It was also customary in gardens of large country houses to move the best clump of violets to a cold frame in late autumn to provide flowers for the winter.

varieties

Viola odorata
Sweet Violet
Also known as garden violet. Hardy perennial. Ht 7cm, spread 15cm or more. Sweet-smelling white or purple flowers from late winter to early spring. Heart-shaped leaves form a rosette at the base, from which long-stalked flowers arise. *Viola odorata* is one of the few scented violets. It has been hybridised to produce Palma violets, with a single or double flower, in a range of rich colours. A recent revival in interest in this plant means it is being offered again by specialist nurseries.

Viola reichenbachiana
Wood Violet
Hardy perennial. Ht 2–20cm, spread 15cm or more. Pale lilac/blue flowers in early spring. Leaves are green and heart-shaped. The difference between this plant and the common dog violet is the flowering time; there is also a slight difference in flower colour but it is difficult to discern.

Viola riviniana
Common Dog Violet
Also known as blue mice, hedging violet, horse violet and pig violet. Hardy perennial. Ht 2–20cm, spread 15cm or more. Pale blue/lilac flowers in early summer. Leaves are green and heart-shaped. This violet does not grow runners.

cultivation

Propagation

Seed
The small seed should be sown in early autumn in prepared seed or plug trays. Use a soil-based seed compost; I have found violets prefer this. Water in and cover with a layer of compost, and finally cover with a sheet of glass or polythene. Put the trays either in a

VIOLET

corner of the garden or in a cold frame (because the seeds germinate better if they have a period of stratification, though it will still be erratic). In the spring when the seedlings are large enough to handle, prick out into pots. If grown in cells allow a period of hardening off. Plant out as soon as temperatures have risen at a distance of 30cm.

Cuttings
These can be taken from the parent plant, with a small amount of root attached, in early spring and rooted in cell trays, using a standard seed compost mixed in equal parts with composted fine bark. Harden off and plant out into a prepared site in the garden in late spring when they are fully rooted. Water in well.

When using runners to propagate this plant, remove them in late spring and replant in a prepared site in the garden, 30cm apart. Plant them firmly in the ground, making sure that the base of the crowns are well embedded in the soil; water in well. Runners can be grown on in pots in early autumn. Remove a well-rooted runner and plant in a pot of a suitable size. Over-winter in a cool greenhouse, watering from time to time to prevent red spider mite. Bring into the house in the spring to enjoy the flowers. After flowering, plant out in the garden into a prepared site.

Division
Divide well-established plants as soon as flowering is over in early summer. It is a good idea to plant 3 crowns together for a better show and as an insurance policy against damage when splitting a crown. Replant in the garden in exactly the same way as for runners.

Pests and Diseases
The major pest for container-grown violets in warm weather is red spider mite. A good way to keep this at bay is to spray the leaves with water. If it is persistent, use a liquid horticultural soap as per the manufacturer's instructions. In propagating violets, the disease you will most probably come across is black root rot, which is encouraged by insufficient drainage in the compost. Young plants can also be affected by damping off root rot, which is caused usually by too much water and insufficient drainage.

Maintenance
Spring Take cuttings from established plants. Remove runners, pot or replant in the garden.

Viola riviniana

Summer Divide well-established plants, and replant.
Autumn Sow seed. Pot up root runners for wintering under cover.
Winter Feed the garden with well-rotted manure.

Garden Cultivation
Violets thrive best in a moderately heavy, rich soil in a semi-shaded spot. If you have a light and/or gravelly soil, it is a good idea to add some texture – a mulch of well-rotted manure – the previous autumn. In spring dig the manure in.

Plant out in the garden as soon as the frosts have finished, allowing 30cm between plants. When they become established, they quickly create a carpet of lovely sweet-smelling flowers. There is no need to protect any of the above-mentioned violets, they are fully hardy.

Harvesting
Pick the leaves in early spring for fresh use or for drying. Gather the flowers just when they are opening, for drying or crystallising.

Dig up the roots in autumn to dry for medicinal use.

container growing

Violets make good container plants. Use a soil-based compost mixed in equal parts with composted fine bark. Give them a liquid feed of fertiliser (following the manufacturer's instructions) after flowering. During the summer months, place the container in partial shade. In winter they do not like heat, and if it is too warm they will become weak and fail to flower. So, it is most important that they are in a cool place with temperatures no higher than 7°C. There must also be good air circulation, and watering should be maintained on a regular basis.

culinary

The flowers of sweet violet are well known in crystallised form for decorating cakes, puddings, ice-cream and home-made sweets. They are also lovely in salads, and make an interesting oil – use an almond oil as base. The flowers of common and dog violet can also be added to salads and used to decorate puddings. Their flavour is very mild in comparison to sweet violet, but they are just as attractive.

medicinal

Only sweet violet has been used medicinally. Various parts are still used, most commonly, the rootstock. It is an excellent, soothing expectorant and is used to treat a range of respiratory disorders, such as bronchitis, coughs, whooping cough and head colds. It also has a cooling nature and is used to treat hangovers.

Made into a poultice, the leaves soothe sore, cracked nipples. Also they have a reputation for treating tumours, both benign and cancerous. Strong doses of the rhizome are emetic and purgative. The flowers have a reputation for being slightly sedative and can be so helpful in cases of anxiety and insomnia.

other uses

The flowers of sweet violets are used in potpourris, floral waters and perfumes.

Sweet violet perfume

Vitex agnus-castus

CHASTE TREE

Also known as Monks Pepper, Lygos, Hemptree, Agnus Castus, Abraham's Tree and Chaste Berry. From the family *Verbenaceae*.

This aromatic shrub, indigenous to the Mediterranean and Central Asia, has been used medicinally for thousands of years. The first known records of its medical use were made in the fourth century BC by Hippocrates; he used it to treat female disorders, particularly the diseases of the uterus. Historically the Christian monks used to chew the leaves and grind the dried berries over their food to reduce their libido. *Agnus castus* translates as 'chaste lamb', which is the Christian symbol of purity. In Germany in the twentieth century Dr Gerhard Madaus conducted scientific research into the plant's effects on the female hormonal system. He subsequently developed a medicine made from the dried berries called 'Agnolyt', which is still available today.

varieties

Vitex agnus-castus

Chaste tree

Deciduous shrub. Ht and spread 2.5m. Upright panicles of fragrant, tubular violet blue flowers in late summer until mid-autumn, which are followed by small round orange/red fruit. Dark green, aromatic leaves are divided into 5 or 7 lance-shaped leaflets.

cultivation

Propagation

Seed

The best time to sow the seeds is in the late autumn when they are fresh. Sow into prepared plug module trays or small containers using a standard seed compost mixed in equal parts with perlite. Once the seedlings are large enough to handle, pot up using a standard potting compost mixed in equal parts with composted fine bark. Place the container in a frost-free environment for the winter. Do not think it is dead when all you are left with is a twig for the first winter; new shoots will appear in the spring if you do not over water. It will flower in the 4th or 5th summer.

Cuttings

The easiest and the most reliable method of propagation is by softwood cuttings taken in the late spring, early summer from non-flowering shoots. Prepare module plug trays or a small pot using a standard seed compost mixed in equal parts with perlite. Place the cuttings in a sheltered, warm environment; they do not need bottom heat. Once rooted, pot up in exactly the same way as the seedlings and again winter in a frost-free environment. It will flower in the 2nd or 3rd summer.

Pests and Diseases

As the whole plant is aromatic, it is rarely attacked by pests. Over-watering of young container plants can cause rotting in winter, especially in cold and cool climates.

Maintenance

Spring Prune back last year's growth to 5cm.
Summer Take cuttings in early summer.
Autumn Sow seeds.
Winter Protect from excessive wet.

CHASTE TREE

Garden Cultivation

This most attractive aromatic shrub is worthy of a place in the garden: it smells good, looks good and does you good – what more can you ask of a plant? Plant in a fertile soil; it will tolerate, dry soils, moist soils but not cold, heavy clay soils. When living in cold areas, plant against a warm, sunny wall; the wall will help cut down the rainfall by 25 per cent and give added protection and warmth in winter. Allternatively, plant in full sun in a sheltered position. It will not tolerate shade.

Harvesting

The leaves are picked in early summer for use fresh or for drying. Stems are cut in late summer or autumn and are dried. Roots are lifted, from plants over 5 years old in late summer and autumn and dried. The fruits are harvested in autumn for use fresh or for drying. However this plant rarely sets fruit in cool or cold climates.

companion planting

Good late-nectar plant for butterflies, which is therefore beneficial for late pollination.

container growing

From experience I know that this plant adapts happily to being grown in a container. Use a soil-based compost mixed in equal parts with composted fine bark. Feed the container throughout the growing season with a liquid feed following the manufacturer's instructions.

In winter allow the plant to drop its leaves, then clear away any fallen debris to prevent disease. Place the container in a sheltered spot. In wet conditions raise the pot on bricks so it does not become waterlogged.

medicinal

Vitex agnus-castus is one of the most important herbs for treating menstrual and menopausal problems and infertility. The key part of the plant used is the berries, which are taken in tablet or tincture form. It is used to regulate the hormones, for increasing female fertility, for regulating irregular periods and during the menopause to balance the hormones. It is also used to relieve spasms of pain, especially PMS, and to treat migraines and acne associated with the menstrual cycle.

other uses

The flowers are used in the making of perfume. A yellow dye is obtained from the leaves, the seed and the roots.

warning

Do not take excessive doses. Do not take *Vitex agnus-castus* during pregnancy. Do not take when taking any other product or drug that affects the female hormone system, such as HRT or the contraceptive pill.

Vitex agnus-castus

culinary

The small aromatic fruit when dried is often used as a pepper subsitute; the flavour is milder and spicier. Today, it is rare to find chaste tree berries in recipes, although they can appear in a Moroccan spice mixture called *ras el hanout*, which translates as 'top (or head) of the shop', referring to the best combination of spices the seller can provide.

Ras el Hanout

No two recipes for this combination of spices are the same.

1 tablespoon dried chaste berries
2 cinnamon sticks, broken into several pieces
1 tablespoon cloves
1 tablespoon coriander seeds
1 tablespoon cumin seeds
1 tablespoon fenugreek
1 tablespoon fennel seeds
25g dried Damask rose petals; these are available in Middle Eastern food stores

Place all the ingredients in a heavy-based metal frying pan and place over a low heat. Do not let them burn, but cook until the seeds begin to pop in the pan. As soon as they start, shake or toss gently, cook for a further minute, then allow to cool slightly. Remove all the ingredients from the frying pan and grind the mixture in a coffee grinder, mini food processor or pestle and mortar. Store in an airtight container, in a cupboard, and use this mixed spice within two weeks with meat, vegetable and rice dishes.

Zingiber officinale

GINGER

Also known as Sweet Ginger, Ginger Root, Shunthi, and Adrak. From the family *Zingiberaceae*.

Ginger is a very ancient herb. Confucius (551–479 BC) was known to have eaten fresh ginger with every meal as a digestive and carminative, and it has been used in India from the Vedic period (1500 BC) when it was called Maha-aushadhi, which means 'the great medicine'. Together with black pepper, ginger was one of the most commonly traded spices during the 13th and 14th centuries. Arabs carried the rhizomes on their voyages to East Africa and Zanzibar to plant at coastal settlements. During this time in England, ginger was sought after, and one pound in weight of ginger was equivalent to the cost of a sheep.

Zingiber officinale roots

varieties

Zingiber officinale

Ginger

Tropical, subtropical, deciduous perennial. Ht 1.5m, spread indefinite. The yellow/green flowers, with a deep purple and cream lip in summer, grow on a single stem and are followed by red fleshy fruits, each having three chambers containing several small black seeds. Commercially cultivated plants are often sterile. Aromatic long, mid-green, lance-shaped leaves. The root is a thick branching rhizome.

cultivation

Propagation

Seed

Rarely grown from seed as root cuttings are very simple and quick.

Cuttings

Unless you live in the tropics, the best source for fresh ginger root is from a good greengrocer. However, be aware that the quality can be variable. For example, air-freighted rhizomes can be often killed by the cold temperatures or they may have been treated with chemicals to inhibit sprouting. Choose a fresh plump, juicy-looking root with a stout horn-like growing bud. Fill a pot with seed compost mixed in equal parts with vermiculite. Slice the root 5cm below the bud, place in the container with the bud facing up, gently push the root into the compost, being careful not to break the bud. The compost should just cover the bud. Place the pot in a plastic bag and seal, put in a warm place or on a propagator at 20°C. In about 3 weeks, maybe longer depending on the warmth, shoots will emerge. As soon as they do, remove the plastic bag. Keep the container warm and out of direct sunlight until the plant is fully established. Then pot up, one size, using a soil-based potting compost mixed in equal parts with composted fine bark.

Division

Divide established plants in the spring. Use a spade to remove a clump, which includes some rhizomes, from the established plant in the garden, replanting into a prepared site in the garden. Alternatively, for container-raised plants, ease off some rhizomes either by hand or using two small forks, repotting into a container that is just big enough. Use a soil-based potting compost mixed in equal parts with composted fine bark.

Pests and Diseases
Red spider mite can be an occasional problem in older plants: regular misting and keeping the leaves well-washed will reduce this. If it gets out of hand, use a horticultural soap following the manufacturer's instructions. If you live in the northern hemisphere you might find it difficult to over-winter plants due to the low light levels.

Maintenance
Spring Take cuttings from fresh roots. Divide established plants.
Summer Feed container-grown plants regularly.
Autumn Harvest roots, cut back on watering container-raised plants.
Winter Protect from frost. Minimum temperature 20°C.

Garden Cultivation
When grown outdoors in the tropics it needs a minimum annual rainfall of 150cm, temperatures of 30°C or over, a short dry season and a deep, fertile soil. It usually takes 5–9 months to produce a crop.

Zingiber officinale

In the northern hemisphere it is best grown as a container plant either indoors, in a conservatory or a heated greenhouse. However, it rarely flowers outside of the tropics.

Harvesting
Outside the tropics one can only produce a small amount of fresh rhizome. The rhizome can be harvested in late autumn. The leaves can be used as a flavouring; pick as required throughout the growing season. A fresh rhizome can be stored in the refrigerator for up to 2 weeks. Freshly grated ginger can be frozen.

container growing

Container-raised plants do not produce much useful rhizome, although you can use the leaves in the kitchen for flavouring foods. Grow your plant in a soil-based potting compost mixed in equal parts with composted fine bark. In summer keep the compost damp and warm, place the container in a slightly shaded position and feed during the growing season weekly with a general-purpose liquid fertiliser. In dry weather plants will benefit from being lightly misted with rainwater daily. In the autumn cut back on the watering, keeping the plant fairly dry over the winter at a minimum of 20°C.

medicinal

Ancient Indian and Chinese herbalists used this herb for many aliments. It has a wide range of demonstrated health-giving properties; it both stimulates the heart and settles the stomach. It will improve your digestion and circulation, reduce any inflammatory process in your body and improve the absorption of anything you eat. Ginger works best when treating post-operative nausea and morning sickness, although its effectiveness in treating sea-sickness and other forms of motion sickness is questioned by some scientists, and I personally can vouch that it did not help me.

Fresh rhizomes are said to help to reduce inflammation in conditions like osteo- and rheumatoid arthritis. In Ayurvedic medicine it is used as a cure for cholera, anorexia and 'inflamed liver'.

culinary

When you buy fresh ginger, choose roots, which are often called 'hands', that are plump, not shrivelled. Fresh ginger, peeled and ground or grated into a pulp, is used in many types of curries and spicy foods. In India it is used in curries, in China it is often used with seafood and mutton, in Japan it is pickled and is known as *gari* and *beni-shoga* and eaten with sushi. In European cooking, ginger is mainly used in sweet preparations, spiced breads, biscuits and cakes. Many countries make liqueurs and drinks from the root; the Europeans make ginger beer and green ginger wine, and in China they make a liqueur called Canton, to name just a few.

Salmon, Watercress and Fresh Ginger
Serves 2
I love the combination of fish and ginger. Here is a very simple recipe that does not fail.

2 fillets of salmon, boned
1 tablespoon grated ginger
1 clove garlic, finely chopped
1 tablespoon chopped coriander leaf
juice of ½ lemon
I bunch watercress, washed and chopped
Olive oil and balsamic vinegar dressing (3 tablespoons olive oil, 1 tablespoon vinegar)
2 pieces of foil large enough to encapsulate the salmon fillets.

Preheat the oven to 200°C/400°F/gas mark 6. Place the fillets of fresh salmon onto an individual piece of foil, sprinkle each fillet with the ginger, garlic and coriander and finish with the lemon juice. Wrap loosely with the foil, bake in the oven for about 14 minutes or until the fresh fish is cooked. Serve on the watercress that has been drizzled with an oil and balsamic vinegar dressing.

FRUITS

In this section, fruit is defined as plant flesh that we are induced by the plant to eat, in order to distribute its seeds. Included is the closely allied group of nuts, which are in fact the seeds themselves. Presumably the plants are satisfied if a small percentage of the nut seeds are distributed to grow elsewhere.

The fruits covered in the three chapters Orchard Fruits, Soft, Bush and Cane Fruits and Annual Tender Fruits are determined by the manner in which we most usually grow them in temperate gardens. Those that require some protection or cover, such as the annual, perennial and tender fruits, are mostly fruits that are not hardy enough outdoors for the UK and northern European gardener, but are achievable in southern Europe, Australia and much of the USA. Of course, we commonly grow many of these, such as melons, grapes and even lemons, quite easily with the aid of a greenhouse, in cooler regions. I've also included many exotic tropical and sub-tropical fruits on sale in supermarkets and abroad which may be grown (or eaten!) out of curiosity and interest. Most of these were once grown and fruited in Victorian stovehouses, and can often be fruited at home with a modest heated glasshouse or conservatory. Failing that, most of these make spectacular, educational and decorative houseplants. (Please bear in mind, however, the potential final height and size of your humble date palm seedling before you start dreaming of ever ripening a crop!)

The chapter on Shrub and Flower Garden Fruits includes those forgotten and unsung heroes that are called upon only in times of shortage and famine, and by those country folk who appreciate the sharp, strong flavours these piquant fruits offer. The potential of these fruits has often been overlooked; many of them are worthy of deliberate cultivation, and with only a little breeding and selection they could become sweet and tender attractions for our delectation. The strawberry today is gigantic and succulent compared to those of two centuries ago; we can only imagine what new fruits we may conjure from nature's raw materials in the future.

ORCHARD FRUITS

When fruits are mentioned, these are probably the first that come to mind: apples, pears, plums and cherries – the tree-hard or top fruits, as they are known. They consist of two main groups: the pome fruits, which are the apple- and pear-like members, and the stone fruits, which are the plums, cherries, peaches and apricots. The pomes have small seeds in a core around which the 'stalk' from immediately behind swells, enclosing them with flesh. The stone fruits have a single seed in a hard shell around which the flesh forms. Both of these groups are related, as they are both members of the *Rosaceae* family. Mulberries and figs come from different families. Nonetheless, all are similar in hardiness, size and manner of cultivation to orchard trees.

Most of these fruits have been cultivated since ancient times. They were nearly all known to the Romans, who spread them throughout their empire. However, much knowledge of their cultivation was then lost during the Dark Ages. The monasteries, and a few noblemen, maintained fruit gardens and orchards, but the common people reverted to farming and cropping from the wild, with little interest in fruit cultivation. Indeed, fruits were often seen as poor fare compared to meat, and more suited for animal feed. If it was not for the ease with which many of these fruits could be fermented to make intoxicating beverages, they would probably have been even more neglected. After the Norman Conquest of England the new lords proved to be more interested in fruit than the Saxons they had defeated, bringing many of their own improved varieties with them from France. Orchards became more widely planted, and the wealthy vied with one another in collecting the greatest number and variety and in having the earliest and longest-lasting fruits.

In the sixteenth century, Henry VIII brought many new fruit varieties from the Netherlands and France, and the streets of London became full of home-grown and imported favourites. The arrival of completely different and new fruits from the New World aroused more enthusiasm for horticulture, reviving interest in the old fruits as well as the new discoveries. As old trusted varieties were exported to the colonies, new species and varieties were imported. These produced new varieties, which followed the old abroad at the same time as descendants of the first wave were already returning.

By the Victorian era the number of varieties in cultivation had escalated from a few hundreds to many tens of thousands, if you counted local varieties worldwide. The great cities were served by the immense orchards and market gardens that surrounded them.

Every gentleman aspired to a house with grounds that would include an orchard, at least, if not a grouse moor.

In the twentieth century, after two World Wars, the labour to maintain great gardens and orchards was not available and the land was needed for more basic crops. Orchards were grubbed up for cereals and more exciting fruits became available from abroad. The British orchard all but disappeared from many counties. Houses with grounds and orchards were demolished to make way for executive hutches at a dozen an acre.

However, the green movement, combined with people's increasing awareness of the utility of trees, the value of fruit, and the ecological advantages of permanent culture as opposed to annual crops, have all caused a reawakening of interest in orchard fruits. More are now being planted than for nearly a century.

Malus domestica from the family *Rosaceae*

CULINARY AND DESSERT APPLES

Tree up to 10m. Life span: medium to long. Deciduous, hardy, sometimes self-fertile. Fruits: up to 15cm, spherical, green to yellow or red. Vitamin value: vitamin C.

Malus domestica apples are complex selections and hybrids of *M. pumila* with *M. sylvestris* and *M. mitis*. Thus the shape of the fruit varies from the spheres of **Gladstone** and **Granny Smith** to the flattened buns of **Bramley** and **Mère de Ménage**, or the almost conical **Spartan**, **Golden Delicious** and **Worcester Pearmain**. The colour can be green, yellow, scarlet orange or dark red to almost purple. The texture can vary from crisp to pappy and they may be juicy or dry, acid or insipid, bitter, bland or aromatic. All apples have a dent in the stalk end, the remains of the flower at the other and a central tough core with several brown seeds. These are edible in small amounts, though there is a recorded death from eating a quantity, as they contain small amounts of cyanide.

The trees will often become picturesque landscape features, particularly when seen in an orchard. They frequently become twisted or distorted when left to themselves. They have soft downy or smooth leaves, never as glossy as pear leaves. The flowers are often pink- or red-tinged as well as snow white.

Apples are native to temperate Europe and Asia. They have been harvested from the wild since prehistory and were well known to the ancient Phoenicians. When Varro led his army as far as the Rhine in the first century BC, every region had its apples. The Romans encouraged their cultivation, so although Cato had only noted a half dozen varieties in the second century BC, Pliny knew of three dozen by the first century AD. The Dark Ages caused a decline in apple growing in Britain and only one pomerium (orchard), at Nottingham, is mentioned in the Domesday Book. However, interest increased after the Norman invasion. **Costard** and

Pearmain varieties are first noted in the twelfth and thirteenth centuries, and by the year 1640 there are nearly five dozen varieties recorded by Parkinson. By 1669, Worlidge has the number up to 92, mostly cider apples. *Downing's Fruits*, printed in 1866, has 643 varieties listed. Now we have more than 5,000 named apple varieties, representing about 2,000 actually distinguishable clones. Several hundred are easily obtainable from specialist nurserymen, though only a half dozen are grown on a commercial scale.

This sudden explosion in numbers was most probably due to the expansion of the colonies. The best varieties of apple trees from Europe mutated and crossed as they were propagated across North America, and then later Australia, and these then returned to be crossed again.

Apples are now grown extensively in every temperate region around the world. The first apples in North America, supposedly, were planted on the Governor's Island in Boston Harbour, but the Massachusetts Company had requested seeds in 1629, and in 1635 a Mr Wolcott of Connecticut wrote he had made 500 hogsheads of cider from his new apple orchard.

CULINARY AND DESSERT APPLES

varieties

The oldest variety known and easily available is **Court Pendu Plat** (mid-winter, dessert) which may go back to Roman times and is recorded from the sixteenth century. It is still grown because it flowers late, missing frosts. **Nonpareil** and **Golden Pippin** also come from the sixteenth century and keep till mid-spring. However, they are rarely available, though there are a dozen and a half part-descendants all called **Golden Pippin**. **Golden Reinette** (mid-winter, dessert) is still popular in Europe and dates from before 1650. The large green **Flower of Kent** (1660) has nearly disappeared. This was the apple that prompted Sir Isaac Newton in his discoveries of the laws of motion and gravity. **Ribston Pippin** (mid-winter, dessert) has one of the highest vitamin C contents and superb flavour. It was bred in 1707 and is not happy on wet heavy soils. From 1720 comes **Ashmead's Kernel**, one of the best-tasting, late-keeping dessert apples, but it is a light cropper. **Orleans Reinette** (mid-winter, dessert) is known from 1776. It is juicy, very tasty with a rough skin and is not very good on wet cold sites. 1785 saw the birth of the rare but choice **Pitmaston Pine Apple** (mid-winter, dessert). This has small fruits with a rich, honey-like flavour. **Wagener** is a mid- to late winter, hard-fleshed keeper, which was raised in New York State in 1791. **Bramley's Seedling** (mid-winter, culinary) raised in 1809, has one of the highest vitamin C contents of cooking varieties. It grows large, so have it on a more dwarfing stock than others. The **Cornish Gillyflower** is a very tasty, late-keeping dessert raised in 1813. Unlike many other apples, it will flourish in a mild wet climate. It is unsuited to training or cordon culture. One of the best dual-purpose apples is **Blenheim Orange** (mid-winter), a wide, flat, golden russeted fruit and a large tree. Raised in 1850, **Cox's Orange Pippin** (late autumn) is reckoned the best dessert apple. However, it is not easy to grow as it is disease-prone, hates wet clays and does best on a warm wall. **Sunset**, raised in 1918, and **Suntan**, in 1955, are more reliable offspring. **Beauty of Bath** is one of the best-known earlies, fruiting in late summer with small, sharp, sweet and juicy, yellow fruits stained scarlet and orange. It was introduced in 1864. It is a tip bearer and not suitable for training.

Apples exposed to full sun are highly coloured, most aromatic and sweetest

These were fairly well thinned – but they would have been bigger and better if I'd been even more ruthless

Egremont Russet (late autumn), bred in 1872, is one of the best russets, a group of apples with scentless, roughened skin and crisp, firm flesh, which is sweet and tasty but never over-juicy or acid. Just a century old is **James Grieve** (mid-autumn, dessert). It is prone to canker, but makes a good pollinator for **Cox** and is a good cropper of refreshingly acid, perfumed fruits. The ubiquitous **Golden Delicious**, so much grown commercially in Europe, is a conical yellow. It actually tastes pretty good when grown at home, but must be waxed for keeping as otherwise it wilts. It was found in West Virginia in 1916. **George Cave** is a modern (in horticultural terms), early dessert variety from 1945 much better flavoured than **Beauty of Bath**. **Discovery**, introduced 1962, an early; fairly frost tolerant and scab resistant, crisp scarlet and a heavy cropper. **Esteval(e)** a cross between **Worcester Pearmain** and **Golden Delicious** is a new early much vaunted for its sweet, sharp flesh. Other newer apples proving popular are: **Queen Cox** and **Fiesta**, both have the Cox taste with reliable cropping, **Queen Cox** is also self-fertile, both keep till January. **Jonagold** (cross of **Jonathan** and **Golden Delicious**) produces huge red cartoon apples that keep till January. **Saturn** keeps till February, has good resistance against scab and mildew; crisp, juicy and sweet, it's self-fertile and a reliable cropper. **Braeburn** is a crisp, tasty supermarket favourite, with a tangy flavour it stores till March.

It is interesting to note some varieties have much more vitamin C than others that grow in the same conditions. **Ribston Pippin** typically has 31mg/100g, **Orlean's Reinette** 22.4mg, **Bramley's Seedling** 16mg, **Cox's Orange Pippin** 10.5mg, **Golden Delicious** 8mg and **Rome Beauty** 3.6mg. Maybe the famous saying should go 'A Ribston Pippin a day keeps the doctor away'.

A shady wall will grow apples but, surprisingly, pears might be a better choice

cultivation

Apples are much-abused trees. They prefer a rich, moist, well-drained loam, but are planted almost anywhere and yet often still do fairly well. What they will not stand is being waterlogged, or growing on the site of an old apple tree or near to others that have been long established, and they do not thrive in dank frost pockets. Pollination is best served by planting more than three varieties, as many apples are mutually incompatible, having diploid or triploid varieties with irreconcilable differences in their chromosomes. A **Cox** and a **Bramley** will not fruit on their own, but if you add a **James Grieve** all three bear fruit. Crab apples usually prove good pollinators for unnamed trees.

James Grieve

Growing under Glass

Apples do not like being under glass all the time, as they need a winter chill, and they are more susceptible to pests and diseases, especially if hot and dry, but the usual remedies apply. Some French varieties, such as the classic **Calville Blanche d'Hiver**, can only be grown to perfection under cover in Britain.

container growing

On very dwarfing stocks apples are easily grown in large pots. They need hard pruning in winter and in summer the lengthening shoots should be nipped out, thus tip-bearing varieties are not really suitable. Special varieties have been developed for containers, which supposedly require little pruning.

Ornamental and Wildlife Value

The pink and white blossom is wonderful in late spring. The flowers are valuable to insects and the fruits are important to birds.

Maintenance

Spring Weed, mulch, spray seaweed solution monthly.
Summer Thin fruits, summer prune, spray with seaweed solution, apply greasebands.
Autumn Use poor fruits first, pick best for storage.
Winter Hard prune, add copious compost, remove mummified fruits.

CULINARY AND DESSERT APPLES

Propagation
Apple pips rarely make fruiting trees of value; however, many of our best varieties were chance seedlings. Apples are grafted or budded onto different rootstocks depending on site and size of tree required. Few grow them from cuttings on their own roots or as standards on seedling stocks as these make very large trees only suitable for planting in grazed meadows. Half-standards are more convenient for the home orchard and these get big enough on M25 stock at about 5.2m high and 6m apart. At the other extreme, the most dwarfing stock is M27, useful for pot culture, but these midgets need staking all their lives and the branches start so low you cannot mow or grow underneath them. M9 produces a 2m tree, still needing staking but good for cordons. On such very dwarfing stocks the trees do badly in poor soil and during droughts. M26 is bigger, growing to 2.8m and still needs a stake but is probably the best for small gardens. It needs 3m on each side. MM106 is better on poor soils, and on good soils is still compact at about 4m, needing 4.5m between trees.

Pruning and Training
Apple trees are often left to grow and produce for years with no pruning other than remedial work once the head has formed. They may be trained and hard pruned summer and winter, back to spur systems, on almost any shaped framework, though rarely as fans. For beauty and productivity, apples are best as espaliers; to achieve the maximum number of varieties, as cordons; for ease and quality, as open goblet-pruned small trees. Stepovers are low single-tier espaliers designed for edging beds or paths; these require pruning hard to spurs, like cordons. Some varieties, especially many of the earliest fruiters, are tip bearers. These are best pruned only remedially as hard pruning will remove the fruiting wood. Minarettes are slender columnar relatively dwarfed varieties; these require minimal pruning and are most suited to tub and patio culture. As important as the pruning is the thinning. Removing crowded and congested, damaged and diseased apples improves the size and quality of those remaining and prevents biennial bearing. Thin after the June drop occurs and again twice after, disposing of the rejects to destroy any pests.

Once the first is damaged the rest will soon follow – don't delay; gather them NOW

An early then a second thinning means these **Discovery** are near perfect

Pests and Diseases

Apples are the most commonly grown fruit tree in much of the temperate zone. They have thus built up a whole ecosystem of pests and diseases around themselves. Although they have many problems they still manage to produce enormous quantities of fruit for many years, in often quite poor conditions. Vigorous growth is essential as this reduces many problems, especially canker. The commonplace pests require the usual remedies (see pages656–58), but apples suffer from some annoying specialities. Holes in the fruits are usually caused by one of two pests. Codling moth generally makes holes in the core of the fruit, pushing frass out at the flower end. They are controlled by corrugated cardboard band traps, pheromone traps, permitted sprays as the blossom sets, and hygiene. The other hole-maker is apple sawfly, which bores narrow tunnels, emerging anywhere. They may then eat into another or even a third. They are best controlled by hygiene, removing and destroying affected apples during thinning. Permitted sprays may be used after flower set, and running poultry underneath an orchard is effective. Many varieties are scab-resistant. If it occurs, it affects first the leaves then the fruits and, like brown rot and canker, is spread by mummified apples and dead wood. It is worst in wet areas. All of these problems, and mildews, are best controlled by hygiene, keeping the trees vigorous, well watered and mulched, and open pruned. Woolly aphis can be sprayed or dabbed with soft soap. Sticky non-setting tree bands control many pests all year round, especially in late summer and autumn. Apples also get damaged by birds, wasps and occasionally earwigs, so for perfect fruits, protect them with paper bags.

An old fridge makes a good mouse-proof insulated store

Harvesting and Storing

Early apples are best eaten off the tree. They rarely keep for long, going pappy in days. Most mid-season apples are also best eaten off the tree as they ripen, but many will keep for weeks if picked just under-ripe and stored in the cool. Late keepers must hang on the trees till hard frosts are imminent, or bird damage is getting too severe, then if they are delicately picked and kept cool in the dark they may keep for six months or longer. Thus apples can be had most months of the year, provided early- and late-keeping varieties and a rodent-proof store are available. They are best picked with a cupped hand and gently laid in a tray, traditionally padded with dry straw. (This may taint if damp so better to use shredded newspaper.) Do not store early varieties with lates nor near pears, onions, garlic or potatoes. The fruits must be free of bruises, rot and holes and the stalk must remain attached for them to store well. If apples are individually wrapped in paper they keep longer. Apples can be puréed and frozen, or juiced and frozen, or dried in thin rings, or made into cider.

Apple surpluses can easily be dried as rings or pieces

CULINARY AND DESSERT APPLES

culinary

Apples are excellent raw, stewed and made into tarts, pies and jellies, especially with other fruits which they help set. The juice is delicious fresh and can be frozen for out-of-season use, and made into cider or vinegar. Cooking apples are different from desserts, much larger, more acid and less sweet raw. Most break down to a frothy purée when heated and few retain their texture, unlike most of the desserts. **Bramley's**, **Norfolk Beauty** and **Revd Wilke's** are typical, turning to sweet froths when cooked. **Lane's Prince Albert**, **Lord Derby** and **Encore** stay firm and are the sorts to use for pies rather than sauces.

Flying Saucers
Per person

1 large cooking apple
Approx. 3 dessertspoons mincemeat
Knob of butter
7g sesame seeds
Cream or custard, to serve

Wash, dry and cut each apple in half horizontally. Remove the tough part of the core but leave the outside intact. Stuff the hollow with mincemeat, then pin the two halves back together with wooden cocktail sticks. Rub the outside with butter and roll in sesame seeds, then bake in a baking tray in a preheated oven at 190°C/375°F/gas mark 5 for half an hour or until they 'lift off' nicely. Serve the saucers immediately with cream or custard.

companion planting

Apples are bad for potatoes, making them blight-prone. They are benefited by Alliums, especially chives, and penstemons and nasturtiums nearby are thought to prevent sawfly and woolly aphis. Stinging nettles nearby benefit the trees and, dried, they help stored fruits keep.

Malus pumila **from the family** *Rosaceae*

CRAB AND CIDER APPLES

Kingston Black

Trees up to 10m. Life span: long. Deciduous, often self-fertile. Fruits: 1–7cm, spherical, yellow, green or red. Value: make tonic alcoholic beverages.

Crab apples grow wild in hedgerows and have smaller, more brightly coloured fruits than cultivated varieties. The fruits vary from being unpalatable to completely inedible raw, though they make delicious jellies. Cider apples are more like eating and cooking apples, with fruits in-between in size but similarly bitter and astringent. Selected crab apples are grown ornamentally and are useful pollinators for other apples, so they may be found on semi-dwarfing stocks as small trees. Cider apples were always grown on strong or seedling stock, gaining immense vigour, which they needed to tower above animals grazing underneath.

Crabs have been used since prehistory and doubtless cider has been made for as long. It was probably brought to Cornwall initially by Phoenician tin traders.

varieties

Many crabs are well-known ornamentals, such as **John Downie**, which has long golden orange fruits, and **Golden Hornet**, with bright yellow. Most are mixed hybrids of *Malus pumila* with the native *M. sylvestris*, which has sour, hard, green fruits and is sometimes thorny, and *M. mitis*, from the Mediterranean region, which has softer leaves and sweeter, more coloured fruits. *M. baccata*, the Siberian crab, and *M. manchurica*, from China, are widely planted for their bright red fruits. Cider apples are more improved and closer to *M. domestica* hybrids. Every local variety is rated the best. In trials, **Yarlington Mill** was thought outstanding. **Sweet Coppin**, **Kingston Black**, **Tremlett's Bitter** and **Crimson King** are also excellent. **Dabinett** is still widely grown in the West Country; ripening in November it makes a bitter-sweet cider. **Michelin**, raised 1872, is self-fertile, popular in the Midlands and makes a medium sweet cider.

cultivation

Tougher than the finer apples, these may be grown almost anywhere not waterlogged or parched.

Growing under Glass

Neither crab nor cider apples are happy permanently under glass and do better in the open. They need a winter chill, otherwise they crop badly. They are hardy enough for most places.

Ornamental and Wildlife Value

The crabs are very attractive in flower and many also in fruit. Cider apples are bigger and less pretty. Both are valuable to insects when in flower and to birds, rodents and insects with their fruit.

John Downie

CRAB AND CIDER APPLES

Maintenance
Spring Weed, mulch and spray seaweed solution.
Summer As spring.
Autumn Pick fruit.
Winter Prune, and spread the compost.

Propagation
Although some crabs can be grown from seed they are not reliable. Named varieties are grafted on to rootstocks suitable for the size and site intended. Neither crab nor cider apple varieties are usually available on the most dwarfing rootstocks, but on more vigorous stock they make large trees.

Pruning and Training
Crab apples are usually worked as half-standards on semi-dwarfing stock and are only pruned remedially in winter. Cider apples are worked as standards on strong growing stocks and make big trees. Likewise, cider apples should only be pruned remedially after forming a head.

Pests and Diseases
Although these can suffer from the same problems as dessert and culinary apples, the crops are rarely badly affected. Crab apples are usually remarkably productive whatever care they get. For cider apples many problems such as scab are also mostly irrelevant.

Harvesting and Storing
Crabs can be picked under-ripe for jellying, but you can hang on, as the birds do not go for them as fast as softer fruits. Cider apples are left as long as possible to get maximum sugar and to soften. If they are shaken down they bruise, so the apples are then best pressed immediately, not stored in heaps to soften further before pressing.

companion planting

The plants that benefit dessert and culinary apples (see page 451) also associate well with these varieties. Both crab and cider apples are often grown in hedgerows and grazed meadow orchards. They seem content with grass underneath and old trees often have mistletoe growing in their boughs, to no obvious detriment.

Malus baccata

A grape press for extracting juice from apples works better if they're pulped first

container growing

Crab apples can be fruited in pots; indeed, they often do, even in the small pots in which garden centres sell them. Cider apples are shy and less likely to produce heavy crops.

other uses

Pressed apple pulp can be dried and stored till late winter for wild bird and livestock food.

culinary

Crab apples make delicious tart jellies by themselves, or mixed with other fruits which have less pectin and so do not set so easily. Cider apples are used solely to make cider. They are cleaned, crushed and pressed and the juice is fermented, often with the addition of wine yeast and sugar. After fermentation, cider may be flat, cloudy, sweet or sparkling, green or yellow, depending on local taste. Ciders are made from several varieties of apple, to give a blend of acidity, sweetness and tannin. (Palatable cider can be made from a mixture of dessert and cooking apples.) Some cider is made into vinegar, deliberately.

Crab Apple Jelly
Makes approx 3.5kg

2kg crab apples
Approx.1.5kg sugar

Chop the apples, then simmer them in water to cover until soft. Sieve or strain through a jelly bag and weigh the juice. Add three-quarters of its weight in sugar. Return to the heat and bring to boil, stirring to dissolve the sugar. Boil until setting point is reached, then skim and pour into warm, sterilised jars. Cover while still hot.

Pyrus communis from the family *Rosaceae*

CULINARY AND DESSERT PEARS

Tree up to 20m. Life span: very long. Deciduous, hardy, rarely self-fertile. Fruits: up to 8x18cm. Value: potassium and riboflavin.

Pears very closely resemble, and are related to, apples, but there are no known natural hybrids between them. Pears have a fruit that elongates at the stalk end, which stands proud, whereas an apple's stalk is inset in a dent in the top of the fruit. Some pears such as Conference and Bartlett will set fruit parthenocarpically (i.e., they fruit freely without pollination). However, these fruits are usually not as good as fertilised ones, being misshapen and, of course, lacking seeds. Pear trees resemble apples but have shiny leaves, more upright growth and the fat stems usually have a glossier brown hue and more angled buds than apple. Pears on their own roots make very big trees, too big to prune, spray or pick, and the fruit is damaged when it drops. Pears are thus usually worked on quince roots, which makes them smaller, more compact trees.

Pears are native to Europe and Asia. The first cultivated varieties were selected from the wild in prehistory. The ancient Phoenicians, Jews and pre-Christian Romans grew several improved sorts; by the time of Cato there were at least a half dozen distinct fruits, Pliny records 41 and Palladius 56. A list of fruits for the Grand Duke Cosimo III, in late medieval Italy, raises the number to 209, and another manuscript lists 232. In Britain in 1640 only five dozen were known. This had risen to more than 700 by 1842. In 1866 the American author T. W. Field catalogued 850 varieties. The rapid increase in numbers and quality, from poor culinary pears to fine desserts, was mainly the work of a few dedicated breeders in France and Belgium at the end of the eighteenth century, who selected and bred most modern varieties.

varieties

As pears are used more for eating raw than in cooking, and dessert pears can be cooked, but not vice versa, it is not worthwhile growing purely culinary pears. **Doyenne du Comice** (late season) is by far the best dessert pear. None other approaches it for sweet, aromatic succulence, and the fruits can reach a magnificent weight. These are very choice and deserve to be espaliered on the best warm wall. **Bartlett/Williams' Bon Chretien** (early/mid-season) is widely grown for canning, but is an excellent table fruit – if a bit prone to scab. It is also parthenocarpic as is **Conference** (early/mid-season). The latter is a reliable cropper on its own and is scab-resistant. Raised in 1770 in Berkshire, it is now very widely grown. **Jargonelle** is an old variety first recorded in 1600. It crops for me in early August in East Anglia, about the same time as **Souvenir de Congress**. These are the earliest with good flavour. **Improved Fertility** (mid) is very hardy and crops heavily and regularly. Other superb pears are **Clapp's Favourite**, **Dr Jules Guyot** (both early/mid season); **Glou Morceau** and **Durondeau** (both late) keep into the New Year. The recently introduced **Beth**, ripening in September is a very tasty early. **Concorde**, ripe from October may store till January. These last two are improved offspring from **Conference**, and **Concorde** is similarly fairly self-fertile. There are hundreds of other pear varieties, many of which are cooking, not dessert, and several species which have more ornamental than edible value. The Birch-leaved Pear, *P. betulifolia*, comes from fourteenth-century China. Almost all parts were eaten, flowers and leaves as well as the small fruits. The Asian Pear or Nashi, such as **Kumoi** or **Shinseiki**, is the most apple-like pear, brown green, russeted and crunchy, juicy, but a bit insipid, easy to grow, partly self-fertile and productive. Similar is the **Chinese Sand** or **Duck Pear**, *P. sinensis*. In Syria the **Three-lobed-leaved Pear**, *P. trilobata*, is popular. This may be the same as the native Turkish *Malus trilobata*.

cultivation

Dessert pears need a rich, well-drained, moist soil, preferably light and loamy. They can be cropped in the open in southern England, but need a wall further north. The blossom and fruitlets need protection from frosts as they flower early in spring. Pollination is best ensured by planting a mixture of varieties. Do not plant deep as pears are prone to scion rooting, which allows them to make big, less fruitful trees.

Growing under Glass

Pears appreciate a warm wall, but are not easy under glass as they do not like to get too hot and humid; they are thus difficult to grow in the tropics. Ideally plant in containers, keep indoors for flowering and fruiting, and put outdoors for summer and most of winter as they need some chill. They are hardy and can stand light frost, but the roots are most susceptible in pots.

Ornamental and Wildlife Value

Why plant an ornamental flowering tree when you can plant a pear? They are just smothered with blossom and buzzing with bees in early spring. The fruits are splendid, but soon eaten in autumn by insects and birds.

Pears should not be left till they drop; pick before that point to colour and ripen indoors

Kumoi

Maintenance

Spring Weed, mulch, spray with seaweed solution monthly; protect flowers against frosts.
Summer Thin fruits.
Autumn Pick fruit – better early, not late.
Winter Prune hard or not at all; compost heavily.

Propagation

Pears do not come true from seed, reverting to their unproductive forms. They may root from cuttings and can occasionally be layered, but get too big on their own roots. Normally they are grafted on to quince rootstocks, which are generally best. For heavy, damp soils and for big trees, pear seedling stock was always used, but there is little demand for this nowadays. Pears have been grafted on to apple stocks and even on to hawthorn. As some varieties do not bond readily with quince stock they are 'double-worked', or grafted on to a mutually compatible inter-graft on the quince. This takes more work and an extra year in the nursery.

Pruning and Training

Grown as trees or bushes, pears can be left to themselves except for remedial pruning. They tend to throw twin leaders which need rationalising. Alternatively, they respond better than most fruits to hard winter pruning and are almost as amenable to summer pruning; thus they can be trained to an endless variety of forms. As they benefit from the shelter of a wall, they are commonly espaliered, and are more rewarding for less work than a peach in the same position. There are many specialised training forms for pears in addition to the cordon, espalier and fan, and equally intricate and specialised pruning methods. These include Pitchforks and Toasting Forks, with two or three vertical stems on short arms, the Palmette Verriers, with long horizontals that turn vertical at the ends, and the multi-curved L'Arcure. Pears will take almost any shape you choose, and you will not go far wrong if you then shear almost all long growths off by three-quarters in summer and then again by a bit more in winter. Take care not to let them root from or above the graft, which destroys the benefit of the rootstock!

Espaliered **Beurre Hardy** is happy on this sunny wall where it pollinates a **Doyenne du Comice** close by

Pests and Diseases

Pears suffer fewer problems than apples. Providing the flowers miss the frosts and have a warm summer they usually produce a good crop of fruits despite any attacks. However, ripening is poor in very cool or hot weather and results in hard or mealy fruits. Leaving the fruit too long on the tree or in storage causes them to rot from the inside out. Pear midge causes the fruitlets to blacken and drop off; inspection reveals maggots within. These are remedied by the hygienic removal and disposal of affected fruits; running poultry underneath in orchards is as effective. Fireblight causes damage resembling scorching. It usually starts from the blossoms. Prune and burn damaged parts immediately, cutting back to clean wood. Scab is a problem in stagnant sites. It affects the fruits before the leaves, which is the opposite to apples. Sprays are usually unnecessary with the more resistant varieties such as **Conference**. Good open pruning and healthy growth, not overfed with nitrogen, reduce attacks. Always remove all mummified fruits and dead wood immediately. Leaf blistering is usually caused by minute mites. These used to be controlled with lime and sulphur sprays before bud burst. Modern soft-soap sprays have also proved to be effective.

Harvesting and Storing

Early pears are best picked almost, but never fully, ripe. They should come off when lifted to the horizontal. Left on the tree they go woolly. Watch them carefully while they finish ripening. They will slowly ripen if kept cool, faster if warm. Late pears should be left until bird damage is too great, and picked with a stalk. Kept in a cool, dark place, late varieties last for months, ripening up rapidly if brought into the warm. Do not wrap them with paper as one does with apples, and take care not to store the two fruits near each other as they will cross-taint.

container growing

Pears have often been fruited in tubs. They are amenable to hard pruning, and so respond better than most other fruits to this and to the cramped conditions of a pot.

other uses

The bark contains a yellow dye and arbutin, an antibiotic. The leaves have been used medicinally for renal and urinary infections. The wood is hard and uniform, so loved by carvers. It is beautifully scented and good for smoking foods.

companion planting

Pears are hindered by grass, so this should not be allowed near them in their early years. However, the pears may be grassed down later, especially if they are situated in rather too heavy and/or damp conditions.

Pear gluts are easily turned into sweet syrup!

CULINARY AND DESSERT PEARS

culinary

Pears are exquisite as dessert fruits and may be used in much the same way as apples, sliced and used in tarts, baked, stewed or puréed. They are delicious pickled with onions and spices in vinegar.

Windfall and over-ripe pears can be chopped and simmered, then sieved and boiled down to a maple like syrup. Nashi Pears such as the **Kumoi** dry to delectable rings, chewier and sweeter than dried European pears and far better than apple rings.

Pear Islands
Serves 4

4 large pears
90g caster sugar
Half bottle sweet Muscat wine
2 level tablespoons cornflour
Splash of milk
Dark chocolate, to taste

Peel, core and halve the pears. Dissolve two-thirds of the sugar in the wine over a gentle heat and poach the pears in this syrup. Place the pear halves on an oiled baking tray, dredge with the rest of sugar and pop under a hot grill for a minute or two until the tops caramelise. In the meantime, blend the cornflour with a little milk and stir it into the syrup. Return to the heat and cook gently, stirring, until thick. Pour the sauce into a serving bowl, grate on some dark chocolate and set the pear halves into this. Serve piping hot or chilled.

Pyrus from the family *Rosaceae*

PERRY PEARS

Blakeney Red

Tree up to 20m. Life span: very long. Deciduous, rarely self-fertile. Fruits: up to 10 x 7cm, variable, pear-shaped, yellow, brown or red. Value: make a healthy tonic beverage.

Perry pear fruits are smaller and hardier than their culinary and dessert cousins, while the trees are generally enormous. As these fruits were wanted in quantity for pressing, they were selected for large trees that would stand well above grass and the depredations of stock. The fruits are bitter and astringent, containing a lot of tannin, even though they may often look very appealing. They are pressed for the juice, which ferments to an alcoholic beverage.

Perry pears may have been brought to Britain by the Romans, but they were introduced in force by the Normans. Normandy was by then well established as a perry-growing region and the Normans found parts of the West and Midlands of England equally suitable. Many of the original pear orchards would probably still have been planted for perry without the Norman influence, as early pears tended to be unpalatable raw anyway. The colonists of North America certainly did not start off without perry pears. By 1648 a booklet entitled *A Perfect Description of Virginia* describes a Mr Kinsman regularly making forty or fifty butts of perry from his orchard. Whereas cider has remained a popular drink, perry waned in the nineteenth century. It is now coming back into fashion as large cider makers have started selling it.

varieties

There are some 200–300 perry pear varieties known, though these are called by twice as many names, each district having its own local variation. The **Thorn Pear** is recorded from 1676 and grows on a very upright tree. Other early varieties such as **Hastings** and **Brown Bess** could be eaten or made into perry. Later varieties such as **Holmer** are more single purpose and almost inedible. In the nineteenth century, breeders thought dual-purpose pears would be useful and produced **Blakeney Red** and **Cannock**. The **Huffcap** group all make very big trees with fruits of a high specific gravity; **Rock** varieties make even stronger perry, but are still often called **Huffcaps.**

cultivation

Perry pears are less demanding than culinary or dessert pears and will do quite well in fairly poor soils. They do not like shallow or badly drained sites.

Growing under Glass

These are fairly hardy and will make big trees, so are generally not likely to thrive under glass.

Perry pears are not good eating

PERRY PEARS

Perry pears are smaller, more acidic and often more brightly coloured than dessert pears

Ornamental and Wildlife Value

Perry pears make a mass of flowers and are tall attractive trees that live for several hundred years, making them very suitable as estate and orchard trees, but too big for most gardens except on modern dwarfing stock. The flowers are good for insects and the fruits are eaten by birds in winter.

Maintenance

Spring Cut grass, spray with seaweed solution.
Autumn Cut grass, collect fruits.
Winter Prune and spread compost.

Propagation

Perry pears were often grown from seed or grafted on to stocks grown from pips in order to get large trees. Now they can be had on dwarfing stocks, such as quince, and are more manageable.

Pruning and Training

Once the initial shape is formed these are only pruned remedially, though care should always be taken to ensure that such large trees are sound.

Pests and Diseases

These suffer the same few problems as dessert and culinary pears which make little impression on the immense crops. Fireblight is a risk, but there is little one can do with such large trees.

Harvesting and Storing

Immense crops are produced; a ton or even two per tree is possible, after a wait of several decades. One tree in 1790 covered three-quarters of an acre and produced six tons per year. Perry pears do not keep and rot very quickly. Perry is usually pressed from one variety only, with sugar and yeast added. Dessert and culinary pears do not make a perry of any value, but can be made into a pear wine.

companion planting

Young trees were usually cropped between cereals or hops and then grassed down as they matured, often grazed by geese.

container growing

Although they would resent it by being short-lived and light-cropping, it should be possible to grow these in large tubs.

other uses

The wood is useful for carving and, as firewood, scents the room.

culinary

Perry pears are not normally very good for eating even when they are cooked, though **Blakeney Red** was once considered a good baker.

Pressless Perry Wine

3kg pears
1.5kg sugar
4.5 litres boiling water
Approx. 1 teaspoon wine yeast

Chop the pears, peel and all, into the boiling water. Stir in half the sugar, and bring back to the boil, then allow to cool to 22°C before adding the yeast. Seal with a fermentation lock and keep warm for one week, then strain and add the remaining sugar to the liquor. Reseal with the lock and ferment in a warm place until all action has stopped. Siphon off the lees and store in a cool place for 3 months, then bottle and store for a year before drinking.

Cydonia vulgaris/C. oblonga **from the family** ***Rosaceae***

QUINCES

Tree up to 6m. Self-fertile, deciduous, hardy, long-lived. Fruits: 7 x 12cm, yellow and fragrant.

These are small, bushy trees, often twisted and contorted. There are two forms: a lower mounded one with lax branches, more suited to ornamental and wildlife use, and a stiffer, erect type, which bears larger fruits and is better for orchards. The leaves are downy underneath, resembling apple leaves more than pear. They turn a gorgeous yellow in autumn. The decorative pink and white flowers resemble apple blossom, but appear singly on short shoots, up to 12cm long, that grow before the flower opens, so they are rarely bothered by late frost. In some varieties the unfurled flower bud looks like an ice-cream cone with strawberry stripes, or a traditional barber's pole. The quinces are hard fruits somewhat resembling pears in shape and colour, often covered with a soft down when young, inedible when raw, but delicious and aromatic when cooked. Because they have long been used as rootstocks for pears and other fruits, they are sometimes found as suckers, surviving long after the scion has passed away.

Quinces are old fruits, still much grown in many parts of Europe though originally from Persia and Turkestan. Dedicated by the ancients to the Goddess of Love, they were promulgated by the Roman Empire as one of their favoured crops and were well known to Pliny and Columella. In 812 AD Charlemagne encouraged the French to grow more and Chaucer refers to them by the French name as *coines*. Their pulp makes *Dulce de Membrillo* or *Marmelo*, still popular in Portugal and Spain, and the origin of the marmalade we now make from citrus fruit.

varieties

The **Portuguese** is pear-shaped, vigorous, but slow to crop. **Vranja** (**Bereczki**) is from Serbia, large-fruited, pear-shaped and with erect growth. **Meech's Prolific** is also pear-shaped, early to bear and late keeping. **Champion** is rounder and mild-flavoured. Also available are **Ispahan** from Persia and **Maliformis** (apple-shaped). Also in the USA: **Orange**, **Pineapple** and **Smyrna**.

The Chinese Quince, *Pseudocydonia sinensis*, has a similar large yellow edible fruit but is usually planted for its gorgeous mottled bark and autumn leaf colour.

Cydonia quinces may be confused with *Chaenomeles*

cultivation

Quinces need a moist soil and flourish as waterside specimens. They prefer a warm site, doing rather badly in cold or exposed places. Plant them at least 3m apart. Normally they will only need staking in their first years of life. They are self-fertile.

Ornamental and Wildlife Value

Quinces make excellent small specimen trees as the flowers, fruits, autumn colours and the knotted branches give year-round interest. The flowers feed beneficial insects while the fruits are relished by birds and other wildlife after the apples and pears have long gone.

QUINCES

culinary

Quinces can be made into aromatic clear jelly, jam or a pulpy cheese that goes well with both sweet and savoury dishes. Pieces of quince (if you can hack them off) keep their shape when cooked, adding both texture and aroma to apple and pear dishes.

Quince Cheese
Makes approx 2kg

1kg ripe quinces
1 unwaxed small orange
A little water
Approx. 1kg sugar
1 or 2 drops orange flower or rose petal water (optional)

Roughly hack the quinces into pieces. Finely chop the orange and simmer both, with just enough water to cover them, until they are a pulp. Strain the pulp and add its own weight in sugar. Bring to boil and cook gently for approximately 1½ hours. Add the orange flower or rose petal water, if liked. Then pot into oiled, warmed bowls, seal, and store for three months or more before using.

Turn the cheese out of the bowl and slice for serving with cooked meats or savoury dishes.

Maintenance
Spring Spray monthly with seaweed solution; weed and mulch well.
Summer Spray monthly with seaweed solution and weed.
Autumn Remove rotten fruit and pick best for storing.
Winter Prune out dead and diseased wood; add plenty of compost.

Propagation
Quinces can be had from seed or by suckers removed from pear trees as these are usually grafted on to quince stocks. Better fruits will always result from buying a ready-formed tree of an already named variety.

Pruning and Training
Quinces can be trained but the twisted contorted growth makes this difficult. They are best grown as bushes or standards, with pruning restricted to removing dead, diseased and crossing wood.

Pests and Diseases
There are no widespread problems for quinces. Even the birds and wasps will ignore the fruits for most of the autumn.

Harvesting and Storing
Pick the fruits in autumn before they drop and keep them in a cool airy place. Do not store with apples or pears or vegetables as they may taint.

Cydonia quinces are very pear-like, but much firmer

companion planting

Like pears and apples, quinces may be expected to benefit from underplantings of chives, garlic and the pungent herbs.

container growing

There is little to be gained by growing quinces under glass. They can be grown against a wall with some success and could be grown in pots and taken indoors by those living in harsh climates, but the fruit hardly merits such effort.

other uses

The wood is hard and prunings make good kindling. The fruits are excellent room perfumers and can be used as bases for pomanders.

Prunus domestica from the family *Rosaceae*

PLUMS

Tree 6m. Long-lived, deciduous, hardy, slender, thorny, not usually self-fertile. Fruits: ovoid, usually 3–6cm in any colour. Value: rich in magnesium, iron and vitamin A.

Plums come in more variation than most other fruits. They differ in season, size, shape, colour and taste. We have mixed hybrids, descendants of plums originally selected from fifteen or more different wild species. The European plum, *Prunus domestica*, is thought to be predominantly a hybrid between *P. cerasifera*, the cherry plum or Myrobalan, and *P. spinosa*, the sloe. It is a small to medium, slender, deciduous tree with small white blossoms.

The European plum came from Western Asia and the Caucasus. It naturalised in Greece first and then throughout most of the temperate zone. Pliny describes cultivated varieties from Syria coming to Italy via Greece, and it is likely they were spread by the Roman Empire to Britain and Northern Europe. They were reintroduced in the Crusades. Henry VII is recorded as importing a Perdrigon plum and Brogdale Fruit Research Station have a plum grown from a stone salvaged from the wreck of the *Mary Rose*, Henry VIII's splendid warship. Plum stones were ordered in 1629 for planting in Massachusetts, and plums became widely cultivated in the temperate parts of North America. In 1864 more than 150 varieties were offered in nurserymen's catalogues. Some American plums returned to Europe; California, for example, became famous for exporting prunes – late, dark-skinned plums, which are dried on the tree. Although unsuccessful in cool, damp climates, they can be dried with machinery. Fellemberg and Prune d'Agen are from Europe, but are more widely grown in California.

Czar

varieties

There are hundreds of good varieties, which ripen through the season. **Victoria** is fully self-fertile, pollinates many others and is always worth having. It has golden-yellow-fleshed, large, yellow, ovoid fruits, flushed with scarlet. It ripens middle to late August in Sussex, where it was found around 1840. **Opal** is like a smaller, sweeter, earlier, better-flavoured **Victoria** and even partly self-fertile and a good pollinator for **Victoria**. **Coe's Golden Drop** is a shy cropper, but superb. It needs a warm spot or a wall and closely resembles an apricot. **Severn Cross** is a delicious golden seedling from **Coe's** and self-fertile. **Czar** has frost-resistant flowers and is self-fertile; it is for culinary rather than dessert use. **Marjorie's Seedling** is dual purpose, late cropping and self-fertile.

cultivation

Plums like a heavier, moister soil than many other fruits. Even on a shady wall some, such as **Victoria** or **Czar**, do well. This means they may often be relegated to cold, damp sites and heavy soils, which they really do not like.

Growing under Glass and in Containers

Because of their susceptibility to spring frost, brown rot and wasps, plums are worth growing under glass.

PLUMS

Victoria

However, they would be best confined to large pots so they can spend some time outdoors, as they do not relish hot conditions.

Ornamental and Wildlife Value
Plums are wonderful during their short blossoming. The flowers are loved by early insects and the fruits by birds, insects and rodents, who also chew the bark.

Maintenance
Spring Protect blossom from frost; weed, mulch, spray with seaweed solution, prune.
Summer Put out wasp traps.
Autumn Remove mummified plums.
Winter Protect buds from birds with cotton or nets.

Propagation
Graft on **Pixy** and other new dwarfing stocks, unless you have a big orchard and want immense quantities. Plums can be grown from stones, but take years to fruit and do not come true.

Pruning and Training
Plums make good high standards because, eventually, heavy fruiting branches weep, bringing the fruit down to a skirt. Overladen branches will need propping. Plums are usually grown as a short standard or bush. Leave them alone once a head is formed, except to cut out dead and diseased wood. Any work is best done in the growing season. They can be usefully trained on walls if on dwarfing stock, for example **Pixy**. Plums prefer herringbone, not fan shapes.

Pests and Diseases

Plums get a host of the usual pests and a few more besides, but when they avoid the frost and crop at all they are so prolific there is usually a surplus for private gardeners. They are susceptible to silver leaf disease, so should not be pruned in autumn, winter or early spring but in late spring and summer and during dry weather.

Harvesting and Storing
Plums picked under-ripe for cooking will keep for days, but often have poor flavour. Many varieties can be quite easily peeled, and this avoids some of the unfortunate side effects of too many plums.

culinary

Plums can be turned into jam, juice or cheese, or frozen if stoned first, and are Epicurean preserved in plum brandy syrup.

Prunes in Semolina
Serves 4

250g dried prunes
100ml plum brandy
100ml water
900ml milk
Twist of lemon rind
½ teaspoon salt
7 dessertspoons semolina
2 dessertspoons honey
2 small eggs, separated
Grated nutmeg, to taste

Soak the prunes in the brandy and water overnight. Strain off the juice and simmer it down to syrup. Put the fruits and syrup into the base of a pudding basin. Boil the milk, lemon rind, salt and semolina for 10 minutes, stirring continuously.

Cool a little, remove the lemon rind and stir in the honey and egg yolks. Whisk the egg whites and fork them in. Immediately pour the mixture over the back of a spoon on to the fruit and syrup. Grate nutmeg over the surface and bake in a preheated oven at 200°C/400°F/gas mark 6 for 20 minutes, or until the top is browning.

As delicious cold for breakfast as it is hot for dinner.

companion planting

Avoid anemones, which harbour plum rust. In the USA curculios are reportedly kept off by surrounding plum trees with garlic.

other uses

Potent brandy is made from plums in Hungary and in central Europe.

Double Victoria
Serves 6–8

500g Victoria plums, halved and stoned
50g honey
25g flaked blanched almonds

for the sponge:
100g sifted self-raising flour
100g vanilla-flavoured caster sugar
100g softened butter
2 eggs
pinch of salt

Place the plums in a buttered pudding dish and dribble the honey over them. Beat together the sponge ingredients in a mixer until creamy and light in colour. Dollop the sponge mixture over the fruit, smooth and garnish with almonds. Bake in a preheated oven at 190°C/ 375°F/gas mark 5 for 35 minutes or until golden brown and firm to the touch.

Prunus italica from the family *Rosaceae*

GREENGAGES

Tree/bush 4–5m. Hardy, long-lived. Fruits: 2–4cm, green to red.

Greengages are like plums, fruiting in mid-season with sweet, greeny yellow or golden, lightly scented flesh. The fruits are smaller, firmer, more rounded and less bloomed than plums. They have a deep crease down one side and frequently russet spotting. The trees are sturdy, not often thorny, and bushier than most plums, though not quite as hardy.

Wild greengages are found in Asia Minor. Possibly introduced to Britain and Northern Europe by the Romans, they disappeared from cultivation during the Saxon period or Middle Ages and were not reintroduced until 1725. Originally known in France as the Reine Claude, the first greengage was brought to Britain by, and named after, Sir Thomas Gage, who lived in Bury St Edmunds, Suffolk – less than 20 miles from where I now write. He was fortunate to live in East Anglia as the conditions suit greengages, which need a drier, warmer summer than plums. The original greengage almost always came true from seed, but there are some larger-fruited selections, and also some good crosses between gages and plums.

Mirabelle

varieties

The **Old Greengage** is original, but can be unreliable; an improved seedling is **Cambridge Gage**. The **Transparent Gage** is another old variety from France and is honeyed in its sweetness. It has almost transparent, golden-yellow flesh and is heavily spotted with red. It fruits in late summer. The true **Mirabelle** is very similar, smaller fruited and of dwarf growth, but is rarely found except in southern France. (Sometimes the yellow **Myrobalan** plums may erroneously be called **Mirabelles** to make a sale!) **Denniston's Superb** comes from the USA and is close to the original in flavour, but is larger fruited, hardier, regular cropping and most valuable of all, self-fertile. **Jefferson** is similar, later, but not self-fertile.

Reine Claude de Bavay possibly has a plum as one parent. It fruits a fortnight or so after the previous varieties, in early autumn. It makes a most delicious jam, a touch more acid and tasty if the fruit is picked a week or so early. **Golden Transparent**, another hybrid, is a large, round, transparent yellow. It ripens late and needs a wall in most cool areas, but is self-fertile. **Oullin's Golden Gage** is self-fertile. A plum with a richer, sweeter flavour, it flowers late, missing frosts.

cultivation

Greengages need a lighter soil than plums but still need it to be rich, moist and well aerated. They are easiest to tend as standards at least 6m apart. In colder areas they need a wall. Some, such as **Denniston's Superb**,

GREENGAGES

Prunelle liqueur is a warming drink akin to brandy

Early Transparent Gage, **Jefferson's Gage** or **Oullins Golden Gage**, will fruit on a shady wall.

Growing under Glass and in Containers
Greengages, especially the choicer **Mirabelle** or **Transparent** varieties, are worth growing in pots as this is the only way to restrict their growth. They can then be taken under cover for the flowering and ripening periods. Greengages will need a rest in winter.

Ornamental and Wildlife Value
As for plums (see page 463).

Harvesting and Storing
Greengages are loved by the supermarkets green and hard, as in that condition they keep for weeks. Picked fully ripe off the tree they are delectable, but do not last, especially if wet. They can be jammed, turned into cheeses, juiced or frozen if stoned first.

Pruning and Training
They are usually grown as low standards or bushes with remedial pruning in late spring once the shape has formed. Overladen branches need propping, but not as much as for plums. As they also tend to irregular bearing, thinning of heavy crops is sensible, but not as effective as for most fruits. Gages are best worked like plums in a herringbone pattern on a wall, and summer pruned.

Maintenance
Spring Protect blossom from frost; weed, mulch, spray with seaweed solution, then prune.
Summer Put out wasp traps.
Autumn Remove mummified fruits.
Winter Protect buds from birds with cotton or nets.

Pests and Diseases
As greengages are so sweet they suffer particularly from bird damage, and the buds are attacked as well as the fruits! The damage caused also allows dieback to get a foothold, so trees need netting or cottoning over the winter in bird-infested areas. Wasps also make a mess of the crop, so be prepared to be ruthless!

Propagation
Gages are usually grafted on to Myrobalan stocks, but newer dwarfing stocks mean that it is easier to fit them on to walls. Cuttings can be taken in late autumn with some success and the oldest varieties come nearly true from stones.

companion planting

Greengages benefit from being positioned on the sheltered, sunny side of the larger plums.

other uses

Prunelle (left) is a liqueur from Alsace and Angers, made from **Mirabelles**. Slivovitsa, an eau-de-vie, comes from the Balkans.

culinary

They make the best plum jams. True Mirabelle jam is almost apricot flavoured, but even better.

Greengage jam
Makes 2kg

1kg greengages
1kg sugar
A little water

Wash, halve and stone the gages. Crack a few stones, extract the kernels and add these to the fruit. Pour in enough water to cover the bottom of the pan and simmer until the fruit has softened. Add the sugar and bring rapidly to the boil, then carefully skim, jar and seal.

Prunus species* from the family *Rosaceae

DAMSONS

BULLACES AND JAPANESE PLUMS

Zwetsche

Tree/bush up to 5m. Long-lived, some self-fertile. Fruits: 2cm+, some vitamin value.

Bullaces (*P. insititia*) have globular, bluey-black or greeny-yellow fruits, which ripen in late autumn, at least a month later than most other plums. The fruits are small and generally too acid to eat raw, but make good preserves. The trees are sometimes thorny.

Damsons (*P. damascena*) are closely related to bullaces, with larger, blue-black fruit which more closely resemble plums. However, they are more oval, with less bloom, and have a sweet, spicy flavour once cooked. Damson trees are compact and are reasonably self-fertile.

The Cherry Plum or Myrobalan (*P. cerasifera*) can be woven into hedges; it is often used as a windbreak. The self-fertile white flowers open at the same time as the leaves, which are glossier than those of other plums. The fruits are yellow, red or purple, spherical and a little pointed at the bottom. They have a sweet, juicy, if somewhat insipid flesh and make good jam.

Japanese plums (*P. salicina* and *P. triflora*) are large, conical, orangey-red or golden fruits without much flavour. They blossom early and are vulnerable to cold, but they are also more productive and tolerant of a wider range of warm conditions than ordinary plums, so are extensively grown in Australia, South Africa and the USA. They have shiny, dark twigs, white flowers on bold spurs and leaves that turn a glorious red in autumn.

Bullaces are native to Europe and Asia Minor, but the current apparently wild stock has probably been inadvertently selected over the centuries. Damsons come from Damascus, or certainly that region, and were brought to Europe during the Crusades in the twelfth century, supposedly by the Duke of Anjou, after a pilgrimage to Jerusalem. Cherry plums come from the Balkans, Caucasus and Western Asia and were introduced to Britain in the sixteenth century. Some of these went to the New World and were interbred with American native species to cope with the harsher climatic conditions. Japanese plums, originally natives of China, were introduced to Japan about 1500 and only to the USA in 1870. Being rather tender, they have never really expanded into Europe.

varieties

Bullaces now available are the purpley-blue **Black**, the greener yellow **Shepherd's** and the **Langley**, a blue-back jamming variety raised as recently as 1920. The common wild plum of central Europe, **Zwetsche** or **Quetsche**, the source of many good liqueurs, has blue-black berries with golden-yellow flesh on a compact but twiggy tree. Damsons are now found in only three varieties: **Farleigh**, from Kent, is a heavy cropper if well pollinated and a sturdy bush often used as a windbreak; **Merryweather** has slightly larger fruits with greenish- yellow flesh, earlier in autumn; the **Shropshire**, or **Prune damson**, which ripens last, has the better flavour, but is a light cropper. **Cherry plums** are most often called **Red** or **Yellow Myrobalans**. There are several ornamental forms that may fruit; some have pink flowers, many are purple-leaved. Japanese plums, such as **Burbank**, may be available. Similar is **Beauty**, which is red with sweet, juicy, yellow flesh that clings to the stone.

DAMSONS

Farleigh

cultivation

Most of these are more tolerant of soil and site conditions than true plums, except for the Japanese, which require warmer conditions.

Growing under Glass and in Containers
As most of these plum types are hardy and culinary, it seems unprofitable to fruit them under cover. If only small quantities are required then these plums could be tried in pots.

Ornamental and Wildlife Value
The trees are not imposing until they flower, then they froth and foam. Later their displays of fruiting profligacy are quite stupendous. Great masses of fruits continue to feed the wildlife through many months.

Maintenance
As for plums (see page 463).

Propagation
Most types do well on Myrobalan stocks and many on their own roots. Japanese plums can bear in three years or so, the others maybe in five. Those on their own roots can be grown from suckers, and seedlings of most will come nearly true.

Pruning and Training
As for plums, these are best left alone except for remedial work.

Pests and Diseases
These varieties of plums are generally not as susceptible to brown rot as true plums.

Harvesting and Storing
Bullaces and damsons are supposedly mellowed and taste better raw after frosts, but the birds will have eaten them by then.

companion planting

Avoid anemones, which harbour plum rust.

other uses

As hedges and windbreaks the bullaces and cherry plums are excellent. Damsons can be planted in sheltered sections.

Myrobalans

culinary

Almost all of these fruits are now used culinarily only for jamming, jellying, fruit cheeses, winemaking and liqueurs.

Damson Cheese
Makes 2.5kg

1.5kg damsons
About 1kg light brown sugar

Wash the fruit and then simmer till soft with just enough water to prevent burning. Sieve and boil the pulp with three-quarters its own weight of sugar. Skim off any scum. Cook until the scum has finished rising and the jam is clear. Pot the pulp in warmed oiled bowls and seal.

Prunus armeniaca* from the family *Rosaceae

APRICOTS

Tree up to 6m. Hardy, deciduous, self-fertile. Fruits: 3 x 5cm, yellow/orange. Value: rich in vitamin A and potassium.

This small tree has white flowers (occasionally tinged with pink) very early in spring, well before the leaves emerge. These are spade-shaped and often glossy. The young shoots can also appear glossy, as if varnished in red or brown. The leaves are more similar to those of a plum than those of a cherry or peach, and apricot fruits closely resemble some plums, but the stone is more spherical and the flavour distinct. Some think that Moorpark, one of the commonest varieties, is a plumcot (a plum-apricot hybrid).

The first apricots came from China or Siberia, not Armenia, where Alexander the Great found them. The fruits became loved by the Romans, but they never succeeded in transplanting any to northern Europe. Apricots reached Britain in the thirteenth century and were introduced again, more successfully, in the sixteenth. Bredase may be the oldest variety in cultivation; it closely resembles descriptions of Roman apricots.

varieties

New Large Early is earliest, **Alfred** and **Farmingdale** ripen next, the latter is large, good and has some dieback resistance. **Moorpark**, the commonest variety ripens a little later. **Tomcot**, **Flavourcot** and **Goldcot/Gold Cott, Lil Cot** and **Perle Cot** are all recent introductions bred for cool damp conditions and probably the better choices. Also available for enthusiasts are **Hemskerke, Breda, Bredase, Hongaarse, Luizet, Shipley's Blenheim** and **Tross Orange**. **Hunza** wild apricots of Northern India can be grown from stones in dried fruits bought from natural food shops; these make large bushes with more ornamental than cropping potential. The **Japanese Apricot**, *Prunus mume*, has scented flowers and sour fruit usually eaten salted or pickled.

cultivation

Like most stone fruits, apricots need a cold winter period to rest and warm summers to ripen the fruit. The plants themselves are tough, but the flowers are so early that they are always in danger from frost. The soil should not be heavy nor the site wet. In most cool areas they will only crop reliably against a wall.

Growing under Glass and in Containers

Grown under glass, apricots are more sure to crop if allowed a cold resting period, so do not plant in continuously heated greenhouses. A better plan is to confine them in large pots. Keep these outside unless very cold, bring them under glass for flowering and take them back outside again once the weather is warm.

Ornamental and Wildlife Value

Apricots are not noticeable trees, though the flowers are pretty enough. Ornamental varieties and **Hunzas** are attractive but unfruitful. Their earliness makes the flowers useful to bees and beneficial insects.

Pruning and Training

The least work is to grow apricots as trees, only removing dead and diseased wood as necessary. Cut out all the dieback until no discoloration is seen. The wood is brittle, so watch for overladen branches and prop or prune early. On walls, build a fan of old wood with fruiting spurs. Prune the frame and any dieback in late winter, then prune again in summer to restrict the growth.

Pests and Diseases

Apricots are not particularly bothered by pests. Ants may introduce and farm scale insects and occasionally caterpillars and aphids may be seen. Apricots do

Apricots grown in cool climates are still delicious

APRICOTS

New Large Early

suffer from dieback and gummosis, however, the twigs dying back and resiny gum oozing out of the branches or, worse, the trunk. Both these conditions are symptomatic of poor growth and every effort should be made to improve the conditions; fewer weeds, more compost, more mulches, more water or better aerated roots, seaweed sprays and hard pruning should do it.

Propagation
Apricot trees are obtained budded on to suitable rootstocks, such as **St Julien A** in the UK. This is better for wetter, heavier soils; seedling peach or apricot rootstocks suit lighter, drier ones. Successful trees have been raised from stones.

Maintenance
Spring Cover flowers on frosty nights; weed, mulch and spray with seaweed solution monthly.
Summer Thin the fruits early, summer prune, bag fruits against wasps and birds.
Autumn Keep weeded, tie in new shoots.
Winter Protect from very hard frosts, prune out dieback.

Harvesting and Storing
Apricots ripened on the tree are heavenly, but soon go over. Picked young enough to travel, they never develop full flavour. Thus they are best eaten straight away, or jammed or frozen.

culinary

Apricots make scrumptious jam and can be preserved in brandy and syrup. Unlike the majority of *Prunus* fruits, most apricots have kernels that are sweet and edible and can be used to make ratafia biscuits.

Apricot Sponge
Serves 6–8

500g apricots
Knob of butter
Coarse, light brown sugar

For the sponge:
55g each of butter, fine, light brown sugar, white self-raising flour
1 large egg
A splash of milk
1 teaspoon real vanilla extract

Halve and stone the apricots (score or remove the skin if you prefer) and place in a buttered flan dish, cut side up. Sprinkle with coarse sugar. Cream together the butter and sugar, stir in the egg, then fold in the flour, together with the milk, if the mixture is stiff, and the vanilla extract. Spoon the sponge mixture over the apricots and cook in a preheated oven at 190°C/375°F/gas mark 5 for about half an hour. When cold, turn out and serve with whipped cream.

companion planting

Do not grow tomatoes, potatoes or oats near apricots, but they benefit from Alliums nearby, especially garlic and chives.

other uses

The wood is brittle, but of use as kindling.

Apricot halves freeze easily

Prunus persica from the family *Rosaceae*

PEACHES

Tree/bush up to 5 x 5m. Generally short-lived. Self-fertile, deciduous. Fruits: 6–8cm, yellow, orange, red. Value: rich in vitamin A, potassium and niacin.

Peach trees are small and resemble willows, with long, lightly serrated leaves. They bloom before the leaves appear. The smaller the flowers, the darker rose coloured they are, with the largest being lightest pink. Flowering a fortnight later than almonds, they are closely related, the almond having a tough, inedible, leathery skin over a smooth stone. The skin has a partition line along which it easily splits. Most peach stones are ribbed or perforated with small holes in the shell. In varieties known as clingstones the flesh clings to this shell; in others, the freestones, the fruit is free and easier to enjoy without having to tease it off the stone. The flesh may vary from white to yellow; there are even blood peaches with red staining. The texture varies with cultivar, soil and climate. The skin colour can be from dull green through yellows and orange to dark red. The most distinctive feature of peaches is the soft downy fluff on the skin.

The peach was known three hundred years BC to the Greek philosopher Theophrastus, who thought it came from Persia, and so it became named. Early Hebrew writings make no reference to it, nor is there a Sanskrit name, so it seems likely peaches did not reach Europe to any extent until shortly before the Christian period. Dioscorides mentions the peach during the first century, as does Pliny, who states that the Romans had imported it only recently from Persia. Peaches are, in fact, of Chinese origin. They are mentioned in the books of Confucius from the first century BC, and can be traced back to the tenth century BC in artistic representations. The Chinese still have an immense number of varieties and these were initially spread by seed. The stones produce trees with ease, but, of course, do not come true. The variability

may have accounted for the slow spread to Europe. However, this was more likely to have been because peaches were initially tried in hot countries at too-low altitudes. Thus the trees did not get their winter dormancy and would have fruited badly, effectively discouraging further experiments. Pliny indeed mentions that peach trees were taken from Egypt to the island of Rhodes, but this transplantation did not succeed; they were then brought on to Italy.

It took till the middle of the sixteenth century for peaches to reach England and in 1629 a quantity of peach stones was ordered by the Governor of the Massachusetts Bay Colony of New England. The peach found its new home in North America highly suitable and spread abundantly. In fact, it spread so rapidly through the wild that it was thought to be a native fruit. Peaches spread so widely that they had conquered much of South America by Darwin's time. He spotted them on islands in the mouth of the Paraná, along with thickets of similarly fugitive orange trees.

PEACHES

varieties

Amsden June is early with greenish-white flesh, semi-freestone and a month behind its name. **Duke of York** has better-flavoured, creamy-white flesh and is also semi-freestone. **Hale's Early** has pale yellow flesh and is freestone. **Peregrine** is the best choice for a bush tree in the UK and has yellowish-white flesh of excellent flavour. It is freestone and comes mid-season. **Rochester** is my second favourite for a bush; it is not quite so well-flavoured and yellow-fleshed. **Royal George** comes late, has tasty, big, yellow fruits, flushed dark red with pale, yellowish-white flesh and a small free stone, but is prone to mildew. **Saturn(e)**, a Peento or Honey Peach from China, has a flattened fruit, very sweet and juicy and is especially suited for container growing. **Bonanza** is a dwarf peach with full-size fruits, well suited to patio and container culture. A very new introduction is **Avalon Pride**, a seedling found near Seattle in the USA, the first leaf curl-resistant variety. Also available are: **Bellegarde**, **Alexandra Noblesse**, **Barrington** and **Dymond**.

Peaches ideally need a well-enriched, well-aerated, but moist piece of soil. They prefer open gravelly soils to heavy, and need to be planted at least 6m apart. If they take (peaches do not always), they establish quickly and need no staking after the first year. They need copious quantities of compost annually and mulches are obligatory to ensure the constant moisture they demand. Where they are planted against walls, great care must be taken to ensure adequate water constantly throughout the season or the fruits will split. However, they are most intolerant of any waterlogging. Peaches also should not be planted near to almonds, as the two fruits may hybridise, resulting in bitter nuts.

Growing under Glass and in Containers

Peaches are often grown under glass, where the extra efforts of replenishment pruning and tying in are repaid by gorgeous, succulent, early fruits free from the depredations of birds. The greenhouse must be unheated in winter to give peaches a dormant rest period. More problems occur under cover – the red spider mite can be especially troublesome unless a high humidity is maintained. Peaches are good subjects for large pots as they can take heavy pruning if well fed and watered. Pots enable them to be kept under cover during the winter and through flowering and then brought out all summer, thus avoiding peach leaf curl and frost damage. The flowers must be protected from frosts and so must the young fruitlets. The flowers are more susceptible to frost damage after pollination and the fruitlets likewise for a further fortnight.

It's worth thinning Peento or flat peaches to get bigger fruits

One – or better two – of these must go, or all will be small

Ornamental and Wildlife Value

The peach is a pleasure to have. The willowy leaves are held well into autumn and the sight of a good crop of fruits is magnificent. The blossom is wonderful too; peaches in bloom are a joy. The wildlife value of the tree is rather low in wet areas as the trees succumb to leaf curl, but in some drier areas peaches become weeds and are appreciated by birds, insects and rodents.

Thinning to wide spacing gives bigger, more luscious fruits

Pruning and Training
Peaches fruit on young shoots, thus it is essential to have plenty of these growths. They are best obtained by a partial pollarding operation late each winter. This makes the pruning more akin to that of blackcurrants than to that of most other tree fruits. Basically the top ends of the higher branches are removed to encourage prolific growth from the lower branches and stubs. This also serves to keep the peach bushes lower and more manageable.

On walls and under cover, peaches are usually fan-trained. Selected young shoots are allowed to spring from a main frame and then tied in to replace the previous growths, once those have fruited.

Fortunately, if the pruning of peaches is temporarily neglected, healthy bushes respond to being cut back hard by throwing plentiful young growths. More important than pruning is thinning. Peaches are prone to overcropping, breaking branches and exhausting themselves. Thin the fruits hard, removing those touching or anywhere near each other. Do this very early and then again later.

Maintenance
Spring Protect tree from leaf disease with Bordeaux sprays before buds open and blossom. Protect blossoms from frost. Hand-pollinate. Weed, mulch and spray with seaweed solution monthly.
Summer Thin the fruits early, and often. Protect fruits from birds and wasps.
Autumn Remove mummified fruits from the trees.
Winter Prune hard. Spray with Bordeaux mixture.

Propagation
Peaches can be raised from stones, but the results are haphazard and take years to fruit. Budded on to suitable stocks, **St Julien A** in the UK, for example, peaches will normally fruit in their third year. Plum stocks are more resistant to wet, but for warmer, drier conditions seedling apricot or peach rootstocks are better.

Pests and Diseases
Peaches suffer most losses from birds and wasps. Small net or muslin bags will protect the crop. Earwigs can get inside the fruits and eat the kernel out, but are readily trapped in rolls of corrugated paper around each branch. Protect the bark from animals such as rabbits and deer. Peaches may get minor diseases such as peach scab, but their main problem is peach leaf curl. This puckers and turns the leaves red and yellow and they cease to function properly. Severe attacks cripple the tree and can even kill it. Keeping the plant dry under cover or under a plastic sheet prevents the disease almost completely. Spraying with Bordeaux mixture prevents the disease if done several times as the buds are opening in late winter. Dieback and gummosis are symptomatic of poor growth and are best treated by heavy mulching and hard pruning in very late winter.

Harvesting and Storing
A truly ripe peach is a bag of syrup waiting to burst. If picked under-ripe the flesh never develops the full gamut of flavour, or the liquidity. A good peach is a feast, drink and all. As with so many fruits, they are best eaten straight off the tree. They can be picked a few days early if handled with absolute care. Kept cool, they may last. The slightest bruising, however, and they decompose.

companion planting

Peaches are benefited by Alliums, especially garlic and chives. Clover or alfalfa leguminous green manures give the richness they need. Nettles nearby are reputedly helpful at preventing the fruit from moulding.

other uses

Peach stones are used for making activated charcoal for filters. The wood is brittle, but the prunings make good kindling. In some countries, gluts of peaches are used for feeding the local livestock.

Pick just before they drop, peel and eat sun-warm

PEACHES

Home-grown peach preserves are exotic fare for the table

culinary

Peach jam, which is potentially more aromatic than plum, is easily over-cooked and the perfume lost. Make it set more easily and sooner by adding apple purée. Fruits that are ripening but starting to rot are best prepared then cooked in syrup. This, if not consumed immediately, can be frozen for winter use. Over-ripe fruits can be juiced, which is the nectar of the gods, and unripe peaches that never look like ripening make excellent chutney.

Peach Macaroon Cheesecake

Serves 4–6

125g macaroons
125g digestive biscuits
100g butter, melted
1kg peaches
15g gelatine
250g cream cheese
125g natural yogurt
100g honey or peach jam
Few drops vanilla essence
125g alpine strawberries

Crush the macaroons and biscuits and mix with the butter. Press into a flan dish. Peel, slice and chill the peaches. Simmer the skins and stones in minimal hot water, sieve and dissolve the gelatine in the warm liquid. Beat the cream cheese, yogurt, honey or jam and vanilla, then stir in the gelatine. Immediately pour over the biscuit base, chill and leave to set. Top with peach slices and alpine strawberries.

Prunus persica from the family *Rosaceae*

NECTARINES

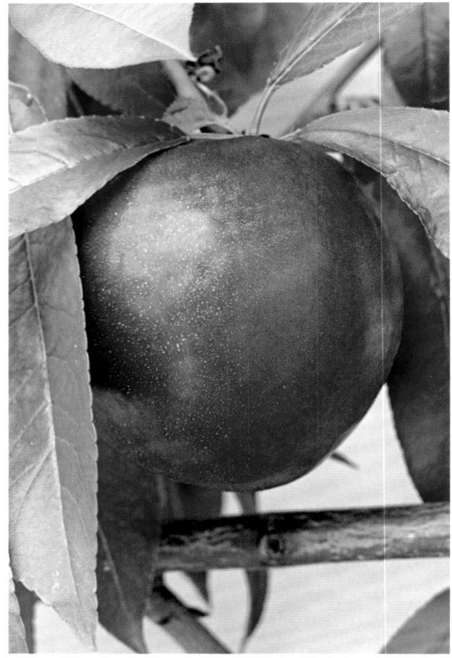

Early Rivers

Tree up to 5m. Generally short-lived, self-fertile, deciduous. Fruits: up to 8cm, greenish yellow/orange and red. Value: rich in vitamin A, potassium and niacin, as well as riboflavin and vitamin C.

In almost every way nectarines are just varieties of peach. However, there are several subtle and fascinating differences. Nectarines are more difficult to grow and are less hardy. The fruits are smaller, on average, than peaches. The flesh is firmer and less melting than a peach and more plum-like, less prone to falling apart while you eat it. Nectarines have a definite, almost peculiar, rich, vinous flavour quite distinct and in addition to that of a well-ripened peach. The colour of some nectarines also makes them distinguishable from peaches, as many of the older varieties have a greenish or sometimes even a purplish hue over a quite yellowish or greenish ground. Most noticeably, nectarines do not have that downy fuzz on the skin so typical of peaches, but instead are smooth and shiny; indeed they closely resemble a very large, plump plum.

Darwin noted how peach trees occasionally spontaneously produced nectarines, and also the converse; he even noted the case of a nectarine tree that produced a fruit that was half peach, half nectarine and then reverted to peaches. Despite the peach's long history of cultivation, however, rather strangely, no mention is made of nectarines by pre-Christian authors. Pliny mentions an unknown fruit, a duracina, but the first, if indirect, reference is by Cieza de Leon, who lived in the early sixteenth century and described a Caymito of Peru as being 'large as a nectarine'. They were seen growing amongst peaches in Virginia in 1720 and A. J. Downing listed 19 varieties in the USA by 1857. Many dozens of nectarine varieties are now in cultivation, and travellers also report local varieties of nectarine in most of the world's peach-growing areas, so their spontaneous emergence is not really a rare phenomenon.

varieties

Early Rivers, white-fleshed is the earliest, just before **John Rivers**, then comes **Lord Napier** a tasty white-fleshed mid-season and the most common. These are followed by **Elruge** and the new and very popular **Queen Giant**. **Pineapple** is too late and better container grown under cover. **Nectarella** is a new dwarf variety ideal for growing in containers on warm patios.

cultivation

Nectarines need even better conditions than peaches do. Good water control is critical to prevent the fruit splitting, so thick mulches are obligatory. They are less likely to set crops as bushes than peaches, so in any cool region they must be given a wall or extra shelter.

Growing under Glass and in Containers

Nectarines are more suited than many plants to culture under glass, preferring warmer conditions than peaches. However, they need a cool period of rest each winter or they become unfruitful, so they are best grown in large pots moved outside after cropping and brought back into a cool greenhouse in mid-to-late winter. Being dry under cover prevents peach leaf curl, keeps the frost off the flowers and fruitlets and gets them an early start. But they will only fruit successfully if religious attention is given to watering and ventilation. Hand pollination is desirable.

NECTARINES

Pruning and Training
Exactly as for the peach (see page 472).

Propagation
Nectarines sometimes come from peach stones and vice versa. More often they occur when a peach bud produces a sport. Most varieties are budded on suitable stocks as for peaches.

Pests and Diseases
In the USA nectarines are not much grown away from the Pacific coast as a common pest, plum curculios, can easily do much damage to their smooth skins. Spray against peach leaf curl.

Ornamental and Wildlife Value
Nectarines are almost exactly the same as peaches and are attractive, with the long, glossy leaves being held late into the season. The pink flowers are beautiful and appreciated by early insects; the lack of fruits reduces their wildlife interest. Nectarines need spraying to prevent peach leaf curl in many areas, so cannot be left to themselves.

Maintenance
As for the peach (see page 472).

Harvesting and Storing
Also as for the peach (see page 472).

Lord Napier

culinary

Nectarines can be used in similar ways to peaches, but as their skin is not fuzzy it is less of a barrier and makes them more toothsome to eat fresh. The firm flesh is also less melting and reduces one's need for a bib to catch half the juice as with a ripe peach. This makes the nectarine a civilised fruit for eating at the dinner table, and this is only enhanced by the succulent vinous flavour.

Nectarine Melba
(all quantities according to taste)

Nectarines, honey, vanilla pod
Vanilla ice-cream (the real sort!)
Frozen raspberries
Bitter chocolate, nutmeg

Halve, peel and stone, say, one ripe nectarine per person. Poach half of them, those that are the least decomposingly ripe, in a syrup made by gently warming the vanilla pod in the honey. Poach the nectarines very gently until they are tender but not breaking down, then drain and chill well. Create a bed of vanilla ice-cream, interspersed with the chilled ripest halves. Lay the poached halves on this base. Crush the frozen raspberries in the drained honey syrup. Chill, then pour over the nectarines, followed by a generous grating of bitter chocolate and a hint of nutmeg.

companion planting

Again as for the peach.

other uses

As for peaches.

Prunus avium from the family *Rosaceae*

SWEET CHERRIES

Tree up to 10m. Long-lived, deciduous, not self-fertile. Fruits: 2cm, yellow, red, black, rich in riboflavin

Although described as sweet cherries, some varieties, and particularly wild ones, are not actually sweet. They are tall trees, often reaching over 9m, with white blossoms, occasionally pink, which are a massive display at about the same time as the peach. The leaves are plum-like with a lengthening and thinning at the tip. Dangling in pairs on long pedicels, the near-spherical fruits hang in groups along the fruiting branches. The flesh is cream or yellow, sweet or bitter, but once ripe is rarely acid.

The wild form native to Europe, known as the bird cherry, gean or mazzard, is of little value as fruit except in Central Europe, where it is used for liqueurs. Mazzards, little improved from the wild version, still survive and have richly flavoured black fruits only slightly larger than the wild. They make good seedling stocks for better varieties, though if left to themselves they can grow to 20m or more high. Gean fruits were thought to have softer more melting flesh and varieties called Bigarreau had a crisp texture. Years of continuous selection and cross breeding have given us sweet cherries of mixed parentage. One group, the Duke or Royal cherries, have some sour cherry (Morello) blood, which serves to make them tasty and lightly acid as well as generally more hardy.

varieties

There are countless varieties new and old but finding combinations compatible for pollination is difficult so the most useful cherries for small gardens are at least partly self-fertile. One is **May Duke** which is very early, though not May, more like mid-June; it may even fruit (later) on a north wall. **Merton Glory**, a red-flushed, white-fruited variety with very tasty flesh, ripens later in June. **Stella** is a vigorous and upright tree with sweet dark red fruits in early July. **Sunburst** is another new cherry with very dark red sweet cherries in mid-July. **Summer Sun** is probably the best sweet cherry for tougher positions; it has dark red cherries in mid-July. Older varieties need matching: **Governor Wood** and **Napoleon Bigarreau** are my favourite mutually compatible pair. The former has red fruits a week or so before the yellow of the latter, which makes the bigger tree. One alternative is that almost any late-flowering cherry or cherries will be pollinated by a **Morello** sour cherry planted with them.

cultivation

Sweet cherries are very particular to soil and site. They need plentiful moisture at the roots but loathe waterlogging and need a very rich but well-aerated soil. Poor soil conditions can be helped by growing through grass, as was the traditional manner.

Merton Glory

SWEET CHERRIES

Sweet cherries rarely ripen unless protected from birds

Growing under Glass and in Containers
As the trees get so big and are particular to site and weather, pot cultivation is the most effective way to get cherries. In this way you avoid damage to the flowers from frost or rain and likewise protect the fruits. Do let the trees have a cool dormant period in winter.

Ornamental and Wildlife Value
Sweet cherries are quite staggering to behold in blossom, as if festooned with snow. The fruits are one of the most addictive to birds, and they will come despite all discouragement. The number of pests suggests that cherries must benefit some wildlife immensely.

Propagation
Normally budded on to strong-growing rootstocks, which make for big trees, they are now offered on dwarfing stocks such as **Colt**. This makes them smaller and easier to constrain. Stones grow, but the trees take years to fruit and get tall by then.

Pruning and Training
Prune as little as possible after the initial head formation. Any pruning must be done early during the growing season and then only to remove any dead and diseased growth.

Pests and Diseases
Cherries really suffer from bird damage, but at least escape the wasps by fruiting early in summer. Bacterial canker can kill the tree, and there are several virus diseases, but these are treatable only by improving the conditions. Warm drizzle during flowering can cause the flowers to mould and a sudden heavy rain during ripening invariably succeeds in splitting the fruits.

culinary

Sweet cherries often do not make as good culinary dishes as sour cherries because many lack acidity. Their jams and jellies are better combined with redcurrant or whitecurrant juice for this reason.

Pickled Cherries
Makes about 1.5kg

1kg cherries
500g sugar
300ml vinegar
6 cloves
15g fresh ginger, peeled and chopped
Hint of ground cinnamon

Wash and stone (if desired) the cherries. Dissolve the sugar in the heated vinegar and add the spices. Simmer the cherries in the spiced vinegar for a few minutes, then pack them into warmed jars, and cover with the vinegar. Seal and store.

Eat with pâtés, cold meats and especially quiches.

Cherry Jam
2.25kg cherries
3 lemons, squeezed
1.25kg sugar

Wash the fruit thoroughly; remove stalks and stones. Put the fruit into a pan with the lemon juice and simmer gently for 30–35 minutes. Warm the sugar and add it to the cherries over a gentle heat and allow the sugar to dissolve. Then bring the jam to a rapid boil and continue boiling until setting point is reached; this will take about 15 minutes. Test for setting point, remove from the heat and leave for 5 minutes. Pot and cover. Store in a dry place.

Maintenance
Spring Protect the flowers, weed, mulch and spray seaweed solution monthly. Prune if necessary.
Summer Protect the fruits from birds with netting or muslin sleeves.

Harvesting and Storing
Cherries can be picked and kept for several days if absolutely dry. Leave them on the sprigs and do not pack them deep. They can be frozen, but are fiddly as they are better stoned first.

companion planting

They are normally grown in grass sward, include clover and alfalfa, to give more fertility. Cherries supposedly suppress wheat and make potatoes prone to blight.

other uses

Sweet firewood.

Stella

Prunus cerasus from the family *Rosaceae*

MORELLO CHERRIES

Tree up to 8m. Long-lived, deciduous, self-fertile. Fruits: 2cm, crimson to black. Value: rich in vitamin A.

Morello cherries are similar to sweet cherries, except they are not sweet, so are useful only for culinary purposes. The trees are smaller than sweet varieties, with slightly lax, more twiggy branches and greener foliage that does not have such a red tinge when young. The fruits have shorter stalks, tend to have darker colours and are more acid.

Sour cherries were selected from *Prunus cerasus* (also known as *P. acida*), which grew wild around the Caspian and Black Seas. In about 300 BC sour cherries were known to the Greek Theophrastus and proved so popular with the Romans, who developed at least half a dozen different varieties, that by the time of Pliny, in the first century AD, sour cherries had already long reached Britain. However, during the Dark Ages, the art of their cultivation was lost and the trees had to be re-introduced to England in the sixteenth century by Henry VIII, who had them brought from Flanders. They were soon adopted by the growers of Kent and by 1640 they had over two dozen varieties. The first cherries to reach the New World, the Kentish Red, were planted by the Massachusetts colonists.

Sour cherries get sweeter – if given time

varieties

Before the Second World War more than 50 varieties of Morello and sour cherry were in cultivation, and almost every country had its own wide choice. Now few are grown commercially and only the generic **Morello** is offered by most suppliers. This last favourite, namesake for all its disappearing brethren, is deep crimson to black in colour with a richly bitter, slightly sweet flavour. Late flowering, it misses more frosts than sweet cherries so is more reliable and ripens in mid- to late summer, towards the end of August in southern England. It is self-fertile, and would pollinate almost any other cherry of any type if the flowering period was not so late in the season.

Red Morellos, **Amarelles** or **Griottes** are closely related, slightly more stubby trees with red fruits that are probably of mixed origin. The **Kentish Red** has endured since at least 1700. It is self-fertile with a scarlet skin, soft flesh and is only slightly bitter. It ripens in midsummer. The **Flemish** is a similar cherry variety. A new, larger-fruited variety, **Crown Morello**, is claimed to have the largest fruits, at half an ounce apiece.

cultivation

Much the same conditions are needed as for pears, with an increased demand for nitrogen and even more

MORELLO CHERRIES

water than needed by sweet cherries. Though the trees will do badly if they are waterlogged, they are more tolerant of poor drainage than sweet cherries.

Growing under Glass and in Containers
Although this would be possible, it would usually be more worthwhile to nurture sweet cherries, especially as Morellos will even crop on a cold wall in cool areas.

Ornamental and Wildlife Value
Not as elegant or floriferous a tree as the sweet cherry, so of less ornamental value. However, Morellos are well-loved by birds and the early flowers are good for insects.

Maintenance
As for sweet cherries (page 477).

Propagation
As for sweet cherries.

Pests and Diseases
Morellos and sour cherries are particularly unbothered by pests or diseases other than losses to birds.

Pruning and Training
Morellos fruit on younger wood than do sweet cherries and thus they can be pruned harder. However, it is usually more convenient to stick to removing dead, diseased and congested growths in spring or summer. Usually grown as standards, they are ideal as low bushes for picking and bird protection, although they can also be trained as fans. They will even crop well on cold walls.

Harvesting and Storing
Cut the cherries off the tree rather than risk damage by pulling the stalks. Morellos were one of the first fruits to be stored frozen and are one of the best. They can be frozen without sugaring and retain their flavour superbly.

companion planting

As for sweet cherries.

other uses

The prunings make good kindling.

culinary

They are primarily a culinary fruit and make fabulous pies, tarts, jams and cakes.

Black Forest Gâteau
Serves 6–8

500g Morello cherries
125g honey
4 tablespoons kirsch
Little water

For the sponge:
100g butter
100g fine light brown sugar
100g white flour
2 large or 3 small eggs
1 heaped teaspoon baking powder
1 heaped teaspoon cocoa powder

For the filling:
600ml double cream
1 heaped teaspoon sugar
1 heaped teaspoon cocoa powder
1 teaspoon vanilla extract
100g dark chocolate, grated

Wash and stone the cherries. Dissolve the honey and kirsch in sufficient hot water just to cover the stoned fruits. Simmer till they soften, then strain (keeping the syrup) and chill. Mix the sponge ingredients and divide between three oiled or lined sandwich tins. Bake in a preheated oven at 170°C/325°F/gas mark 3 for about half an hour. Turn out and cool, then soak the sponge cakes in the syrup. Whip the cream, then stir in the sugar, cocoa powder and vanilla. Build up alternate layers of sponge, cream filling and fruit. Finish off with a covering of filling and grated chocolate.

Morus nigra and *Morus alba* from the family *Moraceae*

MULBERRIES

Tree up to 9m. Long-lived, hardy, deciduous, usually self-fertile. Fruits: 2–3cm, purple, also red through pink.

Mulberry trees are medium-sized trees with a round domed crown, formed by their habit of having no terminal bud on their overwintering twigs. They can become big, gnarled and picturesque in advanced old age. The leaves are heart-shaped, toothed and occasionally lobed, lightly downy underneath but rough on top. The black mulberry, *Morus nigra*, fruits are like loganberries, of a blackish dark red through purple when ripe and stain all they touch. The white mulberry, *Morus alba*, fruits are red or pinkish white, and slightly inferior to eat. There is a red mulberry (which can be almost black), native to North America, and several similar species in hotter climates.

Black mulberries are thought to be native to western Asia. Known at least since the time of the Greeks, they failed to become popular until the Roman Emperor Justinian encouraged them for the production of silk. The trees appeared all over the empire and *M. nigra* was not superseded by *M. alba*, the more productive variety for feeding silkworms, until the sixteenth century. Exceptionally long-lived, the black mulberry may attain some stature; giants found in many parts of the world were planted during the seventeenth and early eighteenth centuries in vain attempts to foster local silk industries. Some mulberries, such as *M. alba* var. *tatarica*, the Russian mulberry, seed freely and become weeds. The red mulberry was native to the Northern Missouri region and along the Kansas river system, where it was an esteemed fruit.

MULBERRIES

varieties

Chelsea may be identical with **King James I** and was once the only available variety, though now joined by the remarkably similar **Charlton House** and **Large Black**. The **Red Mulberry** (*M. rubra*) has red/black fruits and lovely autumn colour with bright yellow leaves; it is said to have the preferable fruit. The **White Mulberry** (*M. alba*) has wider leaves, up to 15cm across, and is grown for silkworm fodder, though the fruit is quite edible, often red or pinkish and sweet. *M. alba* is also popular as an ornamental in many varieties. Cultivated mulberries are mostly self-fertile with unisexual flowers, though the species tend to be dioecious when growing in the wild.

cultivation

Mulberries succeed in any well-drained moist soil, but the roots are brittle so these need care during planting, which is best done in early spring. They prefer a warm site, as typical of southern Britain, to fruit happily in the open, and are rarely bothered by late frost damage as they are tardy coming into leaf and flower. Normally they are grown in a fine grass sward to enable the fruit to be collected reasonably cleanly as it drops from the tree.

Growing under Glass and in Containers
This may be feasible, but I've not heard or read of it.

White mulberries

culinary

The fruits can be made into jams or jellies, though they are usually combined and bulked out with apples, and the wine is an old country favourite.

Mulberry No Fool
Per person

Clotted or thick fresh double cream
Some honey
A pot of tea
Biscuits
A mulberry tree in fruit and a sunny day off

Take a bowl with a large portion of clotted or thick cream, honey and the tea and biscuits. Sit peacefully under the mulberry tree, savouring the tea and biscuits while waiting for enough fruits to fall to fill your bowl. Eat them with the cream and honey. In emergencies, tinned or fresh fruit of any sort and a patio umbrella can be usefully substituted for the mulberries and tree!

Ornamental and Wildlife Value
Mulberries become gnarled and attractively grotesque as they age. The traditional situation is in the middle of a lawn with a circular wooden seat around the trunk. The berries are much appreciated by birds and small children.

Pests and Diseases
No significant pests or diseases bother the trees, other than the usual hazards of birds and small children.

Maintenance
Spring Cut grass and spray seaweed solution monthly.
Summer Pick the fruits as they ripen and drop.
Autumn Cut grass and enjoy the seat occasionally.
Winter Thin and remove dead and diseased growths.

Propagation
25–30cm cuttings of newly ripened growth with a heel taken in December are ideal. Layering is possible as are (supposedly) whole branch cuttings.

Pruning and Training
Mulberries can be pollarded hard to give the maximum foliage needed for feeding silkworms. If left to them-selves, they make congested heads, so thin out twiggy, dead and diseased growths. Their narrow forks, big heads and brittle wood mean old trees should be carefully inspected and excess weight removed by a competent tree surgeon.

Harvesting and Storing
Mulberries are aromatic and mouth-watering when fresh and ripe, but decompose rapidly, so are really only of use as they drop, and will not store or travel, though they can be frozen.

companion planting

Mulberries must be grown in a grass sward if there is to be any hope of getting clean fruit. They were one of the traditional trees to support grapevines in Classical times.

other uses

Of course you can try silkworm production and the foliage is palatable to other animals. The fruits will certainly make dye!

Ficus carica **from the family** *Moraceae*

FIGS

Brown Turkey

Tree/bush up to 9 x 7m. Hardy, self-fertile. Fruits: pear-shaped, 6 x 10cm, browny green.

Fig leaves are large and distinctive, but vary in exact shape with each variety. Although figs are deciduous, young plants tend to be almost evergreen and are then more tender. Figs can be grown as trees or bushes, but are most often trained on walls. Almost all fig cultivars set the fruit parthenocarpically, without actual fertilisation. However, some old and mostly inferior varieties in hot climates have female fruiting plants and male caprifigs that are separately and specifically cultivated to produce female fig wasps. These wasps will crawl through the minute end hole into the fruits to pollinate them.

Figs are indigenous to Asia Minor and were one of the first fruits to be brought into cultivation. They thus became an intrinsic part of the diet of the Mediterranean basin long before Classical times, though the Greeks claimed they were given to them by the goddess Demeter. Cato knew of six different figs, and two centuries later, in about 60 AD, Pliny notes no fewer than 29 varieties! Figs were certainly brought to England by the Romans as remains have been found, though the plants were not officially introduced until the early sixteenth century. There are more than 600 fig species. Many of the varieties that we encounter today are ornamentals such as the India Rubber Plant (*Ficus elastica*), the Weeping Fig (*F. Benjamina*) and the Fiddle Leaf Plant (*F. lyrata*).

varieties

In the UK the main crop of figs ripens from late August, but fruits that make it safely through the winter may crop in late June or July. **Brunswick** produces large fruits that are most tasty a couple of days after picking; **Brown Turkey** is reliably prolific; **White Marseilles** has pale fruits. Also: **Angelique**, **Black Ischia**, **Bourjasotte Grise**, **Castle Kennedy**, **Negro Large**, **Osborne's Prolific**, **Rouge de Bordeaux**, **St John's**, **Violette Dauphine**, **Violette Sepor**, **White Ischia** and, in the USA, **Adriatic**, **Celeste**, **Kadota, King**, **Magnolia (Brunswick)** and **Mission**. **Violetta**, a new Bavarian-bred variety, is the most cold resistant and therefore probably the best for tougher positions.

cultivation

Varieties vary in hardiness, but even if the tops are lost many will regrow from the roots if these are protected from the frost. Thus figs may be planted a little deep to encourage stooling. Often grown against a wall, they can be cropped in the open in southern Britain, their main enemy being the excessive wet, rather than the cold. It is traditional to confine the roots of figs grown against walls to promote fruiting. This appears unnecessary if excessive feeding is avoided, but too much nitrogen will promote abundant undesirable soft growth.

You rarely see them this big in the shops....

FIGS

.... Okay, well, sometimes you do

 culinary

Figs are delicious fresh and they can be made into jams, jellies, cheeses and chutneys.

Savoury and Sweet Figs
Serves 4 as a snack or canapé

12 dried figs
90g mild hard cheese, broken into chunks
Pinch of celery seeds
1 shallot, thinly sliced
25g dark cooking chocolate
50g marzipan
25g each sultanas and raisins
A little Cointreau or sweet liqueur
Honey

Stuff half the figs with small chunks of cheese, a sprinkling of celery seeds and a thin slice of shallot.

Grate the chocolate and marzipan. Mix in the dried fruit and liqueur, then stuff the remaining figs with this mixture. Smear all the figs with honey and put them in an oiled dish in a preheated oven at 180°C/ 350°F/gas mark 4 for 10–15 minutes. Serve the figs hot or cold.

Growing under Glass and in Containers
Figs fruit and ripen much more reliably under glass, and with heat and care three crops a year are possible. Figs can be grown confined in large pots and can make excellent foliage plants for house or patio, though they will need careful watering and training.

Ornamental and Wildlife Value
The very attractive foliage adds a luxuriant touch to a garden. Of little use to wildlife in colder climes, figs are much more valuable in hotter countries.

Maintenance
Spring Remove frost protection, weed, spray seaweed solution monthly, and thin the fruits.
Summer Protect the fruits carefully from wasps and birds; make layers.
Autumn Remove all fruits of all sizes before the first frosts arrive.
Winter Prune, take cuttings, protect from frost.

Propagation
Layers can be made during summer. 20cm well-ripened or old wood cuttings taken during the early winter will root, especially if given bottom heat. Seeds produce plants that are unlikely to fruit well.

Pruning and Training
The most fruitful wood is well-ripened, short-jointed and sturdy; long soft shoots are unproductive and better removed. Figs can be pruned any time during dormancy, but are best left until growth is about to start in spring. More importantly, remove any fruit or fruitlet in autumn to prevent these starting into growth, failing and thereby spoiling the second crop that could otherwise succeed.

Pests and Diseases
Fig plants have few problems except birds and wasps. Weeding must be thorough near the trunks to prevent rodents nibbling the bark. Under cover, on walls or in pots they may suffer from red spider mites.

Harvesting and Storing
The fruits are soft-skinned and do not travel or store well once ripe. In hotter regions figs can be dried and will then keep well.

 companion planting

One of the few plants to get on well with rue.

 other uses

Figs are known by many for their syrup, administered as a laxative. Also valuable as a food, they contain nearly half their weight in sugar when dried.

A promising crop of **Rouge de Bordeaux** – needs thinning out with the odd one!

SOFT, BUSH AND CANE FRUITS

I prefer to group these fruits as fruitcage fruits because this accurately describes their most common factor: without a cage most of them cannot effectively produce fruits at all. The reasons for their other names are fairly self-evident: soft, because they are not hard fruits such as apples, bush is self-descriptive and cane refers to the slender branches. (The term vine is applied differently depending on region; in some places it is used solely for grapes, in others to plants with long, flexible branches, even Cucurbits.)

The fruitcage members are grouped together by the gardener because they must be netted or grown in a cage. However, as with orchard fruits, most of them come from the *Rosaceae* family. They are shrubby perennials, not tree-like, and thus fairly compact, or can be kept so, quick to crop, and small-fruited. Most are found growing naturally on the woodland's edge and thrive given moist root runs with plentiful humus-rich mould and thick mulches of leafy material. Most will grow and even crop in light shade, though, of course, they usually produce sweeter, better-tasting fruit given more sun. Grapevines will most obviously require much more sun than do the other plants of this group.

The majority of these fruits have been gathered from the wild native species since prehistory and their cultivation will have occurred inadvertently around sites of human habitation, from waste heaps and primitive latrine arrangements. (Both, of course, afford remarkably well-fertilised ground, and thus select for strains that could use such conditions.) Being fast-growing plants and quick to crop, the proximity and opportunity would have created more, but still inadvertent, selection. Such a process would have produced much improved cultivars, spread by migrating peoples, and these may well have influenced wild populations. Today, some wild fruits such as the blackberry are just such hybrids in most areas, and it may well be that some supposed native species are in reality escaped cultivars from our distant past.

Thus the cropping potential of some wild fruits may have been raised over the millennia by our own unintended selection. In any case, the wild forms of most of these fruits were so productive and available in such vast quantities that they were simply not brought into cultivation until the end of the Middle Ages, with the important exception of grapes, which have been tended since prehistory. Some, such as the strawberry, have undergone intense development and hybridisation and so have improved dramatically, while others, such as aronia, have remained almost unaltered. Many closely related edible species are available, which could be crossed with cultivated forms, so improved fruits may be only an experiment or two away. The Tayberry is the result of one such crossing, and is better by far than all its similar predecessors.

Fragaria hybrids from the family *Rosaceae*

SUMMER STRAWBERRIES

Herbaceous. 30cm. Life span: short. Self-fertile. Fruits: up to 5x5cm, red, conical. Value: some vitamin C and half as much iron as spinach.

This most delicious of fruits really needs no introduction! Modern strawberries are hybrids based on *F. chiloensis*, the Chilean Pine, and *F. virginiana*, the Scarlet Virginian. The former contributed larger fruit with a pineapple tang and was brought to Europe in 1712; the latter, first mentioned in Massachusetts in 1621, provides the superior flavour, and is still grown commercially as Little Scarlet for jam-making. These two were first combined in the nineteenth century, with intensive development since. Every region has its favourite varieties.

varieties

There are far too many to list and new improved ones are continually becoming available. My favourites are: **Royal Sovereign** (mid-season), beautifully flavoured if light cropping, originally introduced in 1892; **Silver Jubilee** (mid) is tasty with some disease resistance; **Cambridge Vigour** (early) is a commercial variety, but has rich flavour; **Late Pine** is also simply delicious. **Mae** is exceptionally early and earlier under cloches. **Gariguette** is a French strawberry from the 1930s, small but very well flavoured and early. **Alice**, mid-season, and **Florence**, mid-season-late, **Pegasus**, late, are all modern with good pest and disease resistance and fair taste. **Cambridge Favourite** is remarkably popular with a sweet flavour and reliable habit. **Marshmello**, mid-season, has superb flavour. **Maxim**, late, produces very big fruits, which are surprisingly tasty, and does well in dry conditions. **Sonata** is said to be the most weather resistant and reliable cropper of perfect fruits if flavour is not the priority.

cultivation

All preparation is well repaid. Strawberries need a very rich soil, full of humus, and benefit from slow-release phosphates, such as bone meal. Well-rotted manure or compost, and/or seaweed meal dressings, will be recouped with better cropping. The more space you give them, the better they will do and the less work they will be! 60cm each way is the minimum.

Growing under Glass

They do not like the hot dry conditions under cover, and these also encourage red spider mite. However, the advantages of protection from birds, prevention of rain damage and earlier crops mean that they are widely grown under glass and plastic. For winter cropping, the new Californian day-length indeterminate types are far superior.

Growing in Containers

Many containers, often tower shaped, are sold specifically for strawberries. While a good idea, the small amount of root run, hot in an above-ground position, and usually dry because of the low water-holding capacity, make these poor growing conditions for strawberries. The plants respond with low yields, all of which occur in a short period, which makes them neglected for much of the year.

Straw is essential to stop such low berries getting muddy

SUMMER STRAWBERRIES

Extra early crops indoors planted in the tubs of other fruits such as citrus and grapevines

However, strawberries can be grown successfully in pots and containers. They need regular feeding as well as copious automatic watering!

Ornamental and Wildlife Value
Rather too straggly to be considered decorative, they would make effective but short-lived ground cover and would not be very productive. By the losses, one must assume that they are valuable to birds!

Maintenance
Spring Weed, root first runners, spray seaweed monthly.
Summer Weed, straw, protect from birds and mould; pick.
Autumn Establish new plants, tidy up the best, eradicate old.
Winter Rake up old leaves and mulch and compost them well.

Propagation
Some special varieties can be grown from seed, but otherwise seed generally produces poor results. Runners are the obvious replacement, usually available to excess. These are best taken from quality plants that have been reserved for propagation and deflowered; half a dozen good ones, or many more poor, can be had thus. The first plantlet on an early runner is chosen, pinned or held to the ground – or preferably rooted into a pot of compost for an easier transplantation. Start new beds in late summer or early autumn to allow them to establish; they can then crop well the next summer. Late autumn or spring plantings should be deflowered the first summer to build up their strength for a massive crop the next.

Pruning and Training
Strawberries will become relatively unproductive after three or four years, so in practice a continual annual replacement of a quarter or a third of the plants is best. Replacing all every four years or so means years of gluts and years of shortages, as they are all the same age. Strawing up is essential for clean crops. Do not straw too early – only when the first green fruits are seen swelling. Remove surplus runners regularly.

Pests and Diseases
Major losses are from birds and mould. Individual bunches can be protected from both with jam jars. Netting is usually obligatory. Mould can be decreased by strawing up and removing fruits that rot, preferably before the mould goes 'fluffy'. After fruiting has finished, tidy the plants, shearing back surplus runners and dead leaves, and in winter tidy them again, removing old straw as well for composting. Aphids spread the dreaded virus diseases, so you should buy in fresh clean stock every ten years or so.

Late Pine

culinary

Wimbledon Fortune
Fresh strawberries, sprinkled with sugar, cream, a shortbread biscuit or two – and a small mortgage, unless enjoyed at home!

Harvesting and Storing
Strawberries must be used quickly as they keep only for a day or so – they will last better picked with a stalk. They can be juiced and made into syrup, or jammed and jellied – the addition of apple, red- or whitecurrant juice will help with the setting. If frozen they will lose their delicate texture, but are still delicious.

companion planting

Before establishing a new bed, dig in a green manure crop of soya beans to help prevent root rots. Growing borage, any of the beans or onions nearby reputedly helps strawberry plants.

other uses

As for perpetual strawberries (see page 489).

Fragaria hybrids from the family *Rosaceae*

STRAWBERRIES
PERPETUAL AND REMONTANT

Herbaceous. 30cm. Life span: short. Self-fertile. Fruits: up to 5 x 2cm, red, conical. Value: some vitamin C and half as much iron as spinach.

The remontant or perpetual strawberries are just as much a mixture of, and very similar to, summer strawberries, except that they fruit continuously through the autumn. In fact, they start to set early crops even before the summer varieties, but these are normally removed to ensure bigger crops later, when the summer crowd has finished. Some have extremely good flavour, probably due to some *F. vesca* ancestry.

This particular group of strawberries has been improved far more in Continental Europe than in America or the UK. Their development has been parallel to that of the summer strawberries, but with different mixes of species used in the hybridisation. Most of the strawberry family are greatly affected by the day length, and this group perhaps most of all. Most strawberry varieties respond to a change from their native latitude by producing primarily runners if they move north and fruits if they move south. It would thus seem that any variety would be a perpetual fruiter if cultivated at the 'right' latitude, and that perpetual fruiters are really just summer croppers enjoying a warmer latitude than nature intended.

varieties

My favourite and one of the finest flavoured is **Aromel**; this also does well under cover. **Mara des Bois** is a new French variety and is extremely well flavoured and crops well. Both of these varieties runner well. **La Sans Rivale**, an old French variety, and **Hampshire Maid** produce few runners. The Dutch have some runner-forming, climbing, remontant varieties. (They do not climb, but have to be pushed!) **Ostara** and **Rabunda** are good autumn croppers, especially if the first flowers are disbudded in the early summer months. **Flamenco,** a new supermarket everbearing variety that is very sweet, has good flavour and is reliable.

cultivation

Remontants need the same rich, moist conditions as summer strawberries. As they crop for longer and can give higher total yields they deserve even better treatment.

Growing under Glass
These are better value grown under glass than summer strawberries, as they use the space productively for longer. They benefit most from such cover at the end of autumn. They must have good ventilation or they mould rapidly, and they are prone to red spider mite attacks. Either use the commercial predators or suffer.

Growing in Containers
These perpetuals are often recommended for growing in containers, as they are productive for far longer than summer fruiters and make better use of limited space. However, they will do much better in the ground, especially if erratic watering is a possibility!

Ornamental and Wildlife Value
Climbing varieties can be trained well over trellis and can also be quite decorative, especially when in fruit. Their long flowering period makes these plants valuable to insects, and the fruits themselves are enjoyed by both birds and rodents!

Maintenance
Spring Weed, mulch, spray with seaweed solution and deflower.
Summer Deflower, root some runners, remove rest.

STRAWBERRIES

Autumn Straw up and pick the fruits.
Winter Tidy bed, de-runner, plant out in late winter.

Propagation

The perpetual fruited varieties cannot be reliably propagated by seed as they are such mixed hybrids. I have tried! Runners can be rooted in pots in summer and detached. They crop significantly more easily the first year than summer fruiters. This is because when planted out they have longer to establish before the fruiting commences, whether this happens in the preceding autumn, late winter or even early spring.

Pruning and Training

Remontants are grown similarly to summer strawberries, but it pays to give them a sunny spot so their later fruits can ripen.

Pests and Diseases

Because they have such a long season, early mouldy fruits must be rapidly and hygienically removed, or later fruits will suffer exponentially. They are more prone to moulds because of the humidity in autumn and benefit from cloches. Bird damage is less as there are plenty of other attractive fruits around.

Pop a fresh strawberry or three in a glass of white wine for pleasure

Mara des Bois

culinary

They can be used just like summer strawberries, but are conveniently fresh from summer till the frosts.

Baked Strawberry Apples

Serves 4

4 cooking apples
A little sugar and butter
Punnet of strawberries
A few raisins

Wash, dry and core the apples and smear with butter. Cut a thin slice off one end to ensure they sit flat in a greased baking dish. Setting aside the four best strawberries, eat some and pass the rest through a sieve. Fill the holes in the apples with strawberry purée. Push in raisins to bring the level over the top. Finish with the reserved fruits and a sprinkling of sugar. Bake at 190°C/375°F/gas mark 5 till the apples are bursting – for approximately 20 minutes. Serve hot with custard, cream or yogurt.

Harvesting and Storing

Because the remontants crop late into autumn they are juicier than summer varieties, but not always so sweet, and often rot before ripening. They can be picked green if there is a hint of colouring, and are then fine for culinary use.

companion planting

As with other strawberries, they do well with beans and are benefited by onions and borage. Most of all they love a mulch of pine needles.

other uses

Eating strawberries will supposedly whiten your teeth. Strawberry leaves have been used as a tea substitute.

Looks humble, so the strawberry centre is a surprise!

Fragaria **hybrids from the family** ***Rosaceae***

STRAWBERRIES
ALPINE AND WILD EUROPEAN

Herbaceous. 30cm. Life span: short! Self-fertile. Fruits: up to 1 x 2cm. Value: some vitamin C.

Alpine strawberries differ from the common garden strawberries in two distinct ways. The fruits and plants are smaller and they do not form runners. The alpines form neat clumps about 30cm across, with lighter green leaves, and flower all season almost from last to first frost. The other wild (European) strawberries, *F. vesca*, and Hautbois, *F. elatior/moschata*, are more like miniature versions of garden strawberries. With smaller fruits than even the alpines, they do make runners. Indeed some of the wild woodland forms only produce runners and rarely fruit.

These wild forms were the earliest strawberries cultivated and are native to Europe, but only north of the Alps. They were thus unknown to the Ancient Greeks, and although passing reference is given to them by Roman and early medieval writers, it is as wild, not cultivated fruits. In England the fruits are mentioned during the thirteenth century in the Countess of Leicester's Household Roll. By the reign of Henry VIII the fruit was highly esteemed and cost four pence a bushel. At the same time the Hautbois strawberry, *F. elatior/moschata*, was also popular, especially on the Continent. It was one of the most fragrant of all and made few runners. The species from the Americas arrived during this period, but were considered not as good as these native wild varieties. It was not until the nineteenth century that the natives became superseded by 'modern' hybrids.

varieties

Alpines are grown from seed. They are little improved on the true wild form, though yellow- and white-fruited versions are available. **Baron Solemacher** is a slightly larger-fruited selection; it needs nearly 60cm of space each way. **Alexandria** is tastier, juicier and does better in moist half-shade. Wild, runnering strawberries, known as **Fraises des Bois**, are obtainable on the Continent as seed or plants. The **Green Strawberry**, *F. collina/ viridis*, has sadly vanished. Wild American strawberries, *F. virginiana*, which were crossed with *F. chiloensis* to give the first modern strawberry are presumably most happy at home in Virginia. The **Beach Strawberry**, *F. chiloensis*, can be grown for the unique flavour though benefitting from pollination by other species such as *F. moschata*. The **Indian Strawberry**, from another genus entirely, *Duchesnea indica*, is a low runner forming strawberry-like plant with a strawberry-like fruit except it's evergreen and the fruit is bland.

cultivation

Alpines are easier to grow, needing less richness and moisture than other types of strawberry, though they will do much better in improved conditions. They can be spaced at 30cm or so apart. Wild strawberries are better grown as ground cover in moist partial shade and allowed to run.

Strawberry flowers are beneficial for insects too

STRAWBERRIES

The fruits are small but fragrant

Pests and Diseases
Alpine and wild species are much tougher plants than conventional varieties and rarely suffer from pests or diseases. The fruits are also less appealing to birds, so they can often be cropped without protection.

Pruning and Training
As with other varieties and species, it is best to replace the entire stock in stages, normally running over a three- or four-year period.

Growing under Glass and in Containers
Alpines can be cropped under glass to extend the season, but become more prone to red spider mite. Other wild species resent being under cover more. In pots they are easier than more conventional varieties, however, as they need less water and do not run.

Maintenance
Spring Weed, mulch, spray thoroughly with seaweed solution monthly. Sow seed, divide runners.
Summer Pick regularly all through summer.
Autumn Remove old, worn-out plants. Divide runners if they have become invasive.
Winter Sow seed in pots in coldframes in late winter.

Ornamental and Wildlife Value
Alpines make good ground cover and are more decorative than other strawberries, as they form neat mounds. The long flowering period makes them beneficial to insects and the almost evergreen clumps are good shelter and hibernation sites.

Harvesting and Storing
They can be picked under-ripe for culinary purposes. Pick and freeze them until sufficient quantities are gathered, then shake the frozen fruits in a dry cloth and many seeds can be removed. The fruits can be made into jams, jellies, sauces, syrups and compôtes.

Propagation
As true alpines make no runners, they are grown from seed. Rub the seeds off fruits that have been left in the sun to shrivel. Sow in late winter, early spring for plants to put out in spring. Do not cover. A bottom heat of 15°C (60°F) is helpful. Later in spring the seeds can be sown without heat.

Sometimes the crowns can be successfully divided. The new plantlets can be replanted where required during the growing season.

culinary

Both alpine and wild strawberries are superlatively fragrant and delicious raw, if fully ripe. Cooking brings out even more flavour, from under-ripe fruits as well. These are less moist than ordinary strawberries, so require some water or redcurrant juice to make jam or jelly. They are also firmer after freezing and go well in compôtes.

Alpine Strawberry Tarts

Make individual sweet shortcrust pastry tart cases. Smear the insides with butter, then fill each tart with a mixture of both alpine strawberries and strawberry jam (alpine or otherwise). Bake in a preheated oven for ten minutes or so at 190°C/375°F/gas mark 5. Cool and top each with clotted cream before serving.

companion planting

Sixteenth-century poet Thomas Tusser said:

'Gooseberries, raspberries, roses all three
With strawberries under do trimly agree.'

In the wild strawberries are sometimes found growing with vervaine.

other uses

The leaves and fruit of wild strawberries can be used for medicinal purposes.

Vaccinium species from the family *Ericaceae*

BLUEBERRIES AND BILBERRIES

Bush. 30cm–4m. Life span: medium to long. Deciduous, self-fertile. Fruits: 1cm, spherical blue-black. Value: some vitamins C and B and full of antioxidants and anthocyanins.

Bilberries, blaeberries or whortleberries, *Vaccinium myrtillus*, are low shrubs native to Europe, found on heaths and moors in acid soils. They have slender, green twigs with myrtle-like leaves and spherical pink flowers followed by blue-black fruits in late summer. They are fiddly to pick and rarely cultivated. Highbush Blueberries, *V. corymbosum*, and many near relations come from North America. The Highbush is a tall shrub up to 4m, so lower-growing cultivars for garden use have been bred, though commercial growers still prefer the tall.

Many similar species are used as ornamentals because in autumn the leaves turn to some amazing reds and pinks. The Rabbiteye Blueberries, *V. virgatum/ashei* are similar. The Lowbush Blueberry, *V. angustifolium*, is different, only about 30cm high, and is much hardier than the Highbush. The berries are large, sweet and early so this has been crossed with the Highbush.

Bilberries were once highly popular. They would be picked from the wild and taken to market in the towns where they were esteemed for tarts and jelly. They were a staple food to the Scots Highlanders who ate them in milk and made them into wine. Bilberries went into oblivion when the better fruiting blueberries from America became available. Blueberries are a traditional American fruit, with different regions favouring different species, such as Highbush, or Swamp, and Rabbiteye Blueberries in the warmer areas and the Lowbush, Early or Low, Sweet Blueberry further north. The last was particularly useful as it was easily dried during preservation for winter. The dried berries were beaten to a powder and made into cakes with maize meal. In the north-west they even smoke-dried them for extra flavour.

Conveniently for the home gardener they do not ripen all together but over weeks

varieties

Most of our garden varieties are of mixed American origin. Many of the better ones were bred in Maine, where the climate is similar to that of Britain. The early **Earliblue**, **Goldtraube** and **Jersey** are fairly compact bushes. **Berkeley** is a spreader, **Blue Crop** more upright and with good drought resistance. **Ozark Blue**, **Toro, Brigitta** and **Chandler** are taller later croppers. **Herbert**, mid-season, is considered one of the best flavoured. **Nui** is early, very light blue and very large-berried. All are commonly available and need planting about 1–1.5m apart but more dwarf varieties such as **Top Hat** and **Blue Pearl** can be planted closer at 0.6–1m. Many of the species have edible berries.

Vaccinium ovatum, the **Huckleberry** or **Box Blueberry** is a small, attractive, evergreen shrub with tasty berries and can be used to make a hedge. The **Red Huckleberry, Red Bilberry**, *V. parvifolium,* is a large deciduous shrub which has red fruits. *V. membranaceum*, the **Big Huckleberry**, has big berries, is fairly drought resistant and is one of the tastiest. *V. hirsutum*, the **Hairy Huckleberry**, is not very hardy and has hairy fruits.

BLUEBERRIES AND BILBERRIES

cultivation

An acid soil suitable for heathers or rhododendrons is essential; a substitute of peat and leaf mould will do. The tall species prefer wetter sites, the dwarfer ones suffer drier, but all crop better with moister positions. They prefer sunny sites though they will grow in partial shade. Partly self-fertile, they do better if several varieties are grown together.

Growing under Glass
They are sufficiently hardy almost anywhere, but glass protection may be worthwhile temporarily while they fruit to prevent losses to birds.

Growing in Containers
Blueberries have to be grown in containers in many areas as they die on lime soils. Ericaceous compost or a mixture of peat, sand and leaf mould is essential as is regular, copious watering with rain or acidic water. In limy areas avoid tap water!

Ornamental and Wildlife Value
Stunning colours in autumn. There are countless ornamental species and varieties. All berries and are valuable to birds.

Maintenance
Spring Weed and mulch.
Summer Make layers, pick fruit.
Autumn Take suckers and transplant once leaves fall.
Winter Prune if necessary.

Propagation
The species come true from seed but better varieties are layered in summer. Suckers can be detached in winter.

Pruning and Training
Bilberries and blueberries need little pruning except to remove dead or diseased growth, best done in winter. Older bushes should have congested twiggy branches removed, retaining the younger stronger ones.

Pests and Diseases
Apart from the usual losses to birds, this is a remarkably pest- and disease-free family. Any distress will probably be due to an alkaline soil or lime in the water supply.

Harvesting and Storing
The berries should be picked when fully ripe and easily detached or they are too acid. They may be jammed, juiced, jellied or frozen; commercially they are obtainable dried.

companion planting

As these are ericaceous they grow well near heathers. They can also have cranberries or lingonberries underplanted with them as ground cover. I grow wild (English, not French, American or alpine) strawberries with them.

other uses

The leaves were used medicinally. Chewing dried bilberries was a cure for diarrhoea and mouth and throat infections.

culinary

Blueberries and bilberries can be used in pies, tarts, jams, jellies and syrups. Blueberry cheesecake and blueberry muffins are very popular American dishes.

Blueberry Grunt
Serves 4–6

500g blueberries
50g sugar
1 teaspoon allspice
1 small lemon
Maple syrup, to taste
125g white flour
Pinch of salt
1 ½ teaspoons baking powder
50g butter
300ml single cream

Simmer the washed blueberries gently with sugar, allspice and the lemon's juice and grated rind. Add maple syrup to taste. Meanwhile rub the flour, salt, baking powder and butter into crumbs and blend in enough cream to make a smooth creamy dough. Carefully spoon the dough on top of the blueberries, cover the pan and simmer till the crust puffs and sets. Serve with the rest of the cream and more maple syrup.

Vaccinium species from the family *Ericaceae*

CRANBERRIES AND COWBERRIES

Bush. Prostrate to 60cm. Life span: medium to long. Evergreen, self-fertile. Fruits: up to 2cm, reddish-orange. Value: some vitamin C.

Cranberries are very similar and closely related to blueberries and bilberries, the most noticeable differences being that cranberries have red berries and are evergreen.

***V. oxycoccus*, the cranberry, is a native of most northern temperate countries and is found on bogs and moorlands. The low-growing, evergreen shrub is tough and wiry with long, sparsely leafed stems. The leaves are longer and thinner than those of the blue-berried *Vaccinium* species. The flowers are tiny, yellow and pink, in early summer and are followed by the round, red fruits, which are pleasantly acid to taste. The American Cranberry, *V. macrocarpon*, is much the same, but larger in size and berry. The Cowberry, Crane or Foxberry, *V. vitis idaea*, is also similar, more densely leafed, with rounded ends and clusters of berries, which are more acid and less agreeable than cranberries. The Lingonberry or Lingenberry, or Mountain Cranberry *V. vitis-idea*, is very similar but even more drought resisting; resembling a small box hedge, it can be clipped likewise, the berries can be eaten, pretty sour from August and better later, after the frosts have touched them. The native cranberry has been gathered from the wild in most cool regions by native peoples throughout the northern hemisphere.**

The cowberry has not been enjoyed so widely as it is usually too acid to eat raw, though it is excellent after cooking. Strangely the British never much liked it either, although both the cranberry and cowberry were very popular in Sweden. When better North American cranberries came as a sauce to accompany the new festive dish of roast turkey, cowberries were suddenly in demand – as now they could be sold to the unwary in London as 'cranberries'. *Arctostaphylos* species, Manzanitas, Bearberries, Kinnikinnik, are small trees and shrubs, needing similar conditions and producing similar fruits used in similar ways.

varieties

Pilgrim has large berries and is less mositure demanding than the wilder forms. **Early Black** has darker berries and is claimed to be most frost resistant. Apart from those already mentioned, there are several other *Vacciniums* that fall between cranberries and blueberries and have edible berries; *V. floribundum*, the **Mortinia**, is the least hardy and comes from Ecuador but will survive in southern England. It is an attractive, evergreen shrub with heavy racemes of rose-pink blooms followed by masses of red berries. From East Asia and Japan comes *V. praestans*, a prostrate, creeping, deciduous shrub with sweet, fragrant, glossy, red berries. *V. nummularia* is one of the choicest little evergreens for an alpine house. It resembles a prettier cowberry, with arching, hairy stems and small black berries.

cultivation

These really need moist boggy conditions, in lime-free soil and water. The best sites are made on the edge of a river or pond by slowly building up a layer of stones covered with a thick layer of peaty, humus-rich, acid soil. They must be moist but not drowned; they need to stand above the water! Most are small, needing as little as 60–90cm each way and even tolerating some light shade.

CRANBERRIES AND COWBERRIES

The commercial cranberry harvest, honest. It's an amazing corner of horticulture, or is it aquiculture, indeed is it fishing?

Growing under Glass
As the usual varieties are hardy this is only worthwhile for bird protection. *V. floribundum* and *nummularia* benefit from cool cover, such as that afforded by an alpine house, and are very beautiful small shrubs.

Growing in Containers
Cranberries have to be grown in containers in many areas as they will die on lime soils. Ericaceous compost or a mixture of peat, sand and leaf mould is essential, as is regular and copious watering with rain or acidic water. Avoid tap water in limy areas!

Ornamental and Wildlife Value
On acid soils cowberries make excellent ground cover. Cranberries are not as dense, so they suppress weeds less well. The berries are loved by wildlife and, being evergreen, the plants will provide good shelter.

Maintenance
Spring Weed, mulch, make the layers.
Summer Keep well watered.
Autumn Divide plants, pick fruit before frost.
Winter Protect less hardy species from frost.

Propagation, Pruning and Training
The species can be grown from seed, or by dividing in autumn, or they can be layered in spring. Only remedial tidying is required.

Pests and Diseases
Cowberries and most of the denser-growing species suppress weeds well. They are all pest- and disease-free in most gardens. Grown under glass they need to be kept cool or they suffer. Any problems are most often due to lime in the soil or water. Watering on sequestrated iron chelates will help but these are not available to organic growers.

culinary

Invariably used for the jelly, but also in tarts and pies and added to many other dishes.

Cranberry Jelly
Makes about 3.75kg

500g cranberries
750g apples
Approx. 2kg sugar

Wash the fruits, chop the apples and simmer both with enough water to prevent burning. Once the apples are soft, strain and add 500g sugar to each 600ml of liquid.

Bring the liquid back to the boil, stirring to dissolve the sugar. Cook briefly, skim and pour into sterile warmed jars. Seal at once. Serve with roast turkey.

Harvesting and Storing
Cowberries are made sour by frosts so must be gathered promptly. In Siberia they were kept under water through the winter, so that they gradually became less acid, and were then eaten in spring.

companion planting

They are ericaceous and enjoy similar conditions, root bacteria and fungi as rhododendrons and azaleas so can be used as ground cover between these.

other uses

The leaves and fruits of most of these berries have been used medicinally, Cowberries have been eaten as a cure for diarrhoea.

Ribes nigrum **from the family** ***Grossulariaceae***

BLACKCURRANTS

Bush 1.5m. Life span: short. Deciduous, self-fertile. Fruits: 1–2cm, black, spherical. Value: very rich in vitamin C.

Blackcurrants are quite different to the other types of *Ribes*, though they are often bundled in with redcurrants. They fruit on young wood, not old, and have dark purple, almost black, berries with a most distinct and unforgettable aroma, which is similar to that of the aromatic foliage and stems.

These, like other *Ribes*, seem to have been unknown to the Ancient Greeks or Romans and were used only medicinally, as quinsy berries, for curing colds and throat problems until the sixteenth century. Then they became more popular as a garden crop and are now very widely grown commercially in Europe, but not so much in the USA. Their rise in fame was due to their very high vitamin C content, and probably also to the fact that sugar, needed to make this naturally sour fruit palatable, became available more cheaply. The native plants can occasionally be found in wild wet areas of northern Europe and Asia, but are now more likely to be garden escapes. Improvement has been done mostly by selection rather than producing hybrids with other *Ribes*, though the Josta is a good example of what is possible.

Blackcurrants can be huge and sweet enough to eat fresh

varieties

New varieties such as **Ben Sarek** (mid-season), **Ben Lomond** (late) and ***Ben More*** (very late) are numerous and generally more productive, with better disease resistance, than old favourites. **Laxton's Giant** (mid) is still one of the biggest, and **Seabrook's Black** (mid) is supposedly resistant to big bud. **Ben Connan**, a smaller bush with big well-flavoured currants, is also resistant to mildew and leaf curling midges. **Ben Hope** has been bred to resist most pests and diseases including Big Bud mite, which means it should not get Reversion disease; it ripens late, just after **Ben Lomond**. **Ebony**, an introduction from Eastern Europe, has particularly sweet (plus 15 per cent) currants, making it the first dessert variety. The **Jostaberry** is a much larger hybrid, more like a thornless gooseberry, with heavy crops of large blackcurrant-flavoured berries. Some American species have been esteemed, such as the fragrant, bright-yellow-flowered **Buffalo Currant** or **Golden Currant**, *R. aureum/odoratum*, which is also used as the stock for standard gooseberries. *R. americanum*, **American Blackcurrant**, has yellowish flowers and inferior fruit, but turns glorious colours in autumn.

cultivation

Blackcurrants revel in rich, moist ground, the richer the better, and similarly respond to heavy mulching. They do not mind light shade. Late varieties are usually chosen to avoid frost damage during their flowering period.

BLACKCURRANTS

Growing under Glass and in Containers
The bushes prefer to be cool so they are not happy for long under cover. They can be grown and fruited successfully in large pots.

Ornamental and Wildlife Value
Rather dingy plants of little decorative appeal, though the smell of foliage and stems is most pleasing. However, the currants are as valuable to birds as to us and thus good subjects for wild gardens.

Maintenance
Spring Weed, mulch, spray seaweed solution monthly.
Summer Protect and pick fruits.
Autumn Take cuttings.
Winter Prune back hard and compost heavily.

Propagation
There are no easier cuttings! As blackcurrants are best grown as a stool, the main requirement is for multiple shoots from ground level. Thus cuttings have all buds left on and new bushes are planted deeper than is the standard practice for almost every other subject.

Pruning and Training
In order to provide as much young fruitful wood as possible the optimum pruning is to remove annually all branches over two or three years old, allowing new ones to come up from near ground level. Or annually remove a one-third segment of the stool – like removing a slice of pie from above. The laziest though less effective way is just to cut back totally, down to near ground level once every three years. (Better done to one in three of your bushes each third year in turn or you will have years with no crop.)

Pests and Diseases
Weeds must be kept from encroaching on the stool, but rarely germinate there because of the intense shade. Birds bother these less than most other fruits but still need to be prevented. Mildew is aggravated by stagnant air and dry roots; hygienic pruning and vigorous growth is usually sufficient redress. Big bud is obvious. It is caused by microscopic pests that also carry virus diseases such as Reversion. The simple solution is to replace old infected stock with new clean material after ten or fifteen years or when yields have dropped too far.

Harvesting and Storing
Blackcurrants will keep for several days once picked as they are so firm and tough-skinned. They can be frozen, jammed, jellied and turned into delicious syrups and juices.

They hang on little wiry stems, sprigs, which need careful detaching

culinary uses

The currants have too little liquid to simmer down on their own, so they need water or other juices. Add redcurrant juice to make blackcurrant jams and dishes more pleasantly acid. Jelly is easier work than jam as de-sprigging the berries is tedious.

Bob's Cunning Blackcurrant Jam
Makes about 3kg

1.5kg blackcurrants
500g redcurrants
A little water
Approx. 2kg sugar

De-sprig best quarter of the blackcurrants and set aside. Simmer the rest with the redcurrants, and enough water to just cover them, till they break up. Strain the juice, and repeat the process (if frugal and caring little for the quality). Weigh the (combined) juice, add the reserved blackcurrants and bring to boil simmering till these just start to break up. Then add half the juice's weight in sugar, bring back to boil, skim and pour into sterilised jars. Put the lids on immediately.

companion planting

Nettles nearby benefit blackcurrants. In some parts of the USA blackcurrants may not be grown as they are host to white pine blister rust.

other uses

The leaves have been used as tea for medicinal and tonic purposes and dried currants likewise, especially for throat infections.

Ribes sativum from the family *Grossulariaceae*

REDCURRANTS AND WHITECURRANTS

Shrub 2m. Life span: long. Deciduous, self-fertile. Fruits: 1–2cm, globular, glossy red or white.

Redcurrants are reliable, productive plants we rarely notice except when translucent, glossy red berries festoon the branches. Otherwise they are insignificant, similar to flowering redcurrant, *Ribes sanguineum*, which has such prolific pink tassels in early spring. (This sets inedible fruits occasionally.) Whitecurrants are varieties selected without the colour and with their own flavour. I find that the whites crop slightly less extravagantly than reds and are culinarily useful.

Redcurrants, *R. sativum* and *R. rubrum*, are native European fruits. They were not cultivated by the Romans. They gained garden notice in the sixteenth century, when they rapidly became a stalwart of the cottager's garden. Long-lived, they often survive, neglected and unproductive, whereas, given a good site and bird protection, they produce prodigiously. Surprisingly little known in much of Europe and the USA, they are popular in Scandinavia and Russia.

White Versailles

varieties

Earliest of Four Lands is my favourite, but all are very amenable and productive, varying little in taste, acidity and season – though this may cover three months in a good year. **Raby Castle** has *R. rubrum* blood so is most hardy of a really tough group, **Wilson's Longbunch** are crimson pearls, **White Grape** and **White Versailles** the equivalent whitecurrants. **Rovada** is a recently introduced improved redcurrant but cropping rather later than most older sorts. **Jonkheer van Tets** is much earlier ripening. **Blanka** is a new heavier cropping whitecurrant ripening in August.

cultivation

Redcurrants respond best to a cool, well-mulched soil. They do not need as rich conditions as blackcurrants or raspberries and will grow in partial shade and quite happily on cold walls, although the berries are sweeter in the sun. The main requirement is protection from birds. Without thorough netting, all else is useless, as these are the bird food supreme.

Pests and Diseases

The only major threat to the crop is the birds. Minor sawfly caterpillar attacks sometimes occur but do little damage. Berries left to ripen fully mould in damp weather unless protected. The leaves regularly pucker with red and yellow blotches from the leaf blistering aphis, though surprisingly this does not affect the yield. Indeed these leaves are removed wholesale, with aphids on board, during summer pruning.

Pruning and Training

These are the most easily trained and forgiving of plants. No matter how you misprune them, they respond with new growth and ample fruit. Redcurrants can be made to take any form – cordon, goblet, fan or espalier – and moreover are quick to regrow. They

REDCURRANTS AND WHITECURRANTS

will fruit best on a permanent framework with spurs and need every shoot, except leaders, cut back in summer and again harder in winter. Growing in good conditions, redcurrants can be trained over large walls, including north-facing ones. Moderately pruned as goblet bushes, they need to be nearly 2m apart. As cordons they can be grown two to the metre.

Propagation
The easiest of all plants to root from autumn cuttings.

Maintenance
Spring Weed, mulch and spray seaweed solution monthly.
Summer Pick fruit as required. After mid-summer, prune.
Autumn Take cuttings once leaves fall.
Winter Cut back hard to spurs on frame.

Growing under Glass and in Containers
This can be worthwhile if you want pristine berries over a longer season or have no space for bushes. They do not want to get too hot!

Ornamental and Wildlife Value
Redcurrants have little value in the ornamental garden as the berries disappear so rapidly. Redcurrants are the most palatable bird snack and so they will be popular in wildlife areas, briefly.

Harvesting and Storing
Currants can be picked from early summer as they colour, for use as garnishes, adding to compôtes and for the most acid jellies. Ripening continues into late autumn in dry years when the fruits become less acid and tasty raw. Being seedy, the fruit is conveniently stored juiced and frozen. Similarly, currants are better as jelly rather than jam. Dry, cool berries keep very well.

companion planting

Redcurrants may benefit from nettles nearby, but the fruit picker will curse. *Limnanthes douglassii* is the best companion and ground cover once the bushes are well established.

other uses

Once popular with apothecaries, as they could be stored for months packed fresh and dry into sealed glass bottles and were thus available as 'vitamin pills' in bleak late winter and spring when fresh fruit and vegetables become scarce. The juice of redcurrants must be the best edible red dye you could want.

Raby Castle

culinary

Redcurrants are immensely useful because of their colour and acidity. They add deep red to everything and their juice makes other fruit jellies set. Redcurrant juice adds tartness and flavour to other juices and can be used in many sweet and savoury dishes.

Whitecurrant juice is a substitute, if not an improvement, for lemon juice and makes even more delicious jellies.

Mint Sauce Supreme
Makes about 1kg

500g whitecurrants
Approx. 500g sugar
125g fresh mint leaves, finely shredded

Simmer the whitecurrants till soft, then strain, or simply juice. Add the same weight of sugar to the juice and slowly bring to boil. Skim and remove from heat. Stir in the finely shredded mint and bottle in warm sterile jars. Cover immediately. This is the sauce for spring lamb roasts, for yogurt dips and for barbecue glazes.

Ribes uva-crispa (Ribes grossularia) **from the family** ***Grossulariaceae***

GOOSEBERRIES

Bush up to 1.5m. Life span: long. Deciduous, self-fertile. Fruits: up to 3cm, oval, green to purple. Value: some vitamin C.

Gooseberries you find in the shops are green bullets, for culinary use, nothing like the meltingly sweet, well-ripened dessert varieties. Compared to other *Ribes*, gooseberries have bigger, hairy berries and sharp thorns. They are easy to grow but often handicapped by being grown as a stool. Given attention and good pruning, large, succulent berries can be had in almost any colour and with a delicious flavour, which can range from a clean, acid-sweet taste to vinous plumness.

Unnoticed by Classical writers but a European native, gooseberries, *Ribes grossularia*, are first mentioned in purchases for the Westminster garden of King Edward I in 1276. They became popular almost solely in Britain, and by the nineteenth century there were hundreds of varieties, and countless clubs where members vied to grow larger fruits, achieving berries the size of bantam eggs. One variety, London, an outstandingly large, not so hairy, red was the biggest exhibited every year from 1829 to 1867, 37 years unbeaten champion! The wild relation is found in rocky terrain as a small shrub, variable in berry colour, size and in habit of growth, with some being inconveniently lax. American gooseberries, Worcesterberries, are derived from *R. divaricatum*. They have smaller berries and are resistant to the American mildew disease that can damage European varieties.

varieties

London (mid-season) is biggest, and dark red. I love **Langley Gage** (mid-season) which has divine, bite-sized, syrupy sweet, translucent white globes hanging in profusion. **Early Sulphur** (very early) has golden yellow, almost transparent, tasty, medium-sized berries. For a substantial, dark olive green, strongly flavoured, large berry choose **Gunner** (mid), though it's not a heavy cropper. **Leveller** (mid) is, and has delicious yellow-green fruits. **Whinham's Industry** has become the most widely grown dual-purpose red. **Careless** is a heavy cropper of greeny white, large and smooth-skinned culinary fruits. **Invicta** is a heavy cropping pale green-berried, disease-resistant, culinary gooseberry. **Pax**, a red dessert, is fairly mildew resistant and has few if any thorns. **Xenia** and **Captivator** are newer, almost spine-free red desserts ripening in June and July. **Hinnonmaki Yellow** and **Hinnomaki Red** are not from these shores and have sweet aromatic yellow green or red berries, the former claimed to taste like apricot. Both are resistant to American mildew. The **Worcesterberry** is even meaner-thorned and is really an American species with smaller black berries more like blackcurrants. **Pixwell** is an improved, green-fruited form. *R. hirtellum*, the **Currant Gooseberry**, is another edible American species with small reddish fruits.

Think of dessert gooseberries as hairy grapes and skin them for delight

cultivation

They love rich, moist, loamy soil and don't like hot, dry, sandy sites or stagnant air, doing better with a breeze.

GOOSEBERRIES

Growing under Glass and in Containers
Gooseberries are so hardy they need no protection, as well as being too thorny. They can be potted, but are easier in the ground.

Ornamental and Wildlife Value
The bushes are drab, the flowers inconspicuous and the berries not brightly coloured – the ideal landscaping plant to go with modern buildings! The flowers benefit early insects and the berries disappear.

Maintenance
Spring Weed, mulch, spray seaweed solution monthly.
Summer Thin and pick fruit, watch out for sawfly and mildew.
Autumn Take cuttings.
Winter Prune hard.

Propagation
Gooseberries are propagated by 30cm long cuttings. Disbud the lower end to prevent suckers.

Pruning and Training
Often misgrown as a stool with many shoots direct from the ground, gooseberries are better hard pruned to spurs on a goblet-shaped frame with a short leg. I leave the pruning till late winter so that the thorns protect the buds from the birds, which perversely delight in disbudding gooseberries. To get larger berries or more varieties in a confined space, gooseberries may easily be grown as vertical cordons, fans or even standards.

Although delicious in themselves, the reds, even if picked green, will turn your green gooseberry jam red

My favourites are the small whites, such as **Langley Gage**

Pests and Diseases
American mildew is the worst problem, burning tips and felting fruits with a leathery coat that dries them up. Hygiene, moist roots, hard pruning and good air circulation reduce the damage. Sodium bicarbonate sprays and sulphur-based ones (which burn some varieties) are available to organic growers. Occasionally, often in the third year or so after planting, gooseberries suffer damage from sawfly caterpillars. First appearing as a host of wee holes in a leaf, they move on to stripping the bush. However, vigilance and early action prevent serious damage.

Harvesting and Storing
Gooseberries do not have as much bird appeal as many fruits and can even be got unripe without protection. Birds and wasps do steal them once they're ripe, otherwise the fruits mellow and hang on till late summer if protected from such pests and damp.

companion planting

Tomatoes and broad beans nearby are reputed to aid them and I always grow them with *Limnanthes douglasii* as ground cover.

other uses

Gooseberries make a powerful wine much like that of the grape.

culinary

Picked small and green, they make the most delicious acid jams and tarts – which turn red if overcooked. As they ripen they become less acid and fuller flavoured for dessert purposes. Ripe fruits for cooking combine well with redcurrants to keep up the acidity and are often jellied to remove the tough skins and seeds.

Gooseberry Fool
Serves 4

500g ripe green gooseberries
Approx. 75g light honey or sugar
300ml thick cream
Dark chocolate and grated nutmeg, to garnish

If the gooseberries are soft, press them through a sieve. If not, warm very carefully till soft first, or freeze and defrost first. Add sweetening to the purée to taste, and cool. Immediately before serving, whip the cream and fold in the purée. Garnish with grated dark chocolate and nutmeg.

Aronia melanocarpa from the family *Rosaceae*

CHOKEBERRIES

Bush up to 2m. Life span: medium. Deciduous, self-fertile. Fruits: 2cm, spherical, black. Value: very rich in vitamin C.

With a name like chokeberry you can be sure the fruits are astringent and sour raw, though fine cooked and sweetened. They are hard, red, ripening to purple or almost lustrous black. They closely resemble blackcurrants in appearance and even in taste, though are more acid and almost pine-flavoured, making them a useful substitute where blackcurrants may not be grown, such as in some parts of their native USA. The bushes are easy to grow, reliable, highly productive, and compact with a height and spread of about 1.5m. White, hawthorn-like flowers and brilliant autumn leaf colours make this a most decorative fruit bush.

Distantly related to the pear and *Sorbus* genus, these berries came from eastern North America in 1700. They were relished by Native Americans who would mix the dried fruits with others to make 'cakes' for winter storage, but it was the autumn colouring that recommended them to European plantsmen. The Royal Horticultural Society Award of Merit was eventually granted in 1972, but still rather for their ornamental appeal than their taste. They are a fruit with great potential. I'm sure they would do better if called the tastyberry!

varieties

Aronia melanocarpa, **Viking** is available as bushes for fruit production and is self-fertile. Another similar cultivar of *A. melanocarpa*, **Brilliant**, is available but for ornamental plantings because it has exceptionally good autumn leaf colouring. *A. arbutifolia*, the **Red Chokeberry**, also has good autumn colour and produces red berries that were eaten by Native American children for their aroma rather than for their taste. There is also a more erect form. Although completely unrelated, *Rhus glabra*, **Scarlet Sumach, Vinegar tree**, *Anacardiaceae* is very similar. This small easily grown hardy shrub was also introduced from North America in 1622 for its spectacular scarlet autumn foliage. It also produces masses of fruits, if both male and females are grown, and these were likewise eaten by the Native Americans and small boys. They also have a sour taste and were used as a substitute for vinegar; dried and crushed, they were sprinkled over meat and fatty dishes as a seasoning. Some other members of the *Rhus* family have apparently had their fruits or foliage eaten but as they are closely related to Poison Ivy, *Rhus toxicodendron*, great caution is advisable!

Chokeberries are soon robbed by the birds – ignore the seakale leaves underneath, the Aronia's are those small ones nicely colouring

CHOKEBERRIES

cultivation

Chokeberries are easy and do well on any reasonable soil other than very shallow chalk or in very boggy ground. Naturally they will respond to better conditions by becoming larger and more prolific, and are happier with well-mulched peaty conditions. The bushes need to be 2m apart for effective cropping, but possibly closer for displays of berries and autumn colour.

Maintenance

Spring Weed, mulch and spray with seaweed solution.
Summer Net to keep birds off and pick the fruit.
Autumn Prune out dead and diseased growths, and take cuttings.
Winter Mulch with compost and straw, leaf mould or bark.

Propagation

The species come true from seed but named varieties are best reproduced from early autumn cuttings or by division.

Pruning and Training

They can tend to sucker, turning them into a stool, but cultivation is easier if they are kept to a single stem. Pruning is mostly remedial, removing suckering, low-growing, congested and diseased growths. I suspect chokeberries would be good trained on wires or a wall – they would certainly be most decorative.

Pests and Diseases

Other than the usual hazards of choking weeds and losses to the birds, these plants are remarkably free from problems. One reason chokeberries are coming into cultivation is that they are as good a source of vitamin C as blackcurrants, but also more productive, with none of their potential problems of big bud, reversion or mildew.

Growing under Glass and in Containers

There seems no need to grow chokeberries under glass as they thrive outdoors and are of little value fresh, only tasty once preserved. However, it is worth growing them in a container if you need a rich source of vitamin C and have no garden space available.

Ornamental and Wildlife Value

The tough reliability, the profusion of spring flowers, immense quantities of glossy black berries and the colour of the autumn leaves make this an essential plant for any area, ornamental or wild, especially if you like birds.

companion planting

No good or bad companions are yet recognised as they have been little cultivated. I grow mine in a bed with rhubarb and seakale.

other uses

The berries can be used for an edible dye and I can vouch for their being a high-vitamin self-service food for my poultry, who head for them whenever they get out.

culinary

They can be used in the same way as blackcurrants and indeed taste not dissimilar, if more piney and aromatic. However, their preserve goes better with savoury dishes in the manner of cranberry jelly. Jams, jellies and preserves only set well if apple juice or purée is combined with the berries; without it you get berries in syrup.

Chokeberry preserve

1kg chokeberries
500g whitecurrants, or redcurrants if white unavailable
1 small lemon
A little water
Approx. 1kg sugar

Thoroughly wash the berries and currants, chop the lemon and simmer all three with just sufficient water to cover. Simmer till soft then sieve out the skins and pips, weigh the juice and return it to the pan with three-quarters of its weight in sugar. Bring to the boil. Skim well, then pot in small jars. Store for six months before use. Serve with gammon steak, new potatoes and peas.

Rubus idaeus from the family *Rosaceae*

RASPBERRIES

Bush/vine. Life span: short. Deciduous, self-fertile. Fruits: up to 2.5cm, red, yellow, black, conical. Value: valuable amounts of vitamin C, riboflavin and niacin.

Although many have described the strawberry as the finest fruit, others consider the raspberry to be as good, if not better. However, these exquisite berries are not as common by far, because the fruits perish so rapidly. Strangely they are also little grown as garden fruits, although they are amongst the easiest to care for and cultivate.

Raspberries vary considerably in size and are usually red, with black and yellow sorts also sometimes encountered. In good varieties, the conical fruit pulls off the plug easily, leaving a hole. Some are less easy to pick and the berries may be damaged in the extrication as they are very soft and thin skinned.

Their young shoots are usually green but soon go red or brown as they grow. Eventually the canes reach 1–3m high. They are often bristly and occasionally thorny. Individually the canes are short-lived, springing from a suckering root system one year, to die the following year after fruiting. The root systems could spread perpetually, though they most often fade away from virus infections. The flowers are small, usually white, with the fruitlet visible in the middle once the petals fall. While the flowers are usually unscented, the leaves have a slight fragrance. This is more pronounced in the North American wild species *Rubus odorata*, which has a resiny smell to the stems, though the pale reddish-purple flowers remain scentless. (Another, *R. deliciosus*, has large, rose-like flowers which are deliciously perfumed.)

R. idaeus raspberries are native to Europe and Asia in hilly areas, heaths and on the edge of woodlands, especially those with acid soils. They are found growing wild in Scandinavia as far north as 70° and have long been gathered. Seeds and debris from the plants have been found preserved in the remains of the prehistoric lake villages of what is now Switzerland. Strangely, like other northern temperate fruits, raspberries went unregistered by the Classical writers. The Romans were such ravenous gourmands that it seems unlikely they did not eat these delightful fruits when colonising cooler, wetter lands than their own. Perhaps because these were gathered from the wild in such profusion they needed no mention, being taken as commonplace. Raspberries are included in the practical poetry of Thomas Tusser and noted by Gerard in the sixteenth century. It seems that at that time the fruit of the closely related bramble was considered far superior and raspberries were used more for medicinal and tonic purposes. There are many similar and wild species which are enjoyed in other temperate countries.

RASPBERRIES

Galante

varieties

There are many *Rubus* species closely resembling raspberries, and others more like brambles or blackberries. The latter are dealt with as a group with Japanese wineberries. In North America they cultivate red and yellow raspberries descended from *R. strigosa*, similar to European raspberries. They also have Black Raspberries, **Blackcaps**, *R. occidentalis*, which are less hardy and with fewer stouter canes more given to branching. Their **Rocky Mountain** raspberry, *R. deliciosus*, is large fruited, delicious but a shy fruiter especially in the UK.

Raspberries can be summer or autumn cropping, though many cultivars vary as much with the pruning method as inherently. Those that readily crop on old wood, last year's canes, are now called floricane varieties while those that crop on this year's new shoots, effectively in autumn, are primocanes. **Autumn Bliss** was long the best primocane for autumn though **Polka**, a Polish variety, has exceptional cropping and flavour and is set to replace it. Another newcomer, **Galante**, is very well flavoured, reliable with large berries and **Joan J** has similar claims made. These last three fruit on and on from mid-summer till the frosts.

As to choice, new varieties of today are soon replaced by others in profusion and no list is ever complete. I still like the almost obsolete but very tasty **Malling Jewel**, an early that crops lightly but holds on well and is tolerant of virus infection, but may infect others. **Glen Moy** is a heavier cropping early, resistant to greenfly. **Glen Ample** is rapidly replacing the very popular **Glen Prosen** and **Malling Joy** as the most reliable mid-season. **Glen Fyne** has by far the best flavour and is aphid resistant and heavy cropping. Lates ripening in August include the superbly flavoured **Malling Admiral**, **Glen Doll**, **Autumn Treasure**, **Octavia** and the Canadian bred and very tasty **Tulameen**. Closely related to the last is **Cascade Delight**, which has very large fruits, is late cropping but very resistant to root rot so best for wet areas. I love the yellow raspberries: these originated alongside the reds and are less vigorous, with paler leaves and naturally tend towards autumn fruiting. **Golden Everest** is superb with richly flavoured soft sweet berries lacking the sharpness of many reds. **Allgold** may be identical with **Fall Gold,** primocane autumn fruiters much like **Golden Antwerp**, which had the better flavour; one of the oldest varieties it's now rare. **Red Antwerp** is lost; **Norfolk Giant**, another legendary old variety is also no longer obtainable.

cultivation

Preferring cool, moist conditions, raspberries do wonderfully in Scotland. Although they can be grown elsewhere, in most soils, they do considerably better given plentiful moisture, a rich neutral or acidic soil, or at least copious quantities of compost and very thick mulches. With hot, dry summers and wetter autumns the autumn-fruiting varieties can be much more productive, especially on drier sites, and these also suffer less from maggots. Summer-fruiters need a moist site, and do not mind quite heavy shade. They do not like the dry conditions against walls, but can be grown on cool, shady ones with a moist root run.

Growing under Glass

Being mostly very hardy, raspberries need no protection. Under cover, however, provided they are kept cool and airy, they can produce good crops and benefit immensely from the bird protection. Growing them under glass also spreads the season. *R. niveus*, the **Mysore** black raspberry, comes from Asia and needs to be grown under glass.

Fall Gold

Two to pick now, one when you've finished, one for tomorrow and another for the day after

Growing in Containers
I have fruited raspberries in large pots. They resent it and do not crop well or flourish as they really need a bigger, cooler root run.

Ornamental and Wildlife Value
Cultivated raspberries are not very ornamental themselves and the fruit does not last long enough to be called a display! Some of the species are more decorative, and scented, so are worth considering, but are nowhere near as productive as modern fruiting varieties. The birds adore raspberries and will get to them anywhere, so they are good in wild gardens. However, you have a duty to others to eradicate the berries when they become virus-infected. The flowers are very beneficial to bees and other insects.

Maintenance
(Summer Fruiters)
Spring Weed, mulch heavily, spray seaweed solution at monthly intervals.
Summer Protect and pick fruit, thin shoots.
Autumn Cut out old canes, tie in the new.
Winter Add copious quantities of compost.

(Autumn Fruiters)
Spring Weed, mulch heavily, spray seaweed solution monthly, thin the shoots as they emerge.
Summer Thin young canes, shorten or tie in the best.
Autumn Protect and pick the fruit.
Winter Cut all canes to the ground, add copious compost to soil.

Propagation
Varieties are multiplied by transplanting any piece of root with a bud or young cane in the autumn. Pot-grown ones may be planted in spring. Summer fruiters should not be cropped the first year but built up first; autumn fruiters may be cropped if they were well established early in the previous autumn. I have found seed can provide very vigorous and productive, if variable, plants, but it is a better plan to buy new, named cultivars.

Pruning and Training
Too lax to be left free, they are best restrained by growing between pairs of wires or winding the tips around horizontal ones. Alternatively, they can be grown as tripods, with three well-spaced stools being joined to an apex. Pruning for summer raspberries is done in autumn; remove all old canes that have fruited or died and fix the new ones in place, selecting the biggest and strongest at about 12cm apart. (It helps to pre-thin these when the shoots emerge in early summer.) Autumn fruiters are easier still: just cut everything to the ground in late winter. (Pre-thinning the canes in spring is, again, quite advantageous.)

Pests and Diseases
Weeding must be done carefully because of their shallow roots. Thick mulching is almost essential. Birds are the major cause of lost crops – no protection, no fruit! The raspberry beetle can be controlled with hygiene and mulching (methodically rake thick mulches aside in winter to allow birds to eat the pupae) or the use of permitted sprays if necessary. These maggots are

Rubus ideaus

rarely a problem with autumn fruiters. Virus diseases may appear, mottling the leaves with yellow and making the plants less productive. Replacing the stock and moving the site is the only practical solution, but wait till the yields have dropped. Interveinal yellowing is a reaction to alkaline soils; seaweed solution sprays with added magnesium sulphate are a palliative. Compost and mulching provide the most effective cure.

Harvesting and Storing
Pick gently, leaving the plug; if it won't come easily, do not force it! They do not keep for long if they are wet, and less still if warm. If you want to keep them longest, cut the fruiting stalks with scissors and do not touch the fruits. They must be processed or eaten within a matter of hours as they are one of the least durable or transportable fruits. Those sold commercially are the toughest – and therefore obviously also the least meltingly sumptuous!

companion planting

They reputedly benefit from tansy, and garlic or marigolds and strawberries may be grown close by but not underneath them. Do not grow potatoes nearby as they will then become more prone to blight.

other uses

Raspberry canes are bristly if not thorny. They have little strength or heat value but can be useful for wildlife shelters. I find short lengths, bundled together, then 'Swiss rolled' in newspaper and jammed into a cut-open plastic bottle, will make superb dry but airy ladybird hibernation quarters which I hide in evergreen shrubs. Raspberry leaves and fruits have been used medicinally, and have often traditionally been used as a tea.

culinary

Raspberries make wonderful juices, jellies, drinks and sorbets. They are often combined with redcurrant juice to add tartness. Their wine is delicate and beautifully coloured. Raspberries can be frozen, but have poor texture afterwards.

Kitty Topping's Raspberry Conserve
Makes about 2kg

1kg freshly picked raspberries
1kg caster sugar

Pick fresh raspberries and hurry straight to the kitchen. Wash them and immediately heat them in a closed pan, rapidly but gently, swirling the pan to prevent sticking and burning. Once most berries have softened, but before they break down totally, add the same weight of pre-warmed caster sugar. Stir while heating. One minute after you are absolutely sure all the sugar has completely dissolved, pour into small, heated jars and seal. Keep in the cool and use quickly once opened as the aim of this recipe is the stunning flavour, not keeping quality.

Rubus fruticosus from the family *Rosaceae*

BLACKBERRIES

Bush/vine. Life span: medium to long. Deciduous, self-fertile. Fruits: up to 2cm, black, drupe. Value: rich in vitamin C.

Even without its glistening blackberries, the native bramble is known to everyone for the long, arching and scrambling stems armed with vicious thorns. What may be appreciated more by the picker than the walker is that the fruits vary widely from plant to plant. In fact, there is no one native bramble or blackberry, but hosts of them. These have occasionally been crossed deliberately, and often inadvertently, with one another and then again with other garden escapes. Although in some remote areas the stocks remain as several distinct but variable species, they may nevertheless have been altered in prehistory by unconscious human behaviour, as has also been suggested for raspberries (see p504). Certainly now, any blackberry found near enough to human habitation to be picked, is highly likely to be a hybrid. And those nearby will probably be different. Just by looking at the fruits or the flowers in any area you will usually see great diversity.

The common brambles have fern-like leaves, thin purple or green stems, white or light purple flowers and small hard berries. A better and recognisable type of wild blackberry is *Rubus ulmifolius*, which has strong-growing, plum-coloured branches with five-lobed leaflets, which are very light on the underside. It is not self-fertile.

An improved form of this blackberry, *bellidiflorus*, is grown ornamentally for its pink double flowers. *Rubus caesius*, the dewberry, is another distinctive species, with three-lobed leaflets on long, thin, creeping, almost tendril-like stems, which can form large mats. The dewberry is smaller than most blackberries, with fruits containing fewer drupelets, which break up as you pick them. These come earlier in summer than the blackberries and do not have the same shiny glossiness, but are more matt, with a whitish bloom similar to a plum's.

Blackberry remains have been found in many of the earliest European habitations and they have been an important autumn crop since before recorded history. They were known to the Ancient Greeks, as much for the herbal properties of their leaves as for the fruits. They have always grown in great profusion in woods and hedges, on heaths and moorlands and indeed on every site as soon as it has been vacated by us. So much fruit has always been available free that brambles have never been cultivated on a large scale and even the markets were satisfied by gleaning the wild crop. More recently there has been some breeding, with improved varieties such as **Bedford Giant** and **John Innes**. Most work has gone into producing thornless and large-fruited varieties, in effect neglecting flavour, so many people prefer to pick the wild berries rather than use cultivated sorts. Certainly there has been far more interest in the development of hybrids with other *Rubus*.

The North American dewberries, which derived from *R. alleghaniensis*, were introduced to Europe. They are less vigorous and larger-fruited than the natives, but have unfortunately also proved more tender. Other introductions have been more successful. The **Himalayan Giant** is an exceptionally vigorous variety which has encroached on the wild populations all over. Strangely enough, the thornless **Oregon Cutleaf** blackberry is not American, as used to be thought, but an old English variety of *R. laciniatus*, or **Parsley-leaved Blackberry**. It has bigger fruits, is nearly evergreen and comes almost true from seed. It was discovered in Surrey in 1770.

BLACKBERRIES

Blackberries never ripen all at once

varieties

Helen, a new thornless sort, is very early ripening in early July. **Silvan** is recommended for heavy soils, dry situations and for disease resistance and is also very early. **Black Butte** produces really huge well-flavoured berries in mid-July. **Karaka Black, Waldo** and **Fantasia** also make similar claims of producing the largest tastiest berries. **Veronique** and **Loch Maree** are thornless, serious croppers but effectively sold for their ornamental pink flowers. **Loch Tay** and **Loch Ness** are a new sort, with short upright thornless canes so these can be trained more like raspberries, as can the similar USA bred **Chester**; all crop from mid-August. In late August and September ripens another American bred thornless variety, **Triple Crown**, reckoned to be very productive and well flavoured.

cultivation

The whole bramble family are gross feeders and love rich, moist soils. They will crop in a light shade but are sweeter in the sun. Their cultivation and control is more like a pitched battle – if you give way they will take over your garden! Each stem arches over, grows down and roots from the tip to form a new stool of stems. Their sheer size and exuberance makes them too much for most small, modern gardens, though they do crop handsomely. Heavy dressings of compost and thick mulches will keep up the yields. They do well grassed up underneath, and left to themselves will exclude weeds and anything smaller than large trees.

Growing under Glass and in Containers

They are so easy outside it would be bizarre to grow them under cover. Blackberries do not like the cramped conditions pots afford, and as most of them are thorny they are seldom grown this way. The thornless ones are still too vigorous to thrive in any reasonable pot.

Ornamental and Wildlife Value

The fruiting species are delightful in flower, with a mass of blossom in early summer, and there are many ornamental varieties and species, though most are too vigorous for modern gardens. Their vast quantity of blossom is valuable for bees and other insects and the fruit is an immense feast for wildlife, fattening up the

Oregon Thornless

Pick fruits with shiny skins; matt ones are old

bird population for the winter. The thicket of the bushes makes a snug, dry home for many small creatures, from ladybirds and beetles up to small rodents and birds.

Maintenance

Spring Weed, mulch, spray seaweed solution.
Summer Tie in new shoots.
Autumn Root tips or cut off, pick fruit.
Winter Prune, add copious amounts of compost.

Pests and Diseases

Remarkably tough and reliable, blackberries pose few problems. Even the birds find it hard to eat them as fast as they are produced. Weeds can get into the stool but rarely succeed for long as the brambles are so vigorous a weed themselves.

Propagation

Seedlings come up everywhere, but are variable. The tips readily root and form new plants in the few weeks at the end of summer and into autumn. At this time, make sure the tips go into pots of compost if you want extra plants, or cut them off if not.

Pruning and Training

Blackberries can fruit on wood older than one year old and the canes do not always die, as with raspberries. However the new wood is better and carries fewer pests and diseases so it is best to cut all the old and dead wood out and tie in the new. The canes are much longer and carry heavier loads than raspberries so strong supports are necessary. The young canes need tying in during summer. If new plants are not needed they are best de-tipped in late summer to stop them from rooting wherever they hit the ground.

Harvesting and Storing

Traditionally blackberries are picked as they ripen, from late summer up until Michaelmas, or the first frost, when the Devil was supposed to have spat on them and made them sour. They are unusable red but turn soft to the touch as they blacken. **Bedford Giant** ripens one berry in each bunch way ahead of the others. The berries are fairly tough-skinned, so can travel and last longer than raspberries if picked dry and not so overloaded that they pack down. They are best used as soon as possible or frozen; the spoilt texture when they defrost is no problem if they are to be used in cooking anyway.

companion planting

Blackberries benefit from tansy or stinging nettles nearby and they are a good companion and sacrificial crop for grapevines.

other uses

Brambles make a secure and quick barrier to many four-legged animals and two-legged rats. There are few in the world who will try to come over or through a fence or hedge clothed in any of this thorny bunch. **Himalayan Giant** is so big and tough it will stop almost anything between the size of a rabbit and that of a tank! The prunings are vicious but do burn well.

Waldo

BLACKBERRIES

culinary

Blackberries are often too sour to eat raw but once cooked they are much tastier and do not have such a deleterious effect on one's insides. They make excellent jams, although, as they are rather seedy, the jelly is more often made. Frequently apples are included in blackberry dishes, especially jams and jellies, to aid setting and also because the flavours combine so lusciously. Blackberry wine is made by country folk everywhere and the berries used to be added to wines and spirits to give a distinctive colour, such as with the Red Muscat of Toulon.

Bob's Blackberry and Apple Pancake Supreme
Serves 4

Approx. 300ml pancake batter
250g blackberries
Golden syrup, to taste
30g cornflour
A little milk
2 apples
Knob of butter
Sugar, lemon juice, lots of cream or yogurt

Prepare the pancake batter and set aside. Wash the blackberries and simmer with golden syrup till soft. Strain, reserving the juice, and keep the fruit warm. To the juice add enough water to make it up to 200ml, return to the heat and bring to the boil. Cream the cornflour in a little milk, pour it into the boiling blackberry juice, stirring all the time, and cook until the juice thickens. Set aside.

Peel, core and chop the apples into chunks. Heat them rapidly with a little butter till they start to crumble at the edges, then remove from the heat and keep warm. Next preheat a grill and a frying pan. Oil the pan. Once it is smoking, pour in all the pancake batter. As the bottom sets, but while the top is still liquid, take the pan off the heat, rapidly spoon in apple chunks, blackberry blobs and stripes of sauce. Swirl slightly so that the liquid batter blends a little but does not mix or cover completely. Sprinkle sugar generously over the top, then add the lemon juice and put under the red-hot grill. Serve immediately the top has caramelised. This goes well with yogurt or cream.

Rubus hybrids from the family *Rosaceae*

LOGANBERRIES
BOYSENBERRIES AND TAYBERRIES

Bush/vine. Life span: medium. Deciduous, mostly self-fertile. Fruits: 6 x 2cm. Value: some vitamin C.

The loganberry resembles a blackberry in manner of growth, but the fruits are more like raspberries: cylindrical, dull red and firm, with a more acid flavour than either, making them sour raw but exquisite cooked. Boysenberry fruits are sweeter, more blackberry-like, larger and reddish-purple. They can be savoured raw with cream but also make the celebrated jam. The tayberry is bigger and sweeter than either, with an aromatic flavour. When fully ripe, the enormous loganberry-like fruits are dark wine red or purple and nearly three times the size of most other berries.

Loganberries were reputedly a hybrid of American dewberry and raspberry, raised by a Judge Logan of California in 1882. Introduced to Britain in 1897, loganberries have remained the supreme culinary berry for nearly a century. The boysenberry has a similar history. It is believed to be another hybrid dewberry but in fact is probably a youngberry x loganberry. It is not as hardy as other hybrids and does better in a warmer site than the rest. The Medana Tayberry was developed by the Scottish Crop Research Institute, who crossed the Oregon blackberry Aurora with a tetraploid raspberry to produce this excellent fruit. It is outstandingly the best of all hybrids so far.

Tayberry

Thornless loganberries

varieties

There are several varieties and many other similar hybrids. The **Thornless Loganberry 'LY59'** is often thorny; the fruit does not pull off the plug easily, but it is a very heavy cropper and has a good flavour. The totally thornfree **'LY654'** is better picking but is not quite as productive. I similarly find the **Thornless Boysenberry** not as good a cropper as the thorned. There is now a thornfree **Tayberry**. The **Tummelberry** is like a Tayberry but the very tasty berries ripen red not purple. The **Marionberry** and the **Youngberry** have the habit and appearance of blackberries but the fruits have more flavour. The Youngberry, sometimes called the **Young Dewberry**, is available as a thornless version. The **Black Loganberry** is a New Zealand variety; it has cylindrical, tapering fruits, and is slow to establish and crop. The **Laxtonberry** has a round, raspberry-like fruit and is not self-fertile. The **Veitchberry** is a blackberry crossed with an autumn-fruiting raspberry, more mulberry-like, with a later fruiting season than that of other hybrids and large sweet fruits. Tayberry **Buckingham Thornless** is a new thorn-free variety that retains both the vigour and flavour of the superb original.

cultivation

The hybrids all need much the same, cool conditions and rich moist soils. The boysenberry will cope with drier sites and indeed prefers some shelter. They are generally quite happy on a cool shady wall if they have a moist root run.

LOGANBERRIES

BOYSENBERRIES AND TAYBERRIES

Growing under Glass and in Containers
They prefer cool, shady positions so find the dry heat under glass too much and become prone to pests such as red spider mite. Generally needing a bigger, cooler root run than can be afforded in a pot, they will resent the confinement, sulk and crop poorly.

Ornamental and Wildlife Value
Not very useful ornamentally, and thorny, but their flowers and fruits are valuable in the wild garden, bridging the gap between raspberries and blackberries well.

Maintenance
Spring Weed, mulch and spray seaweed solution at monthly intervals.
Summer Protect and pick fruit, tie in new canes.
Autumn Root tips in pots or cut them off.
Winter Cut out old canes and tie in the new; add copious compost.

Propagation
As these are hybrids; they will not come true from seed, though interesting results may be had. Tips can be rooted in late summer and early autumn and occasionally the roots can be successfully divided.

Pruning and Training
They grow much like blackberries but can have more brittle canes, like raspberries, so care must be taken when bending them. They mostly fruit on young wood which dies and is cleared completely after the second year. Canes may produce again a third year, as blackberries might, but are usually unproductive, so annual replacement of all the old by new is generally considered a better policy.

Pests and Diseases
The only common major problem is bird losses, which are high as these plants mostly crop after the summer fruits but before the wild blackberries. Weeds choking the stool can reduce their vigour more than with blackberries.

companion planting

Tansy, marigolds and alliums are all beneficial.

other uses

These canes will make good additional barriers with fences and hedges.

Buckingham Thornless

culinary

Generally best flavoured when fully ripe, these fruits are not very acid and benefit from the addition of redcurrant juice to many recipes. Varieties from which the plug is not easily removed, or which even detach a thorny stalk with the fruit, are best used by straining or juicing them first. These fruits make excellent jams, jellies, tarts and pies. The juices are delicious as drinks and make wonderful sorbets.

Summerberry Squash
Makes about 2.5 litres

1kg mixed berries
500g redcurrants
1 litre water
1kg sugar

Wash the fruits and simmer them down with half the water till soft, then strain. Cover the fruit pulp with the rest of the water, bring almost to boiling point, and strain again. Combine the strained juices and the sugar, heating gently if necessary to make sure the sugar dissolves completely. Cool, then pour into plastic bottles when cold and freeze till required. Defrost and dilute with water to taste.

Rubus species from the family *Rosaceae*

JAPANESE WINEBERRIES AND RUBUS SPECIES

Rubus spectabilis – Salmonberry

Bush/vine. Life span: medium to long. Deciduous, self-fertile. Fruits: variable. Value: some vitamin C.

***Rubus phoenicolasius*, the Japanese wineberry, is the best of the vast raspberry/blackberry clan. This delicious and highly ornamental cane fruit resembles a vigorous raspberry covered with russet bristles and thorns. Unlike those of blackberries, these prick rather than jab so are more pleasant to handle and pick. The fruits are smaller than blackberries, orange to cherry red, and they are generally far more palatable.**

Japanese wineberries are not a hybrid but a true species coming from North China and Japan. They certainly do come true, as you will find as they appear all over the garden once the birds spread the seed. Introduced to Britain around 1876, Japanese wineberries were considered worth cultivating and won a First Class Certificate from the Royal Horticultural Society in 1894. During the century since, however, they have not proved popular except with children of all ages who are lucky enough to find them.

varieties

There are no named varieties of any of these species, though there is some variation in leaf and fruit colour so there is scope for improvement. The **Black Raspberry** is close to the **American Black Raspberry**, *R. occidentalis*, in habit and fruit. *R. leucodermis*, **Blackcap**, has thorny, bluish stems with light green leaves, white underneath, on a medium-size bush, small white flowers and purple-black sweet fruits with a plum-like bloom. Yellow- and red-fruited forms occur in its native Northwest America. Another, *R. parviflorus*, the **Thimbleberry**, has large, fragrant, white flowers on strong, thornless stems and large, flattened, insipid, red berries. *R. parvifolius*, the **Australian Bramble**, was fruited in England in 1825 and has small, pink, tasty, juicy berries. The **Salmonberry**, *R. spectabilis*, has maroon-red flowers on prickly erect stems and acid orange-yellow fruits. Apparently Native Americans ate the cooked young shoots. *R. arcticus*, the **Arctic** or **Crimson Bramble**, has amber-coloured fruits that are said to taste of pineapple. Very unusual is the **Rock** or **Roebuck Bramble**, *R. saxatilis*, which grows just like strawberry plants, and is eaten in much the same way. The Russians used to distill a spirit from the berries.

cultivation

Although they will grow almost anywhere, the biggest berries come from plants growing in rich, moist soil well enriched with compost and leaf mould. They will grow in moderate shade or full sun and are self-fertile. Best grown on a wire fence or wired against a wall, after the manner of blackberries, they need a spacing of at least 3–4m apart and wires to at least 2m in height. Like most fruits, they do best when grown in well-mulched, clean soil, but will still produce when grassed down around.

Growing under Glass

These are mostly so easy to grow outside that there is little advantage to having them under glass except to extend the season.

Growing in Containers

Growing them in pots will shorten their life and give greatly reduced yields, but it may be well worthwhile for both their snack and their garnishing value.

JAPANESE WINEBERRIES

Very ornamental and tasty, too

Ornamental and Wildlife Value
Japanese wineberry leaves are a striking light green, with russet bristled stems, bright orangey-red fruits and a star-shaped calyx left afterwards. They are highly decorative – probably the best fruiting plant to train against a whitewashed wall or up a pole for all-year-round interest and colour. Their value to wildlife is as immense as that of the whole clan.

Maintenance
Spring Weed, mulch, spray seaweed solution.
Summer Protect and pick fruit, tie in canes.
Autumn Pick the fruit and tie in canes.
Winter Cut out old canes, tie in the new, add copious quantities of compost.

Propagation
These are species, so they come true from seed and the tips can be layered in late summer and early autumn. Remove the old canes and tie in the new each autumn. Plants have a long life if cared for. Prune out any infections early.

Pests and Diseases
Choking and climbing weeds such as nettles and bindweed must be well controlled. There are no major problems other than the birds.

Harvesting and Storing
Japanese wineberries are one of the most delicious of all fruits eaten fresh and also in quantity, though they will keep for a while in the cool of a refrigerator.

companion planting

Tansy, garlic and French and pot marigolds are all potential good companions.

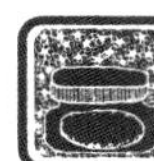

other uses

Their dense growth and prickly bristles make them attractive but impenetrable informal boundaries.

culinary

Very valuable as garnishing for sweet and savoury dishes and simply eating off the plant. Some berries can be frozen to add to mixed fruit compôtes. Japanese wineberry jelly does not set, but forms a treacly syrup, ideal to accompany ice-cream.

Wineberry Ripple
Serves 6

1kg Japanese wineberries
Approx 500g sugar
1kg superb vanilla ice-cream

Freeze a few berries for garnishing. Simmer the rest till soft with just enough water to prevent sticking. Strain and weigh the juice. Thoroughly dissolve three-quarters of the juice's weight in sugar in the warm juice. Leave to cool completely. Once cooled, interleave scoops of ice-cream with the syrup, pressing it all down into a new container. Freeze the new rippled block and then scoop as required, garnishing with the frozen berries.

Vitis vinifera from the family *Vitaceae*

GRAPEVINES

Vine up to any height. Life span: long. Deciduous, self-fertile. Fruits: 2–3cm, ovoid, white, black, red. Value: generally beneficial.

These scrambling vines have smooth, peeling, brown stems, large lobed leaves and bunches of grapes in autumn. The flowers are so insignificant they are rarely noticed but are white and sweet scented. The leaves colour well in autumn; red-berrying varieties tend to go red and white ones yellow.

Grapes have been with us since Biblical times; Noah planted a vineyard. The Egyptians show full details of vineyards and wine-making in their relics from 2440 BC. The Romans spread vines all over Europe until, in the first century AD, Emperor Domitian protected his home market and ordered the extirpation of the grape from Britain, France and Spain. Two centuries later, Emperor Probus restored the vine and long after the Roman Empire collapsed the monasteries kept vineyards going.

By the time of the Domesday Book, in the eleventh century, there were still thirty-eight vineyards in Britain. But the climate was cooling and the last UK vineyards disappeared in the eighteenth century. During the nineteenth century grapes were widely grown in glasshouses and the Victorians raised grape cultivation to perfection, almost year round, in hothouses. Without the cheap labour and even cheaper fuel of Victorian times these hothouse grapes disappeared – though some survived, unproductive, on the sunny walls left when the glass had long gone. Since the Second World War there has been a revival of British viticulture and the vineyards are returning. Of course in Europe the grape has remained part of life, and in 1494 was already being grown in the New World. Over the last centuries the vine spread to almost every part of the world. Now most grapes are grown on American roots to prevent Phylloxera root aphids.

varieties

Specific grapes best for growing under glass or outdoors for wine making are recommended in the next sections. The most reliable and therefore best two grapes for dessert use I have found for growing outside, ideally on a warm wall, in a cool English climate, from many dozens tried, are **Boskoop Glory** a delicious dark purple grape, remarkably problem free and ripening in October, and **Siegerrebe** which has a rose fruit with a sweet Muscat flavour though bitter pips. **Siegerrebe** comes so early, outdoors it can ripen in August; the wasps bother it whereas most other outdoor varieties miss them. Others worth trying in a warm spot are: **Phoenix**, a pale yellow grape with a hint of Muscat, and fairly disease resistant and **Lakemont** which is similar with more of the Muscat flavour and is seedless. **Perlette** is seedless, lacks the Muscat flavour but ripens earlier than most others, save **Siegerrebe**. **Polo Muscat** is not seedless but has that Muscat flavour, is fairly disease resisting and also early cropping. **Regent** is a blue black, fleshy grape ripening later in October as the leaves turn a deep red. **Muscat Bleu** ripens a few weeks earlier and does have that delicious Muscat flavour. **Dornfelder** is a heavy cropping red with some disease resistance and a fair flavour. **Flame** is a red seedless grape, with a unique crunchy texture. It is quite more-ish but will only ripen well on a warm wall. If you just want sheer quantity plant the **Strawberry Grape**, which most years produces rose purple bunches in abundant profusion with a hint of that American foxiness. It has given me more than eight gallons of juice from a single vine. If left to hang the scent of these last grapes perfumes the whole garden.

GRAPEVINES

Little is more pleasing than ripe grapes hanging aplenty

cultivation

Grapevines are very easy to grow. They are usually too vigorous and do not need rich conditions to crop well. They need a hot, dry autumn to ripen well and thus are usually best grown on walls in cooler regions.

Ornamental and Wildlife Value

Grapevines are quick to climb over and hide objects and turn bright colours in autumn, so they are valuable where space allows them to ramble. In the wild garden grapevines are useful in both flower and fruit.

Growing under Glass and in Containers

These are so advantageous there is a further section on pages 520–521.

Maintenance

Spring Weed, mulch and spray with seaweed solution at monthly intervals.
Summer Tie in and later nip out ends of shoots.
Autumn Protect and pick the fruit.
Winter Prune back hard.

Propagation

In the UK we have no Phylloxera so we can grow grapes on their own roots; any ripe cutting will root in autumn. Otherwise they are budded on resistant rootstocks.

Pruning and Training

There are many ways to prune grapes and many sub-variations, enough to fill a book on their own. Left to themselves vines often produce rank growth and exhaust themselves with overcropping; see sections on pages 519, 521 and 523.

Pests and Diseases

Birds are the main cause of losses! Mould can be common in damp ripening seasons and mildews in dry ones. There are organic sprays to combat these but often the more resistant varieties crop unaided. The vine weevil may appear and is best treated with traps and by applying a parasitic nematode solution.

Harvesting and Storing

Kept cool and dry, the grapes hang on the vines well. Cut bunches with a stalk, place the stalk in a bottle of water and keep the grapes in a cool, dry cellar for weeks. The less they are handled, the better they will keep!

companion planting

Traditionally grown over elm or mulberry trees, grapevines are benefited by blackberries, sage, mustard and hyssop growing nearby and inhibited by cabbages, radish, Cypress spurge and even by laurels.

culinary

The best are excellent dessert fruits, are easily juiced and the juice can be frozen for year-round use. They make good jellies and any surplus used for wine.

Love Nests

Per person

Approx. 6 large dessert grapes, preferably Muscat
Marzipan, apricot conserve, clotted cream and dark chocolate, to taste
Individual meringue case

Peel and seed all but one grape per portion and fill each with a pellet of marzipan. Smear the meringue bases with apricot conserve, then a layer of cream. Press in the filled grapes, cover with more cream, top with grated black chocolate and the perfect grape. Serve the nests immediately.

Vitis species from the family *Vitaceae*

GRAPEVINE SPECIES

Vine. Life span: long-lived. Deciduous, sometimes self-fertile. Fruit: up to 2cm, in bunches, red or black.

There are hundreds of true grapes, or *Vitis* species, which are mostly ornamental climbing vines. (Very closely resembling them are the *Ampelopsis* and *Parthenocissus*, of which the most common are known as Virginia creepers. These resemble grapevines and even form bunches, which are not edible.) These *Vitis* species are much grown for their covering capacity as they soon hide eyesores, and for their spectacular autumn colour. The fruits are often considered a bonus, but offer different and exciting flavours.

Although the *Vitis vinifera* varieties are almost exclusively used for commercial purposes and wine-making there are countless other *Vitis* species grapes eaten throughout the world, and have been since time immemorial. Their blood has also influenced the *V. vinifera* varieties on many occasions. These are but a few.

varieties

Vitis aestivalis, the **Summer, Bunch** or **Pigeon Grape** is from North America. It has heart-shaped leaves, downy underneath, scented flowers and early black grapes. It was first seen in Europe in 1656. *V. labrusca*, the **Plum, Skunk** or **Fox Grape** is another first brought to Europe in 1656. Its young shoots are covered in down and the leaves are thick, dark green on top, ageing to pink underneath. The fruits are rounded, blackish purple and have a distinctive musk or fox flavour which some dislike. I enjoy it but not in wine! It has given rise to several good cultivars such as **Concord**. *V. vulpina* is similar, with glossy leaves. Both have sweet-scented flowers. *V. rotundifolia* is the **Muscadine** or **Southern Fox Grape**, widely used for wine in southern USA. It produces only half a dozen large fleshy grapes per cluster, usually black, though there are cultivated local white varieties. The **Winter, Chicken** or **Frost Grape berry**, *V. cordifolia* is very hardy and does well on lime soils. The dark purple fruit has to be frosted before it is edible and is used for wine. Some local cultivated varieties have sweeter, tastier, red or black fruits. *V. riparia* is the high climbing **Riverbank Grape** from North America, with large, glossy, deeply lobed leaves and big panicles of male flowers that smell most distinctly and sweetly of mignonette. The fruits are black or amber and very acid. *V. coignetiae* comes from Japan and Korea. It has enormous leaves up to 30cm across. It is strong-growing and the leaves turn crimson and scarlet in autumn so it is much used ornamentally, but the black grapes with a bloom are not very tasty. *V. davidii*, once called **Spinovitis** because it has spines on the shoots, stems and leaves, was brought from China for its glorious, rich crimson, autumn colouring. It also has edible black fruits. *V. californica* was originally cultivated by the native American Pueblo Indians. It must have been good because, to quote Sturtevant, 'The quantity of the fruit that an Indian will consume at one time is scarcely credible'.

cultivation

Most species require much the same treatment as *V. vinifera* grapes. Over-rich conditions should be avoided. For most, a warm site is better, but they do not like warm winters and do best with some chilling in winter.

GRAPEVINE SPECIES

Some varieties colour beautifully in autumn

Growing under Glass
The freedom from frost, rain and birds is valuable but vines grown under glass become more susceptible to pests and mildew.

Growing in Containers
Like *vinifera* grapevines, the species do not enjoy cramped conditions, but they can be grown in large pots. This shortens their life and gives small crops but conveniently controls their vigour, allowing many to be grown in a small area.

Ornamental and Wildlife Value
These are of the highest value as ornamentals and are useful for quick screens and coverings, though they can be too vigorous for small gardens unless hard pruned. Their bountiful flowers and fruit make them very good for wildlife gardens.

Maintenance
Spring Weed, mulch and spray seaweed monthly.
Summer Tie in shoots, protect fruit, cut off tips.
Autumn Pick fruit.
Winter Prune well.

Propagation
As these are species they can be grown from seed. Ripe wood cuttings, taken in late autumn, are best, and budding or grafting is possible.

Pests and Diseases
As these are species, they are generally resilient to most of the common grape pests and diseases. However birds are still as much, if not more, of a problem.

Pruning and Training
See also sections on Grapes for Dessert and for Wine. There are many ways to prune grapes and many sub-variations, enough to fill a book on their own. The species are best treated much as regular vines which, left to themselves, often produce rank growth and nearly exhaust themselves with overcropping. For ornamental purposes that is no problem, but for fruit they are best hard pruned and trained on a wall on wires about 50cm apart, and on walls under cover for the more tender varieties. The main framework is formed the first years, covering the wires with stems furnished with fruiting spurs. Thereafter these shoot each spring and a flower truss appears between the third and fifth leaf. After another three or four leaves, each shoot is tipped, as are any replacements as they come. Thinning the number of bunches is recommended, leave no more than three per metre run of cane. In winter new canes are all cut back hard to two buds on a stub on each spur. Many pruning and training methods are available.

companion planting

Species grapevines are probably benefited by blackberries, sage, mustard and hyssop growing nearby and they are inhibited by cabbages, radishes, Cypress spurge and laurels.

other uses

As with other grapes, the grapevine prunings make great kindling.

culinary

Most of these grapes are too small and pippy or sour to be used raw as dessert. They are best juiced or turned into jellies. The strong flavour of some, such as the Fox Grape, can make them unsuitable for wine.

Grape Jelly
Makes approx. 2kg

1kg grapes
Approx. 1kg sugar

Simmer the grapes with only just enough water to stop them sticking. When they are soft, strain and weigh the liquid. Add the same weight of sugar to the juice and bring to the boil, skim until clear, bottle into hot jars and seal.

Vitis vinifera from the family *Vitaceae*

DESSERT GRAPES UNDER GLASS

Vine. Life span: exceptionally long-lived. Fruits: up to 3cm, oval or round, any colour.

Although most dessert grapes are grown in hot areas, the very best flavour, size and succulence come from grapes grown under glass in cooler regions. The varieties selected make bigger grapes with thinner skins than outdoor varieties. They are hardy, but depend on protection and warmth to ripen in time, so they will not crop outside except in favourable years.

Growing dessert grapes under glass was raised to an art by the Victorians, who could afford the heat and labour to produce perfect bunches of fine grapes almost every day of the year. Now cheaper greenhouses and plastic-covered tunnels are within the reach of most gardeners and many varieties can be grown with little or no extra heat, so these gourmet fruits are once again achievable.

Don't spurn small grapes; they are often sweetest and seedless

varieties

Muscat Hamburg is by far the best. It has oval, black bunches of firm, sweet grapes with a superb flavour. It is superior to **Black Hamburg**, though this too is a fine old variety, with bigger grapes in larger bunches. Prize berries could measure 10cm in circumference! The **Black Hamburg** planted at Hampton Court Palace in 1796 still produces hundreds of bunches every year. Neither will crop outdoors save in an exceptional situation. **Buckland Sweetwater** has small, white, sweet and long-keeping grapes. **Chasselas Doré/ Golden Chasselas** is early, with translucent yellow fruits. It is most reliable and will even crop outdoors on a warm wall. **Perle de Czaba** and **Siegerrebe** are both speedier still and may crop outdoors, but they are better and reliable only on a hot wall or indoors. They both have the spicy muscat flavour; the **Perle** is yellow and **Siegerrebe** a rosé. The latter is a shy cropper with very choice berries, but dislikes limy soils. Also recommended for growing under cover in pots are any other allegedly outdoor dessert grapes listed on page 516, especially **Flame** and **Boskoop Glory**. Two more grand old indoor dessert varieties that I have found superb are the big black Muscat variety **Madresfield Court** and the sweet white **Foster's Seedling**. If you wish to make your own currants grow **Zante**, which is nearly seedless, introduced in 1855 it helped make the Californian fruit industry but is unsuitable outdoors in cooler regions.

DESSERT GRAPES

cultivation

Amenable to almost any soil, they do not require rich conditions. The very best varieties need a long season with an early start. With Victorian heating they were started into growth in early spring. Heat was used to keep them frost-free through spring, and again in autumn to finish off a late crop. However, in a good season, many varieties can be cropped under glass just with the extra natural warmth afforded, the more so if they are grown in pots and brought in after chilling outside.

Growing in Containers
This is the only sensible way with grapes. They resent the confined root system and need careful watering and pruning, but become controllable. Several varieties can go in a greenhouse too small for one planted in the ground. In tubs they are also conveniently moved outside for winter chilling then brought under cover for an early start. This ripens them months sooner, allows several to be cropped in turn AND their sojourn outdoors after harvest until the following spring keeps them clean of pests and diseases.

Ornamental and Wildlife Value
Well-pruned vines in pots or trained on walls are very decorative. The framework is easily manipulable so almost any form can be achieved as long as all fruiting wood is kept at roughly the same level and in the light.

Maintenance
Spring Bring in or heat greenhouse, spray with seaweed solution.
Summer Prune back tips, thin bunches, spray with seaweed solution.
Autumn Pick the fruits.
Winter Prune and put outside or chill in a glasshouse.

Pruning and Training
In pots and tubs vines are grown vertically or wound as spirals around a central supporting post or cane about 2m tall. Once the canes reach the top all further growth and sideshoots are removed. Each autumn the canes are ruthlessly reduced to two buds on a stub of new growth, on three or four spurs on the stump. In spring the best four or five canes are chosen when the flowers have set and all other canes broken off and the remainder tied to the support. I now do not advise planting in the ground for vines under cover but if you must then the vine is best planted outside, trained in through a hole and then treated much as on a wall. This means wires about half a metre apart and nearly the same distance from the roof to allow for the annual growth. The vine framework is formed over the first year or three, covering the wires with main stems soon furnished with fruiting spurs. Thereafter these spurs shoot each spring, a flower truss appears between the third to fifth leaf, the best placed shoots are left, the surplus broken off. After three leaves beyond each flower truss each remaining shoot is tipped, as are any other further growths. Thinning the number of bunches is essential; leave no more than two per metre run of cane. Thinning the grapes in the bunch is tedious and only for show. In autumn the canes are all cut back to two buds out from each spur.

Pests and Diseases
Grapes can suffer many problems but usually still produce. If the air is too humid when grapes are ripening they may mould. If they are kept too dry before then they get mildew and red spider mite. However, the permitted sprays and usual remedies work well with most problems. Vine weevils can be excluded from vines in pots by making a lid that fits snugly around the stem.

Harvesting and Storing
Under cover they ripen early and hang longer as they are less threatened by pests or weather. Late varieties protected with paper bags may keep almost until the New Year.

Chasselas D'Or sets the standard for reliable sweet transparent yellow desserts

culinary

The dessert fruit *par excellence*, their juice is delicious and freezes well. For wine-making, the flavour and high sugar content go well in combination with outdoor grapes, which have lower sugar and higher acidity.

Recipe
Just eat them as they come, sun warmed.

companion planting

Grow French marigolds underneath the vines to help deter whitefly.

other uses

Pieces of old vine, detached when pruning, make good supports for climbers in pots.

Vitis vinifera from the family *Vitaceae*

GRAPES
OUTSIDE AND FOR WINE

Vine. Life span: long-lived. Deciduous, self-fertile. Fruits: up to 2cm. Some vitamin value.

Vine grapes are as sweet, or more so, than dessert varieties. They have been bred to produce many small bunches rather than large berries, and this suits the vine's natural habit. The grapes are every bit as tasty, just smaller, and if you don't want the wine the juice is still valuable. They are hardier than dessert grapes and many are cropped commercially.

Although wine is predominantly produced in the Mediterranean region and areas with a similar climate, vineyards have been and are successful in many cool regions. The wines are usually light whites, but new hybrids now produce reds as well. Unfortunately European legislation does not permit commercial plantings of the new, high-yielding, disease-resistant hybrids, but they are still available to the amateur.

varieties

The classic wine grapes such as **Cabernet, Chardonnay/ Pinot Blanc**, **Pinot Noir** and even **Riesling** are not suitable for cooler regions. **Mueller Thurgau/Riesling Sylvaner** is much planted for its excellent white wine, but mildew can be a problem, **Seyve Villard 5/276** is a white hybrid. It is more reliable but lacks the character, so often both are grown. **Siegerrebe** is a light cropper of rosé berries, which add flavour to blander grapes. The hybrids are by far the best croppers and most disease-resistant; **Triomphe d'Alsace**, **Leon Millot**, **Seibel 13053** and **Marshall Joffre** produce masses of dark black bunches, which make good red wine or juice. The **Strawberry Grape** and **Schuyler** produce well and easily but their flavour is not to everyone's taste. **Boskoop Glory** is the best outdoor dessert grape, a consistent producer of large, sweet black grapes. Any other of the outdoor dessert grapes from page 516 may be used to make wine, or add flavour or just to juice. Indeed this is really their better use as a good juice can be enjoyed from grapes not quite ripe enough to eat.

Ripening Boskoop Glory

cultivation

Rich soils should be avoided as they will grow excessively. Obviously the warmer and sunnier the better; wires and supports should ideally run north–south, to give sun on both sides of each row.

Growing in Containers

Although crops are light they can be grown in pots, see page 519.

Ornamental and Wildlife Value

A vineyard is quite an accessory to any estate. Quote in bottles per year – it sounds bigger. The aim is two bottles per square metre/yard. The wildlife are certainly going to like your vineyard unless you invest in netting against rabbits, rodents and birds.

Reds are more disease-free

Maintenance

Spring Weed, spray with seaweed solution monthly.
Summer Thin and tie in new shoots, tip after flowering.
Autumn Protect fruit and pick when ripe.
Winter Prune back hard.

Pests and Diseases

The major losses are due to bird damage. Wet summers cause mouldy crops with little sweetness – little can be done about this. Mildew is best avoided by growing the more resistant varieties. Vine weevils are controlled by having clean cultivation and keeping chickens underneath, except when the fruit is ripening.

Whites are more prone to rots

GRAPES

Pruning and Training

There are many different ways of treating vines outside. They can be grown with a permanent framework, as with indoor grapes or those on a wall, and this can be high or low, with benefits from air circulation or heat from the ground. You pays your money, and more for the higher methods! Strong posts and wires up to shoulder height are probably best, so the fruit is borne high enough up to avoid soil splash. Thus the bottom wire should not be less than 30cm high. A modified form of Guyot pruning is often used instead of spur pruning. A short leg reaches to the bottom wire and supports a strong shoot, or two, of last year's growth tied down horizontally to fruit from the buds along its length. Two replacements are allowed to grow from the leg and all other new shoots there are removed. Fruiting shoots are nipped out a few leaves after the flower truss, as with other methods.

Harvesting and Storing

When the fruit has finally ripened enough, but before losses to the birds and mould have mounted, pick the bunches. A dry day after a rainy period gives cleaner bunches. Cut out mouldy bits as you go and press as quickly as possible for juice for drinking or for white wine. White wines can be made from black grapes; only certain Teinturier grapes have red juice. With most varieties the colour only comes from fermenting the skin. Red wines are fermented entire, with the grapes merely mashed. The juice is pressed from the pips and skins later. Extra pips and skins of those squeezed for juice can be added with benefit to the red wine mix as they increase the tannins and sweetness.

companion planting

Asparagus is sometimes grown with the vines in France.

other uses

The vine prunings make good kindling. Big stems are often made into corkscrew handles.

culinary

The juice is one of the most satisfying drinks. It can be used from the freezer throughout the year and is useful as a sweetener. The wine may be even better.

Fruit Salad Soup

Serves as many as you like

Chop and slice finely as many fruits as available and serve in copious grape juice with cream or yogurt and macaroon biscuits.

Triomphe d'Alsace

TENDER FRUITS

This section covers a huge range of fruits, from annual tender fruits, such as melons, to perennial tender ones, which are a much more diverse group, and include citrus, olives and kiwis. All need warm conditions in which to ripen, so in cool, temperate conditions benefit from some protection for at least part of the year, if not all year round. A few tender perennial fruits also need a winter period of cooling so that they will fruit the following year. These can survive outside in temperate areas, though their fruit may not ripen.

The annual tender fruits are amazing. They come from tropical regions, so their natural requirement is for frost-free conditions, yet they are grown in almost every country across the world – including cold ones, with only short summers. In culinary terms, they are used as often, if not more often, for savoury dishes as for sweet.

In their native lands these fruits are mostly very short-lived perennials and are grown as annuals in cultivation. They were brought to Europe and then America by the exploratory voyages of the sixteenth century. Hybridisation and development during the next centuries in both continents slowly introduced better, more hardy and productive varieties. Most still take longer to fruit than the short growing season of temperate climates permits, so we gain extra weeks at the beginning by starting them under glass, with heat. The availability of myriad varieties has made it possible to grow them with only an amateur greenhouse and a sheltered garden or warm wall.

With the exception of Actinidia – kiwi fruit – all the perennial tender fruits have long been in cultivation and their use from wild stocks predates history, so early on they spread well beyond their native lands. This is probably due to the fact that most of them possess thick, rind-like skins and thus have some ability to travel and store well, as compared to other fruits anyway. The human desire for new tastes and flavourings made each of these fruits important items of commerce as they are all distinctly different to other fruits. Once sampled, they became desired, and as world trade started to increase in the seventeenth century, most of these fruits became expensive luxuries and thus indispensable for the developing European, and later the American, markets. They had become common by the Victorian era, as increased production in the warm temperate countries, such as those around the Mediterranean basin and in California and Florida, displaced longer distance imports. These areas became so competitive that few of these fruits have ever become worth growing commercially under glass in the colder regions, despite the ease with which they can be cultivated.

Some of the tropical and sub-tropical fruits are the tastiest and most luscious in the world. Strong sunlight and hot conditions produce sweeter, stronger flavours than can be achieved in temperate zones. The natural conditions change little during the year, in contrast to the fluctuating heat and light and winter chilling of cooler regions. Tropical plants therefore often set fruit several times a year, or continuously throughout it. They were grown on large estates in colder countries from the seventeenth century, but the advent of the stovehouse in Victorian times – a large greenhouse with a massive stone keeping the temperature tropical – allowed more success with fruiting these plants. Many were successful but others could not be persuaded to fruit, even when given extra heat. Now we are more fortunate: with electric light to replicate sunlight, we can give these plants the brightness and day length they need. We also have automatic heat and humidity control, so it is easy to grow many exotic plants ourselves. And if they will still not fruit, they always make attractive houseplants!

Physalis species from the family *Solanaceae*

CAPE GOOSEBERRIES AND GROUND CHERRIES

Herbaceous. Life span: annual or short-lived perennial. Fruits: up to 6cm, yellow to purple in papery husk. Value: rich in vitamin C.

The *Physalis* are distantly related to tomatoes and potatoes. Their best known member is the old garden perennial **Chinese Lanterns**, *P. franchetii*, and the incredibly similar if slightly less vigorous and smaller-lanterned **Bladder Cherry**, *P. alkekengi*. Both have straggling stems, heart-shaped leaves, small pale flowers and bright orangey scarlet papery lantern calyces surrounding a red fruit. Both are often said to be poisonous – even though *P. alkekengi* has been eaten at least since the Greek Dioscorides in the third century AD, and many also think *P. franchetti* is but another variety, not a species! However, the calyce, foliage and even the fruits of some uncommon species are poisonous. ALL are dangerously similar – do not eat unknown unusual varieties!

The **Ground Cherry**, **Strawberry Tomato** or **Cossack Pineapple**, *P. pruinosa*, is low-growing, up to knee level, and has small, green fruits ripening to dirty yellow. Sweet and acid, they are vaguely pineapple-flavoured. *P. peruviana*, the **Cape Gooseberry**, **Ground** or **Winter Cherry**, is taller (about 1m) with yellower fruits. Both fruits are enclosed in similar, though duller, papery husks to those of the **Chinese Lantern**. Another similar fruit is the **Tomatillo** or **Jamberry**, *P. ixocarpa*, which is perennial, with much larger green or purplish berries filling the husk.

The annual *P. pruinosa*, which grows wild in North America, was popular with Native Americans and, apparently, also with Cossacks. It was introduced to England in the eighteenth century, but never caught on. The perennial **Cape Gooseberry**, *P. peruviana*, comes from tropical South America and became an important crop for the settlers on the Cape of Good Hope at the beginning of the nineteenth century. *P. ixocarpa*, the **Tomatillo** or **Jamberry**, comes from Mexico, but has become popular in many warm countries as it fruits easily and reliably and makes good sauces and preserves. Improved versions are now being offered for greenhouse culture elsewhere.

Physalis peruviana

varieties

Golden Berry, *P. edulis* seems much like Cape Gooseberries, *P. peruviana*, while **Little Lanterns** and **Pineapple**, claimed as improved Cape Gooseberries, look as if they have more *P. pruinosa* in them. *P. ixocarpa* has some improved forms such as **Large Green** and **New Sugar Giant** with fruits up to 6cm across and yellow or green instead of the usual purple. Many other *Physalis* species are cultivated locally in warm countries but seed or plants are rarely available.

cultivation

The genus are all best started under cover and planted out. The perennials can ripen outdoors in a good season in southern England, but are far more reliable under cover. *P. pruinosa* is tough and may crop outdoors without protection. They all prefer a rich, light, warm soil and a sunny position. Little support is really necessary, though they do flop.

CAPE GOOSEBERRIES
AND GROUND CHERRIES

Physalis ixocarpa

Growing under Glass
All the family give better and sweeter fruits grown under glass and present no major problems; indeed they seem to be designed for it.

Growing in Containers
P. pruinosa is easily grown in pots. The larger *Physalis* can be grown likewise, but do not do quite as well, preferring a bigger root run. The decorative *P. alkekengi* is worth having in a pot just for the show it provides.

Ornamental and Wildlife Value
Most of the productive species are nowhere near as attractive as their more ornamental cousin, the **Chinese Lantern**. They have small value to wildlife, though the flowers are popular with insects.

Maintenance
Spring Sow indoors, pot up and plant out once hardened off.
Summer Support lax plants.
Autumn Pick fruits once fully ripe, discard husks.
Winter Protect the roots of perennials well for another season.

Propagation
Normally grown from seed, the perennial varieties can be multiplied by root cuttings or division in the spring. Start them off early and pot up regularly to build up a large root system.

Pruning and Training
They need little attention other than tying in the lax growths and clearing away the withered stems after cropping. The roots of perennial varieties can be got through mild winters under protection for earlier crops the following year.

Pests and Diseases
They are remarkably pest- and disease-free. The **Tomatillo** is especially useful as it can be used much like a tomato, but can ripen as early in cool conditions and does not suffer blight as tomatoes may.

Harvesting and Storing
The fruits must be fully ripe to be edible. They can hang on the plant till required as they are rarely attacked by pest, disease or bird. The husk is inedible and must be removed.

companion planting

There are no known companion effects.

other uses

The ornamental Chinese Lanterns can be dried for winter decoration and have been used medicinally.

Pretty in a pot but not good eating

culinary

Most *Physalis* berries are relatively tasteless and insipid raw but make delicious preserves, sauces and tarts. The Cape Gooseberry often tastes best on first acquaintance and may rapidly lose its appeal after the initial elusive strawberry flavour. It was once imported in vast quantities from South Africa, when it was known as Tippari jam or jelly.

Tippari Jelly
Makes approx. 2kg

1kg Cape Gooseberries
A little water
Approx. 1kg sugar

Remove the husks from the fruit and boil them with just enough water to prevent the fruit from sticking. Strain the juice and add its own weight in sugar. Simmer till fully dissolved, skim off scum, then jar and seal.

Cucumis melo from the family *Cucurbitaceae*

MELONS

Herbaceous vine. Life span: annual. Self-fertile but requires assistance as separate male and female flowers. Fruits: 5–25cm spheres, whiteish cream to green, netted or smooth. Value: rich in vitamins A and C, niacin and potassium.

Melons belong to a very wide family of tender, trailing annual vines, much resembling cucumbers in habit. They have broad leaves, softly prickled stems and small yellow flowers, followed by fruits that can be any size from small to very large, round or oval. Melons are characterised by a thick, inedible rind covering succulent, melting flesh, which encloses a central cavity and a battalion of flat, pointed oval, whitish seeds.

The tasty, sweet, aromatic melons we know were apparently unknown to the Ancients. They certainly grew similar fruits, but these seem to have been more reminiscent of the cucumber. Pliny, in the first century AD, refers to the fruits dropping off the stalk when ripe, which is typical of melons, but they were still not generally considered very palatable. To quote Galen, the philosopher-physician, writing in the second century AD, 'the autumn (ripe) fruits do not excite vomiting as do the unripe'. By the third century they had become sweeter and aromatic enough to be eaten with spices, and by the sixth and seventh centuries they were distinguished separately from cucumbers.

The first reference to really delicious, aromatic melons comes in the fifteenth and sixteenth centuries, probably as the result of hybridisation between many different strains. The seeds were left wherever humans ventured. Christopher Columbus returned to the New World to find melons growing aplenty where his previous expedition had landed and eaten the odd meal of melons, liberally discarding the seeds. Likewise, both deliberately and inadvertently, melons have reached most warm parts of the globe and are immensely popular crops for the home garden in many countries. The Victorians developed reliable varieties that were successfully cropped year-round under glass, though most melons are now imported to Britain from warmer countries.

MELONS

varieties

There are countless varieties, literally hundreds if not thousands, and many more go unrecorded worldwide. Most of those that are available, either as seed or commercially, fall into three or four main groups. **Cantaloupe** varieties usually have orange flesh. The fruits tend to be broadly ribbed, often with a scaly or warty rind, but not netted. The flesh is sweet and aromatic. A good typical variety is **Charentais**. They are the hardiest – well, least tender – of the melons and **Sweetheart** is one of the most reliable. **Ogen** melons are an Israeli strain. These resemble a much improved but more tender **Cantaloupe**. The fruits are smooth, broadly ribbed and yellow when ripe, with very sweet, green, aromatic flesh. **Musk** melons are netted and **Nutmeg** melons have distinct netting, are lighter in colour and raised from the yellow or green rind. These are the typical hothouse melons – large oval or round fruits with very sweet and aromatically perfumed flesh from green to orange. A good old variety is **Blenheim Orange**. **Winter** melons are round to oval, yellow or green, smooth or with a leather-like surface and hard yellow flesh that is not very sweet or perfumed. Often called **Honeydews**, these are long keeping, up to a month or so, thus they are popular in commerce.

cultivation

Melons are not difficult if their particular requirements are met. They need continuous warmth, greater than that needed for tomatoes, peppers or aubergines, and must have a much higher humidity. This makes them difficult to accommodate with the Solanum glasshouse crops. They grow most happily with okra and other Cucurbits such as cucumbers, as these all prefer similar humidity and heat and likewise will thrive in slightly less light than the Solanums. Indeed, they do not like bright light and prefer diffused to direct sun. The soil must be rich, very well drained and, like the air, kept continually moist. Melons really do best on hotbeds or heating-up compost heaps.

Growing under Glass

This is almost essential, though in sheltered parts of Southern England in a good year the new hybrid varieties such as **Sweetheart**, **Romeo** and **Emir** may be grown in the open with some hope of success. Otherwise a coldframe is only just enough and needs to be sat on a hotbed or compost heap to be really worth considering. Cold greenhouses can produce light crops, heated ones far more. A coldframe or curtained area in a greenhouse is better still and the ease of providing extra warmth and humidity more than makes up for the diminished light. A coldframe sat on a hotbed, or with soil warming cables, in a greenhouse or polytunnel, is the best attainable by the average gardener and can produce an impressive crop. Indoors, hand pollination is advisable as a precaution.

Yellowing is a sign of approaching heaven

Cantaloupe

Growing in Containers

Melons are one of the easiest plants to crop well in a large pot, provided they are kept warm, well watered and fed regularly with a liquid feed. (I even grow them successfully in bags of fresh grass clippings topped off with a bucketful of sieved garden compost to seal in the heat and smell. The seed is sown direct in a mound of sterile compost set on top of that and covered with a plastic bottle cloche. The bag stands in my polytunnel, a self-contained mini hotbed. When the plant is growing, the bottle is reversed to make a useful watering funnel.)

Ornamental and Wildlife Value

These are not really very decorative plants, but can impress with their luxurious growth. The scent of ripening melons is heavenly. The fruits are well liked by rodents and birds, making these useful for the wild garden in warmer countries.

Maintenance

Spring Sow as soon as warm conditions can be maintained, pot up, nip out tip after four true leaves, pollinate, reduce excess number of fruits, spray with seaweed solution weekly and mist frequently.
Summer Spray with seaweed solution weekly, mist frequently and support swelling fruits.
Autumn Continue as for summer.
Winter Melons can be grown throughout the year if sufficient heat is maintained.

Propagation

Melons are normally started from seed, which does not come true when self-saved unless you are very careful, as Cucurbits are promiscuous cross-pollinators. The seed needs warm, moist conditions to start and the plants need continuous warm, rich, humid conditions. Avoid the roots making a tight ball in the pot, but do not over-pot. The stems can be layered or even taken as soft cuttings to continue the season.

Pests and Diseases

Provided the growing conditions are just right, melons usually suffer no major problems. The slightest drop in humidity and red spider mite may need controlling, as this can check the plants. It is worth introducing the commercially available predator *Phytoseuilis persimilis* if red spider mites are spotted. Aphids and whiteflies sometimes appear, requiring the usual remedies. Melons do suffer occasionally from neck rot where the stem enters the soil, usually during cold conditions. Sterile compost, warmth and clean, carefully applied water usually prevent any occurrence. Victorian gardeners always grew melons on little mounds to keep the neck dry. If neck rot appears, rub with sulphur dust and then earth up with moist, gritty compost to encourage rooting from the base of the stem. Rodents and slugs attack the fruits.

Harvesting and Storing

Their heavily perfumed, aromatic sweetness and luscious, melting texture make them divine when ripened to perfection – though too often they are taken young to travel and are then not sweet, but woody and never well perfumed. For sybarites, they really must be ripened on the vine until they are dropping – the nets are not just there to support! Once they are ripe enough to scent a room, the fruits should be chilled before eating to firm the flesh and then removed from the refrigerator a short while before serving to allow the perfume to emerge fully.

At this stage, beware of slugs creeping in

A tile slipped underneath prevents a yellow bottom

companion planting

Melons like to ramble under sweetcorn or sunflowers, enjoying their shelter and dappled shade, even in English summers! They also get on with peanuts, but do not thrive near potatoes. Morning Glory seed sown with melons is said to improve their germination. Most of all, melons need the same hot, humid conditions as cucumbers and there seems little problem with their pollinating each other. However, there might be if you want to save seed.

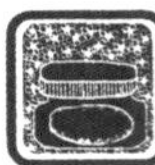

other uses

Melons accumulate a great deal of calcium in their leaves, making them especially useful for worm compost. The empty shells make good slug traps. *Cucumis melo dudaim*, **Queen Anne's Pocket** melon, is grown for its strong perfume, but the flesh is insipid.

MELONS

Melon chutney

culinary

Quintessentially a dessert fruit, melons are nevertheless most often served as a starter in the place of savoury dishes. They may be combined with savoury or sweet dishes and are exquisite as chunks combined with Dolcelatte cheese and wrapped in Parma ham. Melons can be made into jam or chutney, added to compôtes, and used as bowls for creative cuisine.

Melon Sundae
Quantities to taste

Ripe melon
Vanilla ice cream
Toasted flaked almonds
Sultanas
Honey
Dark chocolate
Glacé cherries

Remove balls of melon with a spoon. Layer these in sundae glasses with scoops of vanilla ice-cream, almonds and sultanas. Then pour over the melon juice thickened with honey. Top with grated dark chocolate and a glacé cherry.

Citrullus lanatus/vulgaris from the family *Cucurbitaceae*

WATERMELONS

Herbaceous vine. Life span: annual. Self-fertile. Fruits: variably large, green. Value: rich in vitamins and minerals; a serving contains more iron than spinach.

Watermelons are scrambling, climbing vines. Their leaves are darker, more bluey-green, hairy and fern-like than melon leaves; the flowers are similar, small and yellow. Watermelon fruits vary in size from small to gigantic, from light green to dark green or yellow in colour. The thin, hard rind is packed with red flesh, embedded in which are small, dark seeds. Eating the very juicy flesh is like drinking sweet water.

This productive and nutritious, thirst-quenching fruit comes from Africa and India, but was first mentioned by botanists and travellers in the sixteenth century. The fruit became widely cultivated, but it appears to have been little improved until it reached North America. There it was developed to produce examples weighing over 45kg and many varieties with different coloured flesh, rind or seed. These included a sub-group with ornamental 'painted', 'engraved' or 'sculptured' seeds.

varieties

The flesh is always exceedingly juicy and sweet and red, but once it varied in colour, with black, white, cream, brown, purple and yellow-fleshed forms. Now few varieties other than the reds are grown on any scale. **Charleston Gray** is long, oval and light-green-skinned with crisp red flesh. **Sugar Baby** is round, darker green and with very sweet red flesh, but needs an early start for maximum sweetness. **Yellow Baby** is similar but with yellow flesh rated very highly and I agree. In desperation for seed I've tried those from supermarket fruits with occasionally good results.

cultivation

Warm (70°F plus) well aerated soil is necessary. Pre-warm their site with opaque plastic sheet laid flat with another clear sheet raised above on sticks or whatever. Start in warmth and do not plant out in cold or roots rot. Although copious water is needed at the roots to swell the large fruits, watermelons prefer rather less humid conditions than melons or cucumbers. They also do not need the same high degree of fertility and thrive in any reasonable sandy soil, as they benefit from watering as much as from feeding. In temperate areas with hot summers, the watermelon can be grown outdoors as it produces more quickly than the melon.

Growing under Glass

This is absolutely essential anywhere other than in a country with hot summers. The vines do not mind light shade but need warmth and prefer a drier atmosphere to melons or cucumbers – although they do better with them than with the Solanums. They need plenty of space; grow them a metre apart. Provide support and train them up if space is limited, but keep the fruits well supported.

Growing in Containers

Watermelons can be grown in large pots although they resent the restricted root run and lack of aeration. Be sure to use an open, gritty compost and ensure that religious watering and feeding are maintained throughout the season.

WATERMELONS

Ornamental and Wildlife Value
The vines are much more decorative than those of melons. The fruits are appreciated by wildlife in hotter countries.

Maintenance
Spring Sow as soon as warm conditions can be maintained, pot up, pollinate, reduce excess number of fruits, spray with seaweed solution weekly.
Summer Spray with seaweed solution weekly and straw under swelling fruits.
Autumn Continue as for summer.
Winter Save the seeds for spring sowing.

Propagation
These are readily started from seed, but need regular potting up. Watermelons are not as easy to layer as melons. They prefer clay pots to plastic ones and react very badly to over watering or to compaction of the compost.

Pruning and Training
Watermelons do not need stopping like melons, but it is a good idea to limit each vine to one fruit to ensure a decent size. The plants are best allowed to ramble. If they are trained up anything, provide support for the immensely heavy fruits. On the ground they are best laid on straw or a tile to keep them clean.

Pests and Diseases
Very prone to attack by red spider mite under glass, so commercial predators should be introduced early, before a major attack starts. Although watermelons like it as warm as melons, a similar degree of humidity does not suit them.

Harvesting and Storing
Watermelons are ripe when they sound taut and 'hollow' to a tap from the knuckle. If they are really ripe and well grown they will split open as soon as the knife bites. However, if left intact, they will keep for a week or more.

companion planting

Watermelons do not object to potatoes, as melons do, and may run amongst the plants to advantage in warm countries.

Although small when grown in the UK, they still get luscious and very sweet

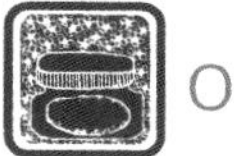

other uses

The closely related *Citrullus colocynthis*, **Colocynth**, **Bitter Gourd**, is like a small, intensely bitter watermelon. It is used medicinally and occasionally pickled or preserved after many boilings.

culinary

Watermelons are best eaten fresh. Their seeds may be eaten – they are oily and nutritious. The pulp can be made into conserves, or reduced to make a sugar syrup.

Watermelon Ices
All quantities to taste

Watermelon
Melted chocolate
Grated desiccated coconut

Freeze bite-size cubes of watermelon on a wire tray. Once they are frozen solid, dip each quickly in cooling melted chocolate, sprinkle with grated coconut and freeze again. Serve them nearly defrosted, but still just frozen, with piped cream and macaroon biscuits, if liked.

Citrus species **from the family** *Rutaceae*

LEMONS, ORANGES
AND OTHER CITRUS FRUITS

Tree/bush up to 8m. Life span: medium to long. Evergreen, self-fertile. Fruits: variable size, orange, green or yellow. Value: rich in vitamin C.

Citrus fruits of the *Rutaceae* family are small, glossy-leaved evergreens with green stems that are occasionally thorny, especially in the leaf axils. Typical of this family, the leaves have glands which secrete scented oil. The small, white, star-shaped flowers are intensely and similarly perfumed, and are followed by the well-known fruits, which take up to a year to ripen. These swell to a size which ranges from that of a cherry to a human head, depending on the species. They are yellow or orange with light-coloured flesh inside a tough, bitter and scented peel. The flesh is sweet or sour, always juicy, and segmented. Each piece may contain a few small seeds.

Originally from China and Southeast Asia, some species and closely inter-related cultivars have been in cultivation since prehistory. They moved slowly westward to India and then on to Arabia and thence to the Mediterranean countries. The Ancient Greeks seem not to have been aware of any citrus. The Romans knew the citron, which is recorded in Palestine in the first century AD, but probably arrived several centuries before. They were widely planted in Italy in the second and third centuries, becoming especially popular near Naples.

The Romans were such gourmands that they would hardly have failed to notice a delight such as an orange. These did not reach Arabia until the ninth century. It was recorded as growing in Sicily in the year 1002 and was grown in Spain at Seville, still famous for its oranges, while it was occupied by the Moors in the twelfth century. It is said St Domine planted an orange in Rome in the year 1200 and a Spanish ship full of the fruits docked at Portsmouth, England, in 1290; the Queen of Edward I received seven. These were probably bitter oranges, as many believe the sweet orange did not reach Europe till later. First seen in India in 1330, the sweet sort was planted in 1421 at Versailles; another planted in 1548 in Lisbon became the 'mother' of most European sweet orange trees and was still living in 1823.

The lemon reached Egypt and Palestine in the tenth century and was cultivated in Genoa by the mid-fifteenth century. The new fruits were soon spread around the warmer parts of Europe, and then further afield, with the voyagers of the fifteenth and sixteenth centuries. Columbus must have scattered the seeds as he went, for they are recorded as growing in the Azores in 1494 and the Antilles in 1557. They had reached orchard scale in South America in 1587 and by then Cuba was covered in them. They are now mainly grown in Florida, California, Israel, Spain and South Africa, though every warm to tropical area produces its own and more.

LEMONS, ORANGES
AND OTHER CITRUS FRUITS

varieties

The various types are of obscure parentage and were probably derived by selection from a distant common ancestor. *Citrus aurantium* is the **Seville, Bitter** or **Sour Orange**. Too sour to eat raw, this is the best for marmalade and preserves and was the first sort to arrive in Europe. *C. sinensis* is the **Sweet Orange**, often known by the variety such as **Valencia Late, Jaffa**, which is large, thick-skinned and seedless, or the nearly seedless and finest quality **Washington Navel**. Blood Oranges, such as the **Maltese**, are sweet oranges with a red tint to the flesh.

C. limon is the lemon. The fruits are distinctly shaped yellow ovoids with blunt nipples at the flower end and the characteristic acid taste. The commonest are **Lisbon, Eureka** and **Villafranca**; the hardiest and most convenient for a conservatory is the compact **Meyer's Lemon**.

C. aurantifolia is the lime. This makes a smaller tree of up to 3.5m. The small, green fruits do not travel well and are mainly consumed locally or made into a concentrate. Limes offered for sale are often small, unripe lemons, given away by the nipple, which a true lime does not have. Alternatively, they may be the similar *C. limetta*, **Sweet Lime**, which is insipidly sweet when ripe. True limes are grown mostly from seed and will require near tropical conditions.

Citrus under cover usefully crop from mid-winter

The scent of their blooms alone make them worth having

C. paradisi is the grapefruit. Not as acid as a lemon, this is relished for breakfast by many. They may be a hybrid of the **Pomelo** or **Shaddock**, *C. grandis*, which is similar but coarser. **Marsh's Seedless** is the commonest variety of grapefruit, with greenish-white flesh; but some prefer the Texan varieties with pink flesh. Grapefruit are nearly hardy.

C. reticulata is the **Mandarin, Satsuma, Tangerine** or **Clementine**. These names are confused and interchanged for several small, sweet, easily peeled and segmented sorts of small orange. *C. medica* is like a large, warty lemon and is now mainly produced in a few Mediterranean countries for making candied peel. There are many other citrus species and hybrids, **Uglis, Ortaniques** and **Tangelos**, to name but a few. The **Kumquat** is not a citrus, but belongs to the similar genus *Fortunella*. The fruits are very like small, yellowish, tart oranges, and are especially good for making preserves. *Citrus/Poncirus trifoliata* is hardy and the musky, fuzzy small orange fruits unpalatable. Hybrids with others look interesting for the future. *Aegle marmelos*, **Bael fruit**, is another relation with hard shelled 'oranges' used fresh, for sherberts and marmalade.

cultivation

Citrus need a warm, rich, moist soil, well aerated and never badly drained. They are all tender, though lemons and oranges have, despite the odds, been grown successfully outdoors in favourable positions on warm walls in Southern England, and even cropped in some years. In warm countries they are spaced about 5–6m apart each way and are in their prime at ten years old. Trees with fruits of orange size will give a crop of over 500 each winter; small fruits, such as lemons, will also crop more than 500; while big fruits, such as grapefruits, will crop less.

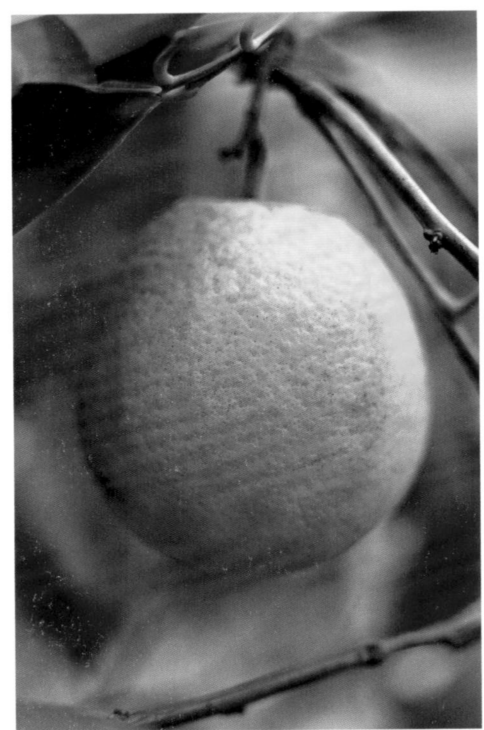

If the fruits feel soft, the plants need more water

Growing under Glass

Frost-free protection in winter is necessary in northern countries. Citrus do not like being under glass all year round. They are much happier outdoors in summer and enjoy a rest in autumn. Thus they are best grown in large containers and moved under glass only for the frosty months, which conveniently are the months that they flower and crop. Limes are among the least hardy sorts and need more heat than the others.

Growing in Containers

This is ideal for citrus as it keeps them compact and makes it easy to give them winter protection indoors. They must have a well-aerated, well-drained, but rich compost. Avoid plastic pots or give them extra perforation. The 'old boys' always reckoned that diluted, stale urine was the best feed for citrus.

Ornamental and Wildlife Value

Very decorative in leaf, flower and fruit, all of which have a wonderful scent, these are ideal subjects for a conservatory and for a warm patio in the summer. The flowers are loved by insects; the fruits are less use to wildlife – which is fortunate for us.

Maintenance

Spring Prune, spray with seaweed solution weekly, move outdoors.
Summer Prune, spray with seaweed solution weekly.
Autumn As for summer.
Winter Prune, move indoors, pick the fruit.

Propagation

Commonly, commercial plants are grafted or budded, often on *Poncirus* stock to dwarf them. Cuttings can be taken. I find some succeed quite easily in every batch. Seedlings are slow to bear fruit and may not be true; however, citrus seeds occasionally produce two seedlings, one being a clone copy of the original plant and the other normal. Seedlings are often more vigorous and longer lived than worked plants, which may offset their slow development.

Pruning and Training

Once these have grown tall enough they are best cut back hard to form a neat cone or globe shape. Regularly remove and/or shorten straggly, unfruitful, diseased and long shoots, cutting back hardest before growth starts in spring. They generally need no support until they are in fruit, when heavy crops may even cause the branches to bend severely and tear.

If the fruits don't yellow well, they're too hot

Pests and Diseases

All manner of pests bother these plants under glass, but when they go out for the summer most of the problems disappear. The usual remedies work and soft soap sprays may also be useful against scale insects, which can particularly bother citrus, especially if ants are about to farm them. Bad drainage will kill them more rapidly than cold weather will!

Satsumas give very good value for their space

LEMONS, ORANGES
AND OTHER CITRUS FRUITS

These containers make it easy to move the citrus outside to the orchard for summer and autumn

Harvesting and Storing

Usually picked too young so they can travel, they are of course best plucked fresh off the tree and fully ripe. They do not all ripen at once and picking may continue over many weeks. The rind contains the bitter oil which can be expressed to give a zest to cooking.

companion planting

In warm countries citrus are benefited by growing aloes, rubber, oak and guava trees nearby. However, they are also said to be inhibited by *Convolvulus* or possibly by the *Ipomoea* species.

other uses

The leaves, flowers and fruits, especially of *C. bergamia*, the Bergamot (NOT the herbaceous *Monarda didyma*), are used in perfumery. The empty shells of the fruits make slug traps and firelighters if dried. *Citrus/Poncirus trifoliata* is hardier than the others and heavily thorned; it is used as a hedge in mild regions.

culinary

The fruits can be juiced – much of the world's crop is consumed this way – or jammed or jellied and made into marmalade. The peel is often candied or glacéed. Small amounts of lemon juice prevent freshly prepared fruits and vegetables oxidising and give a delightful, sharp, clean taste to most things, savoury or sweet. The Victorians grew the seeds for the young tender leaves to add to salads. The peels are much used for liqueurs and flavourings.

Orange Sorbet

Serves 4

4 large unwaxed oranges
Mace
Sugar
Egg white
Parsley sprigs

Cut the tops off the oranges and scoop out the contents. Freeze the lower shells to use as serving bowls. Strain the juice from the pulp and weigh and measure it. Simmer the chopped tops of the oranges with the pulp and a small piece of fresh mace in half as much water as you have juice, then strain out the bits and add half the juice's weight in sugar. Once it has dissolved, mix this sweetened water and the juice and partially freeze. Take it from freezer and beat vigorously, adding one beaten egg white per 450g of mixture, refreeze, then repeat the beating. Serve, partially thawed, in the reserved shells with a garnish of parsley.

Orange marmalade

Olea europaea **from the family** ***Oleaceae***

OLIVES

Tree up to 10m. Life span: long. Evergreen, self-fertile. Fruits: up to 2.5cm, ovoid, green to black. Value: rich in oils.

Cultivated olive trees are gnarled and twisted with long, thin, dark leaves, silvered underneath, though the wild species are bushier with quadrangular stems, rounder leaves and spines. The inconspicuous, sometimes fragrant, white flowers are followed by green fruits that ripen to brown or bluey-purplish-black, and occasionally ivory white, each containing a single large stone.

Found wild in the Middle East, olives have long been cultivated. They were amongst the fruits promised to the Jews in Canaan. According to Homer, green olives were brought to Greece by Cecrops, founder of Athens. They were certainly the source of its wealth. By 571 BC the olive had reached Italy and in the first century AD Pliny records a dozen varieties grown as far as Gaul (France) and Spain. These are still the major producing areas; olives are also grown in California, Australia and China.

varieties

There are several dozen varieties grown commercially in different regions but usually unnamed 'olive' trees are offered. Recently **Picholine, Sativa**, and the newly bred for cool climates **Veronique** have been introduced. **Queen Manzanilla** produces the biggest green pickling variety.

cultivation

Olives grow well in arid sites that will not support much else. They prefer a well-drained, light, lime-based soil. They will grow, but rarely fruit well, outside Mediterranean climatic regions. Small strong trees, they need little support.

Growing under Glass

Olives are almost hardy but it is worth growing them in big pots and brining them in for winter and putting outside in summer. If kept indoors, it must not be too hot or humid.

Growing in Containers

Olives make good subjects for containers, though they are unlikely to be very productive. They must have a free-draining compost and then are fairly trouble-free.

Ornamental and Wildlife Value

Very attractive shrubs, these are worth having even if they never fruit, and they probably won't. The flowers are beneficial to insects, and some are fragrant. They are dense evergreens, making good shelter, and the fruits are rich in oils, so these are useful plants for wild gardens in warmer climes.

Maintenance

Spring Prune, spray with seaweed solution monthly.
Summer Spray with seaweed solution monthly.
Autumn As for summer.
Winter Take indoors or protect from frost.

Propagation

Seeds may not come true, though they are often used, and the resulting plants can be slow to come into fruit. Cuttings with a heel can be taken in late summer but, like the seed, need bottom heat to ensure success. Grow seedlings on for a year or two in large pots before planting out – they get tougher as they get bigger.

Pruning and Training

As olives bear on the previous year's growth, they must have only remedial pruning to remove the dead and diseased or crossing branches. This is best done in late

Not the South Downs yet…

OLIVES

Green olive crops come before the ripe black and are more attainable in the UK

winter or early spring. They can be trained as fans on walls for the extra protection. If the tops are frosted, they can still come again from the root and can be cut back very hard or pollarded. Old trees often throw suckers as replacements.

Pests and Diseases
Olives have very few problems in private gardens, though scale insects can bother them occasionally. Protection from frost and good drainage are more important factors.

Harvesting and Storing
In the Mediterranean region the trees bear when they are about eight years old. They produce about 25kg of fruit each, which reduces to about a quarter to half that weight in oil. The green fruits are ones picked unripe and pickled; the black fruits are ripe and ready for pressing for oil or preserving. The oil that is squeezed out without heat or excess pressure is called extra virgin (an interesting concept). Cheaper grades are produced by heating or adding hot water to the mass.

companion planting

Oaks are thought to be detrimental to olive trees.

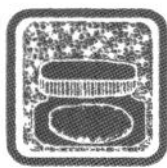

other uses

The oil has many industrial as well as culinary uses. Much is used in cosmetics and perfumery and it was once burned in lamps.

culinary

The oil is used in many ways – in Mediterranean countries it is used in preference to animal fats – and the fruits are added to various dishes. The fruits are also eaten as savoury accompaniments pickled in brine, often stuffed with anchovy or pimento, or dried. If beaten to a paste, olives will make a delicious savoury spread.

Olive Bread
Makes 1 loaf

500g strong white flour
1 packet dried yeast
Water
60g black olives, stoned
Olive oil
Poppy seeds

Mix the flour and yeast with enough water to form a dough. Knead and allow to rise until it is half as big again. Knead again and work in the olives. Rub the dough with olive oil; place it in an oiled tin to rise again with a sprinkling of poppy seeds. Once it has risen to half its size again, put it in a preheated oven at 220°C/425°F/gas mark 7 for 20 minutes or till brown on top. Serve as an entrée with a crisp green salad and a sharp dressing.

Actinidia chinensis/deliciosa from the family *Actinidiaceae*

KIWI FRUIT OR CHINESE GOOSEBERRY

Vine, up to 10m or more. Life span: medium. Deciduous, some not self-fertile. Fruits: up to 5cm, flattened ovoid, brown, hairy. Value: very rich in vitamin C.

The kiwi fruit or Chinese gooseberry is the best-known member of a small number of deciduous clambering and twining shrubs closely related to camellias. The kiwi has large, hand-sized, heart-shaped leaves, downy underneath, on softly bristled stems. The flower is like a small, poorly developed rose, off-whitish and fragrant. The brown, furry fruits are thin-skinned and firm with luscious green pulp containing many tiny black seeds around the centre.

These were not known in the West till the end of the nineteenth century, when they were introduced from Japan and East Asia, more for their use as decorative climbers than for their fruits. Though now commonly known as kiwi fruit, they are not native to New Zealand, but were introduced there in the early years of this century and became more popular as a fruit when greenhouse growers needed to look for alternative crops to their oversubscribed tomato market. Recent breeding has developed self-fertile varieties. A dwarf, shrubby, form would be handy.

varieties

A. chinensis/deliciosa **Hayward** is the most widely available but is not self-fertile so needs planting with a male. **Tomari** is similar. **Jenny** and **Oriental Delight** are self-fertile. **Blake** is very high-yielding (up to 90kg per vine is claimed). It is self-fertile and needs protection for the lower stems in cold regions. *A. arguta*, the **Siberian Kiwi** or **Tara Vine**, is very vigorous with deeply toothed leaves and fragrant, triple, white flowers with purple anthers, followed by green, sweet, slightly insipid fruits. The copious sap is drinkable. Improved self-fertile varieties, **Issai** and the newer **Arguta** are available with small hairless fruits. *A. kolomikta* is very attractive with large, oblong leaves which start green then go cream and pink. They are tapered at the end and hang from white, downy stems. The single, white flowers are sweetly perfumed and followed by long, yellowish, sweet berries. This is less vigorous than the other species and is thus more useful for cramped, modern gardens. It tolerates shade well. *A. polygama* is a native of Japan and has large, heart-shaped leaves that open bronze, maturing to green on red stems, with white, fragrant flowers and yellow fruits. There are many more sweet edible species.

cultivation

Kiwis require a warm site, preferably against a wall, as their young growths and flowers are easily damaged by late frosts. They need rich, loamy soil and strong supports. They will grow, but not crop, in shade. Some of the early varieties, such as **Hayward**, are female only and require one male to be planted to every half dozen females. Modern varieties are self-fertile, though they may perform better if planted with others to cross-pollinate. Some species similarly need pollinators, so it is often best to plant several (more than three) to be sure of getting one male for several female plants.

Growing under Glass

If crops are wanted, kiwis have to be grown under glass to give them the warmth and long season they require. However, they take up a lot of space. They like similar conditions to tomatoes, so the two could possibly be grown together.

KIWI FRUIT

Growing in Containers
These are too vigorous to be confined for long in a pot and are never likely to be very productive in one.

Ornamental and Wildlife Value
Kiwis are very attractive climbers. The young shoots are particularly pretty as they are covered in many fine, red bristles. Kiwis are hardy enough to be used to cover eyesores, but they are not likely to crop unless they are given a warm site. The species are equally useful.

Maintenance
Spring Prune and tie in, spray with seaweed solution monthly.
Summer Tie in new shoots.
Autumn Cut back sideshoots to shorter stubs.
Winter Pick the fruit – if you are lucky.

Propagation
The species can be grown from seed, but varieties are grown from half-ripe summer or from hardwood autumn cuttings rooted with bottom heat in a frame.

Pruning and Training
The fruits are borne on sideshoots. Unless these are left there are no fruits, so these must not be hard pruned, but may be shortened. Kiwis must be well trained or they form an unruly thicket; therefore give them plenty of space. Old shoots die, so tie in replacements in spring. Good supports are necessary for these vigorous plants.

Their fruits swell early but hang on late

Long after the leaves fall the crop hangs on, making picking easier

Pests and Diseases
There seem to be no major problems with these fruits, save dieback in cold winters. Give them extra protection if you are after fruits.

Harvesting and Storing
The fruits ripen late; hang well with protection against frosts from the elements, ideally under glass.

companion planting

There are no known companion effects.

other uses

These plants are useful for covering old trees and other eyesores.

culinary

Kiwis can be eaten raw or made into juices, jellies and jams and are much liked for decorating other dishes, usually sweet but occasionally savoury. They contain an enzyme that breaks down gelatine, so should not be used to make dessert jellies. This enzyme can also tenderise meat.

Kiwi Meringue Pie
Serves 4

4 individual meringue bases
Green gooseberry jam or preserve
Clotted cream
4 kiwi fruits
Toasted chopped nuts
Glacé cherries

On each meringue base build a thick layer of jam, then a layer of clotted cream covered with very thin, overlapping slices of peeled kiwi. Top with nuts and a cherry or two.

Passiflora from the family *Passifloraceae*

PASSION FRUIT

Herbaceous vine up to 10m. Life span: short. Semi-deciduous, self-fertile. Fruits: from 2.5cm to 10cm, spherical to cylindrical, yellow, orange, red, brown, black or green. Value: rich in vitamin C.

Passion fruit are a family of perennial climbers with tendrils, deeply lobed leaves, amazing flowers and peculiar fruits. These vary in size from that of a cherry to a coconut, and in colour, coming in almost any shade from yellow to black. They are usually thick-skinned with a juicy, acid, fragrant, sweet pulp inside, almost inseparable from smooth, black seeds. The passion fruit, *Passiflora edulis*, is the most widely grown species. It has white or mauve flowers fragrant of heliotrope and purple-black fruits that are best when 'old' and wrinkled. A yellow-skinned form, *P. edulis flavicarpa*, is also popular in Brazil.

Passion fruit are native to America and were first recorded in Europe in 1699. The flowers caused quite a stir in European society, with many contemporary Christians claiming that they were a sign of Christ's Passion (the Crucifixion). The three stigmas represented the nails, the central column the scourging post, the five anthers the wounds, the corona the crown of thorns, the calyx the halo, the ten petals the faithful apostles, and the tendrils the whips and scourges of His oppressors. These delightful climbers, with their stunning flowers and delicious fruits, have now become popular in most warm countries. They are often grown in conservatories and in pots on patios for the flowers rather than the fruits.

varieties

Passiflora edulis is the tastiest variety and hardy enough to survive in a frost-free greenhouse. Improved varieties such as **Crackerjack** are available. The **Giant Granadilla**, *P. quadrangularis*, needs more tropical conditions, but can produce fruits weighing many pounds; these are often used as vegetables in their unripe state. *P. laurifolia* is the **Water Lemon** or **Yellow Granadilla**, much esteemed in Jamaica. *P. incarnata* **Maypops**, comes from eastern North America. It has attractive creamy flowers, ornamental, three-lobed leaves and tasty yellow fruits. It is not self-fertile and spreads by underground runners. It can survive outdoors, as it comes again from the roots. *P. caerulea* is the hardiest, with blue flowers and orange fruits which may be edible, but are not palatable, even after boiling with sugar. Dozens of edible-fruited passion flowers are grown locally, such as *P. antioquiensis*, with yellow, banana-shaped fruits, *P. foetida*, the goat-scented passion flower, *P. mixta Curuba di Indio* and *P. ligularis*, said by connoisseurs to be the most delicious.

cultivation

A humus-rich, moist soil and a sheltered position on a warm wall suit the hardier varieties, with thick mulches to protect their roots. However, most of these plants are happier under cover.

Old and wrinkly is better!

PASSION FRUIT

Big crops of tasty tender *P. edulis* can be had under cover

Growing under Glass
For edible fruit production, and to grow most of the more tender varieties, glass is essential. Fortunately, they can also be grown in pots, making them less rampant and allowing some to be put outside for the summer.

Growing in Containers
Surprisingly, such rampant climbers take fairly well to pot culture, but require a lot of watering. They will crop in pots and, indeed, this is one of the better ways of growing them, so they can be taken indoors for winter. They need an open, free-draining, rich compost and regular feeding.

Ornamental and Wildlife Value
Amongst the most attractive of all climbers in flower, foliage, or when festooned with fruits. The flowers are enjoyed by many insects and the fruits contain plentiful seeds for the birds.

Maintenance
Spring Tie in new growth, spray with seaweed solution monthly.
Summer As for spring.
Autumn Pick fruits after a long, hot summer.
Winter Prune before frosts and protect roots or take pots inside.

Propagation
Passion flowers can all be grown from seed, which can give good results for they are still relatively unimproved and most are true species. Propagate good varieties from heel or nodal cuttings in mid-summer if they are given bottom heat.

Pruning and Training
Pruning is remedial, removing dead and surplus growth. Strong wires are needed, as these are quite vigorous and productive.

Pests and Diseases
Passion flowers tend to form rather dense stools, which means they are prone to weed infestations which make the crown damp and short-lived. Occasionally they fail to thrive, but generally they are no problem save for frost damage: protect the roots and lower stems.

Yellow Granadilla

Harvesting and Storing
Best ripen the fruit on the vine till they drop, though picked young for transport they keep well. As they ripen they will shrivel, appearing old and wrinkled, and the flavour is then at its best.

other uses

The empty shells make good slug traps for the garden.

culinary

Thirst-quenching raw passion fruit are made into juice which is a popular drink in many countries, much as orange juice and squash are in others. They can be made into jams, liqueurs and sorbets.

Passion Fruit Sorbet
Serves 4–6

12 ripe passion fruit
Approx. 120g sugar
Mint sprigs

Scoop out the fruit pulp and seeds and sieve. Discard the seeds. To the juice, add half its weight in water and the same of sugar, stir till the sugar has dissolved and freeze. Partially thaw and beat vigorously, then refreeze. Serve partially thawed, scooped into glasses and decorated with mint sprigs.

Punica granatum **from the family** *Lythraceae*

POMEGRANATES

Bush up to 4m. Life span: medium. Deciduous, self-fertile. Fruits: up to 8cm, spherical, orange. Value: good source of vitamin C.

Pomegranates have coppery young leaves that yellow in autumn, glorious orange or red, camellia-like blooms, and orange fruits with a rough nipple and thin, leathery rind. Inside they have bitter yellow pith and are stuffed with seeds embedded in pink, sweet pulp. They are infuriating to eat. Natives of Persia, the pomegranates were cultivated in Ancient Egypt and other Mediterranean countries. Some are now grown in California. They are easily transported, so were widely known in early times, even in most of the colder countries where they would not fruit.

varieties

Iraq is said to have the best pomegranates, large with perfumed flesh and almost seedless; such varieties have also been known in Kabul and Palestine. **Wonderful** is a Californian variety.

Pomegranates outdoors will do best not far but sheltered from the sea

cultivation

They will grow and flower outdoors but seldom set a crop in England, even on a warm wall. They prefer well-drained, limy soil.

Growing under Glass

This is the only way to have a crop. They need intense heat in summer to ripen and are more manageable in pots.

Growing in Containers

Pomegranates are easily grown in pots, especially the dwarf variety, **Nana**. This variety is also hardier than taller types.

Ornamental and Wildlife Value

They are as ornamental as any rose and several improved flowering varieties are available. The flowers are good for insects.

Maintenance

Spring Spray with seaweed solution, prune outdoor specimens.
Summer Keep hot to ripen the fruits.
Autumn As for summer.
Winter Prune indoor specimens in early winter.

Propagation

By seed, layering or half-ripe summer cuttings using bottom heat.

Pruning and Training

They can be fan trained but are easiest allowed to grow out as a bush from the wall. Old, diseased and weak wood is best pruned out in late spring. Prune indoor specimens in early winter and give them a reduced temperature for a few weeks afterwards.

Harvesting and Storing

They travel very well, keeping for weeks, so are easy to distribute.

Pests and Diseases and Companion Planting

Pomegranates suffer from few problems and no companion effects have been noted.

culinary

The juice is used for drinks, syrup, conserves and fermenting.

Pomegranate Pastime

Sit down under a shady tree with your ripe pomegranate. Cut open the rind, pick out the seeds individually with a pin and eat the pink pulp. The seed may be swallowed or rejected.

Diospyros kaki from the family *Ebenaceae*

PERSIMMONS

Tree/bush, up to 6m. Life span: medium. Deciduous. Fruits: 5–8cm, round, orange-red. Value: rich in vitamin A and potassium.

Persimmons have lustrous, dark green leaves. The flowers are drab and the fruits look like large, orange tomatoes. Japanese Persimmons were first seen in 1776. Extremely popular in Japan and China, they are now grown on the French Riviera and in California. Other persimmons are eaten locally in hot countries.

varieties

Diospyros kaki is the **Japanese Persimmon**. There are many varieties from the East, some of which are parthenocarpic. However, you should plant several to ensure pollination. *D. virginiana* is the **American Persimmon**, which has smaller, redder fruits and crops as far north as the Great Lakes. *D. lotus*, the **Date Plum**, is grown in the East and Italy, and many other species are eaten worldwide. *D. blancoi*, **Mabolo/Butter-fruit**, resemble small orangey winter squashes smelling of cheese – but once you get the skin off, their creamy white flesh is surprisingly sweet and tasty. *D. digyna*, the **Black Persimmon** a.k.a **Chocolate Pudding Fruit**, from Mexico has thick finger-sized dark greeny-black, soft fruits which, mixed with spices and sugar, resemble real chocolate pudding and are thus exceedingly popular.

cultivation

Persimmons are hardy, but need a well-drained soil and warmth to produce passable fruit. They grow better under glass, though the American species may crop well outdoors.

Growing under Glass
They need glass protection in northern Europe to extend the season long enough to ripen good fruits.

Growing in Containers
Persimmons are better in borders than pots, though these will do. Thin out the fruits to 25cm apart, otherwise they overcrop and will not swell.

Ornamental and Wildlife Value
These are very attractive small trees, grown for their autumn colouring and ornamental fruits. The fruits are of value to birds.

Maintenance
Spring Weed, spray with seaweed solution.
Summer Disbud if over-vigorous.
Autumn Cut off fruit and store till ripe.
Winter Dress with compost and with copious amounts of straw.

Propagation
By seed or grafting for the better varieties.

Pruning and Training
The branches split easily and are not easily trained. Only remedial pruning is needed, so plant at about 6m apart. Thin the fruits to get bigger ones.

Pests and Diseases
There are few common pests or diseases, though a fungus has been known to kill wild trees in America.

Harvesting and Storing
Cut, rather than pick, to retain the short pedicel. This way they keep better. Store cool and dry until ripe – up to four months, as they are best when somewhat shrivelled. When soft, they lose the astringency that makes unripe persimmons mouth-puckering and rather disgusting. They can be frozen like tomatoes and similarly de-skinned as thawing.

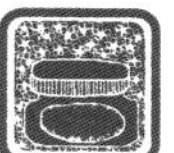

other uses

The fruits are successfully used as pig food. Pounded persimmons dropped into a pond stupefy the fish.

culinary

Persimmons are eaten fresh, dried or candied, but rarely cooked. They can also be made into a yummy ice cream.

Percinnammons
Serves 4

4 soft, ripe persimmons
Cinnamon
Apricot conserve
Cream
4 glacé cherries

Peel the fruit and cut in half. Dust each half with cinnamon, cover with conserve and top with cream and a cherry.

Eriobotrya/Photinia japonica **from the family** ***Rosaceae***

LOQUATS

JAPANESE MEDLARS OR PLUMS

Tree/bush, up to 10m. Life span: medium to long. Evergreen. Fruits: up to 5cm, pear-shaped, orange. Value: minor.

Loquats have very large, leathery, corrugated leaves, woolly white underneath, and fragrant, furry, yellowish flowers. The fruits are orange and pear-shaped with one or more big, brown-black seeds and sweet, acid, chewy pulp.

First reported in 1690, these were imported from Canton to Kew Gardens in London in 1787. Widely cultivated in the East, they are now popular in the Mediterranean countries and in Florida.

varieties

In the UK named varieties are unavailable though improved sorts are found abroad such as **Advance, Champagne, Gold Nugget, Mammoth** and **Oliver** – surely some of these may suit the conditions here…

cultivation

They grow outdoors well enough in northern Europe, but do not fruit. Any reasonable soil and a warm, well-drained site will suffice. They do best on a warm wall, but it needs to be a big one. They will crop only under glass or in countries with warm winters.

Growing under Glass
As they flower in autumn and the fruits ripen in late winter and spring, loquats need to be grown under glass if they are to fruit in any country that does not have a warm winter.

Growing in Containers
Loquats make big shrubs, so they are usefully confined in large pots.

Propagation
Loquats can be grown from fresh seed, or layers, or soft wood cuttings taken in spring with bottom heat.

Ornamental and Wildlife Value
Very architectural plants with a lovely scent, they will also make good shelter for birds and insects.

Pests and Diseases
Few problems with these.

Maintenance
Spring Prune if needed, spray with seaweed solution, pick fruit.
Summer Move outdoors for the summer.
Autumn Bring indoors.
Winter Protect outdoor plants from frosts.

Pruning and Training
Only remedial pruning is needed. They are best trained on a wall and allowed to grow out from it or grown as bushes in pots. Trim back any dead and diseased growths in spring.

Harvesting and Storing
The fruits need warmth and protection to ripen in late winter/early spring, so they must be grown under glass in cold countries.

companion planting

These are dense evergreens that will kill off any plants grown underneath.

other uses

These shrubs make tall and attractive screens in countries with warmer climes.

culinary

Loquats are eaten raw, stewed, jammed or jellied. They are made into a liqueur in Bermuda.

Loquat Jam
Makes approx. 2.4kg

1.4kg loquats
Approx. 1kg sugar

Wash and stone the loquats, then simmer till soft with just enough water to prevent burning. Weigh and add three-quarters of the weight in sugar. Stir to dissolve the sugar, then bring to the boil. Skim and pot in sterilised jars. Store in a cool place.

Opuntia ficus indica/dillenii from the family *Cactaceae*

PRICKLY PEARS

BARBERRY FIGS

Herbaceous up to 2m. Life span: medium to long. Evergreen, self-fertile. Fruits: 5–9cm, ovoid, red, yellow or purple. Value: minor.

These are typical cacti, with round or oval, thick, fleshy pads covered with tufts of long and short spines. The flowers are large, 5–7cm, yellow, with numerous petals, stamens and filaments. These are followed by red, yellow or purple, prickly, oval cylinders which are the fruits. Under the skin the flesh is very acid and sweet.

These are natives of the Americas, where they have long been used. They have naturalised in the Mediterranean basin and almost every hot, dry country, even flourishing on the lava beds of Sicily.

varieties

Burbank, the great American breeder, raised a spineless prickly pear. Many other *Opuntia* species have edible fruits: *O. compressa/fragilis/goldhillea/humifusa/macrorisa/maxima/phaeacantha* and *rutila*. Other cacti with edible fruits are: *Brachycereus* spp. **Rositas**, a trailer with red flowers and small fruits; *Borzicactus aequatorialis*, **Zoroco**, which is a fragrant white-flowered cacti with white-fleshed fruits; *Cereus peruvianus*, **Peruvian Apple Cactus**, **Pitaya**, are very good white-fleshed, red-skinned fruits; *Echinocereus triglochidiatus*, is a mound-forming cactus with red flowers and fruits rated as strawberry like; *Espostoa lanata*, **Cacto lanudo**, has red flowers with purple sweet fruits; *Pachycereus pectin-aboriginum*, **Hairbrush Cactus**, gets to the height of a house and has orange-size spiny fruits with sweet flesh, good fresh or cooked; *Trichocereus pachanoi*, **San Pedro**, is a huge cylindrical cacti that can reach tree size, covered with fragrant flowers and edible fruits; *Hamatocactus hamatacanthus*, **Lemon cactus, Turk's Head** are spherical tender cactii with red and yellow flowers followed by lemon-flavoured fruits used raw, in beverages and cakes.

cultivation

Opuntia need a well-drained, open, limy soil and a warm position. They are remarkably hardy, for cacti. Several survive outside at the Royal Botanic Gardens at Kew in England, and I have had them for many years in Eastern England.

Growing under Glass and in Containers

Growing under glass is vital if fruits are to be produced, and you will also need extra artificial light. In large and free-draining pots, prickly pears are happy enough.

Ornamental and Wildlife Value

Very decorative and quite a talking point in a garden. They are more reliable under glass and just as attractive. The flowers are good for insects.

Pruning and Training

No pruning is required. When the pads become heavy they may need propping up.

Pests and Diseases

Weed control needs to be good, as these are nasty to weed between. Slugs and snails may develop a taste for the pads.

Maintenance

Spring Keep weed-free.
Winter Protect outside plants during coldest weather.

Propagation

These can be grown from seed, but are slow. Detached pads or pieces root easily and are much quicker.

Harvesting and Storing

This is a thorny task. Wrap a piece of bark round to pick the fruit, which will keep for several days. The peel is best skinned off completely before the fruit is eaten.

other uses

After ensilaging or pulping with salt, prickly pears make a useful animal feed.

culinary

Prickly pears are usually eaten raw in place of drink and are occasionally fried or stewed. The red varieties will stain everything.

Tasty they may be, but the picking and peeling…

Monstera deliciosa from the family *Araceae*

CERIMANS

SWISS CHEESE PLANTS

Herbaceous vine, may ramble or climb to over 12m. Life span: appears perpetual and invulnerable. Fruits: about 2.5 x 22cm, green. Value: some vitamin C.

The Swiss cheese plant is one of the commonest and most enduring house plants, somehow surviving hostile conditions in dark, dry rooms the world over. The leaves are dark green, large, scalloped and uniquely and curiously, have natural holes in them, presumably to let tropical winds pass with less damage. In its native habitat it is an epiphytic climber, rambling on the forest floor and climbing up vigorously, clothing the trees and throwing down masses of aerial roots. It is a close relation of the arum lily and the flowers are similar. The long, cone-like spadix fruit, or ceriman, is green, cylindrical and leathery with tiny, hexagonal plates for skin. The flesh is sweet and richly flavoured, resembling a cross between a pineapple and banana. The fruits are as delicious as the name suggests, but the spicules make unripe fruits unpleasant, so they have never become widely popular.

Swiss cheese plants are native to Central America, but have been spread worldwide for their attractive leaves and amazing durability. This fruit was discovered in Mexico and was originally known as the Mexican bread fruit. It became known as the shingle plant and classed as *Philodendron pertusum*, then *Monstera acuminata* and now as *M. deliciosa* but is known worldwide as the Swiss cheese plant. In 1874 the fruits were exhibited before the Massachusetts Horticultural Society.

'Swiss cheese' originates from holes in the leaves

cultivation

One of the most enduring and robust plants discovered, but will fruit only if given warmth and moisture. It does not need as much bright light as most tropical fruits, and has been grown successfully under glass in most countries.

Ornamental and Wildlife Value

Superb ornamental value almost anywhere frost-free.

Growing under Glass and in Containers

Ideally suited to almost any treatment! For fruits, give better conditions, e.g., copious watering and syringing.

Propagation, Pruning and Training

They can be air-layered or cuttings will root easily. They are happiest climbing up a stout, rough-barked tree or log, or wired on a wall.

Harvesting and Storing, Culinary

When the fruit is ripe the inside appears to swell and the leathery skin plates loosen up; they can be eased off like tiny buttons. Then the flesh can be eaten off the stem. Try it with care – the tiny spicules of calcium oxalate irritate some people's throats but appear harmless. Do not worry about the spicules if you eat only ripe fruits – many people regularly enjoy them. Cerimans are only used as dessert fruit and are widely popular in native markets.

other uses

The vines make a quick-growing screen in warm regions.

Cyphomandra betacea **(syn.** ***C. crassicaulis*****) from the family** ***Solanaceae***

TAMARILLOS
TREE TOMATOES

Shrubby tree up to 5m. Life span: generally short. Fruits: 4 x 6cm, ovoid, purple. Value: some vitamin C.

The tree tomato is an evergreen, semi-woody shrub from the same family as the tomato. The leaves are large and lightly felted and smell muskily aromatic. Greeny-pink, fragrant flowers are followed by copious fruits similar to tomatoes but more pointedly egg-shaped and purple. Each fruit is thick-skinned with two lobes containing about a hundred seeds. They are tasty raw only when well ripened, and then eaten with salt to bring out the flavour. Less well ripe and they are too acid and require cooking. Native to Peru or Brazil, they are widely cultivated in warm zones. They fruit conveniently during much of the year.

varieties

In the UK only one species is easily available, and is rather variable anyway. Grown in many countries, often up country in the cooler regions of the hottest, the tree tomato has spawned a host of varieties. Frequently grown from seed, these vary from the type and are given convenient names by the grower to sell the fruits. The Fruit, Berry and Nut Inventory lists no varieties as such, though the fruits are widely grown in the warmer states and in tubs in the northern states. There are two main strains: the original reddish purples and the selected golden ones, the latter having yellow skins and flesh. A visit to your local supermarket may provide the former but you will have to search ethnic markets for more variety.

cultivation

Tamarillos resemble aubergines in many ways and prefer medium to high altitudes for their coolness in the tropics. They demand deep, well-manured soils to produce good dessert fruits. However, for culinary purposes they can be grown anywhere frost-free and are then fairly obliging as to soil, preferring well-drained sites. They are easy to propagate and quick to bear in under two years so can be tried with no great loss. Where replacements are home grown and with a little protection they have been grown outdoors for a few years or so in sheltered spots in the UK.

Pests and Diseases

More prone than almost any other plant to the usual pests and so requiring all the usual remedies. I find these really do need growing with French marigolds to help discourage the whitefly and it is worth introducing the predatory wasp *Encarsia formosa* as soon as things warm up in early spring. Keeping the humidity high will discourage the red spider mite but introduce the predatory mite *Phytoseuilis persimilis* at the first sign of infestation (tiny yellowish pin pricks). Aphids simply require soft soap sprays to be kept under control. I have not noticed vine weevil damage to the leaves but they probably eat the roots.

Growing under Glass and in Containers

These plants need a lot of light and heat to produce tasty dessert fruits but then usefully give high yields of fresh fruits for cooking purposes during the winter months, with the occasional tastier one.

culinary

The fruits are fairly robust until fully ripe and keep for several days. When ripe they are purple and can be eaten raw, but tend to be sour so they are better stewed like tomatoes or plums.

Tree Tomato Jam

Makes approx. 2kg

1kg tamarillos/tree tomatoes
Approx. 1kg light brown sugar
1 small lemon

Stew the tamarillos in a little water until soft. Sieve and return to the heat. Add the same weight of sugar and the lemon's grated rind and juice. Bring to the boil and then bottle in warm, sterilised jars and seal.

Ornamental and Wildlife Value

These resemble *Daturas/Brugmasias* in many ways, including the mawkish but strangely appealing smell of the foliage and the delicious scent of the pinkish, though small, flowers. Much loved by whitefly, aphids and red spider mite these are veritable bankers if you want sources of these pests to keep your predators alive. Otherwise use soft soap sprays routinely! In the wild where they have self-seeded the fruits are soon eaten by birds and rarely ripen.

Propagation, Pruning and Training

Tree tomatoes are easily grown from seed or cuttings, often fruiting in their second year. Best trained as short standards, pruning is then mainly nipping out growing points to keep the bush compact. Otherwise, they reach 3 metres and then bend over, throwing cascades of fruit dripping from the slender branches, if you're lucky. Or cascades of honeydew dripping from the hordes of pests – I don't want to be negative but these are legion and legendary.

Harvesting and Storing

The fruits are fairly robust till fully ripe, which conveniently occurs in mid-winter, and once ready will keep for several days or a week or so if kept cool. When ripe, the fruits lift off readily and the red ones become more purple and can be eaten raw, but more often the fruits tend to be sour and so are better stewed.

companion planting

Plant French marigolds alongside to deter whitefly.

TROPICAL AND SUB-TROPICAL FRUITS

Some of these fruits are the tastiest and most luscious in the world. Strong sunlight and hot conditions produce sweeter, stronger flavours (indeed, often excessively so) than in temperate zones. The natural conditions change little during the year, usually being hot, bright and dry, followed by hot, bright and wet, or by more of the same, rather than the fluctuating heat and light and winter chilling of cooler regions. Tropical plants therefore often set fruit several times a year, or continuously throughout it. Thus tropical fruits are a more reliable source of food than those of cooler areas and less reliance has to be put on storage till the next harvest.

The original species of many tropical fruit have disappeared, to be replaced, even in the wild, by our partially selected stock. This happened far back in prehistory; there is no wild date or banana that corresponds to the cultivated varieties. Many of these long-cultivated plants have been turned into clones, superior varieties propagated vegetatively, often without seed, or not coming true from seed. This handicaps the gardener as only inferior forms can be had from seed; to get better sorts, living plant material has to be obtained. With international restrictions this can be difficult.

However, some tropical fruits are still species, little changed from the original, and useful to the gardener, for if the fruits can be obtained, it may be possible to grow the plant. We can find some fruits in ethnic shops and even occasionally in more adventurous supermarkets. Modern refrigeration and air travel now permit us to enjoy a tremendous range of fresh exotic fruits. The availability and range is constantly extended, but, of course, our predecessors discovered and enjoyed these fruits before us, and they also wanted them back home after travelling abroad.

The first tropical fruits probably reached northern Europe in Roman times, but were then forgotten until rediscovered by late medieval travellers. To be enjoyed fully, many needed to be ripened longer on the plant than early long-distance transport allowed. The colonisation of the Canary Islands during the Renaissance enabled exotic fruits to be cultivated within reach of the European market, and this was given further impetus by the newly discovered New World fruits.

These new, expensive, luxury fruits were so desirable that owners of large estates in colder countries encouraged their gardeners to try them even though they needed more than a frost-free greenhouse or simple winter protection to grow, let alone fruit. Surprisingly, just as they had succeeded with orangeries, they met with more success; for example, pineapple plants were grown in Britain as early as 1690. However, it required the Industrial Revolution to provide the iron, glass and heat for serious home production.

The stove house was just that – a large glass greenhouse with a massive stove keeping the temperature tropical; different sections were arid and dry or steamy moist. With steam- or water-heated pipes, the stove could be moved to a separate boiler room. The Victorian age saw British-grown pineapples, bananas and mangosteens gracing many a table. Nothing seemed impossible to these horticultural pioneers.

Along with private extravagance, public botanic gardens were built, and specimens were obtained of many plants. Of course when kept in pots, some grew far too big to be considered. Many were successful, but others could not be persuaded to fruit even in extra heat. Now we are more fortunate: with electric light to replicate sunlight we can give these plants the brightness and day length they need. We also have automatic heat and humidity control, so it is easy to grow many exotic fruits ourselves. And if they still will not fruit, they always make attractive house plants.

Ananas comosus from the family *Bromeliaceae*

PINEAPPLES

Herbaceous, 1 x 1m. Life span: short-lived, perennial. Fruits: average 10 x 20cm, dull orange or yellow. Value: rich in vitamins C and A.

Pineapples need little description; they are the most distinctive of fruits – there is nothing else like them. They are Bromeliads, like many houseplants. They resemble common garden yuccas, being nearly cylindrical with a tuft of narrow, pointed leaves emerging from the top. The skin of the fruit is green to yellow, with many slightly raised protuberances. Wild species have serrated, thorny-edged leaves and set seed. Modern cultivars are seedless, with smoother leaves and smaller fruits; those of traditional varieties weighed up to 8kg.

Cultivated and selected from the wild by the people of Central America for thousands of years, the fruits were sensational in 1493 to the crew of Columbus. The first fruit, surviving the voyage back, was regarded as nearly as great a discovery as the New World itself. Pineapple motifs appeared, sometimes distorted, throughout European art – often as knobs on pew ends. By 1550 pineapple was being preserved in sugar to be sent back to the Old World as an exotic, and profitable, luxury. By the end of the sixteenth century pineapples had been spread to China and the Philippines and were naturalising in Java, and soon after were colonising the west coast of Africa. An enterprising M. Le Cour of Holland succeeded in growing them under glass in 1686 and was supplying plants to English gardeners in 1690. Within a few years there was a craze for pineapples, with noblemen's gardeners growing them under glass on deep hotbeds of horse dung and leather wastes as far north as Scotland. British-grown pineapples were sold in the markets at a guinea apiece. The Victorians raised the cultivation of pineapples to a high level with the regulated heat from steam boilers. A photograph of English pineapples from the turn of the century is a humblingly impressive display.

varieties

There have been hundreds of varieties, but most have disappeared as a few commercial cultivars monopolised trade. **Smooth Cayenne**, or **Kew Pine**, was widely grown for many years, but connoisseurs preferred **Queen** and **Ripley**. Hard to find now are Victorian favourites **Montserrat** and **Black Jamaica,** averaging 5 to 6 pounds apiece, the bigger Envilles from 7 to 9 pounds and the huge **White Providences** that weighed 10 to 12 pounds or more apiece. Even bigger varieties were grown for show but not esteemed and rated coarse.

cultivation

Growing under Glass and in Containers

A tropical plant, the pineapple is happiest in Hawaii. It needs very high soil and fairly high air temperatures, high humidity during the growing season except when ripening, as much light as possible and a very rich, open, fibrous compost in quantity. In plantations

PINEAPPLES

The juice acts as high fibre!

pineapples have about a square metre/yard of ground each, so generous pots are required! Never let them chill below 21°C or 'cook' above 32°C. Dry plants may survive cold down to near freezing while damp ones succumb very quickly, then rot. Keep their centre from getting wet in the cold as this is first lost.

Ornamental Value
They are easy to root and grow as houseplants, but require lots of heat and care to fruit well.

Maintenance
Spring Pot up all plants, start to mist, water and feed.
Summer Mist frequently and water lightly, plus seaweed in spray, and fish emulsion in water, both weakly weekly. Root suckers, gills and stem cuttings.
Autumn Tidy and dry off plants for winter.
Winter Keep warm!

Propagation
Suckers from healthy plants are best, fruiting in a year and a half or so. Gills, the shoots produced at the base of the fruit, seeds and even stem cuttings can be used, but take longer to grow and fruit. The crown from a fruit is easily rooted; cut it off whole with a thin shoulder, prise off the shoulder and pull off the lower withered leaves individually. Root over heat in a gritty compost. Once growing, keep foliage moist with regular misting from spring till autumn. From autumn till spring keep dry unless preferred temperatures can be continuously maintained.

Pruning and Training
Young plants rooted in summer need potting up in spring, growing on for a year and repotting the following spring to fruit that summer or autumn. Pot up annually, and remove superfluous shoots and fruited stumps if you have any. Bury the plants progressively deeper each year – relatively tall but narrow pots are needed. If happy, pineapples may grow and produce for five to ten years.

Pests and Diseases
The main pests are white scale and mealybugs, controlled by sprays of soft soap or by the use of commercial predators.

companion planting

Once the plants are big enough a dressing of banana skins may encourage them to flower, and again later to help ripen the fruit. A good smoking can also induce flowering.

other uses

Pineapple leaves have a strong fibre used for thread and fishing lines and pinya cloth to make the national dress shirt of Filipino men.

culinary

Picked underripe and kept cool, they can be stored for several weeks. They make tasty jam, delicious juice and are best-known canned. Although a fruit, they add sweetness and texture to many savoury dishes, from Chinese to curry, and can even be fried with gammon. Pineapples cause egg whites and jellies not to set and turn milk products bitter unless cooked to above 40°C/104°F. If a pineapple is too acid try eating it with salt – honest, this works!

Pineapple-baked Ham
Serves 4

1 very large tin of ham
1 larger pineapple
Garlic (optional)

Carefully empty the meat from the tin in one piece. Using a very sharp knife, carefully extract the pineapple from its skin, first slicing off the top shoulder. Leave the skin intact. (Slice and chill the pineapple flesh for dessert, cutting out the tough central core.) Insert the ham in the hole created, pin the shoulder back on to seal, and bake for at least an hour in a preheated oven at 200°C/400°F/gas mark 6. (Garlic lovers may rub the ham over before insertion.)

To serve, turn the pineapple on its side to cut circles of ham with a pineapple rind. Serve with puréed sweetcorn and baked sweet potatoes.

Musa species from the family *Musaceae*

BANANAS

Tree/herbaceous plant, 3–9m and nearly as wide. Life span: perpetual as vegetative clones but sterile. Fruits: up to 30cm, green to yellow or red. Value: nutritious and rich in starch.

Banana plants form herbaceous stools of shoots like small trees. Each enormous shoot unfurls sheaths of gigantic, oblong leaves up to 4.5m long. A mature shoot disgorges one flower stalk, which hangs down under its mighty bunch of many combs or hands of bananas. The hands point upwards, sheltered by succulent, purple bracts the size of plates, along the length of the stalk, and a mass of male flowers adorns the end. One bunch can hold about a dozen combs, each with over a dozen fingers. Bananas are the most productive food crop, giving about forty times the yield of potatoes. The fruits are green, ripening yellow, sweet, notoriously shaped and unforgettably scented.

The Ancient Egyptians had the culinary Abyssinian banana (*Musa ensete*), but early travellers from Europe soon discovered sweeter, more edible bananas in most tropical and semi-tropical regions. The hardiest, the smaller Chinese banana, was brought from the East Indies for cultivation in the Canary Islands by 1516 – and, it has been suggested, was introduced to the Americas from there, but evidence of indigenous bananas discounts this. Most cultivated bananas are probably descendants of *M. sapientum*, *M. acuminata* and *M. balbisiana*, wild bananas thought to come from eastern Asia and Indo-Malaysia. *Musa maculata* and *M. rosacea* descendants are popular in parts of Asia and worldwide there exist many other species. In prehistoric cultivation, seed-bearing species were replaced by selected hybrids with big, seedless fruits propagated vegetatively. These superior sorts multiplied and by the nineteenth century there were countless local varieties of bananas in most hot countries. Many of these have now been lost and replaced by a few high-yielding commercial cultivars.

varieties

Common cultivars are **Robusta**, **Lacatan** and **Gros Michel**, the latter pair being old varieties that survived the Panama disease which destroyed many. However, they are tall and need hot, moist conditions such as those of Jamaica, where **Gros Michel** was introduced in 1836. The **Chinese Banana** (*M. cavendishii*), also known as **Dwarf** or **Canary Island Banana**, is hardier, smaller and particularly tasty with shorter, delicate fruits, making it widely grown for domestic use, and well suited for greenhouse culture. A new, medium-sized and more productive hybrid, **William**, is taking over commercially. Plantains are separate varieties, possibly descendants of *M. paradisiaca*, larger and slightly horn-shaped. Plantains are most often used for cooking, not dessert purposes; thus cooking bananas are sometimes referred to as plantains.

BANANAS

No pollination required or wanted

cultivation

In warm climates bananas need a rich, quite heavy, deep, well-drained soil with copious moisture. Freedom from strong winds is also essential. The smallest varieties are usually planted about 3m apart, the larger proportionately more.

Growing under Glass and in Containers
In cooler climates the **Canary/Chinese/Cavendish** is best, as it is very compact and fruits at less than 3m high. Keep the temperature above 19°C in winter and under 30°C in summer. Potted plants can go out for summer.

Maintenance
Spring Pot up or top dress, thin and root suckers.
Summer Top dress, thin suckers.
Autumn Cut fruits, remove old shoot, thin suckers.
Winter Keep warm.

Propagation
Some ornamental and inferior varieties grow from seed. The best edible plants are bits of rhizome with a bud or sterile hybrids, propagated by suckers. Viable buds will resemble enormous, sprouting bulbs. They root easily in a gritty compost over heat in a moist atmosphere.

culinary

Bananas are best matured off the tree and kept in a warm room. Most are at their best when yellow. They can be dried, made into flour or fermented to produce a sweet liqueur.

In hot countries, bananas are a staple food, being used as a vegetable and a source of flour, as well as a fruit.

Banana Custard Pie
Serves 4

175g digestive biscuits
75g butter, melted
30g strawberry jam
300ml milk
60g honey
15g cornflour
Dash of vanilla essence
3 or 4 bananas, sliced and chilled
Nutmeg

Crush the biscuits and mix with most of the butter. Press into a pie dish and freeze. Spread the frozen surface with the remaining butter, paint with the jam and refreeze. Heat the milk and honey, mix the cornflour with a drop of water and the vanilla, then pour the hot milk on to the flour mixture. Return to the heat, stirring continuously, till thickened, then allow to cool. Put chilled banana slices into the cooled custard, mix together and pour into the biscuit case. Top with grated nutmeg and chill till required.

Pruning and Training
Bananas fruit continuously if growing happily. A stem will flower and fruit in about a year and a half. Only one main shoot and a replacement are allowed; all others should be removed, and once the main shoot has fruited it is cut out. The clump or stool may live as long as a human being, but is replaced commercially every dozen years or so.

Pests and Diseases
No major problems for private gardeners.

other uses

Often used in beauty preparations and shampoos. The foliage will provide good animal fodder.

My plastic tunnel is not quite tall enough

Phoenix dactylifera from the family *Arecaceae*

DATE PALMS

Tree 25 x 6m. Life span: as long as a human being. Fruits: just under 2 x 5cm, blue, brown or yellow. Value: rich in vitamins A, B1 and B2 and some B3.

A tall tree with ferny leaves, the russet dead leaf bases protecting the trunk. Separate male and female plants are required for fruiting, which is prolific. Each tree annually bears several bunches of three or four dozen strings, each string carrying two or three dozen dates, the total weight being up to 68kg.

Dates have been cultivated in the Middle East for at least 4,000 years and all wild forms have disappeared. They have always been one of the staple foods of the Arab peoples, much of whose lives were centred on the oases where the palms were found.

varieties

Numerous local varieties exist. **Deglet Noor** is from north Africa and is the world's most popular variety. **Saidi** is the common date; **Fardh** is favourite in Arabia, **Weddee** a feed date for donkeys and camels, **Farayah** a choice, long blue. *P. pusilla/zelanica*, **Ceylon Date Palm** is smaller and possibly easier to crop. **Peach nut** and **Borassus** – *Bactris/Guillielma utilis*, is a native of Central America, similar to date palms with similar fruit, usually cooked in salty water before eating they taste of chestnuts. The best varieties are seedless. Other relations are **Prickly Palm**, *B. major*, and **Tobago palm**, *B. minor. Borassus flabellifer*, **Borassus** or **Palmyra palm**, **Sea Apple** or **Lonta**, like a tall date palm with shorter fan-shaped leaves. The white flesh is used as a vegetable or a fruit and even canned, the fruits also contain refreshing sap drunk like coconut water. This, and the true sap tapped from the trunk, are boiled down into sugar, Jaggery, or fermented into Toddy.

cultivation

One of the few crops to revel in hot, even scorching, dry places, date palms still need irrigation to fruit well but can use brackish water.

Growing under Glass and in Containers

They are well suited to indoor life, though they may find it too humid with other fruits, and they get too big and need too much heat to produce good fruit. Date palms are easy to distribute as the small rootballs transplant with ease.

Ornamental and Wildlife Value

Choice ornamentals, not quite hardy enough for planting outside in Britain, they are popular in hotter climates both with people and wildlife.

Propagation

For choice fruit, offshoots or suckers have to be potted up; however, for ornamental use, seedlings are easily germinated but grow slowly. Large trees are easily transplanted with small rootballs. One male will be needed to every fifty females.

other uses

The palms have their sap extracted for sugar or fermenting. They may also be used as lumber.

culinary

Soft dates are partly dried, stoned and pressed into cakes, then exported all over the world. Semi-soft dates are those we find in packs at Christmas. Dry dates are hard, ground to a flour and commonly found only in Arab markets. Dates can be used in cakes, biscuits and confectionery and with curries and savoury dishes. They are also made into wine.

Stuffed Date Chocolates

1 box of dates
1 pack (225g) real marzipan
150g bar of dark chocolate, melted
Icing sugar

Stone the dates and stuff with marzipan. Dip in melted chocolate and cool on a tray dusted with icing sugar.

Persea americana from the family _Lauraceae_

AVOCADOS

ALLIGATOR PEAR, VEGETABLE MARROW

Tree up to 9 x 6m. Life span: short, self-fertile. Fruits: 15 x 8cm, green. Value: rich in fats, proteins, vitamins A, B and C and thiamine.

The avocado is a medium-sized bushy tree with large, green, often rough but shiny, pear-shaped fruits. Each is filled with an enormous stone and a morsel of greeny-yellow flesh next to the leathery skin. (Large avocados are better value than small ones!) The small, greenish-cream, hermaphroditic flowers are especially fragrant. Natives of tropical America, avocados soon spread to Australia, India and other tropical areas. They were naturalised on Mauritius in 1758, but also as north as Florida, California and even Madeira.

varieties

Each tree can yield 500 fruits a year weighing up to 1kg, but smaller ones are preferred in supermarkets. Mexican or Canary varieties generally withstand the lowest temperatures, so use these to grow from stones. **Fuerte** and **Hass** are the commonest varieties, though if you want to grow your own it would be best to try as wide a selection as obtainable. Ethnic shops may offer some of the less common sorts.

cultivation

This is one tropical tree that surprises gardeners everywhere by its near hardiness. Grown by children on windowsills these eventually become planted out once they get leggy and unattractive. And some survive. There are full-size trees alive in the open all over the British Isles, mostly in sheltered gardens, of course, but thriving. I have seen some myself and been amazed. Of course one hard spell kills off the leafy tops but they seem to re-sprout much as Citrus usually manage. So avocados could be a garden crop of the future given more selection. Pollination is a potential problem probably most easily obviated by growing several varieties. Although many are planted out both survival and cropping will be more likely with pot or tub culture as with Citrus; i.e., out in summer but take under cover for winter. Rich soil, full of humus, and good drainage with plentiful moisture are necessary. Be generous with pot size, but avoid excessively nitrogenous composts.

Growing under Glass and in Containers
If fruits are produced in northern areas, they ripen late, through autumn and into winter, so are best kept under glass. They can be grown in large pots, surviving easily, but are harder to fruit than, say, Citrus.

Ornamental Value
Avocados are attractive and often grown as houseplants. They do tend to get very leggy – the answer is to nip out the tip early on and then repeat this with each shoot making a bushier shape. However it's usually left too late and the legginess has to be dealt with; try winding it slowly down and around in a circle and up again – the stems are fairly pliable. If it bends; tie it in place. If it breaks off; trim neatly and hope for re-sprouting.

Maintenance
Spring Repot containerised plants and move outdoors once last frosts over.
Summer Feed, harvest and water as necessary.
Autumn Bring containerised plants indoors in cool temperate zones.
Winter Reduce watering and stop feeding avocados grown under glass.

Propagation, Pruning and Training
Easily grown from fresh stones kept warm and moist. The best fruiting varieties are layered, or grafted on seedling stock. Little pruning is needed; nip out the centre shoot if you want a more squat shape!

Pests and Diseases
Under glass they may get attacks of common pests, but are usually robust. In the tropics avocados are one of the first trees to be attacked by termites – secreting fresh logs of their wood about your dwelling and inspecting fortnightly indicates whether any termite nests are hidden about and enables these to be quickly dealt with before they get well established.

culinary

Bruised and blackened is not good, but preferable to hard and under-ripe! More savoury than sweet, an avocado is best with vinegar, pepper and salt, but is usually overwhelmed with prawns and mayonnaise. Containing up to 25 per cent fat, it adds body (and calories) to dips. Chunks or slices go very well with or in salads and rather than vinegar use lemon or lime juice for a different taste. Likewise try avocado with walnut oil and balsamic vinegar, wow.

Avocado Dip/Guacamole
Serves 6

3 avocados
1 large or 2 small hard-boiled eggs
1 plum tomato, skinned and seeded
1 small sweet red pepper (I prefer red)
3 spring onions
1 or more garlic cloves (optional)
1 tablespoon lemon or lime juice
Dribble of olive oil
Salt and pepper in moderation
A little ground coriander
As much chilli powder as you like

Scoop out the flesh of the chilled avocados and blend with all the other ingredients. Chill and serve with crudités.

Mangifera indica from the family *Anacardiaceae*

MANGO

Tree, variable size. Life span: as long as a human being. Fruits: variable in size, flat ovoids, green/yellow or red. Value: rich in vitamins A, B and C.

Medium to large trees, with luxuriant masses of narrow leaves, these carry fruits that weigh anything from a few grams to a kilogram. They have an inedible, tough skin and an enormous, flat stone to which the fibrous flesh adheres.

Mangoes are native to India and exist there in countless variety. Doubtless early Europeans came across them, but they were first recorded by a Friar Jordanus in about 1300. Mangoes have now spread to most hot regions, including Florida.

varieties

India's favourite varieties are **Alphonso**, large and of fine quality and flavour, and **Mulgoa**, medium-large, green and blotchy. The West Indies prefer **Bombay** (**Peters**), round, flat, yellow when ripe and very juicy, and **Julie**. Cuba once exported one rated very highly called **Biscochuelo**.

cultivation

Mangoes want a hot, dryish climate and deep, well-drained, rich soils. Excessive rain spoils pollination and drought spoils the quality of the fruit. They want exceptionally deep and wide planting holes dug about 10m apart.

Growing under Glass and in Containers
As they are large when fruitful, the glasshouse would have to be big. They may be grown in pots for ornamental use but fruiting success seems unlikely.

Propagation
Mangoes raised from seed are often polyembryonic, giving several seedlings. Some are near-clones of the original, and some produce poor fruits with stringy texture and a turpentine taint! The best varieties are grafted or layered and will bear in about four years.

Young fruits are often pulled – but not this young

Pruning and Training
Remedial pruning of thin and poor growth is necessary once a head has formed, plus root pruning if the tree persists with strong, unfruitful growth.

other uses

The seeds have been boiled and eaten in famines. The wood is poor, but used for packaging crates and firewood.

harvesting, storing and culinary

Mangoes soften and turn yellow or red as they ripen and do not keep. Picked unripe, they travel well and are fine for culinary use. They are very messy to eat raw! They are widely used for chutneys, jams, tarts, pickles and preserves.

Mango Chutney
Makes approx. 1kg

1kg green mangoes
175g salt
600ml vinegar
75g each peeled chopped garlic cloves; sultanas; chopped dates
50g each chopped fresh ginger; chopped blanched almonds
2 teaspoons hot chilli powder
500g brown sugar

Peel, stone and chop the mangoes, sprinkle them with salt and keep cool overnight. Rinse and drain thoroughly. Mix all the ingredients except the sugar and simmer for several hours till soft. Add the sugar, bring to the boil and bottle in clean sterilised jars

Store for six months before use.

Carica from the family *Caricaceae*

PAWPAWS

PAPAYA

Herbaceous, up to 6m. Life span: very short. Fruits: up to 30 x 15cm. Value: rich in papain, calcium and in vitamins A and C.

Pawpaws (also known as papaya) are small herbaceous 'trees' that resemble palms as they are unbranched with ornate, acanthus-like foliage clustered on top and up to fifty green 'melons' underneath. There are male, female and hermaphrodite plants. The fruits uncannily resemble melons, turning yellow-orange as they ripen. The flesh is usually pink with a central hole full of small round seeds and can weigh up to 2kg.

Although it is indigenous to Central America, the pawpaw has rapidly spread to every warm country. Seeds were sent to Nepal as early as 1626 from the East Indies.

varieties

Solo is a commercial dwarf, but most pawpaws are local varieties selected from seed and varying considerably. Grow them from fruit you like. *Carica candamarcensis*, or the **Mountain Papaw**, is hardier with coarser leaves and smaller fruit, which have blunt ridges and an apple-like aroma. They are too acid to eat raw, but good cooked or as jam. *Asimina/Anona triloba*, **Northern Pawpaw**, *Anonaceae*. The **Custard Banana** is a bottle-shaped fruit with fantastic potential. Closely related to the papaya, it is much hardier and in its Native America is found as far north as Michigan and New York. A large attractive long-leafed suckering bush, pest-resistant with fragrant purple flowers, it prefers moist not wet soils. Male and female plants are needed: pollination is by flies. The fruits vary from 5 to 15cm long, are green, ripening yellow to bronze with yellow pulp, big brown seeds and a resinous flavour. Occasionally tasty raw, they are usually best cooked. Slow growing, slow bearing and long living, unlikely to fruit save under glass, they were introduced to Britain in 1736.

cultivation

Pawpaws prefer deep, humus-rich soil, and to be about 3m apart. They need support while young. You should eliminate most males and replace the whole lot every five years.

Growing under Glass and in Containers
They are practical for tall, heated greenhouses, and highly ornamental, so also well worth growing as pot plants, but they do not fruit well easily.

Ornamental and Wildlife Value
Their foliage is very attractive and they're so easy to grow they are worth trying anywhere frost-free and elsewhere as summer bedding!

Propagation
Variable from seed; usually grown by sowing several to a hole and eliminating the poorest seedlings. (Only one male is needed to fifty females, but in fact they can be differentiated only when flowering.) Fruiting occurs within a year.

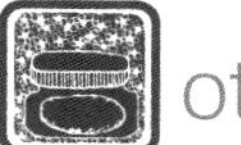

other uses

Papain, the active ingredient, is used medicinally, and for chewing gum.

culinary

Pawpaws are in season all year, and unripe ones keep for many days. Eaten as dessert or cooked as a fruit or vegetable, pawpaw is delicious, especially the first time. However, most importantly, the fruit and leaves contain papain, which tenderises meats cooked with them.

Pawpaw Breakfast Juice
Serves 2

1 ripe pawpaw
1 small lime
honey or sugar to taste

Scoop out the flesh and sift out the seeds of the pawpaw. Add lime juice and sweetener and liquidise. Serve immediately in frosted glasses with sugared rims as breakfast starters.

Psidium and *Acca* from the family *Myrtaceae*

GUAVAS AND FEIJOAS

Tree 3–9m. Life span: short. Fruits: 5–8cm, yellow or red. Value: very rich in vitamin A and C, pectin, iron, potassium and calcium.

Guavas are small trees or spreading shrubs with leathery leaves. The bark peels off the smooth, ruddy branches in flakes. The fruits are round, ripening from green to yellow or red, full of acid yellow or red pulp and many hard, round seeds.

Native to tropical America, guavas soon became popular the world over and were grown in orangeries.

varieties

There are many species of guavas. *P. guajava* is the commonest, with yellow fruits. Many varieties of it are pyriferum – bearing somewhat pear-like fruits, which are a little acid and better for cooking. Some varieties are pomiferum, bearing apple-shaped fruits which are thought better. For dessert, gourmets choose *P. cattleianum*, the **Strawberry Guava**. This has a reddish-purple, plum-size, sweeter fruit on a hardier, shaggy-barked tree. *P. araca, P. montanum, P. pigmeum* and *P. polycarpon* are all reputed more delicious still. *Acca sellowiana* a.k.a. **Feijua** is so similar as to be a variety of guava. It is smaller, with crimson and white flowers and fragrant fruit. It is almost hardy and not self-fertile, and grown in California. Feijoas are rich sources of vitamin C.

cultivation

Any good soil and a sunny site is all they require. In a warm climate they are normally planted at about 4–5m apart. They are thirsty plants!

Growing under Glass and in Containers
Acca is most reliable, the **Strawberry Guava** next, but all guavas are easily grown under glass and/or confined in pots. They can be put outdoors during the warmer months of the year.

Ornamental and Wildlife Value
Attractive pot plants even if they never fruit, in the wild they are loved by birds, which spread the seeds.

Propagation, Pruning and Training
Propagate by seeds for the species but also by suckers, layers or grafts for better varieties. Some plants occasionally produce seedless fruits: note these and propagate from them. Nip out top shoots to promote bushiness; otherwise prune only remedially.

other uses

The heavy wood is used for agricultural implements. The leaves and bark are a native cure for dysentery. The burning wood and leaves add a delicious aroma to barbecue and jerked meats.

harvesting, storing and culinary

Guavas are best fresh off the tree, or picked early, as they soften. They are good for dessert, and cooked as tarts, jam and, of course, the jelly.

Guava Jelly
Makes approx. 2kg

1kg guavas
1 large lemon
Approx. 1kg sugar

Simmer the washed, chopped guavas for two hours with a little water. Strain, weigh and bring back to the boil. Add the same weight of sugar and the juice of the lemon, reboil, pour into warmed sterilised jars and seal.

Strawberry guava – it really does taste of strawberry!

Artocarpus from the family *Moraceae*

BREADFRUIT AND JACKFRUIT

Trees up to 28m. Life span: medium. Fruits: up to 20cm diameter, leathery balls. Value: mostly starch.

Both breadfruit and jackfruit are attractive, tall trees with large, deeply incised leaves. From the branch ends hang green, round to ovoid fruits, which have a thin, warty rind and are white and starchy within. Some have about 200 fleshy edible seeds or more, some have none.

Breadfruit are native to the Pacific and East Indies, jackfruit come from the Asian mainland and Indian sub-continent. They were first noted in the voyages of the sixteenth century and soon taken to other hot regions, but never proved really popular. Breadfruit plants being taken to the West Indies played a part in the famous 1787 mutiny on *The Bounty*. When the water supply ran low, the valuable cargo was given water in preference to the crew.

varieties

Artocarpus communis (*incisa* or *altilis*) is the **Breadfruit** proper, bearing a remarkable resemblance, once cooked, to bread. From the West Indies comes the **Bread-nut Tree**, which, when cooked, is claimed to rival a new loaf in both taste and texture. *A. integrifolia* (*heterophylla*) is the **Jackfruit** or **Jakfruit**. The fruit is much bigger, weighing up to 30kg. These fruits strangely spring from older branches and direct from the trunk. When ripe they have a strong odour of very ripe melon.

A. odoratissima, the **Johore Jack** is smaller and esteemed for its sweetness and flavour. *A. integer*, **C(h)empedak** is another similar fruit, even sweeter and pongier eaten fresh and with more seeds which are also much prized.

cultivation

Any reasonable soil and site in a hot and moist climate.

Growing under Glass and in Containers
The size of the fruiting tree makes it impractical to grow these in pots or under glass except for ornamental value.

Ornamental and Wildlife Value
Very handsome trees, the fruits are valuable to wildlife.

Propagation, Pruning and Training
They can be propagated by seed, but the best varieties come only by root suckers or by layering. Little pruning is required except remedial.

Harvesting and Storing
Breadfruit are eaten fresh after cooking, but used to be stored in pits where the pulp was fermented to make a nauseous soft 'cheese', which would keep for several years. Breadfruit are traditionally stored under water! Jackfruit are said to become the best if a small one on a bent-down low branch is heavily earthed over before it starts to swell.

companion planting

The jackfruit tree is often used to support pepper (*Piper nigrum*) and as a shade tree for coffee plantations.

other uses

Jackfruit wood is like mahogany, valuable and useful. A yellow dye for clothing is extracted from the wood in India and the East. Breadfruit wood is light and used for box manufacture, and in Hawaii for surfboards.

culinary

Breadfruit are usually eaten roasted, boiled or fried as a vegetable. The edible seeds are often preferred, when they occur. Jackfruit are eaten in the same way and have a stronger flavour. Breadfruit can be dried and ground to a flour offensive to some and may also be eaten raw, in soups or even ice cream.

Baked Breadfruit
Serves 2

Bake a breadfruit in a preheated oven at 190°C/375°F/gas mark 5, until you can push a knife through it easily. Extract the pulp, seeds and all, and serve with curry or a savoury sauce.

Durio zibethinus from family *Malvaceae/Bombacaceae*

DURIANS

CIVET FRUIT

Tree 30m. Life span: fairly long. Fruits: about 25x20cm, ovoid, green to yellow. Value: a little protein, a little fat; one quarter to a third is fat and starch.

This is an infamous fruit, banned from airlines and loathed by most people on first acquaintance. It has an aroma similar to that of an over-ripe Gorgonzola cheese in a warm room. The flavour was long ago described as 'French custard passed through a sewer pipe'. The texture is like that of blancmange or custard, and the sweet flavour delicious and addictive. Once bravely tasted, the durian is unforgettable and 'the sensation is worth a voyage to the East'.

The trees are very large and upright with leaves not dissimilar to those of a peach. The fruits are round to ovoid, very large, weighing up to 4–5kg, green initially, yellowing as they ripen, with a long stalk. They are covered in short, sharp spikes and resemble some brutal medieval weapon. The pulp is white with up to a dozen big seeds.

Durians come originally from Malaysia and spread to Southeast Asia in prehistoric times. Widely grown all over the region, they have never managed to become more than a curiosity elsewhere.

varieties

Local varieties exist with variation in size, shape and flavour, but no widespread commercial clones are currently available.

cultivation

The large trees need a deep, heavy soil to secure them, but they are not too difficult to please.

Propagation, Pruning and Training
Seeds, if fresh, germinate in about a week and come nearly true. Only remedial pruning work is necessary.

Growing under Glass and in Containers
Their size and their need for tropical heat and moisture make them difficult. They may make good pot specimens if fresh seed can be obtained.

Ornamental and Wildlife Value
Attractive trees in their native climate.

other uses

Durians help get railway compartments to oneself. They are also reputed to be an aphrodisiac.

harvesting, storing and culinary

Durians must be eaten fresh and they quickly spoil due to a chemical change (and the aroma gets worse!). They are best eaten raw but may be made into ice cream or jam, or juiced and drunk with coconut milk. The large, fleshy seeds are boiled or roasted and eaten as nuts.

Durian Delight
Serves 6

225g each unsweetened durian purée, honey, natural set yogurt, cream
Dash of vanilla essence

Mix the ingredients and beat till smooth. Partially freeze, beat again and repeat this two or three times before freezing firm. Keep well sealed until immediately before eating!

They can be located by the nose alone

Annona **from the family** *Annonaceae*

CHERIMOYAS
CUSTARD APPLES AND SOUR SOPS

Tree up to 6m. Life span: short. Fruits: variable in size, green and scaly. Value: all varieties rich in vitamins.

A family of small trees, known as custard apples or sour sops, from the flavour and texture of the better fruits. Many have aromatic leaves and/or fragrant flowers. Despite the varying appearance of each species, the common names are often swapped or confused in different countries. The flesh is usually white, sweet and acid, with up to 30 black seeds embedded in it and covered with a thin rind, which breaks off like scales when ripe.

Natives of the Americas, they are most popular there but have spread to other tropical and warm zones. They are grown in Madeira and the Canaries for the European trade.

varieties

All varieties have leaves with some scent if rubbed, often downy underneath; the odd green tri-petalled flowers are often fragrant, the fruits resemble artichokes with a banana-cum-pineapple flavour and lots of hard raisin-sized seeds. *Annona squamosa* the **Sugar Apple** or **Sweet Sop**, is small fruited and knobbly and most common in the West Indies and is always eaten fresh. *A. cherimolia* is the **Cherimoya**, this is deciduous, more hardy, growing at high elevations in hotter areas and best for growing under glass in cool countries. *A. muricata* or **Sour Sop**, is evergreen with large (up to 2–3kg) green fruits rich in vitamins B and C with soft spines in ridges and the best flavour, but fibrous. Eaten raw as a vegetable it is made into delicious beverages when ripe. *A. reticulata* is the **Bullock's Heart** called because of the ruddy colour, it has firm, sweet, yellow pulp rich in vitamin A. Numerous other species are grown all over the Americas and especially good are the huge rugby ball-sized **Rollinias**.

cultivation

They will thrive in poor soils but give better crops with better treatment and prefer dryish, hilly conditions.

Growing under Glass and in Containers
The cherimoya is worth growing even if it never fruits and custard apples are probably worth trying. Sour sops may not crop, but all are good value as pot specimens.

Propagation, Pruning and Training
They grow from seed, but best varieties are budded. Only remedial pruning and nipping out is required.

Harvesting and Storing
Picked under-ripe they keep for up to a week or so.

culinary

They are usually eaten raw or used to flavour drinks and ices. Ripen wrapped in brown paper, yellowing and scent indicate time to use – black bits on skin are not rot! They make excellent ice cream and a really weird cider.

Sour Sop Sorbet
Serves 6

500g sugar
1 large ripe sour sop
60g crystallised ginger

Dissolve the sugar in 450ml boiling water and chill. Squeeze the juice from the sour sop pulp and strain. To each cupful add half a cup of sugar syrup. Partly freeze, beat and refreeze. Repeat twice. After the final beating, mix in the chopped ginger.

Averrhoa carambola* from the family *Oxalidaceae

CARAMBOLAS

Tree up to 11m. Life span: medium. Fruits: up to 5 x 15cm. Cylindrical, star-shaped in cross section, yellow. Value: vitamins A and C.

Averrhoas are small trees with delicate, pinnate (walnut-like) foliage. A profusion of sprays of little white or pinkish flowers is followed by huge quantities of yellow, cylindrical, star-shaped fruits with five prominent angles, which weigh down the branches.

Natives of Indonesia and the Moluccas, averrhoas are still mostly grown in Southeast Asia and the Indian subcontinent. However, because of their decorative value, and being quite robust, they are now exported and appear almost everywhere.

varieties

Carambolas are variable in taste. Some are much better than others, though they are all juicy. I have eaten them from European supermarkets, when they were like yellow soap, and also fresh, when they were amber and a joy. *Averrhoa bilimbi* or **Billings** are similar, resembling gherkin cucumbers, but are used more as vegetables and in curries, and for the popular billings jam.

cultivation

Adaptable to most warm, moist climates, and any reasonable, well-drained soil, they fruit prolifically with little attention.

Ornamental and Wildlife Value
Burdened with yellow fruit, they are impressive.

Growing under Glass and in Containers
Carambolas are pretty plants, so make good specimens. If they fruit they will put on a terrific show.

Propagation
Usually they are grown from seed, but better varieties are grafted. Only remedial pruning is needed.

other uses

Carambola juice removes stains from linen and can be used for polishing brass. Billings also polishes brass.

harvesting, storing and culinary

They travel fairly well if picked just underripe and then keep for a week or two. The narrow-ribbed carambolas are sourer, the wider fleshier-ribbed ones sweeter. Carambolas are as often used for refreshing drinks as dessert fruits. Their star shape makes excellent decorative garnishes in compotes and fruit salads.

Camaranga or Carambola Jam
Makes approx. 2kg

1kg carambolas
1kg white sugar

Cut the fruit up into finger-thick pieces and discard the sharp edges. Do not discard the seeds, as they improve the flavour of the jam. Add water to cover them and boil till the pieces are softening – about 15 minutes should be sufficient. Add the sugar and bring to the boil again for another 15 minutes, then bottle and seal.

Achras sapota/Manilkara zapota* from the family *Sapotaceae

SAPODILLAS

OR SAPOTAS, NASEBERRIES, BULLY TREES, CHIKKUS

Tree 4–16m. Life span: medium. Fruits: about 6cm, rounded, brown.

Sapodilla makes a medium to big tree with glossy leaves. The fruit is a large, round berry with a rough, brown skin over luscious pulp similar to a pear's, containing a core with up to a dozen seeds much like an apple's – black, shiny and inedible. The tree's milky sap can be tapped in the same way as rubber. Once collected, it is coagulated with heat and the sticky mass produced is strained out and dried to form chicle gum.

Still mostly grown in its native region of Central America, sapodilla is found wild in the forests of Venezuela. Sapodillas were more extensively planted when chicle gum started to be used for a booming new commodity – chewing gum. Now it is an important crop for Mexico and Central American countries. The naseberry is Jamaica's national fruit.

varieties

Local varieties are cultivated for fruit, but there is little commercial demand, as it must be eaten absolutely ripe. However, there is a big demand for chicle gum; varieties selected for sap production can give up to 3kg of gum per year.

Pouteria/Calocarpum sapote, **Mamey sapotes** or **Marmalade Plums** are similar *Sapotaceae*, with red fruits used fresh and preserved, and the seed can be made into a bitter chocolate, nearly hardy and grown in many local varieties.

cultivation

Sapodilla prefers very hot, moist climates with rich soil.

Propagation, Pruning and Training
Propagate by seed or preferably by grafting for better varieties. Only remedial pruning work seems necessary.

Growing under Glass and in Containers
The trees are variable in size and unlikely to crop well under glass, but they are attractive so should make good specimen pot plants.

Ornamental and Wildlife Value
Sapodilla has very attractive foliage and the fruits are enjoyed by wildlife.

Harvesting and Storing
Best finally ripened off the tree, when the fruit softens and mellows in a few days to a treacly, gummy consistency. Left on the tree, the fruits become veined with milk, which makes them too acid until bletted like medlars (see page 577).

One of the great unknowns

companion planting

Sapodilla trees are normally grown for the first five years with underplanted legume crops.

other uses

Sapodilla wood is hard and durable, so it is used for handles and tools.

culinary

Perfectly ripened, they are considered superb dessert fruit. The fruits keep up to a month or more in a cold refrigerator. Sapodillas also make a good ice cream and, believe it or not, even an interesting milk pudding with garlic!

Chewing Gum

120g each of chicle gum, glucose, caramel paste, icing sugar
225g sugar
Spearmint or mint flavouring, to taste

Melt the gum carefully in a bain-marie. Meanwhile boil 100ml water, the sugar and glucose to exactly 124°C. Remove from the heat, add the caramel and boil again. Off the heat, mix the syrup into the melted gum, beating steadily and briskly. Add the flavouring and pour on to a cold surface thickly coated with icing sugar. Roll flat. When cold, cut into strips, wrap and label.

Nephelium lappaceum and *N. chinensis/litchi* from the family *Sapindaceae*

RAMBUTANS AND LITCHIS

Tree up to 18m. Life span: medium to long. Usually not self-fertile; male and female flowers often on separate trees. Fruits: 2.5–5cm, round, reddish or yellow.

Rambutan or ramtum trees are large and spreading, with pinnate leaves, and festooned with hairy, chestnut-like conkers. These fruits are apricot-sized, covered with red or orange-yellow soft spines like tentacles. Underneath the skin the flesh is wrapped around the single, inedible, brown seed. The flesh is sweet, acid, almost like pineapple with a hint of apricot – shame there is so little, as it is one of the best fruits I've ever tried.

Very similar yet more perfumed are litchees, lychees or litchis. These have a prickly, crackly shell and grow on a smaller tree. Originally from the Malay archipelago, rambutans are greatly appreciated in Southeast Asia where they are often grown in gardens, but, surprisingly, have never proved popular anywhere else. Litchis have long been a Chinese speciality, so they followed their people to many other suitable areas, such as Florida.

varieties

Many different varieties of **Rambutan** and **Litchi** are grown in their regions. The **Pulassan**, *Nephelium mutabile/chryseum*, is another species; native to Java it is similar but covered with fleshy spines instead of tentacles. *N. longana/Dimocarpus longan*, the **Longan**, is popular in Southern China; it is smaller, browny yellow and nearly smooth skinned with similar chewy flesh. Very similar is *Euphoria malaiense*, **Cat's Eye** or **Dragon's Eye**, though the seeds are huge relative to the flesh. The **Keppel** is an almost mythical fruit that allegedly turns body odours to the sweet smell of violets – should be compulsory eating before all public events, then.

cultivation

They need tropical conditions. Litchis prefer lower humidity, rambutans more.

Growing under Glass and in Containers
Sadly they are too large. They may possibly do as pot specimens for foliage, but are unlikely to fruit.

Ornamental and Wildlife Value
Handsome trees, they are often planted in gardens and parks in Southeast Asia and the Indian subcontinent. The fruit is exceptionally attractive to birds and also bats.

Propagation, Pests and Diseases
They come nearly true from seed, but the best varieties are budded. Bird and bat damage is a severe problem.

harvesting, storing and culinary

Both will ripen if picked early off the tree, so they are often found in temperate country shops. Litchis are often preserved in syrup or dried to perfumed 'prunes'. All the fruits are superb desserts. Litchis are often used to close a Chinese meal.

Litchi Sundae
Serves 4–6

1kg litchis
600ml real vanilla ice cream
60g blanched toasted almonds
Nutmeg
30g coarse brown sugar
4–6 glacé cherries

Peel and stone the litchis, then chill them. Layer the fruit in tall glasses with ice cream and almonds and top with a flourish of nutmeg, sugar and a cherry.

Garcinia mangostana from the family *Clusiaceae (Guttifereae)*

MANGOSTEEN

Tree 14m. Life span: long. Fruits: 6cm, round, brownish. Value: small amounts of protein, mineral matter and fat; approx. one-seventh sugar and starch.

The trees are small to medium-sized, cone-shaped and have large, leathery leaves somewhat like a lemon's. The fruits are round, purplish-brown, about apple size, with a rosette of dead petals around the stalk and an odd 'flower'-shaped button on the other end. If you cut the rind around the fruit's circumference, the top can be lifted off to reveal about half a dozen kernels of melting white pulp, tasting between grape and strawberry, tart and sweet, almost syrupy and chewy. There is a seed contained in many kernels, which is not eaten.

Mangosteens are natives of Malaya and were described by Captain Cook in 1770 in detail, and with delight. They were introduced to Ceylon (now Sri Lanka) in 1800 and were successfully fruited in English greenhouses in 1855. Widely held to be the world's most delicious fruit, they may be found in gardens in every tropical area, but are nowhere grown on a commercial scale.

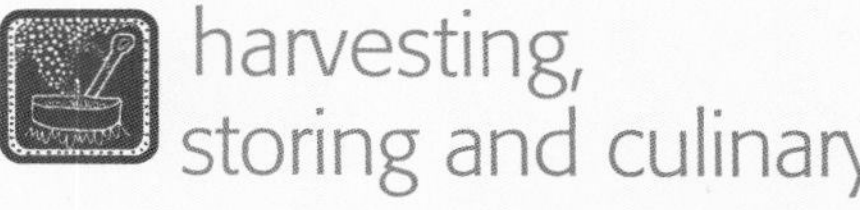

harvesting, storing and culinary

Use a ladder to climb up and pick mangosteens as they bruise if they fall. They can be picked unripe and kept for a few days or may be made into conserves.

Mangosteen Ecstasy
If you are fortunate enough to have a mangosteen, just eat it!

Not all segments have seeds

varieties

There are many local varieties and also close relations, most of which are found in the East Indies. *Garcinia cambogia* has a smaller, yellow-pulped, yellow fruit. *G. cowa* is the **Cowa-Mangosteen**. Bigger, ribbed and apricot-coloured, it is generally too acid for dessert but makes good preserves. Another, *G. indica*, the **Cocum, Conca** or **Kokum**, has a sour, purple pulp used to make a vinegar, and the seeds are pressed for cocum oil. *G. dulcis* is a yellow-fruited variety found in the Moluccas. *G. morella* is common in Southeast Asia and the plant also provides an orange-red resin, gamboge.

cultivation

Mangosteens need deep, rich, well-drained soil, a sheltered site, shade when young, a hot, moist climate.

Growing under Glass and in Containers
If it can be done in England in the mid-nineteenth century, it can be done now.

Propagation
They are slow and unreliable from seed. They are then slow-growing, reaching only to the knee after two or three years. The best varieties are layered.

companion planting

Mangosteens benefit from light shade, especially when young, preferring tropically bright but not direct light, so they are planted in the shade of taller trees.

TRAVELLERS' TALES TROPICAL FRUITS

There are many more fruits the inveterate traveller may come across in tropical and semi-tropical countries. Some of these are of more practical or commercial interest, such as the spices, while others are of such purely local interest they are rarely if ever recorded. Many may be unknown, as they are unpalatable by 'modern' standards, or they may be delicious but hard to cultivate. As we move into a homogenised world of mass consumption, the numbers of varieties of even the most popular fruits is declining. Quaint, difficult and unusual fruits have already disappeared from all but local native markets and botanic and private gardens. If you travel far off the beaten track, you may come across the following, and others – but, when choosing to taste them, do not rely on my identification.

Aberia gardneri,
Ceylon Gooseberry, *Bixinea*
Native to Sri Lanka, this is a small, shrubby tree with large, purple-brown, round berries mostly used for making jams and preserves. Closely related is *A. caffra*, the **Kai**, **Kau** or **Kei Apple** of South Africa, which is yellow and so acid it is used as a pickle, omitting the vinegar.

Baccaurea dulcis/Pierardia motleyana,
Rambeh or **Rambei**, *Euphorbiaceae*
Found in Malaysia, especially Sumatra, and China, this has long, hanging bunches of large, yellow berries that are reputedly sweet-tasting, juicy and luscious. *B. griffithii*, **Tampoi**, a similar fruit, which has buff berries with yellow flesh, is considered better eating than the **Rambeh**/**ei**.

Blighia sapida,
Akee, *Sapindaceae*
A medium-sized tree from West Africa, grown in the West Indies, especially Jamaica, where the fruit, fried with saltfish, is considered excellent fare. The fruits are bright red, heart-shaped pods (see image above), which burst to reveal three glossy, black seeds the size of peas sitting in a yellowish cup, which is the tasty bit. The seeds are inedible, the pink flesh highly poisonous and even the edible bit is poisonous if under- or over-ripe. One wonders how this ever became popular!

Carissa grandiflora,
Natal Plum, *Apocynaceae*
This and *Carissa carandas* are large, thorny shrubs used as hedges in Natal. They have purple, damson-like fruits tasting of gooseberry which are widely used for tarts and preserves. *C. carandas*, which is also used for pickling, prefers drier areas. *Carissa congesta*, **Kerandang** is preferred in Malaya.

Ceratonia siliqua,
Carob Tree, *Leguminosae*
The **Locust Bean** or **St John's Bread** is an edible, purple-brown, bean-like pod that tastes so much like chocolate it can be ground up and used as a substitute. The seeds are not eaten, but are so exact in size and weight they were used for weighing gold and were the original 'carat'. This tree is native to the Mediterranean region, and preserved pods have been found at Pompeii. Nowadays, the pods are used as animal feed. Remarkably similar are the closely related **Honey Mesquite**, *Prosopis juliflora*, and **Honey Locust**, *Gleditsia triacanthus*, often planted as ornamental trees; their pods have a very sweet pulp around the inedible seeds.

Chrysophylum cainito,
Star Apple, *Sapotaceae*
Noted by Cieza de León in Peru in 1532–50, this is a large, evergreen tree with purple-brown 'apples', which, when cut through, have a star shape in the middle with about half a dozen shiny, brown seeds in a sweet/acid pulp.

Coccoloba uvifera,
Seagrape, *Polygonaceae*
A coloniser of tropical shores, this is a very salt-tolerant, small, evergreen shrub or tree found from Florida to Venezuela. The 'grapes' are up to half an inch across, mild and sweet. They are eaten raw or jellied. The wood is hard and takes a polish well. It is often used as hedging or for good windbreaks.

Eugenia caryophyllus

TRAVELLERS' TALES TROPICAL FRUITS

Coccoloba uvifera

Eugenia caryophyllus,
Cloves, *Myrtaceae*
Cloves are the dried flower-buds of this Indonesian tree. The **Brazilian Cherry**, **Surinam Cherry** or **Pitanga Cherry**, *E. uniflora* has fruits varying from yellow to red and purple: unusually, the darker are the more acid. Other *Eugenia* species have bigger, juicier fruits, though often rather bland, such as the Jambus, the **Malay Apple**, *Syzygium/E. malaccensis*, the **Water Apple**, *E. aquea*, and the **Rose Apple**, *E. jambos*.

Hovenia dulcis,
Japanese Raisin Tree, *Rhamnaceae*
I remember having this as a child. Resembling brown, candied angelica, the 'fruit' is the dried, swollen flower stalk from behind the pea-sized seed of a small, attractive, Asian tree with glossy foliage. Eaten in China as a hangover cure, where they also make sugar from seeds, leaves and shoots.

Mimusops elengi,
Sapotaceae
This large East Asian tree has fragrant flowers. The 2.5cm yellow berries are eaten when ripe and an oil is expressed from the seed. Other Mimusops are also grown for their similar fruits. *M. elata* of Brazil is the **Cow Tree**. Its apple-sized fruits contain a milk-like latex that resembles milk, when fresh, and is drunk with coffee, but soon congeals to a glue.

Pimenta dioica, Allspice,
Pimento, *Myrtaceae*
This small, evergreen West Indian tree has pea-size berries that are dried and valued unripe for their mixed spice flavour.

Spondias,
Spanish Hog or **Brazilian Plum**, *Anacardiaceae*
Distantly related to cashews and pistachios, the *Spondias* have edible fruits, most of them only when made into preserves, but some are eaten raw. They are purple to yellow, resembling a plum with a central 'stone'. The stone of the **Spanish Plum** or **Otaheite** or **Golden Apple**, *Spondias purpurea*, is eaten by some people.

Syzygium paniculatum,
Australian Bush or **Brush Cherry**
It has rose-purple fragrant berries used for jellies. *Eugenia dombeyi/brasiliensis, Grumichama*, is a red-purple cherry-sized berry, very popular in Brazil, fresh and cooked in many ways.

Syzygium paniculatum

Tamarindus indica

Tamarindus indica,
Tamarind Tree, *Caesalpiniaceae*
This large, handsome tree has brown pods containing very acid pulp used for beverages and in chutneys, curries and medicine.

Vanilla planifolia,
Vanilla, *Orchidaceae*
Vanilla is the fruit of an orchid that is native to Central America but is mostly grown in Madagascar. The bean-like pods are cured and dried for flavouring.

Zizyphus jujuba,
Jujube, **Chinese Date**, *Rhamnaceae*
An East Indian native, this reached China and was improved to dessert quality. In China it is popular dried or preserved in syrup. Jujubes resemble large, yellowish or reddish cherries with thick, tough skin, a hard kernel and a pithy, acid pulp, rich in vitamin C. The thorny, shrubby trees survive in cooler climates and have been in Mediterranean countries since biblical times. *Z. vulgaris*, a native of the Middle East, is similar but less agreeable. *Z. lotus* is like a sweet olive and is thought to be the lotus Odysseus had trouble with.

Adansonia digitata
Baobab, **Monkey Breadfruit,** *Bombacaceae*
This large African tree has a swollen trunk often hollowed by age. The leaves are eaten as a vegetable and the fruits, up to 30cm long, are fat, cylindrical, dark and hairy. They have a floury white pulp that reputedly tastes like gingerbread, with small black seeds that have been ground and eaten in times of famine.

Antidesma bunius
Bignay
Coming from the Far East and Northern Australia, this tropical evergreen tree – originally from India – produces grape-like bunches of multicoloured berries which are eaten raw, made into jams or eaten with fish.

Billardiera longiflora
Appleberry, *Pittosporaceae*
An acid-loving, nearly hardy climber from Australasia, which has oblong dark blue berries with a pleasing taste. Many similar species are eaten by the indigenous Australians.

Bixa orellana
Annatto, **Lipstick tree**, *Bixaceae*
This is a small evergreen tropical shrub-cum-tree that has heart-shaped leaves with peachy flowers followed by heart-shaped pods with seeds embedded in a coloured flesh used as a lipstick, for dyeing hair and clothes and commercially as the colouring Annatto.

Bouea macrophyla
Gandaria
This is just like a small mango gone very wrong. However, many in the Far East become addicted to its odd, though very juicy, taste.

Butyrospermum parkii
Shea Butter Tree, *Sapotaceae*
A native of central Africa, this small stout African tree has white fragrant flowers followed by plum-like fruits with a thick pericap that bleeds when green. Once fully ripe, this has some pulp, which is sweet and perfumed, but the fruits are gathered more for their seeds which have a high fat content and are used to make shea butter. The wood is resistant to termites.

Canarium commune
Chinese olive, **Java almond**
This is a closely related, rather attractive tree from Southeast Asia, which is grown for the tasty kernel.

Canarium edule/Dacryoides edulis
Safu, *Burseraceae*
The fruit is a large violet drupe that is too bitter to be eaten raw, but is fatty and nutritious once cooked. It is popular in West Africa.

Casimiroa edulis
Casimiroa, *White Sapote or Zapotl*
From South America (but not a Sapota or Sapodilla), this plant looks pretty much like an apple crossed with a mango, buttery custardy, tasting a bit like pear flesh, with white seeds and a bitter papery skin that must be avoided. Chill before eating and they will not keep long, going rock hard to too soft in days. Claimed to have medicinal values as well as vitamins A and C.

Cecropia peltata
Trumpet Tree, **Monkey's Paw**, *Urticaceae*
This stinging nettle relation is the tropical equivalent of elderberry as a weed; from South America, it takes over old human habitations. The small apple-sized fruits are watery and fig flavoured.

Cicca acida
Malay Gooseberry, *Euphorbiaceae*
A small tree resembling a Carambola but with bunches of small greeny yellow berries, each with a hard seed and sour acid flesh eaten fresh but more often made into pickles.

Bixa orellana

Cynometra cauliflora
Nam-Nam, *Fabaceae*
Popular in Malaysia, this tree relation of the beans has odd, wrinkled, kidney-shaped, small apple-sized pods concealing juicy yellow, sour, flesh wrapped around a big seed.

Dillenia indica
Chulta, **Elephant Apple**, *Dilleniaceae*
This is a tree with huge leathery leaves and strange fruits a bit like globe artichokes with acidic pulpy bits and is rather stringy. It is used for sherbets, curries, jelly, making vinegar and, apparently, as treats for elephants.

Emblica officinalis
Indian Gooseberry, *Euphorbiaceae*
The fruits are small marbles of yellow or green. They are too acid to be eaten raw, but are amazingly rich in vitamin C, containing three hundred times as much as orange juice. They are often jammed and pickled and used to treat scurvy.

Euterpe oleracea
Acai Palm, *Palmaceae*
From the Amazonian basin, this plant has acid berries that are claimed to be incredibly health giving and so these are now appearing in smoothies and other tonic beverages.

Gnetum gnemon
Gnemon, *Gnetaceae*
A primitive plant from Java, this has what are technically cones, though not resembling a pine's; from these swell green 'fruits' that are really seeds. When these turn orangey-red the skin or rind is edible and most often used in cakes and confectionery.

Grewia asiatica
Phalsa, **Pharsa**, *Tiliaceae*
This is a shrubby reddish tree from India and the East Indies, which has many small red berries that are too acid and dry to eat many of. However, they make the most delicious sorbets and syrups, and are very popular in northern India.

Harpephyllum caffrum
Kaffir Plum
A dark red fruit, used for jelly, on a house-high tree with glossy leathery leaves.

TRAVELLERS' TALES TROPICAL FRUITS

Heteromeles arbutifolia
Christmas Berry, **Tollon**, *Rosaceae*
This is an easy-to-grow, attractive pot plant closely resembling a holly with white scented flowers and red, holly-like berries. It is a unique plant, found only in California, where it makes a shrubby tree 9m in height. It may also be hardy enough to grow on a wall in southern Britain.

Irvingia gabonensis
Duika, **Wild mango**, *Irvingiaceae*
This is a large central African tree which has an inferior mango pulp, but the seed is oily and is used for cooking and soap-making. It is also used in Gabon chocolate or *pain de dika*.

Lansium domesticum
Langsat, **Dokong**, **Longkong** or **Duku**
Meliaceae
These are a tribe of small yellowish berries with parchment-like skins hanging like slim bunches of grapes from large Far Eastern trees. The Duku is the preferred choice, though has a bitter seed to be avoided; the kongs are not big ape relatives but bigger sweeter Dukus.

Lardizabala biternata
Zabal Fruit, Aquiboquil, *Lardizabalaceae*
These are sold in Chilean markets: a climber with dark evergreen leaves, clusters of purple black flowers are followed by finger-length purple sausages full of sweet pulp. Nearly hardy, this is a superb conservatory or cold greenhouse plant or even for the warmest garden.

Malpighia glabra/ punicfolia/ emarginata Acerola
Barbados Cherry, *Malpighiaceae*
Big evergreen tropical shrubs with pink or red flowers followed by thin-skinned juicy red berries with a very high vitamin C content that are mouth-puckeringly acid, so are used in drinks and tonics with much sweetening.

Mammea americana
Mammey Apple, *Guttifereae*
This is a large evergreen tropical tree with an orange- to grapefruit-sized russeted yellowish fruit; this contains one to four big seeds in firm juicy flesh which is best eaten stewed, though some eat them raw if very ripe.

Morinda citrifolia
Dog Dumpling, *Rubiaceae*
This is a handsome glossy foliaged, small tropical tree which almost continuously flowers, attracting hummingbirds whilst dropping strange, off-whitish, macabrely pineapple-like fruits which make ripe Camembert seem like fresh air. Yet these are believed to be very healthy medicine, especially if fermented first. The leaves are also used as poultices for pain, fever and headaches and the roots widely used for dye.

Owenia acidula
Australian Native Nectarine, *Meliaceae*
This is a small, tender, ornamental tree with pinnate foliage, white flowers and bluish black fruits with very acid red pulp, which is good for drinks, juices and jellies. The large stony seed is used for jewellery.

Parinarium curatefolium
Mupunda, *Chrysobalanaceae*
There are a number of other closely related fruits found in Africa. This is considered the best, with reddish-brown or greyish plum-sized drupes on a shrubby bush.

Parmentiera aculeata
Cucumber Tree, *Bignoniaceae*
A Central American small tree, this bears flowers directly from the trunk which turn into very cucumber-like fruits used raw, roasted and cooked.

Pereskia aculeata
Rose Cactus, **Barbados Gooseberry**, *Cactaceae*
These very floriferous plants closely resemble small trees with woody stems and deciduous leaves and are the least succulent of the cacti family. The highly perfumed flowers vary from orange to white through yellow and the fruits resemble orange or yellow cherries. They are popular for preserving in the West Indies.

Randia formosa
Raspberry Bush or **Blackberry-jam Fruit**, *Rubiaceae*
This South American tender shrub can be grown in a frost-free conservatory to fruit in a large pot. The white flowers are followed by woody shelled fruits with a jam-like sweet, sticky centre.

Salacca zalacca
Snakefruit, Salak, *Palmaceae*
Another palm tree fruit; coming in bunches of a couple of dozen, the 'nuts' are an inch or so across and covered in what looks just like snakeskin with flesh tasting like a mixture of banana and pineapple all wrapped about a brown seed.

Sicana odifera
Musk Cucumber, **Casabanana**, **Curuba** or **Coroa**
Cucurbitaceae
This South American climbing vine resembles a lushly grown and prettier cucumber plant, but its tendrils really serve to glue themselves at the tips as well as twine. A perennial in Brazil and Ecuador, it can be grown very easily in more northern climates as an annual or conservatory subject. A very vigorous plant, it has yellow flowers and yellowish-red fruits with a sweetish flavour, but which are strongly fragrant – too strong for many. The fruits can be eaten young as vegetables and ripe as fruits, most commonly in jams.

Theobroma cacao
Cocoa, *Sterculiaceae*
The nibs are not eaten raw but fermented, dried and ground to make our chocolate. The trees are small natives of the Americas but now mainly grown in West Africa. The melon-like pods are green, ripening to red or yellow, and spring directly out of the trunk and main branches of the tree, following delightful fuschia-like pink flowers. Their pulp is sweet and edible also. Closely related is **Pheng Phok**, **Chinese Chestnut,** *Sterculia monosperma*, of which the nuts are much esteemed by the Chinese, who eat them boiled or roasted.

Yucca baccata
Eve's Date, *Liliaceae*
This is closely related to our **Adam's Needle**, the spiky garden perennial which flowers but rarely sets fruit. *T. baccata* is too tender, or rather too loathing of damp, so needs to be grown under cover. The flowers need hand pollinating, then purple peach-sized, stubby cylindrical fruits form which have an aromatic bitter-sweet taste. The native Americans were fond of them fresh or dried and also ate the flowerbuds roasted or boiled.

SHRUB AND FLOWER GARDEN FRUITS

WILD GARDEN FRUITING TREES AND SHRUBS

Many of our familiar garden and countryside plants also bear fruit. Some of these are edible and can add variety and nutritional range to the diet. But first, I must insist that you never eat anything you are not sure of. Have it identified as safe by an expert on the spot. Also you must realise that although the fruits of some plants, such as yew, may not themselves be harmful, the foliage and seeds are deadly if ingested in quantity. Various parts may also be an irritant to some people.

However, such warnings aside, there are many familiar plants that have edible, if not actually delicious, fruits. Although they might not seem at first glance very attractive or be popular with many of us today, some of these were once greatly esteemed by native peoples, and others were part of the country fare of our not-so-distant predecessors. With a little care and attention, these plants can provide us with fresh and unusual dishes, far exceeding in vitamins and flavour than those made from the tired and flabby fruits usually offered for sale.

The plants at the beginning of this section are all excellent garden subjects, worthy of anyone's attention, which incidentally bear edible fruit. Some of the other plants with edible fruits are not attractive enough, or grow too big, for most gardens, but are of interest or value to insects or birds, and are more often planted in larger, public or wildlife gardens.

The ideal picturesque garden, often called the cottage garden, is typified these days by extravagant species at flower shows with odd mixtures of flowers, half of them out of season, grown elsewhere in pots and jammed full, with a camouflage of bark. The true cottager's garden was indeed a mixture of plants, but all with a purpose – to provide medicines, herbs, flavourings, fruits and, last of all, flowers.

Fruiting trees and shrubs can be easily cultivated with other plants underneath, and were the backbone of a true cottage garden. A mixture of plants that were found to grow happily together for both production and ornament was sensible, as it produced a mixed ecology, so there were rarely pest or disease problems. In addition, the plants grown in flower, shrub and wild gardens are often innately more reliable than those especially cultivated for fruits, as they are closer to the wild forms, with natural pest and disease resistance.

The biggest handicap for some of these plants has probably been their very attractiveness. If they had been a little less pretty, they might have been developed further for their fruits and have remained part of our diet.

Few of them are palatable raw, at least not to most people's taste, and all are better made into jams, jellies and preserves, but they have appealing and interesting flavours and are of inestimable value to the adventurous gourmet or those wishing to expand their dietary range. And for those interested in breeding, they offer plenty of opportunity for rapid improvement towards bigger, tastier and better fruits.

Amelanchier canadensis from the family *Rosaceae*

JUNEBERRIES
SNOWY MESPILUS, SHADS, SWEET/GRAPE PEARS

Tree/bush up to 6–9m. Life span: medium. Self-fertile. Fruits: 1cm, round, purple-black. Value: rich in vitamin C.

The Amelanchiers are small, deciduous trees or shrubs tending to suckering growth, most noticeable when absolutely covered with white blossoms. The fruits are purplish, spherical and about pea-size, but can be larger. *A. canadensis* is the best species and a native of North America. Though there are relatives in Asia and Europe, these are not as palatable. *A. vulgaris* grows wild in European mountain districts and was long cultivated in England, as much for the flowers as the fruits.

varieties

Amelanchier canadensis was a favourite fruit of Native Americans. It was adopted by French settlers and became **Poires** in Canada and **Sweet** or **Grape Pear** in the USA. It has small, purple berries which are sweet. *A. alnifolia*, or **Western Service Berry**, is larger and found wild in the states of Oregon and Washington. All cultivars are supplied for ornamental rather than fruiting purposes. *A. spicata*, grown in Europe since the eighteenth century but not in the wild, is a suckering shrub that has blue black berries good for jam or wine.

cultivation

Amelanchiers do best in moist, well-drained, lime-free soil. They are slow-growing, tending to sucker.

Ornamental and Wildlife Value

Neat, compact, floriferous with good autumn leaf colours, they are excellent shrub border plants. The berries are much liked by birds.

Propagation, Pruning and Training

Sow seeds fresh for the species, but graft choice varieties in April on to *Sorbus aucuparia* stock. They may need to have suckers removed, otherwise prune only remedially.

Growing under Glass and in Containers

They are so hardy that it's not worth the space under cover. They could be delightful small specimens in pots.

Pests and Diseases

No particular problems affect these tough plants.

Frequently grown as a flowering shrub on lime-free sites

harvesting, storing and culinary

They can be eaten raw, but are better as jams or tarts, or dried like raisins.

Snowy Mespilus Sponge Cakes

Makes about 10

60g each butter, powdered sugar and white self-raising flour
1 large or 2 small eggs
Dash of vanilla essence
Splash of milk
120g dried amelanchier berries

Cream the butter and sugar, beat in the eggs and vanilla, fold in the sifted flour, then add enough milk to make a smooth mixture. Stir in the berries; pour into greased paper cups. Stand the cups on a metal tray and bake in a preheated oven at 190°C/375°F/ gas mark 5 for 20 minutes or till firm.

Myrtus communis from the order *Myrtaceae*

MYRTLE

Tree up to 4.5m. Life span: medium. Fruits: 1cm, usually black-blue.

Myrtle is one of sixty small evergreen shrubs which are densely and aromatically leafed, with white flowers.

***Myrtus communis* has long been grown in southern Europe and western Asia. It was highly revered and much enjoyed by the Ancient Greeks, whose mythology held myrtle sacred to Aphrodite, the goddess of love. As all other myrtles are native to South America or New Zealand, common myrtle may indicate a prehistoric link between the continents.**

culinary

Myrtle berries are eaten raw or in tarts and jams or dried. The jam has an aromatic flavour which goes well with savoury dishes. The fresh flowers can be added to a salad. Mediterranean, especially Tuscan, cooking uses the dried fruits and flower buds as a spice. In Chile, **Temo**, *M. molinae*, seed was used to make a coffee.

Myrtle Jam
Makes approx. 2kg

1kg myrtle berries
1kg sugar

Wash the berries and prick them with a darning needle. Simmer them until soft with just enough water to prevent burning. Add the sugar and bring to the boil then bottle in warm, sterilised jars and seal.

varieties

Myrtus communis is grown for fruit and has ornamental forms such as *microphylla* and *tarentina* which are small, rarely reaching more than 60cm high. Their foliage is aromatic when crushed, making them good for small patios or sunny windows, but they rarely fruit well. *M. ugni* is the **Chilean Guava Myrtle**, which has delicious, mahogany red, pleasantly fragrant berries which are used to flavour water. These apparently fruit well in English greenhouses. *Luma apiculata,* **Arrayan**, nearly hardy, from Chile, this small evergreen has beautiful leathery leaves, white flowers and then juicy aromatic sweet purple berries, good raw or as jam.

cultivation

Myrtles need full sun and a well-drained soil. They are not averse to lime and are also good by the seaside. *M. communis*, *chequen* and *nummalaria* are the hardiest.

Myrtles come in several species and varieties, though rarely selected for cropping

Ornamental and Wildlife Value
Very attractive evergreens for mild regions. The flowers and berries are attractive to wildlife.

Pests and Diseases
As myrtles are very tough there are no real problems.

Propagation and Training
The species can be grown from seed, or half-ripe cuttings can be taken in late summer with some success. Myrtle is best grown as a bush against a wall for the shelter and warmth. In early summer, clipping is usually preferred to pruning.

Growing under Glass and in Containers
These are very good subjects to have under glass and/or in containers, especially *M. ugni* or *M. communis tarentina*.

other uses

Used in medicine. The dried leaves and wood are fragrant. The flowers are made into perfume and toilet water.

Berberis vulgaris **from the family** *Berberidaceae*

BARBERRIES

MAHONIAS, OREGON GRAPES

Bush up to 4 x 4m. Life span: short. Fruits: under 1cm, white, yellow, scarlet, purple or black.

The Berberis family contains hundreds of small- to medium-sized, spiny shrubs that have masses of berries in many colours. The leaves of most deciduous varieties turn bright shades in autumn and the wood is usually yellow.

Various species are found all over the world. Our common barberry is now seldom relished, but was once widely popular. Indeed, the settlers in Massachusetts grew so many that in 1754 the province had to forbid further planting.

varieties

Berberis vulgaris is the **Common Barberry**, which once existed in a host of local forms and colours. One found in Rouen was seedless. Other species are enjoyed all over the world: *B. darwinii* is popular; *B. buxifolia*, the **Magellan Barberry**, is large and said to be the best raw or cooked. **Mahonia** is now a separate genus, but is very similar in many ways, only lacking the spines and having pinnate leaves. The flowers are yellow, usually scented, and the blue-black berries of *M. aquifolium* were made into preserves as Oregon grapes.

cultivation

They grow almost anywhere not actually dark, bone-dry or waterlogged, and are even fairly tolerant of salt spray.

Ornamental and Wildlife Value

Some ornamental varieties are very attractive, though not as productive of berries, which are well liked by birds. Most varieties are excellent when used for wildlife gardens.

Propagation, Pruning and Training

The species grows from seed, layered or grafted. They can usually be cut to the ground and will recover.

Beautiful in flower and berry, and nice tart jelly

Growing under Glass and in Containers

They are hardy enough not to need protection, but do make good plants in containers.

Pests and Diseases

Barberries are an alternate host for wheat rust, so care should be taken not to plant near wheat.

other uses

They make good hedges and game cover. A dye was made from the **Yellow Barberry** wood and roots.

culinary

The berries can be pickled in vinegar, preserved in sugar or syrup, candied or made into jam. The leaves were once used as a seasoning.

Colonel Flowerdew's Bengal Chutney

Makes approx 1.4kg

1kg grated apples
120g each of the following:
dried barberries, soft brown sugar, Demerara sugar, mustard seed, golden syrup
60g each of the following:
chopped onions, chopped garlic, chopped fresh ginger and salt
15g cayenne pepper
600ml vinegar

Mix all the ingredients and simmer till soft, say, 2–3 hours. Bottle in small sterilised jars for six months.

Mespilus germanica from the family *Rosaceae*

MEDLARS

Tree up to 9m. Life span: medium. Self-fertile. Fruits: 2.5–5cm, green to russet.

Medlars resemble pear trees but are smaller, have bigger leathery leaves, single, large, white flowers and fruits like giant, distorted rose hips. The brownish-green fruits have a rough 'leafy' end, which shows the seed chambers. Medlars seldom ripen fully on the tree in cool regions and were eaten 'bletted' – stored till at the point of decomposition. The taste is like that of rotten pear and they are now disdained. Originally from Persia, medlars became naturalised over much of Europe. Theophrastus mentions them in Greece in 300 BC and Pliny refers to the Romans having three sorts. Once very popular, they are now planted infrequently.

varieties

There are only a few left. Generally the bigger the tree, the larger the fruit tends to be. **Dutch** and **Monstrous** are the largest, **Royal** and **Nottingham** the tastiest and smaller. The seedless **Stoneless** has disappeared.

cultivation

Medlars are obliging and will grow in most places, generally preferring a sunny spot in a lawn or grass.

Growing under Glass and in Containers
The tree is hardy, so it is a waste to grow it under cover, but the twisted framework and general attractiveness make it a good specimen plant for pot growing.

Ornamental and Wildlife Value
Pretty flowers, large leaves and gorgeous autumn golds make this a delightful specimen tree, with a twisted and contorted dark framework for winter interest. The fruits are useful to the birds late in winter.

Propagation, Pruning and Training
Medlars can be grown from seed, but are usually grafted on pear, quince or thorn stock. They are best pruned only remedially, as they fruit on the ends of the branches.

Pests and Diseases
Medlars rarely suffer from any problem.

A beautiful small tree in foliage, flower and fruit

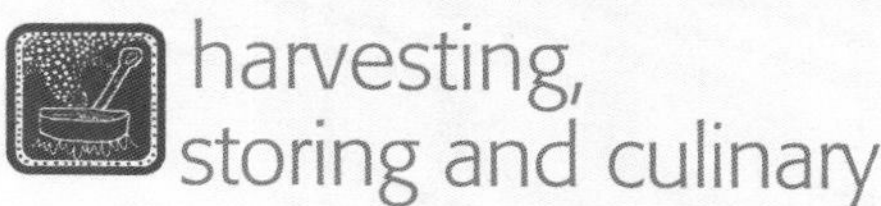

harvesting, storing and culinary

The fruits should be left on the tree till winter and then stored in a cool, dry place till they soften (blet). The pulp was once popular raw, mixed with liqueur and cream, but is better jammed or jellied.

Medlar Fudge
Makes about 2kg

1kg medlars
1 large or 2 small lemons
3 cloves
600ml apple cider
Approx. 1kg light brown sugar
Honey or maple syrup to taste, cream and macaroons to serve

Wash and chop the fruit, add the cloves and cider and simmer until the fruit is soft. Sieve and weigh the pulp. Add three quarters of its weight in sugar and bring back to the boil, then bottle and seal. When required, whip the fruit cheese with honey or maple syrup till soft, spoon into bowls and top with whipped cream and broken macaroons.

Chaenomeles japonica from the family *Rosaceae*

JAPONICA QUINCES

Shrub up to 3 x 3m. Life span: short. Self-fertile. Fruits: 2.5–5cm, round, green/red/yellow.

A tangled mass of dark, occasionally thorny branches covered with red, white, orange or pink blossom in early spring. This is followed later in the year by hard, roundish fruits, green flushing red or yellow.

Often just called Japonica, this shrub arrived in Europe from Japan as late as 1800. It was rapidly accepted and is now widely planted in many varieties and hybrids, for its flowers, not the fruits. There is opportunity for developing a better culinary or even a dessert form.

varieties

The botanical *C. japonica* has orange flowers and is not common. The true **Japonica** is *C. speciosa*. There are many ornamental hybrids and varieties that also fruit. **Boule de Fer** is my favourite and is a heavy cropper. A related species, *C. cathayensis*, is larger, with green fruits up to 15cm long, and is a thorny brute!

cultivation

Hardier than Cydonia quinces, these are easy almost anywhere, even on shady walls.

Growing under Glass and in Containers

Tough and unruly plants, they are better outdoors. However, I've found them dependable in pots to force for early flowers.

Ornamental and Wildlife Value

The displays of early flowers and long-lived fruit are exceptional, making these valuable shrubs. The flowers come in late winter, benefiting early insects, but the hard fruits often rot before the birds take them.

Propagation, Pruning and Training

Unlikely to come true from seed. Better varieties are easily layered or can be grafted. Winter cuttings may take; softwood cuttings are better but trickier. The shrubs have a congested form and are best left alone or tip-pruned in summer to control their size. I weave young growths like basket work to produce a tight surface that can be clipped.

Pests and Diseases

Other than weeds and congestion, they have no common problems.

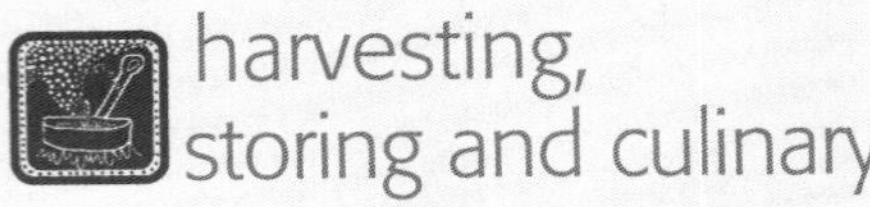

harvesting, storing and culinary

Inedible – well, impenetrable anyway – until cooked, when they have an aromatic scent similar to, but different from, Cydonia quinces. They can be used for tarts, baked or stewed or made into cheese or jelly.

Japonica Jelly
Makes approx. 4kg

2kg Japonica quinces
Approx. 2kg sugar

Chop the fruit and simmer in 3 litres water till tender, then sieve and weigh the pulp. Add 500g of sugar per 600ml of pulp and return to the heat. Bring to boil, bottle and seal. Store for three months before use.

Arbutus unedo from the family *Ericaceae*

STRAWBERRY TREE

CANE FRUIT, ARBUTE

Tree up to 6m. Life span: long. Fruits: about 2cm, spherical, red.

***Arbutus unedo* is a small, evergreen tree often with gnarled, shedding bark, rich brown underneath. It has white, heather-like flowers in late autumn as the previous year's crop of spherical fruits ripens. These are stubbily spiky, resembling litchis. The pulp is really no good raw, especially in cool regions, but may be better in warmer climes. *Arbutus unedo* is native to the Mediterranean. The Ancient Greek Theophrastus knew it was edible, but three hundred years later, the Roman Pliny did not regard it as worth eating. Ever since, it has been planted for its beauty, yet not really for the fruit, which have remained undeveloped. I gather the descriptive name *unedo* means 'I eat one only'.**

Rubra (above); **Killarney** (top right)

varieties

The ordinary **Killarney Strawberry Tree** is commonest and in few varieties; *Rubra* has pink flowers and abundant fruits. Other species are *A. canariensis*, whose berries are made into sweetmeats, and *A. menziesii*. **Madrona**, from California, has cherry-like fruits that are said once to have been eaten.

cultivation

A. unedo is, surprisingly, not particular as to soil, but prefers a mild climate or a warm site. Although ericaceous, it does not mind some lime, but does better in a leaf-mould-rich woodland or acid soil. The other species are not as compliant.

Ornamental and Wildlife Value

One of the most highly prized, small, ornamental evergreens. Of slight value to wildlife, though the flowers are handy for insects late in autumn.

Pests and Diseases

Few problems occur.

Growing under Glass and in Containers

So ornamental it is worth having under cover, but only if confined in a pot. Be careful to use rainwater.

Propagation, Pruning and Training

Seed of the species comes true, layers are possible and winter cuttings may take. The strawberry tree needs little pruning and usually recovers if cut back hard.

other uses

The hard, tough wood is turned into carved souvenirs of Killarney, especially small cudgels.

harvesting, storing and culinary

Harvested a year after the flowers, the fruits are made into sweets, confections, liqueurs and sherbets, but rarely eaten raw.

Killarney Strawberry Surprise

Serves 6

225g wheatmeal biscuits
120g butter
120g strawberry jam
225g strawberries
30g gelatine
1 large cup water
Killarney strawberries (arbutus *fruits) to garnish*

Break up the biscuits and mix them with most of the butter. Press into a tart dish and chill. Once set, rub the biscuit crust with the rest of the butter, line with half the jam, then fill with strawberries. Dissolve the gelatine and the rest of the jam in a cup of hot water, then cool. Pour the jelly over the fruit, swirl and chill. Garnish with Killarney strawberries and serve with cream.

Viburnum trilobum from the family *Caprifoliaceae*

HIGHBUSH CRANBERRIES

Shrub 4m. Life span: relatively short. Fruits: 2cm, round, scarlet.

A large, spreading shrub with maple-like leaves that give good autumn colour, and scarlet, glossy, translucent berries in mid-summer, very similar to the European Guelder Rose.

***Viburnum* is an immense genus of many species, some of which have berries that were once considered highly edible. The Guelder Rose (Snowball Tree, Whitten, Water Elder, *V. opulus*) is now forbidden fruit and as children we are warned how dangerous it is. Though the foliage is poisonous, the fruit was cooked and eaten by the poor for millennia. A native European fruit, it thrives in wet soils and damp hedgerows. Similar conditions suit most species.**

varieties

V. trilobum, **Highbush Cranberry**, which is very similar to *V. opulus*, is one of several edible wild North American species; *V. lentago* (**Sweet Viburnum, Nannyberry** or **Sheepberry**) has sweet black berries; *V. nudum* (**Naked Viburnum** or **Withe Rod**) is similar, with a deep blue berry; *V. prunifolium* (**Black Haw**) is sometimes good enough even to eat raw.

V. lentago

cultivation

Very easy to please. Most species do best in moist, humus-rich soil; many, except *V. nudum*, like lime.

Growing under Glass and in Containers
Tough individuals, these are mostly hardy and need no glass, though other viburnums may be grown in pots so that their scented flowers may be enjoyed.

Ornamental and Wildlife Value
Most viburnums are grown for their pretty or scented flowers. The berrying sorts are often as decorative and both fruit and flowers are very useful for attracting insects and birds.

Propagation, Pruning and Training
Species come true from seed but are slow. Most varieties can be layered and some will strike from hardwood cuttings taken in winter.

Pests and Diseases
No common problem, just weeds and bird losses.

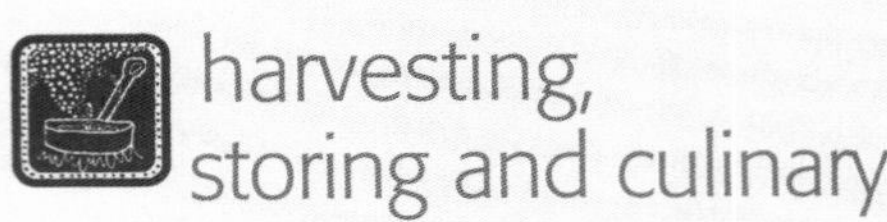

harvesting, storing and culinary

The berries are not eaten raw, save occasionally after frost, but can be made into sharp-tasting jams and jellies.

American Highbush Chutney
Makes approx. 1.25kg

680g highbush cranberries
300ml distilled vinegar
120g each currants, sultanas, raisins, sugar
15g salt
2 teaspoons each cinnamon, allspice
Pinch of nutmeg

Wash the fruits and then remove each of the stalks. Simmer, with enough water to prevent burning, until soft. Add in the vinegar and the other ingredients, simmer until the chutney thickens. Store in sterilised jars for three months.

Fuchsia from the family *Onagraceae*

FUCHSIA

Semi-herbaceous shrub to 2 x 2m. Life span: short. Fruits: 1cm, round/ oval purple-black.

This gloriously flowered plant needs little description, and most of us must have noticed the roundish oval, purple fruits it occasionally sets. As I have always searched for new fruits from far places, I was amused when I first saw fuchsia jelly and learned that I had overlooked these berries, which are so close at hand, often edible and as delicious as many from distant shores.

The first fuchsia was recorded in 1703. Over the following century a few species arrived in Britain from South America with little remark, but in 1793 James Lee astutely launched his *F. coccinea*, the first with impressive flowers, and took the world by storm. Other species were introduced from New Zealand and now there is an amazing range of colours and forms of hybrids of many species. James Lee claimed he found this new Fuchsia in a sailor's window but it was actually supplied to him by Kew (labelled **Coccinea**) who got it from a Captain Firth in 1788. However, Pere Plumier wrote of *Fuchsia triphylla flore coccinea* in Plantarum Americanum Genera, in 1703. What was so 'astute' about launching *F. coccinea* was the way he did it – putting two on display, saying he would only sell one of the pair, for a guinea, which he did, and then replacing it with one of the three hundred he had propagated as soon as the client had left.

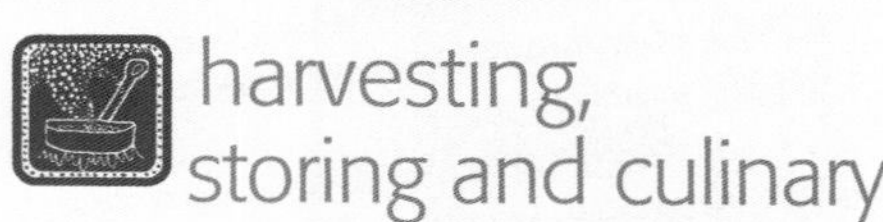

harvesting, storing and culinary

Some are tasty raw, but all are best jellied or in tarts.

Fuchsia Jelly
Makes approx. 2kg

1kg fuchsia berries
Approx. 1kg sugar
Juice of 1 lemon

Simmer the berries with sufficient water to cover. Once they have softened, sieve and weigh. Add the same weight of sugar and the lemon juice. Bring back to boil, then transfer to steralised jars.

varieties

Fuchsia species *corymbiflora* and *denticulata* were eaten in Peru and *F. racemosa* in Santo Domingo. Fuchsia clubs and societies often have jelly competitions. There appears to be no known poisonous variety – and I have tried many. As they are all bred for flowers, the berries are neglected and should be easily improved – there is enough variety to start with! *F. magellanica* is the hardiest, but rarely fruits. The **California Dreamer** series can produce thumb-sized fruits!

cultivation

Even many less hardy species can be grown in cold regions if the roots are well protected, as fuchsias usually spring again from underground to flower and fruit. Make backup plants to be safe and keep these indoors. Fuchsias are amenable to almost any soil but prefer a sunny site.

Propagation, Pruning and Training
Seed produces mixed results; cuttings are easy to train to any shape. Control growth by nipping out tips in summer; cut back in winter.

Ornamental and Wildlife Value
Fuchsias are appealing. They can support large numbers of unwanted wildlife.

Growing under Glass and in Containers
An ideal plant for a cool or heated greenhouse or conservatory. Can be grown for years in large pots.

Pests and Diseases
They suffer from common pests requiring the usual remedies.

Rosa from the family *Rosaceae*

ROSE HIPS

Clambering shrub up to 9m. Life span: short to medium. Usually self-fertile. Fruits: up to 2.5cm, ovoid, red, yellow or purplish black. Value: very rich in vitamin C.

There can be none of us who does not know roses and few who have never nibbled at the acid/sweet flesh of a rose hip. These are so rich in vitamin C that they were collected on a massive scale during the Second World War for rose-hip syrup for expectant mothers and babies. Rose stems are commonly thorny with a few exceptions such as the divine **Zéphirine Drouhin**. The deciduous leaves vary from glossy to matt, the flower colour is any you want save black or blue, though wild roses are almost all white or pink. The Eglantine rose leaves smell of apples after rain.

Rose hips and flowers are eaten in countries all over the world. The **Brier** or **Dog Rose**, *Rosa canina*, and **Eglantine** or **Sweetbrier**, *R. rubiginosa*, are natives of Europe and temperate Asia. Their fruits have been eaten by country folk since time immemorial, but are now regarded with some disdain. However, eglantine sauce was made at Balmoral Castle from sweetbrier hips and lemon juice and was considered good enough for Queen Victoria. Roses are bred for flowers, not for their hips, so most varieties have small hips. However, a little selective breeding could produce hips as large as small apples within a few generations.

A little breeding and these could be as large and sweet as apples

varieties

The **Brier** and **Eglantine** rose hips are the most common varieties used for hips, though *R. rugosa*, the **Rugosa Rose**, offers larger hips so is more rewarding. *R moyesii* has large flask-shaped hips and *R. omiensis* has pear-shaped, yellow and crimson fruits that ripen early. *R. spinosissima/pimpinellifolia*, the **Scotch** or **Burnet Rose**, is another European native, often found in maritime districts. It has a sweet, purplish-black fruit.

cultivation

Most roses are accommodating, but they prefer heavy soil, rich in organic matter, and need to be kept well mulched with their roots cool and moist, yet not wet.

Propagation

Some species can be grown from seed. Cuttings taken in early autumn are reliable for many varieties and most species.

ROSE HIPS

Ornamental and Wildlife Value
Roses are *the* garden plant. At least one or more can fit into almost any garden anywhere to good effect. Single-flowered roses are valuable to insects and the hips are choice meals for winter birds and rodents.

Growing under Glass and in Containers
Only tender roses such as *R. banksiae* are happy under glass. Hardier varieties tend to become soft and drawn and suffer from pests. They must be kept moist at the roots, well-ventilated and shaded against scorch.

Maintenance
Spring Weed, mulch heavily and spray with seaweed solution often.
Summer Deadhead regularly, watch for aphids.
Autumn Take cuttings.
Winter Cut back or tie in.

culinary

Hips can be made into jellies, preserves and the famous syrup. The flower petals may be used as garnishes, preserved in sugar or syrup, used for rosewater flavouring, honeys, vinegars and conserves, pounded to dust for lozenges and they make good additions to salads. The leaves of *R. canina* have been used for tea (in desperation, one suspects).

Rose Hip Tart
Serves 6

for the pastry:
225g self-raising flour
150g butter
1 large egg yolk
30g fine brown sugar
Pinch of salt
Splash of water

for the filling:
500g rose hips
60g each fine brown sugar, honey, chopped stem ginger preserved in syrup
1 teaspoon cinnamon
Sprinkling of sugar and grating of nutmeg

Rub together the ingredients for the pastry, making it a little on the dry side, and use it to line a tart dish. Wash, top, tail and halve the rose hips, extract every bit of seed and hairy fibre, rinse and drain. Mix with the sugar, honey, cinnamon and ginger and spoon on top of the pastry. Decorate with pastry offcuts and sprinkle with sugar and nutmeg before baking at 180°C/350°F/gas mark 4 for half an hour or until the pastry is light brown on top.

Pruning and Training
This requires a chapter on its own just to list the methods! Basically, for most roses, plant them well apart, prune as little as possible and wind in growths rather than prune. Reduce tall hybrid bushes by a third to a half in height with hedge trimmers annually in late winter.

Pests and Diseases
Roses suffer from a host of common diseases, but provided they are growing reasonably well the only real threat to flower and hip production is aphids. Control these with jets of water and soft soap.

Harvesting and Storing
The berries need to ripen fully on the bush before being eaten raw, but are best taken before they soften for culinary use. All seed hairs must be removed!

Don't forget the petals for salads and garnish

companion planting

Underplantings of Alliums, especially garlic or chives, help deter blackspot and pests. Parsley, lupins, mignonette and lavender are beneficial; catnip and *Limnanthes douglassii* are good ground cover underneath roses.

other uses

Strong-growing roses such as *R. rubiginosa, R. spinosissima* or **The Queen Elizabeth** make excellent stock- and people-proof hedges. The flower petals are dried for potpourri and used in confectionery, medicinally and in perfumery. The seeds are covered in hairs that cause itching when put down the back of other children's collars...

Cornus mas* from the family *Cornaceae

CORNELIAN CHERRY, SORBET

Tree/bush up to 8m. Life span: medium to long. Deciduous, self-fertile. Fruits: 1cm, ovoid, red.

Cornelian cherry fruits resemble small, red cherries but are generally too sour to eat raw, except for the occasional better one. The trees or large bushes are tall, deciduous, densely branched and suckering, common in hedgerows and old grasslands. The stems are greyish and the leaves are oval, coming to a point with noticeable veins. The flowers are primrose yellow and appear in small clusters early in spring before the leaves.

The Cornelian cherry is one of a genus of about a hundred, mostly small, shrubby plants up to 3m high, but ranging from creeping sub-shrubs to small trees. One of the two species native to Europe and western Asia, *Cornus mas* was once widely cultivated and rated very highly, though now it is rarely eaten even by country folk. There are species from North America and the Himalayas and it seems a shame they have not been cross-bred for better fruits. The Cornelian cherry really is a fruit that has stalled in development and probably would not take much further work to improve immensely.

varieties

The variety *C. mas macrocarpa* has somewhat larger fruits and is still available. There used to be many other improved forms, now apparently lost. In France and Germany there are records of several varieties of Sorbets, as they were called – one that had a yellow fruit, some with wax-coloured fruits, white fruits and even one with a fleshy, rounded fruit. Other *cornus* such as *C. stolonifera*, the **Red Osier**, and *C. amomum,* **Kinnikinnik**, are found in North America and are edible. The former was eaten more in desperation than for pleasure; the latter, found in Louisiana, is said to be very good. The berries of *C. suecica* used to be gathered by Native Americans, who froze them in wooden boxes for winter rations. *Cornus kousa chinensis* comes from China via Japan and is a smaller tree than *C. mas.* The flowers are greyish-purple backed by immense pale bracts; the fruits are more strawberry-like and juicy with better flavour. This species needs a moist, acid soil.

Cornus capitata

C. macrophylla and the tenderer *C. capitata* come from Asia and the Himalayas and are eaten raw and made into preserves in India. *C. canadensis (Chamaepericlymenum canadense)*, **Bunchberry** or **Dwarf Cornel**, is a different type altogether. A lime hater more resembling a soft dwarf raspberry in manner of growth, it has white flowers on low, soft shoots with vivid red fruits. These are pleasant enough, if tasteless, and can be added to summer puddings.

cultivation

C. mas is easy to grow almost anywhere but prefers calcareous soil. Some species need acid conditions. Generally they are among the most reliable of shrubs, requiring little attention.

Maintenance

Spring Cut back ornamental stemmed varieties, weed, mulch and spray with seaweed solution.
Summer Make layers.
Autumn Preserve fruit for winter.
Winter Take cuttings or suckers and prune.

CORNELIAN CHERRY, SORBET

Cornus canadensis

Growing under Glass and in Containers
They are hardy, so are hardly worth growing under glass unless you wish to force them for their early flowers. They will grow well in pots.

Ornamental and Wildlife Value
Many species are liked particularly for their autumn colour and some species and varieties are grown for their brightly coloured stems. The flowers are early, benefiting insects, and birds and rodents love the berries. These factors render them one of the best backbone shrubs of a wild garden.

Propagation
The species come true from seed with some variation, but are slow. Suckers taken in autumn are best; hardwood winter cuttings may take, but layering is more sure.

Pruning and Training
Generally only remedial pruning is needed – though, as these plants sucker, some root pruning may become necessary. Ornamental coloured-stem varieties are best sheared to ground level in early spring.

Pests and Diseases
No common problems bother these tough plants.

other uses

Cornus sanguinea, the **Cornel Dogwood**, **Dogberry** or **Pegwood**, is a common European relation, not really edible, though the fruits were once used for oil and in brewing. *C. alba* varieties tolerate wet or dry situations and can be used to reinforce banks. Dogwoods grow stiff and straight, so were used for arrows.

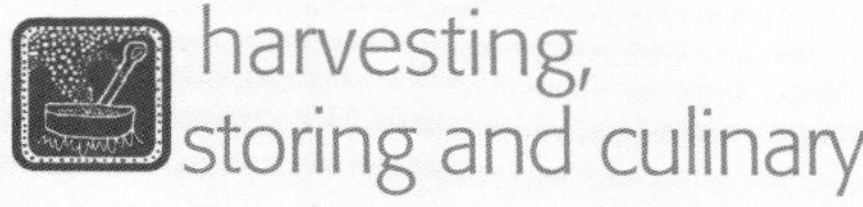

harvesting, storing and culinary

In Germany the fruits were sold in markets to be eaten by children (who presumably liked them). They were widely made into tarts, confectionery and sweetmeats, even used as substitutes for olives. In Norway the flowers were used to flavour spirits and in Turkey the fruits were used as flavouring for sherbets. It is noticeable that fruits vary on different bushes and some are more palatable raw than others.

Sorbet Sorbet
Serves 4

1kg cornelian cherries
1 small lemon
Approx. 600g sugar
2 or 3 small egg whites

Wash the fruits, slit the cherries, chop the lemon and simmer them till soft, just covered with water, in a deep pan. Strain and measure the juice, then dissolve 200g sugar per cup of juice. Bring back to the boil, cool and partially freeze. Remove from the freezer and beat, adding one beaten egg white for every two cups of sorbet. Repeat the freezing and beating one more time before freezing until required.

Crataegus azarolus from the family *Rosaceae*

HAWTHORN AND AZAROLE

Tree/bush up to 8m. Life span: long. Deciduous, self-fertile. Fruits: up to 2.5cm, roundish, usually orange. Value: rich in vitamins C and B complex.

The azarole is a more palatable relation of the well-known hawthorn and is cultivated in many of the Mediterranean countries for its cherry-sized fruits. These are usually yellow to orange, but can occasionally be red or white. They are larger than a hawthorn haw and have an apple-flavoured, pasty flesh with two or three tough seeds. Small, spreading trees or large shrubs, these are typical of the thorn genus, with clusters of large, white flowers that do not have the usual family scent.

The thorn family are remarkably hardy, tough plants for wet, dry, windswept or even coastal regions. They are survivors and various species can be found in almost every part of the world, many of which bear similar small, edible, apple-like fruits. Native to North Africa, Asia Minor and Persia, the azarole, *C. azarolus*, may be the *mespile anthedon* about which Theophrastus wrote. More popular in the Latin countries, it was brought to Britain in 1640. In 1976 it got an award of merit from the Royal Horticultural Society, but it has never really caught on, most probably because other more floriferous varieties and species were readily available.

Crataegus. tomentosa

varieties

The **Azarole** is the more productive of the species and is grown commercially for flavouring liqueurs. It is probably the best choice for a tree for preserves. The Armenian *C. tanacetifolia*, the **Tansy-leaved Thorn** or **Syrian Hawberry**, is another good choice. The berries are almost relishable raw as dessert and have an aromatic apple flavour, which is surprising, as they also closely resemble small yellow apples. They are pale green to yellow, with slight ribs like a melon, and a tassel of 'leaves' at the end. The **Common Hawthorn** or **Quickthorn Haw**, *C. monogyna*, has one seed and is edible but not at all palatable, so is seldom eaten, save by curious children. Reputedly it was eaten raw when fully ripe by Scots Highlanders. The fruits are dark red and hang in immense festoons in autumn. The flowers have a sweet perfume when new, but go fishy as they age – on some trees more than others.

Equally common, *C. oxycantha/laevigata* is very similar, usually with dark red flowers. There are several edible North American species. *C. tomentosa*, **Black Thorn** or **Pear Thorn**, has hard, orange-red, pear-shaped fruits; *C. flava* has yellow fruits; *C. douglasii* is a better species with small but sweet, black berries with yellow flesh. One identified as *C. coccinea* (now *mollis*, sub *mollis, pedicellata* or *intricata*) was very popular with Native Americans, who dried the large scarlet or purple fruits for winter use. Sometimes these fruits were mixed together with chokecherries and service berries before they were dried and pressed into cakes for storage. *C. mollis* also has amazingly big thorns.

cultivation

The thorns are all extremely easy to please and require little skill or attention. They tend to lean in the more exposed situations.

HAWTHORN AND AZAROLE

Common Hawthorn

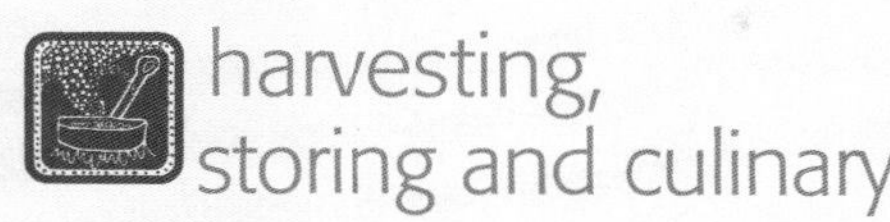

harvesting, storing and culinary

The flowers of the **Common Hawthorn** once made a heady liqueur or wine. The young leaves and buds, known to schoolchildren as bread and cheese, had a nutty taste and made a welcome addition to salads. We are now told that both are slightly poisonous. However, the berries of **Azarole**, **Common Thorn**, and especially the Armenian or Syrian, will make excellent preserves, wines and jellies.

Hedge Jelly
Makes approx. 3kg

1kg haws
500g crab or cooking apples
225g elderberries
Approx. 1.5kg sugar

Wash the fruits, chop the apples and simmer the fruits together, just covered with water, for about 2 hours, till softened. Strain and weigh the juice. Add the same weight of sugar to the juice and bring back to a boil. Skim off the scum, jar and seal.

Growing under Glass and in Containers
Because they are so hardy they do not need protection, but several of the ornamental varieties can be grown in pots for forcing to produce early flowers.

Ornamental and Wildlife Value
The shows of blossom and masses of bright fruits make these an excellent choice for larger shrub borders and informal gardens. The flowers, fruit and foliage are useful to all manner of wildlife. The common thorn flowers attract over 150 different insect species.

Propagation
This is more difficult than for many fruits. The haws need stratifying (burying outdoors in a pot of sand) for a year and a half before sowing properly the following spring. They may produce mixed offspring unless they are from a true species grown far from any others. Cuttings are difficult so choice varieties are best budded in May or grafted in April on to common hawthorn stock.

Maintenance
Spring Weed, mulch and spray with seaweed monthly.
Summer No maintenance.
Autumn Collect fruits before the birds do.
Winter Prune if necessary.

Pruning and Training
Very little is needed. Do not over-thin the branches, as thorns will naturally have a congested head.

Pests and Diseases
Thorns rarely suffer badly from problems, though they are occasionally defoliated by caterpillar attacks. These attacks may be easily avoided by diligent observation and prompt action.

other uses

Thorns make the best and most traditional hedge with a trimmed surface like fine tweed. The wood is heavy and hard and will burn with a good heat.

The May blossom is cheerful, too

Sorbus aucuparia from the family *Rosaceae*

ROWAN, WHITEBEAM AND SERVICE BERRIES

Tree, up to 15m. Life span: short. Deciduous, self-fertile. Fruits: up to 1cm, spherical, scarlet, in clusters. Value: very rich in vitamin C and pectin.

Rowans are most attractive, small trees with distinctive, pinnate leaves, dark green above, lighter underneath. They have big heads of foamy, cream flowers like elderflowers, but smell unpleasant. In autumn the branches bend under massive clusters of bright red to scarlet berries, which would hang through the winter if the birds did not finish them so quickly. The Latin name *aucuparia* means 'bird catching' and refers to the fruit's early use as bait.

The *Sorbus* family is large and includes dwarf shrubs and large trees. They are spread all over the world and the majority are quite hardy. They colour richly in autumn and are widely grown for their attractive shows of fruits, also in yellow and white. Many new ornamental species were introduced from China during the nineteenth century, but little advance has been made in fruit quality since *S. aucuparia edulis* (*moravica* or *dulcis*) was first introduced in about 1800.

Rowan

varieties

Sorbus aucuparia, the **Rowan** or **Mountain Ash**, has scarlet berries that birds love and which are generally too sour and bitter for our tastes. There are many other ornamental species and varieties, for example *S. aucuparia xanthocarpa*, which has yellow fruits. However, the common rowan is still the frequent favourite choice from the genus and much planted in metropolitan areas. The best variety by far for the gourmand is *edulis* which has larger, sweeter fruits carried in heavy bunches. *S. aria*, the **Whitebeam**, has similar red berries to a rowan and they were once eaten and used for wine. *Sorbus domestica*, the **Service Tree**, is a native of Asia Minor and was also once widely liked, but is now less common. It has smaller clusters of larger fruits of a brownish green, resembling small pears. They need to be bletted like medlars (see page 577) before they are edible. There were pear-shaped and apple-shaped versions; the flavour and texture were improved after a frost and they were commonly sold in London markets. In Brittany they were used to make a rather poor cider. The **Wild Service** or **Chequer Tree**, *S. torminalis*, has smaller, still harder fruits, that would pucker even the hungriest peasant's mouth, but were once eaten by children. Note the seeds contain cyanides and should not be swallowed.

cultivation

Mountain ashes are tolerant of quite acid soils and do not like chalky or limy ones, protesting by being short-lived. They generally prefer drier to wetter sites but, strangely, are commonly seen growing naturally by mountain streams and in wet, hilly country.

ROWAN, WHITEBEAM AND SERVICE BERRIES

Growing under Glass and in Containers
They are hardy and rather too large for growing inside. They can be grown, and fruited, in pots if only a small crop is desired.

Ornamental and Wildlife Value
Almost all the species and varieties are very attractive to us in flower and fruit, turning glorious shades of crimson in autumn, but most unfortunately do lack a sweet scent. Rowans are valuable to the birds and insects and are pollinated by flies and midges.

Maintenance
Spring Weed, mulch and spray with seaweed solution.
Autumn Pick fruit before the birds get them.
Winter Prune as necessary.

Service tree

Whitebeam

Propagation
Seed from the species may come true if the tree is isolated, but it must be stratified over winter first. Cuttings are rarely successful, so selected forms are budded in mid-summer or grafted in early spring onto seedling rootstocks. The fruits of rowans are thought to get larger if the trees are grafted on to Service stock.

Pruning and Training
Pruning should be only remedial once a good head has formed, but watch for overladen branches and prop in time. The strongly erect growth of young trees can be bent down and into fruitfulness: pull them down gently with weights tied near the ends.

Pests and Diseases
Apart from bird losses, and being rather short-lived on alkaline soils, the members of this genus need little care and rarely suffer problems. They are thus well liked for amenity planting.

other uses

Rowan bark was used in dyeing and tanning. The wood is strong and was used for handles. Service wood is tough and resists wear well. The berries of the **Wild Service** tree, *S. torminalis*, were also used medicinally.

culinary

Rowan berries make a delicious jelly, almost like marmalade, which goes well with venison, game and fatty or cold meats. They can be used in compôtes, preserves and syrups, and are sometimes added to apple dishes to liven them up. They have been used for fermenting and distilling liquor, and in emergencies the dried berries have been ground into meal to make a substitute for bread. Service fruits were used traditionally in quite similar ways.

Mountain Ash Jelly
Makes approx. 2kg

1kg firm ripe rowan berries
1 small lemon
Approx. 1kg sugar

Wash and de-stem the berries, add the chopped lemon and 425ml water and simmer till soft (for about an hour). Strain and add 450g sugar to each 600ml. Bring back to the boil, skim and jar.

Prunus species **from the family** *Rosaceae*

SLOES, BIRD CHERRIES, BEACH PLUMS

Tree/bush up to 9m. Life span: medium. Deciduous, self-fertile. Fruits: up to 2.5cm, ovoid, single stone, red to black. Value: rich in vitamin C.

Closely related to orchard plums and cherries, the wild *Prunus* species remain very much tough, hardy alternatives for difficult spots or wild gardens. Sloes are the fruits of the blackthorn, *P. spinosa*, which is a medium-sized shrub, many-branched, very thorny, with blackish bark. The fruits are black ovoids with a bloom that resembles plums, and hard, juicy, green flesh that is usually far too astringent to eat raw. However, like many children, I forever searched for a sweeter one.

Sloes are native to Europe, North Africa and Asia. The stones have been found on the sites of prehistoric dwellings, so they have long been an item of diet. They may be one of the ancestors of the damson and some of their 'blood' has probably got into many true plums. The Bird Cherry is a native of Europe and Asia, especially grown in the north of England. The sixteenth-century herbalist Gerard claimed it was in 'almost every hedge'. Many ornamental *Prunus* species were later introduced in the eighteenth and nineteenth centuries; *P. maritima* was introduced by R. J. Farrer in 1800.

varieties

Prunus padus, the **Bird Cherry** or **Hag Berry**, is a small tree with white, fragrant flowers in late spring; the double-flowered form, *P. padus plena*, is more heavily almond-scented. The leaves and bark smell of bitter almonds and contain highly poisonous prussic acid. *P. maritima*, the **Beach Plum**, comes from the eastern seaboard of North America, from Maine down to the Gulf of Mexico. It is a small, compact shrub with masses of white flowers and red or purple juicy fruits up to 2.5cm across. These can be eaten raw, but are better if preserved. Another North American species is *P. virginiana*, the **Choke Cherry**. This is a tall shrub with glossy, green leaves and variable red to purplish-black berries. The **American Red Plum, Goose, August, Hog** or **Yellow Plum**, *P. americana*, was the Native Americans' favourite and is good for eating raw, stewing or jamming. It is reluctant to fruit in the British Isles, apparently preferring a more Continental climate. *P. simonii*, the **Apricot Plum** from China, has large, attractive, red and yellow, scented fruits.

Bird Cherries

SLOES, BIRD CHERRIES, BEACH PLUMS

cultivation

Sloes and Bird Cherries are very hardy and good for making windbreaks. They like an exposed site and thrive on quite poor soils, even one packed with chalk, while *P. maritima* does best in coastal regions, and is happy with salt winds and sandy soils. Do mulch these heavily; water well while establishing on very sandy soils as these hold little water.

Growing under Glass and in Containers
Most of these are hardy and not tasty enough to merit space under cover. *P. americana* may be more fruitful if pot-grown, wintered outside and brought in for the spring through to autumn.

Ornamental and Wildlife Value
All Prunus are good at flowering time, but few are worth space in most small, modern gardens. They are more use to the wild garden, as the flowers are early, benefiting insects, while the fruits are excellent winter fare for birds and rodents.

Pruning and Training
Minimal pruning is required and is best done in summer to avoid silver leaf disease. As hedges, blackthorn should be planted at forty-five degrees, staggered in two or three close rows, each laid in opposite directions. Most species can be trained as small trees, but leave beach plum bushes alone.

American Red Plum

Choke Cherries

Maintenance
Spring Weed, mulch, spray with seaweed solution.
Summer Prune if any pruning is needed.
Autumn Protect fruits from the birds.
Winter Pick fruits.

Propagation
The species usually come true from seed. Choicer varieties have been developed for ornamental use and must be budded in summer or grafted in spring. Cuttings do not take.

Pests and Diseases
These wildest members of the Prunus family suffer least from pests or diseases.

other uses

Sloes are good hedging plants. Their leaves were formerly used to adulterate tea. Bird Cherries do well in a hedge. Their wood is hard and used for carving and especially liked for rifle butts. Sloe bark was once used for medicinal purposes.

harvesting and storing, culinary

Sloe berries are much used for liqueurs, especially gin-based ones, and to add colour to port-type wines. All over Europe they are fermented for wine or distilled to a spirit. In France the unripe sloes are pickled like olives. They can be made into juice, syrup or jelly. Bird Cherries have been used in much the same way but are inferior. The American species are used similarly. The Apricot Plum seems strangely under-rated.

Sloe Gin

Sloes
Brown sugar or honey
Peeled almonds
Gin, brandy or vodka

Wash and dry the sloes and prick each several times. Pack the sloes loosely into bottles. To each bottle add 120g brown sugar or honey and a couple of almonds, then fill with gin (or brandy or vodka). Leave for several months before drinking.

Pinus pinea* from the order *Pinaceae

PINE KERNELS

PIGNONS, PINONS, PINOCCHI

Tree up to 25m. Life span: long. Evergreen, self-fertile. Fruits: up to 1cm long, ivory-white seeds inside a cone. Value: high in minerals and oils.

Pine kernels are more nuts than fruits as we eat the seed and not the surrounding part (in this case, the cone), but they are softer than true nuts. Most pine kernels come from the stone or umbrella pine. It is an attractive, mushroom-shaped tree with glossy, brown cones which expand in the sun, and drop the seeds. Each is in a tough skin which needs to be removed before they are eaten.

The stone pine is indigenous to the Mediterranean region. Loved by the Ancient Greeks, it was dedicated to the sea god Poseidon. The kernels of other species are eaten almost anywhere they grow.

varieties

Pinus pinea grows happily in northern regions, but without enough sun the cones do not ripen. *P. cembra*, the **Arolla Pine**, is native to Central Europe and Asia and the seeds are a staple food in Siberia.

Gerard's Pine

P. gerardiana, **Gerard's Pine**, comes from the Himalayas. *P. cembroides* is a native of North America and has pea-sized kernels that taste delicious roasted. *P. edulis*, from Mexico, is the best variety, but unlikely to fruit in Britain. Likewise the **Araucaria Pine** or **Monkey Puzzle Tree**, *A. aurucana*, which has edible kernels and grows but rarely fruits in cool climates. The **Parana Pine**, *A. angustifolia*, from South America, is not hardy and equally is too big for glass-house culture.

cultivation

Stone pines will need shelter to fruit in cold regions. They prefer to grow in sandy soil and acid conditions.

Growing under Glass and in Containers
Most pines grow too large unless confined and are unlikely to crop. *P. canariensis*, however, is a particularly pleasing indoor pot plant.

Ornamental and Wildlife Value
Very attractive if the space is available. Pine kernels will provide excellent food for many species of bird.

Pests and Diseases
Pines need companion bacteria and fungi, so do better if soil from around another pine is used to inoculate new sites.

P. edulis

Pruning and Training
Pines are best left well alone.

Propagation
They can be grown from seed or grafted (with skill).

other uses

Pines are a source of turpentine and resin.

culinary

Pine nuts are roasted and salted in the same way as peanuts, and also used in marzipan, confectionery, salads and soups.

Piñon Truffle Salad
Quantities to taste

Pine kernels
Butter
Truffles
Walnut oil
Vinegar
Lettuce

Gently fry the kernels in butter till light brown. Remove from the heat; add finely sliced truffles, oil and vinegar. Cool the mixture and toss it in clean, dry lettuce leaves.

VERY WILD FRUITS

There are many other edible, if not palatable, fruits in gardens and parks, just waiting for breeders to improve them. The following are a few of the many that are occasionally eaten by some, but are not widely used. Please remember not to eat anything unless you are 100 per cent sure of it and you have had it identified on the spot by an expert!

Carpobrotus edulis
Hottentot Fig *Aizoaceae*
This resembles the *mesembryanthemums*, with which it was once classed. Native to South Africa, it is a low-growing succulent, found wild in maritime south-west England and southern Europe. Large, magenta or occasionally yellow flowers are followed by small, figlike fruits, which can be eaten raw or cooked, pickled or preserved. The fleshy, triangular leaves can be eaten as saladings as can those of the similar *Mesembryanthemum crystallinum*.

Cephalotaxus fortuni, wilsoniana, harringtonia, drupacea
Chinese and **Japanese Plum Yews** or **Cow's-Tail Pines**
Very much resembling yews, these grow in moist conditions in heavy shade, even on chalky soils; usually dioecious, the females bear small plum-size khaki brown olives which bizarrely taste like butterscotch – could be developed into a superb new fruit!

Chiognes hispidulum,
Snowberry (NOT the poisonous *Symphoricarpus* Snowberry) is very similar and closely related to Gaultheria with delicious whitish mini-gooseberries with an oil of wintergreen aroma.

Decaisnea fargesii, Lardizabalaceae
A deciduous shrub from the Himalayas grown as an ornamental for the amazing metallic blue pods, which surprisingly, have an edible white pulp which is actually sweet and pleasant. Really ought to be improved!

Elaeagnus umbellata
Autumn Olive *Elaeagnaceae*
A strong-growing, spreading, deciduous shrub from Asia with fragrant, yellow flowers and small, orange or red berries used like redcurrants or dried to 'raisins'. Many species in this genus bear edible fruits, and most of these have fragrant flowers. *E. commutata*, the **Silver Berry**, has pasty, silver berries and silver leaves. *E. angustifolia*, **Oleaster** or **Wild Olive**, has sweet berries, and is still popular in southeastern Europe.

E. multiflora/edulis, **Goumi**, from the Far East has very acid, orangey-red silver-speckled berries good for conserves, beverages and jellies. *E. orientalis*, **Trebizond Date**, was once sold commercially as the fruit was considered very tasty. The plant was more famous for its aphrodisiacally perfumed flowers.

Gaultheria procumbens,
Checkerberry, Teaberry, *Ericaceae*
A low-growing, evergreen native of North America with white flowers and red berries, needing moist, acid soil and partial shade. The berries are odd raw, but can be cooked for jellies and tarts and were once popular in Boston. Ironically, the leaves were once used as 'tea'. *G. humifosa* was also used, as was *G. shallon,* another taller, shrubbier version with great clusters of purple berries which were eaten dried by Native Americans.

Hippophae rhamnoides
Sea Buckthorn *Elaeagnaceae*
Often grown as an ornamental for its silver leaves, this tall, thorny shrub produces acid, orangey-yellow fruits if both sexes are planted. The fruits are too sour for most tastes, but have been eaten in famines and by children and are apparently widely collected in Russia, as they are very rich in vitamin C. They are used as a sauce with fish and meat in France, and in Central Europe they are made into a jelly that is eaten with fish or cheese.

Lonicera caerula var. *Kamtschtica*
Honeyberry
The first fruit to ripen, so immensely valuable. The edible berried honeysuckle comes from Siberia and, unlike its poisonous relations, is a small shrub only a metre each way that resists cold and drought really well. The more cylindrical but very blueberry-like berries are effectively seedless and full of anthocyanins. Two plants are required for polination. *L. angustifolia*, *L. involucrata* and *L. ciliata*, the **Fly Honeysuckle** are other relations with berries once eaten by native peoples – but be careful, most other honeysuckle berries are toxic.

Lycium barbarum
Goji
Also known as the **Wolf berry**, the **Box Thorn**, or the **Duke of Argyll's Tea Plant**. A nearly hardy *Solanaceae* this has gone native in the south-west UK and is found from Eastern Europe through to northern India and Tibet. The red berries follow small violet flowers on lax, often spiny, sprawling stems. 'More vitamin C than oranges, more beta carotene than carrots and more iron than steak' claims one vendor! Much vaunted as the ultimate healthy fruit, these are not awfully full of mouth appeal but easy enough to grow under cover or in a sheltered garden.

Maclura pomifera
Osage Orange, *Moraceae*
Hardy, small thorny tree with inedible orange-like fruits – needs much improvement!

Mitchella repens
The Partridge Berry, Squawberry, Twinberry
Rubiaceae
From North America, this resembles cranberries in many ways but is unrelated; the berries come in pairs, persisting all winter as even birds find them a bit bland.

Podocarpus macrophyllus
Kusamaki, Japanese Yew *Taxaceae*
Whereas we scare our children away from our native yew, apparently Japanese children enjoy these small red to purple fruits raw and cooked. *P. totara* is a dwarf evergreen from New Zealand with a sweet cherry-like fruit.

Podophyllum peltatum
May Apple *Podophyllaceae*
Often grown as an ornamental, this poisonous herbaceous perennial has waxy white flowers followed by small lemon-like fruits that are edible and not unpleasant. *P. emodi* from the Himalayas has larger edible scarlet fruits. Hybridisation could maybe make much improved fruits from these.

Prinsepia sinensis
Cherry Prinsepia *Rosaceae*
From Manchuria comes this rather graceful spiny shrub with yellow flowers and juicy acid fruits much resembling red or purple cherries. I suspect you need two or more different plants for crops as mine never sets.

Shepherdia argentea
Buffalo berry *Eleaegnaceae*
A deciduous hardy shrub with spines, producing scarlet or yellow small currant-like berries that are tart raw but good for jellies or drying, both male and female plants needed for pollination. *S. canadensis*, the **Soapberry**, is smaller with sour orange berries that can be made into 'Indian ice cream', a froth made by beating them with sweetened water.

Smilacina racemosa
Treacle Berry *Convallariaceae*
This delectable, herbaceous garden plant has scented, foamy, white flowers followed by apparently edible, sweet, red berries. One of the best of unknown fruits!

Tilia species
Limes, Lindens *Tiliaceae*
These enormous, well-known trees have sweet sap which was formerly boiled down to sugar. The fruits were once ground into a 'chocolate', but this never caught on, as it kept badly.

NUTS

Nuts are different from fruits. We eat the seeds of nuts and usually not the coverings, though we often find uses for these as well. With fruits, plants are giving us, the animals and birds who consume them, a sweet pulp so that we will, inadvertently, help distribute their seeds. This trade-off is easy for the plant, as the seeds are the expensive items to manufacture and the sugary pulp takes little resource.

However, when we eat nuts, we eat big seeds that are very expensive for the plant to make, as they are rich in oils, minerals, proteins and vitamins. They have a high dietary value to us, but of course this does not serve the plant well. The trade-off is that the plant 'hopes' that if it produces a lot of nuts, some will escape the slaughter and be trampled underfoot or carried elsewhere and hidden but never recovered and thus start a fresh territory.

Squirrels are well known for assisting this process by burying nuts, but many other small mammals, particularly rodents, are also involved. Birds similarly hide nuts; there are plausible stories of birds that have filled attics with nuts, popping them in singly throughout the autumn through a small space such as a knot hole.

Because of their high oil content and nutritional value, nuts have long been gathered from the wild. The rise of industry created greatly increased demand for nuts as a source of oils for lighting and lubrication and for turning into margarines, soaps and cosmetics. The pulps that remained were rich animal feeds and helped to fuel the expansion of farming in the nineteenth century. Nuts were no longer wild crops, but had become cultivated crops on a vast scale.

Although nut trees generally require little work, they mostly grow too big for the garden and are best grown agriculturally. (Of course they then suffer from a build-up of pests and diseases – problems associated with all crops if they are grown as monocultures.)

They are slow to come into production, though on the plus side they are mostly long-lived and make good timber. Many nut trees produce very hard or oily wood that lasts well. Walnut is one of the most prized of all timbers, so precious that it is mostly used only as veneer.

Nuts suffer different pests from those that attack fruits. Bigger birds and rodents are more of a threat and, as nuts are larger, they are harder to protect or grow under cover (indeed, except as bonsai, many are almost impossible to keep small). Also, they are unfortunately more tender and susceptible to frost damage than many fruits. This, plus the need for a hot summer and autumn to ripen the nuts, means that most nut trees are best grown in countries warmer than Britain.

One interesting connection between many of the nut trees covered in this chapter is how many of them have catkins and are wind-pollinated, even though these nut trees belong to entirely different families. This also means that they do not generally have scented flowers and give little nectar to insects, but they are, of course, a rich source of pollen.

From a commercial point of view, it is curious how almost all retail nut sales take place at Christmas, a period that serves to outsell the rest of the year put together.

Prunus dulcis/amygdalus from the family *Rosaceae*

ALMONDS

Tree up to 6m. Life span: short. Deciduous. Fruits: 5cm, pointed oval, in brown skin. Value: rich in protein, calcium, iron, vitamins B2 and B3 and phosphorus.

Almond trees resemble and are closely related to peaches, with larger, light pink blossoms appearing before long, thin leaves. Flowering a fortnight earlier than peaches, they are often affected by frost. The wild varieties sometimes have spiny branches. The almond fruit has tough, inedible, leathery, greenish-brown, felted skin with a partition line along which it easily splits. The skin peels off a smooth, hard stone, full of small holes that do not penetrate the shell, containing the single, flat, pointed oval seed. Originally from the Middle East, almonds were known to the Ancient Hebrews and Phoenicians. Long naturalised in southern Europe and western Asia, they are now widely grown in California, South Africa and south Australia. Almond trees were introduced to England in 1548.

Almond orchards must be well away from peach trees

varieties

Prunus dulcis dulcis is the **Sweet Almond**, *P. dulcis amara,* the **Bitter Almond**. The former is the much-loved nut; the latter is used for producing oil and flavourings and is too bitter for eating. It contains highly poisonous amounts of prussic acid. In Spain, a light-cropping old variety, **Jordan**, is still grown and commercial varieties are available in some regions. The variety **Texas** is especially tasty roasted and salted with the skin on. Ornamental varieties rarely bear fruit. **Robijn** is a new introduction, pink flowered, soft shelled and self-fertile.

cultivation

Almonds want well-enriched, well-aerated, light soil, and to be at least 3m apart. They need no staking after the first year. They want copious quantities of compost and mulches. Usually grown as a bush, they can be planted against walls. They should not be planted near peaches, as they may hybridise, resulting in bitter nuts. Hand-pollination is recommended. They are not completely self-fertile, so several trees should be planted together.

ALMONDS

Growing under Glass
Almonds are rarely grown under glass because the necessary extra efforts of replenishment – pruning and tying in – are not usually well rewarded. The greenhouse must be unheated in winter to give them a dormant rest period. More problems occur under cover, and red spider mite can be troublesome unless high humidity is maintained.

Growing in Containers
Almonds can be grown in large pots, as they can take heavy pruning if well fed and watered. Pots enable them to be kept under cover during winter and through flowering and then brought out all through the summer, thus avoiding peach leaf curl and frost damage. The flowers must be protected from frosts and so too must the young fruitlets.

Ornamental and Wildlife Value
The almond is a real beauty – the fruiting tree as much as the many ornamental varieties. In hot, dry countries they are useful to birds, insects and rodents.

Propagation
Almonds can be raised from stones, but take years to fruit and may produce a mixture of sweet and bitter fruits. Budded on to **St. Julien A** in the UK, for example, almonds will normally fruit in their third year. Seedling almond or peach rootstocks are better.

In the almond the outer flesh is leathery and peels away

Bitter Almond

Maintenance
Spring Protect blossoms from frost, hand-pollinate, weed, mulch and spray with seaweed solution monthly.
Summer Thin fruits early.
Autumn Remove ripe and mummified fruits from the trees.
Winter Prune hard, spray with Bordeaux mixture.

Pruning and Training
Almonds fruit on young shoots, like peaches, but need to carry more fruits, so the trees are not pruned as hard. Commercially, every third year or so a few main branches are cut back hard and the top end of the remaining higher branches is removed to encourage prolific growth from the lower branches and stubs. This is best done in late winter, though it may let silver leaf disease in. It also keeps the bushes lower and more manageable. On walls and under cover, almonds may be fan trained, as for peaches (see page 472). Thinning the fruits is not essential, but prevents biennial bearing.

Pests and Diseases
Earwigs can get inside the fruits and eat the kernel, but are readily trapped in rolls of corrugated paper around each branch. Protect the bark from animals such as rabbits and deer. Almonds' main problem is peach leaf curl; see peaches (page 472) for treatment. Rainy weather in late spring and summer may cause fruits to rot. Dieback and gummosis are symptomatic of poor growth and are best treated by heavy mulching and hard pruning.

Harvesting and Storing
Once the nuts start to drop, knock them down, peel and dry. Commercially they are hulled, but the nuts will keep better intact. They can be stored in dry salt or sand for long periods.

other uses

The wood is hard and makes good veneers; the oil is used in cosmetics.

companion planting

Almonds are benefited by Alliums, especially garlic and chives. Clover or alfalfa and stinging nettles are also reputedly helpful.

culinary

The nuts are left with the brown skin on, or are blanched to provide a cleaner-tasting product. They may be eaten raw, cooked or turned into a milk.

Sirop d'Orgeat Milkshake
Quantities to taste

Sirop d'Orgeat
Ice-cold full-cream organic milk
Grated nutmeg

To one part Sirop d'Orgeat add approximately seven parts milk, mix well and top with the grated nutmeg.

Carya species from the family *Juglandaceae*

PECAN AND HICKORY NUTS

Tree up to 30m. Life span: medium to long. Deciduous, partly self-fertile. Fruits: up to 5cm, green-skinned, hard-shelled nuts.

Pecans are very large, fast-growing trees with pinnate leaves, male catkins and insignificant flowers followed by pointed, rounded, cylindrical fruits. These have a leathery skin that peels off to reveal a reddish, smooth shell enclosing the walnut-like kernel. Hickories are similar, though not as large in tree or fruit, with non-aromatic leaves and peeling bark, while the pecan has grey, resinous leaves. Hickories also prefer more humid conditions than pecans.

These are natives of North America and have been long enjoyed by Native Americans. They are grown in Australia, but rarely crop well elsewhere, though they are widely grown for their timber. The first trees were introduced to Britain in 1629.

Pecans

varieties

Carya illinoensis is the pecan, a fruit much like a walnut in most ways save that the reddish shell is smooth, not embossed, and is also more cylindrical. The trees grow up to 21m high, so are not very suitable for most modern gardens. Burbank developed new varieties of pecan with very thin shells, but these are not widely available. The hickories are similar to the pecan, with more flattened nuts. The trees generally grow up to half the pecan's height again, which is pretty big anywhere. The **Shellbark Hickory**, *C. laciniosa/alba*, and *C. ovata*, the **Shagbark Hickory**, are the more popular and productive sorts. *C. tomentosa*, the **Mocker** or **Square Nut**, has a tasty nut, but is very difficult to shell. *C. cordiformis* is the **Bitternut Hickory**, and *C. porcina/glabra* is the **Pignut**. As their names suggest, these are suitable only for pigs and, of course, hungry children. *C. sulcata* is the **King Nut**, considered the best variety by Native Americans, but not yet developed commercially. Many other minor species are also occasionally eaten from the wild.

cultivation

They will grow in the British Isles, but rarely fruit here. The pecan prefers a hotter, drier, more sub-tropical climate; the hickories prefer one that is warmer and wetter, so may produce crops in a favourable site in certain western regions.

Growing under Glass and in Containers

There is little practical possibility of getting these huge trees under cover. They resent being confined in pots

PECAN AND HICKORY NUTS

Bitternut Hickory

and are unlikely to crop, though they may make good bonsai subjects.

Ornamental and Wildlife Value
Given a suitable setting, these are large, attractive trees with decorative foliage that turns a rich yellow in autumn. Their nuts are of significant wildlife value.

Maintenance
Spring Weed, mulch, spray with seaweed solution.
Summer Prune, but only when needed remedially.
Autumn Collect nuts if it has been a long, hot summer.
Winter Cut out coral spot if any is seen.

Propagation
Normally grown from seed, they are best pot-grown and then planted out as soon as possible in their final site, as they do not like to be transplanted. They are slow to establish and then fast-growing. Improved varieties are grafted or budded on seedling stock.

Pruning and Training
Minimal pruning and training are required. Large specimens need staking for the first few years, as they are slow to take.

Harvesting and Storing
Ripe fruits are knocked down from the tree, peeled (if the peel has not dropped off already) and dried. They do not store as well as walnuts, though they may keep up to a year if they are stored in a cool, dry place. Commercially they are hulled before storage and will keep up to two years at −15°C.

Pests and Diseases, Companion Planting
Very few pests or diseases are problems for the hickories in European gardens. The pecan finds Britain too cool to ripen its wood and suffers from coral spot, which needs pruning before it spreads. In America pecans suffer from many pests and diseases. No companion effects are known about for either tree.

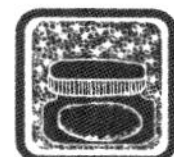

other uses

Hickories are planted for their tough, elastic timber and are renowned as fuel for smoking foods. Hickory bark was used for a yellow dye.

Square Nut

culinary

Pecan nuts taste much like mild, sweet walnuts and to my taste are preferable. They are used raw or cooked in savoury and sweet items, especially cakes and ice cream. Pecan pie is a legendary dessert. Hickory nuts are used similarly and can be squeezed to produce nut milk or oil.

Pecan Pie
Serves 4–6

175g shortcrust pastry
60g shelled pecans
90g brown sugar
3 small or 2 large eggs
225g golden syrup
30g maple syrup
½ teaspoon vanilla essence
¼ teaspoon salt

Roll out the pastry and line a wide, shallow pie dish. Bake blind, weighed down with dried peas or similar, at 190°C/375°F/gas mark 5 for 20 minutes or until cooked. Cool and fill the case with pecans, arranged aesthetically. Beat together the other ingredients till the sugar is dissolved and pour over carefully without disturbing the nuts, which may float. Bake at 220°C/450°F/gas mark 8 for 10 minutes, then reduce the temperature to 180°C/350°F/gas mark 4 and cook for another 30 minutes. Cool and chill well before serving in slices with lashings of cream.

Juglans species from the family *Juglandaceae*

WALNUTS

Tree 40m. Life span: long. Deciduous, partially self-fertile. Fruits: up to 5cm, green sphere enclosing nut. Value: rich in oil; the husks contain much vitamin C.

Walnuts are slow-growing, making massive trees up to 40m eventually, with aromatic, pinnate foliage, silvery bark, insignificant female flowers and male catkins. All parts have a distinct sweet, aromatic smell. The fruits have a green husk around the nut, enclosing a kernel wrinkled like a brain.

***Juglans regia*, the common or Persian walnut, is native to western Asia. Introduced to the Mediterranean basin before the first century BC, it became an important food in many regions and was also grown for timber. Walnuts reached Britain in the sixteenth century, if not in Roman times. The black walnut, *J. nigra*, comes from north-east America and was introduced to Britain in 1686. It is even bigger than the common walnut and widely grown for timber. The nuts are large, very hard to crack and a valuable dietary source of phosphorus.**

varieties

Named varieties of common walnuts such as **Franquette** are hard to find in Britain, though may be available elsewhere. France had one called the **Titmouse**, because the shell was so thin that a titmouse could break in to eat the kernel! Black walnuts are usually offered as the species here, but there are several named varieties available from the USA. They can grow half as high again as common trees, up to 45m. **Broadview** comes into cropping earlier than most after just 3–4 years. **Lara** is a French-bred producer of big nuts. **Majestic** is a heavy cropper of big nuts on a very vigorous tree. **Rubis** produces attractive, different red-skinned kernels. Another American species, the **White Walnut** or **Butternut**, *J. cinerea*, introduced in 1633, is grown for timber and ornamental use. The nuts are half as big again as common nuts, strong-tasting and oily. *J. sieboldiana cordiformis* is the **Heartnut** from Japan. Fast-growing, fruiting after only five years or so, it has leaves up to 1m long and small, easily shelled nuts which hang on strings.

Pterocarya fraxiniflora, **Caucasian Wing-nut** is native to the Caucasus and Persia, and was introduced to Britain in 1782. These are strong-growing relations of walnuts, succeeding in damp places. They have large non-aromatic pinnate leaves, catkins and small edible nuts surrounded with semi-circular wings. They are hardy though sustaining some dieback after hard frosts and could be improved, perhaps by crossing with other species such as the **Japanese Wing Nut**, *P. rhoifolia* or the large-fruited Chinese *P. stenoptera*.

cultivation

Walnuts prefer a heavy, moist soil. They should not be planted where late frosts occur. As pollination is difficult, it is best to plant several together. If the walnut has a damaged bark, it produces a more valuable distorted grain in the timber.

Growing under Glass and in Containers

Huge trees, these can hardly be housed. I have a fifteen-year-old bonsai walnut, but I doubt that it will ever fruit!

Ornamental and Wildlife Value

Very attractive and sweetly aromatic trees, they grow too big for most small gardens. They are not in general very valuable to wildlife, save to rodents and squirrels.

Butternut

WALNUTS

Japanese

Maintenance

Spring Weed, mulch and spray with seaweed solution.
Summer Take young fruits for pickling.
Autumn Collect nuts, prune.

Propagation

The species can be grown from seed, but are slow. Improved varieties are grafted or budded, and still take a decade to start to fruit, finally maturing around the century. They are best started in pots and moved to their final site while still small, as they resent being transplanted.

Pests and Diseases

There are few problems. Late frosts damage them, otherwise they are slow reliable croppers, each averaging 68kg annually.

Pruning and Training

Walnuts must be pruned only in autumn, as they bleed in spring. Minimal pruning is required, but branches become massive, so remove badly positioned ones early. They need no stake after the first years.

Harvesting and Storing

The nuts are knocked down and the sticky staining peel is removed before drying. Walnuts can then be stored for up to a year. For pickling, pick the nuts green when a skewer can still be pushed through.

companion planting

Varro, in the first century BC, noted how sterile the land near walnut trees was. Walnut leaves and roots give off exudates that inhibit many plants and prevent their seeds from germinating. The American species are more damaging than the European and they are particularly bad for apples, *Solanaceae*, *Rubus* and many ornamentals.

other uses

The foliage and husks have long been used as brown dyes and the oil as a hair darkener and for paints. The wood has always been valued for veneers and gun stocks, the more gnarled the better. Walnut trees were often planted near stables and privies, as their smell was thought to keep away flies. The walnut sap has traditionally been boiled to produce sugar.

Black Walnut

culinary

Walnuts may be eaten either raw or cooked, often in confectionery or cakes, and associate particularly well with coffee or chocolate. They yield an edible, light oil. The young fruits may be pickled before the stone forms.

Walnut Aperitif or post-prandial liqueur

Young green walnuts
Brandy
Red wine
Sugar, as required

Wash and prick enough nuts to fill a wide-necked bottle. Fill with brandy, seal and store in a cool, dark place. After a year, decant the brandy into another bottle, refill the original bottle with red wine and reseal. After another year, decant the wine into the brandy, refill the walnut bottle with wine and reseal. After another year decant again, fill the walnut bottle with white sugar and reseal. After a year (making four in total,) discard the nuts and add the sugar syrup to the wine and brandy. Serve in sherry glasses before meals.

Corylus species from the family *Corylaceae*

HAZELS, COBS AND FILBERTS

Tree/bush up to 6m. Life span: medium to long. Deciduous, partially self-fertile. Fruits: up to 2.5cm, pointed round, or oblong oval, brown nuts. Value: rich in oils.

These are shrubby trees, typical of woods and thickets, with dark stems, leaves rounded to a point and magnificent, yellow catkin male flowers in early spring. The gorgeous, carmine-red, female flowers are tiny, sea-urchin-like tentacles that protrude on warm days. The nuts are a pointed round or oblong oval. Hazels and cobs have a husk around the base, while filberts (full-beards) are completely enveloped by the husk. The shell is thin and the kernel sweet. Wild hazels are *Corylus avellana*, but, being wind-pollinated, these have often been influenced by *C. colurna*, cobs, also known as Turkish or Barcelona nuts, and *C. maxima* or filberts.

Hazelnuts of wild species were known in ancient times and filberts were introduced by the Romans from Greece. Pliny claims they came there from Damascus. The Romans may have brought filberts to Britain, but they were not noticed officially till introduced in 1759. Cob nuts were introduced earlier, in 1582, and the American hazelnut, *C. americana*, a similar, smaller nut with a thicker shell and heart-shaped leaves, arrived in 1798. Any appellation no longer signifies true breeding, as these all became interbred during the nineteenth century, giving us most of our current varieties.

Hazel catkins

varieties

I adore **Red-skinned Filberts**; they are certainly small and fiddly, with tight, russet husks, but they are so delicious. **Cosford Cob** is thin-shelled and a good pollinator of others. **Kentish Cob** (**Lambert's Filbert**) is a prolific cropper of large nuts if pollinated by **Cosford** or **Pearson's Prolific** (**Nottingham Cob**). This last is compact, a good pollinator and has large nuts. For the best flavour, however, you can't beat the wild hazelnut.

cultivation

Hazels thrive in stony, hilly ground. A well-drained, loamy soil will do, but heavy, damp, rich soils cause too much rank growth and few female flowers. Hazels need no support and are best planted severally to ensure pollination. They are an immensely easy crop for the lazy gardener, requiring even less effort than most.

Growing under Glass and in Containers

There seems no reason to grow them under cover and I suspect they would not like it anyway. They survive in pots quite well, looking attractive but seldom cropping.

Ornamental and Wildlife Value

These are not generally noticed, save when their catkins make a welcome display. The **Twisted Hazel** is attractively distorted and deformed and still crops well. The hazels will support many life forms, both large and small, and are also excellent plants for wild or native gardens.

HAZELS, COBS AND FILBERTS

Maintenance
Spring Weed, mulch and spray with seaweed solution at monthly intervals.
Summer Cut close underneath if grassed.
Autumn Clear mulch or cut the grass close to disclose any fallen nuts.
Winter Prune, mulch well if not under grass.

Propagation
These can be grown from seed but do not come true. Layering or grafting is possible, but root suckers are best, detached in autumn and potted up or planted in situ or a nursery bed for a year before their final move.

Pruning and Training
Traditionally hazels were grown on a low, flat, cartwheel frame. They are probably best trained to spurs on goblets, but are often left to be bushes or thickets. It is worth keeping them on a single trunk, uncongested, and removing the suckers, to prevent losing any of the nuts.

Pests and Diseases
Weedy growth underneath makes it hard to find nuts, so hazels are best planted in grass or mulched. In gardens they suffer few problems on a scale sufficient to damage crops, other than the attentions of birds, rodents, children and especially squirrels.

Kentish Cob

Kentish Cob

Harvesting and Storing
The nuts can be eaten a little unripe, but have to be fully ripe to keep. Ideally they should be allowed to fall off, but many are stolen by wildlife. They need to be dehusked and dried to keep well, though I never dehusk my Red-skinned Filberts. They will keep best, and for years, if packed in salt.

companion planting

They seem to associate naturally with bluebells and primroses, and truffles can be grown on their roots.

other uses

Hazels make good hedges and windbreaks. The foliage is eaten by many animals, including cows. The stems are tough and flexible, so are good for baskets and hurdles, and forked branches make divining rods. The wood is used for smoking fuel and I smoke my cheese with the shells.

culinary

Hazelnuts of all varieties are used in savoury dishes, but more often in sweet dishes and confections. They are used for liqueurs and can be squeezed to express a light, edible oil.

Hazelnut Macaroons
Makes about 12

120g hazelnuts
120g light brown sugar
1 egg white
Drop of vanilla extract
Whipped cream, to serve

Grind three-quarters of the nuts in a food processor, then add the other ingredients and cream them together. Pour rounds of the mix on to rice paper on a baking tray, place the remaining nuts on top and bake for 15 minutes at 180°C/350°F/gas mark 4. Serve sandwiched with thick or clotted cream.

Castanea sativa from the family *Fagaceae*

SWEET OR SPANISH CHESTNUTS

Tree up to 37m. Life span: long. Deciduous, rarely self-fertile. Fruits: 5cm, prickly burrs containing two or three nuts. Value: rich in oils.

Sweet chestnuts make massive and beautiful trees. They have very large, serrated-edged leaves. The male flowers are long, yellow catkins, different from walnut or hazel catkins, as they are divided like pearls on a string. The fruits are brownish-russet, softly spiny burrs usually containing three brown nuts with thin, tough, leathery shells, flattened on one side and pointed.

Sweet chestnuts are native to the Mediterranean region. They were highly valued by the Romans for food and timber and became widely distributed. They fruit well only after hot summers in Britain, but are still capable of reaching a large size, so they have been planted for timber and were often coppiced. The major exporter of these nuts has always been Spain, though most southern European countries have their own production, as chestnuts have become a staple food. Madeiran nuts are said to be the biggest and were traditionally served to sustain the peasants for months each year.

varieties

Castanea sativa is usually available only as the species or ornamental selections, though better varieties exist, such as *C. sativa macrocarpa*, **Marron de Lyon**, from France. **Marigoule** is an early fruiting variety similar but not as dwarf as **Regal,** which is claimed to get no more than 4.5m after 10 years; both start cropping at 2 or 3 years and have very fragrant flowers. *C. dentata* was the **American Sweet Chestnut**, with smaller, richly flavoured, sweeter nuts, but has been almost wiped out by a fungus, chestnut blight. Now American growers are breeding hybrids from the resistant *C. mollissima*, the **Chinese Chestnut**. *C. pumila*, the **American Chinquapin**, is rare in cultivation, and unlikely to fruit in Britain, but the nut is said to be very sweet. Other *Castanea* species are widely grown and the delicious nuts eaten with appreciation all around the world.

Castanea sativa – not for the smaller garden

cultivation

Sweet chestnuts are far too big for most gardens, rapidly reaching 30m or even more. They do not like thin chalky soils, but are not calcifuges and will grow on an alkaline soil if it is also a light, well-drained loam or light, dry, sandy soil. They need no staking after the first year or so. If nuts are required rather than timber, plant on the sunny side of woodlands or windbreaks, ideally of yew or holm oak.

Maintenance

Spring Weed, spray with seaweed solution.
Summer Hope for hot weather.
Autumn Collect nuts if following a hot summer.
Winter Enjoy the nuts toasted over the fire.

Ornamental and Wildlife Value

Sweet chestnuts make statuesque trees and the large leaves colour well in autumn. The nuts are often rather too useful to wildlife.

Growing under Glass and in Containers
Far too big to grow under glass, they resent being confined in a pot and they are unlikely to ever fruit in one, but there's a challenge.

Pests and Diseases
Generally they are problem-free in Britain and will set crops in the south following hot summers. In America, chestnut blight wiped out their best species; so far this is no problem elsewhere. Rodents, birds (especially rooks and pheasants) and squirrels soon take the nuts.

Propagation
They can be grown from seed and are fast-growing but still slow to fruit. The best varieties are budded or grafted, but hard to find in the UK.

Pruning and Training
Minimal pruning is required and is best tackled in winter. Chestnuts get very large, so care must be taken to remove unsound branches.

Harvesting and Storing
The nuts are beaten down, the husks then removed and dried. They will keep for a year in cool, dry conditions; longer if totally dried first.

companion planting

Chestnuts are considered healthier when they are grown near oak trees.

The nuts come well protected

culinary

Chestnuts are not eaten raw, but are delicious roasted. They are made into marrons glacés (crystallised chestnuts), ground into flour and then made into porridge, puddings, breads, cakes, muffins and tarts and they are also much used for savoury dishes such as pâtés and in stuffing for meats. Sweet chestnuts are even made into liqueurs.

Chestnut Amber
Serves 4

225g chestnuts
300ml milk
1 lemon
1 vanilla pod
60g breadcrumbs
30g butter
60g caster sugar
2 eggs, separated
120g shortcrust pastry

Roast the chestnuts for 20 minutes, cool and remove the skins. Simmer gently, with sufficient water to cover, until tender, then drain and sieve to a purée. Simmer the milk with the lemon peel and vanilla pod for 15 minutes, then strain on to the breadcrumbs. Blend the butter and half the sugar, mix in the egg yolks and lemon juice and stir in the puréed chestnuts, breadcrumbs and milk. Line a deep dish with pastry and fill with the mixture. Bake at 200°C/400°F/gas mark 6 for 25 minutes, or until firm and brown. Whisk the egg whites to a stiff froth, add a teaspoon of sugar, whisk again and spoon on top of the pie. Sprinkle with more sugar and then return to the oven until the meringue turns a delicious amber.

other uses

Sweet chestnut has long been used for cleft paling. The wood is durable, but has 'shakes' or splits in it. Often used for rough or external timber, such as coffins and hop poles, it makes a poor firewood but superior charcoal. The nuts were traditionally esteemed for the self-service fattening of swine.

Castanea dentata

Pistacia vera from the family *Anacardiaceae*

PISTACHIOS

Tree, up to 9m. Life span: medium. Deciduous, not self-fertile. Fruits: 2.5cm, pointed oval nuts. Value: rich in oil.

Pistachios are small, tender trees with grey bark and grey-green, slightly downy, pinnate leaves. The male and female flowers are borne on separate trees in the wild, but some cultivated varieties may be unisexual. It is hard to tell, as the flowers are inconspicuous though carried in panicles, which later become masses of small, pointed red fruits. These have a thin, green husk covering the thin, smooth, twin-shelled nut, housing a small, oval, green kernel that has a most appealing taste, especially after roasting and salting. Pistachios are usually the most expensive variety of nuts to buy.

Pistachios are natives of the Middle East and Central Asia and, with some close relations, have been cultivated since ancient times. They were grown in Italy in the late Roman period but were not introduced to Britain till the sixteenth and seventeenth centuries and then never proved hardy enough. The seed was distributed by the American Patent Office in 1854 and the trees proved successful in many southern states, where the nuts are now produced in quantity, rivalling the output of the Mediterranean and Turkey.

varieties

Pistachios, *Pistacia vera*, are not hardy enough for outdoors in Britain except in a very favoured site, for example against a hot wall or under glass. The variety **Aleppo** is an old favourite, but probably unobtainable. *P. terebinthus*, the **Cyprus Turpentine Tree** or **Terebinth**, is similarly not hardy. The whole plant is resinous, with nuts which start coral red, ripening to brown, and have edible, oily kernels, also green and pointed but smaller. It was used as a pollinator for pistachios as these shed pollen too early to fertilise themselves. Thus most seedling pistachios became hybrids of these two species. This has prevented much improvement of the pistachio nut. *P. lentiscus* is the evergreen **Mastic Tree**. The nuts yield edible oil, but this tree is grown mainly for the resin that is obtained from bark incisions, used in turn for chewing gum and medicines.

cultivation

Pistachios are not particular as to soil, but will require a protective wall in Britain. In warmer countries they are useful for growing on poor, dry, hilly soil where other nuts will not thrive. Their greater value commercially also encourages this. One male staminate tree is needed for six female pistillate trees, but as the pollen is shed early it needs to be saved in a paper bag until the pistils are receptive.

Mastic Tree

PISTACHIOS

Pistacia vera

Growing under Glass and in Containers
Pistachios have been grown under glass and they make very attractive shrubs in large pots. However, good provision must be made for their pollination.

Ornamental and Wildlife Value
In warm countries or on a warm wall these are not unattractive, small trees. The unproductive **Chinese Pistachio**, *P. chinenis*, is a small, very pretty and hardy shrub. Pistachio nuts are readily taken by maurauding birds and mammals.

Maintenance
Spring Put out pots or uncover trees; weed, mulch, spray with seaweed solution.
Summer Water plants in pots regularly but minimally.
Autumn Gather nuts if the summer has been hot.
Winter Bring in pots or protect trees.

Propagation
Although better varieties could be grown, pistachios seem to lack development. Continual cross-hybridisation with terebinths has prevented improvement, but means that nuts come as true as their parents, which is not claiming a lot. Better varieties can be layered, budded or grafted.

Pruning and Training
Only remedial pruning is required and this is best done in mid-summer. They can be trained against walls and are best fan-trained initially, then allowed to grow out to form a bush.

Pests and Diseases
No common pests or diseases are a problem for pistachios in garden cultivation, though under glass and on a warm wall they may suffer red spider mite attacks.

Harvesting and Storing
As these hang severally on panicles, they can be cut off to be dried and husked. They are best stored in their shells, which open automatically if they are roasted.

other uses

P. terebinthus was long ago cultivated for producing first terebinth, the resin which oozes from the tree, and later turpentine. The wood is dark red and hard and used in cabinet making.

Pistacia terebinthus

culinary

Pistachios can be eaten raw, but are most commonly roasted and salted, the shelling being left to the purchaser. They are much used as a colouring and flavouring for a wide range of foods, some savoury and many sweet, including nougat and ice cream.

Pistachio Ice Cream
Serves 6

3 small or 2 large egg yolks
175g honey
1 teaspoon vanilla extract
600ml cream
120g shelled pistachios
Natural green food colouring
Candied lemon slices and glacé cherries to taste

Whisk the egg yolks, honey, vanilla and half the cream. Scald the remaining cream in a bain-marie, add the egg mixture and stir until the mixture thickens. Remove from the heat, cool, chill, then partially freeze. Remove from the freezer, beat vigorously and return. Repeat, but after the second beating mix in the nuts and colouring, then freeze again. Partially defrost before serving the ice cream scooped into glasses, garnished with lemon slices and cherries.

Anacardium occidentale from the family *Anacardiaceae*

CASHEWS

Tree up to 12m. Life span: medium. Semi-evergreen. Fruit: up to 8cm long, unusual. Value: kernels are nearly half fat and one-fifth protein.

Cashews are medium-sized, spreading trees with rounded leaves, related to pistachios. The cashew comes attached underneath the bottom of the much larger and peculiar fruits, cashew apples, which are juicy and astringent. The nut is grey or brown, ear-shaped and contains a white kernel within the acrid, poisonous shell.

Indigenous to South America, cashews were planted in the East Indies by the sixteenth century and are now grown in many tropical regions, especially India and eastern Africa.

varieties

Other cashews are eaten: *A. humile*, the **Monkey-nut**, and *A. nanum* are from Brazil and have similar nuts. *A. rhinocarpus* is the **Wild Cashew** of Colombia and British Guyana.

cultivation

These are best grown by the sea in moderately dry tropical regions and will thrive in any reasonable soil.

Growing under Glass and in Containers

Plants dwarfed by large pots could probably be grown in a hot greenhouse or conservatory, if the seed could be found.

Ornamental and Wildlife Value

Cashews make interesting subjects for a collection or botanical garden.

Propagation

They are normally grown from seed, but this is difficult to obtain.

Pruning and Training

Only remedial pruning is necessary.

Pests and Diseases

No problems are known. The trees exude a gum obnoxious to insects, which was used in book-binding.

Harvesting and Storing

Once the nuts are picked from underneath the fruits, they have to be roasted and shelled, which, despite mechanisation, is labour-intensive. This is because all the shell must be removed, as it contains an irritant in the inner membrane around the kernel, though this is rendered harmless by heat.

other uses

The shells of the nuts contain an oil used industrially. The 'apples' are then fermented to make a liquor. The sap makes an indelible ink.

culinary

Cashew nuts are popular raw, i.e., already partially roasted, or roasted and salted. They are used in many sweet and savoury dishes and can be liquidised to make a thick sauce. Cashews are fermented to make wine in Goa. The cashew apple is the more valued part and is eaten fresh, preserved in syrup, candied and fermented to wine.

Cashew Tarts
Makes 12

225g marzipan
A little icing sugar
120g cashew nuts
60g honey
1 teaspoon vanilla extract
A little milk
Glacé cherries

Roll the marzipan as pastry and form individual tart cases in a tray dusted with icing sugar. Liquidise the other ingredients, adding just enough milk to ensure success. Pour into the marzipan cases, set a cherry in each and chill them to set.

Macadamia integrifolia (Macadamia ternifolia) **from the family** *Proteaceae*

MACADAMIAS
OR QUEENSLAND NUTS

Tree up to 18m. Life span: medium. Semi-evergreen. Fruits: up to 2.5cm, grey-husked nuts. Value: over 70% fat.

Macadamia trees are densely covered with narrow, glossy, holly-like, dark green leaves. The tassels of whitish flowers are followed by strings of small, hard, roundish, pointed nuts in greyish-green husks.

The kernel is finely flavoured and of exquisite texture.

These nuts, despite the Greek-sounding name, are natives of north-eastern Australia. Not widely appreciated, they are mostly consumed in the United States from plantations in Hawaii. They were introduced to Ceylon, now Sri Lanka, in 1868.

varieties

No species or varieties are available in Britain.

cultivation

Macadamia nuts prefer tropical or sub-tropical, moist conditions. They are not particular as to soil and do best on the volcanic slopes of Hawaii. They thrive at medium elevations.

Growing under Glass and in Containers
If the seed could be obtained they might be grown under glass, in pots to constrain growth, though it is doubtful they would crop.

Ornamental and Wildlife Value
They are attractive, glossy, dark trees, but too tender for growing successfully outside sub-tropical zones.

Propagation
They are propagated by seed, but as the nuts are usually sold roasted and salted this may be difficult to find.

Pruning and Training
Only remedial pruning is necessary and they will form bushy trees.

Harvesting and Storing
The shells are very hard to crack, so the bulk crops are collected mechanically and taken to factories to be de-husked, shelled, roasted and salted before packaging and storing, when they will keep for up to a year or so.

culinary

Most macadamia nuts are eaten roasted and salted, but they are also used in certain baked goods and confectionery.

Macadamia Slice
Serves 8–10

120g macadamia nuts
Icing sugar
225g marzipan
1 dessertspoonful apricot jam
30g chopped candied peel

Rinse and dry the macadamia nuts, if they are salted. Dust a rolling board with powdered sugar and roll out the marzipan thickly. Coat thinly with jam and cut into two equally shaped pieces and an approximate third. On one piece spread a layer of nuts and peel, then place one-third of marzipan on top, sticky-side down. Smear the top with jam and add another layer of nuts and peel. Then put the last third on top (also sticky-side down). Carefully press and roll this sandwich flatter and wider until the nuts almost push through. Trim, cut into small portions and sprinkle the mixture with powdered sugar before presenting.

Cocos nucifera from the family *Arecaceae*

COCONUTS

These attractive palms, so typical of dreamy, deserted islands, are spread by their floating, oval-husked nuts. The thick, fibrous husk is contained in a rind and itself encloses a thick-shelled, oval nut with a hollow kernel that is full of milk when under-ripe.

Palm, up to 28m. Life span: medium to long. Evergreen, not usually self-fertile. Fruits: 30cm plus, green-brown, oval husk containing the nut. Value: 65% oil.

Venerated in the islands of the Pacific as a sacred emblem of fertility, coconuts are distributed and known around the world.

Each tree holds a huge weight of small head-crushing bombs

varieties

The **King Coconut** of Ceylon is esteemed for its sweet juice. The **Dwarf Coconut**, *Nyiur-gading*, of Malaysia has small fruits, but crops when young and at only about a metre high. The **Maldive Coconut** is small and almost round; the **Needle Coconut** of the Nicobar Islands is triangular and pointed.

Definitely not Dorking

cultivation

Coconuts thrive by the sea in moist, tropical heat and rich, loamy soils and are planted about 10m apart.

Growing under Glass and in Containers

These are very attractive, easy plants to start with, rapidly outgrowing most places. The **Dwarf Coconut** may fruit, given good conditions, in only four years – thus, while still small enough to stay indoors!

Ornamental and Wildlife Value

Very attractive trees, these can be used for indoor display until they grow too large.

Propagation

Ripe nuts that are laid on their side and barely covered with compost will germinate readily in heat.

Harvesting and Storing

The nuts are used as they drop, but for milk, processing or cooking are picked by climbing or by using trained monkeys.

companion planting

Coconuts are often grown in alternate rows with rubber trees, and with cacao while young. Climbing peppers, *Piper nigrum*, are grown up the coconut trunks.

culinary

The milk is drunk fresh or fermented. The nut is eaten raw or cooked, often in the form of desiccated or shredded coconut.

Coconut Biscuits

Makes approx. 10

1 egg white
150g powdered sugar
75g dried coconut
Rice paper
Glacé cherries
Crystallised angelica

Beat the egg white until stiff, then beat in the sugar and coconut. Spoon blobs of the mixture on to rice paper on a baking tray. Garnish each with a cherry and angelica and bake at 180°C/350°F/gas mark 4 for 15 minutes or until they are browning.

Bertholletia excelsa* from the family *Myrtaceae

BRAZILS
OR QUEENSLAND NUTS

Tree up to 30m. Life span: long. Semi-evergreen. Fruits: up to 15cm; brown, spherical shell containing many nuts. Value: 65% fat and 14% protein.

These are tall handsome trees found on the banks of the Amazon and Orinoco Rivers. They have large, laurel-like leaves and panicles of white flowers, which drop brown, spherical bombs with thick, hard cases. These need to be smashed to reveal inside a dozen or more nuts shaped like orange segments, each with its own hard shell enclosing the oval, brown-skinned, sweet, white kernel. Natives of Brazil, these are still mainly produced there and also in Venezuela and Guyana. They are grown ornamentally in other countries such as Sri Lanka, but rarely on a commercial scale.

varieties

Many consider the **Sapucaya Nut** superior; it is similar, though it comes from a different tree, *Lecythis zabucajo*.

You really don't want to climb a Brazil nut tree so you wait till they drop

cultivation

Brazil trees thrive in deep, rich, alluvial soil in tropical conditions.

Ornamental and Wildlife Value
Very attractive trees, but too large and requiring too much heat and warmth for widespread use.

Growing under Glass and in Containers
These can be grown from seed and kept dwarfed in containers, making interesting specimens, but are unlikely ever to fruit.

Harvesting and Storing
The individual nuts are obtained by cracking the spherical containers, which are sealed with wooden plugs. Inside their shells the nuts will keep for up to two years.

Pruning and Training
These need no special attention, but are slow. They are often not cultivated, but are gathered from the wild, as they take fifteen years to start fruiting.

Propagation
The nuts can be started off in heat, but actually take months to germinate.

other uses

The oil expressed from the kernels is used industrially; bark once caulked ships.

culinary

Most often Brazil nuts are eaten raw at Christmas time, but are also widely used in cooking, baking and confectionery.

Treacly Brazil Pie
Serves 6

225g shortcrust pastry
75g Brazil nuts
60g breadcrumbs
75g golden syrup
Juice and grated rind of 1 lemon
Cream, to serve

Roll out three-quarters of the pastry and line a pie dish with it. Use dried peas to weigh it down, and bake blind at 190°C/375°F/gas mark 5 for 10 minutes. Remove the peas and put a layer of nuts around the base of the case. Mix the other ingredients together and pour on top. Decorate with strips of pastry and then bake for 20 minutes at 190°C/375°F/gas mark 5. Serve the pie with lashings of whipped cream.

Arachis hypogaea **from the family** *Leguminosae*

GROUNDNUTS OR PEANUTS

Herbaceous, 60cm. Life span: annual. Self-fertile. Fruits: 1cm small oval seeds. Value: rich in oil, protein and vitamins B and E.

Peanuts are always known and used as nuts, although they are in fact the seeds of a tropical, pea-like, annual plant. After pollination of the yellow 'pea' flower, the stalk lengthens and pushes the seed pod into the ground, where it matures. The light brown husks shell easily to reveal a few red-skinned, whitish-yellow seeds.

Natives of tropical America, peanuts were brought to Europe in the sixteenth century and remained curiosities until the nineteenth. Useful for oil, animal feed and 'nuts', they are now grown worldwide.

varieties

Spanish Bunch or **Virginia** varieties average fewer kernels than the better **Valencia** varieties, which have up to four. Mauritius peanuts are believed to be of superior quality. Similar in several ways are **Tiger Nuts**, *Cyperus esculentus*, *Cyperaceae*. Also called **Ground-Almond** or **Chufa** these are also not a nut at all but edible underground rhizomes of a small perennial grass-like sedge grown in dry sandy soils mostly in Western Asia and Africa.

cultivation

Peanuts prefer loose, dryish, sandy soil. Cool, wet, British summers are unfavourable and they do better under glass, but can be raised indoors and planted out, and will then produce light crops. They are normally grown on ridges to make digging the crop easier.

Ornamental and Wildlife Value
Too pea-like to be attractive, they have curiosity value. The seeds are only too valuable to wildlife!

Pruning and Training
These require no care. The old, runnering varieties were more difficult.

Pests and Diseases
On a garden scale these are problem-free, save for rodent thefts.

Harvesting and Storing
The pods are dug in autumn, thoroughly dried and shelled before storing. Commercially the bulk is pressed for oil and feedcake.

Propagation
Sow in pots in heat for growing on in large pots to fruition, or planting out in favourable areas. In warm countries they are grown outdoors, sown 8cm deep, about 60cm apart each way.

Growing under Glass and in Containers
Peanuts are good subjects for greenhouses, or in large pots, which can be kept under cover at the start and end of the season and put out on the patio in summer.

culinary

Peanuts are commonly roasted and salted, and are much used in baking and confections, savoury sauces and for their butter and edible oil. They should be kept dry until required.

Roast Peanuts
Makes 1kg

1kg peanuts
30g garlic
1 x 50g can anchovies
7g oregano

Boil the peanuts for 2 minutes and slip off their skins, then dry the nuts. Blend the garlic, anchovies and oregano. Coat the peanuts with this mixture and roast at 180°C/350°F/gas mark 4 for 10 minutes or so. Stir and cool.

companion planting

Peanuts have been grown with rubber and coconuts.

other uses

Peanut oil is used industrially.

Grow them in the ground, not pots

OTHER NUTS

Araucaria araucana,
Monkey Puzzle or **Chile Pine**
Auracariaceae
These well-known trees, with spiny, overlapping, dark green leaves festooning the long, tail-like branches, rarely fruit in the UK, it seems, as they are usually planted singly. Where they have been planted severally, as at a school in Sussex, they reportedly set seed and the nuts were shed most years. In Chile the seeds are eaten raw, roasted or boiled. Closely related trees are also grown in Brazil and in Australia.

Brosimium alicastrum,
Maya Breadnut is a tall Caribbean tree with round yellow fruits containing a large edible seed. Not to be confused with the Breadnut which is the seedy form of the normally seedless Breadfruit, *Artocarpus*, see p 561.

Canarium ovata,
Pili Nut
Burseraceae
This is the legendary and rare Pili nut frustratingly rated as the best tasting of all nuts by those fortunate enough to have found it.

Castanospermum australe,
Moreton Bay Chestnut
Leguminosae
These poisonous Australian nuts are relished by native Australians, who leach them in water before drying and roasting the nuts to render them edible.

Coffea arabica,
Coffee
Rubiaceae
Small, evergreen trees, which once grew wild in Arabia and are now cultivated in most hot countries. Coffee 'beans' are seeds from the cherry-like berries, roasted to oily charcoal, then leached with hot water.

Terminalia catappa,
Geranium tree, **West Indian Almond**
Has red geranium-like flowers followed by small flattened walnut-like fruits containing a very hard-to-get-at edible kernel much resembling almond.

Cordia sebestena,
Geiger tree
This is another Caribbean tree with hairy leaves, orange-red tubular flowers and sweet sticky fruits that can be
eaten raw or cooked with sugar or even made into cough medicine.

Fagus sylvatica,
Beech
Fagaceae
A well-known tree that can reach 30m and chokes out everything underneath with heavy, dry shade. Although parts of the tree are poisonous and have been used medicinally, an edible oil can be extracted from the seeds and they have been eaten raw and roasted to make 'coffee'. In sheer desperation, beech sawdust has been boiled, baked and mixed with flour to make 'bread'. A worthy subject for parks on acid or alkaline soil, but far too large for most gardens! The **American Beech**, *F. grandiflora*, is similar.

Myristica fragrans,
Nutmeg
Myristiceae
These nuts are only ever used as a spice. They are natives of the Moluccas Islands in Indonesia and are commercially grown in few other places save Grenada in the West Indies. The trees reach 18–21m and have fruits resembling apricots or peaches which split, like almonds, revealing a nut surrounded by a reddish yellow aril. This is the spice, mace. Inside the thin shell is the brown nutmeg kernel which rattles when ripe. If still alive, they will germinate in heat after three months or so.

Quercus species,
Acorns
Fagaceae
Although acorns are not edible raw these have been ground, leached and cooked in times of hardship, some varieties were apparently relished, and with huge numbers of species these have great potential for breeding a new tasty nut or oil source. Acorns fatten swine but can kill horses! *Q. macrocarpa*, **Bur Oak**, has the biggest acorns, they also have a low tannin content and are easily made edible. The **Cork Oak**, *Q. suber* also bears more edible acorns.

Simmondsia chinensis,
Goat Nut, **Jojoba**
Grown mainly for the oil which is odourless and does not go rancid, the small nut-like fruit can be eaten if desperate.

Xanthocerus sorbifolium,
Bob's Nuts
Sapidaceae
From China comes this pretty slowgrowing, very hardy tree much resembling a rowan with sprays of almost orchid-like flowers followed by green, smooth, small apple-sized capsules with thick hard walls that split and peel back to drop about a dozen or so black-shelled nuts. These much resemble macadamias when roasted. Very good!! Apparently the flowers and foliage were also cooked and eaten.

PRACTICAL GARDENING

PLANNING YOUR VEGETABLE GARDEN

Several factors determine the planning and layout of a vegetable garden and the varieties that can be grown:

- the locality and climatic conditions
- the size and shape of the plot
- the number of people to be supplied with vegetables
- the duration of cropping
- the skill of the gardener, and the time available for maintaining the plot
- whether the vegetables are intended for use when fresh, stored or both

It is a good idea to list the vegetables you like, then decide how and where they can be grown to achieve the best results. In large gardens a vast range and volume of tasty vegetables can be produced using crop rotation and protected cropping to extend the growing season. Smaller sites allow fewer opportunities for self-sufficiency, but it is still possible to grow a good selection of high-value crops like asparagus and try unusual varieties. Even the tiniest gardens, balconies or patios are suitable, particularly with the use of containers, while areas surrounded by buildings sometimes create favourable microclimates for tender vegetables and early crops. Protected cropping provides further opportunities to defy the cold weather.

Before preparing a planting plan it is important to be aware of the advantages and disadvantages of the site, considering everything from aspect and shelter to soil quality and drainage. Choose a design that suits your site and taste: traditionally, plots were planned in beds and rows, but you may prefer the potager (or 'edible landscape') developed by the French, or the raised bed system, which is ideal for intensive small- or large-scale cropping. It is essential to provide the best possible growing conditions for optimum production and the old saying, 'the answer lies in the soil', is particularly relevant to vegetable growing.

Accurate timing is also a prerequisite. A cropping timetable should make full use of the ground all year round. It is advisable to plan backwards from the intended harvesting date to work out when crops should be sown. The aim is to provide home-grown vegetables throughout the year, even if you have to rely on frozen or stored produce.

Top: Kitchen garden
Bottom: Is there a more glorious site for a garden than the foot of the Alps?

Site and Soil

Most vegetables are short-term crops, which are harvested before they reach maturity. To achieve the necessary rapid growth, the ideal site is warm and light with good air circulation. This is particularly important for the fertilisation of wind-pollinated crops such as sweetcorn, and to discourage pests and diseases, which flourish in still conditions. However, it is worth noting that strong winds can reduce plant growth by up to 30 per cent.

On a *gently sloping site* facing the sun, the soil warms faster in spring than in other aspects, making such an area perfect for early crops.

It is more difficult to work the soil on *steeper slopes*, especially if machinery is being used: crops should be planted across, rather than down the slope to reduce the risk of soil erosion during heavy rain. On *very steep slopes* the ground should be terraced.

A sloping site is ideal for early crops

There's plenty of cropping potential here

THE ORNAMENTAL VEGETABLE GARDEN

The sight of a neat, well-tended vegetable garden burgeoning with healthy crops is immensely satisfying. Conditioned to believe that vegetables are functional and flowers beautiful, many fail to appreciate the beauty and bounty of a vegetable plot with its contrasting colours, forms and textures, neatly framed by well-kept paths.

The traditional design for a vegetable garden is based on a system of rows. This gives ease of access and maintenance, enables crops to be rotated regularly and for maximum production while preventing the build-up of pests and diseases. Large kitchen gardens were formerly attached to a 'great house', with the gardeners' task to cultivate a wide range of crops and supply the household with vegetables throughout the year.

By way of contrast, most gardeners now have smaller gardens, and the basis for what they grow and how they grow is more a matter of choice than necessity. This provides them with the opportunity to grow vegetables for their ornamental and culinary values so you can savour the flavour. For the same reasons, it may also be preferable to grow small quantities of a wider range of cultivars.

Colour

Vegetables are not only eaten for their flavour and nutritional value but add welcome colour to a meal. Consider the range of colours and tonal variation in a salad; it is generally greater than the range of flavours.

There are tones of green in lettuce, endive and cucumber, yellow in peppers, purple in beetroot and leaf lettuce and red in tomatoes and peppers. For centuries, only the French potager capitalised on this display of colours, but vegetables are at last more commonly used as design elements within ornamental planting schemes. The sight of vegetables dotted individually or grouped in the flower border has in recent years become a familiar sight. To the open mind and eye, all vegetables are ornamental, even if their attraction is more obvious in some than in others. The contrasting red and green foliage of ruby chard is sumptuous, as is beetroot 'Bull's Blood', while purple- or yellow-podded French beans and the mauve leaves of cabbage 'Red Drumhead' create their own exotic magic. Red-skinned onions like 'Red Baron' and the red-leaved lettuce 'Lollo Rosso' are equally stunning. Squashes impress with their bold shapes and colour (think of 'Turk's Turban') while tomatoes like the golden 'Yellow Perfection' and striped 'Tigerella' delight the eye. Flowers, fruits and companion plants add other colour accents.

Form

While you may not consider the shape of certain vegetables to be interesting (swedes or potatoes, for instance), a good many have appealing foliage or a distinctive habit. The bold architectural leaves and sculptured form of globe artichokes are prized by garden designers, a 'wigwam' of bicoloured runner beans or trailing cucurbits makes a dramatic focal

Look closely and see a bud's true beauty

point, while the compact growth of parsley and chives is a neat edging for borders. Members of the onion family, leeks, and chives have spiky, linear leaves, which contrast well with the rounded shapes of lettuce and the feathery, arching growth of carrot tops. Celery and Brussels sprouts naturally have a distinguished upright habit. Perhaps the most unusual of all, e.g.,, a purple Brussels sprout, 'Rubine', which is planted as a single specimen in a container, looks like an angular, alien sculpture.

Texture

Texture is rarely considered a design feature, yet the different colours and forms of vegetable foliage are underlined by their texture. Consider the solidity of a compact cabbage head, wreathed in glaucous, puckered leaves, or the soft billowing effect created by the dissected leaves of fennel or asparagus foliage. Exploiting such contrasts provides a foliage display as interesting as a bedding scheme or herbaceous border. Even after harvest the impact remains; the tall dried stems of sweetcorn look wonderful when frosted on a winter's day and the rustling sound as the wind blows through their dead leaves brings life to a desolate garden.

Vegetables as Bedding Plants

Many vegetables are grown as annuals and, as each crop matures and is harvested, the appearance of the vegetable garden changes. Often, the only constant elements are the framework of paths and hedges and a few long-lived perennial crops.

Such intensively grown crops are easier to maintain when grown in a formal pattern as this provides the ideal opportunity to arrange crops in brightly coloured, bold patterns or emphasise their subtle qualities. Brightly coloured chard, lettuce or kale or the fern-like foliage of carrots can be used to provide a foliage display rivalling any bedding plant.

Vegetables which have run to seed look particularly spectacular; lettuces are upright and leafy, beetroot display their red-veined leaves and bold flower spikes while onions and leeks produce symmetrical globes of flowers which are invaluable for attracting pollinating insects. Some gardeners allow a few plants to go to seed to enjoy such effects, but would you be daring enough to create a planting scheme with the specific purpose of featuring vegetables in their later stages?

One approach to creating a bedding scheme is to group together plants with a similar life span, for a long-term display. Alternatively, with skilful planning, it is possible to plant crops taking different times to reach maturity, filling any gaps with suitable vegetables after harvest, to retain the impact of the design. The time scale can vary considerably, with radishes taking a mere five weeks, and sprouting broccoli and winter cabbages remaining in the ground for several months. The rapid changeover of these crops means that colours and flavours are constantly changing within the scheme; there is a fine example at the Eden Project.

Perennial vegetables such as asparagus, globe artichoke and rhubarb can be grown in separate beds or used as permanent feature plants within the design. In smaller or irregularly shaped gardens, crop rotation allows a wide range of annual design patterns and planting arrangements. Growing vegetables purely for their ornamental value is an option which gardeners of such gardens should consider.

The Potager

Their love of food and appreciation of aesthetics motivated French gardeners to create the potager, or ornamental vegetable garden, with a more obvious visual appeal than the English kitchen garden. Simple, formal, geometric shapes such as four square-shaped raised beds dissected by straight paths form a permanent structure, enlivened with a succession of vibrantly coloured leafy crops chosen for their culinary and visual qualities. The main paths are wide enough for a wheelbarrow, but narrower ones, and stepping stones, allow access for maintenance and picking. Vegetables are placed according to their height and habit, larger vegetables forming centre pieces and smaller ones making up the rows. Climbing vegetables growing over ornamental tripods, arches or even canes provide height, and the whole area can be screened by trellis and can include vegetables in containers as additional features.

To maintain the symmetry, harvest plants with an eye to pattern: work evenly from both ends and from the centre, or cut every other plant to keep the coverage balanced as far as possible. Make full use of successional cropping to ensure that the soil is always utilised and the pattern maintained.

Random Systems

Scattering vegetables individually or in groups among ornamental borders is becoming more popular. They should be planted according to their ultimate height with lower plants near the front of the border, using brightly coloured vegetables strong enough to stand up to their brightly coloured neighbours. Grow purple-pod climbing French beans through shrubs instead of clematis, runner bean 'Painted Lady' instead of sweet peas and purple-leaved cabbage alongside nasturtiums. Mini vegetables can be grown in window boxes or hanging baskets; try tomato 'Tumbler' or grow climbers as trailing plants.

The only requirement for successful growth of vegetables among ornamental plants is adequate soil or compost fertility, which can easily be maintained with well-prepared soil and careful feeding.

PLANNING YOUR HERB GARDEN

Herbs are so versatile that they should appeal to anyone, be they a cook, a lover of salads, or someone just wanting to enjoy the rich scents of plants and watch the butterflies collecting nectar from the flowers. And there are herbs for every space; they will grow in a window box or in a pot on a sunny window ledge; and some can be grown indoors as houseplants as well as outside in gardens, small or large. The best way to grow herbs is the organic way. Quite apart from the fact that if you use natural products, the soil remains clean and free from chemical pollutants, in organic herb gardens there is no chance of contaminating a plant before you eat it. Organic methods also encourage bees and other insects to the garden, which in turn helps maintain the healthy natural balance of predator and pest.

Conditions

As herbs are basically wild plants tamed to fit a garden, it makes sense to grow them in conditions comparable with their original environment. This can be a bit difficult, for they come from all over the world. As a general rule, the majority of culinary herbs come from the Mediterranean and prefer a dry sunny place. But herbs really are adaptable and they do quite well outside their native habitat, provided you are aware of what they prefer.

Choosing the Site

Before planning your site, it is worth surveying your garden in detail. Start by making a simple plan and mark on it north and south. Show the main areas of shade – a high fence, a neighbouring house and any high trees, noting whether they are deciduous or evergreen. Finally, note any variations in soil type – wet, dry, heavy etc. Soil is one of the most important factors and will determine the types of herb you can grow. For different soil types, see page 633.

Use

Next, decide what you want from your herb garden. Do you want a retreat away from the house? Or a herb garden where the scents drift indoors? Or do you want a culinary herb garden close at hand to the kitchen door?

Assessing the position of your herbs before planting

Style
Then think about what shape or style you want the garden to take. Formal herb gardens are based on patterns and geometric shapes. Informal gardens are a free-for-all, with species and colours, all mixed together. Informal gardens may look un-planned, but the best have been well planned. This is worth doing even if it is just to check the final height and spread of the plant. It can be misleading buying plants from a garden centre – they are all neat, uniform and fairly small, and it is well worth investigating and trying to visualize their mature size.

The plants have to be accessible, either for using fresh or to harvest, so paths are a good idea. They also introduce patterns to the design and can help to define its shape. For convenience, herbs should be no more than 75cm (30in) from a path, and ideally the beds no more than 1–1.2m (3–4ft) wide. If they are more than 1.2m (4ft) wide, insert stepping stones to improve access.

laying paths

There are a number of choices of materials you can use for paths.

Grass
A grass path is quite easy to achieve and looks very attractive. Another plus point is the minimal cost. Make it at least as wide as your lawn mower, otherwise you will be cutting it on your hands and knees with shears. It is a good idea to edge the path either with wood, metal or, more attractively, with bricks, laid on their side end to end. Disadvantages to this kind of path are that it needs mowing and will not take heavy traffic.

Gravel
Gravel paths really do lend themselves to being planted with herbs. Be sure to prepare them well otherwise a water trap will form, and the plants would be better off being aquatic.

First remove the topsoil carefully. Then dig out to a depth of 30cm (12in), putting the soil to one side. It is advisable to put a wooden edge between the soil of the garden or lawn and the new path. This will stop the soil falling into the path, and keep the edge neat. Fill the newly formed ditch with 14cm (6in) hard core. Mix the topsoil with peat, bark and grit in the ratio 3:1:1:1, and put this mix on top of the hard core to a depth of 8cm (3in). Finish with 8cm (3in) gravel or pea shingle. There are many colours of gravel available, most large garden centres stock a good range, or if you have a quarry nearby it is worth chatting them up.

Roll the path before planting. Herbs that will grow happily in gravel are creeping or upright thymes, winter savory, and pennyroyal. The disadvantage of gravel is that you will find that weeds will recur, so be diligent.

Bricks
Paths made out of bricks have become very fashionable. There is now a subtle range of colours available, and varied and original patterns can be created. For standard-size bricks, dig out to a depth of 10cm (4in). As with the gravel path, include a wooden edge. Spread 5cm (2in) of sharp sand over the base; level and dampen. Lay the brick on top of the sand in the desired pattern, leaving a 2–5mm (⅛–¼in) gap between the bricks. Settle them in, using a mallet or a hired plate vibrator.

If you wish to plant the path with herbs, it is as well at this juncture to leave out one or two bricks, filling the gaps later with compost and planting them when the path has settled. When the whole path has been laid, spread the joints with a mix 4:1 of fine dry sand and cement, and brush it in. The mixture will gradually absorb the moisture from the atmosphere, so setting the brick.

Paving stones
These can take up a lot of space and are expensive to lay over a large area. Garden centres now stock a large range in various colours and shapes. Also try builders' merchants, you may get a better deal, especially if you require large quantities. Paving stones are ideal for the classic chequer-board designs and the more formal designs.

If you want the paving stones to lie flush with the ground, dig out the soil to the depth of the slab plus 5cm (2in). Put 5cm (2in) of sharp sand on to the prepared area, level off and lay the slabs on top, tapping them down, and making sure they are level. If the chosen area is already level and you want the slabs to be proud, then lay them directly in position with only a small layer of sand underneath.

alternative planting

Raised beds
If your soil is difficult, or if you wish to create a feature in the garden, raised beds are a good solution. Also, plants in raised beds are easier to keep under control and will not wander so much around the rest of the garden. Finally, they are more accessible for harvesting and stand at a good height for those in wheelchairs.

The ideal height for a raised bed is between 30cm (12in) and 75cm (30in). If you raise it over 1m (3ft) high you will need some form of foundation for the retaining walls, to prevent them keeling over with the weight of the soil. Retaining walls can be made out of old railway sleepers (which are not as cheap as they used to be), logs cut in half, old bricks, or even red bricks – leave the odd one out and plant a creeping thyme in its place.

For filling a 30cm (1ft) raised bed, the following ratios are ideal. First put a layer of hard core (rubble) on top of the existing soil to a depth of 8cm (3in), followed by an 8cm (3in) layer of gravel, and finally 4cm (6in) of topsoil mix – made up of 1 part peat, 1 part bark, 1 part grit or sharp sand, with 3 parts topsoil.

Raised bed: ideal for confinng plants that tend to roam

Lawns
Many herbs are excellent ground cover and can make a fragrant lawn but, as already mentioned under chamomile, beware of planting too large an area to begin with. It can be an error costly in both time and money. Small areas filled with creeping herbs give great delight to the unsuspecting visitor who when walking over the lawn, discovers a pleasant aroma exuding from their feet!

It may sound repetitious, but it is worth saying that preparing your site well is the key to a good garden. Given a typical soil, prepare the site for the lawn by digging the whole area out to a depth of 30cm (12in) and then prepare in exactly the same way as for the raised bed: 8cm (3in) hard core, 8cm (3in) gravel, 14cm (6in) topsoil mix – this time, 1 part peat, 2 parts sharp sand, 3 parts top soil. Apart from chamomile, other plants that can be used for a herb lawn are Corsican mint, *Mentha requinni*, planted 10cm (4in) apart, or creeping thymes – see pages 424–426 for varieties – and plant them about 23cm (9in) apart.

HERB GARDENS

The gardens I have designed can be followed religiously, or adapted to meet your personal tastes, needs and of course space. It is with this last requirement in mind that I have specifically not put the exact size into the design and concentrated on the shape, layout and the relationship between plants. I hope these plans give you freedom of thought and some inspiration.

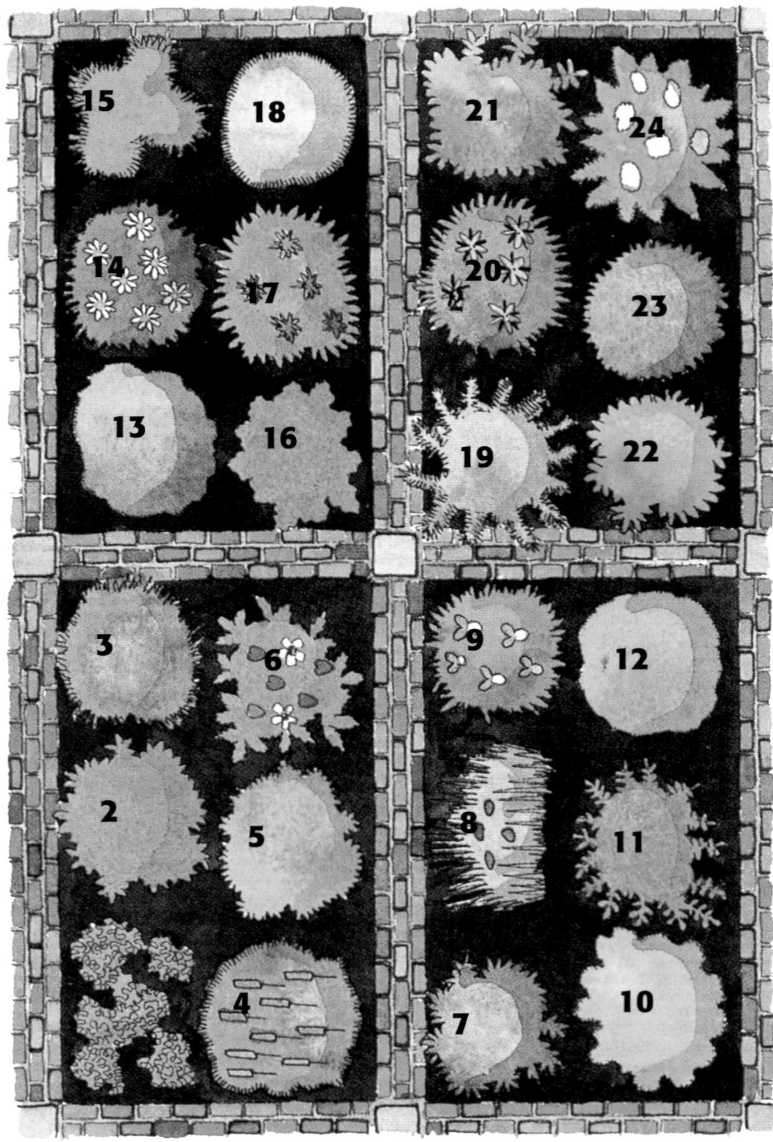

first herb garden

When planning your first herb garden, choose plants you will use and enjoy. I have designed this garden in exactly the same way as the one at my herb farm. Much as I would love to have a rambling herb garden, I need something practical and easy to manage, because the nursery plants need all my attention.

It is also important that the herbs are easy to get at, so that I can use them every day. By dividing the garden up into four sections and putting paving stones round the outside and through the middle, it is easy to maintain and provides good accessibility.

For this garden, I have chosen a cross-section of herbs with a bias towards culinary use, because the more you use and handle the plants, the more you will understand their habits. There is much contradictory advice on which herb to plant with which but many of these are old wives tales.

There are only a few warnings I will give: Do not plant dill and fennel together because they intermarry and become fendill, losing their unique flavours in the process. Equally, do not plant dill or coriander near wormwood as it will impair their flavour. Also, different mints near each other cross-pollinate and over the years will lose their individual identity. Finally, if you plan to collect the seed from lavenders, keep the species well apart. Aside from that, if you like it, plant it.

1 Parsley *Petroselinum crispum*
2 Pineapple Mint *Mentha suaveolens* 'Variegata'
3 Fennel *Foeniculum vulgare*
4 Lavander Munstead *Lavandula angustifolia* 'Munstead'
5 Greek Oregano *Origanum vulgare* subsp. *hirtum* 'Greek'
6 Alpine Strawberry *Fragaria vesca*
7 Purple Sage *Salvia officinalis* 'Purpurascens'
8 Chives *Allium schoenoprasum*
9 Heartsease *Viola tricolor*
10 Golden Curly Marjoram *Origanum vulgare* 'Aureum Crispum'
11 Salad Burnet *Sanguisorba minor*
12 Lemon Thyme *Thymus citriodorus*
13 Garden Thyme *Thymus vulgaris*
14 Roman Chamomile *Chamaemelum nobile*
15 Rock Hyssop *Hyssopus offinialis* subsp. *aristatus*
16 Buckler-Leaf Sorrel *Rumex scutatus*
17 Bergamot *Monarda didyma*
18 Dartington Curry Plant *Helichrysum italicum* 'Dartington'
19 Rosemary *Rosmarinus officinalis*
20 Borage *Borago officinalis*
21 Variegated Lemon Balm *Melissa officinalis* 'Aurea'
22 Apple Mint *Mentha suaveolens*
23 Winter Savory *Satureja montana*
24 Chervil *Anthriscus cerefolium*

herb bath garden

This may seem a bit eccentric to the conventially minded but when my back is aching after working in the nursery, and I feel that unmentionable age, and totally exhausted, there is nothing nicer than lying in a herb bath and reading a good book.

The herbs I use most are thyme, to relieve an aching back, lavender, to give me energy, and eau-de-cologne to knock me out. Simply tie up a bunch of your favourite herbs with string, attach them to the hot water tap and let the water run. The scent of the plants will invade both water and room. Alternatively, put some dried herbs in a muslin bag and drop it into the bath.

Remember when planting this garden to make sure that the plants are accessible. Hops will need to climb up a fence or over a log. Again, quite apart from the fact that the herbs from this garden are for use in the bath they make a very aromatic garden in their own right. Position a seat next to the lavender and rosemary so that when you get that spare 5 minutes, you can sit in quiet repose and revel in the scent.

1 Lavender Seal *Lavandula* x *intermedia* 'Seal'
2 Lemon Verbena *Aloysia triphylla*
3 Benenden Blue Rosemary *Rosmarinus officinalis* var. *angustissimus* 'Benenden Blue'
4 Gold Sage *Salvia officinalis* icterina
5 Valerian *Valeriana officinalis*
6 Bronze Fennel *Foeniculum vulgare* 'Purpureum'
7 Tansy *Tanacetum vulgare*
8 Golden Lemon Thyme *Thymus citriodorus* 'Golden Lemon'
9 Double-Flowered Chamomile *Chamaemelum nobile* 'Flore Pleno'
10 Houseleek *Sempervivum tectorum*
11 Black Peppermint *Mentha* x *piperita*
12 Orange-scented Thyme *Thymus* 'Fragrantissimus'
13 Meadowsweet *Filipendula ulmeria*
14 Fennel *Foeniculum vulgare*
15 Pennyroyal *Mentha pulegium*
16 Creeping Hops *Humulus lupulus*
17 French Lavender *Lavandula stoechas*
18 Bay *Laurus nobilis*
19 Roman Chamomile *Chamaemelum nobile*
20 Lemon Balm *Melissa officinalis*
21 Porlock Thyme *Thymus* 'Porlock'
22 Prostrate Rosemary *Rosmarinus officinalis* Prostratus Group
23 Golden Marjoram *Origanum vulgare* 'Aureum'
24 Eau de Cologne Mint *Mentha* x *piperita* f. *citrata*
25 Comfrey *Symphytum officinale*
26 Lady's Mantle *Alchemilla mollis*
27 Yarrow *Achillea millefolium*
28 Lavender Grappenhall *Lavandula* x *intermedia* 'Pale Pretender'
29 Silver Posie Thyme *Thymus vulgaris* 'Silver Posie'

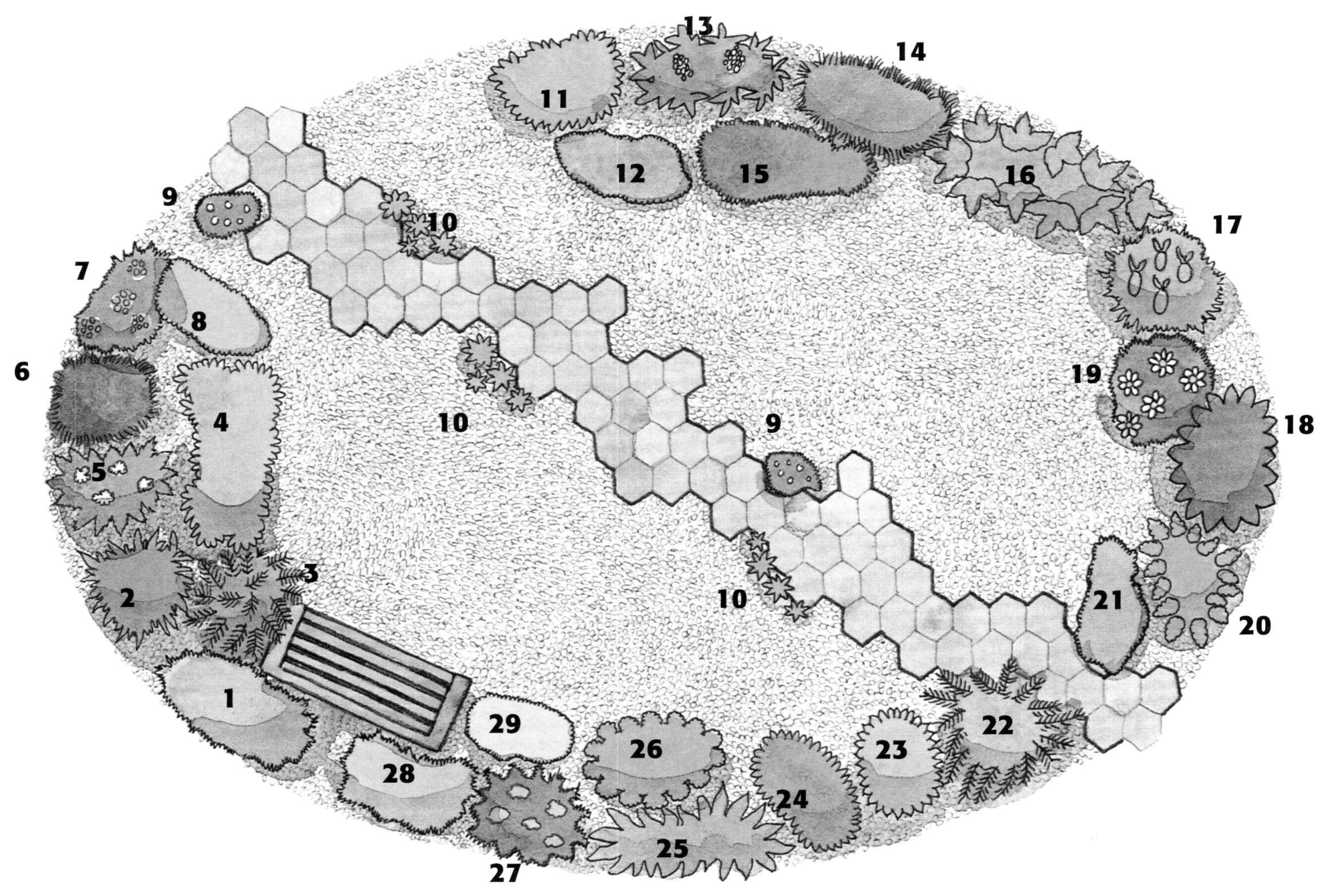

white herb garden

This garden gave me great pleasure to create. For me, it is a herb garden with a different perspective.

It has a row of steps going from the road to the front door of the house. Either side of the steps is a dwarf white lavender hedge. In spring before the lavender, and just before the lily of the valley, are in flower, the sweet woodruff gives a carpet of small white flowers. This is the start of the white garden, which then flowers throughout the year through to autumn. It is a most attractive garden with a mixture of scents, foliage and flowers.

This planting combination can easily be adapted to suit a border. Even though it is not a conventional herb garden, all the herbs can be used in their traditional way. The garlic chives with baked potatoes, the horehound for coughs, the chamomile to make a soothing tea, and the lavender to make lavender bags or to use in the bath.

The great thing about a garden like this is that it requires very little work to maintain. The hedge is the only part that needs attention – trim in the spring and after flowering in order to maintain its shape.

1 Dwarf White Lavender *Lavandula angustifolia* 'Nana Alba'
2 Sweet Woodruff *Galium odoratum*
3 Bergamot Snow Maiden *Monarda* 'Schneewittchen'
4 Jacob's Ladder (white) *Polemonium caeruleum* subsp. *caeruleum* f. *album*
5 Yarrow *Achillea millefolium*
6 Foxgloves (white) *Digitalis purpurea* f. *albiflora* (POISONOUS)
7 Roman Chamomile *Chamaemelum nobile*
8 Lily of the Valley *Convallaria majalis* (POISONOUS)
9 White Thyme *Thymus serpyllum* var. *albus*
10 White Hyssop *Hyssopus officinalis* f. *albus*
11 Snowdrift Thyme *Thymus serpyllum* 'Snowdrift'
12 Garlic Chives *Allium tuberosum*
13 Pyrethrum *Tanacetum cinerariifolium*
14 Sweet Cicely *Myrrhis odorata*
15 Valerian *Valeriana officinalis*
16 White Horehound *Marrubium vulgare*
17 Prostanthera *Prostanthera cuneata*

cook's herb garden

The best site for a culinary herb garden is a sunny area accessible from the kitchen. The importance of this is never clearer than when it is raining. There is no way that you will go out and cut fresh herbs if they are a long way away and difficult to reach. Another important factor is that the sunnier the growing position, the better the flavour of the herbs. This is because the sun brings the oils to the surface of the leaf of herbs such as sage, coriander, rosemary, basil, oregano and thyme.

The cook's herb garden could be made in the ground or in containers. If it is to be in the ground, make sure that the site is very well drained. Position a paving stone near each herbs so that it can be easily reached for cutting, weeding and feeding, and also to help contain the would-be rampant ones, such as the mints, which might otherwise take over.

Alternatively, the whole design could be adapted to be grown in containers. I have chosen only a few of the many varieties of culinary herb. If your favourite is missing, either add it to the design or substitute it for one of my choices.

1 Ginger Mint *Mentha* x *gracilis*
2 Chervil *Anthriscus cerefolium*
3 Coriander *Coriandrum sativum*
4 French Parsley *Petroselinum crispum* French
5 Chives *Allium schoenoprasum*
6 Corsican Rosemary *Rosmarinus officinalis* var. *angustissimus* 'Corsican Blue'
7 Garden Thyme *Thymus vulgaris*
8 Angelica *Angelica archangelica*
9 Fennel *Foeniculum vulgare*
10 Winter Savory *Satureja montana*
11 Greek Basil *Ocimum minimum* 'Greek'
12 Buckler-Leaf Sorrel *Rumex scutatus*
13 Bay *Laurus nobilis*
14 Sweet Cicely *Myrrhis odorata*
15 Garlic *Allium sativum*
16 Greek Oregano *Origanum vulgare* subsp. *hirtum* 'Greek'
17 French Tarragon *Artemisia dracunculus*
18 Lovage *Levisticum officinale*
19 Garlic Chives *Allium tuberosum*
20 Lemon Balm *Melissa officinalis*
21 Moroccan Mint *Mentha spicata* var. *crispa* 'Moroccan'
22 Dill *Anethum graveolens*
23 Parsley *Petroselinum crispum*
24 Lemon Thyme *Thymus citriodorus*
25 Sweet Marjoram *Origanum majorana*

salad herb garden

Herbs in salads make the difference between boring and interesting; they add flavour, texture and colour (especially the flowers).

Included in the design is a selection of salad herbs and salad herb flowers. There are two tall herbs in the middle, chicory and red orach (blue and red), which are planted opposite each other. Also, I have positioned the only other tall plant – borage – on the outside ring, opposite the chicory so that the blue flowers together will make a vivid splash. To make access easy, there is an inner ring of stepping stones.

The herbs chosen are my choice and can easily be changed if you want to include a particular favourite. Remember to look at the heights; for instance, do not plant angelica in the outside circle because it will hide anything in the inner circle. Equally, in the inner circle make sure you do not plant a low growing plant next to a tall spreading herb because you will never find it.

This whole design can be incorporated in a small garden or on the edge of a vegetable garden to give colour throughout the growing season. As the majority of these herbs are annuals or die back into the ground, the autumn is an ideal time to give the garden a good feed by adding well rotted manure. This will encourage lots of leaves from the perennial herbs in the following season, and give a good kick start to the annuals when they are planted out in the following spring.

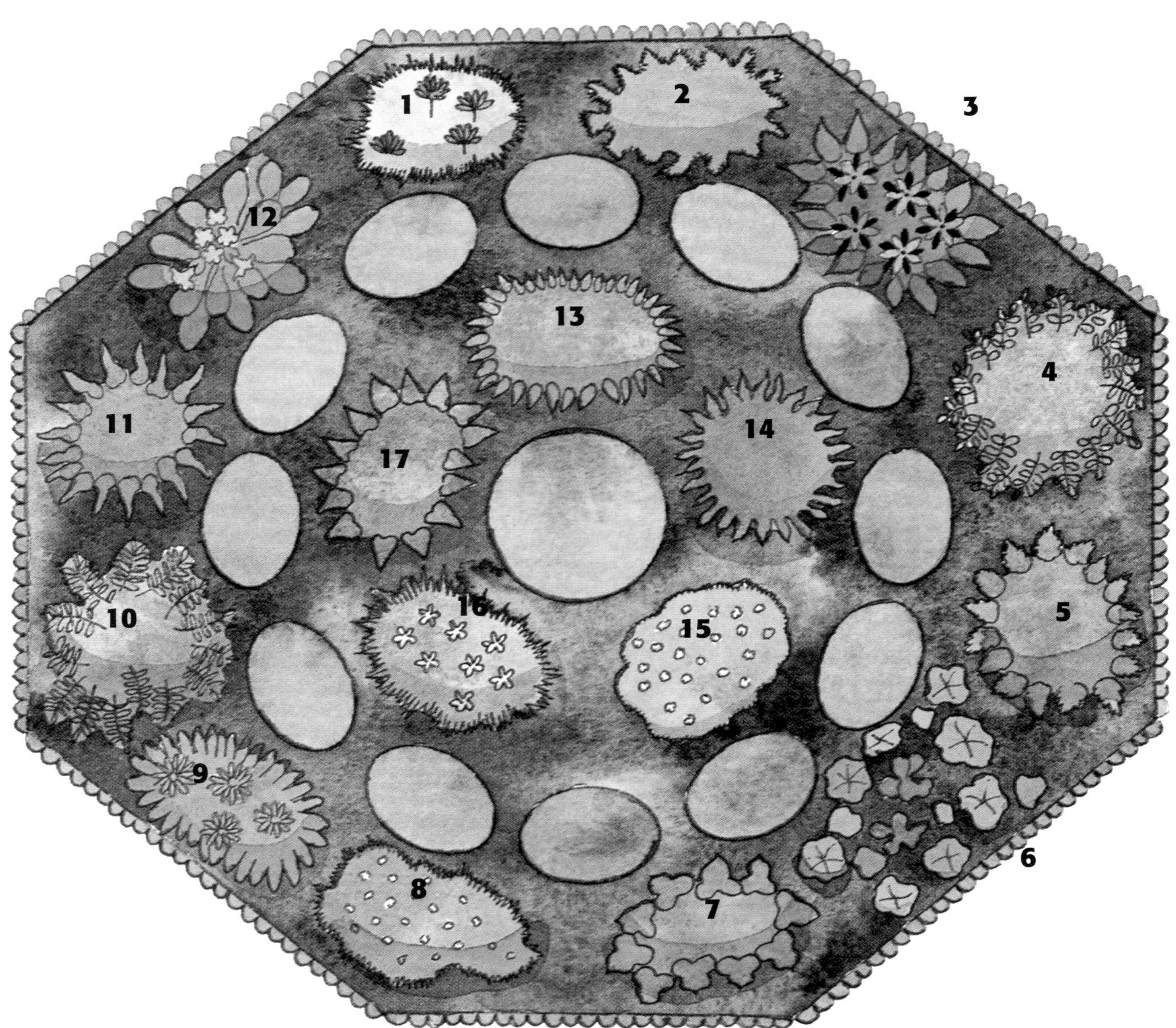

1 Chives *Allium schoenoprasum*
2 Caraway *Carum carvi*
3 Borage *Borago officinalis*
4 Salad Burnet *Sanguisorba minor*
5 French Parsley *Petroselinum crispum* French
6 Nasturtium *Tropaeolum majus*
7 Buckler-Leaf Sorrel *Rumex scutatus*
8 Hyssop *Hyssopus officinalis*
9 Pot Marigold *Calendula officinalis*
10 French Tarragon *Artemisia dracunculus*
11 Salad Rocket *Eruca vesicaria* subsp. *sativa*
12 Cowslips *Primula veris*
13 Spearmint *Mentha spicata*
14 Chicory *Cichorium intybus*
15 Lemon Thyme *Thymus citriodorus*
16 Garlic Chives *Allium tuberosum*
17 Red Orach *Atriplex hortensis* var. *rubra*

medicinal herb garden

I would like this garden, not just for its medicinal use, but for the tranquillity it would bring. The choice of herbs is not only for internal use but for the whole being. I can imagine sitting on the seat watching the dragonflies playing over the pond. Some of the herbs included are certainly not for self-administration, for instance blue flag iris, but this is a beautiful plant and would look most attractive with the meadowsweet and the valerian. Chamomile, peppermint, dill and lemon balm are easy to self-administer with care as they all make beneficial teas. One should not take large doses just because they are natural, as some are very powerful. I strongly advise anyone interested in planting this garden to get a good herbal medicine book, see a fully trained herbalist, and always consult your doctor about a particular remedy.

1 Blue Flag Iris *Iris versicolor*
2 Meadowsweet *Filipendula ulmaria*
3 Valerian *Valeriana officinalis*
4 Horseradish *Armoracia rusticana*
5 Sage *Salvia officinalis*
6 Lady's Mantle *Alchemilla mollis*
7 Rosemary *Rosmarinus officinalis*
8 Dill *Anethum graveolens*
9 Chamomile *Chamaemelum nobile*
10 White Horehound *Marrubium vulgare*
11 Comfrey *Symphytum officinale*
12 Feverfew *Tanacetum parthenium*
13 Heartsease *Viola tricolor*
14 Lemon Balm *Melissa officinalis*
15 Garlic *Allium sativum*
16 Black Peppermint *Mentha* x *piperita*
17 Fennel *Foeniculum vulgare*
18 Pot Marigold *Calendula officinalis*
19 Lavender Seal *Lavandula* x *intermedia* 'Seal'
20 Garden Thyme *Thymus vulgaris*
21 Houseleek *Sempervivum tectorum*

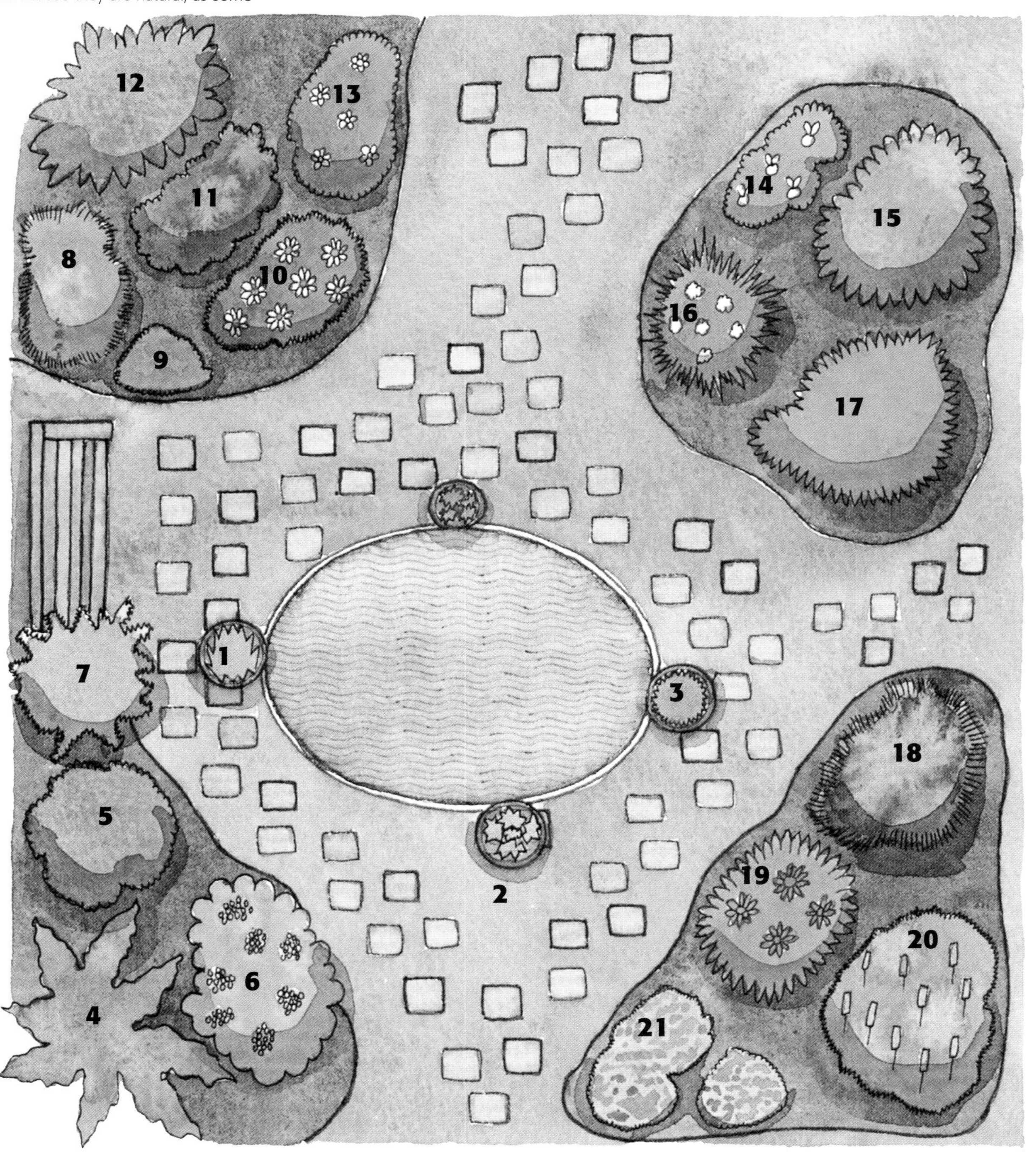

PLANNING THE FRUIT GARDEN

Planning your fruit garden means deciding your priorities – what do you want most? A little thought beforehand can save you a lot of wasted effort and ensure you actually get what you are after. With vegetables and bedding plants we have the luxury of burying our mistakes annually; with our trees and bushes we need to be more certain.

Although most of us get our garden fortuitously with our house, we usually have quite a wide choice of what we actually do with it, though the tendency is rarely to make radical changes. However, if we spend as much time and effort planning and remaking the garden as we do on decorating and furnishing the rest of our home it will turn out a mighty fine place!

Obviously the soil, climate, large trees, buildings and the rest of the hard landscape have to be worked around. But with skill and cunning, and modern materials, we can have almost any fruit we desire. Which fruits we actually choose to grow must depend on our budget as much as our climate. Obviously growing fruit in the open garden is easiest and cheapest, but a heated greenhouse allows for growing many more.

I think the first criterion for choosing fruit must be taste. After all, if you are growing for yourself, there is no point having poorly flavoured varieties or ones that are widely available commercially. Go for those with flavour and sweetness, even if they are poor croppers. If you find you especially like a particular kind, you can always grow more.

The Greater Plaited in his native surroundings

Vine arches add a productive aesthetic

Freshness is invaluable and one's own fruits are the most truly fresh. It makes sense to choose fruits and varieties that are best eaten straight off the plant and thus rarely found in shops. Likewise dessert are preferable to cooking varieties as we eat them with all their vitamins and flavour while culinary fruits lose some in the cooking.

Of course, growing your fruit yourself guarantees freedom from unwanted chemical residues, and applying plentiful compost will ensure a good internal nutritional balance in the fruit. However, when choosing fruits, bear in mind that their dietary value can vary as much with variety as with type or growing conditions. For example, **Golden Delicious** apples contain a third or less vitamin C than **Ribston Pippin**, while **Laxton's Superb** has only one sixth!

Economy must always be considered. Fruit growing requires higher investment initially than vegetables but running costs are lower. Similarly, soft fruit plants are cheaper than tree fruits individually, but require netting from birds in many areas. For maximum production, tree fruits produce as much weight from fewer plants per acre and usually require less maintenance, but are slower to crop, and live longer. Likewise vine fruits and cordons require more posts, ties and wires than trees or bushes. Any form of greenhouse or cover is costly, requiring upkeep, and of course heating uses much expensive energy.

The seasonal implications, and the time taken to maintain different fruits need to be considered, although in general fruit requires much less labour per yield than vegetable production. Initially the preparation and planting are heavy demands on time and energy but afterwards the workload is light, for the amateur if not for the professional. Fruit trees and bushes generally need mulching, thinning, picking and pruning, which are all light tasks and can be done upright in pleasant conditions. Growing the fruit is only half the battle, though. After picking we need to process and store the fruit and this takes more time than the growing! Don't plan to grow fruits that mature just when you go away on your annual holiday!

ORCHARDS

Orchards are devoted to the production of top or tree fruits, most frequently, privately and commercially, for apples, pears and plums. Other fruits are less common, only being widely grown in suitable areas – for example, cherries are little grown in wetter maritime regions but are common in drier zones such as Kent, in south-east England.

Orchards were always grassed down for convenience and soil preservation, though recent practice has been for bare cultivation, but this has been shown to be irresponsible. Initially weed-free conditions must be maintained in the orchard, but in general the preservation of bare soil is counterproductive. Most home orchard fruits will be best with a heavy mulch, ground cover or companion plants, or a grass sward.

For comfort and least cost, private orchards are still customarily planted with standard or half-standard trees on strong rootstocks. Though more intensive plantings with more dwarfing stock are more productive, they also require much more pruning and training and are difficult to mow underneath!

Grass clippings are an excellent mulch if put on in thin layers, and are a good fertiliser if returned to the sward. However, orchards, and indeed most other swards, should not be composed solely of grasses, as these compete strongly for resources in the topmost layers and do not contribute much to the mineral levels. Include clovers, alfalfa and chicory seeds in your sowing mixture. Special blends are now available ready mixed. Traditionally orchards were combined with grazing livestock. Though no longer done on a commercial scale, the practice is still useful for amateurs. Running chickens underneath adds interest and fertility, and almost guarantees freedom from most pest problems. Cynics might add, 'especially if you do not overfeed them!' Ducks control slugs and snails better than hens and do not scratch and damage so much, and the drakes do not crow so they are preferable from that point of view. Geese are superb lawnmowers, converting grass into fertiliser and eggs as well as being the most noisy watchdogs. Be warned, though, as geese may severely damage young plants with thin bark if they are hungry.

Other forms of livestock are more dangerous to the home orchard. I suggest that no four-legged herbivorous animals are allowed anywhere near valued plants, unless each is individually and securely fenced and protected.

Orchards are also an enticement to two-legged rats and it has long been established best practice to surround them with a thick, impenetrable hedge of thorny plants such as quickthorn or blackthorn. This, and a narrow verge of long grass and native plants, simultaneously provides a good background ecology to help control all the other pests of the orchard. A fence or wall does not contribute in the same way, and is usually more expensive and surmountable.

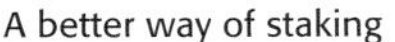

A better way of staking

ORNAMENTAL FRUIT GARDENS

A Doyenne du Comice pear espalier is also decorative

Trained fruit trees and bushes which can be bought ready-made in a host of interesting and architectural forms, such as espaliers and fans, create interest out of season as the framework of the plants becomes revealed. Attention to the appearance of the supports is essential as they are also disclosed for much of the time. Neatness and uniformity are thus of the greatest moment. Pergolas can look as beautiful clothed in grapevines as they do with any climber and even a fruit cage can be fashioned ornamentally from the right materials.

It is important in a fruit garden to allow for plentiful light and air – more than perhaps might be granted with a shrub garden. Wide paths aid such design and can be of grass sward where wear is light. Gravel is next choice for practicality and economy, and concrete or stone flags where the area is small or budget large.

More colour and interest, with benefit to the main planting, is obtained by having suitable companion plants to provide shelter, ground cover, flowers for nectar and pollen for the beneficial insects, and sacrificial plants that give up their fruits that others may not be eaten. However, vegetables are not easily mixed in. They do not grow well surrounded by vigorous competitors such as fruiting plants, but most of the culinary herbs can be grown to advantage and use.

Wildlife Fruit Gardens

Although these may be as stylised and neat as any purely decorative garden, an accurate description of many so-called wild gardens is 'unkempt'. Indeed, the term is often used to justify total neglect.

However, if the aim is truly to provide more and better habitats for endangered native flora and fauna, then neglect is not enough. A wild garden needs to be managed so that we maximise the number and forms of life supported. The more fruiting and berrying plants we include, the more wildlife we attract, and we need to ensure other basic necessities for wild creatures.

The fruit helps immensely but shelter for nesting and hibernation, water and peace are also required. Dense brambles, shrubs and evergreens are mandatory, but do ensure the gardener can still gain access. Paths should be maintained to permit various tasks, but of course excess traffic will soon drive away most creatures.

Many garden soils are too rich for the more appealing wild flowers. To establish these it is frequently necessary to start them in pots and plant them out into sites prepared by removing the turf. Thus they should be kept away from the fruiting plants which require richer conditions.

Fruits do well with flowers underneath

SITE, SOIL PREPARATION AND PLANTING

Most old gardening books started off with instructions to make a garden on a well-drained, south-facing slope of rich loamy soil. If only we had such a choice! We must often take what comes. And as we get small gardens with modern houses we rarely have much choice of positioning within them.

site

Shelter is the most important aid we can give our plants – good hedges, fences, windbreaks, warm walls, cloches, plastic sheets and even old curtains on frosty nights. But be careful not to overdo it and make the area stagnant. Drainage is occasionally necessary to prevent waterlogging, as few plants survive for long with drowning roots. But for most gardens, water is more often a problem in its absence. Growing on raised mounds is ideal for draining away the water in areas with wet summers.

soil

For most crops, a neutral to slightly acid soil with moderate fertility and organic matter content is ideal. There are basically four types of soil.

Chalk or Limestone Soil

This soil tends to be light and very well drained. But its inability to hold moisture can cause problems in a hot summer. It is alkaline in character and it is sometimes difficult to lower its pH level, so some plants become stunted and leaves go yellowish in colour, because the minerals, especially iron, become locked away. If you find the plants are not thriving, try a raised bed where you can introduce the soil they require.

Clay Soil

This soil is made up of tiny particles that stick together when wet, making the soil heavy. When dry, they set rock hard. Because it retains water and restricts air flow around roots, it is often known as 'a cold soil'. It may have a natural reserve of plant food, but even so it is better to work compost, sharp sand, gypsum (often sold as clay breaker) and horticultural grit into the top layer as this will help get the plants established and improve drainage. If you continue to do this every year, it will gradually become easier to cultivate, and extremely productive. Your efforts will be well rewarded.

Loam Soil

This soil is a mixture of clay and sand. It contains a good quantity of humus and is rich in nutrients. There are various types of loam: heavy, which contains more clay than loam and becomes wet in winter and spring; light loam, which has more sand than clay; and medium loam, which is an equal balance of clay and sand.

Sandy Soil

This is a very well-drained soil, so much so that plant foods are quickly washed away. A plus point is that it warms up quickly in the spring so is ideal for early crops. For hungry plants you will need to build the soil up with compost to help retain moisture and stop the leaching of nutrients. Plants such as the Mediterranean herbs and carrots will thrive on this soil.

Soil Acidity and Alkalinity

The natural acidity of the soil must be taken into account. Find out whether your soil is acid or alkaline using a reliable soil-testing kit, available from any good garden centre. Soil acidity and alkalinity is measured on a pH scale ranging from 0 to 14; 0 is the most acid and 14 the most alkaline.

Acid soils (0–6.5 pH), on a sliding scale from acidic to nearly neutral, include sphagnum moss peat, sandy soil, coarse loam soil, sedge peat and heavy clay. Alkaline soils (7.7–14 pH) tend to contain chalk or lime and may be thin or shallow.

A reading that approaches either end of the pH scale indicates that the soil will tend to lock up the nutrients necessary for good growth. If it is very acid, you will need to add lime in the late autumn to raise the pH. Fork the top 10cm (4in) of the soil and dress with lime. Clay soils need a good dressing, but be careful not to over-lime sandy soils. It should not be necessary to do this more than once every 3 years unless your soil is very acid. Never add lime at the same time as manure, garden compost or fertilizer, as a chemical reaction can occur that will ruin the effects of both. As a general rule, either add lime 1 month before, or 3 months after manuring, and 1 month after adding fertilizers. If the soil is alkaline, dress it every autumn with well-rotted manure to a depth of 5–10cm (2–4in), and dig over in the spring.

Soil pH also influences the number and type of beneficial soil-borne organisms and the incidence of

Testing your soil's acidity

pests and diseases; worms dislike a low pH, but leather-jackets and wireworms are usually more common in acid conditions, as is the fungal disease clubroot.

Some plants and plant families have quite specific pH requirements; others are more tolerant. On the whole, we have unwittingly selected plants that grow happily in the average, mildly acid to slightly alkaline soils most of us have. Herbs and vegetables generally do best in a soil between 6.5 and 7.5 pH (fairly neutral). A few fruits such as blueberries need acid conditions. They resent lime in the soil and need to be cultivated in pots of ericaceous compost and watered with rain, not ground water. Brassicas benefit from a slightly alkaline soil to reduce the risk of clubroot.

preparation

The more you plan and the more carefully you prepare the site, the better the results and the more pitfalls avoided. Always work out schemes on paper. Draw a map of existing features and plan how you will fit in the new. For vegetables, plan your rotation (see page 637). For fruits and perennial herbs, imagine walking round after five years when all the plants have grown up.

A good soil structure can be easily damaged by cultivating when it is excessively wet or dry and by using heavy machinery – even walking on wet soil causes compaction and smearing. To reduce the impact, lay a plank on the soil to disperse the weight. Once the soil is under cultivation, aim to stick to the path rather than tread on the soil.

Eliminating Weeds

If you don't weed thoroughly before you start, it will become much harder once the plants go in. Most weeds do not emerge until the spring, and continue well into the season. If you wish to get rid of them successfully, it is best to cover the plot with black plastic the previous autumn. Dig a shallow trench around the area and bury the sides of the plastic in it to keep it secure. Starved of light, the weeds will eventually give up. This will take until the following spring, so if you do not like the sight of the black plastic, position a few paving stones (taking care not to make any holes because the weeds will find them and come through), cover the plastic with bark, and place terracotta pots planted up on top of the stones.

Alternatively, fork through the plot and remove weeds by hand. If you do not have the patience to use a plastic mulch, you can dig over the area, pulling out weeds by the roots. This works most easily when the soil is just moist, which is most often in autumn. With less patience still, some may resort to a herbicide to kill all the weeds initially. There are no organic herbicides. We disapprove, but do understand the dilemma. If you consider you really need to, then purchase whatever a reputable supplier recommends and follow the instructions religiously! Even so, results will not be instant and may need repeat applications. There is no short cut to eradicating weeds.

Soil Cultivation

The soil structure essential for successful growth is created by incorporating different materials into the soil:

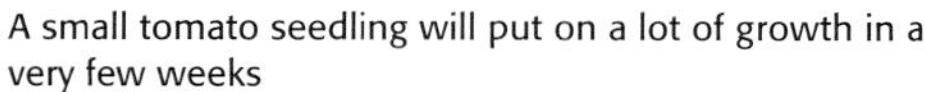

A small tomato seedling will put on a lot of growth in a very few weeks

well-rotted compost improves moisture and nutrient retention, creates an open structure and aerates the soil; sharp sand helps drainage; and 'green manures' improve the structure.

Digging is the most important method of cultivation prior to planting. In the vegetable garden it is an annual event. On heavier soils in cooler climates autumn digging is beneficial as the winter frosts that follow help to break down these wet, sticky soils into a workable tilth. Sandy soils can be lightly forked over and it is advisable to cover them with compost or green manure to reduce the loss of nutrients during the winter rains, or to leave the digging until spring. No matter what type of soil you have, it is beneficial to lay a good layer of well-rotted manure or garden compost over the surface in winter and dig it in the following spring. You can also use the 'no-dig' method for vegetables: dig over the soil, remove weeds, then add organic matter on the surface annually and let nature build the structure.

Single digging consists of cultivating the soil to about 30cm, a good spade's depth and the average rooting zone of many plants. Double digging is used on poorly drained and previously uncultivated land. It encourages deep rooting and promotes rapid drainage. 'Double' refers to the depth of cultivation, some 45–60cm deep or approximately two spades' depths. When double digging it is important to ensure that the topsoil and subsoil are not mixed.

There should be no need to add extra fertilizers to the soil if it has been well-prepared. Too much fertilizer causes lank, soft, disease-prone growth.

If necessary, in the vegetable garden you can gradually improve the soil by planting a ground-breaking crop such as potatoes or Jerusalem artichokes. Potatoes are so good for this because they need a number of cultivations to 'earth up' and cover the tubers, so preventing weeds from becoming established. Potatoes also produce large, leafy tops, blocking the light that is so essential for the germination of many weeds.

planting

If this is a new garden, and you are planting fruit trees and bushes, herbs or perennial vegetables, it is worth laying the plants out on top of the prepared ground first, walking round and getting an overall view. Make any changes to the design now, rather than later. For details on buying plants, see page 647.

Any stakes or supports must be in place beforehand, and for trees and bushes they must be strong enough to do the job and must last for a reasonable number of years. It is false economy to be mean on these, as it will be hard to correct when the plants are grown.

Remember that once the planting is done, the future is determined. So don't skimp on digging or preparation. Dig a wide planting hole: bigger is better! Don't mix in compost, manure or fertiliser: apply later once plants have established. Don't bury plants too deeply. Almost all wish to be planted at the same depth as they have grown. For grafted fruit trees, don't bury the rootstock.

Keep roots in their respective and different layers. Don't force them doubled up into a cramped hole, and never pack them all down in a flat layer unless they grew like that. Gently pack soil around roots, filling and firming as you go. It is better to over-firm! Then attach support if needed.

aftercare

For fruit trees and bushes, herbs and perennial vegetables, use a mulch to retain moisture and suppress weeds. Try a thick layer of organic matter or a plastic sheet or carpet mulch. In the fruit garden, whatever the final intention, do not allow a weed or grass within a circle as wide as the tree or bush is high for three years.

Good watering in dry spells is absolutely crucial! For long-term plants, water regularly for three years.

Add plenty of well-rotted organic matter to the soil

CROP ROTATION

Rotation is a system whereby groups of vegetables are grown on a different section of the plot each year, so maintaining the balance of soil nutrients for successive crops. Growing crops in this way avoids the build-up of pests and diseases, assists in weed control and prevents the soil from deteriorating.

For crop rotation to be effective, a large area of ground is required, particularly when controlling soil-borne pests and diseases. White rot (which attacks the onion family), clubroot (which damages members of the cabbage family) and potato cyst eelworm remain dormant in the soil for many years and can survive on any weeds that are relatives. This makes good husbandry just as vital as crop rotation.

Planning

Before planting, first list the vegetables you want to grow and group them together according to their botanical relationship. Allocate each group to a plot, then compile a monthly cropping timetable for each space.

Rotational groups	*Legumes*	*Onion family*	*Carrot and tomato families*	*Brassicas*
	Broad bean	Bulb onion	Carrot	Cabbage
	French bean	Garlic	Celery	Cauliflower
	Pea	Leek	Pepper	Radish
	Runner bean	Salad onion	Parsnip	Swede
		Shallot	Potato	Turnip
			Tomato	

Next, allocate each rotation group to a plot of land ('plot A' etc), and draw up a month-by-month timetable for each space.

Plot A	*Year 1*	*Year 2*	*Year 3*	*Year 4*
	Legumes	*Onion family*	*Carrot and tomato families*	*Brassicas*
	Broad bean	Bulb onion	Carrot	Cabbage
	French bean	Garlic	Celery	Cauliflower
	Pea	Leek	Parsnip	Radish
	Runner bean	Salad onion	Pepper	Swede
		Shallot	Potato	Turnip
			Tomato	

Plot B	*Year 1*	*Year 2*	*Year 3*	*Year 4*
	Onion family	*Carrot and tomato families*	*Brassicas*	*Legumes*
	Bulb onion	Carrot	Cabbage	Broad bean
	Garlic	Celery	Cauliflower	French bean
	Leek	Parsnip	Radish	Pea
	Salad onion	Pepper	Swede	Runner bean
	Shallot	Potato	Turnip	
		Tomato		

Plot C	*Year 1*	*Year 2*	*Year 3*	*Year 4*
	Carrot and tomato families	*Brassicas*	*Legumes*	*Onion family*
	Carrot	Cabbage	Broad bean	Bulb onion
	Celery	Cauliflower	French bean	Garlic
	Parsnip	Radish	Pea	Leek
	Pepper	Swede	Runner bean	Salad onion
	Potato	Turnip		Shallot
	Tomato			

Plot D	*Year 1*	*Year 2*	*Year 3*	*Year 4*
	Brassicas	*Legumes*	*Onion family*	*Carrot and tomato families*
	Cabbage	Broad bean	Bulb onion	Carrot
	Cauliflower	French bean	Garlic	Celery
	Radish	Pea	Leek	Parsnip
	Swede	Runner bean	Salad onion	Pepper
	Turnip		Shallot	Potato
				Tomato

This timetable fully utilises the land and provides continuity of cropping. As space becomes available, plant the crops due to follow immediately. Crops may come from different groups, which means that rotation from one plot to another is a gradual process, rather than a wholesale changeover on a set date. Wind-pollinated sweetcorn should be sown in blocks in any remaining space on the plot, bearing in mind it casts shade. Ensure it is not planted in the same place in successive years. Courgettes, squashes, cucumbers and marrows can be treated similarly.

GROWING SYSTEMS

Rows
Traditionally, vegetables are grown in long straight rows, with plants close together within the rows, and paths to allow access. This has some drawbacks. Competition between plants for space in a row means that much of the plants' extension growth is into the pathways, causing leafy vegetables like cabbages, cauliflowers and lettuces to produce oval 'hearts' rather than round ones.

Beds
These are effectively multi-row systems, with equidistant spacing of plants in and between the rows. Plants may be set in staggered rows, creating a diagonal pattern. Major paths between beds should always be wide enough to accommodate a wheelbarrow; access paths can be narrower. Pathways between the beds are slightly wider than those on the row system, but closer plant spacing means that more plants are grown per square metre and their growth and shape are more uniform. Close spacing ensures that weed growth is suppressed in its later stages and the soil structure remains intact, because there is less soil compaction when pathways are further apart.

Raised beds should be about 90cm wide so that it is easy to reach into the centre without overbalancing. After initial double digging, and the addition of organic matter, they will not be trodden on, allowing a good soil structure to form. Layers of well-rotted compost added annually will be further incorporated by worm activity. Where natural populations are low, they can be bought from specialist suppliers. You can't have enough!

Bed system, with paths separating them

Random Planting
Vegetables can be grown as 'edible bedding plants' scattered among flowering plants and borders in the ornamental garden. Swiss chard, cabbage 'Ruby Ball', beetroot 'Bull's Blood', fennel and Brussels sprout 'Rubine' are particularly pleasing. (See also pages 618–619, 'The Ornamental Vegetable Garden'.)

The Potager
This is the French tradition of planting vegetables in formal beds; a practical and visually pleasing display.

Long-term Crops
Perennial vegetables such as asparagus are difficult to incorporate into a crop rotation programme. For this reason, they are often grown on a separate, more permanent site for several years before being replaced.

What shall I harvest today?

Any space can be utilised

Sowing for Continuity

The greatest challenge is to have vegetables ready for harvest throughout the year. Some, such as asparagus, have only a short harvesting period, while vegetables like cabbage, cauliflower and lettuce are available at almost any time, provided that there is successional sowing. Careful selection of cultivars also helps to achieve continuity; growing both rapidly maturing and slow-growing types extends the cropping period. The season for many root vegetables such as parsnips can be lengthened by growing a portion of the crop for eating fresh, and the remainder for storage. These can slowly be eaten over the winter period.

Successional Sowings

It is usually quick-maturing crops that are prone to gluts and gaps, often bolting and becoming inedible. To avoid this, sow sufficient seed for your needs a little and often, so that if something goes wrong, the problem is only a small one. A good guide is to sow a batch of each seed when the first 'true' leaves of the previous sowing begin to emerge.

Intercropping

Spaces between slow-growing crops can be used as a seedbed for vegetables such as brassicas that will later need to be transplanted into permanent positions at wider spacings. This is only successful where the competition for moisture, nutrients, light and space is not too great. This can be partially solved by growing them between rows of deep-rooted crops, so that the roots are in different levels of the soil. To achieve this, it may also be necessary to space crops a little further apart than the usual spacings. Radishes sown on the same day as parsnips will have germinated, grown and been harvested before the parsnips have fully developed, achieving the maximum possible yield, covering the soil with vegetation and suppressing weed growth. This intercropping requires some flexibility in the cropping plan.

Catch Crops

These are rapidly-maturing vegetables and include radishes, lettuce and salad onions, all of which can be grown before the main crop is planted. Catch cropping is useful when growing tender crops such as tomatoes, sweetcorn and courgettes, which cannot be planted until the risk of late frost is well and truly over.

Extending the Season

The season for vegetables can be extended by growing them under cover at the beginning or end of their natural season; sowing early under cover hastens maturity while protection later in the season extends the cropping period. (For more detail, see pages 342–643, 'Container Culture' and 653, 'Protected Cropping'.)

Pot-raised Plants

Raising plants in pots or modules is useful for those that are frost tender and allows plants to mature while the ground is cleared of earlier crops. This saves time and space and gives more control over the growth and development of young plants. Brassicas, lettuces, leeks and onions can be grown in this way.

'The Hungry Gap'

Few vegetables mature from late winter to early spring – a period often referred to as 'the hungry gap'. Careful planning allows this gap to be bridged; spring cauliflower, winter cabbage, celeriac, kale, leeks and oriental vegetables can all be grown for harvesting during this period.

Raised beds are highly productive

POLLINATION AND COMPANION PLANTING

Almost all our fruits require their flowers to be pollinated. Many are self-fertile and can set by pollinating themselves, but crop better if cross-pollinated. Some of these – for example figs – have varieties that will produce fruit parthenocarpically, without pollination; thus they are often seedless. A few fruits, such as Conference pears, are partly parthenocarpic: if not pollinated by another variety, their fruits are different and oddly shaped.

Some fruits do not have male and female flowers on the same plant and so we have to grow a non-fruiting male to pollinate every half dozen or so females. Wild kiwis are good examples. For convenience, we have bred varieties that carry both sexes.

Some fruits and many nuts are wind-pollinated, others by bees and other insects. Under cover, neither of these natural pollinators exists and we have to assist. A rabbit's tail, a cotton ball or piece of wool lightly touched on each flower – moving pollen from flower to flower – should suffice. Lure more insects to help by growing attractant companion plants and this can add a bit of colour and life as well.

Outdoors early in the year there are few insects about and for the earliest flowerers pollination is risky. Again, hand pollination is effective, but tedious. It is better to ensure insect pollination by increasing the numbers. Planting companion plants as attractants helps; taking up bee-keeping makes an enormous improvement.

To pollinate each other, not only do varieties have to be compatible, but they also have to flower at the same time. All reputable catalogues have a choice of suitable cross-pollinators indicated. When in doubt, choose the wild species, which is often the best pollinator for most of its varieties. Alternatively, simply plant more varieties. This is especially true for a few fruits which are not good pollinators themselves. For example, the apples Cox's Orange Pippin and Bramley's Seedling never crop together, but if you add a James Grieve, all three fruit.

Companion planting is of immense benefit to most fruiting plants as it brings in and supports pollinating and predatory insects, and maintains them throughout the rest of the year. We must aim at having a continuity of flowers throughout the year as it is these that provide nectar and honey for the bees, hoverflies and other beneficial insects.

Of particular usefulness is *Limnanthes douglasii*, the poached egg plant, which is a low-growing, self-seeding, weed-suppressing, hardy annual. I find it especially

There are three water butts hidden here

beneficial under soft fruit and gooseberries. *Phacelia tanacetifolia*, *Convolvulus tricolor*, pot marigolds and clovers are all very good for beneficial insects.

Other plants can be good companions to our fruiting plants by repelling pests. French marigolds are one of the strongest and their smell will keep whitefly out of a greenhouse. Planted about the fruit garden, they disorientate pests sniffing for their quarry. The aromatic herbs such as rosemary, thyme, sage, southernwood and lavender are all excellent companions as they not only camouflage the air with their perfume but they are long-flowering, also benefiting the various pollinators and predators.

Alliums are valuable companions. Their smell deters many pests, yet their flowers attract beneficial insects. They help protect the plants they grow with from fungi and are even claimed to improve the scent of roses. Garlic and chives are the easiest to grow and use in quantity at the base of the most fruit trees and bushes.

Many companion plants help access nutrients and make them available for our fruiting plants. Clovers, lupins and the other leguminous plants are especially good as they fix nitrogen from the air and the surplus

Strawberries hitch a ride with the grapes and peaches

Hand pollinating should be called brush pollinating, really

will feed our crops. Alfalfa/lucerne is exceptionally deep-rooted and brings up minerals from depths other plants cannot reach. It should be included in the seed mixture for orchard and wild garden swards. Similarly a few thistles and docks can be tolerated in those areas for the same purpose, but cut off flowers before they set seed. Clovers and chicory should be included whenever grass is seeded as the sward that is produced will be both richer and lusher.

In grass at the base of trees, dense shrubs and hedges, bulbous, spring-flowering plants can be fitted in with no difficulty. These areas can remain uncut till the bulbs' foliage dies down and their flowers will be a rich source of pollen and nectar for the earliest pollinators and predators. Some ivy in the hedges will provide late flowers to feed the insects in autumn before they hibernate.

Ground-cover plants provide habitats for ground beetles and many other useful creatures, but may also harbour slugs, snails and over-wintering pests. On the whole it is better to have such habitats kept as far as possible from soft fruit and seed beds – after all, the beetles, frogs and hedgehogs can walk further than slugs and snails!

Some plants are specifically helpful to others and a few hinder. These are noted with the entry for that particular fruit. For example, the Romans noticed that oaks were bad for olives, and cabbages were not good companions for vines, but rue benefited from the presence of figs.

CONTAINER CULTURE

Growing in containers is an ideal technique for fruits and vegetables that might otherwise be difficult, enabling you to control the size of the plant and conditions in which it is kept. Almost all herbs can be grown in containers. The individual plant entries give details of the best size of container for each crop and where to place it to receive the ideal growing conditions. This section provides more general information and advice on container growing.

Advantages of Containers

With attention to feeding and watering, almost any plant can be coaxed to grow in a container. Where room is limited, containers offer an alternative to ground space. They can also offer different condtions to those found in the ground. Tomatoes and cucumbers have long been grown under glass in containers to prevent any contact with disease-infected soil. If you have a chalky soil, lime-hating plants can be grown in containers of ericaceous compost and watered with rain water.

Portable containers allow you to respond to weather conditions, moving plants outdoors once spring frosts end, and putting them under shelter again at the onset of colder weather.

For large fruit plants, a container cramps the root system, preventing them growing too large or too quickly. Allied with pruning, this allows us to dwarf plants we would find too large to handle. Often this restriction is resented by the plant and, perversely, may result in earlier fruiting. The fruits may not be prolific, but a selection of containers can give several different varieties in a space otherwise occupied by one, allowing greater variety and a longer season.

Vegetables most suited to container culture are either rapidly maturing crops such as mini beetroot, carrots, lettuces, radishes and salad onions, or dwarf varieties of bush or climbing vegetables, including aubergines, beans, cucumbers, peas, peppers and tomatoes, as these need little support. Deep-rooted vegetables like sprouts, maincrop carrots and parsnips can be grown in containers at least 45cm (18in) deep.

Choosing a Container

Choose the container to suit the plant. If it is a tall plant, make sure the container has a base wide enough to prevent it toppling over, even outside in a high wind. For vegetables, the chance of successful cropping is improved with large containers; they allow for a greater rooting depth, dry out slowly and provide adequate anchorage for crops that need staking.

Plastic containers (with drainage holes) are more moisture-retentive than those of wood or terracotta, but the latter are preferred by some plants, such as citrus.

If using unconventional containers – old watering cans, sinks, a half beer barrel – make sure they have drainage holes, and gravel or broken pots in the bottom of the container to stop the holes clogging up.

Herbs can suit hanging baskets, but the position is crucial. They dislike high wind and full sun all day. Also, they are mostly fast growers and if too cramped or over- or under-watered they will drop their leaves.

Composts

Choosing the right compost is essential for healthy plants. Composts for containers are either loam-based (soil-based) or loamless.

John Innes

This is loam-based and includes chemicals. On the plus side, it stays richer in nutrients longer, making feeding less critical; it holds water well and is a stable compost for a large, long-term plant. If it does dry out, it takes up water easily. The major watch point is that it is easy to over-water.

Peppers do well in containers and need little support

There are usually three different grades of compost, though some manufacturers combine No.1 and No.2. No.1 is for growing rooted cuttings, No.2 for seedlings and No.3 for final potting. The numbers indicate the amount of nutrients. You must choose the correct one for the job. Loam-based composts are generally ideal for vegetables and fruits in containers, and organic matter can be added if required. This is not a suitable compost for hanging baskets, because it is heavy when wet.

Own Soil-based Mix
If you wish to make a soil-based compost of your own this recipe is fairly reliable.
4 parts good weed-free garden topsoil
3 parts well-rotted garden compost
3 parts moist peat
1 part horticultural sharp sand

Multipurpose Potting Compost
This is usually peat-based with added nutrients. The bags are light and the compost is clean and easy to use. On the negative side, you will need to feed regularly as nutrients are soon depleted; also the compost is light, so watch out that large plants do not fall over. The most frequent problem is that it takes up water poorly if ever it dries out.

Own Mix Bark, Peat, Grit
This is ideal for many herbs as well as some fruits and vegetables. The open mix helps prevent over-watering; the bark retains water, which protects against under-watering and keeps the compost open to help absorb water if ever it dries out completely. It is suitable for containers and hanging baskets alike. Another plus is that you know what nutrients are in it, so will be able to feed in a balanced way.

Alternative Composts
Peat-free composts are increasingly available as we become aware of the need to conserve our diminishing peat fields. They are generally made of coir, a by-product of coconuts, or composted bark. Coir is very free-draining and so some composts include a jelly that retains water, releasing it gradually. Coir compost is light, so there may be a stability problem with tall or large plants. A final minus point – the nutrients are soon depleted so you will need to feed from the start.

Composted bark has similar drawbacks if used straight, with no peat or soil: watering and nutrient loss. If the bark has not been composted for long enough, it can leach the nutrients and starve the plants. However, mixed with peat or soil, bark is a great asset.

Apricots can be cropped in big buckets

Maintenance

Most plants need moist soil but drown if waterlogged and wilt if dry. Water stress due to drought, even for short periods, can greatly reduce the potential productivity of plants. If you cannot maintain consistent watering, install an automatic system of irrigation. It is best to water containers in the early morning or the evening, especially in summer, as this reduces the amount lost through evaporation.

Repotting into a larger container helps provide new nutrients, but this may become difficult with larger plants. The alternatives are top-dressing with an enriched mix of compost and organic fertilizer, or feeding little and often with a diluted liquid feed such as seaweed, or comfrey and nettle extract.

Liquid seaweed contains small amounts of nitrogen, phosphorous, potassium, and it is also rich in trace elements. It not only makes a good soil feed but, as the elements are easily taken in by the plant, it can also be sprayed on as a foliar feed.

Maintenance Calendar

Spring Pot on perennial plants if necessary (look for roots protruding from the bottom of the container). Use a pot the next size up. Carefully remove the plant from its old pot. Give it a good tidy up, removing any weeds and dead leaves. Place gravel or other drainage material in the bottom of the container and keep the compost sweet by adding a tablespoon full of granulated charcoal. As soon as the plant starts producing new growth or flowers, start feeding regularly with liquid feed. Prune fruits as needed. Plant pots of ornamental vegetables.

Summer Keep a careful eye on the watering; make sure the pots do not dry out fully. Move some plants out of the midday sun. Deadhead any flowers on herbs. Feed with liquid feed, on average once a week. Remove any pest-damaged leaves. Prune fruits as needed. Harvest vegetables.

Autumn Cut back the perennial herbs. Weed containers and at the same time remove some of the top compost and re-dress. Bring any tender plants inside before the frosts. Start reducing the watering. Prune fruits as needed.

Winter Protect all container-grown plants from frosts. If possible move into a cold greenhouse, conservatory or garage. If the weather is very severe, cover the containers in a layer of sacking. Keep watering to a minimum. Prune fruits as needed.

PROPAGATION AND BUYING PLANTS

One of the great joys of gardening is propagating your own plants. Success is dependent on adequate preparation and the care and attention you give in the critical first few weeks. This section provides general instructions for the main propagation methods. Individual guidelines are given under each plant entry. Where space and time are short, or for larger perennial plants such as fruit trees and bushes, it may be easier to buy your plants.

seed

Most vegetables, annual herbs and annual fruits are usually grown from seed. There is a huge range of seed available from seed catalogues, or you can save your own. Seed is not always advisable for perennials, particularly fruits; a fruit seedling will not fruit until mature, which may be quick for some but takes many decades for most trees. Until the fruit is produced there is usually no way of telling what it will be like – and it may not be very good. Most of the best varieties will not come true from seed.

Seed can be sown outside, either directly where it is to mature or into a seedbed for transplanting. Alternatively, starting off the seeds in a greenhouse or on a windowsill gives you more control over the warmth and moisture they need, and enables you to begin propagating earlier in the season.

Types of Seed
Seeds are usually bought as packets of individual 'naked' seeds; but vegetable seeds in particular may be available in other forms.

Pelleted seeds are individual seeds coated in a ball of clay, which moistens and disintegrates in the soil. They are easy to handle and sow at precise spacings, reducing the need for thinning. Sow at a depth of about twice their diameter and keep the soil moist but not waterlogged until germination.

Seed tapes and sheets are individual seeds encased at the correct spacing in tapes or sheets of tissue paper or gel. These substances are soluble when placed in moist soil. Tapes and sheets are quick to use and are precisely spaced so there is no need for thinning.

Primed seeds are in the first stages of germination and are used for those that are hard to germinate or need

Seedlings almost ready to be planted out

high germination temperatures, such as cucumbers. They arrive by post in plastic packets to prevent moisture loss and are pricked out into pots or trays on arrival.

Preparation of Seed
Most seeds need air, light, particular temperatures and moisture to germinate. Some have a long dormancy, and some have hard outer coats and need a little help to get going. Here are two techniques. See individual plant entries for when to use these techniques.

Scarification
If left to nature, seeds that have a hard outer coat would take a long time to germinate. To speed up the process, rub the seed between two sheets of fine sandpaper. This weakens the coat of the seed so that moisture needed for germination can penetrate.

Stratification (vernalization)
Some seeds need a period of cold (from 1 to 6 months) to germinate. Mix the seed with damp sand and place in a plastic bag in the refrigerator or freezer. After 4 weeks sow on the surface of the compost and cover with perlite.

Sowing Outside
In an average season the seed should be sown in mid to late spring after the soil has been prepared and warmed. It is simplest to sow direct where the plants are to mature, and this is best for plants that resent root disturbance, such as many root vegetables. Alternatively you can use a seedbed if you need the space seedlings are to occupy for something else, and to save space as small plants are not at their final spacing until they mature.

Before starting, check your soil type (see page 633), making sure that the soil has sufficient food to maintain a seedbed. Dig the bed over, mark out a straight line with a piece of string secured tightly over each row, draw a shallow drill, 6–13mm deep (this will vary according to seed size), using the side of a fork or hoe, and sow the seeds thinly. For larger seeds, sow individually. If your soil is sticky clay, give the seeds a better start by adding a fine layer of horticultural sand along the drill. Do not overcrowd the bed, otherwise the seedlings will grow leggy and weak and be prone to disease. For more precise details of seed spacing see individual plant entries

Protected Sowing

Sowing under cover is expensive and often labour-intensive, yet it allows seeds to be sown whatever the weather. It is most often used in cooler climates for tender crops that cannot be planted out until there is no longer any danger or frost.

Start with a thoroughly cleaned container. Old compost also provides ideal conditions for damping off fungi and sciarid flies, so remove any spent compost from the greenhouse or potting shed. Always use fresh compost.

Compost

It is best to use a sterile seed compost. Ordinary garden soil contains many weed seeds that could easily be confused with the germinating seed. The best compost for most seed sowing is 50 per cent propagating bark: 50 per cent peat-based seed compost. However, for plants that prefer a freer-draining compost, or for those that need stratification outside, a 25 per cent peat-based seed compost: 50 per cent propagating bark: 25 per cent horticultural grit mix is ideal. And if you are sowing seeds that have a long germination period, use a soil-based seed compost or buy ready-mixed.

Sowing in Seed Trays

Fill a clean seed tray with compost to 1cm below the rim and firm with a flat piece of wood. Do not press too hard as this will over-compress the compost and restrict drainage, encouraging damping off disease and attack by sciarid fly.

The gap below the rim is essential, as it prevents the surface-sown seeds and compost being washed over the edge when watering.

Water the prepared tray using a fine rose on the watering can so as not to disturb the seed. Do not over-water. The compost should be damp, not soaking. After an initial watering, water as little as possible, but never let the surface dry out. Once the seed is sown, lack of moisture can prevent germination and kill the seedlings, but too much excludes oxygen and encourages damping-off fungi and root rot.

Sowing Methods

There are three main methods, the choice dependent on the size of the seed. They are, in order of seed size, fine to large:

1 Scatter on the surface of the compost, and cover with a fine layer of perlite.
2 Press into the surface of the compost, either with your hand or a flat piece of wood the size of the tray, and cover with perlite.
3 Press down to one seed's depth, cover with compost.

The Cardboard Trick

When seeds are too small to handle, you can control distribution by using a thin piece of card cut to 10cm x 5cm, and folded down the middle. Place a small amount of seed into the folded card and gently tap it over the prepared seed tray. This technique is especially useful when sowing into plug trays (see below).

Sowing in Plug (Module) Trays (Multi-cell Trays)

These plug trays are a great invention. The seed can germinate in its own space, get established into a strong seedling, and make a good root ball. When potting on, the young plant remains undisturbed and will continue growing. This is very good for plants like coriander, which hate being transplanted and tend to bolt if you move them. Another advantage is that the problem of overcrowding is cut to a minimum, and damping-off disease and sciarid fly are easier to control. Also, because seedlings in plugs are easier to maintain, planting out or potting on is not so critical.
Plug trays come in various sizes; for example, you can get trays with very small holes of 15mm x 15mm up to trays with holes of 36.5mm x 36.5mm. To enable a reasonable time lapse between germination and potting on, the larger are recommended.

Prepare the compost and fill the tray right to the top, scraping off surplus compost with a piece of wood level with the top of the holes. It is better not to firm the compost down. Watering in (see above) settles the compost enough to allow space for the seed and the top-dressing of perlite.

The principles of sowing in plug trays are the same as for trays. Having sown your seed, label the trays clearly with the name of the plant, and the date.

Sowing in Pots

Multi-sowing in pots speeds the growth of root and bulb vegetables. Sow up to six seeds (two for beetroot) into a 7.5–10cm pot, leave the seedlings to develop and then transplant the potful of plants, allowing extra space within the rows. Thinning is not needed. This technique works well for beetroot, cauliflower, turnips, kohlrabi, leeks and onions.

Teasing two plantlets apart

Taking a hardwood cutting

Fluid Sowing

This is useful when weather and soil conditions make germination erratic. Seeds germinated under ideal conditions are sown, protected by a carrier gel. To germinate the seeds, place some moistened kitchen towel in the base of a plastic container. Scatter the seeds evenly on the surface, cover with a lid or cling film and keep at 21°C (70°F). When the rootlets are about 5mm (¼in) long they are ready to sow. Wash them into a fine mesh strainer. Mix carrier gel from half-strength fungicide-free wallpaper paste, scatter the seeds in the paste, and mix. Pour into a clear bag, cut off a corner and force the mixture through the hole. Cover the seeds with soil or vermiculite.

Seed Germination

Seeds need warmth and moisture to germinate. In a cold greenhouse, a heated propagator may be needed in early spring for seeds that germinate at warm to hot temperatures. In the house you can use a shelf near a radiator (never on the radiator), or an airing cupboard. Darkness does not hinder the germination of most seeds, but if you put your containers in an airing cupboard check them daily. As soon as there is any sign of life, place the trays in a warm light place, not in direct sunlight.

Hardening Off

When large enough to handle, prick out seed tray seedlings and pot up individually. Allow them to root fully. Test plug tray seedlings by giving one or two a gentle tug. They should come away from the cells cleanly, with the root ball. If they do not, leave for another few days.

When the seedlings are ready, harden them off slowly by leaving the young plants outside during the day. Once weaned into a natural climate, usually after 10–14 days, plant them directly to where they will mature.

Seed Storage and Viability

Once seed is harvested from the plant, it begins to deteriorate; even ideal storage conditions can only slow down the rate of deterioration. Viability (the ability to germinate) usually declines with age. Always use fresh seed, or store in cool, dark, dry conditions (definitely not the corner of the greenhouse) in airtight containers.

cuttings

Taking cuttings is the best way to propagate many fruiting plants and non-flowering herbs. This may be the only way to reproduce a particular variety or cultivar that will not come true from seed. For successful softwood cuttings it is worth buying a heated propagator, which can be placed in a greenhouse or on a shady windowsill. For successful semi-ripe, hardwood and root cuttings, a shaded coldframe can be used. For specific details, see under the individual entry.

Softwood Cuttings

Softwood cuttings are usually taken between spring and mid-summer, using the new, lush, green growth. To produce successful rooting material from herbs, prune the plant vigorously in winter to encourage new growth, and take cuttings as soon as there is sufficient growth.

Prepare a pot, seed tray, or plug tray with cutting compost – 50 per cent bark, 50 per cent peat. Firm the compost to within 2cm of the rim. Collect the cuttings in small batches in the morning. Choose sturdy shoots with plenty of leaves. Best results come from non-flowering shoots with the base leaves removed. Cut the shoot with a knife, not scissors. Place the cutting at once in the shade in a polythene bag or a bucket of water; softwood cuttings are extremely susceptible to water loss.

To prepare the cutting material, cut the base of the stem 5mm below a leaf joint, to leave a cutting of roughly 10cm long. If the cutting material has to be under 10cm, take the cutting with a heel. Remove the lower leaves and trim the tail which is left from the heel. Trim the stem cleanly before a node, the point at which a leaf stalk joins the stem. Remove the leaves from the bottom third of the cutting with a knife, leaving at least 2 or 3 leaves on top.

Make a hole in the compost and insert the cutting up to its leaves. Do not overcrowd the container or include more than one species, because quite often they take different times to root. Label and date the cuttings clearly, and only water the compost from above if necessary. Keep out of direct sunlight in hot weather. If it is very sunny, heavy shade is best for the first week.

Place in a heated or unheated propagator, or cover the pot or container with a plastic bag supported on a thin wire hoop (to prevent the plastic touching the leaves), or with an upturned plastic bottle with the bottom cut off. If you are using a plastic bag, turn it inside out every few days to stop excess moisture from condensation dripping onto the cuttings. Spray the cuttings with water every morning for the first week. Average rooting time is 2–4 weeks. The cutting medium is low in nutrients, so give a regular foliar feed when the cutting starts to root. Harden off the cuttings gradually when they are rooted. Bring them out in stages to normal sunny, airy conditions. Pot them on using a prepared potting compost once they are weaned. Label and water well after transplanting. About 4–5 weeks after transplanting, when the plant is growing away, pinch out the top centre of the young cutting. This will encourage the plant to bush out, making it stronger as well as fuller. Allow to grow on until a good-size root ball can be seen in the pot, then plant out.

Semi-hardwood or Greenwood Cuttings
These are usually taken from shrubby herbs such as rosemary and myrtle towards the end of the growing season (from mid-summer to mid-autumn). Use broadly the same method as for softwood cuttings, but the compost should be freer-draining. Make the mix equal parts peat, grit and bark. Once the cuttings have been inserted in the compost, place the pot, seed tray or plug tray in a cold greenhouse, cold frame or cool conservatory, not in a propagator, unless it has a misting unit. Average rooting time for semi-hardwood cuttings is 4–6 weeks. If the autumn is exceptionally hot and the compost or cuttings seem to be drying out, spray once a week. Begin the hardening off process in the spring after the frosts. Give a foliar feed as soon as there is sufficient new growth.

Hardwood Cuttings
Taken mid- to late autumn in exactly the same way as softwood cuttings, but with a freer-draining compost of equal parts peat, grit and bark. Keep watering to the absolute minimum. Winter in a cold frame, greenhouse or conservatory. Average rooting time can take as long as 12 months. For hardwood cuttings of hardy fruits, push cuttings into a slit trench lined with sharp sand in moist ground, firm well and keep them weed-free and protected from drying winds – a cloche is usually advantageous.

Root Cuttings
This method of cutting suits plants with creeping roots, such as bergamot, comfrey, horseradish, lemon balm, mint, soapwort and sweet woodruff. Dig up some healthy roots in spring or autumn. Fill a container with cutting compost – 50 per cent bark, 50 per cent peat, firmed to within 3cm of the rim. These cuttings lend themselves to being grown in plug trays. Water well. Cut 4–8cm lengths of root that carry a growing bud. For comfrey and horseradish, simply slice the root into sections, 4–8cm long, using a sharp knife to give a clean cut through the root. Make holes in the compost with a dibber. If using pots or seed trays these should be 3–6cm apart. Plant the cutting vertically. Cover with a small amount of compost, and a layer of perlite level with the top of the container. Label and date. Average rooting time is 2–3 weeks. Do not water until roots or top growth appears. Then apply liquid feed. Slowly harden off the cuttings when rooted. Pot on cuttings in seed trays and pots in a potting compost once they are hardened off. Label and water well after transplanting. About 2–3 weeks after transplanting, when you can see that the plant is growing away, pinch out the top centre of the young cutting. This will encourage the plant to bush out, making it stronger as well as fuller. Allow to grow on until a good-size root ball can be seen in the pot. Plant out in the garden when the last frosts are over.

layering

Layering is a process that encourages sections of plant to root while still attached to the parent. Bay, rosemary, sage, and other evergreens suit this method. Blackberries root their tips anywhere they can in autumn, and many plants will root where they touch the ground to form natural layers, which easily detach with roots in autumn.

To layer a plant, cultivate the soil around it during winter and early spring by adding peat and grit to it. In spring trim the leaves and side shoots of a young, low vigorous stem for 10–60cm below its growing tip. Bring the stem down to ground level and mark its position on the soil. Dig a trench at that point. Roughen the stem at the point where it will touch the ground, and peg it down into the trench, then bend the stem at right angles behind the growing tip, so that it protrudes vertically. Return the soil to the trench to bury the stem. Firm in well and water. Keep the soil moist, especially in dry periods. Sever the layering stem from its parent plant in autumn if well rooted, and 3–4 weeks later nip out the growing tip from the rooted layer to make plant bush out. Check that the roots are established before lifting the layered stem. If necessary, leave for a further year. Replant either in open ground or in a pot. Label and leave to establish.

grafting

Some fruit plants are only propagated by grafting or budding small pieces onto more easily grown rootstocks. Often this is done with special rootstocks simply to influence growth or to get the maximum number of plants from limited material. Budding and grafting techniques, although essentially simple, are profoundly difficult to master without much practice and are beyond the scope of this book.

buying plants

Buying fruit, vegetable and herb plants has the advantage of ease and speed over growing them yourself, but is relatively costly and incurs a high risk of importing weeds, pests and diseases along with the plants, especially with pot-grown specimens.

To fill an average garden with plants does not require an immense investment but it is still quite enough to warrant care and budgeting. For fruit, specialist mail-order nurseries usually provide a greater choice, and are often cheaper than most local suppliers. Get several catalogues and compare them before ordering, and do so early.

For fruit trees and shrubs bare-rooted plants are easier to inspect than pot-grown ones for signs of health or disease. The root systems on well-grown, bare-rooted trees are usually more extensive than those of pot-grown plants. Furthermore, for larger growing trees, there is the danger of pot-grown specimens being root bound. However, for most smaller subjects, containerized plants from reputable suppliers are convenient and give good results.

Vegetables may be available pre-grown as seedlings, at the stage when they require pricking out; as plugs with three or four true leaves, which may need growing on; or as plants that are quite large (sold in strips or multi-packs, like bedding plants) and should be transplanted directly on arrival.

Misting unit

MAINTAINING THE GARDEN

To ensure that plants are productive and the crops are of high quality, vegetables, herbs and fruit need careful nurturing and diligent husbandry. Vegetables require more constant attention than either herbs or fruit. However, all plants benefit from regular checking so that any problems can be spotted early. For watering, see pages 651; for pests and diseases, see pages 656–661.

planting out

Plants that have been germinated or grown as cuttings in a greenhouse will need to be planted out after hardening off. The timing depends on the prevailing weather and soil conditions. All plants suffer a check in their growth rate after transplanting, caused by inevitable root disturbance and by the change in environment to somewhere cooler. Younger plants tend to recover rapidly while older ones take longer. To reduce stress, ensure there is plenty of water available, transplant on an overcast day into moist soil and provide shelter from strong sunshine until plants are established.

If you must prune, be ruthless

plant supports

These should be positioned before they are required by the plants. Plants growing tall without a support can be damaged or suffer a check in growth. Select a support according to the plant being grown. Peas climb with tendrils, preferring to twine round thin supports such as chicken wire or spindly twigs, while runner beans hug poles or canes. Vegetables without climbing mechanisms need to be tied to stakes with garden twine in a loose figure-of-eight loop, positioning the stake on the windward side of the plant. Fruit trees trained against a flat surface need strong wires to which the branches can be tied.

pruning fruit trees

We prune for two reasons: to remove diseased, damaged and ill-placed growths, and to channel growth into fruit production. We may also prune to reduce the size of a plant as it becomes too large for the space available.

Excessive pruning, especially at the wrong time, is counter-productive. In general, pruning even moderate amounts from a tree or bush in autumn and winter stimulates regrowth, proportionately as much as the amount removed. This is useful when the plant is young and we wish to form the framework by stimulating the growth of young, vigorous replacement shoots.

However such autumn and winter pruning is not so suitable for more mature fruiting plants whose structure is already formed and from which we wish to obtain fruit. Most respond much better to summer pruning, which is cutting out three-quarters of every young shoot, bar the leaders. This redirects growth and causes fruit bud production on the spurs or short side shoots formed. These may be further shortened and tidied in the winter but then, as only a little is removed, vigorous re-growth is avoided.

Although it may seem complicated, pruning is easy once you've done it a few times. Use clean secateurs, which have been sterilized with alcohol, or a sharp saw, and cover large wounds with a proprietary sealant to prevent water getting in.

Training the Basic Shape

The majority of our perennial woody fruiting plants can be trained and pruned to make a permanent framework which carries spurs, preferably all over. Growing just one such stem, branch or cordon on a weak rootstock allows us to squeeze many varieties into the same space as one full-sized tree. Such single-stemmed cordons do not produce very much fruit, especially as they are hard to support if they reach more than head height. Sloping these cordons at 45° serves to make them longer without going too high.

Growing two, three, or more branches is a better compromise. These can be arranged as espaliers (in tiers), fans (radiating from the centre), or gridirons of almost any design. However, for the vast majority of trees and bushes, the shapes most commonly used are the expanding head, and the open bowl or goblet arrangement, on top of a single stem or trunk.

This bowl shape maximizes the surface area of fruiting growth exposed to the sun and air. To achieve it the main leader is removed from the middle and the branches trained as a bowl with a hollow centre open to the sky. The number of main stems is customarily about five or six which divide from the trunk and redivide to form the walls of the goblet. The stem or trunk may be short, as is common with gooseberries, or taller, which is often more convenient.

Where the branches of a tree divide from the trunk on a short trunk they are termed bushes, at waist to shoulder height they are called half-standards, and standards where they start higher still. Bushes, especially those on the more dwarfing stocks, tend to be too low to mow underneath but can always be mulched instead. Half-standards grow large, depending on stock, and are tall enough to mow underneath. Full standards make very big specimens and are only usually planted in parks and meadows.

Young unformed plants, maidens, can be bought more cheaply than those with a good shape or form already trained. It is very satisfying to grow your own espalier or gridiron from a maiden, but the result will depend on your skill and foresight.

Renewal Pruning

Some fruits need pruning on the renewal principle. Whole branches or shoots are removed at a year or two old, after they have fruited. Summer raspberries are typical, the old shoots being removed at ground level as the young are tied in. Grapevines can be constrained to two young branches emerging from the trunk, replaced each year. Peaches on walls also have young shoots tied in and old fruited ones removed. Blackcurrants have a third of growths from the ground removed annually.

Timing

A few plants are pruned only at certain times of year. Hollow-stemmed, tender and evergreen plants are pruned in spring once the hardest weather is over and Prunus (members of the cherry family or stone fruits) are pruned in summer, as otherwise they are liable to disease.

Working on steps requires two people, or beware…

FERTILITY AND WATER MANAGEMENT

Good stuff for the compost heap

Plants have varying nutrient requirements according to the species. All need sufficiently fertile soil and water to grow and yield well. Fruits are ordinarily not as demanding as many vegetable crops; this is because they are mostly perennial. Growing in open ground, they make extensive root systems, which find the water and nutrients required. Annual and short-lived crops generally need much more attention to soil fertility and water provision because they have limited root systems. Plants in containers similarly require much care, and because of the confinement also readily suffer from any excess.

soil fertility

Soil fertility is the source of essential nutrients needed for healthy growth. There is no precise definition of a fertile soil, but ideally it has the following characteristics: a good crumb structure; plenty of humus and nutrients; good drainage, moisture-retentive; a pH that is slightly acid to neutral. Nutrients are absorbed in solution by the roots and transported through the plant's system. The major elements include nitrogen, phosphorus, potassium, calcium, magnesium and sulphur. Minor, or trace, elements include iron, manganese, boron, zinc, copper and molybdenum. These are needed in minute quantities yet are still essential.

Nutrients are present in the soil as a result of the weathering processes on the mineral particles and the chemical breakdown or decay of organic matter and humus by bacteria and other micro-organisms. These release nutrients in a form available to plants. To function effectively they must have an open, well-drained soil with adequate supplies of air and water. Extra nutrients in the form of compost and fertilizer may be needed to provide sufficient nutrients to sustain healthy growth.

Fertility is best provided organically from materials that slowly convert to a usable form in the soil. Well-rotted farmyard manures, good compost, seaweed meal, hoof and horn meal, blood, fish and bone meal, bone meal, and fish emulsions are all eminently suitable.

If mineral shortages are suspected, then ground rock dusts provide cheap, slow-release supplies; potash, phosphate, magnesium (dolomitic) limestone, calcified seaweed and lime are all widely available, cheap and pleasant to apply. Wood ashes are extremely valuable to most fruiting plants. Seaweed extracts sprayed on the foliage can give rapid relief of mineral deficiencies.

Bulkier materials such as well-rotted manure and compost also provide much humus, which is essential for the natural fertility of the soil, its water-holding capacity and its buffering action (preventing the soil being too acid or alkali). The humus content is conserved by minimal cultivation and refraining from soluble fertilizers.

More humus can be provided by growing green manures. These occupy the soil when other plants are dormant or if parts of the plot are empty throughout winter. Nutrients and water that would have leached away are combined with winter sunlight to grow dense covers of hardy plants. These also protect the soil surface from erosion and rain impaction. When the weather warms up, green manures are incorporated in situ by digging in or composting under a plastic sheet, or are removed and added to the compost heap.

Organic mulches such as well-rotted manures, composted shredded bark, leafmould, mushroom compost, straw or peat are all advantageous to most plants. They rot down at the soil surface aiding fertility and humus levels and suppress weeds if they are thick enough, but most importantly they conserve soil moisture.

Careful rotation in the vegetable plot can aid fertility. Plant 'hungry' crops such as celery, leeks and members of the cabbage family, which require plenty of nitrogen, after peas, beans and other legumes, which fix nitrogen from the atmosphere using nitrogen-fixing bacteria in swollen nodules on their roots and release nitrogen into the soil.

watering

Vegetables and fruits crop to their full potential only when water is plentiful. Yields are reduced when plants suffer from drought stress. The soil is a plant's reservoir and different soils are able to retain different levels of water. Clays retain water very efficiently but dry out in summer, loams have a good balance between drainage and water retention, and sandy soils are very free-draining.

Water is lost from both soil and the leaves of plants and this loss is most rapid on hot, sunny days with drying winds. It is vital to reduce the loss of water from soil by:

- regularly incorporating well-rotted organic matter into the soil, to increase its moisture-holding capacity.
- deep cultivation to open up the soil and encourage deeper rooting. This allows plants to draw water from a greater soil depth.
- covering the soil surface with mulches such as well-rotted manure or black plastic to prevent moisture loss from the surface and upper soil layers.
- improving drainage on waterlogged soil; a high water table restricts root development, making plants prone to stress during drought. By improving the drainage system and lowering the water table, the roots are encouraged to penetrate to a greater depth, thereby leaving them more drought-tolerant.
- removing weeds, which compete with other plants for water. They should be removed at seedling stage.
- reducing plant density, particularly on dry, exposed sites, to allow each plant to draw water from a greater volume of soil.
- providing shelter, especially on exposed sites. Reducing wind speed reduces moisture loss from the soil and plants.
- being careful with accurate timing and frequency of application. Most plants have critical periods when adequate water supplies are very important for successful development and cropping.

Critical Watering Periods

Water before it is too late; if a plant is wilting it is already suffering badly!

Seeds and Seedlings

Water is essential for seed germination and rapid seedling development. Seedbeds should be moist, not waterlogged. Water seedlings using a fine rose on a watering can to provide a shower of tiny droplets and prevent plant damage.

Having easy access to water makes it less of a chore

Transplants

Plants often suffer from transplanting shock due to root damage after being moved, and the effects are more pronounced if the soil is dry. Water plants gently after moving them to settle moist soil around the roots. Liquid seaweed is a useful boost. Water woody plants until they have been established for up to three years, particularly evergreens, which cannot drop their leaves when under stress and so survive dormant.

Fruiting Crops

Fruits and vegetables, such as peas, beans and tomatoes, with edible fruits have two critical periods for water availability once they have established: when flowering (to aid pollination and fruit set) and when the fruit begins to swell.

Leaf and Root Crops

These need a constant water supply throughout their cropping life.

Methods of Watering

It is vital to maintain rapid growth in the growing season, particularly in annual plants, otherwise plant tissues can harden and growth can be affected. Growth check in cauliflowers can cause young plants prematurely to form a small, poorly developed curd, while other annuals will 'bolt'.

Always water thoroughly, soaking the soil to a reasonable depth. It is bad practice merely to wet the soil surface. This draws the roots up to form a surface mat while much is wasted by evaporation. A good soaking descends and draws the roots after it, thus they become deeper and more able to find other soil moisture on their own. Water in the evening if possible.

If you bury a length of hosepipe with the roots of a tree you can inject the water right where it is needed during the critical first few years. Likewise a pot or funnel pushed in nearby is a useful aid.

When time is limited and cash more abundant it is wise to invest in automatic watering equipment. Drip feeds and seeping hoses allow a constant and even supply of water close to the roots without the gardener's constant attention, and soon repay the investment. Overhead sprinklers are extremely inefficient as only about 20 per cent of the water ever reaches the plant.

Mulching

Once the winter rains have drenched the soil a mulch prevents it evaporating away again. Any mulch helps, but the looser and thicker the better. Less than 5cm is ineffective, and initial applications always pack down, so be prepared to add more after a few months. When organic material for mulching is in short supply, inorganic ones may be used. Sharp sand and gravel make excellent moisture-retaining mulches and are cheap and sterile.

Grass clippings are an excellent mulch, if applied in thin layers – though if applied too thickly in wet conditions they may make a nasty claggy mess. Clippings provide a rich source of nitrogen and encourage soil life. They soon disappear and need topping up. Continuous applications slowly make the soil less alkaline and more acid, which is often advantageous.

Weeds being kept down (neatly) with old carpet

GROWING IN GLASSHOUSES

There is no substitute for walk-in cover, which is more valuable if heated and frost-free, and even more so if kept warm or even hot. If maintained as warm as a living room all year round, many exotic fruits and vegetables become possible from oranges to aubergines. It is not only summer crops that benefit. Lettuce, spinach and other hardy winter salads produce more tender, better-quality crops in an unheated greenhouse than outside.

Exotics such as passion fruits are best grown under cover as they would not crop otherwise. The extra heat ripens the fruit and also the wood, which often cannot ripen outdoors. Some plants take many months to crop each year and without cover they would never ripen before frosts come. Extra warmth and shelter allows us to have tender or delicate plants that would not survive in the open, but it also means we can have the same crop earlier than outdoors – for example strawberries or tomatoes.

Often it is best to grow plants in containers that are to go under cover as this allows for more variety in the same area, though each plant will necessarily yield less than if it was in the ground. Containers not only control the vigour of the plants but also make them portable so they can be moved under cover and outside as convenient. This suits many fruits such as the citrus, which really prefer to be outside all summer but need frost protection in winter. By contrast, many varieties of grapevine are best under cover for an early start and for ripening, but need to be chilled in winter to fruit well.

This is why some plants are difficult. They need hot summers and warm autumns and also cold winters to go dormant. Without dormancy they do not ripen wood or fruit well and often just fade away. It is quite easy to chill a greenhouse in a cold area for the couple of months required, but not of course if other tender plants are kept there, too.

The other requirement may be for extra light. Many plants need more light than can be had in winter through dirty glass. The physical barrier reduces light intensity by half. Fortunately, artificial lights are cheap to fit and run – compared to heating anyway. Some plants are very demanding; not only do they want more light and heat but they want enough hours of complete darkness every night as well. This requires the fitting of blinds to exclude daylight and light from any nearby streetlights, too! Similarly some plants find bright light too intense and need shading.

Fortunately most exotics are remarkably easy to grow. The biggest problems are usually the cost of heating and the eventual size of the plants.

Hygiene in the Glasshouse

Glasshouses should be cleaned annually in winter: wash thoroughly with disinfectant to kill overwintering eggs and fungal spores. If you have space, before introducing a new plant to the glasshouse, keep it in quarantine for a few weeks to prevent new problems being introduced.

Tomatoes are reliable under cover

PROTECTED CROPPING

Many crops need protecting at some point in their life. In some cases they need protecting from the elements, to provide conditions more conducive to growth; they may also need protecting from predators such as birds and rodents.

Protection from the Elements

A few simple structures can extend the growing season in spring and autumn, accelerate growth and increase productivity.

Coldframes are ideal for raising seedlings or hardening off plants before transplanting outdoors. The lack of height limits the crops that can be grown in them. Cloches of glass or plastic are easily moved to provide shelter to those individual plants or small groups that need it most. Low polythene tunnels are like small cloches, are cheap, portable and versatile, but are of limited use because of their flimsiness and lack of height. Crop covers or floating film are flexible covers that can be laid immediately over the crop or suspended on hoops as 'floating mulches'. As the plants grow they are forced upwards. Fleece is lightweight, soft-textured spun polypropylene, which lasts for a year if it is kept clean, or sometimes longer if cared for. It is ideal for frost protection and for forcing early crops. Many crops, especially vegetables, can be grown under this, from sowing to harvesting, and it is an effective barrier against pests and diseases. More light and air can penetrate through this than through plastic film. Plastic film is usually perforated with minute slits or holes for ventilation, but the crop may overheat on sunny spring days. The film is laid directly over the plants; as they grow, the flexible material is pushed upwards, splitting open the minute slits and increasing ventilation. It is often used in the early stages of growth. Plastic netting has a very fine mesh, filtering the wind and providing protection, but has little effect on raising temperature. It lasts for several seasons.

As many of these materials are manufactured in long, narrow rolls, the most effective way to use them is to grow crops in long, narrow strips. Insect-pollinated crops like courgettes, marrows, cucumbers and tomatoes should have the covering removed or opened as the crop develops.

Protection from Predators

Fruit cages are covered with netting. They are necessary to protect soft fruit in areas with many birds. The most troublesome birds, such as blackbirds and pigeons, can be excluded with coarse 2cm (1in) mesh, while smaller, insectivorous birds such as wrens and bluetits can still gain access. If all the net is wire, squirrels and rodents can also be stopped. If a finer mesh, down to about 1cm (½in) is used then bees can still enter but large moths and butterflies are excluded.

Fruit cages have other uses. The netting itself makes a more sheltered environment enjoyed by the plants. Light frosts are kept off, and chill winds are reduced. In very hot regions, denser netting gives welcome shade and cooler conditions.

A fruit cage can even be reversed in principle to enclose all of a small garden and to confine ornamental seed- and insect-eating, but hopefully not inadvertently fruit-eating, birds. On a more prosaic level, it is practical to grow plants on a tall leg and then run chickens underneath for their excellent pest control. If you have chickens, they can be running in the cage during most of the year when the plants are not in fruit.

Fruit cages can be hand-crafted from second-hand materials or easy custom-made ones are available. In areas of high wind, obviously, the most substantial materials have to be used. Permanent sides of wire mesh and a light net laid on wires for the roof are most practicable. Make sure you can remove the roof net easily if snows are forecast as a thick layer on top will break most cages.

Most fruit cage plants are natives of the woodland's edge and do not mind light shade, but few of them relish stagnant air so do not overcrowd them.

Cabbage plants protected with garden netting

WEED CONTROL

Any plant that grows in a place where it is not wanted is described as a weed. These compete with crops for moisture, nutrients and light, acting as hosts to pests and diseases, which can then spread to crops.

types of weed

Knowledge of a weed's life cycle enables the gardener to control weeds effectively.

- Ephemeral weeds germinate, flower and seed rapidly, producing several generations each season and copious quantities of seed.
- Annual weeds germinate, flower and seed in one season.
- Biennial weeds have a life-cycle spanning two growing seasons.
- Perennial weeds survive for several years. They often spread through the soil as they grow, producing scores of new shoots and setting seed. New plants sprout from tiny fragments of root, rhizome or bulbils in the soil.

methods of control

There are four main methods of weed control: manual, mechanical, mulching and chemical. Wherever possible, keep the garden weed-free and if germination occurs, remove weeds immediately, before they flower and produce seed. The old saying, 'One year's seed is seven years' weed' is, unfortunately, scientifically proven.

Complete elimination of weeds from an area before planting is invariably worthwhile and is much easier than trying later to weed among the plants (see page 634).

Manual Weeding

Digging, forking, hoeing and hand weeding, are often the only practical ways to eradicate weeds in confined spaces.

Digging cultivates the soil to a depth of at least 30cm (12in). Soil is turned over, burying surface vegetation and annual weeds. Perennial weeds should be removed separately and not buried among the surface vegetation.

Forking cultivates the soil around perennial weeds before they are lifted by hand. If this is undertaken carefully, roots can be removed without breaking into sections and forming new plants.

Hoeing prevents weed seeds from germinating, reduces moisture loss and minimizes soil disturbance. Hoeing is best undertaken in dry weather and the blade should penetrate no deeper than 1cm (½in).

Hand-weed in dry weather when the soil is moist so that weeds are easily loosened from the soil; try to remove the whole plant. Remove weeds from the site to prevent them from rerooting.

Mechanical Weed Control

Machines with revolving rotary tines, blades or cultivator attachments are useful for preparing seedbeds and can also be used to control annual weeds between rows of vegetables. They should be employed carefully to avoid root damage. When clearing the ground of perennial weeds, the blades simply slice through the roots and rhizomes. While this increases their number initially, several sessions over a period of time should gradually exhaust and eventually kill the plants. This is only viable when there is plenty of time.

Biological Weed Control

Mulching is the practice of covering the soil around plants with a layer of organic or inorganic material to suppress weeds, reduce water loss (see also page 651) and warm the soil. Choose your material carefully; straw harbours pests such as vine weevil and flea beetle and contains weed seeds, but is effective as an insulating layer over winter. It also draws nitrogen from the soil in the early stages of decay. Using rotted material like well-rotted farm manure or spent hops avoids this problem

Organic mulches are also limited where crops need to be earthed up – a process that disturbs the soil. Inorganic materials like plastic sheeting, though effective, are unattractive but can at least be hidden beneath a thin layer of organic mulch.

Inorganic mulches suppress weeds and prevent seeds from germinating. They also conserve soil moisture by slowing evaporation; plastic films are the most effective for this purpose. They keep trailing crops clean. White plastic mulches hasten growth and ripening by reflecting light on to leaves and fruit; black plastic mulches warm the soil in spring. Sheet mulches are most effective when applied before the crop is planted, and the young plants are then planted through small holes in the sheet.

Organic mulches improve soil fertility and conserve its structure by protecting the surface from heavy rain and being trodden on, encouraging earthworm activity and adding organic matter and nutrients to the soil. They insulate the soil, keeping it cooler in summer and warmer in winter. Weeds that push their way through are easily removed. For effective weed control organic mulches should be about 10cm (4in) deep, to block out the light.

In general, organic mulches improve soil fertility as they decay, while inorganic mulches are more effective against weeds as they form an impenetrable barrier.

Chemical Weed Control

Choose and use herbicides with extreme care and ensure they are suitable for the weeds to be controlled. **Follow the manufacturer's instructions carefully.**

• Residual herbicides form a layer over the soil, killing germinating weeds and seedlings.
• Contact herbicides only kill those parts of the plant that they touch. They are effective against annual weeds and weed seedlings, but are not recommended against established plants.
• Systemic or translocated herbicides are absorbed by the foliage, travelling through the sap system to kill the whole plant, including the roots. Glyphosate is often the active ingredient. Individual weeds can be 'spot treated' using such herbicides in gel or liquid form.

Several chemical weed-control methods have been developed for use with vegetables and other edible plants but no single chemical is suitable for all crops. This gives the grower three options: stock a range of chemicals and change the type regularly; use chemicals only as a last resort; or grow organically. For edible plants the latter is almost invariably the best.

Using and Storing Weedkillers Safely

Always wear adequate protective clothing, such as face mask, goggles, rubber gloves and old waterproof clothes when mixing weedkillers. Dilute weedkillers according to the manufacturer's instructions, and never mix different chemicals together. Never dilute chemicals in a confined space, as they may give off toxic fumes.

As for mixing, always wear the specified protective clothing when applying weedkillers. Follow the manufacturer's instructions carefully and use only as recommended on the product label. Do not apply weedkillers in windy conditions; nearby plants may suffer serious damage. Do not apply weedkillers in very hot, still conditions when there is a high risk of spray travelling on warm air currents. If you are using a watering can to apply chemicals, ensure that it is obviously labelled and at no times use it for any other purpose.

Store chemicals in the original containers with the labels well secured so that the contents can be identified. Mark the container with the date of purchase, so that you know how long it has been stored. Never store dilute weedkillers for future use. Always store in cool, frost-free, dark conditions out of the reach of children and animals, in a locked cupboard in a workshop or shed.

Thoroughly wash protective clothing, sprayers, mixing vessels and utensils after use. Never use the same sprayer, mixing vessels and utensils for other types of chemical such as fungicides. The results could be disastrous!

grass management

Under woody perennial plants such as fruit trees and bushes, the most convenient ground cover is usually grass sward. Grass competes vigorously, especially if kept closely mowed as a lawn. However grass sward is simple to maintain, ornamental, hygienic and prevents worse weeds. The clippings themselves are a free source of mulching material and can contribute much fertility to our plants.

Grass sward is most productive of useful clippings when the grass is cut often and not too closely. Long grass grows more quickly, is more drought-resistant and suppresses turf weeds better than closely cropped. Cut your grass regularly and frequently with slightly greater height of cut and the longer, more vigorous grass soon chokes out most weeds. When grassing down, it is sensible to include clovers in the grass seed mix as they have the advantage of fixing nitrogen from the air to aid the grasses. A clover grass mixture also stays greener for longer in hot dry summers and, if left to flower, the clovers attract bees and other beneficial insects.

Providing you are not growing ericaceous plants or fine bowling-green grasses you should lime all your turf every fourth year, even if you have lime in the soil. Most swards slowly become acid in the topmost layer; this encourages mosses and acid-loving weeds. If such weeds, say daisies, are increasing you need to add more lime, up to a couple of handfuls per square metre (square yard) per year. You probably also need to raise the height of your cut. If weeds that like wet acid soils such as buttercups appear, then lime is desperately needed and probably better drainage as well.

If they're in the wrong place, take them away!

PESTS AND DISEASES

The control of pests and diseases is done to preserve the yield or appearance of our crops, but we must ensure that the costs incurred do not outweigh the gains. It is ironic that commercial apples are sprayed to prevent scabby patches on their skin, which is now peeled and discarded to avoid the very residues left by many such sprays. Herbs in general suffer from few pests and diseases and in fact many can be used to protect other crops (see pages 586–7).

For the home gardener the main causes of loss of crops most years is the weather. There is very little we can do to make the sun come out, the wind stop or the rain fall.

The second cause of loss is probably the gardener. We all leave action too late, skimp preparation and routine tasks, put plants in less than their optimum positions, and then over-crowd them as well. What we must remember is that our plants 'want' to leaf, flower and crop. They are programmed as tightly as any computer, If we give them the right inputs they must produce the right output.

preventing problems

Organic gardeners avoid the use of artificial chemicals, relying on good management and natural predators to control pests and diseases, while working to create an environment in which they are unlikely to occur. If we give our plants the right conditions then they are also healthy and vigorous enough to shrug off most pests and disease. Healthy plants grow in a well-suited site and soil, with shelter and water when they are small, and are not overfed. Excess fertilizer makes plants soft and prone to problems. Keep them lean and fit with the correct compost and mulches.

Working with Natural Defences

The most effective way to control pests is to persuade others to do it for you. The natural ecology always controls them in the long run. We can help it quickly reach a balance of more ladybirds, thrushes and frogs with fewer aphids, snails and slugs. Our allies require shelter, nest sites, food out of season, water and companion plants. If we provide these they increase in number and thus the pests decrease.

Timing can be of service. For example late raspberries usually escape the depredations of their fruit maggot. Having many fruits that ripen at the same time also reduces damage. When successive plants ripen they are picked clean by small numbers of pests which are overwhelmed by a glut, as may be the gardener!

Check plants daily if possible. Instant eradication is often the best remedy: either remove the affected part of the plant or wipe off the pest or disease as with aphids.

General Hygiene

Plant debris is a logical host for pests and diseases, particularly over winter, allowing them to survive in preparation for reemergence and reinfection the following year. Do everything possible to keep plants healthy by cleaning your glasshouse and equipment regularly. Dispose of organic matter immediately; compost healthy material and get rid of infected plants at once by burning or placing in the dustbin. When buying plants, ensure they come from a reputable source and, if possible, keep them in quarantine for a short while before introducing them to the glasshouse.

Winter is the perfect time to disinfect plant supports, particularly bamboo canes, which are hollow and often split with age, creating cracks and crevices which form suitable sites for fungi and insect eggs to overwinter. These can be sprayed or dipped in disinfectant.

Cultural Practices

Always use sterilized compost for seed sowing and potting. Seedlings are particularly vulnerable, especially when sown under cover. Sow seeds thinly, do not irrigate with stagnant or cold water, ventilate the glasshouse, but avoid chilling plants or seedlings and do not allow compost to become waterlogged. Promote rapid germination and continuous growth to reduce the risk of infection at the time when seedlings and young plants are at their most vulnerable. Inspect plants regularly, particularly if a variety is prone to certain problems. Grow pest- and disease-resistant cultivars where possible. Rotate vegetable crops and always keep the garden weed-free. Handle plants with care, particularly when harvesting, as this is a common cause of infection. Regularly check stored crops as rot spreads rapidly in a confined space.

Yellow sticky trap

Copper rings

control methods

Barriers and Traps

The simplest method of protection is to use a barrier to prevent pests or disease reaching the plant. For example, netting stops birds and horticultural fleece stops many insects.

On fruit trees, traps of corrugated cardboard, carpet or old sack can be made by rolling strips around trunk or stem. These attract pests by replicating creviced bark. They hide within and can be dislodged in winter. Non-setting sticky bands applied to aluminium foil wrapped tightly around trunk, branch or stem stop pests climbing up and down. Do remember to paint the support as well.

Sticky traps can be bought to lure flying pests to them with pheromone sex attractant 'perfumes'. Others use the smell of the fruit or leaf to entice the pest to a sticky end, or simply use colours like yellow. Jars of water with lids of foil containing pencil-sized holes trap wasps when baited with fruit juice or jam.

Organic Sprays

Soap spray suffocates most small insect pests and is remarkably safe for us and the environment. People used to employ ordinary household soap flakes but improved soft soaps are now sold for pesticidal use. Sprays based on plant oils or plant and fish oils are also available, alongside blends of surfactants and nutrients.

Biological Controls

Many predators are now available commercially. Apart from *Bacillus thuringiensis*, the bacterial control for caterpillars, few are effective outside the closed environment of a greenhouse. Most pesticides kill predators with the exception of insecticidal soaps and

Soft soap spray

pirimicarb (against aphids) and so should be avoided. Particularly useful are *Phytoseiulus persimilis*, the predatory mite that controls two-spotted or red spider mite and *Encarsia formosa*, a parasitic wasp, controlling glasshouse white-fly. All biological control methods are relatively expensive when compared with chemical alternatives, and nearly all depend on warm temperatures to work efficiently.

Chemical controls

Chemicals offer a quick and simple solution to many pest and disease problems but, before using them, it is worth considering the following. Many chemicals are toxic and can be harmful to humans, pets and wildlife. They should be applied with extreme care. Chemicals may remain in soil and plant tissue for long periods, affecting predators. Many pests and diseases, particularly whitefly and red spider mite, can develop a resistance to chemicals that are used persistently for long periods, so rendering controls less effective. Applying a range of chemicals can overcome this. Government legislation may make some chemicals unavailable in the future.

Always follow instructions with regard to safety, mixing and application. Apply chemicals in still, overcast conditions, usually in the early morning or evening. Protect surrounding plants from the chemical solution. Thoroughly clean equipment after use. Dispose of unwanted chemicals according to the manufacturer's recommendations. DO NOT POUR THEM DOWN THE DRAIN.

soil-borne pests

Cutworms

These grey-green caterpillars, approximately 5cm (2in) long, appear from late spring until early autumn. They burrow into root crops including beetroot, carrot, parsnip and potato, and any plant with a soft taproot, eating through plant stems at soil level.

Plants grown under cover seem less prone to damage, and heavy watering in early summer often kills young caterpillars. To control them, encourage birds by hanging feeders and bird boxes; cultivate the soil in winter to expose them to predators; encourage ground and rove beetles and use the biological control *Steinernema carpocapsae*.

Cabbage Root Fly

These white larvae, approximately 8mm long, are found from late spring to mid-autumn. They feed on the roots of brassicas just below soil level, stunting growth and causing plants to wilt, seedlings to die and tunnels to appear in root crops. As a preventive measure, grow under horticultural fleece. To control them, place a 12cm collar of cardboard, plastic, carpet underlay or similar around the plant bases. Grow under horticultural fleece or mesh.Use biological control nematodes and practise crop rotation (see pages 634–5).

Carrot Root Fly

These creamy-white maggots, approximately 8mm long, appear from early summer until mid-autumn. They are a common problem with carrots, but can also affect celeriac, parsnip, celery, parsley and other umbelliferous plants. Larvae burrow into the roots, causing stunted growth and reddish-purple coloration of the foliage.

Sow carrots late, after late spring, and harvest before late summer. Use resistant varieties and biological control. Place a barrier 60cm high around crops or cover with horticultural fleece. Where plants are badly affected, pull them up and destroy them. Large herbs should overcome attacks so just pick off dead leaves and boost with a liquid seaweed feed. Water after thinning, or thin when rain is forecast.

Leatherjackets

These grey-brown, wrinkled grubs, can be up to 3.5cm long. They eat through roots and through the stems of young plants just below soil level. Most crops are vulnerable, including Brussels sprouts, cabbage, cauliflower and lettuce. To discourage them, ensure the soil is well drained, dig over the ground in the early autumn, especially if it has been fallow in the summer. The problem is often worse on new sites and usually diminishes with time. Use biological control nematode *Steinernema carpocapsae*, water ground, cover with polythene, then collect or leave for predators.

nematodes (eelworms)

Potato Cyst Eelworm

These white to golden-brown, pinhead-sized cysts are found on roots in mid-summer. Each can contain up to 600 eggs which can remain dormant in the soil for up to six years. Potatoes and tomatoes are affected, with weak and stunted growth, small fruits and tubers, yellow foliage, with the lower leaves dying first followed by premature death of the plants.

Rotate crops and grow resistant varieties like 'Pentland Javelin', 'Pentland Lustre', 'Maris Piper' or 'Kingston'. Do not grow potatoes or tomatoes on infected soil for at least 6 years.

Stem and Bulb Eelworm

These are microscopic pests that live inside the plant and move around on a film of moisture; most active from spring to late summer. The symptoms are weak, stunted growth, swollen bases on young plants, and later stems that thicken and rot. They affect a wide range of young plants from spring until late summer, particularly leek, onion, shallot and chives.

Rotate crops, practise good hygiene and management, and destroy infected crops.

Do not grow potatoes or tomatoes on infected soil for at least 6 years, and any other vulnerable crops on the site for at least 4 years.

A friend indeed

Hens soon get rid of weevil infestation

slugs

Slimy, with tubular-shaped bodies, up to 10cm (4in), from creamy-white through grey to jet-black and orange, slugs are voracious feeders, making circular holes in plant tissue; damaged seedlings are usually killed. They tend to feed at night. Most soft plant tissue is vulnerable. Slugs are more of a problem on wet sites and clay soils; keep the soil well drained and weed-free. Remove all crop debris. Grow less susceptible cultivars. Remove slugs by hand at night and destroy them. Create a barrier of grit, sand, eggshell or soot around plants or cut a 10cm (4in) section from a plastic bottle and place it as a 'collar' around plants. Make traps from plastic cartons, half-buried in the ground and filled with beer or stout, or lay old roof tiles, newspaper or old lettuce leaves or other tempting vegetation on the ground and hand pick regularly from underneath. Attract natural predators like toads to the garden or use biological controls.

Chemical controls include applying aluminium sulphate in the spring as the slug eggs hatch and alternatives such as wool pellets or bran.

wireworms

These are thin, yellow larvae, pointed at each end, about 2.5cm long. They make holes in the roots and tubers, and can cause foliage to collapse; seedlings are usually killed. A wide range of plants are attacked, particularly root crops and potatoes. Symptoms appear from spring until mid-summer. More of a problem on newly cultivated soil, wireworms usually disappear after 5 years; avoid planting root crops until the sixth year. Lift root crops as soon as they mature. Cultivate the ground thoroughly well before planting.

soil-borne diseases

Clubroot

This is a soil-borne fungus that can remain in the soil for up to 20 years. It causes leaves to yellow and discolour, wilting rapidly in hot weather; the roots display swollen wart-like swellings, and become distorted. All members of the Brassicaceae family are vulnerable, including brassicas and weedy relatives. Clubroot is worse on acid soils, so lime to raise the pH and improve the drainage. Practise strict crop rotation, good management and hygiene, and grow your own plants instead of buying in, or at least purchase from a reputable source. Raise plants in pots. Swede 'Marian' has good resistance to clubroot, as do Kale 'Tall Green Curled', Brussels Sprout 'Chronos' and 'Crispus', Calabrese 'Trixie' and Cabbages 'Kilaton' and 'Kilaxy'. Lift and dispose of affected plants immediately, and do not compost.

Common Scab

This fungus causes superficial damage to the underground parts of affected plants. Rough brown lesions appear on the surface of roots and tubers. It is more severe on alkaline soils and in dry summers. Common scab affects mainly potatoes, but can be found on radishes, beetroot, swede and turnips. Avoid liming, do not grow potatoes on soil recently used for brassicas, apply organic matter and irrigate thoroughly in dry weather. Grow resistant potato cultivars like 'Arran Pilot' and 'King Edward'. Avoid 'Majestic' and 'Maris Piper', which are very susceptible. Rotate crops.

Root Rot

This fungus is difficult to eradicate. It attacks beans and peas, causing shrivelled, yellowing leaves, stems and pods; roots rot and stems rot at the base. Plants collapse. Practise a minimum 3-year crop rotation; remove and burn infected plants; or grow in pots or raised bed, avoiding overcrowding and improving the soil with compost for better drainage and fertility.

Stem Rot

This fungus is difficult to eradicate. It attacks aubergine and tomato, causing yellow-brown cankerous lesions with black dots in and around the lesions, occurring at soil level. Remove and burn infected plants and crop debris, disinfect anything that comes into contact with diseased material, including hands and equipment. Improve growing conditions like drainage and feeding, avoid waterlogged soil, and always use fresh compost in pots.

Onion White Rot

This fungus is almost impossible to eradicate. Common on onions and leeks, it also attacks chives and garlic, causing discoloured, yellowing leaves, which die slowly. Roots rot and plants fall over, becoming covered with a white felt-like mould. Do not grow susceptible crops for 8 years after an infection. Remove and burn infected plants immediately; do not compost any plant debris. Try biofumigation with chopped Caliente mustard 199 or water the soil with garlic solution.

air-borne pests

Aphids

These are dense colonies of winged and wingless insects, from pale-green through pink to greeny-black. They attack soft tissue on a wide range of plants, including brassicas, beans and peas, lettuces, potatoes and root crops. Shoot tips and young leaves become distorted; they leave a sticky coating on lower leaves, often accompanied by a black sooty mould. Pinch the growth tips from broad beans, destroy alternative hosts, and take care not to overfeed with high nitrogen fertilizers. Treat small infestations by squashing or, on sturdy plants, washing off with a jet of water; spray with soft soap, derris, pyrethrum-based insecticide; encourage natural predators like lacewings, ladybirds and their larvae and small birds like blue tits. On larger woody perennials these are more of an inconvenience than a real problem.

Cabbage Caterpillars

These are small, hairy, yellow and black caterpillars up to 5cm long. Voracious feeders, the caterpillars eat

Greenfly

holes in the leaves and in severe cases leave plants totally defoliated. They attack all brassicas, including horseradish, and are a problem from late spring to autumn. Inspect plants regularly, squash eggs or caterpillars as they are found, pick off caterpillars and grow crops under horticultural fleece or fine netting. Alternatively, spray regularly with pyrethrum when eggs are first seen, or use the biological nematode *Steinernema carpocapsae*.

Cabbage Whitefly
These small, white-winged insects are usually found on the leaf underside. Clouds of insects fly up whenever the leaves are brushed. The young leaves pucker and there is a sticky coating, often accompanied by a black mould. Whitefly attack all leafy brassicas, usually the outer leaves, so infestations can be tolerated. They can survive severe winter weather and are seen all year round. Remove and burn badly infected plants immediately after cropping, or add to local-authority recycling bin. Spray with sprays based on plant oils, fatty acids or pyrethrum.

Pea Thrips
These small, white to black-brown or yellow-bodied insects, usually on the leaf underside, are commonly known as 'thunder flies'. They cause distorted pods with a silvery sheen; peas fail to swell in the pod. The insects are often found on peas and also on broad beans. Remove and burn badly infected leaves or plants. Spray with sprays based on plants oils or pyrethrum, or use blue-coloured traps or fleece.

Powdery mildew

Vine Weevils
The adult weevil is dark grey, beetle-like 1–2cm long, with a very long snout. It takes rounded bits out of leaves, but most harm is done by the grubs, which are up to the same size with a grey-pink body and brown head. They destroy the roots of many plants. Adults can be trapped in rolls of corrugated cardboard, in bundles of sticks or under saucers, where they hide in the daytime. The grubs can be destroyed by watering on the commercially available predatory nematode.

Leaf Miners
These grubs are sometimes a problem on lovage, wild celery, certain sorrels and various mints. They eat through the leaf, creating winding tunnels in the leaves that are clearly visible. Pick off the affected leaves as you see them. If left, the tunnels will extend into broad dry patches and whole leaves will wither away.

air-borne diseases

Blight
This fungus infects leaves, stems and tubers and is worse in wet seasons. Severe epidemics led to the Irish potato famine. In mid-summer brown patches appear on the upper and lower leaf surfaces, ringed with white mould in humid conditions; leaves become yellow and fall prematurely. In humid climates the fungus spreads rapidly over the foliage and stems which quickly collapse. Tubers show dark, sunken patches. The fungus attacks potatoes, tomatoes and closely related weeds. Remove and burn badly infected plant tops in late summer, avoid overhead watering and use resistant cultivars like 'Cara', 'Romano' and 'Wilja' potatoes, plus 'Sarpo' varieties. Do not store infected tubers. Early potatoes are more likely to escape infection. 'Earth up', use a 4-year rotation, and remove all infected material from the soil and destroy.

Botrytis
This fungus infects flowers, leaves and stems and usually enters through wounds. Discoloured patches appear on stems, which may rot at ground level; flowers and leaves become covered with woolly, grey fungal growth. It causes shadows on unripe tomato fruit called 'ghost spot'. Soft-leaved plants are particularly vulnerable including broad beans, brassicas, lettuces, potatoes and tomatoes.

Scale insect

Maintain good air circulation, handle plants with care to avoid injury and remove and burn badly infected plants. Use a greenhouse fan in winter, avoid high humidity or damp conditions, and don't water or spray leaves.

Downy Mildew
This is a fungus that infects leaves and stems, overwintering in the soil or plant debris. It causes discoloured, yellowing leaves, which have grey/white mouldy patches on the lower surface; plants often die slowly in the autumn. Attacks brassicas including weed species, lettuce, spinach, peas and onions. Avoid overcrowding, remove and burn badly infected plants and maintain good air circulation. Rotate crops, choose resistant varieties like Lettuce 'Chartwell' and Cabbage 'Harmony', and maintain good garden hygiene.

Powdery Mildew
This common fungal disease can occur when the conditions are hot and dry, and the plants are overcrowded. Prevent it by watering well during dry spells, following the recommended planting distances, and clearing away any fallen leaves in the autumn. Adding a mulch in the autumn or early spring also helps. If your plant does suffer, destroy all the affected leaves.

Rust
If only small numbers are affected by pale spots or pustules, remove the leaves by hand. Avoid excess nitrogen, remove all diseased material and dispose of it away from the garden. Use resistant cultivars.

pests and diseases under cover

Red Spider Mite

The spider mites like hot dry conditions and can become prolific in a glasshouse. Look out for speckling on the upper surfaces of the leaves. Another telltale sign is cobwebs. At first sight, raise the humidity and use either a spray of horticultural soap, or the natural predator *Phytoseiulus persimilis*, but not both.

Scale Insects and Mealy Bugs

These are often noticeable as immobile, waxy, brown/ yellow, flat, oval lumps gathered on the backs of leaves or on stems. These leaves also become covered with sticky black sooty mould. Rub off the scales gently before the infestation builds up, or use a horticultural liquid soap or biological controls.

Whitefly

Act as soon as you see the small white flies, either by introducing the natural predator *Encarsia formosa*, or by spraying with horticultural soap or plant invigorator.

Red spider mite

A bell jar can be used to protect crops

other pests

Birds

Birds damage many fruits and some vegetables. Netting or fleece is the answer. If the whole plant cannot be enclosed or moved under cover then protect each fruit or bunch with waxed paper or netting bags. Things that flash, such as pieces of foil and humming tape, all work as bird scarers but only for a short time. Scarecrows rarely work at all!

Ants

These farm aphids, which produce sticky secretions on which sooty mould forms. Use biological control *Steinernema feltiae,* or baits and sprays on hard surfaces.

Wasps

Wasps are valuable allies early in the year when they control caterpillars and other pests. Later they turn to fruit and need trapping with jars (see page 657). Locate nests and call in pest controller. Protect fruit in muslin bags. Use wasp traps.

Rabbits

Only netting the area will keep rabbits out. In case they get in, have a plank ramped against the fence so they are not trapped inside and eat even more. If the perimeter cannot be secured, surround or wrap each plant in wire netting.

DISEASE-RESISTANT VEGETABLES

This list shows vegetables (and varieties) that are either resistant or tolerant to disease. In most years, resistant plants are unaffected, but they will suffer in 'epidemic' years when conditions are ideal and the disease thrives. Tolerant varieties continue to grow and crop despite being infected with little effect on their performance. Whether a variety is resistant or tolerant, the effect is the same: they will crop. This list covers a wide range of vegetables, but they won't necessarily overlap with the varieties mentioned in the A–Z of vegetables. The internet can be a good source of more information on disease resistance.

variety	*resistant or tolerant to*
Asparagus	
Jersey Knight	rust, crown and root rot
Aubergine	
Bonica	cucumber mosaic virus
	tomato mosaic virus
Dusky	tomato mosaic virus
Elondo	tomato mosaic virus
Bean (Dwarf/French/Bush)	
Aiguillon	common bean mosaic virus
Allegria	anthracnose
	common bean mosaic virus
Blauhilde	common bean mosaic virus
Cantare	anthracnose
	common bean mosaic virus
Copper Teepee	anthracnose
Daisy	common bean mosaic virus
Delinel	anthracnose
	common bean mosaic virus
Eva	cucumber mosaic virus
	bean yellow mosaic virus
Forum	anthracnose
	common bean mosaic virus
	halo blight
Green Arrow	anthracnose
	common bean mosaic virus
Helda	common bean mosaic virus
Hildora	common bean mosaic virus
Jersey	anthracnose
	common bean mosaic virus
Laguna	anthracnose
	common bean mosaic virus
Maxi	common bean mosaic virus
Modus	anthracnose
	common bean mosaic virus
	halo blight
Necktar Gold	common bean mosaic virus

variety	*resistant or tolerant to*
Necktar Queen	common bean mosaic virus
Nerina	common bean mosaic virus
Nomad	anthracnose
	common bean mosaic virus
Opera	anthracnose
	common bean mosaic virus
	halo blight
Purple Teepee	common bean mosaic virus
Safari	high disease resistance
Scuba	anthracnose
	common bean mosaic virus
Scylla	anthracnose
	common bean mosaic virus
Sonate	anthracnose
	common bean mosaic virus
Sonesta	anthracnose
	common bean mosaic virus
Speedy	common bean mosaic virus
Venice	anthracnose
	common bean mosaic virus
Bean (Runner)	
Armstrong	good disease tolerance
Galaxy	good disease tolerance
Red Rum	halo blight
Beetroot	
Kestrel	good disease resistance
	including eelworm
Solo	downy mildew
	powdery mildew
	leafspot
Broccoli (Sprouting)	
Express Corona	downy mildew
Ironman	crown rot
	downy mildew
	fusarium wilt
Shogun	downy mildew
Brussels Sprouts	
Bosworth	ringspot
Brigitte	good disease resistance
Brilliant	good disease resistance
Cascade	powdery mildew
Cavalier	light leaf spot
Cor Valiant OGC	ringspot
Cromwell	powdery mildew
Nautic	powdery mildew
Nautic	ring spot

variety	*resistant or tolerant to*
Odette	powdery mildew
Rampart	powdery mildew
	ringspot
Revenge	good disease resistance
Saxon	powdery mildew
	turnip mosaic virus
Sheriff	powdery mildew
Trafalgar	powdery mildew
Wellington	powdery mildew
Cabbage	
Brigadier	fusarium wilt
Kilaton	clubroot
Sherwood	leaf spot
Stonehead	powdery mildew
Cabbage (Chinese)	
China Express	downy mildew
China Pride	downy mildew
	soft rot
	virus
Harmony	clubroot
	downy mildew
	soft rot
Kiansi	virus
Calabrese	
Belstar	downy mildew
Emperor	black rot
	downy mildew
Gitano	downy mildew
Marathon	black rot
Matsui	downy mildew
Samson	downy mildew
Shogun	black rot
	downy mildew
Trixie	clubroot
Typhoon	downy mildew
Zen	downy mildew (moderate)
Carrot	
Artemis	cavity spot
Fly Away	carrot fly
Nantes Express	cavity spot
Nelson	cracking
	greentop
Resistafly	carrot fly
Senior	leaf spot
	powdery mildew

variety	resistant or tolerant to
Sytan	carrot fly
Cauliflower	
Candid Charm	clubroot
Clapton	clubroot
Celeriac	
Prinz	leaf spot
	hollowness
Celery	
Granada	celery leaf spot
Tango	stalk rot
Corn Salad	
Jade	powdery mildew
Vit	downy mildew
	powdery mildew
Courgette	
Afrodite	good all round disease resistance
Astia	powdery mildew
Defender	cucumber mosaic virus
Firenza	powdery mildew
Optima	cucumber mosaic virus
	powdery mildew
Pasqualine	cucumber mosaic virus
	powdery mildew
Supremo	cucumber mosaic virus
Sylvana	good all round disease resistance
Tarmino	cucumber mosaic virus
Zucchino	cucumber mosaic virus
Cucumber	
Aidas	leaf spot
	scab gummosis
Bali	angular leaf spot
	scab
Bella	leaf spot
	scab gummosis
Birgit	gummosis
	leaf spot
	scab gummosis
Boyce	leaf spot
	scab gummosis
Burpee Hybrid	cucumber mosaic virus
	downy mildew
Burpless Tasty Green	downy mildew
	powdery mildew
Bush Champion	cucumber mosaic virus
Carmen	gummosis
	leaf spot
	powdery mildew
Crispy Saladin	cucumber mosaic virus
Cumlaude	powdery mildew
El Toro	gummosis
	leaf spot
Femdan	leaf spot

variety	resistant or tolerant to
Femdan	scab gummosis
Femspot	gummosis
	leaf spot
Flamingo	powdery mildew
Futura	powdery mildew
La Diva	downy mildew
	powdery mildew
Landora	scab gummosis
Marketmore	cucumber mosaic virus
	downy mildew
	powdery mildew
Mistral	powdery mildew
Muncher	cucumber mosaic virus
Passandra	downy mildew tolerant
	gummosis resistant
	powdery mildew
Pepinex	gummosis
	leaf spot
Petita	cucumber mosaic virus
	gummosis
	leaf spot
Silor	cucumber mosaic virus
	gummosis
	powdery mildew
Slice King	angular leaf spot
	downy mildew
	gummosis
	powdery mildew
Styx	powdery mildew
Superator	gummosis
	leaf spot
Swing	powdery mildew
Tiffany	powdery mildew
Tiger Cross	cucumber mosaic virus
Tokyo Slicer	cucumber mosaic virus
Tyria	gummosis
	leaf spot
	powdery mildew
Endive	
Eminence	basal rot
Gherkin	
Diamant	downy mildew
	powdery mildew
Sena	cucumber mosaic virus
	downy mildew
	powdery mildew
Kale	
Tall Green Curled	clubroot
Leek	
Apollo	rust
Autumn Giant	
– Poribleu	rust
Autumn Giant 2	
– Porvite	rust
Autumn Mammoth	
– Tornado	rust (moderate)

variety	resistant or tolerant to
– Verina	rust
Autumn Mammoth 2	
– Walton Mammoth	rust
Bandit	rust
Blue Green Autumn	
– Ardea	rust
– Neptune	rust
– Pandora	virus
Einstein	white tip
Flexitan	rust
	white tip
Kajak	leaf spot
	virus
Oarsman	rust
Porbella	rust
Poribleu	rust
Poristo	rust
Swiss Giant – Zermatt	rust (moderate)
Lettuce	
Action	downy mildew
Amorina	downy mildew
Avoncrisp	downy mildew
Avondefiance	botrytis
	downy mildew
Beatrice	downy mildew
	root aphid
Belize	bolting
	downy mildew
	leaf aphid
	tipburn
Brandon	downy mildew
Bridgemere	downy mildew
Bubbles	downy mildew
Bughatti	downy mildew
Cassandra	downy mildew
	lettuce mosaic virus
Chartwell	downy mildew
Chatsworth	downy mildew
Claremont	downy mildew
Clarion	downy mildew
	virus
Columbus	downy mildew
Corsair	downy mildew
	lettuce mosaic virus
Cosmic	downy mildew
Counter	bremia
	tip burn
Crestand	downy mildew
	virus
Crispino	lettuce mosaic virus
Debby	downy mildew
	lettuce mosaic virus
Diana	downy mildew
Dynasty	downy mildew
	lettuce mosaic virus
Estelle	downy mildew
	leaf aphid
Fatima	bremia
	mosaic virus

variety	*resistant or tolerant to*
Fristina	downy mildew
	tipburn
Iceberg	big vein
Jewel	downy mildew
Lakeland	downy mildew (moderate)
	root aphid
Lilian	downy mildew
Little Gem	root aphid
Lizzy	downy mildew
	lettuce mosaic virus
Marcord	downy mildew
Milan	downy mildew
	mosaic virus
	root aphid
	tipburn
Multy	downy mildew
Musette	downy mildew
	lettuce mosaic virus
Nika	good disease resistance
Novita	downy mildew
Nymans	downy mildew
Pandero	downy mildew
Pinares	tipburn
Pinokkio	high disease resistance
Rachel	downy mildew
Revolution	bremia
Romance	lettuce mosaic virus
Romany	tipburn
Roxy	tipburn
Rushmore	downy mildew
Ruth	downy mildew
Sabrina	downy mildew
Saladin	downy mildew
	tipburn
Saladin Supreme	tipburn
Sierra	good disease resistance
Sigmahead	botrytis
Sonette	downy mildew
Sunny	downy mildew
	lettuce mosaic virus
Sylvesta	downy mildew
	leaf aphid
Targa	downy mildew
TinTin	downy mildew
	leaf aphid
	root aphid
Valdor	botrytis
Vienna	downy mildew
	leaf aphid

Marrow

variety	*resistant or tolerant to*
Badger Cross	cucumber mosaic virus
Tarmino	cucumber mosaic virus
Tiger Cross	cucumber mosaic virus
Zebra Cross	cucumber mosaic virus

Onion

variety	*resistant or tolerant to*
Feast	downy mildew
Golden Bear	botrytis
	downy mildew
	white rot tolerant
Marco	good disease tolerance
Norstar	downy mildew

Parsnip

variety	*resistant or tolerant to*
Arrow	canker
Avonresister	canker
Cobham Improved	canker
Countess	canker
Excalibur	canker
Gladiator	canker
Javelin	canker
Lancer	canker
New White Skin	canker
Palace	canker
Panache	canker
Polar	canker
Tender & True	canker
White Gem	canker

Pea

variety	*resistant or tolerant to*
Ambassador	fusarium wilt
	pea enation virus
	powdery mildew
Avola	fusarium wilt
Cascadia	fusarium wilt
Cavalier	powdery mildew
Delikata	fusarium wilt
	powdery mildew
Green Shaft	downy mildew
	fusarium wilt
Jaguar	powdery mildew
Jessy	fusarium wilt
	powdery mildew
Kelvedon Wonder	fusarium wilt
	powdery mildew wilt
Kodiak	downy mildew
	fusarium wilt
Markana	fusarium wilt
Norli	fusarium wilt
Onward	fusarium wilt
Oregon Sugar Pod	fusarium wilt
	pea enation virus
	powdery mildew
Rondo	fusarium wilt
Spring	fusarium wilt
Starlight	downy mildew
Sugar Snap	fusarium wilt
Sugarbon	fusarium wilt
Twinkle	fusarium wilt
Zuccola	fusarium wilt

Pepper

variety	*resistant or tolerant to*
Anaheim	tobacco mosaic virus
Antaro	tobacco mosaic virus
Bell Boy	tobacco mosaic virus
Canape	tobacco mosaic virus
Gypsy	tobacco mosaic virus
Luteus	tobacco mosaic virus
Midnight Beauty	tobacco mosaic virus
Ringo	tobacco mosaic virus
Unicorn	tobacco mosaic virus

Potato

variety	*resistant or tolerant to*
Accent	eelworm
	scab
Anya	blight (moderate)
	scab
Arran Pilot	blight (moderate)
	scab
Arran Victory	blight
Cara	blight
Catriona	scab
Charlotte	blight
	scab (moderate)
Colleen	blight
Cosmos	blight
Desirée	blight
	scab (moderate)
Duke of York	scab
Estima	blight (moderate)
	scab
Eve	blight
Foremost	blight (moderate)
	scab
Homeguard	scab
King Edward	scab
Lady Balfour	blight
Lady Christl	eelworm
	scab
Majestic	blight (moderate)
	scab
Marfona	blight
	scab (moderate)
Maris Bard	blight
	scab (moderate)
Maris Peer	blight
	scab (moderate)
Maris Piper	blight (moderate)
	eelworm
Markies	blight
Nadine	blight (moderate)
	eelworm
	scab
Orla	blight
Pentland Javelin	eelworm
	scab
Pink Fir Apple	scab
Pomeroy	blight
Premiere	blight
Ratte	scab
Remarka	blight
Rocket	eelworm
	scab (moderate)
Sarpo Axona	blight
Sarpo Mira	blight (100% resistant)
Saxon	blight (moderate)
	eelworm
	scab

variety	resistant or tolerant to
Sharpes Express	blight (moderate)
	scab
Swift	blight (moderate)
	scab
Valor	blight
Verity	blight
Wilja	blight (moderate)
	scab
Winston	fusarium wilt
	nematode
	scab
Spinach	
Bergola	downy mildew
Fiorano	downy mildew
	powdery mildew
Galaxy	downy mildew
	powdery mildew
Mazarka	downy mildew
Palco	downy mildew
Senic	downy mildew
	powdery mildew
Space	downy mildew
Tetona	downy mildew
Tornado	downy mildew
Trinidad	downy mildew
Triton	downy mildew
Squash (Butternut)	
Metro	powdery mildew
Swede	
Marian	clubroot
	powdery mildew
Magres	powdery mildew
Virtue	clubroot
	powdery mildew
Willemsburger	clubroot
Sweetcorn	
Landmark	high disease tolerance
Tuxedo	high disease tolerance
Tomato	
Alexanovas	fusarium wilt
	tomato mosaic virus
Alicante	greenback
	powdery mildew
Amarel	blossom end rot
	tomato mosaic virus
Aviro	tomato mosaic virus
Cherry Wonder	fusarium wilt
	greenback
	leaf mould
	tomato mosaic virus
Craigella	greenback
Cristal	good disease resistance
Cumulus	fusarium wilt
	tomato mosaic virus
Cyclon	leaf mould

variety	resistant or tolerant to
	tomato mosaic virus
	verticillium wilt
Delicate	good all round resistance
Dombito	tomato mosaic virus
Dona	fusarium wilt
	tomato mosaic virus
	verticillium wilt
Estrella	fusarium wilt
	greenback
	leaf mould
	tomato mosaic virus
	verticillium wilt
Eurocross	greenback
	leaf mould
Falcorosso	fusarium wilt
	leaf mould
	tobacco mosaic virus
	verticillium wilt
Favorita	good all round disease resistance
Ferline	blight
	fusarium wilt
	verticillium
Golden Boy	alternaria stem canker
Golden Sweet	fusarium wilt
	rootknot nematode
Grenadier	fusarium wilt
	leaf mould
Herald	leaf mould
Husky Gold	fusarium wilt
	verticillium wilt
Ida	fusarium wilt
	leaf mould
	tomato mosaic virus
	verticillium wilt
Incas	blossom end rot
	fusarium wilt (high)
	verticillium wilt
Legend	blight
Libra	fusarium crown and root rot
Matador	greenback
Matina	greenback
Moravi	fusarium wilt
	leaf mould
	tobacco mosaic virus
	verticillium wilt
Nimbus	fusarium wilt
	tomato mosaic virus
	verticillium wilt
Olivade	good disease resistance
Piccolo	leaf mould
	tomato mosaic virus
Piranto	fusarium crown and root rot
Pixie	fusarium wilt
	tomato mosaic virus
	verticillium wilt
Primato	fusarium wilt
	leaf mould
	tomato mosaic virus
	verticillium wilt

variety	resistant or tolerant to
Royal des Guineaux	good all round disease resistance
Sakura	fusarium wilt
	tomato mosaic virus
Seville Cross	leaf mould
Shirley	cladosporium
	fusarium wilt
	greenback
	tomato mosaic virus
Sonatine	fusarium wilt
	leaf mould
	tomato mosaic virus
Sungold	fusarium wilt
	tomato mosaic virus
Typhoon	fusarium wilt
	leaf mould
	tomato mosaic virus
	verticillium
Turnip	
Oasis	turnip mosaic virus

Companion planting in action

GROWING SPROUTING SEEDS AND MICRO-GREENS

Onion sprouts

Sprouting seeds and micro-greens are eaten in salads, sandwiches or as a garnish for other dishes. Containing high levels of minerals, protein, fibre and enzymes, they are not only delicious but also nutrient-rich. The difference between the two is that sprouted seeds are harvested when the seed leaves are present and micro-greens when the first true leaves appear.

Suitable plants include: alfalfa, aduki beans, amaranthus, beetroot, basil, broad beans, broccoli, buckwheat, cabbage, carrot, chard, coriander, celery, clover, fennel, kale, leek, lovage, *Mesembryanthemum chrystallinum*, mizuna greens, mustard salad, onion, Oxalis, peas, radish, rape salad, red mustard, rocket, salad cress, salad mixes, shiso, soya bean, wheatgrass.

sprouting seeds

Seed sowing

They can be sown in a 'seed sprouter', a clear glass jar with a muslin 'lid' or a similar receptacle. Sterilise the equipment before use and after every 3 or 4 crops of seedlings by soaking for 10 minutes in a solution of 1 tablespoon of bleach or similar to a pint of water, before scrubbing and rinsing thoroughly.

To encourage germination, soak seeds for the time recommended on the seed packet at a 3:1 ratio of water to seed to start the germination process. Stir the seeds with a tablespoon or fork to ensure they are all moistened then place them in the sprouting container. Rinse the seeds thoroughly 2 to 3 times a day, using water under a high pressure and drain thoroughly to avoid disease problems. Keep the seedlings in a bright position, away from scorching sunlight and maintain the temperature between 13–21°C (55–70°F).

Harvesting and Storing

Harvest when the seedling leaves appear. Ensure that sprouts are properly dry before storing in the refrigerator. Around 12 hours after their final watering, drain the seedlings thoroughly, using a salad spinner if possible and store them in a sealed plastic bag to prevent dessication. They will store for several days in the salad drawer of the fridge. When eating large quantities of sprouts, it is advisable to cook them first.

micro-greens

At the time of writing, micro-greens are very trendy and valued by top chefs for the subtle flavour of their stems. They are easy to grow and are the ideal way to make the most of half-used packets of seeds. Seedlings of a wide range of herbs and vegetables can be used as micro-greens, particularly those whose leaves are traditionally eaten at maturity. Do not use seedlings from plants with poisonous leaves like tomato, potato and rhubarb.

Seed sowing

Sow successionally from spring to autumn for a constant supply. Sow in seed trays in a 5cm layer of compost. For larger volumes, make a wooden frame in a heated propagator in spring and fill it with compost to make a large seed-bed. Sow seeds close together over the surface of the compost, but not so densely that they are too congested and prone to disease. Cover with a layer of compost or vermiculite or perlite to reduce moisture loss and keep the stems clean and place the seed tray in a warm, bright place.

When growing seedlings on a windowsill, turn the trays every two or three days to maintain balanced growth and stop them from growing towards the light. Keep the compost moist but not waterlogged.

Harvesting and Storing

Harvest with a pair of scissors when the first true leaves appear, retaining as much of the stem as possible. Use new compost for each crop adding used material to your compost heap. Large seeds like peas can be germinated on several layers of moist kitchen towel.

Water 24 hours before storing as they deteriorate rapidly if refrigerated when wet. Micro-greens with soft stems are more vulnerable to damage through being wet when stored.

Small-scale sprouting is still productive

HARVESTING AND STORING

One of the pleasures of a productive garden is picking your produce and cooking or eating it fresh. However, there are myriad means of storing many crops, thus extending the season, and avoiding unmanageable gluts. See also details under individual crops.

when to harvest

Vegetables

While the majority of vegetables are not harvested until they reach maturity, others, like some lettuces, are harvested while semi-mature and others still, such as rocket, are harvested while juvenile. Some vegetables, again like lettuce, cannot be stored for long and so must be harvested and eaten fresh. Many other leafy vegetables, such as Brussels sprouts and sprouting broccoli, are hardy, surviving outdoors in the ground in freezing conditions. The flavour of Brussels sprouts even improves immeasurably after they have been frosted. Some root vegetables with a high moisture content are easily damaged in winter, even when protected by the soil. This is usually caused by rapid thawing after a period of cold weather. Carrots, parsnip and swede are exceptions; they are very hardy and can be left in free-draining soils until required. In wet soils these crops would suffer winter loss by slug damage and rotting.

Herbs

Herbs can be harvested from very early on in their growing season. This encourages the plant to produce vigorous new growth and allows the plant to be controlled in shape and size. Most herbs reach their peak of flavour just before they flower. Snip off suitable stems early in the day before the sun is fully up, or even better on a cloudy day (provided it is not too humid). Cut whole stems rather than single leaves or flowers. Always use a sharp knife, sharp scissors or secateurs, and cut lengths of 5–8cm (2–3in) from the tip of the branch, this being the new, soft growth. Do not cut into any of the older, woody growth. Cut from all over the plant, leaving it looking shapely. Pick herbs that are clean and free from pests and disease. If herbs are covered in garden soil, sponge them quickly and lightly with cold water, not hot as this will draw out the oils prematurely. Pat dry as quickly as possible. Keep each species separately so that they do not contaminate each other.

Most annual herbs can be harvested at least twice during a growing season. Cut them to within 10–15cm

(4–6in) of the ground, and feed with liquid fertilizer after each cutting. Give annuals their final cut of the season before the first frosts; they will have stopped growing some weeks before.

In the first year of planting, perennials will give one good crop; thereafter it will be possible to harvest two or three times during the growing season. Do not cut into the woody growth unless deliberately trying to prevent growth; again, cut well before frosts. There are of course exceptions: sage is still very good after frosts, and both thyme and golden marjoram (with some protection) can be picked gently even in midwinter.

Pick flowers for drying when they are barely open. Seed should be collected as soon as you notice a change in colour of the seed pod; if, when you tap the pod, a few scatter on the ground, it is the time to gather them. Seeds ripen very fast, so watch them carefully.

Roots of herbs are at their peak of flavour when they have just completed a growing season. Dig them throughout autumn as growth ceases. Lift whole roots with a garden fork, taking care not to puncture or bruise the outer skin. Wash them free of soil. Cut away any remains of top-growth and any fibrous off-shoots. For drying, cut large, thick roots in half lengthways and then into smaller pieces for ease.

Fruit

Without doubt most fruits are best, and certainly are enjoyed most, when they are plucked fully ripe off the tree or vine. Only a few, such as melons, are improved by chilling first. The majority are tastiest fresh and warmed by the sun. Some, such as pears, have to be carefully nurtured till they are fully ripe and need to be picked early and brought to perfection, watched daily, in a warm, not too dry, dim room, Medlars are similarly picked early and are then ripened, or bletted, to the point of rotting.

The best date for picking will vary with the cultivar, the soil, the site and the season, and can only be determined by experience as these factors all vary considerably. Of course, it will generally remain much the same in relation to other fruits nearby, which are also subject to the same conditions, i.e., in a late year, most fruits are late, which is pretty self-evident anyway.

On any tree, the sunny side ripens first. Fruit will also ripen earlier where extra warmth is supplied, so that growing sites next to a wall, window, chimney or vent, or close to the soil are good places for finding early fruits. Likewise, when all the rest have gone, you may find some hidden in the shade.

If you want to store fruits for home use, they need to be picked at just the right stage. Most fruits store best when picked just under-ripe. They may keep much longer if picked even younger, but this is at the cost of flavour and sweetness.

storage conditions

Vegetables

In colder areas, vegetables overwintering in the ground need further protection. This can be provided by spreading a layer of loose straw or bracken over them to a depth of 20cm (8in) and covering with a plastic sheet. While this is labour-saving, crops stored in this way are susceptible to attack by pests and diseases through the winter.

The traditional method for storing root vegetables is in a clamp or 'pie'. Low mounds of vegetables are laid on a bed of loose straw up to 20cm (8in) thick. The top and sides of the mound are covered with a similar layer of straw, and then with a 15cm (6in) layer of soil or sand. Clamps can be made outdoors on a well-drained site or under cover in a shed or outhouse; for extra protection outdoors, they can be formed against a wall or hedge. If crops are to be stored for a long period, find a site that receives as little sunlight as possible during the winter. Although storage conditions are very similar to those in the ground, harvesting from a clamp is much easier. However, losses from rodent damage and rotting can be high.

The length of time that vegetables may be stored depends on the type and cultivar as well as the storage conditions. When traditional methods are used, the main cause of deterioration is moisture loss from plant tissue; for instance, beetroot and carrot desiccate very rapidly. Fungal infection of damaged tissue is a common problem; onions and potatoes bruise very easily. Vegetables for storage should be handled carefully; only store those that are disease-free, check them regularly and remove any showing signs of decay immediately.

Fruits

Although we occasionally store some fruits, such as pears, for a period to improve their condition, predominantly we store fruit to extend the season so that we can enjoy them for as long as possible.

To be stored, all fruit must be perfect. Any blemish or bruise is where moulds start. Choose varieties that are suitable for storing – many early croppers are notoriously bad keepers!

Common, long-keeping fruits such as apples and quinces can be stored at home for months, or even up to a year. The major problems, apart from the moulds, are shrivelling due to water loss and the depredations of rodents and other big pests. A conventional store is too large for most of us and the house or garage is too warm, too cold or too dry. Dead deep freezers and refrigerators make excellent, compact stores. They are dark, keep the contents at the same constant temperature and will keep out night frosts easily. Most useful of all, they are rodent-proof.

Some ventilation is needed and can be obtained by cutting holes in the rubber door or lid seal. If condensa-

Dried herbs will keep for several months and see you tastily through the winter

tion occurs it usually indicates insufficient ventilation, but too much draught will dry out the fruits. The unit can stand outdoors, as it needs no power. In a shed it is out of sight and better protected against the cold, but may then get too warm. In the UK, have it outdoors in the shade or in a cool shed.

When putting fruits in the store it is usually best to leave them to chill at night in trays and to load them into the store in the morning when they have dried off but before they have warmed up again. Similarly it is helpful to chill and dry off the fruits initially by leaving the store open on chill, dry nights and closing it during the day for a week or two after filling.

Most fruits are best removed from the store some time before use, so any staleness can leave them. Care should be taken not to store early and late varieties together or any that may cross-taint. Obviously it is not a good idea to site your store in the same place as strong-smelling things such as onions, paint or bleach! Likewise, although straw is a convenient litter, if it gets damp it taints the fruit. Shredded newspaper is safer, though it also has a slight whiff. Dried stinging nettles are reckoned good but dangerous to handle. Always inspect stored fruits regularly. They can go off very quickly. Remember, if only one in ten goes off every month, you have to start with two trays just to have one tray left after six months. So do not store any fruit long just for the sake of it, but store well what you will use.

Careful storage will cope with seasonal gluts

drying

Vegetables

Peas and beans can be harvested when almost mature and dried slowly in a cool place. Either lift the whole plant and hang it up or pick off the pods and dry them on a tray or newspaper. Beans can be collected and stored in airtight jars until required. They will need soaking for 24 hours before use. Storage areas should be frost-free. Chillies, green peppers, garlic and onions can be hung indoors where there is good air circulation. Peppers, tomatoes and mushrooms can be cut into sections and sun-dried outdoors where conditions allow, otherwise in the gentle heat of an airing cupboard or above a radiator.

Herbs

The object of drying herbs is to eliminate the water content of the plant quickly and, at the same time, to retain the essential oils. Herbs need to be dried in a warm, dark, dry and well-ventilated place. The faster they dry, the better the aromatic oils are retained. Darkness helps to prevent loss of colour and unique flavours. The area must be dry, with a good air flow, to hasten the drying process and to discourage mould.

Suitable places for drying herbs include: an airing cupboard; attic space immediately under the roof (provided it does not get too hot); in the oven at a low temperature and with the door ajar (place the herbs on a brown piece of paper with holes punched in it and check regularly that the herbs are not over-heating; a plate-warming compartment; a spare room with curtains shut and door open. The temperature should be kept at 21–33°C/ 70–90°F.

Herb leaves should always be dried separately from each other, especially the more strongly scented ones. Spread them in a single layer on trays or slatted wooden racks covered with muslin or netting. Place the trays or frames in the drying areas so that they have good air circulation. Turn the herbs over by hand several times during the first two days.

Roots require a higher temperature – from 50–60°C/120–140°F. They dry more quickly and easily in an oven and require regular turning until they are fragile and break easily. Seed should be dried without any artificial heat and in an airy place. Almost-ripe seed heads can be hung in paper bags (plastic causes them to sweat) so the majority of seeds will fall into the bag as they mature. They need to be dried thoroughly before storing and the process can take up to two weeks.

An alternative method for flowers, roots or seed heads is to tie them in small bundles of 8 to 10 stems. Do not pack the stems too tightly together. Then hang them on coat-hangers in an airy, dark room until they are dry. The length of drying time varies from herb to herb, and week to week. The determining factor is the state of the plant material. If herbs are stored before drying is complete, moisture will be reabsorbed from the atmosphere and the herb will soon deteriorate. Leaves should be both brittle and crisp. They should break easily into small pieces but should not reduce to a powder when touched. The roots should be brittle and dry right through. Any softness or sponginess means they are not sufficiently dry and, if stored, will rot.

It is possible to dry herbs by microwave, but easy to over-dry and cook the leaves to the point of complete disintegration. Small-leafed herbs such as rosemary and thyme take about 1 minute, whilst the larger, moist leaves of mint dry in about 3 minutes. Add an eggcupful of water to the microwave during the process. Herbs lose their flavour and colour if not stored properly. Pack the leaves or roots, not too tightly, into a dark glass jar with an air-tight screw top. Label with name and date. Keep in a dark cupboard; nothing destroys the quality of the herb quicker at this stage than exposure to light.

After the initial storing, keep a check on the jars for several days. If moisture starts to form on the inside of the container, the herbs have not been dried correctly. Return them to the drying area and allow further drying time.

Most domestic herb needs are comparatively small so there is little point in storing large amounts for a long time. The shelf life of dried herbs is only about one year so it is sufficient to keep enough just for the winter. If you have large, dark jars, thyme and rosemary can be left on the stalk. This makes it easier to use them in casseroles and stews and to remove before serving.

Fruit

Many fruits can be dried if they are sliced thinly and exposed to warm, dry air. Sealed in dark containers and kept cool and dry, they keep for long periods to be eaten, dried or reconstituted, when required. However, in the UK, and much of maritime Europe and North America, the air is too humid and so drying is not quick enough, not helped by the low temperatures in these regions. Solar-powered dryers – simply wire trays under glass – with good ventilation, allow fruit to be dried to a larger extent, but in the highest humidity regions the fruit may still go mouldy before it dries.

Slicing the fruit thinly and hanging the pieces, separated by at least half their own diameter, on long strings over a cooking range provides the dry warmth and ventilation needed to desiccate most within a day or two, or even just overnight for the easier ones such as apple.

Oven-drying with artificial heat is risky as it can cook the fruit. It is possible, however, if the temperature is kept down and the door kept partly open. It may be convenient to finish off partly dried samples in the cooling oven after you have finished baking. The dying heat desiccates fruit well with little risk of caramelizing.

freezing

Vegetables

Only the best-quality vegetables should be frozen, and they should always be thoroughly cleaned and carefully packed. Vegetables should be fast-frozen. Most freezers have a fast-freeze switch, and keep the freezer door closed during freezing. Do not open the freezer door regularly or leave it open longer than necessary. Many vegetables need to be blanched before freezing, and others need to be shredded, puréed or diced, or frozen when young. See under individual vegetables for details.

Herbs

Freezing is great for culinary herbs as colour, flavour and the nutritional value of the fresh young leaves are retained, and it is quick and easy. It is far better to freeze herbs such as fennel, dill, parsley, tarragon and chives than to dry them.

Pick the herbs and, if necessary, rinse with cold water, and shake dry before freezing, being careful not to bruise the leaves. Put small amounts of herbs into labelled, plastic bags, either singly, or as a mixture for bouquet garnis. Either have a set place in the freezer for them or put the bags into a container, so that they do not get damaged with the day-to-day use of the freezer.

There is no need to thaw herbs before use; simply add them to the cooking as required. For chopped parsley and other fine-leaved herbs, freeze the bunches whole in bags and, when you remove them from the freezer, crush the parsley in its bag with your hand.

Another way to freeze herbs conveniently is to put finely chopped leaves into an ice-cube tray and top them up with water. The average cube holds 1 tablespoon chopped herbs and 1 teaspoon water. The flowers of borage and leaves of variegated mints look very attractive frozen individually in ice-cubes for drinks or fruit salads.

Fruits

Most fruits freeze easily with little preparation. Obviously only the best are worth freezing as few fruits are improved by the process! Most fruits turn soggy when defrosted, but they are still packed full of sweetness, flavour and vitamins so are well worth having for culinary use, especially in tarts, pies, sauces and compôtes. A mixture of frozen fruits is marvellous if they are dechilled but not totally defrosted, so they retain their frozen texture like pieces of sorbet, served with cream.

For most fruits, merely putting them in sealed freezer bags or boxes is sufficient. However they then tend to freeze in a block. If you freeze them loose on open wire drying trays or greased baking trays, they can be packed afterwards and will stay separate. Fruits that are cut or damaged need to be drained first or, if you have a sweet tooth, they can be dredged in sugar, which absorbs the juice, before freezing them.

Stone fruits are best de-stoned before freezing as the stone can give an almond taint otherwise. The tough skins of fruits such as plums are most easily removed after freezing and before use. Carefully squeeze the frozen fruit under very hot water, when the skin will slip off.

Fruits lose value slowly in the freezer. The longer they are frozen the less use they are nutritionally.

Parsley can be frozen stored in a bag, or in ice cubes

juicing

Not all fruits can be juiced, but the majority can be squeezed to express the juice, or heated or frozen to break down the texture then strained. Sugar may be added to taste as it improves the colour, flavour and keeping qualities. When the juice remains unheated, honey may be substituted, though it has a strong flavour of its own. Sweet juices such as apple can be mixed with tart ones like plum.

Fruit juices may be drunk as they are, added to cocktails, drunk as squashes diluted with water, and used in cooking. Grapes are the easiest to press and the most rewarding; they are best crushed first to break the skins. Most of the currants and berries can be squeezed in the same way. Apples and pears must be crushed first and then squeezed; they will go through the same juicing equipment as grapes, but more slowly than the more juicy fruits.

Pulpy, firm fruits such as blackcurrants and plums are best simmered with water till they soften then the juice can be strained off. If you repeat the process and add sugar to the combined juices you also have the basis for jellies. Raspberries, strawberries and fruits with similar delicate flavours are best frozen then defrosted and strained, to obtain a pure juice unchanged by heating.

Suitable equipment for processing large amounts of fruit is widely available if the quantities are too large for kitchen tools.

Commercially juices are passed through microfine filters or flash pasteurized. At home they will ferment rapidly in the warm, last longer if kept cool in the refrigerator, and keep for months or years deep frozen.

HERB OILS, VINEGARS AND PRESERVES

Many herbs have antiseptic and anti-bacterial qualities, and were used in preserving long before there were cookbooks. Herbs aid digestion, stimulate appetite and enhance the flavour of food. I hope the following recipes will tempt you, because a variety of herb oils and vinegars can lead to the creation of unusual and interesting dishes. Herbs can make salad dressings, tomato-based sauces for pasta dishes, marinades for fish and meat, and can act as softening agents for vegetables, introducing myruad new tastes and flavours. They also make marvellous presents.

herb oils

These can be used in salad dressings, in marinades, sauces, stir-fry dishes and sautéing. Find some interesting bottles with good shapes. To start with, you need a sterilised glass jar, large enough to hold 500ml, with a screw top.

Basil oil

This is one of the best ways of storing and capturing the unique flavour of basil.

4 tablespoons basil leaves
500ml olive or sunflower oil

Pick over the basil, remove the leaves from the stalks, and crush them in a mortar. For Greek basil, with its small leaves, simply crush in the mortar. Pound very slightly. Add a little oil and pound gently again. This bruises the leaves, releasing their own oil into the oil. Mix the leaves with the rest of the oil and pour into a small bowl. Put a little water into a small saucepan and place the bowl on the rim of the pan so that the bottom of the bowl does not touch the water. Bring the water to the boil, then turn down to simmer for 15 minutes. Stir the oil occasionally and check that the water does not boil dry. Remove the bowl from the heat, cover and allow to cool before straining into a sterilized bottle. Label and date. This oil will keep in a refrigerator for one month.

Adapt this recipe for dill, fennel (green), sweet marjoram, rosemary and garden or lemon thyme. Garlic also makes a very good oil. Use 4 cloves of garlic, peeled and crushed, and combined with the oil.

Bouquet Garni Oil
I use this oil for numerous dishes.

1 tablespoon sage
1 tablespoon lemon thyme
1 tablespoon Greek oregano
1 tablespoon French parsley
1 bay leaf
500ml olive or sunflower oil

Break all the leaves and mix them together in a mortar, pounding lightly. Add a small amount of the oil to mix well, allowing the flavours to infuse, then proceed as for basil oil above.

Sweet Oils
Good with fruit dishes, marinades and puddings. Use almond oil, which combines well with scented flowers such as pinks, lavender, lemon verbena, rose petals and scented geraniums. Make as for savoury oils above, mixing 4 tablespoons of torn petals or leaves with 500ml almond oil.

Spice Oils
Ideal for salad dressings, they can be used for sautéing and stir-frying too. The most suitable herb spices are: coriander seeds, dill seeds and fennel seeds. Combine 2 tablespoons of seeds with 500ml olive or sunflower oil, having first pounded the seeds gently to crush them in a mortar and mixed them with a little of the oil. Add a few of the whole seeds to the oil before bottling and labelling. Treat as for savoury oils and store.

herbal vinegars

Made in much the same way as oils, they can be used in gravies and sauces, marinades and salad dressings.

10 tablespoons chopped herb, such as basil, chervil, dill, fennel, garlic, lemon balm, marjoram, mint, rosemary, savory, tarragon or thyme
500ml white wine or cider vinegar

Pound the leaves gently in a mortar. Heat half the vinegar until warm but not boiling, and pour it over the herbs in the mortar. Pound further to release the flavours of the herb. Leave to cool. Mix this mixture with the remaining vinegar and pour into a wide-necked bottle. Seal tightly. Remember to use an acid-proof lid (lining the existing lid with greaseproof paper is a way round this). Put on a sunny windowsill and shake each day for 2 weeks. Test for flavour; if a stronger taste is required, strain the vinegar and repeat with fresh herbs. Store as is or strain through double muslin and rebottle. Add a fresh sprig of the chosen herb to the bottle for ease of identification.

To Save Time
1 bottle white wine vinegar (500ml)
4 large sprigs herb
4 garlic cloves, peeled

Pour off a little vinegar from the bottle and push in 2 sprigs of herb and the garlic cloves. Top up with the reserved vinegar if necessary. Reseal the bottle and leave on a sunny windowsill for 2 weeks. Change the herb sprigs for fresh ones and the vinegar is now ready to use.

Seed Vinegar
Make as for herb vinegar, but use 2 tablespoons of seeds to 600ml white wine or cider vinegar. Dill, fennel and coriander seeds make well-flavoured vinegars.

Floral Vinegar
Made in the same way, these are used for fruit salads and cosmetic recipes. Combine elder, nasturtiums, sweet violets, pinks, lavender, primrose, rose petals, rosemary or thyme flowers in the following proportions:

10 tablespoons torn flower heads or petals
500ml white wine vinegar

Pickled Horseradish
As a child I lived in a small village in the West of England where an old man called Mr Bell sat outside his cottage crying in the early autumn. It took me a long time to understand why – he was scraping the horseradish root into a bowl to make pickle. He did this outside because the fumes given off by the horseradish were so strong they made one's eyes water.

Wash and scrape the skin off a good-size horseradish root. Mince in a food processor or grate it (if you can stand it!). Pack into small jars and cover with salted vinegar made from 1 teaspoon salt to 300ml cider or white wine vinegar. Seal and leave for 4 weeks before using.

Pickled Nasturtium Seeds
Poor man's capers!
Pick nasturtium seeds while they are still green. Steep in brine made from 100g salt to 1 litre water for 24 hours. Strain the seeds and put 2 tablespoons of them into a small jar. Into a saucepan add 1 clove of peeled garlic, 1 teaspoon black peppercorns, 1 teaspoon dill seeds, 1 tablespoon English mace leaves and enough white wine vinegar to fill the jar. Slowly bring to simmering point. Then strain the vinegar and pour over the seeds. Seal the jars with acid-proof lids and leave for about 4 weeks. After opening, store in the refrigerator and use the contents quickly.

savoury herb jelly

Use the following herbs: sweet marjoram, mints (all kinds), rosemary, sage, summer savory, tarragon and common thyme.

Makes 2 x 350g jars

1 kg tart cooking apples or crab apples, roughly chopped, cores and all
900ml water
500g sugar
2 tablespoons white wine vinegar
2 tablespoons lemon juice
1 bunch herbs, approx. 15g
4 tablespoons chopped herbs

Put the apples into a large pan with the bunch of herbs and cover with cold water in a preserving pan. Bring to the boil and simmer until the apples are soft, roughly 30 minutes. Pour into a jelly bag and drain overnight.

Measure the strained juice and add 500g sugar to every 600ml fluid. Stir over gentle heat until the sugar has dissolved. Bring to the boil, stirring, and boil until setting point is reached. This takes roughly 20–30 minutes. Skim the surface scum and stir in the vinegar and lemon juice and the chopped herbs. Pour into jars, seal and label before storing.

sweet jellies

Follow the above recipe, omitting the vinegar and lemon juice, and instead adding 150ml water. The following make interestingly flavoured sweet jellies: bergamot, lavender flower, lemon verbena, scented geranium, sweet violet and lemon balm.

preserves

Coriander Chutney
Makes 2 x 500g jars

1kg cooking apples, peeled, cored and sliced
500g onions, peeled and roughly chopped
2 cloves garlic, peeled and crushed
1 red and 1 green pepper, deseeded and sliced
900ml red wine vinegar
500g soft brown sugar
½ tablespoon whole coriander seeds
6 peppercorns tied securely in a piece of muslin
6 all-spice berries
50g root ginger, peeled and sliced
2 tablespoons coriander leaves, chopped
2 tablespoons mint, chopped

Combine the apples with the onions, garlic and peppers in a large, heavy saucepan. Add the vinegar and bring to the boil, simmering for about 30 minutes until all the ingredients are soft. Add the brown sugar and the muslin bag of seeds and berries. Then add the ginger. Heat gently, stirring all the time, until the sugar has dissolved, then simmer until thick; this can take up to 60 minutes. Stir in the chopped coriander and mint and spoon into hot, sterilized jars. Seal and label when cool.

NATURAL DYES

Herbs have been used to dye cloth since the earliest records. In fact, until the 19th century and the birth of the chemical industry, all dyes were 'natural'. Then the chemical process, offering a larger range of colours and a more guaranteed result, took over. Now, once again, there is a real demand for more natural products and colours, which has resulted in a revival of interest in plants as a dye source.

The most common dyeing herbs are listed in the dye chart. You will notice that yellows, browns and greys are predominant. Plenty of plant material will be required so be careful not to over-pick in your own garden (and please do not pick other people's plants without permission! I plead from personal experience). To begin with, keep it simple. Pick the flowers just as they are coming out, the leaves when they are young and fresh and a good green; dig up roots in the autumn and cut them up well before use.

fabric

Any natural material can be dyed; some are more tricky than others. It just takes time and practice. In the following sections I will explain the techniques connected with dying wool, the most reliable and easiest of natural materials. Silk, linen and cotton can also be dyed, but are more difficult.

preparation

First time, this is a messy and fairly lengthy process, so use a utility room, or clear the decks in the kitchen and protect all areas. Best of all keep it away from the home altogether. Some of the mordants used for fixing dye are poisonous, so keep them well away from children, pets and food. The actual dyeing process is not difficult, but you will need space, and a few special pieces of equipment:

1 large stainless-steel vessel, such as a preserving can (to be used as the dye-bath)
1 stainless-steel or enamel bucket and bowl
1 pair of tongs (wooden or stainless-steel; to be used for lifting)
1 measuring jug
1 pair rubber gloves essential for all but Jumblies –
'Their heads are green and their hands are blue,
And they went to sea in a sieve.'
Pestle and mortar
Thermometer
Water; this must be soft, either rainwater or filtered
Scales

Dyeing comprises four separate tasks:

1. Preparation of the material (known as scouring)

2. Preparation of the mordant

3. Preparation of the dye

4. Dyeing process

preparation of the material (scouring)

Prepare the wool by washing it in a hot solution of soap flakes or a proprietary scouring agent in order to remove any grease. Always handle the wool gently. Rinse it several times, squeezing (gently) between each rinse. On the final rinse add 50ml of vinegar.

The flowers of St John's Wort produce a beige dye

preparation of the mordant

Mordants help 'fix' the dye to the fabric. They are available from chemists or dye suppliers. The list below includes some of the more common. Some natural dyers say that one should not use mordants, but without them the dye will run very easily.

Alum: Use 25g to 500g dry wool
This is the most useful of mordants, its full title being potassium aluminium sulphate. Sometimes potassium hydrogen tartrate, cream of tartar, is added (beware! this is not the baking substance) in order to facilitate the process and brighten the colour.

Iron: Use 5g to 500g dry wool
This is ferrous sulphate. It dulls and deepens the colours. It is added in the final process after first using the mordant Alum. Remove the wool before adding the iron, then replace the wool and simmer until you get the depth of colour required.

Copper: Use 15g to 500g dry wool
This is copper sulphate. If you mix with 300ml of vinegar when preparing the mordant it will give a blue/green tint to colours.
WARNING: Wear gloves; copper is poisonous.

Chrome: Use 15g to 500g dry wool
This is bichromate of potash and light-sensitive, so keep it in the dark. It gives the colour depth, makes the colours fast, and gives the wool a soft, silky feel.
WARNING: Wear gloves; chrome is poisonous.

1. Dissolve the mordant in a little hot water.

2. Stir into 20 litres of hot water 50°C.

3. When thoroughly dissolved immerse the wet, washed wool in the mixture. Make sure it is wholly immersed.

4. Slowly bring to the boil and simmer at 82–94°C (180–200°F) for an hour.

5. Remove from the heat. Take the wool out of the water and rinse.

preparation of the dye

No two batches of herbal dye will be the same. There are so many variable factors – plant variety, water, mordant, immersion time.

The amount of plant material required for dyeing is very variable. A good starting ratio is 500g of mordanted wool in skeins to 500g plant material.

1. Chop or crush the plant material.

2. Place loosely in a muslin or nylon bag and tie securely.

3. Leave to soak in 20 litres of soft, tepid water overnight.

4. Slowly bring the water and herb material to the boil.

5. Reduce heat and simmer at 82–94°C (180–200°F) for as long as it takes to get the water to the desired colour. This can take from 1 to 3 hours.

6. Remove the pan from the heat, remove the herb material, and allow the liquid to cool to hand temperature.

dyeing process

1. Gently add the wool.

2. Bring the water slowly to the boil, stirring occasionally with the wooden tongs.

3. Allow to simmer for a further hour.

4. Remove pan from the heat and leave the wool in the dye-bath until cold, or until the colour is right.

5. Remove the wool with the tongs and rinse in tepid water until no colour runs out.

6. Give a final rinse in cold water.

7. Dry the skeins of wool over a rod or cord, away from direct heat. Tie a light weight to the bottom to stop the wool kinking during the drying process.

dye chart

Common Name	Botanical Name	Part Used	Mordant	Colour
Comfrey	*Symphytum officinale*	Leaves and stalks	Alum	Yellows
Chamomile, Dyers	*Anthemis tinctoria*	Flowers	Alum	Yellows
Chamomile, Dyers	*Anthemis tinctoria*	Flowers	Copper	Olives
Elder	*Sambucus nigra*	Leaves	Alum	Greens
Elder	*Sambucus nigra*	Berries	Alum	Violets/Purple
Goldenrod	*Solidago canadensis*	Whole plant	Chrome	Golden Yellows
Horsetail	*Equisetum arvense*	Stems and leaves	Alum	Yellows
Juniper	*Juniperus communis*	Crushed berries	Alum	Yellows
Marigold	*Calendula officinalis*	Petals	Alum	Pale Yellow
Meadowsweet	*Filipendula ulmaria*	Roots	Alum	Black
Nettle	*Urtica dioica*	Whole plant	Copper	Greyish-green
St John's Wort	*Hypericum perforatum*	Flowers	Alum	Beiges
Sorrel	*Rumex acetosa*	Whole plant	Alum	Dirty yellow
Sorrel	*Rumex acetosa*	Roots	Alum	Beige/pink
Tansy	*Tanacetum vulgare*	Flowers	Alum	Yellows
Woad	*Isatus tinctoria*	Leaves	Sodium dithionite, ammonia	Blues

YEARLY CALENDAR

As any gardener knows, you cannot be precise about months and seasons. Each year is different – wetter, windier, hotter, drier, colder than the last. So use this calendar as a general guide.

mid-winter

Look back over the previous year, at successes and mishaps, and plan for the following season. Think about any structural changes as well as ordering seeds and plants. Keep an eye out for any weather damage in the garden, checking especially after heavy frost, snow and gales.

Vegetables

Prepare cropping plants. Lime autumn-dug plots if necessary. Place early seed potatoes in shallow boxes with 'eyes' uppermost, and store in a light, frost-free place. Towards the end of the month, plant out shallots if soil is moist enough.

Under cover, sow radishes and carrots in growbags or in the borders of a cold glasshouse; sow lettuce for growing under cloches, and leeks.

Herbs

This is one of the quietest times. Keep an eye on the degrees of frost and protect tender herbs with an extra layer of agricultural fleece or mulch if necessary.

With a little bit of protection in the garden, bay, hyssop, rosemary, sage, winter savory, thyme, lemon thyme, chervil and parsley can be picked.

Start parsley seed with heat. Outside, if not sown in the autumn, sow sweet cicely, sweet woodruff and cowslip, to enable a period of stratification. Force chives, mint and tarragon in boxes in the greenhouse.

Keep watering of containers to a minimum. Clean old pots ready for the spring 'pot up'.

Fruit

Make a health and hygiene check and examine each plant in your care for pests, diseases and dieback. Check stores, remove and use any fruits starting to deteriorate before they go over and infect others.

late winter

As the days lengthen, and if the weather is not too unpleasant, this is a good time to have the final tidy up before the busy season starts. If you want to get an early start in the garden and you have prepared a site the previous autumn, cover the soil now with black polythene. It will warm up the soil and force any weeds.

Vegetables

Sow broad beans, early peas and spinach. Sow early-maturing cabbage and cauliflowers in pots from the middle of the month on. 'Chit' maincrop potatoes. Sow cabbages, carrots, lettuces and radishes under cloches or in a polytunnel.

Herbs

With a little bit of protection in the garden, bay, hyssop, rosemary, sage, winter savory, thyme, lemon thyme, chervil and parsley can be harvested. Chives start to come up if they are under protection, and mint can be available if forced. Start borage, dill and parsley seeds with heat, and sow chervil in a cold greenhouse.

Herbaceous perennial herbs, including chives, lemon balm, pot marjoram, mints, oregano, broad-leafed sorrel and tarragon, can be divided now, as long as they are not too frozen and are given added protection after replanting.

As the containers have been brought in for the winter, new life may be starting. Dust off, and slowly start watering.

Outside, check that any dead or decaying herbaceous growth is not damaging plants. Check for wind and snow damage.

Fruit

Check stores, remove and use any fruits starting to deteriorate before they go over and infect others. Spread compost or well-rotted manure under and around everything possible and add a good layer of mulch, preferably immediately after a period of heavy rain.

Lime most grass swards one year in four, more often on acid soil, but not amongst ericaceous plants or lime-haters! Once ground becomes workable, plant out hardy trees and shrubs that missed the autumn planting. Sow the very earliest crops for growing under cover. Ensure good weed control, hoeing fortnightly or adding extra mulches on top.

Do major pruning work to trees and bushes missed earlier or damaged in winter (but not stone fruits or evergreens). Prune autumn-fruiting raspberries to the ground.

Spray everything growing with diluted seaweed solution at least once a month, and anything with deficiency symptoms more often. Spray peaches and almonds with Bordeaux mixture to protect against peach-leaf curl.

Examine each plant for pests, diseases and dieback. Apply sticky bands and inspect sacking bands on apple trees, and others if they suffered from many pests.

Check straps and stakes after gales. On still, cold nights protect the blossoms and young fruitlets from frost damage with net curtains, plastic sheeting or newspaper.

early spring

Gradually uncover tender plants outside and look for hopeful signs of life. Give them a gentle tidy.

Vegetables

Plant onion sets in prepared ground. Sow beetroot, cabbage, carrots, parsnip, lettuce and maincrop peas.

Plant lettuce under cloches or in plastic tunnels, sow salad onions in growbags in an unheated glasshouse for early crops. Sow early cabbages, cauliflowers and tomates under glass for transplanting later.

Herbs

Herbs available for picking include angelica, lemon balm, bay, chives, fennel, hyssop, mint, parsley, peppermint, pennyroyal, rue, sage, savory, sorrel and thyme.

Towards the end of the month start borage, fennel, coriander, sweet marjoram, rue and basil seeds with heat. In a cold greenhouse sow chervil, chives, dill, lemon balm, lovage, parsley, sage, summer savory and sorrel. Outside in the garden sow chervil, chives, parsley

Planting seeds

The vegetable patch starts to flourish in spring

(cover with cloches), chamomile, tansy, caraway, borage and fennel.

Check seeds that have been left outside from the previous autumn for stratification. If they are starting to germinate, move them into a cold greenhouse.

Take root cuttings from mint, tarragon, bergamot, chamomile, hyssop, tansy, sweet woodruff and sweet cicely. Divide mint, tarragon, wormwood, lovage, rue, sorrel, lemon balm, salad burnet, camphor plant, thyme, winter savory, marjoram, alecost, horehound and pennyroyal. This is a good time to start mound layering on old sages or thymes.

Tidy up all the pots, trim old growth to maintain shape. Start liquid feeding with seaweed. Re-pot if necessary. Pot up new plants into containers for display later in the season.

In the garden, clear up all the winter debris, fork the soil over and give a light dressing of bonemeal. If your soil is alkaline, and you gave it a good dressing of manure in the autumn, now is the time to dig it well in.

Remove black polythene, weed and place a cloche over important sowing sites a week before sowing to raise the soil temperature and keep it dry.

Towards the end of the month, if the major frosts are over, you can cut lavender back and into shape. This is certainly advisable for plants two years old and older. Give them a good mulch. Equally, sage bushes of two years and older would benefit from a trim, but not hard back. Cut back elder and rosemary, neither of which minds a hard cutting back. Transplant the following if they need it: alecost, chives, mint, balm, pot marjoram, sorrel, horehound and rue.

Fruit

Continue to spread a good layer of compost or well-rotted manure under and around everything possible. Spread wood ashes under and around plants, giving priority to gooseberries and culinary apples. Plant out evergreen and the more tender hardy plants. Protect them from frost and wind the first season. Sow plants grown under cover or for later planting out.

Maintain good weed control by hoeing fortnightly or adding extra mulch on top. Cut the grass at least fortnightly, preferably weekly, returning the clippings or raking them into rings around trees and bushes.

Spray everything growing with diluted seaweed solution at least once a month, and anything with deficiency symptoms more often. Spray peaches and almonds with Bordeaux mixture against peach-leaf curl. Make a health and hygiene check on each plant in your care for pests, diseases and dieback.

Prune back tender plants and evergreens. Protect the new growth against frost afterwards. Pollinate early-flowering plants and those under cover by hand. On still, cold nights protect the blossoms and young fruitlets from frost damage with net curtains, plastic sheet or newspaper.

mid-spring

This is the main month for sowing outdoors, as soon as soil conditions permit. It is also now possible to prune back to strong new shoots the branches of any shrubs that have suffered in winter.

Vegetables

Dig plots occupied by winter greens and prepare for leeks; plant maincrop potatoes, cabbages and cauliflowers (sown under cover in March). Make further sowings of beetroot, radishes, spinach, carrots, cauliflower, maincrop peas, broad beans and parsnips. Sow iceberg lettuce, salad onions and seakale. Sow winter cauliflower, Savoy cabbage, kale and broccoli into seedbeds. Transplant cabbages and cauliflowers sown in nursery beds in late winter. Pinch out the growing tips of flowering broad beans.

Tomatoes sown and pricked out in early spring should be transferred to a coldframe and hardened off. Sow marrows, squashes and sweetcorn at the end of the month. Prepare greenhouse borders or growbags for tomatoes, and plant towards the end of the month.

Herbs

Herbs available for picking include: angelica, balm, bay, borage, caraway, chervil, chives, fennel, hyssop, lovage, pot marjoram, mints, parsley, pennyroyal, peppermint, rosemary, sage, winter savory, sorrel, thyme, tarragon and lemon thyme.

Start basil seeds with heat. In the cold greenhouse sow borage, chervil, coriander, dill, fennel, lemon balm, lovage, pot marjoram, sweet marjoram, sage, summer savory, winter savory, sorrel, buckler leaf sorrel, horehound, rue, bergamot, caraway and garden thyme. Outside in the garden sow parsley, chives, hyssop, caraway and pot marigold. Prick out the previous month's sown seeds, pot on, or harden off before planting out.

Take softwood cuttings of rue, mint, sage, southernwood, winter savory, thymes, horehound, lavender, rosemary, cotton lavender and curry. Take root cuttings of sweet cicely, fennel and mint. Divide pennyroyal, chives, lady's mantle, salad burnet and tarragon.

You should be able to put all containers outside now; keep an eye on the watering and feeding. In the garden, if all the frosts have finished, the following will need cutting and pruning into shape: bay, winter savory, hyssop, cotton lavender, lavenders, rue especially variegated rue, southernwood and thymes.

Fruit

Ensure good weed control, hoeing weekly or adding extra mulch on top. Cut the grass at least weekly, returning the clippings or raking them into rings around trees and bushes. Sow plants grown under cover or for later planting out. Plant out more tender hardy plants under cover or with protection.

Spread a good layer of mulch under and around everything possible, and spread wood ashes under and around fruit trees, giving priority to gooseberries and culinary apples. Spray everything growing with diluted seaweed solution at least once a month, and anything with deficiency symptoms more often.

Water all new plants established within the previous twelve months whenever there has been little rain. De-flower or de-fruit new plants to give them time to establish. Pollinate plants under cover by hand. Tie in new growths of vines and climbing plants. Make a health and hygiene check weekly and examine each plant in your care for pests, diseases and dieback. On still, cold nights protect the blossoms and young fruitlets from frost damage with net curtains, plastic sheet or newspaper. Make layers of difficult subjects.

late spring

Vegetables

Sow French and runner beans, Chinese cabbage, carrots, courgettes, outdoor cucumbers, lettuce, turnips, spinach and parsley. Plant out celeriac, celery, sweetcorn, and summer cabbage. Harvest asparagus, broad beans, cauliflowers, peas, radish and spinach. Earth up potatoes. Under cover, pot on aubergines, outdoor cucumbers, tomatoes, peppers and sweetcorn.

Herbs

Everything should be growing quickly now. The annuals will need thinning; tender and half-hardy plants should be hardened off under a cold frame or by a warm wall. A watch must be kept for a sudden late frost; basil is especially susceptible. Move container specimens into bigger pots or top-dress with new compost.

Nearly all varieties of herbs are available for picking. Outside, keep sowing coriander, dill, chervil, parsley, sweet marjoram, basil, and any other annuals you require to maintain crops. Prick out and pot on or plant out any of the previous month's seedlings.

Take softwood cuttings of marjorams, all mints, oregano, rosemarys, winter savory, French tarragon and all thymes. Containers should now be looking good. Keep trimming to maintain shape; water and feed regularly. In the garden, trim southernwood into shape.

Cut second-year growth of angelica for candying. Cut thyme before flowering to dry.

Fruit

Maintain good weed control. Sow plants grown under cover or outdoors. Plant out tender plants under cover or with protection. Pollinate plants under cover by hand. Cut the grass at least fortnightly, preferably weekly, returning the clippings or raking them into rings around trees and bushes.

Water all new plants established within the last twelve months especially if there has been little rain. De-flower or de-fruit new plants. Spray everything growing with diluted seaweed solution at least once a month, and anything with deficiency symptoms more often.

Examine each plant twice weekly for pests, diseases and dieback. Tie in new growths of vine and climbing plants. On still, cold nights protect the blossoms and young fruitlets from frost damage with net curtains, plastic sheet or newspaper. Make layers of difficult subjects. Protect almost every ripening fruit from the birds.

early summer

Vegetables

Sow French and runner beans, Chinese cabbage, carrots, courgettes, outdoor cucumbers, lettuce, turnips,

Preparing the soil for seed sowing

spinach and parsley. Transplant celery, summer cabbage and tomatoes. Harvest asparagus, broad beans, cauliflowers, calabrese, peas, radish, spinach and turnips. Under cover, transplant aubergines and sweetcorn; pollinate tomato plants.

Herbs

This is a great time in the herb garden. All planting is now completed. Plants are beginning to join up so that little further weeding will be needed. Many plants are now reaching perfection.

All herbs are available fresh. Outside in the garden sow basil, borage, chives, coriander, dill, fennel, sweet marjoram, summer savory, winter savory, and any others you wish to replace, or keep going.

With all the soft new growth available this is a very busy month for cuttings. Make sure you use material from non-flowering shoots. Take softwood cuttings of all perennial marjorams, all mints, all rosemary, all sage, variegated lemon balm, tarragon (French) and all thymes. Also divide thymes and layer rosemary.

Plant up annual herbs such as basil and sweet marjoram into containers to keep near the kitchen. If you must plant basil in the garden, do it now. Nip out the growing tips of this year's young plants to encourage them to bush out.

Trim cotton lavender hedges if flowers are not required and to maintain their shape; clip box hedges and topiary shapes as needed. A new herb garden should be weeded thoroughly to give the new plants the best chance.

Cut second-year growth of angelica for candying. Cut sage for drying.

Fruit

Maintain good weed control. Plant out the tender plants or move them out for summer. Cut the grass at least fortnightly, preferably weekly, returning the clippings or raking them into rings around trees and bushes. Raise the height of cut of your mower. Spray everything growing with diluted seaweed solution at least once a month, and anything with deficiency symptoms more often.

Water all new plants established within the previous twelve months especially whenever there has been little rain. Examine each plant twice weekly for pests, diseases and dieback.

Start summer pruning. This applies to all red- and whitecurrants, gooseberries and all trained apples and pears. From one third of each plant, remove approximately half to three-quarters of each new shoot, except

Autumn's bounty

for leaders. Prune grapevines back to three or five leaves after a flower truss. Tie in new growths of vine and climbing plants.

Also begin fruit thinning. To do this, remove every diseased, decayed, damaged, misshapen, distorted and congested fruitlet. This applies to all apples, pears, peaches, apricots, quality plums, dessert grapes, gooseberries, figs and especially to trained forms. Compost or burn rejected fruitlets immediately. Of course usable ones, such as the larger gooseberries, may be consumed.

Take softwood cuttings if you have a propagator. Make layers of difficult subjects. Protect almost every ripening fruit from birds.

mid-summer

Vegetables

Sow final crops of beetroot, carrots, lettuce, turnips, spinach and parsley. Sow salad onions, spring cabbage and seakale for overwintering. Sow keeping onions in a seedbed for transplanting the following March. In colder districts, plant leeks sown in coldframes in January at the beginning of the month. Remove basal suckers from early trench celery, water well and earth up.

Under cover, harvest cucumbers and tomatoes regularly to encourage further fruiting, and pinch out the growing point when each stem contains about 5 or 6 trusses of fruit.

Herbs

The season is on the wane, the early annuals and biennials are beginning to go over. It is already time to think of next year and to start collecting seeds. Take cuttings of tender shrubs as spare shoots are available.

All herbs can be harvested fresh. Outside in the garden sow chervil, angelica (if seed is set), borage, coriander, dill, lovage and parsley. Take softwood cuttings of wormwood, scented geraniums, lavenders and the thymes. Layer rosemary. Keep an eye on watering of containers as the temperatures begin to rise.

Cut all lavenders back after flowering to maintain their shape. If this is the first summer of the herb garden and the plants are not fully established it is important to make sure they do not dry out, so water regularly. Once established, many are tolerant of drought.

Harvest and dry lemon balm, horehound, summer savory, hyssop, tarragon, thyme and lavender. Use lavender and rosemary for dyeing and potpourris. Harvest seed of caraway and angelica.

Fruit

Maintain good weed control. Cut the grass at least fortnightly, preferably weekly, returning the clippings or raking them into rings around trees and bushes. Raise the height of cut of your mower. Spray everything growing with diluted seaweed solution at least once a month, and anything with deficiency symptoms more often. Water all new plants established within the previous twelve months especially if there has been little rain. Examine each plant for pests, diseases and dieback. Tie in new growths of vines and climbing plants.

Continue summer pruning. For red- and whitecurrants, gooseberries and all trained apples and pears, from the second third of each plant remove approximately half to three-quarters of each new shoot, except for leaders. Prune grapevines back to three or five leaves after a flower truss. Blackcurrants may have a third to half of the old wood removed after fruiting. Stone fruits are traditionally pruned now to avoid silver-leaf disease.

Continue to thin fruits as in June. Take softwood cuttings if you have a propagator, and root tips of the black and hybrid berries. Protect almost every ripening fruit from the birds.

late summer

Vegetables

Sow turnips for spring 'greens' and Japanese onions for overwintering. Sow spring cabbage in nursery rows. To prevent wind-rock in autumn and winter, draw soil around the stems of winter greens, particularly Brussels sprouts, kale and broccoli. Earth up trench celery. Harvest maincrop onions, ensuring that the bulbs' outer skins are well-ripened before storing. Under cover, harvest cucumbers and tomatoes regularly and self-pollinate tomatoes.

Herbs

Traditionally a month for holidays but it is also time to harvest and preserve many herbs for winter use. Collect and dry material for potpourris, and collect seeds for sowing next year.

All herbs are available fresh. In the garden or greenhouse sow angelica, coriander, dill, lovage, parsley, winter savory. Take softwood cuttings of bay, wormwood, rosemary, the thymes and lavenders, scented geraniums, balm of Gilead, pineapple sage and myrtles.

If you are going away, make sure you ask a friend to water your containers for you. Give box, cotton lavender and curry their second clipping and trim any established plants that are looking unruly. Maintain watering of the new herb garden and keep an eye on mints, parsley and comfrey, which need water to flourish. There is no real need to feed your herbs if the ground has been well prepared, but if the plants are recovering from a pest attack they will benefit from a foliar feed of liquid seaweed.

Harvest thyme, sage, clary sage, marjoram and lavender for drying. Pick the mints and pennyroyal to freeze. Gather basil to make a basil oil. Collect the seed of angelica, anise, caraway, coriander, cumin, chervil, dill and fennel.

Fruit

Maintain good weed control. Plant new strawberry plants, if you can get them. Cut the grass at least fortnightly, preferably weekly, returning the clippings or raking them into rings around trees and bushes. Lower the height of cut of your mower. Spray everything that is growing with diluted seaweed solution at least once a month, and anything that has deficiency symptoms more often. Water all new plants established within the previous twelve months especially whenever there has been little rain. Sow green manures and winter ground cover on bare soil that is not mulched; grass down orchards. Check for pests, diseases and dieback, and apply sticky bands and sacking bands to apple trees, and to others if they suffer from many pests.

Finish summer pruning. For red- and whitecurrants, gooseberries and all trained apples and pears, for the last, unpruned third of each plant, remove approximately half to three-quarters of each new shoot, except for leaders. Prune grapevines back to three or five leaves after the fruit truss. Thin fruits as in June. Protect almost every ripening fruit from the birds. Root the tips of the black- and hybrid berries.

early autumn

Vegetables

Order seed catalogues for the following year. Plant out spring cabbages into permanent positions. Sow spinach for harvesting in April. Lift maincrop carrots, beetroot and potatoes, and store in a cool, dark place: later sowings may be left in the ground. Earth up celery before severe frosts. Lift tomato plants with fruits still attached and store; ripen on straw under cloches or in the greenhouse. Wrap green fruits in paper and store in the dark.

Plant thinnings from late-sown salads in frames or under cloches to provide crops during winter. Lettuces reaching maturity should be covered with cloches or frames; sow further crops in a coldframe. Sow broad beans, lamb's lettuce, pak choi, early peas, land cress and spring greens.

Herbs

By mid-month, basil should be taken up and leaves preserved. Line out semi-ripe cuttings of box, cotton lavenders, etc, in coldframes, under cloches or in polythene tunnels for hedge renewal in the spring.

Herbs that can be picked fresh include: lemon balm, basil, bay, borage, caraway, chervil, chives, clary sage, fennel, hyssop, pot marigold, marjoram, the mints, parsley, pennyroyal, peppermint, rosemarys, sages, winter savory, sorrels and the thymes. Sow outside in the garden or greenhouse angelica, chives, coriander, parsley and winter savory. Take softwood and semi-ripe cuttings of rosemary, thyme, tarragon, lavender, rue, cotton lavender, curry plant and box. Divide bergamot.

At the beginning of the month give the shrubby herbs their final clipping (bay, lavender, etc). Do not leave it too late or the frost could damage the new growth. Put basil into a glasshouse or kitchen. Top-dress bergamots if they have died back. If lemon verbena is to be kept outside make sure it is getting adequate protection. Towards the end of this month take in all containers, and protect tender plants like bay trees, myrtles, and scented geraniums. Harvest dandelion (roots), parsley, marigold, clary sage and peppermint for drying or freezing. Collect seed of angelica, anise, caraway, chervil and fennel.

Fruit

Maintain good weed control. Plant out pot-grown specimens and those that can be dug with a decent rootball or moved with little disturbance. Cut the grass at least fortnightly, preferably weekly, returning the clippings and fallen leaves or raking them into rings around trees and bushes. Sow green manures and winter ground cover on bare soil that is not mulched; grass down orchards. Spray everything that is growing with diluted seaweed solution at least once a month, and anything that has deficiency symptoms more often.

Make a health and hygiene check. Apply sticky bands and sacking bands to apple trees, and to others if they suffer from many pests. On still, cold nights protect ripening fruits from frost damage with net curtains, plastic sheet or newspaper. Protect first the tops then the stems and roots of more tender plants before frosts come. Bring indoors tender plants in pots or protect them. Take cuttings of plants as they start to drop their leaves. Prune early fruiting raspberries and hybrids, and blackcurrants and other plants as they start to drop their leaves. Protect almost every ripening fruit from the birds. Root the tips of the black- and hybrid berries.

mid-autumn

Vegetables

Lift potatoes, beet and carrots for storing. Tie onions on to ropes when the skins have thoroughly ripened. Transplant lettuces sown in July to a well-drained, protected site to overwinter. Plant root cuttings of seakale in pots of sand and leave them in a sheltered place until the spring. Cut down asparagus foliage as it turns yellow. Tidy the vegetable plot, removing all plant debris. Add organic matter to plots, and lime if necessary.

Sow lettuce in greenhouse borders or growbags for cutting in the spring. Continue harvesting green tomatoes, storing in a dark, frost-free place to ripen. Clear growbags used for cucumbers, peppers and tomatoes in the summer and replant with winter lettuce.

Herbs

The best time in all but the coldest areas to plant hardy perennial herbs.

Basil, bay, borage, chervil, fennel, hyssop, marigold, marjoram, parsley, rosemary, sage, winter savory, sorrel and the thymes can all be picked fresh. Sow parsley seed with heat. In the garden sow catmint, chervil, wormwood, chamomile, fennel and angelica.

Take softwood and semi-ripe cuttings of bay, elder, hyssop, cotton lavender, southernwood, lavenders, the thymes, curry and box. Take root cuttings of tansy, pennyroyal, the mints and tarragon. Divide alecost, the marjorams, chives, lemon balm, lady's mantle, hyssop, bergamot, camphor plant, lovage, sorrel, sage, oregano and pennyroyal.

Start reducing the watering of containers. Clear the garden and weed it well. Cut down the old growth and collect any remaining seed heads. Cut back the mints, trim winter savory and hyssop. Give them all a leaf mould dressing. Dig up and remove the annuals such as dill, coriander, borage, summer savory and sweet marjoram, and the second-year biennials – parsley, chervil, rocket etc. Protect with cloches or agricultural fleece any herbs to be used fresh in winter, like parsley, chervil, lemon thyme, salad burnet. Dig up some French tarragon, pot up in trays for forcing and protection.

Check the pH of alkaline soil every third year. Dress with well-rotted manure to a depth of 5–10cm and leave the digging until the following spring. Dig over heavy soils; add manure to allow the frost to penetrate.

Fruit

Ensure good weed control. Plant out bare-rooted hardy trees and bushes if soil is in good condition and they are dormant. Cut the grass at least fortnightly, preferably weekly, collecting the clippings with the fallen leaves or raking them into rings around trees and bushes. Spray everything that is growing with diluted seaweed solution at least once a month, and anything that has deficiency symptoms more often.

Check plants for pests, diseases and dieback. Top up the sticky bands and inspect the sacking bands on apple trees, and on others if they suffered from many pests. Check straps and stakes before the gales. On still, cold nights protect ripening fruits from frost damage with net curtains, plastic sheet or newspaper.

Take cuttings of hardy plants as they start to drop their leaves. Prune early-fruiting raspberries and hybrids, and blackcurrants and other plants as they start to drop their leaves. Protect first the tops then the stems and roots of more tender plants before frosts come. Check stores, remove and use any fruits starting to deteriorate before they go over and infect others. Protect almost every ripening fruit from the birds.

late autumn

Vegetables

Sow broad beans, and round-seeded peas in the open: protect with cloches if necessary. Remove dying leaves from winter greens, allowing air to circulate between plants. Check stored vegetables regularly and remove any showing signs of decay. Use those that are slightly damaged immediately. Sow green manure. Lift and store crowns of chicory as well as seakale.

Herbs

The days are getting shorter and frosts are starting. Planting of hardy herbaceous herbs can continue as long as soil remains unfrozen and in a workable condition.

Basil, bay, hyssop, marjoram, mint, parsley, rosemary, rue, sage and thyme are available to pick fresh. Sow the following so that they can get a good period of stratification: arnica (old seed), sweet woodruff, yellow iris, poppy, soapwort, sweet cicely, hops (old seed), sweet violet. Sow in trays, cover with glass and leave outside in a cold frame or corner of the garden where they cannot get damaged.

Cut back on all watering of container-grown plants. Give them all a prune, so that they go into rest mode for the winter. This is the time for the final tidy up in the garden. Cut back the remaining plants, lemon balm, alecost, horehound, and give them a dressing of leaf mould. Give the elders a prune. Dig up a clump of mint and chives, put them in pots or trays and bring them into the greenhouse for forcing for winter use.

Fruit

Keep on top of weeds. Plant out bare-rooted hardy trees and bushes if the soil is in good condition and they are dormant. Cut the grass at least fortnightly, collecting the clippings with the fallen leaves or raking them into rings around trees and bushes. Check plants for pests, diseases and dieback. Top up the sticky bands and inspect the sacking bands on apple trees, and on others if they suffered from many pests.

Check straps and stakes before the gales. Spread a good layer of compost or well-rotted manure under and around everything possible, preferably after a period of heavy rain. On still, cold nights protect ripening fruits from frost damage with net curtains, plastic sheet or newspaper. Protect first the tops then the stems and roots of more tender plants before frosts come.

Take cuttings of hardy plants as they start to drop their leaves. Prune late-fruiting raspberries, hybrid berries, currants and vines, trees and bushes as the leaves fall. Check stores, remove and use any fruits starting to deteriorate before they go over and infect others. Protect almost every ripening fruit from the birds.

early winter

Vegetables

Plan next year's rotation of vegetables, ordering seeds as soon as possible. Prepare a seed-sowing schedule. Lift and store swede and late-sown carrots. If heavy falls of snow or prolonged frosts are forecast, lift small quantities of vegetables such as celery, leeks and parsnips and store under cover in a cool, easily accessible place. Finish digging before the soil becomes waterlogged. On clay soils, spread sand, old potting compost or well-rotted leaf mould on the surface and dig in as soon as conditions are favourable, allowing the frost to break down the soil. Sow green manures on sandy soils or cover with compost to reduce leaching.

Herbs

Bay, hyssop, marjoram, oregano, mint (forced), parsley, chervil, rosemary, rue, sage and thyme are all available for picking fresh. In the garden, remove all the dead growth that falls into other plants, add more protective layers if needed. Wrap and terracotta or stone ornaments in sacking if you live in extremely cold conditions. Bring bay trees in if the temperature drops too low. Keep an eye on the plants you are forcing in the greenhouse.

Fruit

Enusre good weed control, make sure no weeds are getting away, hoe at fortnightly intervals or add extra mulch on top. Plant out bare-rooted hardy trees and bushes if soil is good and they are dormant. Examine each plant in your care for pests and diseases.

Examine plants for pests, diseases and dieback. Top up the sticky bands and inspect the sacking bands on apple trees, and on others if they suffered from many pests. Check straps and stakes before the gales. Spread a good layer of compost or well-rotted manure under and around everything possible, preferably after, but not immediately before, any heavy rain.

Peaches crop well in drier, warmer areas

GLOSSARY

Agricultural fleece A light, woven fleece used to cover crops and protect them from frost, wind, hail, birds, rabbits and other pests that eat or chew the leaves. It allows a light transmission of around 85 per cent and can be permeated by rain.
Analgesic A substance that relieves pain.
Annual A plant that completes its life-cycle from germination to flowering and death in one growing season.
Antidote A substance that counteracts or neutralises a poison.
Apothecary An old term for a person who prepared and sold drugs and administered to the sick.
Aromatherapy The use of essential oils in the treatment of medical problems and for cosmetic purposes.
Astringent A substance that contracts the tissues of the body, checking discharges of blood and mucus.
Atropine An alkaloid obtained from members of the Solanaceae family.

Base dressing An application of organic matter or fertiliser, applied to the soil prior to planting or sowing.
Bed system A method of planting vegetables in close blocks or multiple rows.
Beta carotene The orange-yellow plant pigment and precursor of vitamin A which protects against certain cancers and heart disease.
Biennial A plant that produces roots and leaves in the first growing season, then flowers, seeds and dies by the end of the second.
Blanch To exclude light from leaves and stems and prevent development of green coloration. In the culinary sense, to immerse in boiling water for the removal of skin or colour, often as a preparation for freezing.
Bolt To flower and produce seed prematurely.
Brassica Member of the cabbage family (Brassicaceae), including broccoli, Brussels sprouts, cauliflower etc.
Broadcast To scatter granular substances such as seeds, fertiliser or pesticide evenly over an area of ground.
Bulb A modified plant stem, with swollen leaves acting as a storage organ.
Bulbil A small bulb rising above the ground in the axil of a leaf or bract.

Capping A crust forming on the surface of soil damaged by compaction, heavy rain or watering.
Carminative A substance that allays pain and relieves flatulence and colic.
Catch-crop A rapidly maturing crop sown among slow-growing vegetables to make maximum use of the ground.
Chitting Pre-germination of seeds before sowing. The same term is used for sprouting potatoes.
Clamp A structure made of earth for storing root vegetables outdoors.
Climber A plant which naturally grows upwards covering supports or other plants.
Cloche A small portable structure, often made of plastic or glass, used to protect early crops grown outdoors.
Cold frame A low-lying square or rectangular unheated structure, with a glass or plastic lid.
Columnar Column-shaped.
Compost Decomposed organic material used as soil conditioner, mulch, potting or seed-sowing medium.
Crop rotation A system in which crops are grown in different plots on a three- or four-year cycle, limiting the build-up of pests and diseases and making the best use of soil nutrients.
Cultivar A contraction of 'cultivated variety', a group of cultivated plants which retain desirable characteristics when propagated.
Cultivate The practice of growing plants in soil or compost. Can also be used to describe methods of soil preparation.
Cut and come again crops Seedlings, mature or semi-mature varieties, where several harvests can be taken from one crop.

Damp down To wet the floors and benches in a glasshouse in order to increase humidity and lower high temperatures.
Deciduous Describes a plant that loses its leaves annually at the end of the growing season.
Decoction An extract of a herb (when the material is hard and woody, i.e., root, wood, bark, nuts) obtained by boiling a set weight of plant matter in a set volume of water for a set time.
Diuretic A substance that increases the frequency of urination.
Dormancy Temporary cessation of growth during the dormant season.
Double digging A cultivation technique which penetrates to two spades depth, also known as trench digging or bastard trenching.

Earth up To draw the soil around the base for support or to cover a plant for the purpose of blanching.
Emetic A substance that induces vomiting.
Essential oil A volatile oil obtained from a plant by distillation, having a similar aroma to the plant itself.
Evergreen A plant that retains its leaves throughout the year.

F1 hybrid First-generation plants obtained by crossing two selected pure-breeding parents to produce uniform vigorous offspring.
Fanging A term used to describe the forking of a root vegetable.
Fertiliser A chemical or group of chemicals applied to the soil or plants to provide nutrition.
Fleece Lightweight, woven polypropylene cover used for crop protection.
Floating mulch (floating cloche) Sheets of flexible lightweight material placed over plants to provide protection.
Fluid sow A method for sowing germinated seeds into the soil using a carrier gel.
Folic acid Part of the vitamin B complex, found in leafy vegetables. Deficiency of folic acid causes anaemia.
Friable Used to describe soil with a crumbly, workable texture, capable of forming a tilth.
Fungicide Chemical used for the control and eradication of fungi.

Genus A taxonomic classification used to describe plants with several similar characteristics.
Germination The chemical and physical changes which take place as a seed starts to grow.
Globose Spherical, globe-like.
Green manure A rapidly maturing, leafy crop grown for incorporation into the soil to improve its structure and nutrient levels.
Grow-bag A bag of compost used as a growing medium.

Half hardy A term that describes plants that tolerate low temperatures but not frost.
Harden off To acclimatise plants gradually, enabling them to withstand cooler conditions.
Hardy A term that describes plants that can withstand frost without protection.
Haulm The foliage of plants such as potatoes.

Heart up The stage at which leafy vegetables, like cabbage and lettuce, swell to form a dense cluster of central leaves.
Heavy soil A soil with a high proportion of clay particles, prone to waterlogging in winter and drying in summer.
Herbicide A chemical used to control and eradicate weeds.
Herbaceous Relating to plants that are not woody and that die down at the end of each growing season.
Herbicide A substance that kills plants.
Humus The organic decayed remains of plant material in soils.
Hybrid A variety of plant resulting from the crossing of two distinct species or genera.

Infusion An infusion is made by pouring a given quantity of boiling water over a given weight of soft herbal material (leaves or petals) and left to steep for 10–15 minutes before straining.
Inorganic Term used to describe fertilisers made from refined naturally occurring chemicals, or artificial fertilisers.
Insecticide Chemical used to eradicate insects.
Intercropping The practice of planting fast-growing vegetable crops between slower-growing varieties (see *Catch-crop*).
Inulin An easily digestible form of carbohydrate.

John Innes compost Loam-based growing medium made to standardised formulas.

Knot garden A decorative formal garden popular in the 16th century and normally consisting of very low hedges in geometric patterns.

Leaching The downward washing and loss of soluble nutrients from topsoil.
Leaf mould Partially decomposed leaves.
Legume A single-celled fruit containing several seeds which splits on maturity, e.g., beans and peas.
Lime Calcium compounds used to raise the pH of the soil (i.e., make it less acidic).
Loam The term used for a soil of medium texture.

Maincrop The largest crop produced throughout the main growing season. Also used to describe the cultivars used.
Module A generic term for the containers used for propagating and growing young plants.
Mordant A substance used in dyeing that, when applied to the fabric to be dyed, reacts chemically with the dye, fixing the colour.
Mulch A substance spread around a plant to protect it from weeds, water loss, heat or cold, and in some cases to provide nutrient material. Materials ranging from sawdust and pine needles to black plastic can be used as mulches; leaves and old straw are most commonly employed. Mulches should only be applied to moist soil.

Nematocide Chemical used for the control and eradication of nematodes (eelworms).
Neutral Soil or compost with a pH value of 7, which is neither acid nor alkaline (see pH).
Nutrients Minerals which are essential for plant growth.

Organic Term used to describe substances which are derived from natural materials. Also used to denote gardening without the use of synthetic chemicals and composts, mulches and associated material.

Pan A layer of compacted soil that is impermeable to water and oxygen, and impedes root development and drainage.
Perennial Non-woody plants that die back and become dormant during winter, regrowing tle following spring. Perennials usually survive for 3 or more years, sometimes much longer.
Perlite Expanded volcanic rock. It is inert, sterile and has a neutral pH value (i.e., it is neither acidic nor alkaline).
pH A measure of acidity or alkalinity. The scale ranges from 0 to 14, and is an indicator of the soluble calcium within a soil or growing medium. A pH below 7 is acid and above is alkaline.
Pinch out To remove the growing tip of a plant to induce branching.
Potager An ornamental vegetable garden.
Pot on To move a plant into a larger pot.
Prick out To transfer seedlings, from a seedbed or tray, to a further pot, tray or seedbed.
Propagation The increase of plant numbers by seed or vegetative means.

Radicle A seedling root.
Rhizome A swollen underground stem that stores food and from which roots and shoots are produced.
Root The part of the plant which is responsible for absorbing water and nutrients and for anchoring the plant into the growing medium.
Root crops Vegetables grown for their edible roots, e.g., carrot and parsnip.
Runner A trailing shoot that roots where it touches the ground.

Salve A soothing ointment.
Saponin A substance that foams in water and has a detergent action.
Seed A ripened plant ovule containing a dormant embryo, which is capable of forming a new plant.
Seed leaves or cotyledons The first leaf or leaves formed by a seed after germination.
Seedling A young plant grown from seed.
Sets Small onions, shallots or potatoes used for planting.
Shoot A branch, stem or twig of a plant.
Shrub A perennial plant with woody stems growing from or near the base.
Sideshoot A branch, stem or twig growing from a main stem of a plant.
Single digging A cultivation technique which penetrates to one spade's depth.
Species A plant or plants within the same genus. These can be grown from seed and in the main run true to type.
Spore The reproductive body of a non-flowering plant.
Stale seedbed method A cultivation technique whereby the seedbed is created and subsequent weed growth is removed before crops are sown or planted.
Stamen The pollen-producing part of the plant.
Stigma The part of a pistil (the female organs of a flower) that accepts the pollen.
Subsoil Layers of less fertile soil immediately below the topsoil.
Sucker A stem originating below soil level, usually from the plant's roots or underground stem.
Systemic or translocated A term used to describe a chemical which is absorbed by a plant at one point and tlen circulated through its sap system.

Tap root The primary anchoring root of a plant, usually growing straight down into the soil. In vegetables this is often used for food storage.
Tender Plant material which is intolerant of cool conditions.
Thinning The removal of seedlings or shoots to improve tle quality of those which remain.
Thymol A bactericide and fungicide found in several volatile oils.
Tilth The surface layer of soil produced by cultivation and soil improvement.
Tincture A solution that has been extracted from plant material after macerating in alcohol or alcohol/water solutions.
Tisane A drink made by the addition of boiling water to fresh or dried unfermented plant material.
Top-dressing The application of fertilisers or bulky organic matter to the soil surface, while the plants are in situ.
Topsoil The upper, usually most fertile layer of soil.
Transpiration The loss by evaporation of moisture from plant leaves and stems.
Topiary The art of cutting shrubs and small trees into ornamental shapes.
Transplant To move a plant from one growing position to another.
Tuber A swollen underground stem used to store moisture and nutrients.

Umbel An inflorescence with stalked flowers arising from a single point.

Variety A term used in the vernacular to describe different kinds of plant. Also used in botanical classification to describe a naturally occurring variant of a plant.
Vegetative A term used to describe parts of a plant which are capable of growth.
Vermifuge A substance that expels or destroys worms.
Vulnerary A preparation useful in healing wounds.

Weathering Using the effect of climatic conditions to break down large lumps of soil into small particles.

USEFUL NAMES & ADDRESSES

ORGANISATIONS

Biodynamic Agricultural Association
Painswick Inn
Gloucester Street
Stroud GL5 1QG
www.biodynamic.org.uk

Centre for Alternative Technology
Machynlleth SY20 9AZ
www.cat.org.uk

Friends of the Earth
139 Clapham Road
London SW9 0HP
www.foe.co.uk

Garden Organic
Garden Organic Ryton
Coventry CV8 3LG
www.gardenorganic.org.uk

Horticulture Research International
Wellesbourne
Warwick CV35 9EF
www.hriresearch.org

John Innes Centre
Norwich Research Park
Colney
Norwich NR4 7UH
www.jic.ac.uk

National Council for Conservation of Plants and Gardens
Home Farm, Loseley Park
Guildford GU3 1HS
www.nccpg.com

National Institute of Agricultural Botany
Huntingdon Road
Cambridge CB3 0LE
www.niab.com

National Society of Allotment and Leisure Gardeners Ltd.
O'Dell House, Hunters Road
Corby NN17 5JE
www.nsalg.org.uk

National Vegetable Society
www.nvsuk.org.uk

The Organic Research Centre
Elm FarmHamstead Marshall
Newbury RG20 0HR
www.organicresearchcentre.com

Rothamsted Research
Harpenden AL5 2JQ
www.rothamsted.ac.uk

The Royal Horticultural Society (RHS)
80 Vincent Square
London SW1P 2PE
www.rhs.org.uk

The Royal Horticultural Society of Ireland
Laurelmere Cottage, Marlay Park
Grange Road, Rathfarnham
Dublin 16
www.rhsi.ie

Soil Association
South Plaza
Marlborough Street
Bristol BS1 3NX
www.soilassociation.org

Soil Association of Scotland
Osborne House, 1 Osborne Terrace
Edinburgh EH12 5HG
www.soilassociation.org/scotland

WWOOF (Working Weekends on Organic Farms)
P O Box 2207
Buckingham MK18 9BW
www.wwoof.org.uk

SEED SUPPLIERS

AK Association Kokopelli
Chris Baur
Ripple Farm
Crundale
Canterbury
CT4 7EB
www.kokopelli-seeds.com

B & T World Seeds
Paguignan
34210 Aigues-Vives
France
www.b-and-t-world-seeds.com

Beans and Herbs
The Herbary
161 Chapel St
Horningsham
Warminster BA12 7LU
www.beansandherbs.co.uk

Jennifer Birch
Garfield Villa
Belle Vue Road
Stroud GL5 1JP
Tel/Fax: 01453 750371
(Stock available Sept–Dec only)

J.W. Boyce
Bush Pasture
Fordham
Ely CB7 5JU
www.jwboyce.co.uk

Carroll's Heritage Potatoes
Tiptoe Farm
Cornhill-on-Tweed TD12 4XD
www.heritage-potatoes.co.uk

Charlton Park Garden Centre
Charlton Road
Wantage OX12 8EP
www.charlton-park.co.uk

Chiltern Seeds
114 Preston Crowmarsh
Wallingford, OX10 6SL
www.chilternseeds.co.uk

D.T. Brown & Co.
Bury Road
Kentford
Newmarket CB8 7PQ
www.dtbrownseeds.co.uk

Delfland Nurseries Ltd
Benwick Road
Doddington
March PE15 0TU
www.organicplants.co.uk

Demeter Seeds
Stormy Hall Farm
Botton Village
Danby
Whitby YO21 2NJ
www.stormy-hall-seeds.co.uk

E.W. King & Co. Ltd
Monks Farm
Coggeshall Road
Pantling Lane
Kelvedon
Colchester CO5 9PG
www.kingsseeds.com

Edwin Tucker & Sons Ltd
Brewery Meadow
Stonepark
Ashburton TQ13 7DG
www.edwintucker.com

Exhibition Seeds
'Ridge Lea', 4 Lingrow Close
Runswick Bay
Saltburn by the Sea TS13 5JQ
www.exhibition-seeds.co.uk

G.M. & E.A. Innes
Oldtown
Newmachar AB21 7PR
Tel/Fax: 01651 862333

Heritage Seed Library (HDRA)
Garden Organic
Ryton
Coventry CV8 3LG
www.gardenorganic.org.uk

Medwyn's
Llanor, Old School Lane,
Llanfair P.G.,
Anglesey, LL61 5RZ
www.medwynsofanglesey.co.uk

Moreveg
Shangri-La, Westown
Hemyock
Cullompton EX15 3RP
www.moreveg.co.uk

Mr. Fothergill's Seeds Ltd
Gazeley Rd,
Kentford CB8 7QB
www.mr-fothergills.co.uk

Nicky's Nursery
33, Fairfield Road
Broadstairs CT10 2JU
www.nickys-nursery.co.uk

Organic Gardening Catalogue
52-54 Hamm Moor Lane
Addlestone, Surrey KT15 2SF
www.organiccatalog.com

The Real Seed Catalogue (Vida Verde)
PO Box 18, Newport near Fishguard
Pembrokeshire SA65 0AA
www.realseeds.co.uk

S.E. Marshall & Co Ltd
Alconbury Hill
Huntingdon PE28 4HY
www.marshalls-seeds.co.uk

S M McArd
39 West Road, Pointon
Sleaford NG34 0NA
www.smmcard.com

Seeds by Size
35 Leverstock Green Rd
Hemel Hempstead
Herts HP2 4HH
www.seeds-by-size.co.uk

Seeds of Italy
Unit D2, Phoenix Business Centre
Rosslyn Crescent
Harrow, Middx, HA1 2SP
www.seedsofitaly.com

Select Seeds
58 Bentinck Road
Shuttlewood
Chesterfield S44 6RQ
www.selectseeds.co.uk

Shelley Seeds
5 Speedwell Close
Huntington
Chester CH3 6DX
Tel: 01244 317165

Simply Vegetables
Abacus House
Station Yard
Needham Market
Suffolk IP6 8AS
www.plantsofdistinction.co.uk

Simpson's Seeds
The Walled Garden Nursery
Horningsham
Warminster BA12 7NQ
www.simpsonsseeds.co.uk

Suffolk Herbs
Monks Farm
Kelvedon
Colchester CO5 9PG
www.kingseeds.com

Suttons Seeds
Woodview Road
Paignton TQ4 7NG
www.suttons.co.uk

Tamar Organics
Cartha Martha Farm
Rezare
Launceston PL15 9NX
www.tamarorganics.co.uk

Terwins Seeds
Peppers Hall
Old Hall Lane
Cockfield
Bury St Edmunds IP30 0LH
www.growninengland.co.uk

Thomas Etty Esq.
Seedsman's Cottage
Puddlebridge
Horton
Ilminster TA19 9RL
www.thomasetty.co.uk

Thompson & Morgan
Poplar Lane
Ipswich IP8 3BU
www.thompson-morgan.com

Unwins Seeds Ltd
Alconbury Weston
Huntingdon PE28 4HY
www.unwins.co.uk

W. Robinson & Sons Ltd
Sunny Bank
Forton
Nr. Preston PR3 0BN
www.mammothonion.co.uk

Wallis Seeds
Broads Green
Great Waltham
Chelmsford CM3 1DS
www.wallis-seeds.co.uk

Wisley Plant Centre
RHS Garden
Wisley
Woking GU23 6QB
www.rhs.org.uk

Suppliers of Biological Controls
Biological controls are also available from garden centres and other outlets.

Agralan
The Old Brickyard
Ashton Keynes
Swindon SN6 6QR
www.agralan.co.uk

Defenders Ltd.
Occupation Road
Wye
Ashford TN25 5EN
www.defenders.co.uk

Green Gardener
Chandlers End, Mill Road
Stokesby, Gt Yarmouth NR29 3EY
www.greengardener.co.uk

Scarletts Plant Care / Ladybird Plant Care
Free Spirits Ltd.
The Glasshouses
Fletchling Common
Newick
Lewes BN8 4JJ
www.ladybirdplantcare.co.uk

The Natural Gardener
The Steppes
Hope under Dinmore
Nr. Leominster HR6 0PP
www.thenaturalgardener.co.uk

SELECTED BIBLIOGRAPHY & OTHER USEFUL BOOKS

vegetables

The Allotment Book
Andi Clevely (Collins 2008)

The Allotment Book: Seasonal Planner and Cookbook
Andi Clevely (Collins 2008)

Biodynamic Gardening
John Soper (Souvenir press 1983)

Bob Flowerdew's Complete Book of Companion Gardening (New Edition)
Bob Flowerdew (Kyle Books 2004)

Collins Guide to the Pests, Diseases and Disorders of Garden Plants
Stefan Buczacki & K. M. Harris (Harper Collins 1994)

The Complete Know and Grow Vegetables
J.K.A. Bleasdale & P.J. Salter (Oxford Paperbacks 1991)

Control Pests
Richard Jones (Impact Publishing 2007)

Create Compost
Pauline Pears (Impact Publishing 2007)

Creative Vegetable Gardening
Joy Larcom (Mitchell Beazley 2008)

Crops in Pots: 50 Great Container Projects Using Vegetables, Fruit and Herbs
Bob Purnell (Hamlyn 2007)

Domestication of Plants in the Old World
Daniel Zohary & Maria Hopf (Oxford Science Publications 1994)

Dye Plants and Dyeing
John & Margaret Cannon (Herbert Press 1994)

Encyclopedia of Gardening (RHS) (Revised Edition)
Christopher Brickell (Dorling Kindersley 2007)

The English Gardener
William Cobbett (Cobbett 1833)

Gardening and Planting by the Moon 2008
Nick Kollerstrom (Quantum/Foulsham; yearly)

Growing Fruit and Vegetables on a Bed System the Organic Way (new edition)
Pauline Pears (Search Press Ltd 2004)

The Half-hour Allotment (RHS)
Lia Leendertz (Frances Lincoln 2006)

HDRA: Encyclopedia of Organic Gardening (Henry Doubleday Research Association)
Pauline Pears (Dorling Kindersley 2005)

Jane Grigson's Vegetable Book
Jane Grigson (Penguin 1988)

The Gardeners and Florists Dictionary or a Complete System of Horticulture: Volumes 1 & 2
Phillip Miller (Charles Rivington 1724)

Gourmet Gardener (New Edition)
Bob Flowerdew (Kyle Cathie 2007)

Grow Your Own Veg (RHS)
Carol Klein & Royal Horticultural Society
(Mitchell Beazley 2007)

The History and Social Influence of the Potato
Redcliffe Salaman (Cambridge University Press 1984)

The Kitchen Garden: a Historical Guide to Traditional Crops
David Stuart (Hale 1984)

Maison Rustique, or the Countrie Farme
Charles Estienne & Richard Surflet (1600)

Organic Gardening: The Natural No-dig Way
Charles Dowding (Green Books 2007)

The Organic Salad Garden (New Edition)
Joy Larkcom & Roger Phillips (Frances Lincoln 2003)

Oriental Vegetables
Joy Larkcom (Frances Lincoln 2007)

The Ornamental Kitchen Garden
Geoff Hamilton (BBC Books 1990)

Queer Gear; How to Buy and Cook Exotic Fruits and Vegetables
Michael Allsop & Carolyn Heal
(Century Hutchinson 1986)
RHS Pests and Diseases
Pippa Greenwood & Andrew Halstead
(Dorling Kindersley 2007)

Tropical Planting and Gardening
H.F. MacMillan (Malayan Nature Society 1991)

The Vegetable Garden Displayed
Joy Larkcom (RHS 1992)

Vegetables
Roger Phillips & Martyn Rix (Pan Macmillan 1993)

Vegetables of South East Asia
G.A.C. Herklots (Allen & Unwin 1973)

The Yellow Book: NGS Gardens Open for Charity
(Revised Edition)
Stephen Anderton, Tim Wonnacott & Zac Goldsmith
(The National Gardens Scheme 2008)

herbs

Encyclopedia of Herbs and Herbalism
ed. Malcolm Stuart (Black Cat 1979)

Encyclopedia of Medicinal Plants
Andrew Chevallier (Dorling Kindersley 1996)

Encyclopedia of Medicinal Plants
Roberto Chiej (Macdonald 1984)

Evening Primrose Oil
Judy Graham (Thorsons 1984)

HDRA Encyclopedia of Organic Gardening,
(Dorling Kindersley 2001)

Herbal
John Gerard (1636; Bracken Books 1985)

Home Herbal
Penelope Ody (Dorling Kindersley 1995)

Medicinal Plants of South Africa
Ben-Erik van Wyk, Bosh van Oudtshoorn, Nigel Gericke, (Briza Publications 1997)

Modern Herbal
A.M. Grieve (Peregrine 1976)

Organic Gardening, Month by Month Guide to
Lawrence D. Hills (Thorsons 1983)

RHS Encyclopedia of Gardening
*(*Dorling Kindersley 4th edition 2007)

RHS Encyclopedia of Herbs
Deni Bown (Dorling Kindersley 1995)

RHS Plant Finder
(Dorling Kindersley, published annually)

Seeds
Jekka McVicar (Kyle Books 2001)

Complete Book of Herbs
Lesley Bremness (Dorling Kindersley 1988)

Complete Herbal Culpeper
J. Gleave & Son (1826)

Complete New Herbal, The
ed. Richard Mabey (Penguin 1988)

Complete Cookery Course
Delia Smith (BBC Books 1982)

English Man's Flora, The
Geoffrey Grigson (Paladin 1975)

Herb Book, The
Arabella Boxer & Philippa Black (Octopus 1980)

Herb Book, The
John Lust (Bantam 1974)

Herb Gardening at its Best
Sal Gilbertie with Larry Sheehan (Atheneum/smi 1978)

fruits

These are a few of the many worth looking for. Although they are not all directly concerned with fruit, they contain useful and relevant advice. Some are now out of print and need unearthing from libraries.

The American Gardener
William Cobbett (1821)

Anatomy of Dessert
E. A. Bunyard (1929)

A Handbook of Hardy Fruits
E. A. Bunyard (Picton 1994)

The Diagnosis of Mineral Deficiencies in Plants
HMSO (1943)

Food for Free
Richard Mabey *(*Fontana/Collins 1972/75)

Fruit, Berry and Nut Inventory
Whealy & Thuente (eds.) (Seed Saver Exchange 1993)

The Good Fruit Guide
L. D. Hills *(*Henry Doubleday Research Association)

A Handbook of Hardy Fruits
E. A. Bunyard (Picton 1994)

Handbook of Insects Injurious to Orchard and Bush Fruits
E. A. Ormerod (Simpkin, Marshall, Hamilton & Co., 1898)

Hillier's Manual of Trees and Shrubs
David & Charles

The Miniature Fruit Garden
Thomas Rivers (1860)

Nutritional Value in Crops and Plants
Werner Schupan and C. L. Whittles (Museum Press 1965)

The Orchard House
Thomas Rivers (1859)

Organic Plant Protection
R. B. Yepsen (ed.) (Rodale, 3rd printing 1976)

The Oxford Book of Food Plants
(Peerage Books 1969)

The Physiology of Taste
Brillat-Savarin (trans. M. F. K. Fisher) (Knopf 1949)

Plain and Pleasant Talk about Fruits, Flowers and Farming
Henry W. Beecher (Derby & Jackson 1895)

Plant Physiological Disorders
ADAS, HMSO (1985)

The Plant Finder
Hardy Plant Society (1987)

Plant and Planet
Anthony Huxley (Allen Lane 1974)

Pomarium Britannicum
Henry Phillips (Horticulture Society London 1821)

The Pruning of Trees, Shrubs & Conifers
George E. Brown (Faber & Faber 1972)

Science and Fruit
Long Ashton Research Station (University of Bristol 1953)

Soil Conditions and Plant Growth
Sir John Russell (Longmans, 8th edition 1954)

Sturtevant's Edible Plants of the World
U. P. Hendrick (ed.) (Dover Publications 1972)

Treatise on the Culture and Management of Fruit Trees
William Forsyth (1803)

Trees and Bushes of Britain and Europe
Oleg Polunin (Oxford University Press 1976)

Tropical Planting and Gardening
Macmillan, Barlow, Enoch & Russell (Macmillan 1949)

Weed Control
Robbins, Crafts & Raynor (McGraw Hill 1942)

12 Books on Husbandry
L. Junius Moderatus Columella (Printed for A. Miller 1745)

INDEX

ACKNOWLEDGEMENTS

The authors and publisher wish to thanks the following for permission to quote recipes:

Onion and Walnut Muffins – Hudson Valley Cookbook by Wally Malouf; La Gasconnade – Goose Fat & Garlic by Jeanne Strang; Mrs Krause's Pepper Hash – Pennsylvania Dutch Country Cooking by William Woys Weaver; Risotto with Artichokes – The River Cafe Cookbook by Rose Gray and Ruth Rogers; Pelecing Peria – Indonesian Food and Cookery by Sri Owen; Chayote in Red Wine and Creamed Chinese Artichokes – The Vegetable Book by Jane Grigson; Tzimmes – Russian Cooking by Olga Phklebin.

Photography key:

AL = Alamy; BAL = Bridgeman Art Library; BF = Bob Flowerdew: FY = Fran Yorke; GAP = GAP Photos; GH = Geoff Hayes; GPL = Garden Picture Library; GWI = Garden World Images; JM = Jekka McVicar; LL = Lisa Linder; MG = Michelle Garratt; PC = Pete Cassidy; PG = Photolibrary Group; sf = Stockfood; SPL = Science Photo Library; WH = Will Heap

l = left; r = right; b = bottom; c = centre; t = top

1 David Chapman; 2 Tim Gainey/AL; 4/5 Anneke Doorenbosch/AL; 7 Private Collection © Chris Beetles, London, UK/BAL; 8 Loxton, Margaret/RONA Gallery, London, UK/BAL; 9 Robins, Thomas/© Cheltenham Art Gallery & Museums, Gloucestershire, UK/BAL

Vegetables
10 Mark Bolton/PG; 13 (t) botanica/PG, (b) AGStockUSA, Inc./AL; 14 (t l) Tim Spence/Lightshaft Ltd/PG, (t r) botanica/PG, (b) Michele Lamontagne/PL; 15 (l) PSL Images/AL, (r) botanica/PG; 16 (l) David Sanger Photography/AL, (r) blickwinkel/AL; 17 MG; 18 (l) Natural Visions/AL, (r) Jacqui Hurst/PL; 19 (l) David Marsden/Anthony Blake Photo Library, (r) Steffen Hauser/botanikfoto/AL; 20 (t) blickwinkel/AL, (b) Jacqui Hurst/PL; 21 (l) MG, (r) MG; 22 (l) Iain Bagwell/ALnthony Blake Photo Library, (r) Foodcollection.com/AL; 23 G&M Garden Images/AL; 24 (l) Howard Rice/PL, (r) Maxine Adcock/PL; 25 MG; 26 (l) Tim Hill/AL, (r) JTB Photo Communications, Inc./AL; 28 (l) J S Sira/PL, (r) Robert Harding Picture Library Ltd/AL; 29 (l) botanica/PG, (r) Martin Brigdale; 30 (l) Photofrenetic/AL, (r) Duncan Smith/SPL; 31 MG; 32 (l) Busse Yankushev/PL, (r) Inga Spence/AL; 33 corbis/PG; 34 Jacqui Hurst/PL; 35 MG; 36 Foodpix/PG; 37(t) MG, (b) Jo Whitworth/PG; 38 (l) Bloom Works Inc./AL, (r) Bob Gibbons/AL; 39 (l) Arco Images GmbH/AL, (r) Martin Brigdale; 40 (l) Arco Images GmbH/AL, (r) Michael Grant/AL; 41 Jean-Christophe Novelli; 42 Derek St Romaine; 43 (l) botanica/PG, (r) blickwinkel/AL; 44 MG; 45 (l) Rob Whitrow/PL , (r) Howard Rice/PL; 46 Michael Howes/PL; 47 MG; 48 Martin Brigdale; 49 (l) John Glover/PL, (r) Martin Brigdale; 50 Andy Williams/SPL; 51 (l) Nigel Cattlin/AL, (r) MG; 52 (l) Stephen Shepherd/PL, (r) vario images GmbH & Co.KG/AL; 53 (l) Geoff Kidd/SPL, (r) Martin Brigdale; 54 (l) Kieran Scott/ALnthony Blake Photo Library, (r) JUPITERIMAGES/ Polka Dot/AL; 55 (l) D. Hurst/AL, (r) Malcolm Case-Green/AL; 56 MG; 57 (l) O.D. van de Veer/AL, (r) John Glover/AL; 58 (t) Martin Hughes-Jones/AL, (b) John Daniels/ALrdea.com; 59 (l) MG, (r) MG; 60 John Martin/AL; 61 (l) foodpix/PG, (r) Nic Murray/AL; 62 (l) Adrian Davies/AL, (r) MG; 63 (l) STOCKFOLIO/AL, (r) Bon Appetit/AL; 64 (l) Holmes Garden Photos/AL, (r) MG; 65 (l) Nigel Cattlin/AL, (r) Howard Rice/PL; 66 (l) Marc Hill/AL, (r) Organics image library/AL; 67 MG; 68 (l) Garden World Images Ltd/AL, (r) Marie-Louise Avery/AL; 69 (l) Photo Agency EYE/AL, (r) Sue Atkinson/AL R C Studios/PG; 70 (t) Sue Wilson/AL, (r) Martin Brigdale; 71 (l) David Cavagnaro/PL, (r) Martin Brigdale; 72 MIXA Co., Ltd./AL; 73 (t) John Glover/AL, (b) Westend61 Ohg/PG; 74 Nigel Cattlin/AL; 75 MG; 76 (t) Martin Brigdale, (b) Jeff Morgan food and drink/AL; 77 Simon Fraser/SPL; 78 (l) Shannon99/AL, (r) Martin Brigdale; 79 (l) Shannon99/AL, (r) Martin Brigdale; 80 Nigel Cattlin/AL; 81 (t) blickwinkel/AL, (b) MG; 82 (t) Bon Appetit/AL, (b) Friedrich Strauss/PL; 83 (t) Images Etc Ltd/AL, (b) Maximilian Stock Ltd/Sf; 84 (t) Bon Appetit/AL, (b) Arco Images GmbH/AL; 85 botanica/PG; 86 (t) MG, (b) MG; 87 (l) blickwinkel/AL, (r) Richard Wainscoat/AL; 88 (l) Collections/PG, (r) MG; 89 (l) Kaktusfactory, Ninprapha Lippert/sf, (r) Lemonnier, Nicolas/sf; 90 (l) Philip Smith/AL, (r) Collection/PG; 91 (t) Brian Harris/AL, (b) MG; 92 Nigel Cattlin/AL; 93 Francois De Heel/PG; 94 (l) Christopher Burrows/AL, (r) Chris Burrows/PG; 95 (l) Collection/PG, (r) MG; 96 Piotr Adamczyk shapencolour/AL; 97 (l) Collection/PG, (c) Paul Hart/PL, (r) Carole Drake/PL; 98 (b) Shangara Singh/AL, (t) Foodcollection.com/AL; 99 Mark Winwood/PL; 100 (l) Winkelmann, Bernhard/sf, (r) MG; 101 (l) Collection/PG, (r) Le Scanff Mayer/PG; 102 (l) Collection/PG, (r) MG; 103 (l) Anneke Doorenbosch/AL, (r) Pernilla Bergdahl/PL; 104 (l) Collections/PG, (r) Paul Collis/AL; 105 (t) Lutterbeck, Barbara/sf, (b) FoodPhotogr. Eising/sf; 106 moodboard/AL; 107 (l) Maxine Adcock/PL, (c) Maxine Adcock/PL, (r) David Forster/AL; 108 (l) Mark Bolton/PL, (r) Hoff, Dana/sf; 109 (l) MG, (r) MG; 110 John Heseltine/Fresh Food Images; 111 (l) Michael Warren/PH, (r) MG; 112 Martin Brigdale; 113 (l) Kit Young/PL, (r) MG; 114 FoodStock/AL; 115 (l) Francois De Heel/PL, (r) MG; 116 (l) Bill Barksdale/ALGSTOCKUSA/SPL, (r) D. Hurst/AL; 117 MG; 118 Neil Holmes/PL; 119 (l) Piotr & Irena Kolasa/AL, (r) MG; 120 (l) David Cavagnaro/PL, (r) Michael Warren/PH;

121 MG; 122 (l) David Cavagnaro/PL, (r) Nigel Cattlin/AL; 123 (l) Vietnam Images/AL, (r) Martin Brigdale, 124 (l) David Cavagnaro/PL, (r) Michele Molinari/AL; 125 MG; 126 (l) Jo Whitworth/PL, (r) Howard Rice/PL; 127 (l) Geoff Kidd/SPL, (r) John Swithinbank/PL; 128 (t) Sunniva Harte/PL, (b) botanica/PL; 129 Pernilla Bergdahl/PL; 130 Juliet Piddington/PL; 131 Martin Brigdale; 132 (t) Nigel Cattlin/AL, (b) Dufour Brigette Dit Noun/PG; 133 MG; 134 botanica/PG; 135 (l) Mark Bolton Photography/AL, (r) Bon Appetit/AL; 136 Tim Hill/AL; 137 MG; 138 (r) Itani/AL, (l) Foodpix/PG; 139 (t) MG, (b) AGStockUSA, Inc./AL; 140 John Glover/PL; 141 (l) Stephen Henderson (Burt Latino Com)/PL, (c) Sunniva Harte/PL, (r) David Cavagnaro/PL; 142 (l) botanica/PG, (r) Alec Scaresbrook/PL; 143 (t) botanica/Koenigsberg Diana/PG, (b) Michael Howes/PL; 144 (l) Mark Winwood/PL, (r) Michael Howes/PL; 145 MG; 146 botanica/PG; 147 (t) David R. Frazier PL, Inc./AL, (b) MG; 148 (l) Brian L Carter/PL, (r) Mark Gibson/AL; 149 (l) MG, (r) Joy Skipper/Fresh Food Images/PG; 150 (l) botanica/PG, (r) Anto V Chacko valiyakulam/AL; 151 (l) Henry Westheim Photography/AL, (r) MG; 152 (l) Steffen Hauser/botanikfoto/AL, (r) MG; 153 (l) Martin Brigdale, (r) Martin Brigdale; 154 (l) Bon Appetit/AL, (r) Maxine Adcock/PL; 155 (l) Jacqui Hurst/PL, (r) Steffen Hauser/botanikfoto/AL; 156 (l) David /AL, (r) MG; 157 (l) Foodcollection.com/AL, (r) imagebroker/AL; 158 (t) CuboImages srl/AL, (b) MG; 159 (t) Foodcollection.com/AL, (b) Tim Gainey/AL; 160 (l) NDisc/AL, (r) Neil McAllister/AL; 161 Howard Rice/PL; 162 (l) Niall McDiarmid/AL, (c) Anne Green-Armytage/PL, (r) Tommy Candler/PL; 163 (l) Anne Green-Armytage/PL, (r) MG; 164 (l) Anne Green-Armytage/PL, (r) Howard Rice/PL; 165 (l) Michael Howes/PL, (r) Mark Winwood/PL; 166 MG; 167 (l) botanica/PG, (r) Mark Bolton/PL; 168 (l) Juliette H Wade/PL, (r) Martin Brigdale; 169 (l) botanica/PG, (r) botanica/PG; 170 (l) Mark Turner/PL, (r) botanica/PG; 171 MG; 172 (t) MG, (b) Howard Rice/PL; 173 (l) MG, (r) Maxine Adcock; 174 (l) botanica/PG, (r) Juliet Greene/PL; 175 MG; 176 (l) Juliette Wade/PL, (r) Howard Rice/PL; 177 (t) Ray Roberts/AL, (b) MG; 178 David Cavagnaro/PL; 179 (l) Nic Murray/AL, (r) MG; 180 (l) Mayer/Le Scanff/PG, (r) Maximilian Stock Ltd/sf; 181 (l) Nigel Cattlin/AL, (r) Minowa Studio Co./sf; 182 (l) blickwinkel/AL, (r) imagebroker/AL; 183 (l) blickwinkel/AL, (r) MG; 184 (t) Westend 61/AL, (b) Juliette H Wade (Espalier Media Ltd)/PG; 185 (l) Chris Burrows/PL, (r) MG; 186 Maxine Adcock/PL; 187 (l) botanica/PG, (r) Maxine Adcock/PL; 188 (l) Jacqui Hurst/PL, (r) Jo Whitworth/PL; 189 Maxine Adcock/PL; 190 Maxine Adcock/PL; 191 (t) MG, (b) MG; 192 (l) botanica/PG, (r) Inga Spence/AL; 193 (l) Dufour Brigette Dit Noun/PG, (r) MG; 194 (l) Andrea Jones/AL, (c) Dave Bevan/AL, (r) MG; 195 Arco Images GmbH/AL; 196 (l) Arco Images GmbH/AL, (c) Bon Appetit/AL, (r) FoodPhotography/sf; 197 (l) Nigel Cattlin/AL, (r) blickwinkel/AL; 198 MG; 199 (l) Sunniva Harte/PG, (r) Martin Hughes-Jones/AL; 200 (t) Arco Images GmbH/AL, (b) Foodcollection/sf; 201 (l) Mayer/Le Scanff/PG, (r) Maxine Adcock/PL; 202 PhotoCuisine RM/AL; 203 MG; 204 (l) Rich Iwasaki/AL, (r) Jonah Calinawan/AL; 205 (l) imagebroker/AL, (r) Martin Brigdale; 206 Mark Winwood/PL; 207 (l) Mark Winwood/PL, (r) LWA- JDC/PG; 208 (l) David Cavagnaro/PL, (c) botanica/PG, (r) Carole Drake/PL; 209 MG

Herbs

All photos by Jekka McVicar except:

212 (l) Sally Maltby; 213 (r) MG; 215 MG; 216 (r) Mediacolor's/Alamy, (l) Bickwinkel/AL; 217 Ron Niebrugge/Alamy; 218 (l) John Glover/Alamy, (r) Organica/Alamy; 219 MG; 220 (t) Mediacolor's/Alamy, (b) Torie Chugg; 221 (l) GH; (r) MG; 223 MG; 224 (l) Torie Chugg; 225 MG; 226 (b) Clive Nichols/Garden Picture Library; 227 (l) GH; 228 (l) GH, (b) MG; 229 (l) GH, (r) MG; 233 MG; 235 MG; 237 (l) WH; 239 (both) MG; 241 (b) MG; 243 (b) MG; 245 MG; 246 (b) Sally Maltby; 247 (b) MG; 248 GH; 251 MG; 253 (r) WH; 257 (r) MG; 259 (r) MG; 260 (l) Torie Chugg; (r) GH; 261 (b & r) MG; 262 Pat Behnke/AL; 263 (t) WH, (b) CuboImages srl/AL; 264 (b) MG; 265 MG; 266 Sara Taylor (b); 267 (t) David Hoffman Photo Library/AL; 269 (t) Sally Maltby, (b) MG; 271 (t) WH; 272 (r) GH; 273 (b) GH; 275 MG; 279 (c) John Glover/Alamy, (b) WH; 281 Organics image library/Alamy; 282 MG; 283 (b) WH; 284 Steve Baxter; 285 (t) Photo Dinodia/Garden Picture Library, (b) WH; 287 (t) D. Hurst/Alamy,(b) WH; 288 (b) MG; 289 Sally Maltby; 291 (t) Richard Bloom/GPL, (b) WH; 293 (r) Steve Baxter; 294 WH; 295 (l) Sally Maltby; 297 (b) WH; 298 (b) WH; 300 (b) Organica/Alamy; 301 (b) MG; 302 (t) Rob Walls/AL, (b) WH; 303 (b) MG; 304 (l) GH; 307 (b) Friedrich Strauss/Garden Picture Library; 308 Mediacolor's/Alamy; 309 (b) MG; 310 GH; 311 (b) MG; 313 (t) GH, (b) MG; 314 GH; 315 (t) GH, (b) MG; 317 (r) MG; 318 (b) WH; 319 (t) Bickwinkel/AL; 320 (r) Woodystock/AL; 321 MG; 322 Torie Chugg; 323 (t) GH, (b) MG; 325 (r) Mark Bolton/GPL; 326 (r) GH; 327 MG; 329 MG; 330 GH; 331 (bl) GH, (t & br) MG: 332 (b) MG; 333 (r) WH; 334 (b) MG; 335 (r) GH; 336 MG; 339 Torie Chugg; 340 MG; 342 (br) GH; 343 MG; 345 (t) WH, (b) Fenix rising/AL; 347 MG; 349 (tr) MG; 350 (t) GH; 352 (t & br) MG; 353 MG; 354 (l) GH; 355 (t) WH; 358 (b) MG; 359 (b) MG; 362 MG; 363 Torie Chugg; 364 (r) WH; 366 WH; 367 GH; 368 (l) MG, (r) WH; 369 (bl) GH, (br) MG; 370 (r) Sally Maltby; 372 GH; 373 (t) TH Foto/Alamy; 374 (r) MG; 376 (b) Sally Maltby; 377 (b) MG; 378 (b) WH; 383 MG; 384 GH; 385 (t) MG, (b) GH; 386 (b) GH; 387 (t) GH; 390 (b) GH, (tr) MG; 391 MG; 392 John La Gette/Alamy; 393 (t) MG, (b) Dave Watts/AL; 394 (l) GH, (r) Organics image library/Alamy; 395 MG; 396 (b) GH; 397 (tr & bl) GH, (br) MG; 398 (t) GH, (b) MG; 401 (b) MG; 402 (t) GH; 403 (t) MG, (b) Sally Maltby; 404 (r) GH; 405 (b) WH; 406 Konrad Zelazowski/AL; 407 (tl) WH, (tr) Organica/Alamy, (br) MG; 409 (l) Fenix rising/AL, (r) WH; 411 (t) Chris & Tilde Stuart/FLPA, (b) Howard Rice/Garden Picture Library; 413 (t) GH, (b) MG; 414 Bickwinkel/AL; 415 (t) WH; 417 Sally Maltby; 418 (l) Sally Maltby, (r) MG; 421 MG; 423 (br) MG; 424 (t) GH; 425 (t) Torie Chugg; 427 MG; 429 (r) MG; 430 (t) GH; 431 MG; 432 GH; 433 (b) MG; 434 (r) Sally Maltby; 435 (b) MG; 437 (t) Neil Hardwick/AL, (b) MG; 439 (t) Eising/StockFood; 439 (b) Holmes Garden Photo/AL; 440 Paul Thompson Images/AL; 441 (t) Scherer, Tim/StockFood, (b) Humberto Olarte Cupas/AL

Fruit

442 PC; 444 PC; 446 BF; 447 (t) PC, (b) BF; 448 (l) GH, (r) PC; 449 PC; 450 (t) and (bl) BF, (br) PC; 451 MG; 452 (t) Mark Bolton/GAP, (b) Christina Bollen/GAP; 453 (t) Sally Maltby, (c) Jonathan Buckley/GAP, (b) MG; 454 PC; 455 (l) BF, (r) FY; 456 (l) FY, (r) PC; 457 WH; 458 (t) Dave Bevan/GAP, (b) Juliette Wade/GAP; 459 (t) Juliette Wade/GPL, (b) MG; 460 (t) Emma Lee, (b) Andrea Jones/AL; 461 (t) MG, (b) David Askham/AL; 462 (t) PC, (b) Christina Bollen/GAP; 463 (l) John Glover/GAP, (r) MG; 464 (l) imagebroker/AL, (r) PC; 465 MG; 466 Arco Images GmbH/AL; 467 (l) GPL, (t) (r) blickwinkel/AL, (b) MG; 468 (t) Christie Carter/GPL, (b) FY; 469 (l) Maddie Thornhill/GAP, (tr) MG, (br) BF; 470 PC; 471 (l) FY, (r) BF; 472 (t) BF, (b) FY; 473 MG; 474 (t) Trevor Sims/GWI, (b) BF; 475 (l) Holmes Garden Photos/AL, (r) MG; 476 (t) CuboImages srl/AL, (r) John Swithinbank/

GWI; 477 (l) Sine Chesterman/GWI, (tr) MG, (br) Nic Murray/AL; 478 (t) BF, (b) FY; 479 MG; 480 Dave Bevan/GAP; 481 (l) Frederic Didillon/GAP, (r) MG; 482 (t) Holmes Garden Photos/AL, (b) BF; 483 (tl) Susie McCaffrey/AL, (tr) MG, (br) GardenPixels/AL; 484 PC; 486 (t) Jonathan Buckley/GAP, (b) BF; 487 (tl) PC, (tr) MG, (b) Howard Rice/GAP; 488 BF; 489 (t) Michael Howes/GPL, (bl & br) MG; 450 (t) BF, (b) Organica/AL; 491 (l) FhF Greenmedia/GAP, (b) MG; 492 (r) Botanica GPL, (l) BF; 493 MG; 494 John Glover/GAP; 495 (tl) Phil Degginger/AL, (tr) MG, (b) imagebroker/AL; 496 (t) Juliette Wade/GPL, (b) BF; 497 (t) PC, (b) MG; 498 (t) BF, (b) FY; 499 (t) FY, (bl) BF, (br) MG; 500 (t) PC, (b) BF; 501 (l) Paul Debois/GAP, (t & br) MG; 502 (t) blickwinkel/AL, (b) BF; 503 MG; 504 FY; 505 (t) Paul Debois/GAP, (b) Dufour Brigette Dit Noun/GPL; 506 (l) Paul Debois/GAP, (r) Howard Rice/GPL; 507 MG; 509 (t) BF, (b) Michael Howes/GPL; 510 (t) James Baigrie/GPL, (b) Dave Bevan/GAP; 511 MG; 512 (t) Juliette Wade/GPL, (b) Michael Howes/GPL; 513 (l) Arco Images GmbH/AL, (r) MG; 514 PC; 515 (l) BF, (r) MG; 516 BF; 517 (t) Juliette Wade/GPL, (b) MG; 518 BF; 519 (t) Christie Carter/GPL, (b) MG; 520 PC; 521 BF; 522 BF; 523 (t) MG, (b) Rita Coates/GWI; 524 Inga Spence/AL; 526 (t) BF, (b) Pernilla Bergdahl/GAP; 527 (l) Juliette Wade/GAP, (c) Friedrich Strauss/GAP, (r) WH; 528 Eelco Nicodem/AL; 529 (t) BF, (b) Bon Appetit/AL; 530 BF; 531 MG; 532 (t) BF, (b) Arco Images GmbH/AL; 533 (t) PC, (b) MG; 534 BF; 535 (t) BF, (b) John Glover/GAP; 536 (tl) PC, (tr) FY, (b) BF; 537 (tl) FY, (tr) PC, (b) MG; 538 Emma Lee; 539 (t) GH, (b) MG; 540 PC; 541 (t & bl) BF, (br) MG; 542 (t) BF, (b) Susanne Kischnick/AL; 543 (l) BF, centre Tim Gainey/AL, (r) MG; 544 (t) Arco Images GmbH/AL, (bl) Nigel Cattlin/AL, (br) MG; 545 (t) ImageGap/AL, (b) MG; 546 (t) Arco Images GmbH/AL, (b) MG; 547 BF; 548 (l) blickwinkel/AL, (r) BF; 549 Douglas Fisher/AL; 550 GH; 552 GH; 553 (l) MG, (r) BF; 554 GH; 555 (tl & b) BF, (t) (r) MG; 556 (t) GH, (b) MG; 557 imagebroker/AL; 558 (t) Nigel Cattlin/AL, (c) Photoimagerie/AL, (b) MG; 559 (l) Arco Images GmbH/AL, (r) MG; 560 (l) BF, (tr) MG, (br) PC; 561 (t) GH, (bl) BF, (br) MG; 562 (l) Courtney Turner/GPL, (t) MG, (b) John Lander/AL; 563 (t) imagebroker/AL, (b) MG; 564 BF; 565 (l) Navin Mistry/AL, (r) Photodisc/AL; 566 (l) INSADCO Photography/AL, (r) MG; 567 (tl) Dinodia Images/AL, (tr) Bon Appetit/AL, (b) MG; 568 (l) WoodyStock/AL, (r) David Hosking/AL; 569 (tl) Rob Crandall/AL, (tr) BF, (b) Geraldine Buckley/AL; 570 BF; 571 BF; 572 flowerphotos/AL; 574 (l) All Canada Photos/AL, (r) Martin Hughes-Jones/AL; 575 (tr) MG, centre Martin Hughes-Jones/AL, (b) LL; 576 (t) Javier Etcheverry/AL, centre John Maud/AL, (r) MG; 577 (l) Arco Images GmbH/AL, (tr) MG, (b) BF; 578 (l) Corbis RF/AL, (r) MG; 579 (l) Holmes Garden Photos/AL, (tr) blickwinkel/AL, (b) MG; 580 (tl) Bruce Coleman Inc./AL, (bl) Steffen Hauser/botanikfoto/AL, (r) MG; 581 (l) Dave Zubraski/AL, (r) MG; 582 (l) WoodyStock/AL, (r) Brian Hoffman/AL; 583 (l) Reino Hanninen/AL, (r) MG; 584 (l) Leander/AL, (r) imagebroker/AL; 585 (l) Danita Delimont/AL, (r) MG; 586 (l) John Glover/AL, (r) Niall McDiarmid/AL; 587 (l) John Glover/AL, (tr) MG, (br) Holmes Garden Photos/AL; 588 (l) CuboImages srl/AL, (r) Frank Blackburn/AL; 589 (l) Frank Blackburn/AL, centre Bob Gibbons/AL, (r) MG; 590 (l) Mike Lane/AL, (r) Bob Gibbons/AL; 591 (t) Kris Butler/AL, (bl) WILDLIFE GmbH/AL, (br) Jack Sparticus/AL; 592 (tr) Don Smetzer/AL, (c) blickwinkel/AL, (bl) FLPA/AL; 594 PC; 596 (l) imagebroker/AL, (r) Kirk Anderson/AL; 597 (t) blickwinkel/AL, (bl) flowerphotos/AL, (br) MG; 598 (b) Arni Katz/AL, (t) Nigel Cattlin/AL; 599 (tl) Organica/AL, (b) Grant Heilman Photography/AL, (tr) MG; 600 (t) Eyebyte/AL, (b) Steffen Hauser/botanikfoto/AL; 601 (tl) blickwinkel/AL, (tr) John Glover/AL, (b) MG; 602 (l) blickwinkel/AL, (r) blickwinkel/AL; 603 (t) David Lawrence/AL, (bl) Andrea Jones/AL, (br) MG; 604 (l) Tony Watson/AL, (r) Holmes Garden Photos/AL; 605 (t) MG, (bl) Elizabeth Whiting & Associates/AL, (br) Jim Lane/AL; 606 (t) blickwinkel/AL, (b) Karl Hausammann/AL; 607 (tl) Nigel Cattlin/AL, (tr) MG, (b) Arco Images GmbH/AL; 608 (t) imagebroker/AL, (b) MG; 609 (l) blickwinkel/AL, (r) MG; 610 (tl) Arco Images GmbH/AL, (tr) Nigel Cattlin/AL, (bl) imagebroker/AL, (br) MG; 611 (bl) Leonide Principe/AL, (c) blickwinkel/AL, (r) MG; 612 (l) blickwinkel/AL, (r) Arco Images GmbH/AL

Practical Gardening

614 Howard Rice/PL; 616 (t) Mark Winwood/PL, (b) Nigel Cattlin/AL; 617 (l) Nick Turner/AL, (r) Mark Winwood/PL; 618 Mark Bolton/PL; 619 Angharad Jones/PL; 620; JM; 621 WH; 622–7 artwork by Sally Maltby; 628 (l) PC, (r) Andrea Jones/AL; 629 John Glover/GPL; 630 SM; 631 (t) FY, (b) Stephen Robson/GPL; 632 botanica/PG; 633 WH; 634 PC; 635 (t) FY, (b) Stephen Shepherd/PL; 636 Chesh/AL; 638 (l) Mayer/Le Scanff/PG, (t r) Mark Bolton Photography/AL, (b r) J S Sira/PG; 639 (l) Dave Bevan/AL, (r) Dan Rosenholm/PL; 640 PC; 641 (b) PC, (t) BF; 642 (t) PC, (r) Leroy Alfonse/PL; 643 BF; 644 Dave Bevan/AL; 645 FY; 646 Dave Bevan/AL; 647 Sally Maltby; 648 Jacqui Hurst/PL; 649 (l) FY, (r) BF; 650 PC; 651 (t) Mark Winwood/PL, (b) BF; 652 BF; 653 (t) Jacqui Hurst/PL, (b) Mark Winwood/PL; 654 FY; 655 Maxine Adcock/PL; 656 WH; 657 (l, r) PC; (c) FY; 658 FY; 659 (t) BF, (b) WH; 660 (both) WH; 661 (l) FY, (r) Sunniva Harte/PL; 666 JUPITERIMAGES/Brand X/AL; 667 (l) Creatas/PG, (r) ImageRite/PG; 668 Mark Winwood; 669 MG; 670 botanica/PG; 671 MG; 672 MG; 674 JM; 676 botanica/PG; 677 Lynn Keddie/PL; 678 Howard Rice/PL; 679 BF; 681 BF